GLOBAL CRISIS

Geoffrey Parker is a renowned British historian who taught at the University of St Andrews, the University of Illinois, the University of British Columbia, and Yale University, before becoming Andreas Dorpalen Professor of European History and Associate of the Mershon Center at The Ohio State University. He is a Fellow of the British Academy, the Netherlands Academy of Arts and Sciences, the Spanish-American Academy of Arts and Sciences (Cadiz), the Royal Academy of History (Madrid), and the Royal Society of Edinburgh. He is also a Profesor Afiliado in the División de Historia at the Centro de Investigación y Docencia Económicas, Mexico City. His many books include *The Military Revolution* (1988; winner of the best book prize of the American Military Institute and the Society for the History of Technology), *The Grand Strategy of Philip II*, published by Yale in 1998 (winner of the Samuel Eliot Morison Prize), and, most recently, *Imprudent King: A New Life of Philip II*, published by Yale in 2014. In 2012, the Royal Dutch Academy awarded Parker its biennial Heineken Foundation Prize for History, open to scholars in any field and any period, from any country; two years later, he won a medal awarded by the British Academy for 'a landmark academic achievement ... which has transformed understanding of a particular subject' for *Global Crisis*.

T0327164

GEOFFREY PARKER
GLOBAL CRISIS

**WAR, CLIMATE CHANGE AND CATASTROPHE
IN THE SEVENTEENTH CENTURY**

ABRIDGED AND REVISED EDITION

YALE UNIVERSITY PRESS
NEW HAVEN AND LONDON

For information about this and other Yale University Press publications, please contact:
U.S. Office: sales.press@yale.edu   yalebooks.com
Europe Office: sales@yaleup.co.uk   yalebooks.co.uk

Set in Minion Pro by IDSUK (DataConnection) Ltd
Printed and bound by CPI Group (UK) Ltd, Croydon, CR0 4YY

Library of Congress Cataloging-in-Publication Data

Names: Parker, Geoffrey, 1943- author.
Title: Global crisis : war, climate change and catastrophe in the seventeenth
    century / Geoffrey Parker.
Description: Abridged edition. | New Haven : Yale University Press, 2017. |
    Includes bibliographical references and index.
Identifiers: LCCN 2017005476 (print) | LCCN 2017008357 (ebook) |
    ISBN 9780300219364 (paperback) | ISBN 9780300226355 ()
Subjects: LCSH: History, Modern—17th century. | Military history—17th century. |
    Civil War—History—17th century. | Revolutions—History—17th century. |
    Climatic changes—Social aspects—History—17th century. |
    Disasters—History—17th century. | BISAC: HISTORY / World. | HISTORY /
    Modern / 17th Century.
Classification: LCC D247 .P37 2017 (print) | LCC D247 (ebook) | DDC
    909/.6—dc23
LC record available at https://lccn.loc.gov/2017005476

A catalogue record for this book is available from the British Library.

10 9 8 7 6 5 4 3 2 1

This book is dedicated in admiration to all who fight
multiple sclerosis

'Global Crisis Cruelly Abridged'

And the earth did spin.
An orb far from warming light.
Cast in a dark pall.

From the sun it came;
Wave on wave, chilling absence.
Crops stunted in bloom.

The hardship spread fast.
The golden age did not last.
Rulers toppled first.

Out in barley fields,
Where farmers could scrounge no meals,
Hear the beating drum!

Headless was the king.
And the historians sing:
The weather! Really?

Bah . . .

Richard Parker, January 2014
*Composed upon learning that Gregory C. Johnson, an oceanographer, had
distilled the 2,000-page IPCC Fifth Assessment Report into nineteen haiku*

Contents

Figures

Prologue:
Did Someone Say 'Climate Change'?

Climate change has frequently caused or contributed to widespread destruction and dislocation on Earth. Glaciers around the world advanced and retreated at least five times during the Pleistocene, each one a major climate-related event; the last one, which ended some 12,000 years ago, wiped out most species of large mammals, such as the mammoths. About 4,000 years ago, societies in South and West Asia collapsed amid general drought; between AD 750 and 900 droughts on both sides of the Pacific fatally weakened the Tang empire in China and the Maya culture in Central America; in the fourteenth century, a combination of violent climatic oscillations and major epidemics halved Europe's population and caused severe depopulation and disruption in much of Asia. Then, in the seventeenth century, both the northern and southern hemispheres experienced 'undoubtedly the most pronounced global climate anomaly of the past 8,000 years (until contemporary global warming)' according to Sam White: eight decades of unusually cold weather.[1]

Although climate change can and does produce human catastrophe, until recently few historians included it in their analyses of past events. Even Emmanuel Le Roy Ladurie's pioneering 1967 study *Times of feast, times of famine: a history of climate since the year 1000*, stated that 'In the long term, the human consequences of climate seem to be slight, perhaps negligible'. By way of example, he averred that 'it would be quite absurd' to link 'the adverse meteorological conditions of the 1640s' with the Fronde rebellion that broke out in France in 1648. A few years later Jan de Vries, a distinguished economic historian, claimed that 'Short-term climatic crises stand in relation to economic history as bank robberies to the history of banking'.[2]

Historians are not alone in denying a link between climate change and human affairs. Richard Fortey, a palaeontologist, has observed that 'There is a kind of optimism built into our species that seems to prefer to live in the comfortable present rather than confront the possibility of destruction', with the result that 'Human beings are never prepared for natural disasters'.[3] Extreme climatic events therefore continue to take us by surprise. In 2003, a summer heatwave that lasted just two weeks led to the premature death of 70,000 people in Europe; two years later, Hurricane Katrina killed almost 2,000 people and destroyed property worth over $81 billion in an area of the United States equivalent in size to Great Britain; and in 2012, natural disasters adversely affected 123 million people and caused $157 billion in damage, almost half of which was inflicted on the United States by a combination of droughts, tornadoes, storms and Hurricane Sandy. Yet although we know that the climate caused these and many other catastrophes, and although we also know that it will cause many more catastrophes in the future, we still convince ourselves that they will not happen just yet (or, at least, not to us), and so fail to take appropriate action.

Currently, most attempts to predict the consequences of climate change extrapolate from recent trends, in effect hitting 'fast-forward' on the tape of history. But another methodology exists: we can also hit 'rewind' and study the genesis, impact and consequences of past catastrophes, using two distinct categories of proxy data: a 'natural archive' and a 'human archive'. Both are abundant.

The 'natural archive' on climate change comprises four groups of sources:

- *Ice cores and glaciology*: annual deposits on ice caps and glaciers around the world, captured in deep boreholes, provide evidence of changing levels of volcanic emissions, precipitation, air temperature, and atmospheric composition.
- *Palynology*: pollen and spores deposited in lakes, bogs and estuaries capture natural vegetation at the time of deposit.
- *Dendrochronology*: growth rings laid down by certain types of tree during each growing season reflect local conditions in spring and summer. A thick ring indicates a year favourable to growth, whereas a narrow ring reflects a year of adversity.
- *Speleothems*: annual deposits formed from groundwater trickling into underground caverns, especially in the form of stalactites, reveal annual surface weather.

The 'human archive' on climate change comprises five groups of sources:

- *Narrative* information contained in written texts (chronicles and histories, letters and diaries, judicial and government records, ships' logs, newspapers and broadsheets) and oral traditions.
- *Numerical* information extracted from documents (annual fluctuations in the date when harvesting certain crops began each year, or in food prices; 'Rain fell for the first time in forty-two days').
- *Pictorial* representations of natural phenomena (paintings or engravings of major weather events, or of landmarks, such as the position of a glacier's tongue).

- *Epigraphic* or *archaeological* information (inscriptions that date flood levels, or excavations of settlements abandoned because of climate change).
- *Instrumental data*: regular recordings of weather data (notably precipitation, wind direction and temperature), which exist in some parts of Europe from 1650.

The failure of most historians to exploit these sources for the seventeenth century is particularly unfortunate, because global cooling coincided with an unparalleled spate of revolutions and state breakdowns around the world. In addition, war became the norm for resolving both domestic and international disputes: Europe saw only three years of complete peace during the entire seventeenth century; the Chinese and Mughal empires fought wars almost continuously.

In an influential essay, first published in 1959, Hugh Trevor-Roper popularized the term 'General Crisis' to describe this age of turmoil, which he saw as the gateway to the modern world:

> The seventeenth century did not absorb its revolutions. It is not continuous. It is broken in the middle, irreparably broken, and at the end of it, after the revolutions, men can hardly recognize the beginning. Intellectually, politically, morally, we are in a new age, a new climate. It is as if a series of rainstorms has ended in one final thunderstorm which has cleared the air and changed, permanently, the temperature of Europe. From the end of the fifteenth century until the middle of the seventeenth century we have one climate, the climate of the Renaissance; then, in the middle of the seventeenth century, we have the years of change, the years of revolution; and thereafter, for another century and a half, we have another, very different climate, the climate of the Enlightenment.[4]

Of climate in its literal sense Trevor-Roper said not a word, even though the upheavals he described occurred during a period of prolonged global cooling and extreme climatic events. Temperature readings taken between five and eight times a day between 1654 and 1667 at an international network of observation stations in Southern Europe reveal winters that were, on average, more than 1°C cooler than those of the later twentieth century. Tree-ring series from Scotland reveal that three of the coldest decades (more than 1°C cooler) and three of the coldest years (more than 2°C cooler) recorded in the past eight centuries occurred between 1630 and 1700. The temperatures recorded by an Essex parson suggests a climate 2°C cooler in the 1660s than that of the later twentieth century, and 1.5°C cooler in the 1650s and 1670s. In the Dutch Republic, patents for heating equipment (furnaces, stoves, chimneys and so on) reached their peak in 1660–79.[5]

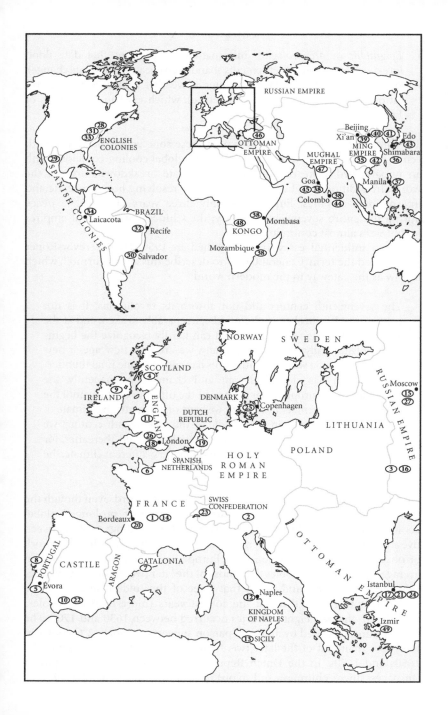

Major Revolts and Revolutions, 1635–66	
EUROPE	**AMERICAS**
1636 1. Croquants' revolt (Perigord)	1637 28. Pequot War
2. Revolt in Lower Austria	1641 29. Mexico revolt (to 1642)
1637 3. Cossack revolt (to 1638)	**30. Portuguese Brazil rebels against Spain**
4. Scottish Revolution (to 1651)	1642 31. English colonies in America take sides in Civil War
5. Évora & S. Portugal revolt (to 1638)	1645 **32. Portuguese colonists in Brazil rebel against**
1639 6. *Nu-pieds* revolt (Normandy)	**Dutch (to 1654)**
1640 **7. Catalan revolt (to 1659)**	1660 33. 'Restoration' in English colonies
8. Portugal rebels (to 1668)	1666 34. Revolt of Laicacota (Peru)
1641 **9. Irish rebellion (to 1653)**	**ASIA AND AFRICA**
10. Andalusia: Medina Sidonia conspiracy	1635 35. Popular revolts spread from northwestern
1642 **11. English 'Great Rebellion' (to 1660)**	China to Yangzi Valley (to 1645)
1647 **12. Naples revolt (to 1648)**	1637 36. Revolt at Shimabara (to 1638)
13. Sicily revolt (to 1648)	1639 37. Chinese (Sangleys) revolt in Manila
1648 14. France: Fronde revolt (to 1653)	1641 **38. Revolt of Portuguese in Mombasa,**
15. Russia: Moscow and other cities rebel (to 1649)	**Mozambique, Goa and Ceylon against Spain**
16. Ukraine revolt against Poland (to 1668)	1643 **39. Li Zicheng declares Shun Era in Xi'an**
17. Istanbul: Ottoman regicide	1644 **40. Li Zicheng takes Beijing and ends Ming rule**
1649 **18. London: British regicide**	**41. Qing capture Beijing and occupy Central Plain**
1650 **19. Dutch regime change (to 1672)**	1645 **42. Qing invade South China; 'Southern Ming'**
1651 20. Bordeaux: Ormée revolt (to 1653)	**resistance (to 1662 in south China; to 1683 in**
21. Istanbul riots	**Taiwan)**
1652 22. Green Banner revolts in Andalusia	1651 43. Yui conspiracy in Edo
1653 23. Swiss Revolution	1652 44. Colombo rebels against Portugal
1656 24. Istanbul riots	1653 45. Goa rebels against Portugal
1660 **25. The 'Danish Revolution'**	1657 46. Anatolia: revolt of Abaza Hasan Pasha (to 1659)
26. 'Restoration' in England, Scotland, and Ireland	1658 **47. Mughal Civil War (to 1662)**
1662 27. Moscow rebellion	1665 **48. Overthrow of Kongo kingdom**
Events listed in **bold** are those that produced a regime change.	49. Shabbatai Zvi proclaimed Messiah at Izmir

1. The Global Crisis.
Although Europe and East Asia formed the heartland of the General Crisis, the Mughal and Ottoman empires, like the European colonies in America, also experienced episodes of severe political disruption in the mid-seventeenth century.

Other records indicate that 1641 saw the third coldest summer recorded over the past six centuries in the northern hemisphere; the second coldest winter in a century experienced in New England; and the coldest winter ever recorded in Scandinavia. The summer of 1642 was the twenty-eighth coldest, and that of 1643 the tenth coldest, recorded in the northern hemisphere over the past six centuries. The Chinese drought of 1627–43 was 'very likely the most persistent severe one in eastern China since AD 500', while the winter of 1649–50 seems to have been the coldest on record in both northern and eastern China. Abnormal climatic conditions in both northern and southern hemispheres lasted until the early eighteenth century – the longest as well as the most severe episode of global cooling recorded in the entire Holocene Era, and the only one known to have affected the entire globe. Climatologists have dubbed this period 'the Little Ice Age'.[6]

This volume seeks to link the Little Ice Age discerned by climatologists with the General Crisis studied by historians – but without painting bull's eyes around bullet holes: without arguing that global cooling must have somehow caused recession and revolution around the world simply because climate change is the only plausible common denominator. Le Roy Ladurie was absolutely correct to insist in 1967 that 'The historian of seventeenth-century climate' must 'be able to apply a quantitative method comparable in rigour if not in accuracy and variety to the methods used by present-day meteorologists in the study of twentieth-century climate', and he regretted that this goal was then unattainable.[7] Happily, written accounts of climatic conditions in Asia, Africa, Europe, and the Americas during the seventeenth century, coupled with millions of measurements of tree-rings, glaciers, ice cores, pollen deposits and stalactite formations, now allow historians to relate variations in the climate to political, economic and social changes with unprecedented precision.

Nevertheless, this data, though abundant and striking, must not turn us into climatic determinists. As early as 1627, Joseph Mede, a polymath with a special interest in astronomy and eschatology who taught at Christ's College, Cambridge, noted a major methodological pitfall: an increase in observed climatic anomalies may simply reflect an increase in the number of observers. When he heard almost simultaneously about an earthquake near Glastonbury and 'another prodigie from Boston [Lincolnshire] of fire from heaven', Mede observed sagely: 'Either we have more strange accidents than was wont, or we take more notice of them, or both.' Subsequent research has corroborated Mede's surmise. Modern astronomy has confirmed that the seventeenth century witnessed an unusual frequency of 'fire from heaven', in the form of comets, but humans also took 'more notice of them' – both because the proliferation of telescopes enabled more of them to be observed, and because dramatic improvements in disseminating news meant that every sighting soon became widely known.[8]

Another obstacle to the accurate assessment of climatic data by historians is the role of contingency. The existence of a well-stocked granary or proximity to a seaport may mitigate the local consequences of colder or wetter weather, whereas war may create a local famine despite a bountiful harvest by destroying or disrupting the food supply. In the aphorism of Andrew Appleby, 'the crucial variable' was often 'not the weather but the ability to adapt to the weather'.[9] This volume therefore examines not only the impact of climate change and extreme weather events on human societies during the seventeenth century, but also the adaptive strategies that evolved to survive the worst climate-induced catastrophe of the last eight millennia.

Introduction:
The Little Ice Age and the General Crisis

In 1638, from the safety of his Oxford College, Robert Burton informed readers of his best-selling book, *The anatomy of melancholy*, that 'every day' he received news of

> war, plagues, fires, inundations, thefts, murders, massacres, meteors, comets, spectrums, prodigies, apparitions; of towns taken, cities besieged in France, Germany, Turky, Persia, Poland, &c.; daily musters and prepa- rations, and such like, which these tempestuous times affoord; battels fought, so many men slain, monomachies [regicides], shipwracks and sea-fights, peace, leagues, stratagems, and fresh alarums.

Six years later, a group of London merchants lamented that 'All trade and commerce in this kingdom is almost fallen to the ground through our own unhappy divisions at home', namely the English Civil War, 'and as the badness of trade and scarcity of money are here, so is all Europe in little better condition, but in a turmoil [of] either foreign or domestic war.' In 1643 the preacher Jeremiah Whitaker warned that '[These] days are days of shaking, and this shaking is universal: the Palatinate, Bohemia, Germany, Catalonia, Portugal, Ireland, England.' Normally, Whitaker argued, God 'shakes all successively', but now it seemed that He planned to 'shake all nations collectively, jointly and universally'. So much simultaneous 'shaking', he concluded, must herald the Day of Judgment.[1]

That same year, in Spain, a tract entitled *Nicandro* [*The victor*] made the same point.

> Sometimes Providence condemns the world with universal and evident calamities, whose causes we cannot know. This seems to be one of the epochs in which every nation is turned upside down, leading some great minds to suspect that we are approaching the end of the world. We have seen all the north in commotion and rebellion, its rivers running with blood, its populous provinces deserted; England, Ireland and Scotland aflame with Civil War.

'What area does not suffer', the *Nicandro* asked rhetorically: 'if not from war, then from earthquakes, plague and famine?'[2]

In Germany, a Swedish diplomat expressed alarm in 1648 at a new bout 'of revolts by the people against their rulers everywhere in the world, for example in France, England, Germany, Poland, Muscovy, the Ottoman empire'. He was well informed: civil war had just begun in France and continued to rage in England; the Thirty Years' War (1618–48) had left much of Germany devastated and depopulated; the Cossacks of Ukraine had just rebelled against their Polish overlords and massacred thousands of Jews; revolts rocked Moscow and other Russian cities; and an uprising in Istanbul led to the murder of the Ottoman sultan. The following year, in Paris, a Scottish exile concluded that he and his contemporaries lived in an 'Iron Age' that would become 'famous for the great and strange revolutions that have happened in it'. In 1653, in Brussels, historian Jean-Nicolas de Parival informed his readers that 'I call this century the "Iron Century"' because so many misfortunes 'have come together, whereas in previous centuries they came one by one'. Rebellions and wars, he continued, now 'resemble Hydra: the more you cut off their heads, the more they grow'. Parival also noted that 'The Elements, servants of an irate God, combine to snuff out the rest of humankind. Mountains spew out fire, the earth shakes, plague contaminates the air', and 'the continuous rain causes rivers to flood'.[3]

Seventeenth-century China also suffered. A combination of droughts and disastrous harvests, rising tax demands and drastic reductions in government programs unleashed a wave of banditry and chaos until in 1644 one of the bandit leaders, Li Zicheng, declared himself ruler of China and seized Beijing from the demoralized defenders of the Ming emperor (who committed suicide). Almost immediately, China's northern neighbours, the Manchus or Qing, invaded and defeated Li, entered Beijing and for the next thirty years ruthlessly extended their authority over the whole country. Several million people perished in the Ming–Qing transition; millions more lost their property and their freedom.

Few areas of the world escaped the consequences of global cooling. North America and West Africa both experienced famines and wars in the mid-seventeenth century. In Japan, following several poor harvests, in 1637–38 the largest rural rebellion in modern Japanese history occurred on the island of Kyushu. Revolution broke out in France, Sicily, Naples, Russia, Ukraine and the Ottoman empire in 1647–48, a year of extreme weather in much of the northern hemisphere. In India, a vicious war of succession in the Mughal empire from 1658 to 1662 intensified the impact of severe drought. In each case natural and human factors combined to create a comprehensive demographic, social, economic and political catastrophe that lasted for two generations.

Those who lived in the mid-seventeenth century believed that they faced unprecedented hardships, leading many of them to record their misfortunes as a warning to others. 'Those who live in times to come will not believe that we who are alive now have suffered such toil, pain and misery', wrote Francesco Voersio, an Italian friar, in his 'Plague Diary'; Nehemiah Wallington, a London craftsman, compiled several volumes of 'Historical notes and meditations' so that 'the generation to come may see what wofull and miserable times we lived in'; and the German Lutheran pastor Johann Daniel Minck started to keep a diary because 'without such records' he feared that 'those who come after us will never believe what miseries we have suffered'.[4]

In 1649, the Welsh historian James Howell speculated that:

God Almighty has a quarrel lately with all Mankind, and given the reins to the ill Spirit to compass the whole Earth; for within these twelve years there have the strangest Revolutions and horridest things happened, not only in Europe but all the world over, that have befallen mankind (I dare boldly say) since Adam fell, in so short a revolution of time ... [Such] monstrous things have happened [that] it seems the whole world is off the hinges; and (which is the more wonderful) all these prodigious passages have fallen out in less than the compass of twelve years.[5]

Two years later, in his book *Leviathan*, the English philosopher Thomas Hobbes provided perhaps the most celebrated description of the consequences of the fatal synergy between natural and human disasters that threatened him and his contemporaries:

There is no place for industry, because the fruit thereof is uncertain, and consequently no culture of the Earth; no navigation, nor use of the commodities that may be imported by sea; no commodious building; no instruments of moving and removing such things as require much force; no knowledge of the face of the Earth; no account of Time; no arts; no letters; no society. And, which is worst of all, continual fear and danger of violent death; and the life of man, solitary, poor, nasty, brutish and short.[6]

When did this fatal synergy commence? In 1652, the Italian historian Majolino Bisaccione traced the sequence of 'popular revolts in my lifetime' back to the rebellion of Bohemia in 1618, which attracted the support of some German Protestants, led by Frederick of the Palatinate, and thus began a civil war in Central Europe. A few years later, the English antiquarian John Rushworth agreed. When he sought to explain 'how we came to fall out among ourselves' in the English Civil War, he too started his account in 1618 because his research convinced him that the conflict originated in 'the causes and grounds of the war in the Palatinate, and how far the same concerned England, and the oppressed Protestants in Germany'. He therefore 'resolved that very instant should be the *Ne Plus Ultra* of my retrospect'.[7]

The human and natural archives both vindicate this chronology. Although Europe had experienced many earlier economic, social and political crises, they remained largely isolated and relatively short-lived. By contrast, the Bohemian revolt began a conflict that lasted three decades and eventually involved all the major states of Europe: Denmark, the Dutch Republic, France, Poland, Russia, Sweden, the Swiss Confederation and the Stuart and Spanish monarchies. The year 1618 also saw a long-running crisis commence in two other parts of the world. In the Ottoman empire, a palace faction toppled the sultan (the first such event in the history of the dynasty), unleashing a series of disasters that a generation later the scholar-bureaucrat Kâtib Çelebi would term *Haile-i Osmaniye*: 'The Ottoman Tragedy'. Meanwhile, in East Asia, Nurhaci, leader of a confederation of Manchurian tribes, declared war on the Chinese emperor and invaded Liaodong, a populous area of Chinese settlement north of the Great Wall. Some observers immediately realized the significance of this step. Years later, Wu Yingji, a gentleman-scholar, recalled 'a friend telling me, when the difficulties began in the eighth month of 1618 in Liaodong, that the state would have several decades of warfare; and my thinking that his words were absurd because the state was quite intact'. Nevertheless his friend had been right: the Manchu invasion initiated almost seven decades of warfare.[8] The natural archive reveals a serious drought in many parts of sub-Saharan Africa from 1614 to 1619; heavy snowfalls in subtropical Fujian in 1618; an intensely cold winter in 1620–21 in Europe; and drought in both the Valley of Mexico and Virginia from 1616 to 1621. In addition, 1618 marked the beginning of a prolonged aberration in the behaviour of the sun, with the gradual disappearance of sunspots, auroras borealis, and coronas. For all these reasons, I follow the lead of Bisaccione, Rushworth, Kâtib Çelebi and Wu Yingji's friend: 1618 is 'the *ne plus ultra* of my retrospect', the point where a fatal synergy between natural and human disasters began.

When did it end? Here the evidence is less consistent. In 1668 Hobbes began *Behemoth*, his account of the English Civil Wars, by observing that:

> If in time, as in place, there were degrees of high and low, I verily believe that the highest of times would be that which passed between the years 1640 and 1660. For he that thence, as from the Devil's Mountain, should have looked upon the world and observed the actions of men, especially in England, might have had a prospect of all kinds of injustice and all kinds of folly that the world could afford.[9]

Yet twenty years later another revolution occurred: William of Orange landed at the head of the largest army ever to invade first Britain and then Ireland, in both of which he created a new regime. On the European continent, the last of the disputes unleashed by the revolt of Bohemia were resolved in 1659–61, but France's invasion of the Palatinate in 1688 generated a new decade of conflict. In the Ottoman empire, the grand vizier Köprülü Mehmed Pasha managed to end the cycle of domestic rebellion in the 1650s and the empire began to expand again; but the defeat of the Turkish army before Vienna in 1683 halted the advance and produced protests and the deposition of another sultan. Elsewhere, however, several conflicts ended. In 1683, Manchu troops finally defeated the last of their opponents, allowing a government inspector to exult that the Qing emperor 'has crushed all the rebels and even the seas are calm. At present, the people have returned to their former lands. Their homes are protected and their livelihood is secure. They will respect and honour Your Majesty's benevolence for generations to come', bringing China's seventeenth-century crisis to an end at last.[10] Three years later, the Eternal Peace of Moscow marked Russia's permanent ascendancy over the Polish-Lithuanian Commonwealth.

Nevertheless, the Little Ice Age continued. In the northern hemisphere, nine of the fourteen summers between 1666 and 1679 were either cool or exceptionally cool, and harvests in Western Europe ripened later in 1675 than in any other year between 1484 and 1879. Climatologists regard the 1690s, with average temperatures 1.5°C below those of today, as the climax of the Little Ice Age – and yet, this time, global cooling did not produce a wave of revolutions. The fatal synergy had ended. This book seeks to explain why.

Writing global history is full of pitfalls. In 2011 Alain Hugon prefaced his admirable study of the revolt of Naples in 1647–48 with a lament that although 'contemporaries clearly stated that no barriers separated the various revolutions of the seventeenth century', they were so numerous that 'we historians of the twentieth and twenty-first centuries dare not study them in their totality, despite our awareness of this synchronicity, of the interdependence, and of the interactions that occurred'. Whenever Hugon 'tried to make historical comparisons appropriate to the mid-seventeenth century', he found that 'the problems that arise from the need to contextualize each historical event render the attempt vain'.[11]

At first sight, this view is very persuasive: we now know about far more upheavals, many of far greater complexity, than previous scholars suspected. Hugon himself uncovered evidence of over one hundred revolts in the kingdom of Naples in 1647–48, and other researchers have revealed that almost half the towns of Portugal rebelled in 1637; over twenty Andalusian towns and cities participated in the Green Banner revolts of 1648–52; and over a million Chinese joined the bandit armies that eventually destroyed

Ming rule. Moreover, contingency complicates the task of explaining the synchronicity, interdependence and interactions of the various revolts. Although most parts of the northern hemisphere experienced both the Little Ice Age and the General Crisis in the mid-seventeenth century, they did so in different ways, for different reasons, and with different outcomes – in part because some structural causes (such as climate change) lay largely beyond human control, while others (such as wars and revolutions) involved so many people that they lay largely beyond the control of any individual. Apparently minor events repeatedly produced consequences that were both unanticipated and disproportionate. As Dr Samuel Johnson observed in 1771:

> It seems to be almost the universal error of historians to suppose it politically, as it is physically true, that every effort has a proportionate cause. In the inanimate action of matter upon matter, the motion produced can be but equal to the force of the moving power; but the operations of life, whether private or publick, admit no such laws. The caprices of voluntary agents laugh at calculation. It is not always that there is a strong reason for a great event.[12]

Historians must identify the precise moment in each historical process when 'the motion produced' ceased to be 'equal to the force of the moving power'. To describe such moments John Lewis Gaddis adopted from physics the term 'phase transitions': the point 'where water begins to boil or freeze, for example, or sand piles begin to slide, or fault lines begin to fracture'. Malcolm Gladwell popularized the term 'tipping point' in his best-selling book of the same name – a more appropriate concept because it implies that such changes, however sudden and dramatic, may one day be reversed. Ice, after all, can easily turn back to water. The subtitle of Gladwell's book, *How little things can make a big difference*, is also appropriate, because the harsher environment of the seventeenth century heightened the role of chance and contingency in human affairs. When historically important outcomes depend on the actions of relatively few people amid fraught conditions, little things can easily 'make a big difference'.[13]

This book studies the Global Crisis of the seventeenth century through three different lenses. Part I presents data from both the human and natural archives that reveal the mechanisms through which the crisis affected humankind. Chapter 1 evaluates how global cooling affects the supply of food, especially of staple crops such as cereals, maize and rice. Chapter 2 examines how early modern states pursued policies that intensified economic hardship and enforced unpopular measures that destabilized societies already under economic stress from climate change, and (more rarely) adopted initiatives that mitigated the consequences of global cooling.

The most intense manifestations of crisis occurred mainly in four zones: composite states, marginal lands, cities and macroregions (densely populated areas that concentrated on producing goods for export rather than for local consumption). Composite states, normally created by dynastic unions, were especially vulnerable because the ruler's authority was normally weaker in peripheral areas, yet in wartime those peripheral areas often experienced enhanced political and economic pressure, and rebelled first (Chapter 2). Marginal lands relied disproportionately on the yield of crops vulnerable to climate change. Cities and macroregions regularly suffered fiscal and military calamities because governments and armies both tended to target places that boasted a large, compact population; they also proved vulnerable to political and military changes both at home and in areas critical for their imports or exports (Chapter 3). Chapter 4 examines the demographic responses by the victims in different regions as they faced a growing imbalance between supply of and demand for resources.

Part II examines a dozen states in Eurasia that experienced the full intensity of both the Little Ice Age and the General Crisis, proceeding geographically from east to west: China; Russia and Poland; the Ottoman empire; Germany and Scandinavia; the Dutch Republic; the Iberian Peninsula; France; Great Britain and Ireland (Chapters 5–12). Each chapter charts the interplay of human and natural forces right up to the tipping point that ended the existing social, economic and political equilibrium; it then analyses the nature of the ensuing crisis; and it documents the emergence of a new equilibrium.

The choice of an east-to-west itinerary, starting with China and ending with Ireland, is arbitrary: it reflects neither chronological differences (in most cases the 'days of shaking' began in or around 1618 and ended in the 1680s) nor the intensity of the crisis (although in terms of physical and personal damage, China and Ireland seem to have suffered worst of all). By contrast, the decision to devote greater space to the experience of Britain and Ireland than to other states gripped by major trauma is deliberate: the upheavals there lasted longer and produced more dramatic changes than anywhere else except China, and the wealth of surviving sources that they generated permit a more detailed understanding of the causes, course and consequences of the crisis than is possible for any other society.

Part III considers two categories of exception to this pattern: areas where at least part of the population emerged from the seventeenth-century trauma relatively unscathed (some European colonies in America; Spanish Italy; South and Southeast Asia; Japan); and regions where the impact of the Little Ice Age remains ambiguous (the Great Plains of North America; sub-Saharan Africa; Australia). Within the first category, the abundant resources of Mughal India and some of its neighbours enabled the state to ride out the crisis (Chapter 13), and in Spanish Italy the government managed to overcome major rebellions by making major concessions (Chapter 14). Elsewhere, notably in Europe's overseas outposts, the prosperity of a few (the European colonists) was achieved at the expense of the many (the indigenous population: Chapter 15). Only Japan seems to have avoided the full effects of the crisis, thanks to human initiatives: although global cooling caused a major famine in the archipelago during the 1640s, a barrage of effective countermeasures first limited and then repaired the damage (Chapter 16).

Despite the diversity of human experience in the states and societies afflicted by the General Crisis, some common denominators emerge, and Part IV considers three of them. First, popular responses to catastrophe exhibited several similar protocols and conventions, ranging from a surprising measure of restraint in violent protests around the world to striking similarities in what James C. Scott termed 'the weapons of the weak': 'foot-dragging, dissimulation, desertion, false compliance, pilfering, feigned ignorance, slander, arson, sabotage' (Chapter 17).[14] Second, an investigation of the individuals and groups in different societies who exploited the mounting instability to produce a tipping point also exposes similarities. In many areas aristocrats played a prominent role, as they had done in many earlier crises; but in the mid-seventeenth century, from China through the Muslim world to Europe, the most prominent 'troublemakers' included men (some clerical, others secular) who had made great sacrifices to acquire advanced education but then failed to find suitable employment (Chapter 18). A third common denominator is the rapid dissemination of radical ideas. Sometimes the rebels themselves spread the word. In 1647 the ease of passage of insurgents between Naples and Palermo synchronized the rebellions in the two capitals; the following year, the return from Moscow of men who had watched spellbound as rioters forced the tsar to make massive concessions triggered copycat rebellions in the provinces. Radical ideas also spread because in much of Asia and Europe the proliferation of printed works and of schools had created a literate proletariat of unprecedented size and sophistication, capable of reading, discussing and implementing new ideas. Thus although the Catholics of Ireland hated and feared the Calvinists of Scotland, they were prepared to learn from them. A few days after the 1641 uprising, when a captured Protestant asked a leading Irish Catholic if '"Yow have made a Covenant amongst yow as the Scotts did?" "Yea", said hee, "The Scotts have taught us our A.B.C."' (Chapter 19).[15]

Part V examines how survivors coped with the aftermath of the crisis. Although the 1690s and 1700s saw further bouts of extreme weather, famine and (in Europe and China) almost continuous war, no revolutions and relatively few revolts occurred. Although the Little Ice Age continued, the General Crisis did not.

Several developments help to explain this paradox. In most parts of the globe, the experience of state breakdown and the 'continual fear and danger of violent death' cooled the general ardour for economic, political and religious change. Most subjects now favoured political stability, economic innovation and religious toleration; many governments switched at least some resources from warfare to welfare and fostered economic regeneration (Chapters 20 and 21). The final chapter examines a variety of intellectual responses devised to cope more effectively with future crises, some (like compulsory universal schooling) imposed by the state, others emerging among subjects (including 'practical knowledge' in China and Japan, the 'new reason' in Mughal India, and the 'scientific revolution' in Europe). These innovations put down deeper roots in the West than elsewhere and formed a crucial ingredient in the 'Great Divergence' that developed between East Asia and Northwest Europe.

The Conclusion considers some implications of the fact that catastrophe is not an aberration but an integral part of human history, and the Epilogue suggests that the current debate on global warming confuses two distinct issues: whether human activity is making the world warmer; and whether or not sudden climate change can occur. Although a few may still deny the first, the seventeenth-century evidence places the second beyond doubt. Since the critical issue is not *whether* climate change occurs, but *when*, the critical question is whether states and societies are willing to invest money now to prepare for future climate-induced catastrophes that are inevitable – hurricanes and typhoons of greater force; more intense heatwaves, droughts and floods; rising sea levels – or would rather wait and pay the far higher costs of inaction that are likewise inevitable.

PART I
THE PLACENTA OF THE CRISIS

In his *Essay on the customs and character of nations, and on the principal facts of history from Charlemagne to Louis XIII*, written in the 1740s for his friend, the marquise du Châtelet (an eminent mathematician who found history boring), the French philosopher and author Voltaire presented the European wars and rebellions of the preceding century as part of a global phenomenon – the first person to do so. After describing the murder of an Ottoman sultan in 1648, Voltaire noted:

> This unfortunate time for Ibrahim was unfortunate for all monarchs. The Holy Roman Empire was unsettled by the famous Thirty Years' War. Civil war devastated France and forced the mother of Louis XIV to flee with her children from her capital. In London, Charles I was condemned to death by his own subjects. Philip IV, king of Spain, having lost almost all his possessions in Asia, also lost Portugal.

Voltaire went on to consider the disruptive careers of Oliver Cromwell in England, Li Zicheng in China and others who had seized power by force, concluding that the mid-seventeenth century had been 'a period of usurpations almost from one end of the world to the other'.[1]

Voltaire's *Essay* repeatedly stressed the global dimension of the crisis: 'In the flood of revolutions which we have seen from one end of the universe to the other, a fatal sequence of events seems to have dragged people into them, just as winds move the sand and the waves. The developments in Japan offer another example ...' Fearing that the marquise might still find his 800 pages of such examples boring, Voltaire summarized his argument in a single sentence: 'Three things exercise a constant influence over the minds of men: climate, government and religion.' Taken together, he proclaimed, this triad offered 'the only way to explain the enigma of this world'. On re-reading his *Essay* two decades later, however, Voltaire added a fourth factor that he believed could 'reconcile what was irreconcilable and explain what is inexplicable' in human history: changes in population size.[2]

Voltaire's vision has attracted few imitators. Although legions of historians have studied individual governments and religions in the seventeenth century, until recently few have tracked population trends and even fewer have considered the impact of climate change. Nevertheless, Voltaire was right: around 1618, just as the human population reached unprecedented size, a phase of global cooling began that produced extreme climate events, disastrous harvest failures, and frequent disease epidemics. Human demographic systems cannot adapt swiftly to such adversity, yet instead of seeking ways to mitigate the natural disasters and save lives, most governments around the globe exacerbated the crisis by continuing their existing policies, above all their wars. These various natural and human factors constituted a 'placenta' (to use the metaphor of the Spanish writer Javier Cercas) capable of nourishing a global catastrophe for two generations, until the global population had diminished by one-third.[3]

CHAPTER ONE

The Little Ice Age[1]

'A Strange and Wondrous Succession of Changes in the Weather'

In 1614 Renward Cysat, botanist, archivist and town historian of Luzern in Switzerland, began a new section of his chronicle entitled 'The seasons of the year'. Because 'the past few years have seen such a strange and wondrous succession of changes in the weather', Cysat decided to

> record the same as a service and a favour to future generations because, unfortunately, on account of our sins, for some time now the years have shown themselves to be more rigorous and severe than in the past, and we have seen deterioration amongst living things, not only among mankind and the animal world but also in the Earth's crops and produce.[2]

Cysat was correct: the 'strange and wondrous succession of changes in the weather' that he reported continued for almost a century. In Africa, Angola and the semi-arid belt of savanna south of the Sahara that stretches from the Atlantic Ocean to the Red Sea, and which is known as the Sahel, faced drought from 1614 until 1619. All Europe experienced an unusually cold winter in 1620–21: the Zuiderzee and many rivers froze so hard that for three months they could bear the weight of loaded carts and, most spectacularly, the Bosporus froze over so that people could walk across the ice between Europe and Asia (apparently a unique climatic anomaly). The summer of 1627 was the wettest recorded in Europe during the past 500 years, and 1628 saw a 'year without a summer', with temperatures so low that many crops never ripened. Between 1629 and 1632, much of Europe suffered excessive rains followed by drought, while northern India suffered a 'perfect drought' followed by catastrophic floods. In China, heavy snow fell in 1618 in subtropical Fujian, and four provinces reported a severe winter in 1620. In the Americas, droughts from 1616 to 1621 afflicted the Valley of Mexico and reduced the crops in the Chesapeake basin so severely that the new Virginia colony almost failed.

After a few years of better weather, drought and frost significantly stunted the growth of trees throughout the western United States from 1640 to 1644, while the Canadian Rockies experienced severe drought from 1641 until 1653. The lack of rain in the Valley of Mexico in 1640, 1641 and 1642 led the clergy of Mexico City to organize processions with the Virgen de los Remedios, an image believed to possess special efficacy in bringing rain, to beg God's intervention before everyone starved to death (the first time the image had ever been used in consecutive years). In Massachusetts, Governor John Winthrop noted early in 1642 that

> The frost was so great and continual this winter that all the bay was frozen over, so much and so long, as the like, by the Indians' relation, had not been so these forty years ... To the southward also the frost was as great and the snow as deep, and at Virginia itself the great [Chesapeake] bay was much of it frozen over, and all of their great rivers.

To the north, English settlers on the coast of Maine complained of the 'most intolerable piercing winter' and found it 'incredible to relate the extremity of the weather'.[3]

Abnormal weather also prevailed on the other side of the Pacific. In Japan the first winter snow of 1641 fell on Edo (as Tokyo was then known) on 28 November, almost the earliest date on record (the average date is 5 January), and both that year and the next saw unusually late springs. According to a 1642 pamphlet published in the Philippines, because of the 'great drought' throughout the archipelago 'a great famine is feared'; and two years later, a resident of Manila recorded that once again 'this year there has been much famine among the Indians [Filipinos] because the rice harvest was a poor one on account of the drought'.[4] In northern China, numerous gazetteers reported drought in 1640 and the following year the Grand Canal, which brought food to Beijing, dried up for lack of rain (another unparalleled event). Between 1643 and 1671 Java fell prey to the longest drought recorded during the past four centuries.

The Mediterranean lands also experienced extreme weather. Catalonia suffered a drought in spring 1640 so intense that the authorities declared a special holiday to enable the entire population to make a pilgrimage to a local shrine to pray for water – one of only four such events in the past five centuries. In 1641 the Nile fell to the lowest level ever recorded and the narrow growth rings laid down by trees in Anatolia reveal a disastrous drought. By contrast, in Macedonia the autumn saw 'so much rain and snow that many workers died through the great cold'; in Spain, the Guadalquivir broke its banks and flooded Seville in 1642; and throughout Andalusia the years 1640–43 remain the wettest on record.[5]

Men and women in England noted 'the extraordinary distemperature of the season in August 1640, when the land seemed to be threatened with the extraordinary violence of the winds and unaccustomed abundance of wet', and in Ireland frost and snow in October 1641 began what contemporaries considered 'a more bitter winter than was of some years before or since seen in Ireland'.[6] Hungary experienced uncommonly wet and cold weather from 1638 to 1641; summer frosts repeatedly devastated crops in Bohemia; and 1641 remains the coldest year ever recorded in Scandinavia. In the Alps, unusually narrow tree rings reflect poor growing seasons throughout the 1640s, and farmsteads and even whole villages disappeared as glaciers reached their maximum extent in historical times, a mile beyond their current positions. In eastern France, each grape harvest from 1640 to 1643 began a full month later than usual and grain prices surged, indicating poor cereal harvests. In the Low Countries, all along the river Maas (or Meuse), floods caused by snowmelt early in 1643 created 'the greatest desolation that one could imagine: the houses all broken open and overturned, and people and animals dead in the hedgerows. Even the branches of the highest trees contained a number of cows, sheep and chicken'. Perhaps most striking of all, a soldier serving in central Germany recorded in his diary in August 1640 that 'at this time there was such a great cold that we almost froze to death in our quarters and, on the road, three people did freeze to death: a cavalryman, a woman and a boy'.[7]

Data from the southern hemisphere reveal similar climatic aberrations. In Potosí (modern Bolivia), the second largest city in the Americas, 'following a long drought came heavy rains' in 1626 that buckled the walls of the great Caricari reservoir, built to supply water to the *ingenios* (engines) used to refine silver. Eyewitnesses described how in a matter of minutes a 'mountain of water' washed away more than thirty engines, destroyed some 2,000 houses and drowned perhaps 4,000 people. In Chile, lack of precipitation from 1615 to 1637 led an inquisitor to apologize to his superiors that he could not send them any proceeds from fines and confiscations because 'we have not collected a penny on account of the drought'.[8] Tree-rings show significantly cooler weather throughout Patagonia in the 1640s, and the glaciers in Bolivia and Peru reached their maximum extent between 1630 and 1680. In Africa, Angolan records show a unique concentration of droughts, locust infestations and epidemics in the second quarter of the seventeenth century.

The 1640s ended with another bout of extreme weather throughout the northern hemisphere. On the Isle of Wight in southern England, a local landowner lamented that 'from Mayday till the 15th of September [1648] we had scarce three dry days together', and when King Charles I (imprisoned nearby) asked him 'whether that weather was usual in our island? I told him that in this forty years I never knew the like before'. In Scotland, 'The long great rains for many weeks did prognosticate famine', and Ireland experienced 'so great a dearth of corn ... not seen in our memory'.[9] The following winter, the barge carrying the corpse of King Charles up the river

Thames to its final resting place after his execution on 30 January 1649 had to contend with ice floes. Other parts of Europe also experienced unusual precipitation – 226 days of rain or snow according to a meticulous set of records from Fulda in Germany (compared with an annual upper limit of 180 days in the twentieth century) – followed by 'a winter that lasted six months'. In France, appalling weather in 1648, 1649 and 1650 delayed the grape harvest into October and drove bread prices to the highest levels in almost a century. In China, the winter of 1649–50 seems to have been the coldest on record. In North America, an epic drought in 1649 created a 'hungry winter' that, according to Kathryn Magee Labelle, 'changed everything' for the Wendat living around Lake Huron: although not destroyed, the Wendat were dispersed for ever.[10]

The 1650s brought no respite. In 1651 the combination of snowmelt and a storm tide caused the worst flooding for eighty years in coastal regions of the Dutch Republic, and snowmelt also caused the Vistula, the Seine and other major rivers to burst their banks. Conversely, the same year saw the longest recorded drought in Languedoc, in southern France: 360 days, or almost an entire year. In the Balkans, in spring 1654 eyewitenesses 'had never before seen such snowstorms and frost, moisture and cold', and even olive 'oil and wine got frozen in the jars', and England experienced an 'unusual drought, which has lain upon us for some years, and still continues and increases upon us'.[11]

The northern hemisphere experienced a landmark winter in 1657–58. Along America's Atlantic coast, deer ran across the frozen Delaware river. In Europe, people rode their horses on the ice across the Danube at Vienna, across the Main at Frankfurt and across the Rhine at Strasbourg; the canal between Haarlem and Leiden remained frozen for two months; and the Baltic froze so hard that the Swedish army with all its artillery marched 20 miles over ice from Jutland to launch a surprise attack on Copenhagen. A diplomat travelling through the Balkans noted in February 1658 that the weather was so cold that even migrating birds turned back, 'causing everyone to wonder'; while John Evelyn judged that he and his compatriots had just lived through 'the severest winter that man alive had known in England: the crow's feet were frozen to their prey; islands of ice enclosed both fish and fowl frozen, and some persons in their boats'. Inevitably, the following spring brought disastrous flooding as the snow and ice melted: the Seine inundated Paris and many other towns, and dikes broke all over the Netherlands. Lieuwe van Aitzema, the official historian of the Dutch Republic, devoted two pages of his chronicle to the extreme climatic events around Europe during 1658, 'a year in which the winter was as harsh and severe at the beginning as at the end'.[12]

Such extreme climatic events often came in unusual concentrations. Of sixty-two recorded floods of the river Seine in and around Paris, eighteen occurred in the seventeenth century. In England (and probably elsewhere in northwest Europe), 'bad weather ruined the harvests of corn and hay for five years from the autumn of 1646 onwards', with five more bad harvests between 1657 and 1661: ten harvest failures in sixteen years. The Aegean and Black Sea regions experienced the worst drought of the last millennium in 1659; in 1675 much of the northern hemisphere experienced another 'year without a summer'; and for six weeks in 1683–84 the Thames at London froze over so hard that enterprising tradesmen built streets and booths on the ice, visited by 'many thousands of people walking sometimes together at once'.[13] In 1686 a military engineer on campaign in what is now Romania complained that 'for three years now, I haven't seen a single drop of rain'. Lakes and rivers dried up, and 'in the swampy soil, cracks were so deep that a standing man could not be seen ... I doubt if there is another example of such a terrible and lasting drought'.[14] In Russia, tree-ring, pollen and peat-bed data show that the springs, autumns and winters between 1650 and 1680 were some of the coldest on record; and, in China, the winters between 1650 and 1680 formed the coldest spell recorded in the Yangzi and Yellow river valleys over the last two millennia. The Turkish traveller Evliya Çelebi encountered snow and hail when he visited Egypt in the 1670s, and complained that 'no one here used to know about wearing furs. There was no winter. But now we have severe winters and we have started wearing fur because of the cold.'[15]

The Search for Scapegoats

So much abnormal weather led some contemporaries to suspect that they lived in an era of climate change. In June and July 1675 (the century's second 'year without a summer'), the Paris socialite Madame de Sévigné complained to her daughter, in Provence, that 'It is horribly cold: we have the fires lit, just like you'; and confided that 'like you, we think the behaviour of the sun and of the seasons has changed'. A generation later the Kangxi emperor, who collected and studied weather reports from all over China, speculated that 'The seasons of Heaven and the physical character [qi] of Earth have undergone some changes' because, in the south,

We remember that before 1671 there would be a new crop of winter wheat by the eighth day of the fourth lunar month [which in 1671 fell on 16 May]. On an earlier tour of Ours to Jiangnan, on the eighteenth day of the third month people were already eating flour from the wheat crop. Now in the middle of the fourth month, the wheat crop has not been harvested ... I have also heard that in the region of Fujian hitherto there was never any snow. Since the time the great armies of this dynasty reached that region [1645], there has been snow.

The emperor also noted with puzzlement that in the north of his empire, 'It used to be that the ice froze to a thickness of eight feet. Now however the weather is milder, not like it was earlier.'[16]

Many contemporaries attributed the extreme weather to divine displeasure. In China, the heavy and prolonged snows in 1641–42 convinced the scholar Qi Biaojia that 'Heaven is extremely angry'; and somewhat later, the Kangxi emperor claimed that 'If our administration is at fault on Earth, Heaven will respond with calamities from above.' A Chinese folk song from the period was blunter:

Old skymaster, you're getting on,
Your ears are deaf, your eyes are gone.
Can't see people, can't hear words.
Glory for those who kill and burn;
For those who fast and read the scriptures,
Starvation.

In 1641, a Jesuit living in the Philippines speculated that the simultaneous eruption of three volcanoes meant that 'Divine Providence wishes to show us something, perhaps to warn us of some approaching catastrophe, which our sins so deserve, or the loss of some territory, because God is angry'. Eight years later a pamphlet published in London, entitled *The way to get rain, by way of question and answer*, included the exchange: '[Question:] Why hath the Earth sometimes too much raine? Answer: Because the Lord is pleased in judgement to send it.'[17]

Such statements reflected the prevailing peccatogenic outlook (from *peccatum*, the Latin word for sin): attributing misfortunes such as military defeats, bad weather and famine to human misconduct, thereby rendering 'intelligible and therefore more bearable a catastrophe in itself unpredictable, while making possible some degree of remedial action'.[18] Examples abound. According to a leading adviser to Philip IV of Spain in 1648: 'The principal cause of the calamities that afflict this kingdom are the public sins and injustices committed', so that punishing the former and 'administering justice with due rectitude and speed are the most important ways to oblige Our Lord to provide the successes that this Monarchy needs so much'.[19] In Germany, in the 1630s, Protestant magistrates in Nuremberg commanded citizens to show moderation in food, drink and fashion and to refrain from sensual pleasure (especially if it involved adultery, sodomy or dancing) in order to avert divine displeasure. Their Catholic neighbour, Maximilian of Bavaria, likewise forbade dancing, gambling, drinking and extramarital sex; limited the duration and cost of wedding festivities; forbade women to wear skirts that revealed their knees; proscribed the joint bathing of men and women; and periodically prohibited carnival and *Fastnacht* celebrations. In the 1640s, the English Parliament 'authorized and required' the magistrates of London 'to pull down and demolish' all theatres, to have all actors publicly whipped and to fine all playgoers, because plays tended 'to the high provocation of God's wrath and displeasure, which lies heavy upon this kingdom'.[20]

The search for scapegoats also fed a witch craze in Europe and its overseas colonies that saw thousands of people tried and executed as sorcerers because their neighbours held them responsible for their misfortunes. As Wolfgang Behringer has noted, not only did the seventeenth century see a notable rise in accusations that blamed witches for unnatural weather, but the peaks of persecution occurred in the coldest decades. Most of the victims were women unable to support themselves unaided, and many lived in marginal areas for crop cultivation (in Lorraine and the Rhineland for vines; in Scotland and Scandinavia for cereals) where the impact of global cooling was felt first and worst. Thus in southern Germany, a hailstorm in May 1626 followed by Arctic temperatures led to the arrest, torture and execution of 900 men and women suspected of producing the calamity through witchcraft. Two decades later, the Scottish Parliament likewise blamed a winter of heavy snow and rain followed by a poor cereal harvest on 'the sin of witchcraft [which] daily increases in this land'; and, to avert more divine displeasure, it authorized more executions for sorcery than at any other time in the country's history. In North America, a witch panic gripped the Wendat during the dearth of 1635–45, and it is no surprise that the Salem witch trials of 1691–92 occurred during a time of unusual cold and scarcity. In China, too, 'To anyone oppressed by tyrannical kinsmen or grasping creditors', a witchcraft accusation 'offered relief. To anyone who feared prosecution, it offered a shield. To anyone who needed quick cash, it offered rewards. To the envious it offered redress; to the bully, power; and to the sadist, pleasure.'[21]

The invocation of sex, stage plays and sorcery to explain catastrophes in the seventeenth century paled when compared with five other suspects: stars, eclipses, earthquakes, comets and sunspots. In Germany, a Swedish diplomat wondered in 1648 whether the spate of contemporaneous rebellions might 'be explained by some general configuration of the stars in the sky'. According to a chronicler in Spain, only 'the malign influence of the stars' could explain the coincidence that 'in a single year [1647–48] in Naples, Sicily, the Papal States, England and France', such 'atrocities and extraordinary events' had occurred. A few years later, Majolino Bisaccione likewise argued that only 'the influence of the stars' could have created so much 'wrath among the people against the governments' of his day.[22]

Others blamed eclipses. In 1640, the author of a Spanish almanac confidently asserted that a recent eclipse of the sun had produced 'great upsets in war, political upheavals and damage to ordinary people'. An English contemporary predicted that the two lunar eclipses and unusual planetary conjunction forecast for 1642 would bring 'sharp tertian fevers, war, famine, pestilence, house-burnings, rapes, depopulations, manslaughters, secret seditions, banishments, imprisonments, violent and unexpected deaths, robberies, thefts and piratical invasions'. In India, even the Mughal emperor Aurangzeb, a devout Muslim, took special precautions during eclipses, staying indoors and eating and drinking little. In Iran, a solar eclipse in 1654 led some 'Persian wise men' to assert that it meant 'that the King had died; others said that there would be a war and blood would be shed.'[23]

Many seventeenth-century people also believed that earthquakes presaged catastrophe. In 1643, a Dutch pamphleteer assured his readers that:

> [The] earthquake not long since felt in the year 1640, was a token of great commotions, and mighty shakings of the kingdomes of the Earth, for a little before and shortly thereupon was concluded the revolt of Cathalonia, the falling-off of Portugal, the stirres in Scotland, the rebellion of the Irish, [and] those civill (uncivill) warres, great alterations, [and] unexpected tumults in England.[24]

Likewise, when severe tremors shook the buildings of Istanbul in 1648, the Ottoman minister and intellectual Kâtib Çelebi solemnly noted that 'when an earthquake happens during daytime in June, blood is shed in the heart of the empire': he was therefore not surprised by the murder of Sultan Ibrahim two months later.[25] Such equations multiplied in the mid-seventeenth century in part because earthquakes became more common. French sources recorded twice as many tremors in the 1650s and 1660s as in any other decade of the century, including one of the most intense ever recorded; and surveys of data from other areas likewise showed a peak of seismic activity around 1650.[26]

Comets also became more frequent. The English astronomer John Bainbridge was apparently the first to suggest that they had become more numerous 'than in many ages before' in one of a multitude of works published in Europe following the appearance of three comets in 1618, predicting dire consequences for humanity, since comets normally brought in their wake 'discord, irritations, deaths, upheavals, robberies, rape, tyranny and the change of kingdoms'.[27] Astronomers in Ming China also interpreted the comets of 1618 as a portent of disasters, and the chronicles of their northern neighbours in Manchuria contain 'an overwhelming number of reports of such heavenly signs'. In Russia, the same comets provoked discussion and doleful interpretations among 'wise men'; in India, a Mughal chronicler claimed that 'no household remained unaffected' by fear, and blamed the comets for both an epidemic of plague and the subsequent rebellion of the crown prince; in Istanbul, writers attributed to their baleful influence not only the extreme weather but also the deposition of one sultan in 1618, the murder of another in 1622, and the provincial revolts that followed.[28] As late as 1683, in Boston, Massachusetts, the Reverend Increase Mather devoted three pages of his *Kometographia, or a discourse concerning comets* to the 'prodigy' of 1618 which, he claimed, had 'caused' not only a major drought throughout Europe, an earthquake in Italy, a plague in Egypt and 'the Bohemian and Germanic war, in which rivers of blood were poured forth', but also 'a plague amongst the Indians here in New England which swept them away in such numbers, as that the living were not enough to bury the dead'.[29]

Some contemporaries blamed the catastrophes that surrounded them on a combination of these natural phenomena. In 1638 Robert Burton assured readers of his *Anatomy of melancholy* that:

> The heavens threaten us with their comets, starres, planets, with their great conjunctions, eclipses, oppositions, quartiles, and such unfriendly aspects. The air with his meteors, thunder and lightning, intemperate heat and cold, mighty windes, tempests, unseasonable weather; from which proceed dearth, famine, plague, and all sorts of epidemicall diseases, consuming infinite myriads of men.[30]

Others, however, doubted such precise links. One Italian historian ridiculed the idea that 'certain celestial constellations have the power to move the spirits of the inhabitants of a country to sedition, tumults and revolutions' in many different places at once; while the comets of 1618 provoked animated debates between astronomers and astrologers over whether or not they were capable of causing 'catastrophes'. Such uncertainty prompted a handful of observers to suggest an alternative natural explanation for the extreme weather of the seventeenth century: fluctuations in the number of sunspots – those dark, cool regions of intense magnetic activity on the solar surface surrounded by flares that make the sun shine with greater intensity. Even though they incorrectly argued that *more* sunspots would produce cooler temperatures on Earth (whereas the reverse is true), unlike comet-watchers and stargazers, the early solar astronomers had identified an important cause of climate change in the seventeenth century.[31]

The development of powerful telescopes after 1609 enabled observers to track the number of sunspots with unprecedented accuracy. They noted a maximum of around 100 spots each year between 1612 and 1614, followed by a minimum with virtually none in 1617 and 1618, and markedly weaker maxima in 1625–26 and 1637–39. And then, although astronomers around the world made over 8,000 observations between 1645 and 1715, the grand total of sunspots observed in those seventy years scarcely reached 100, fewer than appear in even a single year of the twenty-first century. This striking absence, which solar physicists call the Maunder Minimum, suggests a marked reduction in solar energy received on Earth.

Other data confirm this hypothesis. Trees (like other plants) absorb carbon-14 from the atmosphere, and the amount rises as solar energy received on Earth declines; so the increased carbon-14 deposits in many seventeenth-century tree-rings suggests reduced global temperatures. In addition, the *aurora borealis* (the Northern Lights, caused when highly charged electrons from the solar wind interact with elements in the Earth's atmosphere) became so rare that when the astronomer Edmond Halley saw an aurora in 1716 he wrote a learned paper describing the phenomenon – because it was the first he had seen in almost fifty years of observation.[32] Neither Halley nor any other astronomer between the 1640s and the 1700s mentioned the brilliant corona visible nowadays during a total solar eclipse: instead they reported a pale ring of dull light, reddish and narrow, around the moon. All three phenomena confirm that the energy of the sun diminished between the 1640s and the 1710s, a condition normally associated with both reduced surface temperatures and extreme climatic events on Earth.

A further astronomical aberration troubled seventeenth-century observers: the appearance of dust-veils in the sky that made the sun seem either paler or redder than usual. A Seville shopkeeper lamented that during the first six months of 1649, on the few occasions when the sun shone, 'it was pale and yellow, or else much too red, which caused great fear'. Thousands of miles to the east, Korea's royal astronomers reported on several days that 'the skies all around are darkened and grey as if some kind of dust had fallen'.[33] Both the dust-veils and the reddened skies stemmed from an unusual spate of major volcanic eruptions in the mid-seventeenth century, each one hurling sulphur dioxide into the stratosphere, where it deflected some of the sun's radiation back into space and thereby reduced temperatures in the Earth beneath. One day in 1641, a Spanish garrison in the Philippines saw at noon 'a great darkness approach from the south which gradually extended over that entire hemisphere and blocked out the whole horizon. By 1 p.m. they were already in total night and at 2 p.m. they were in such profound darkness that they could not see their own hands before their eyes.' Ash fell on them for twelve hours. They had just witnessed a Force Six eruption, which was heard at exactly the same time 'throughout the Philippines and the Moluccas, and as far as the Asian mainland, in the kingdoms of Cochin-China, Champa and Cambodia – a radius of 900 miles, a wondrous thing which seems to exceed the bounds of the natural world'.[34] The dust veils produced by this and at least eleven other volcanic eruptions around the Pacific from 1638 to 1644, combined with Maunder Minimum, both cooled the Earth's atmosphere and destabilized its climate (Fig. 2).

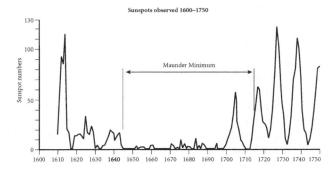

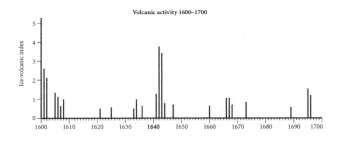

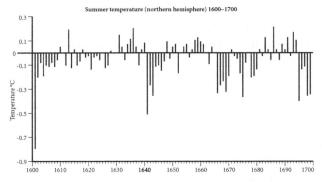

2. Sunspot cycles, volcanic anomalies and summer temperature variations in the seventeenth century.

The number of sunspots observed and recorded by European astronomers (*top*) shows the Maunder Minimum (1643–1715), in which fewer sunspots appeared in seventy years than appear in a single year now. Measurements of volcanic deposits in the polar ice cap (the 'ice-volcanic index') reveal a peak in the 1640s. Both phenomena show a striking correlation with lower summer temperatures in the northern hemisphere.

Blame it on El Niño?

The global cooling caused by reduced sunspot and increased volcanic activity seems to have triggered a change in the phenomenon known as El Niño, which affects much of the world's climate. As the air above the equatorial Pacific warms each spring it creates massive rain clouds: in a normal year, the rain falls on Asia as the monsoon that nurtures the harvest, but in an El Niño year the monsoon weakens and heavy rains fall instead on America, causing floods. Today this reversal – also known as ENSO (El Niño–Southern Oscillation) – happens about once every five years, but in the mid-seventeenth century it happened twice as often: in 1638, 1639, 1641, 1642, 1646, 1648, 1650, 1651, 1652, 1659, 1660 and 1661. The same period saw some of the weakest East Asian monsoons of the past two millennia.

El Niño cannot be blamed for every disaster, because the climates of some regions are El Niño-sensitive whereas others are not (in Africa, the Eastern Cape is susceptible to droughts in El Niño years whereas the Western Cape is not). The global footprint of El Niño normally includes three regions besides the lands adjoining the Pacific: the Caribbean suffers floods; Ethiopia and northwest India experience droughts; and Europe experiences hard winters. In most of the twenty El Niño episodes recorded between 1618 and 1669, and in all twelve from 1638 to 1661, adverse weather afflicted each of these three regions.

The changed climate in the Pacific Ocean during this period emerges starkly from two anomalies recorded in historical sources. The Chinese coastal province of Guangdong suffered more typhoon landfalls between 1660 and 1680 than at any other time in recorded history; and the voyages of galleons sailing from Acapulco to Manila took longer than in any other period. In the first and last decades of the seventeenth century crossing the Pacific took an average of eighty days, but between 1640 and 1670 the average duration rose to over 120 days (and three voyages took over 160 days). Some ships never arrived: of the eleven galleons known to have sunk or run aground before reaching Manila during the seventeenth century, nine did so between 1639 and 1671. The return voyage from Manila to Acapulco also took much longer: the average duration rose from 160 to well over 200 days, and the longest voyages ever recorded (240 days, or eight months) took place in the 1660s. Nothing except a major shift in wind pattern could explain such a dramatic change. Diego de Villatoro, a crown official who had made the round trip twice, saw the connection clearly. In a memorial written in 1676 he noted sadly that 'now we consider a voyage from the Philippines to Acapulco that takes less than seven months to be good', and he perceptively ascribed the longer duration 'to a change in the monsoons'.[35]

Villatoro, of course, lacked the expertise either to blame this change on increased El Niño activity or to associate El Niño with weaker Asian monsoons and increased volcanic activity. But we now know that in normal years, when easterly winds prevail, the Pacific stands some 60 centimetres higher off the Asian than off the American coast, whereas in El Niño years, when westerly winds prevail, those levels reverse. The movement of such a huge volume of water places enormous pressure on the edges of the Earth's tectonic plates around the Pacific periphery, where the most violent and most active volcanoes in the world are located, and this may trigger a spate of eruptions. If this hypothesis is correct, it creates a deadly cycle:

- Reduced solar energy received on Earth lowers global temperatures, which increases the risk of more, and more severe, El Niño events.
- El Niño events may trigger volcanic eruptions around the Pacific that throw sulphur dioxide into the stratosphere, which further reduces the solar energy received on Earth.
- Major volcanic eruptions reduce solar energy received on Earth.

Whatever the exact connections between these natural phenomena (and not all scientists agree), the mid-seventeenth century experienced a marked increase in earthquakes, comets, volcanic eruptions and El Niño episodes, as well as a drastic reduction in sunspot activity, the weakest monsoons and some of the lowest global temperatures recorded in the past few centuries.

Climate and Crops

So what? To a sceptic, global cooling that amounts to a fall of only one or two degrees Celsius in mean summer temperatures may seem insignificant; but that is to think in linear terms. The mean global temperature has shown remarkable stability over the last six millennia: the difference between the Medieval Climate Optimum (the hottest temperatures recorded until the late twentieth century) and the Little Ice Age was probably less than 3°C at the equator. A change of even one degree is thus highly significant. Moreover, in the northern hemisphere, home to the majority of humankind and site of most of the wars and revolutions of the seventeenth century, solar cooling reduces temperatures far more than at the equator, in part because increased snow cover and sea ice reflect more of the sun's rays back into space. The extension of the polar ice caps and glaciers in the mid-seventeenth century would thus have significantly reduced mean temperatures in northerly latitudes.

A recent model of the probable global climate in the later seventeenth century shows significantly colder weather in Siberia, North Africa, North America and northwest India; colder and drier weather in central China and Mongolia; and cooler and less stable conditions in the Iberian Peninsula, France, the British Isles and Germany. As already noted, the states in these same areas – the Russian and Ottoman empires in Eurasia; the Ming and Qing empires in East Asia, and the dominions of Philip IV, Charles I, Louis XIV and Ferdinand III in Europe – reported not only cooler weather in the 1640s and 1650s but also a significant number of extreme weather events. This should cause no surprise: an overall decline in mean temperatures is normally associated with a greater frequency of flash floods, freak storms, prolonged droughts, and abnormal (as well as abnormally long) cold spells. Since each of these anomalies can critically affect the yield of crops that feed the people, each may cause economic, social and political upheavals.

In the temperate zone, which stretches roughly from thirty to fifty degrees of latitude, a cold spell during germination, a drought in the early growing season and a major storm just before harvest affect crop yields disproportionately. In areas of wet rice cultivation, a fall of 0.5°C in the average spring temperature prolongs the risk of the *last* frost by ten days, while a similar fall in the average autumn temperature advances the risk of the *first* frost by the same amount. Either event suffices to kill the entire crop. Even without frosts, a fall of 2°C during the growing season – precisely the scale of global cooling in the 1640s – reduces rice harvest yields by between 30 and 50 per cent, and also lowers the altitude suitable for wet rice cultivation by about 400 metres. Likewise, in cereal-growing regions, a fall of 2°C shortens the growing season by three weeks or more, diminishes crop yields by up to 15 per cent, and lowers the maximum altitude at which crops will ripen by about 150 metres. Drought, too, destroys harvests by depriving crops of the precipitation they required. As a Chinese manual of agriculture, published in 1637, warned: 'All rice plants die if water is lacking for ten [consecutive] days.'[36]

Extreme weather can also destroy crops indirectly. Drought favoured locusts. In 1647 a Moldavian nobleman reported that 'about the time of the year when people pick up their sickles to harvest the wheat', he and some companions were on the road and 'suddenly noticed a cloud towards the south':

> We thought it was a rainstorm until we were suddenly hit by the locust swarm, coming at us like a flying army. The sun disappeared immediately, veiled by the blackness of these insects. Some of them flew high, at three or four metres, while others flew at our level, or even right above the ground . . . They flew around us without fearing anything . . . It took an hour for a swarm to pass, and then after an hour and a half there came another, and then another, and so on. It lasted from noon till dusk. No leaf, no blade of grass, no hay, no crop, nothing remained.

Excessive rain, by contrast, might allow rodents to multiply. In Moldavia in 1670 'myriads of mice' not only ate 'all they found in the vegetable gardens' but also, 'climbing up the trees, ate all the fruit, finishing them up; and to end the job' they consumed all the wheat in the field.[37]

North of the temperate zone, the impact of global cooling increases because the growing season is shorter. In Manchuria, with a total of only 150 frost-free days even in good years, a fall of 2°C in mean summer temperature reduces harvest yields by a stunning 80 per cent. In Finland, where the growing season is the shortest compatible with an adequate harvest even in normal years, a single summer night's frost can kill an entire crop. Seventeenth-century Finland saw eleven crop failures (compared with only one in the eighteenth century).

Global cooling also increases not only the scale but also the frequency of harvest failures:

- In the temperate zone, if early winters or summer droughts occur with a frequency of P=0.1, the harvest will fail once every ten years, and two consecutive harvests will fail once every century. If, however, early winters or summer droughts occur with a frequency of P=0.2, the harvest will fail once every five years (double the risk) while two consecutive harvests will fail every twenty-five years (quadruple the risk).
- In latitudes north of the temperate zone, each fall of 0.5°C in mean summer temperatures decreases the number of days on which crops ripen by 10 per cent, doubles the risk of a single harvest failure and increases the risk of a double failure sixfold.
- For those farming 300 metres or more above sea level, a fall of 0.5°C in mean summer temperatures increases the chance of two consecutive failures a hundredfold.

Climate and Calories

In densely populated parts of the early modern world, most people relied on a single crop, high in bulk and in carbohydrates, known as a 'staple'. Cereals (wheat, rye, barley and oats) formed the principal staple in Europe, northern India and northern China. Rice played the same role in much of Asia, maize in much of the Americas, and millet in upland India and much of sub-Saharan Africa. The economic allure of staple crops is almost irresistible. An acre under cereals feeds between ten and twenty times as many people as an acre devoted to animal husbandry; furthermore, the same amount of money usually bought 10 pounds of bread but only 1 pound of meat. An acre planted with wet rice yields up to 6 tons of food: three times as much as an acre of wheat or maize and sixty times as much as an acre devoted to animal husbandry. Not surprisingly, therefore, according to a Chinese textbook printed in 1637, '70 per cent of the people's staple food is rice'. In Europe, cereals likewise provided up to three-quarters of the total calorie intake of every family (not only in the form of bread but also as a filler for soups and as the basic ingredient for beer and ale).[38]

Steven L. Kaplan has rightly insisted on the tyranny of popular dependence on staple crops in the pre-industrial world. In Europe:

> Cereal-dependence conditioned every phase of social life. Grain was the pilot sector of the economy; beyond its determinant role in agriculture, directly and indirectly grain shaped the development of commerce and industry, regulated employment, and provided a major source of revenue for the state, the church, the nobility, and large segments of the [ordinary population] ... No issue was more urgent, more pervasively felt, and more difficult to resolve than the matter of grain provisioning. The dread of shortage and hunger haunted this society.[39]

Shortage and hunger could arise in three distinct ways. First, since food accounted for up to half the total expenditure of most families, an increase in staple prices soon caused hardship. Second, spending more on food left little or nothing with which to purchase other goods, leading to a fall in demand, which threw many non-agricultural workers out of work and reduced the wages received by the rest. Their income thus fell precisely when their expenditure rose. Third, any shortfall in the harvest reduced the food supply *geometrically* and not arithmetically because the impact of harvest failure on the price of cereals is non-linear. Suppose that:

- In a normal year a European farmer sowed 50 acres with grain and harvested 10 bushels an acre, a total of 500 bushels. Of this, he needed 175 bushels for animal fodder and seed corn and 75 bushels to feed himself and his family – a total of 250 bushels – leaving 250 for the market.
- If bad weather reduced his crop by 30 per cent, the harvest would produce only 350 bushels yet the farmer still needed 250 of them for his immediate use. The share available for the market therefore dropped to 100 bushels – a fall of 60 per cent.
- But if bad weather reduced crops by 50 per cent, the harvest would produce only 250 bushels, all of them needed by the farmer, leaving virtually nothing for the market.

This non-linear correlation explains why a 30 per cent reduction in the grain harvest often *doubled* the price of bread, whereas a 50 per cent reduction *quintupled* it. It also explains why starvation almost always followed when the harvest failed for two or more consecutive years.

Kaplan concluded his study of famines in early modern France by suggesting that this cruel calculus 'produced a chronic sense of insecurity that caused contemporaries to view their world in terms that may strike us as grotesquely or lugubriously overdrawn'. However, a study by Alex de Waal of the Darfur famine of 1984–85 in East Africa rejected the notion of 'overdrawn' because, where harvest shortfalls are concerned, even today failures can 'cross a threshold of awfulness and become an order of magnitude worse'. Not only do large numbers of people die, so does their entire way of life.[40] De Waal identified three characteristics of these landmark famines:

- They force those affected to use up their assets, including investments, stores and goods. Although a family might choose to go hungry for a season in order to preserve its ability to function as a productive unit (for example by keeping back grain to feed its livestock or to use as seed corn), it can rarely maintain that strategy for a second, let alone a third year. Two or three successive harvest failures will leave victims destitute.
- Prolonged starvation also forces those affected to use up their social claims. A hungry family may avoid begging for assistance from other individuals and institutions for a short period, but (once again) it can rarely maintain that strategy for long. If many families suddenly become destitute, the communities where they live may be crippled if not destroyed.
- As communities cease to be viable, some families migrate. Initially, migration may form a reasonable coping strategy in a famine because, although migrants necessarily abandon both their assets and their social claims by leaving their community, those who survive can return to their homes and their previous way of life when conditions improve; but prolonged dearth will sever the links with the world they have left. This will destroy their entire way of life.

Calories and Death

Each day, every human needs to consume around 2,000 calories to maintain her or his basic metabolic rate. Pregnant women and those who earn their living by physical labour require far more: 5,000 calories for those spending eight hours of the day marching or tending their crops; 5,500 calories for those spending eight hours building; 6,500 calories for those harvesting, cutting trees or carrying a heavy load.[41] Few people in the early modern period were so lucky: during the Italian plague of 1630–31, hospital records show that each patient received a daily ration of half a kilo of bread, a quarter of a kilo of meat (probably in a stew), and half a litre of wine – a daily intake of scarcely 1,500 calories (and one seriously deficient in vitamins). Even in normal years during the seventeenth century, the average Frenchman or woman consumed barely 500 calories more than their basic metabolic requirement, and the average Englishman or woman barely 700 calories more.

Nevertheless, as Mirkka Lappalainen pointed out in her study of the Great Finnish Famine of 1695–97, 'Human beings are resilient: it is not easy to starve to death.' We can adjust to a reduced food supply by cutting back on energy demands (working more slowly, resting longer); and, as body weight declines, we can also survive with fewer calories to sustain the basic metabolism (and the reduced physical activity). Lappalainen found that in the 1690s many Finns suffered 'semi-starvation, which people can with-stand for months or longer, although they become increasingly weak and apathetic'.[42] Nevertheless, although a weight loss of 10 per cent reduces energy by about one-sixth, a weight loss of 20 per cent reduces energy by about one-half, and if a woman or man loses 30 per cent of their normal body weight, blood pressure falls and the ability to absorb nutrients fails. In this weakened condition, any additional stress on the body, such as disease, usually proves fatal – and, amid the social disruption normally associated with famine, infectious diseases often spread rapidly – while cold and damp further weaken those who cannot get enough food.

Finnish records studied by Lappalainen revealed that the most striking feature of those suffering from semi-starvation was oedema of face, limbs and abdomen because of protein deficiency: they mentioned people who were swollen and bloated far more often than those who were thin. The commonest cause of death was 'deadly bloody diarrhoea'. A seventeenth-century Chinese official and philanthropist reported much the same:

All beings are physically the same, alike in their intolerance of cold. Those people with old, tattered clothes ... go nearly naked in the dead of winter, their hair dishevelled and feet bare and their teeth chat-tering; crying out and terrified ... Being solitary, they have no place to go ... [and] falling snow covers their bodies. At this point, their organs freeze and their bodies stiffen like pieces of wood. At first they are still able to groan. Gradually they cough up phlegm. Then, their lives are extinguished.[43]

Famines afflicted the young with especial severity. Many infants died because their mothers had no milk to feed them, and even those who survived suffered stunting: the offspring of women who become pregnant during times of dearth are often shorter. Human remains from the Little Ice Age show unmistakable evidence of stunting. When archaeologists exca-vated the skeletons of fifty workers buried in the permafrost at Smeerenburg ('Blubber Town'), a whaling station maintained by the Dutch on the island of Spitsbergen in the Arctic between 1615 and 1670 (when the cold forced them to withdraw), no fewer than forty-three showed evidence of stunting and a corresponding reduction in height. Even more striking, French soldiers born in the second half of the seventeenth century were on average about 3 centimetres shorter than those born after 1700; and those born in famine years were notably shorter than the rest. Stunting reduced the average height of those born between 1666 (when the data first became available) and 1694 to 161 centimetres or less: the lowest ever recorded (Fig. 3).

Malnutrition often impairs the development of major organs as well as long bones, making children more vulnerable to both contagious and chronic diseases, which can further diminish stature. John Komlos, the demographer whose research revealed the reduced height of Louis XIV's soldiers, was surely correct that the seventeenth-century crisis 'had an immense impact on the human organism itself'. His data provide perhaps

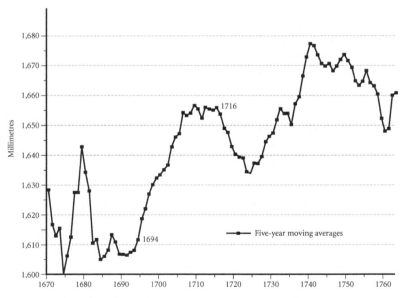

3. Estimated heights of French males born between 1666 and 1770.
John Komlos assembled 38,700 'observations' from the personal records of French males who enlisted in the army born between 1666 and 1770. Even though recruiting officers rejected the shortest volunteers, the 'stunting' effect of global cooling is evident. The average height of Louis XIV's soldiers was 1,617 millimetres, or 5 feet 3 inches.

the clearest – and saddest – evidence of the impact of the Little Ice Age on the human population. The repeated famines not only killed: many of those who survived literally embodied Hobbes's grim assertion that 'the life of man' had become 'solitary, poor, nasty, brutish and short'.[44]

An Overpopulated World?

Although Hobbes and his contemporaries apparently stood somewhat shorter than their grandparents, they were far more numerous. A benign climate for most of the sixteenth century had allowed the human population in most parts of Europe and Asia to increase and in some areas to double, until by 1618 China boasted perhaps 150 million inhabitants, India 116 million and Europe 100 million. In some areas, the number of inhabitants had increased so fast that local resources no longer sufficed to feed them because of another cruel calculus: *population increases geometrically while agricultural output grows only arithmetically*. Just like compound interest, a sustained demographic increase of 1 per cent per year over a century causes a population not merely to double, but to triple; while a 2 per cent increase over a century produces a sevenfold growth. Since crop yields rarely increase at this pace, food shortages can occur very rapidly.

Many people in the early seventeenth century could see that some parts of the world possessed more mouths than could be fed, and they feared the consequences. China's Lower Yangzi Valley, known as Jiangnan, boasted a population of about 20 million by 1618, equivalent to almost 1,200 persons per square mile (by way of comparison, the most densely settled areas of the world today boast 1,000 persons per square mile). According to Alvaro Semedo, a Jesuit long resident in the region, in the 1630s Jiangnan was 'so full of all sorts of people that not only the villages but even the cities can now be seen one from another' and, in some areas, 'settlement is almost continuous'. Indeed, he mused:

> This kingdom is so overpopulated [*eccessivamente popolato*] that after living there for twenty-two years, I remain almost as amazed at the end as I was at the beginning by the multitude of people. Certainly the truth is above any exaggeration: not only in the cities, towns and public places ... but also on the roads there are normally as many people as would turn out in Europe [only] for some holiday or public festival.

Since 'the number of people is infinite', Semedo concluded, 'there can be no capital sufficient for so many, or money enough to fill so many purses'.[45]

Many of Semedo's contemporaries also considered Europe overpopulated. Sir Ferdinando Gorges claimed that England's 'peaceable time affords no means of employment to the multitude of people that daily do increase', and he sent colonists to settle in Maine primarily to reduce population pressure at home. His rivals in the Virginia Company, fearing 'the surcharge of necessitous people, the matter or fuel of dangerous insurrections', likewise sought to remove them from England to their new colony. These and other measures enjoyed such success that by the 1630s thousands crossed the Atlantic each year, which (according to some) promoted England's stability because the colonies 'serve for drains to unload their populous state which else would overflow its own banks by continuance of peace and turn head upon itself, or make a body fit for any rebellion'.[46]

Scarcely had the ink dried on these words than the global population contracted sharply. In China, the victorious Qing believed that in the mid-seventeenth-century crisis 'over half of the population perished. In Sichuan, people lamented that they did not have a single offspring'. In the 1650s, after a decade of sectarian violence and civil war in Ireland, an English eyewitness wrote that 'a man might travel twenty or thirty miles and not see a living creature' except for 'very aged men with women and children' whose skin was 'black like an oven because of the terrible famine'; and a generation later, another English eyewitness estimated that over 500,000 Irish men and women had died 'by the sword and famine and other hardships' in the 1640s. Contemporaries elsewhere made similarly bleak assessments. In southern Germany, one survivor of the Thirty Years' War believed that 'there have been so many deaths that the like of it has never been heard in human history'. A Lutheran minister wrote despondently in 1639 that of his 1,046 communicants a decade earlier, barely one-third remained: 'Just in the last five years, 518 of them have been killed by various misfortunes. I have to weep for them', he continued forlornly, 'because I remain here so impotent and alone'. In France, ravaged by war, famine and disease, Abbess Angélique Arnauld of Port-Royale (just outside Paris) estimated in 1654 that 'a third of the world has died'.[47]

Subsequent research has corroborated each of these striking claims. In China, 'the cultivated area of land decreased by about one-third' during the Ming–Qing transition, while 'the demographic losses were nearly the same'. Sichuan suffered particularly badly, with perhaps a million killed. Ireland's population fell by at least one-fifth during the mid-seventeenth century. In Germany, 'about 40 per cent of the rural population fell victim to the war and epidemics' between 1618 and 1648, and 'in the cities, the losses may be estimated at about 33 per cent'. Many villages in the Île-de-France suffered their worst demographic crisis of the entire Old Regime in 1648–53.[48] These staggering losses were not caused by the Little Ice Age alone, however: it required the misguided policies pursued by religious and political leaders to turn the crisis caused by sudden climate change into catastrophe.

CHAPTER TWO

The General Crisis

'The Century of the Soldiers'

Most of those who lived through the seventeenth-century crisis identified war rather than climate as the principal cause of their misfortunes – and with good reason: more wars took place around the world than in any other era. The historical record reveals only one year entirely without war between the states of Europe in the first half of the century (1610) and only two in the second half (1670 and 1682). Beyond Europe, the Chinese and Mughal empires fought wars continuously for most of the seventeenth century, and the Ottoman empire enjoyed only ten years of peace.

The database of conflicts compiled by Peter Brecke, a sociologist, shows that, on average, wars between states in the seventeenth century lasted longer than in any other period since 1400 (when his survey begins). The 'index of war intensity' proposed by Pitirim Sorokin, another sociologist, rose from 732 in the sixteenth century to 5,193 in the seventeenth – a rate of increase twice or three times greater than in any previous period. Looking only at Europe, Jack S. Levy, a political scientist, considered the sixteenth and seventeenth centuries 'the most warlike in terms of the proportion of years of war under way (95 per cent), the frequency of war (nearly one every three years), and the average yearly duration, extent, and magnitude of war'.[1]

The mid-seventeenth century also witnessed more civil wars than any other period. For six decades, supporters of the Ming and Qing dynasties fought for control of China. The rebellion of large parts of the Stuart and the Spanish monarchies unleashed internal conflicts that lasted over two decades. The states of Germany, with powerful foreign support, fought each other for thirty years. France endured a civil war that lasted five years; the Mughal empire suffered a succession war that lasted two years. Several other countries (including Sweden, Denmark and the Dutch Republic) experienced political upheavals that stopped just short of civil war (see Fig. 4). War, rather than peace, had become the normal state of human society.

Many contemporaries noted the ubiquity of war. The French philanthropist Vincent de Paul (1581–1660) included the word 'war' in his letters 3,200 times, more than twice as often as 'the poor' and four times as often as 'the sick'. The Italian warrior and man of letters Fulvio Testi observed tersely

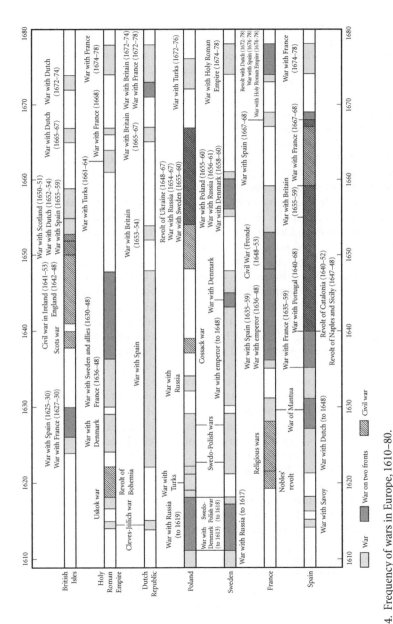

4. Frequency of wars in Europe, 1610–80.

In the six decades between 1618 and 1678, Poland was at peace for only twenty-seven years, the Dutch Republic for only 14, France for only 11, and Spain for only 3. Some states fought wars on several fronts at once. Virtually no European state avoided war during the 1640s.

in 1641, 'This is the century of the soldiers.' Some also noted the destruction wrought by armed conflict. When in the 1690s the Englishman Richard Gough researched the history of his village, Myddle in Shropshire, he found that twenty-one men (one-tenth of the community's adult males) had left to fight in the Civil Wars fifty years earlier, of whom only seven returned; 'and if so many died from Myddle', Gough speculated, 'we may reasonably guess that many thousands died in England in that war'.[2]

Gough was right: the catalogue of 645 'military incidents' fought in England and Wales from 1642 to 1660 compiled by historian Charles Carlton reveals that at least 80,000 men died in action. Nevertheless they formed only part of the destruction caused by war. As Patrick Gordon, a Scottish veteran, noted in the 1650s: 'One can scarse be a souldier without being an oppressor and comitting many crimes and enormityes', because those who failed to take what they needed by violent means 'were sure to be destroyed by vermine, or dy with hunger or cold'. Hans Heberle, a German shoemaker, recorded in his journal how the failure of civilians to appreciate this dynamic could destroy lives and livelihoods. In 1634 the Protestant army of Bernard of Saxe-Weimar approached Heberle's village but, since they, too, were Protestants, the villagers 'did not regard [the prince] as our enemy' and took no precautions. Nevertheless, he wrote bitterly in his journal, Bernard's troops 'plundered us completely of horses, cattle, bread, flour, salt, lard, cloth, linen, clothes and everything we possessed. They treated the inhabitants badly, shooting, stabbing and beating a number of people to death.' As they left, 'they set the village alight and burnt down five houses'. Heberle and his surviving neighbours had learned an important lesson: in the mid-seventeenth century, every soldier was an enemy. Henceforth, 'we were hunted like wild beasts in the forest', and whenever troops approached, he and his family fled with whatever they could carry to Ulm, the nearest fortified city. Their sufferings did not end there, because Ulm lacked the resources to sustain thousands of refugees: water sometimes became so scarce that 'almost everybody drank their own urine, or the urine of their children.'[3]

According to Hobbes, war was like the climate: just 'as the nature of foule weather lyeth not in a showre or two of rain, but in an inclination thereto of many days together: so the nature of warre consisteth not in actuall fighting, but in the known disposition thereto during all the time there is no assurance to the contrary.' This produced widespread insecurity. When Charles II returned to England from exile in 1660, he observed that the Civil War 'had filled the hearts of the people with a terrible apprehension of insecurity ... which keeps the hearts of men awake'. Looking back in 1683 Christina, the former queen of Sweden, wrote that 'In the present century, the whole world is at arms. We threaten each other, we fear each other. Nobody does what they want, or what they could do. No one knows who has lost or who has won, but we know well enough that the whole world lives in fear.'[4]

War became so common in China that a special word emerged to describe military atrocities: *binghuo*, 'soldier calamity' – and according to the eminent sinologist Lynn Struve, 'no locale in China escaped some sort of "soldier calamity"' during the Ming–Qing transition. Some of the worst atrocities occurred when soldiers took a town by storm. The scholar Wang Xiuchu compiled a record of the six-day sack of Yangzhou, a city north of the Yangzi river, by Qing forces in 1645. Although Wang himself escaped by hiding, he noted sadly that whereas before the sack his family numbered eight, 'now there are only three'. Other eyewitnesses claimed that 80,000 inhabitants perished in the sack, and so much lay in ruins that poets began to refer to Yangzhou as 'the weed-covered city'. Its fate in 1645 is still familiar to every Chinese schoolchild.[5] War presented additional dangers for women: the high risk of being raped or abducted. Some suffered through a deliberate effort by the victors to dishonour a community, to prove it could not protect its women. During the sack of Magdeburg, Germany, in 1631, a group of monks watched six Catholic soldiers gang-rape a twelve-year-old girl until she died in the courtyard of their convent. Otto von Guericke, an eyewitness who survived to become a celebrated inventor, reported that 'things went very badly for many of those women, girls, daughters and maids who either had no men, parents or relatives who could pay a ransom on their behalf, or could not appeal to high officers for help or advice. Some were defiled and disgraced, and some were kept as concubines.'[6]

The victims of such violence suffered psychological as well as physical damage. In Germany, soldiers raped Anna Hurter of Hawangen in 1633, and when she died in 1657 her parish priest noted in the burial register that 'for twenty-four years she had not had one sane hour until suddenly she expired'. In Ireland in 1641, as a Catholic officer raped a young servant called Mary:

> To prevent her crying out, one of his souldjers thrust a napkin into her mowth and held her fast by the haire of her head till the wicked act was performed. And the said Mary then complained and sayd that shee had layn sick upon a [bed] for 3 or 4 dayes and was in such a condition that she thought shee should never bee well nor in her right mynd againe, the fact was soe fowle and grievous unto her.[7]

Although wars normally produce personal tragedies like these, the proliferation of conflict in the seventeenth century multiplied them.

'Feeding Mars'

The expense of maintaining armies and navies to fight wars, whether against neighbours or rebels, constantly increased – and not just because of their duration. Europe's Atlantic states built huge fleets of floating fortresses, each one costing £33,000 to build and £13,000 to keep at sea for a campaign. Europe's naval wars during the second half of the century saw these ships deployed in lines that stretched for miles, with thousands of heavy guns firing broadsides at each other, sometimes for several days – an enormous financial outlay. Mediterranean warfare, by contrast, was labour-intensive – each galley carried about 400 men, between oarsmen and soldiers, so that (in the words of a French admiral) 'an infinite number of villages are far from having as large a number of inhabitants' as a single galley – but whether they produced sailing ships or galleys, naval arsenals and dockyards constituted the largest industrial plants in early modern Europe.[8]

Land warfare normally cost even more than naval operations. Apart from raising and maintaining soldiers, most states invested heavily in fortifications. The two largest states in the world, China and Russia, constructed continuous defences along vulnerable frontiers. The Ming emperors extensively rebuilt the Great Wall in stone to take account of gunpowder technology, and although the Manchus breached it in both 1629 and 1642, it still reduced small-scale raiding. The Romanov tsars constructed 'lines' consisting of fortified towns linked by earthen ramparts which by 1658 ran for 800 miles along the edge of the steppe from the Dnieper to the Volga. As in China, although the fortified lines did not bring complete security, they compelled invaders from the south – whether Crimean Tartars or Cossack rebels – to follow certain itineraries where Russian troops could more easily intercept them.

Other European states invested in a network of artillery fortresses, each one a star-shaped complex with extremely thick walls protected by angled bastions, moats and outposts. When in a good state of repair, and defended by sufficient guns and garrison, such positions could seldom be taken by assault, and so artillery fortresses proliferated in many contested areas: the north Italian plain; the France–Spain border; the lands around the Baltic; and the Netherlands, which contained the largest concentration of artillery fortresses in the world. When the Dutch Revolt against Spain began in 1572, twelve towns already possessed a complete set of new-style fortifications and eighteen more had partially updated defences; when the revolt ended in 1648, the same area boasted fifty artillery fortresses and a further sixty towns with partially modernized walls.

Laying siege to these state-of-the-art fortifications formed the largest engineering enterprise of the age – trenches might stretch for 25 miles and operations could last for months – and their outcome dominated most campaigns. The Russo-Polish wars of 1632–34 and 1654–55 hinged on control of the bastions around Smolensk; the turning point in the Habsburg-Turkish wars came with the failed Ottoman siege of Vienna in 1683. 'Battles do not now decide national quarrels, and expose countries to the pillage of conquerors, as formerly', observed a veteran of the Irish wars of the mid-century, 'for we make war more like foxes, than lions; and you will have twenty sieges for one battle.'

'One scarcely talks any more about battles', a German military instructor concurred: 'Indeed, the whole art of war now consists only of cunning attacks and good fortification.'[9] The Spanish and Dutch governments each maintained around 100,000 soldiers in the Netherlands from 1621 to 1648 – and yet they never fought a pitched battle. Instead, each year's campaign consisted of sieges.

Millions of men served in the various armies and navies of seventeenth-century Europe. Philip IV of Spain boasted that 'in the year of 1625 just past, we can count around 300,000 infantry drawing wages and more than 500,000 militiamen'.[10] Some 300,000 troops took part in each campaign fought in Germany between 1631 and 1634; and at least 200,000 men remained in arms when the Thirty Years' War ended in 1648. Over 100,000 soldiers fought the civil wars in England, Scotland and Ireland during the 1640s, and throughout the 1650s over 50,000 soldiers and sailors served the British state. In the 1670s, and again in the 1690s, Louis XIV commanded at least 250,000 troops. All these soldiers required wages, food and clothes, and many also needed expensive equipment and training in how to use it. From the 1620s onwards, up to half of each army in Western Europe fought with muskets in parallel lines, firing repeated broadsides at their enemies. This tactic called for a level of proficiency and discipline from each individual soldier that only prolonged training could provide.

Many non-Western states adopted at least some of these expensive European innovations. Both the Ottoman and the Russian empires recruited infantry armed with muskets: the Janissaries (literally 'new troops') fought for the sultan in ranks with muskets, as did the so-called New Formation Regiments of the tsars. The Mughal and Chinese emperors also made some use of Western firearms and Western military experts, but they normally relied on mobilizing overwhelming numbers rather than investment in new technology. The Mughal emperors normally travelled with an army of 200,000 cavalry and 40,000 infantry, fifty or more heavy guns and numerous war elephants; to repress the Three Feudatories rebellion in China in the 1670s the Qing emperor mobilized over 150,000 bannermen (elite Manchu troops) seconded by 400,000 loyal Chinese troops.

Troops everywhere had to be paid, so that (according to the Italian political theorist Giovanni Botero in 1605) 'wars today are dragged out for as long as possible, and the object is not to smash but to tire, not to defeat but to wear down the enemy. This form of warfare is entirely dependent upon money.' 'The manner of making war at the present time', echoed a Spanish general in the 1630s, 'is reduced to a sort of traffic or commerce, in which he who has most money wins.' Six decades later, an English pamphleteer made exactly the same point:

War is quite changed from what it was in the time of our forefathers, when in a hasty expedition, and a pitch'd field, the matter was decided by courage. But now the whole Art of War is in a manner reduced to money; and nowadays that prince who can best find money to feed, cloath and pay his army, not he that has the most valiant troops, is surest of success and conquest.[11]

This had crushing financial consequences for civilians. In France, the tax burden on a family of four rose from the equivalent of fourteen days' output a year in 1625 to thirty-four days' output by 1675; the Ottoman empire spent 75 per cent of its total budget on war; in Muscovy, 'one-eighth of [all] productive resources went just to pay for the army'.[12]

Devoting so many resources to armies and navies also involved indirect (or opportunity) costs. Rulers who spent so much on war had little left for anything else: official salaries; goods and services; welfare. Philip IV of Spain, who passed every day of his long reign (1621–65) at war, claimed that he lacked the money to set up a national banking system; Charles I of Great Britain, who fought both France and Spain between 1625 and 1630, decided he could not afford to create public granaries for famine relief; and so on. Ming China offers perhaps the most graphic case of the opportunity costs of prodigious military spending. After Manchu raiders broke through the Great Wall in 1629, the emperor's drastic reductions in non-defence spending included the closure of perhaps one-third of all courier and postal stations. Some of those who lost their livelihoods began to prey on those who used the routes they had once serviced: one of them was Li Zicheng, who later became the leader of a bandit alliance and in 1644, briefly, emperor of China.

The Fiscal-Military State

Early modern governments resorted to many expedients to fund their wars. The fiscal history of England, a relatively small country, was both striking and typical. Between 1605 and 1625, the government of James I (much criticized for its extravagance and corruption) raised and spent some £10 million, roughly 25 per cent of which went on military and naval spending; by contrast, between 1642, when the Civil War began, and 1660, when all soldiers and sailors then on foot laid down their arms, the London government raised and spent £34 million on war. The central government's defence spending thus rose twelvefold – from an annual average of £117,000 in 1605–25 to an annual average of £1.5 million in 1642–60. Even so, debts of about £2 million remained unpaid in 1660, leaving subsequent generations of English taxpayers to amortize the cost of the Civil Wars.

Such debts reflected the fact that, then as now, few governments fund their wars from current revenues alone. In Europe, most states raised loans to bridge the gap between income and expenditure, but borrowing created a new set of problems because bankers normally demanded a specific source of revenue as security for each loan, compelling governments to create new taxes. This vicious circle explains the apparently senseless fiscal decisions of so many rulers. Some stifled economic activity by taxing industrial production or exports just when conditions called for economic stimuli and tax breaks; others taxed items in general use, such as foodstuffs, which not only reduced the disposable income of most consumers but also provoked both widespread hardship and resistance. Many revolts began with protests at the point of sale, when a new tax unexpectedly increased the price of everyday items such as a loaf of bread or a basket of fruit. Other revolts started when rulers increased taxes on areas that they believed to be unusually prosperous. When war broke out with Spain in 1635, the French government abruptly doubled the taille (the principal direct tax) payable by the area around the thriving port of Bordeaux from 1 million livres to 2 million. In 1644, although poor harvests had caused grain prices to soar, the government increased the taille to 3 million livres and then in 1648, coinciding with the worst harvest of the century, to 4 million livres. Not surprisingly Bordeaux supported the Fronde revolt that year, and before long considered secession (Chapter 10).

Apart from imposing excise duties and increasing direct taxes, early modern governments at war frequently exploited and extended state monopolies, often known as regalian rights, such as extracting minerals obtained from the sea or underground (including salt and coal, silver and copper), or maximizing the profits from minting coins. Currency manipulation became particularly common in the seventeenth century, with many governments (including those of Spain, Russia and China) either adulterating silver coins with base metals, or issuing copper or paper money with little or no intrinsic value. Forcible devaluation could ruin whole societies. In 1634 Pavel Stránský, a Czech university teacher in exile, recalled devaluation as the most traumatic experience of his life: 'Neither plague, nor war, nor hostile foreign incursions into our land, neither pillage nor fire however atrocious, could do as much harm to good people as frequent changes and reductions in the value of money'. Major revolts broke out after some governments manipulated currency devaluation, notably in Central Europe in 1621–23, in Spain in 1651 and in Russia in 1661–63.[13]

Nevertheless, as the Swedish historian Jan Glete reminds us, in early modern Europe 'Wars were not decided by the existence of resources, but by how these resources were organized'. The key to improved organization, Glete argued, was the fiscal-military state: a polity dedicated to extracting, centralizing, and redistributing resources to finance the use of violence.[14] Only an outstanding ability to mobilize and deploy assets enabled Sweden, with scarcely a million inhabitants, to terrorize 20 million Germans during and after the Thirty Years' War; and allowed the Dutch Republic, also with scarcely a million inhabitants, to defeat the Spanish Habsburgs, with over

30 million subjects. Yet even the Dutch Republic, the most successful fiscal-military state of the early modern world, experienced difficulties in funding its wars. The debt of the federal government rose from 5 million florins in 1618 to 16 million in 1670, and that of the richest province (Holland) from 5 million to 147 million. At the same time taxes, especially sales taxes, rose to dizzying levels both to pay for current wars and to service the debts created during previous wars: in the university town of Leiden, by the 1640s taxes accounted for 60 per cent of the price of beer and 25 per cent of the price of bread. When provinces fell into arrears with their quotas of the overall budget, the federal government imprisoned their citizens as hostages until the shortfall was made good.

Such draconian measures enabled the Dutch government, at war for most of the seventeenth century, to maintain its credit intact while paying its army and navy regularly; few other governments could match this achievement. Calculations in 1633 by the Swedish chancellor, Axel Oxenstierna, responsible for funding all Protestant troops fighting in Germany, show the magnitude of the problem. On paper, the army numbered 78,000 soldiers and each one earned (on average) 125 thalers per year, or almost 10 million thalers for the entire army. Oxenstierna knew that this was far beyond his means, but if he provided each soldier with full wages for just one month, a small cash advance for the other eleven, and a pound of bread every day, then the total annual cost fell to just over 5 million thalers, about half the original cost.[15] But what of the other half? Like other commanders, Oxenstierna expected officers to use their own credit to supply their troops with essential items. He knew that his principal adversary, Albrecht von Wallenstein, had borrowed 5 million thalers (five times his personal fortune) between 1621 and 1628 to sustain his army until a victorious peace would bring reimbursement and rewards; and that most of the 1,500 or so colonels who raised regiments to fight in the Thirty Years' War did much the same (albeit on a lesser scale). Oxenstierna also knew that Wallenstein had introduced a contributions system that forced civilians living near his army to provide it with food and other necessities. His quartermasters worked out with the local magistrates of each community the precise quantities and the exact timetable for delivery, threatening that any shortfall or default would trigger the arrival of a detachment of soldiers who would burn everything to the ground. Oxenstierna expected his quartermasters to do the same.

Although it is hard to assess the precise financial impact of war on civilians, the records of the principality of Hohenlohe in southwest Germany show that in each year from 1628, when hostile soldiers first arrived, until the demobilization of 1650, its inhabitants paid at least double what they had done before the war, and that in some years they paid three, four or even five times as much. Yet even this did not suffice to feed Mars. Just before his regiment was demobilized, one officer complained that he did not 'earn enough daily bread to support my wife and poor children'. In China, Manchu bannerman Dzengšeo recorded in his campaign diary for 1680 that on some days he did not eat, but instead 'wept with sorrow under the blanket'; and on one desperate occasion he 'even sold a woman' whom he had acquired as booty at the capture of a town to buy horses and food. Normally, however, like early modern soldiers elsewhere, when he was hungry Dzengšeo exploited the civilian population: when 'the food supplies for the whole army had been used up', he sent his servants to 'search for food in every single village' and to take what they found by force.[16]

Doing God's Work

Rulers in the mid-seventeenth century could not plead ignorance about the economic hardships that their wars inflicted on soldiers and civilians alike. In China, officials deluged their superiors with memoranda that pointed out that (to quote one missive), 'The present dynasty commands the largest area of land in history. But land without people is worthless, and people without wealth are valueless; and in the present dynasty we find that the poverty afflicting the whole population is unprecedented in the history of China.'[17] Likewise, in 1640, just before the outbreak of revolt in Catalonia and Portugal, a tract published in Madrid perceptively warned Philip IV that 'subjects are more obedient when they are less taxed. A prince who in time of war avoids spending on himself will make the word "tax" acceptable, and avoid being called "ambitious" – but if the king read this, he paid no attention. Likewise, twelve years later, he ignored the protests of his spiritual advisor, Sor María de Ágreda, to 'introduce as few innovations as possible and avoid oppressing the poor, lest their misery leads them to revolt'. Instead Philip loftily replied that, although 'everything possible will be done for the relief of my poor vassals . . . *the requirements of the army pull in the opposite direction*'. His wars therefore continued.[18]

Why did so many seventeenth-century rulers raise taxes to satisfy 'the requirements of the army' instead of taking steps to 'avoid oppressing the poor'? One reason lay in the lack of limits on their power. In China, the emperor claimed to possess the Mandate of Heaven for all his actions, and every aspect of his official life proclaimed unique status and untrammelled authority. At audiences he alone faced south and everyone else faced north; no one else could wear clothes designed like his; he alone used red ink (everyone else used black); the character for 'emperor' received a line of text to itself; no one else could use the character for each emperor's given name or the word he used for 'I' (*chen*). Other Asian rulers also claimed to embody divine power on Earth, which conferred the right to make war at will. Korean kings held that they alone acted with divine sanction to bring the purposes of heaven and of human beings into harmony. In the words of a scholar and minister in 1660: 'The ruler regulates things in place of Heaven, and causes them to find their appointed places.'[19] Political rhetoric in South Asia also presented rulers as endowed with superhuman powers. Successful Buddhist monarchs claimed to be *Chakhravarthi* ('world conquerors'), just as India's Mughal emperors projected themselves as *Sahibkiran* (the 'Shadow of God on Earth'). India's Hindu rulers claimed to be not only the incarnation of one of the gods but also sexual heroes: court poems and dance-dramas, the preferred media for political propaganda in southern India, portrayed the capital as a city of erotic delights and war as a sexual adventure. None of these political visions left room for restraint.

Indonesian rulers likewise acknowledged no limits to their power. According to a foreign visitor, whenever Sultan Iskandar Muda of Acheh (1593–1636) heard about an attractive 'woman, either in city or country, he sends for her to the court. Although she be married, she must come and if her husband seem unwilling or loath to part from her, then [the Sultan] presently commands her husband's prick to be cut off.' Iskandar Muda did not stop at genitalia: he also (according to another foreign visitor) 'exterminated almost all the ancient nobility' in the course of his reign. Therefore no one had the authority to restrain the sultan when in 1629 he decided to lead the entire military and naval strength of his state to attack Portuguese Melaka, or to remind him of the need to fortify his own siegeworks – with the result that a Portuguese relief army destroyed almost all his army, his fleet and his guns.[20]

Most Muslim political writers extolled absolute rulers as both the only alternative to anarchy and the best way to advance the cause of Islam. A treatise of advice presented by an Ottoman palace official to Sultan Murad IV in 1630 ascribed the problems facing the empire to the failure of the ruler to use his divinely sanctioned arbitrary power to the full. To implement this advice, Murad addressed a stream of requests to the chief mufti (*Şeyhulislam*) of Istanbul to certify (usually in the form of a written opinion, or *fatwā*) that a proposed action or edict conformed to Sharia law. In 1638 Murad even took the chief mufti with him on campaign, so that he could ensure that his military as well as his civilian decisions conformed to God's will. Occasionally the chief mufti might defy a sultan – in 1648 one even issued a *fatwā* to legitimize the sultan's deposition (Chapter 7) – but normally the shoe was on the other foot: sultans deposed (and occasionally executed) any chief mufti who challenged their authority.

The tsars of Russia likewise claimed divine status and encouraged writers and artists to portray them as the secular version of the transfigured Christ, as paragons of Old Testament kingship (especially David), the 'image and likeness of God', and their subjects as the Chosen People, their country as the Earthly Paradise, and their capital as the New Jerusalem. The state apartments in the Kremlin sported paintings that interspersed the victories of Moses, Joshua and Gideon with the leading events of Russian history; the main Moscow churches displayed icons in which archangels led the tsar and his troops on campaigns of conquest. When the tsar 'wishes to wage war or make peace with any state', an experienced minister explained in the 1660s, 'or when he wishes to decide any other great or small affairs, it is in his power to do what he wishes'.[21] Like the rulers of Asia, the tsars allowed no discussion or dissent, let alone loyal opposition.

The equivalent of this rhetoric in Latin Christendom was the Divine Right of Kings. Many early modern European rulers claimed that their power was *absolute* (a term derived from Roman Law to describe the authority of one absolved from obeying the laws he had made) and that their actions enjoyed divine approval. In 1609 James I of Great Britain boasted that kings served as 'God's lieutenants upon Earth', and therefore 'exercise a manner or resemblance of divine power upon Earth' because 'they make and unmake their subjects, they have power of raising and casting down, of life and of death, judges over all their subjects and in all causes and yet accountable to none but God only.' A generation later, the funeral oration for a German prince echoed the same sentiments. 'Just as the sun in the heavens above is made and fashioned by God, and is truly a wondrous work of the Almighty, so are kings, princes and lords placed and ordered by God in the secular estate. For that reason, they may themselves be called gods.' In France, a treatise likewise argued that the king's commands must always prevail, because if a measure 'relates to a pressing necessity for the public good ... Necessity knows no law.' Christina, former queen of Sweden, wrote that 'only monarchs must rule: everyone else must obey and execute their orders.' In particular, a royal decision to wage war – even a war of aggression – obliged everyone to obey because sovereigns could discern the true interests of the state better than their subjects.[22]

Most European monarchs received an education crafted explicitly to reinforce these attitudes. They studied history (national, classical and occasionally foreign) primarily 'to examine how each prince acted, well or badly', and to learn how to 'ascertain what our subjects are hiding from us'. During the Fronde revolt of 1648–53, the young Louis XIV read chronicles that described how his predecessors had overcome rebellious nobles; and he, his son and his grandsons all studied Spanish history and literature, and learned to speak Spanish, in case they might one day succeed their ailing relative, Carlos II. They likewise learned the principles of architecture and mathematics explicitly so that they would better understand how to attack and defend fortified towns. Louis XIV even commissioned a set of huge relief models of frontier fortresses so that his son could follow the progress of his wars until he was old enough to participate in person.[23]

Seventeenth-century European rulers believed that 'religion is the most important element that must be taught to a young prince destined to wear the crown' – and this meant not only private but also public devotions.[24] The official *Gazette de France* recorded not only the zeal and humility of Louis XIV during sermons but also his participation in pilgrimages; and it chronicled the occasions on which he 'touched' (and, according to popular tradition, cured) subjects afflicted with scrofula – some 20,000 individuals between 1654 and 1663 – perhaps the most striking public demonstration of divinely delegated power in the early modern world. Catholic and Protestant rulers alike appointed the prelates of their state (the former with papal concurrence) and expected their subjects to follow their theological opinions – or, in the formula that prevailed in the Holy Roman Empire, *Cuius regio, eius religio*: 'Rulers determine religion'.

The overlap of politics and religion influenced foreign as well as domestic policy. In the words of the governor of Louis XIV's heir, the dauphin, Christian princes must not only 'love and serve God' but also 'make others honour Him, avenge His injuries, and take up His causes'; and in 1672 the dauphin, aged eleven, composed a campaign history of the Dutch War that justified his father's invasion of Holland on the grounds that it advanced the Catholic faith.[25] Religion served as a pretext for many wars in Europe in the first half of the seventeenth century. Thus when in 1619 the Bohemians offered their crown to the German Protestant leader, Frederick of the Palatinate, he accepted because, he claimed, it 'is a divine calling that I must not disobey. My only end is to serve God and his Church'. A similar divine calling motivated Frederick's brother-in-law Charles I, who steadfastly refused to negotiate with his rebellious subjects because, as his wife Henrietta Maria put it in a confidential letter of 1642 on the eve of the English Civil War, 'You have testified your gentleness enough; you must show your justice. Go on boldly: God will assist you.' The king complied, assuring his leading general that 'no extremity or misfortune shall make me yield, for I will either be a glorious king or a patient martyr' – and he continued to reject all compromise until in 1649 his victorious subjects made him a martyr.[26]

Most Christian rulers admitted only one restraint on their absolute power: like their Ottoman contemporaries, they submitted controversial decisions to the spiritual advisors and asked for their approval. In Russia, 'when discord and warfare break out with neighbouring powers, the tsar at that time consults with the patriarch, metropolitans, archbishops, bishops and with other hierarchs of the prominent monasteries'.[27] Many Catholic rulers routinely asked their confessors to certify that they might 'in conscience' adopt a controversial policy, and the kings of France and Portugal created permanent Councils of Conscience to advise them on more complex issues. Philip IV created ad hoc committees of theologians to determine whether his sister should marry the Protestant Charles Stuart; whether he could send assistance to the French Protestants; whether he might intervene in the Mantuan succession; and whether he needed to honour concessions made to rebels.

Secular ministers, too, sought to convince their masters that the costly and destructive policies they advocated enjoyed divine favour. In a memorial of 1626 that listed the various successes of his ministry, Don Gaspar de Guzmán, count-duke of Olivares, assured Philip IV triumphantly: 'Sire, God has placed the armed forces of Your Majesty in this situation, with no other assistance or ally. I would be lying to Your Majesty, and a traitor to you, if I claimed that all this stemmed from human provision. No: God alone has done it, and only God *could* have done it.' A quarter of a century later, Philip himself attributed his ability 'to overcome not only my enemies but also storms at sea, epidemics on land and the domestic unrest of the towns of Andalusia' to the support of 'God's most mighty hand'.[28]

Pride and Prejudice

Some early modern rulers may have used such rhetoric and imagery as propaganda without actually believing it. Queen Christina of Sweden asserted as much in 1649, when she debated with her council whether or not to support English royalists after Charles I's execution. Marshal Jakob de la Gardie argued that, since 'such a giddy spirit has arisen' in Europe, rulers of the same faith should support one other; but Christina (still officially a Lutheran) disagreed. 'People use religion as a pretext', she replied, 'and it is used by us against Calvinists and Catholics alike.' 'The Pope, the Spaniards and the rest of the House of Austria have always sought to make use of religion', de la Gardie reminded her. 'Like a raincoat when it's wet', the queen quipped. In similar vein, three decades later (and now a Catholic), Christina noted that although princes should allow their confessors 'to speak freely to us, we must not blindly obey all that they tell us. We must be well aware that it is not always God who speaks to us through them.'[29]

Such overt cynicism was rare in the seventeenth century. More typical was the providential vision of Philip IV of Spain, who in 1629 told a senior minister that 'I seek salvation and want to placate God by obeying His laws and by making sure that all others obey them', because then, 'even if misfortunes rain upon us, you need have no fear that they will harm us.' He still felt the same in 1656. Upon hearing that Britain had joined France, Portugal and the Catalans in making war on him, Philip boasted that although 'The risk is apparent and the distress is greater than any that this Monarchy has ever seen, particularly since we lack the means to withstand even one part of such a great storm', he intended to keep on fighting because 'I have firm faith that, unless our sins make us unworthy, Our Lord will deliver us from this great storm without allowing these kingdoms, so loyal to the Catholic Church, to be brought down by heretics.'[30]

Naturally, a providential vision did not preclude secular motives for waging war. Many seventeenth-century rulers saw waging war as a rite of passage to be performed at the outset of every reign. Three months after his accession, when Charles I asked the English Parliament to vote funds for war with Spain, he said 'I pray you remember *that this being my first action*' as king 'what a great dishonor it were both to you and to me' if it 'should fail for [lack of] that assistance you are able to give me'. A few years later, when Philip IV heard that Louis XIII had just led an invasion of Italy, he scribbled on a ministerial memorandum that 'My intention is to get my revenge on France for its recent behaviour', and to that end 'I shall be there in person. *Fame, after all, cannot be gained without taking personal part in some great enterprise.* This one will enhance my reputation, and I gather it should not be too difficult.' Not only did both monarchs – like their contemporaries Gustavus Adolphus and Charles X of Sweden, Christian IV of Denmark, Emperor Ferdinand III and Tsar Alexei – command their armies in person, they also claimed to relish military life. The same was true of the next generation. Louis XIV personally participated in over twenty sieges, starting in 1650 when he was twelve, and ending in 1692, when he ceded the role to his heir, noting that 'If my son does not go on campaign every year, *he will be totally despised and will lose all respect.*'[31]

Several seventeenth-century rulers also advanced secular strategies to justify starting (or prolonging) their wars. In 1624, despite much public rhetoric about upholding the Protestant cause in Germany, a British diplomat coldly remarked that 'England has no other interest in Germany apart from the Palatinate; it does not matter to them whether all Germany is set in flames, provided they might have the Palatinate.' The diplomat justified this with a domino theory: 'if we lose the Palatinate first, next we lose the Low Countries, then Ireland, and finally ourselves'. Philip IV's ministers constructed a similar domino theory to justify their numerous wars. In 1629 they warned the king that 'once the Netherlands are lost, America and other kingdoms of Your Majesty will also immediately be lost with no hope of recovering them'. Some years later a veteran diplomat extended the argument yet further: 'We cannot defend the Netherlands if we lose Germany'. Spanish troops therefore continued to fight on all fronts until 1648. Sweden's leaders likewise claimed that (initially) invading Germany and (later) occupying large parts of it were essential to Swedish security. 'Pomerania and the Baltic Coast are like an outwork of the Swedish crown; our security against the Emperor depends on them', wrote Chancellor Axel Oxenstierna; Sweden was a fortress 'whose walls are its cliffs, whose ditch is the Baltic and whose counterscarp is Pomerania', echoed Ambassador Johan Adler Salvius. Failure to retain any territorial gains would imperil national security, and so Sweden also continued to fight until 1648.[32]

Such arguments threatened to eternalize armed conflicts. Ending the Thirty Years' War required sixty months of negotiations, with scarcely a break; talks to end the Eighty Years' War between Spain and the Dutch took

twenty months; those to end the Thirteen Years' War between Russia and
Poland required thirty-one sessions over the course of a year – and even
then produced only a truce. Such longevity also reflected factors that tend
to prolong all conflicts. First, rulers found it easier to start than to finish
wars. The nobles of Russia made this point in 1652, when Tsar Alexei sought
their approval to attack the Ottoman sultan: 'It is indeed easy to pull the
sword from the scabbard, but not so easy to put it back when you want,
since the outcome of war is uncertain.' Second, objectives changed. The
English cleric and mathematician John Wallis noted wryly that the ultimate
objectives of the Civil War 'proved very different from what was said to be
at first intended, as is usual in such cases – the power of the sword frequently
passing from hand to hand, and those who begin a War not being able to
foresee where it wil end.' A generation later, another English cleric noted
that 'the ends that those who begin a war designe to themselves are seldome
obtained, but oftentimes that which is contrary thereunto and dreaded by
them is brought to pass'.[33] Third, in the seventeenth century (as today) the
more resources invested, and the more lives sacrificed, the more total victory
seemed the only acceptable outcome. As Arthur Hopton, Britain's ambassador in Madrid, observed in 1638 concerning Spain's struggle against
France:

> The end of all these troubles (unless they outlast Time) must be peace,
> which nevertheless comes on so slowly as I cannot say appearances
> thereof are visible . . . They would be glad of any good occasion to treat
> of a peace; but they are so entangled on both sides, partly out of jeal-
> ousy and partly out of avarice, being unwilling to forgo what they have
> gotten (and indeed have dearly enough paid for), as I find the way to
> begin a treaty to be very difficult.

At this point, the war was only three years old. Eight years later, when
France's diplomatic delegation in Westphalia asked Louis XIV's chief
minister, Cardinal Jules Mazarin, to permit them to conclude a settlement,
he made much the same point as Hopton. 'After the many expenditures that
the war has involved', the diplomats must 'find pretexts for delaying the
signature of a peace treaty' in order to 'profit from the remainder of the
campaigning season'.[34]

Negotiating while hostilities continued also made peace more elusive
because, as Mazarin put it, each state's demands reflected 'how much the
military situation has changed in our favour recently'. The count of
Peñaranda, the chief Spanish negotiator at the Westphalian peace congress,
agreed. In June 1647 he informed a colleague: 'You believe that the war will
last many years, but you are entirely mistaken . . . My lord, the vassals of
both kings (of France and Spain) find themselves so exhausted that asking

them for more could lead either monarch to complete ruin . . . Whether we win or lose, we both must have peace.' Yet one month later, Peñaranda hailed news that the French siege of Lleida had failed as 'the most important and pleasing news I have ever had in my life because it shows that Our Lord in his mercy smiles upon us and wishes to remove the scourge from us'. He therefore urged the king to fight on. Philip duly obliged.[35]

Minorities and Tanistry

Succession struggles increased the frequency of civil wars. Early modern states often experienced anarchy whenever a ruler died leaving no capable and universally recognized successor – and minorities proved unusually common in the seventeenth century. France experienced civil war soon after the accession of Louis XIII at the age of nine in 1610, and again after his death in 1643 when he left an heir aged only five. Civil War broke out in Russia when Alexei died in 1676 leaving three young sons. In Sweden, on the death of Charles IX (1611), of Gustavus Adolphus (1632) and of Charles X (1660), each of whom left a minor to succeed them, the nobility swiftly reduced the powers of the crown. Those of Denmark did the same in 1648 when Christian IV died before parliament had sworn to accept his heir (Chapter 8).

Instability was also endemic in elective monarchies. Although the House of Habsburg retained the title of Holy Roman Emperor throughout the seventeenth century, the Electoral College chose Ferdinand II in 1619 only after a bitter contest that unleashed the Thirty Years' War, and voted for his grandson Leopold in 1658 only after a year of intrigue and bargaining for concessions. Likewise, although Sigismund Vasa and his two sons occupied the Polish throne for almost a century, the death of each monarch gave rise to an interregnum while the federal Diet bargained for concessions before choosing a successor. The Dutch Republic suffered a constitutional crisis in 1650 when, on the death of William II of Orange, the States-General denied the title of stadtholder to his posthumous son; and in Japan the following year, the death of the autocratic Shogun Tokugawa Iemitsu, leaving only a child to succeed him, unleashed plots to overthrow the dynasty. China's Ming dynasty also experienced succession difficulties in the first half of the seventeenth century. The Wanli emperor (1573–1620) refused to recognize his eldest son as his heir, instead intriguing to gain recognition for the claims of another of his offspring; and although the eldest son eventually succeeded, he died shortly afterwards, leaving a young heir who suffered from what today would be called attention deficit disorder. Disputes among ministerial factions racked the central government for a generation (Chapter 5).

This instability paled in comparison with the succession disputes that characterized some other Asian dynasties. The distinguished historian

Joseph Fletcher observed that nomadic peoples such as the Mongols, as well as dynasties like the Qing, the Mughals and the Ottomans who claimed descent from Mongolian forebears, determined each transition of power through a practice he called 'tanistry', after the Celtic practice in which each ruler had a recognized heir (the *táinste*), who nonetheless had to prove himself by defeating, and often killing, all challengers before assuming the full powers of his predecessor.[36] Tanistry created serious political instability because everyone took part in succession disputes (nomadic societies have no civilians, only warriors), and so everyone had to guess which of the various potential successors would emerge victorious in the next generation and position themselves accordingly – knowing that supporters of the victor would monopolize the spoils. Perhaps tanistry made sense on the steppe of Central Asia, where the principal requirement for each clan leader was military talent, but the practice periodically brought more complex states to the brink of extinction. On the death in 1626 of Nurhaci, 'great ancestor' of the Qing, his relatives fought among themselves for several years before a clear successor emerged: his eighth son, Hong Taiji. When he died in 1643, another bitter struggle broke out between his brothers and uncles until the survivors agreed to recognize the late emperor's ninth son as emperor. In the end, half of the sons of Nurhaci who survived to adulthood were executed, forced to commit suicide or posthumously disgraced.

The Ottoman sultans sought to avoid such chaos by confining their male relatives within the Istanbul palace in sealed apartments known as the cage (*kafes*). Immediately after his accession the crown prince normally killed all his male relatives – both siblings and younger sons – in an attempt to avoid succession disputes; but this changed in 1603, when Sultan Mehmed III died leaving only two young sons, and the Ottoman elite allowed them both to survive. His grandson Murad IV murdered most of his brothers and when he died in 1640, only one male member of the House of Osman survived: Murad's youngest brother Ibrahim, who had never left the cage. After eight years of erratic rule, the Ottoman elite murdered him, but once again allowed all his young sons to survive because they were the only surviving male members of the dynasty (four of them would reign; the last of them would be deposed in 1687.) These succession protocols may have been slightly less disruptive than the civil wars among male members of the Qing, but they destabilized the Ottoman state all the same.

The Mughal emperors also faced repeated succession disputes. When Jahangir's heir rebelled in 1606, the emperor impaled 300 of his supporters alive to form an avenue through which his son had to pass to beg forgiveness, but when his heir once more conspired against him Jahangir had the young man blinded and entrusted to his younger son, Shah Jahan – who himself rebelled in 1622. Once again the revolt failed, and once again Jahangir pardoned the miscreant. Shah Jahan therefore survived to succeed his father when he died a few years later – and promptly had all other male members of his family killed or blinded. This left the new emperor and his four sons as the only surviving male members of the dynasty. Each prince built up a powerful following until 1658 when, believing that their father was on the point of death, three of them rebelled and fought a civil war that lasted two years. Aurangzeb, the eventual victor, murdered all his rivals just as his father had done. Later he also tried to partition his inheritance, hoping to avoid another succession war, but his ambitious sons refused to accept anything less than the whole empire – and fought each other after he died in 1707.

The Curse of the Composite State

Although tanistry never took root in early modern Europe, over half the major seventeenth-century revolts there occurred in a 'composite state': a well-integrated core territory linked by loose and often contested bonds to more autonomous regions, some of them far away. These composite states included Denmark (whose monarch also ruled Norway, Greenland, Iceland, Holstein and many Baltic islands); Sweden (which included Finland, Estonia, Ingria and several Polish enclaves); and Russia (with several areas containing distinct religious and ethnic groups annexed by treaties, notably Ukraine and Siberia). The Ottoman empire also incorporated territories that contained disparate religious and ethnic groups (Shiites, Christians of various creeds and Jews) as well as several provinces with separate legal codes and local traditions (notably the Crimea, the Balkan principalities and the North African states).

The three most volatile composite states had been created by dynastic unions: the Stuart monarchy and the lands of the Spanish and Austrian Habsburgs. The political instability of the last two in part reflected repeated endogamy, which reduced the dynasty's gene pool and therefore the viability of its offspring, producing more minorities and more disputed successions. Intermarriage over several generations meant that Philip IV of Spain had only eight great-grandparents (instead of the normal sixteen), and after he married his niece in 1649 he became the great-uncle as well as the father of their children, whose mother was also their cousin. This created the same genetic inheritance as that of the child of siblings, or of a parent and child. Only two of the couple's six children survived infancy, and although their son Carlos II lived to be thirty-nine, he was physically deformed, mentally challenged and sterile. His death unleashed a prolonged succession war between the various claimants that resulted in the dismemberment of the Spanish monarchy.

The second weakness of the Stuart and Habsburg monarchies lay in the fact that many constituent territories preserved their own institutions and collective identity, sometimes reinforced by a separate language or a distinct religion. An English pamphlet of 1641 noted this vulnerability: the Stuart monarchy, it argued, stood on the brink of disintegration 'because there was not heretofore a perfect union twixt England and Scotland, incorporating both into one body and mind'; in the Spanish monarchy, the same 'reason has caused Portugal and Catalonia to revolt from the king of Spain'. Diversity created instability because every political organism has a distinct boiling point – or, as Francis Bacon put it in an essay entitled 'Of seditions and troubles': 'discontentments' are 'in the politique body like to humours in the natural, which are apt to gather a preternaturall heat, and to enflame'. When natural disaster (such as famine) or human agency (such as war) enflamed the parts of a composite state, it became unstable and its 'discontentments' soon emerged. Not only was the political boiling point in composite states unusually low, it was lowest on the periphery, making them the least stable component.[37]

At first sight this may seem surprising, because global cooling, failed harvests and lethal epidemics affected the core as well as the periphery of each composite state. Indeed cores often endured more intense government pressure – taxpayers in England and Castile paid far more than their neighbours in peripheral territories – and yet England was the last of Charles I's kingdoms to rebel against him, while (except for Andalusia) Castile never rebelled against Philip IV. This paradox has three explanations. First, the core of each state often escaped the worst consequences of war, and thus the full synergy between human and natural disasters. Castile contributed soldiers and taxes to the wars, and also suffered from extreme weather, poor harvests and high food prices, but most areas remained safe from the devastation of war. Their inhabitants were rarely robbed or raped by soldiers; troops seldom burnt their property or spread disease; and normally they escaped billeting. By contrast, at a time of dearth and epidemics, the need to feed garrisons as well as the local population created a crisis in frontier regions like Catalonia long before enemy troops wrought further devastation. Second, several parts of Europe's composite states retained not only their own institutions and identity, but also their own economic, defensive and strategic agendas. The priorities of the local elite in Barcelona (like those in Brussels, Lima, Manila, Mexico, Milan, Naples and Palermo) often differed from the priorities of the imperial government in Madrid (just as the priorities of Edinburgh, Dublin, Jamestown and Boston often differed from those of the central government in London). Third, diversity frequently led to what we might call 'sub-imperialism'. The peripheral parts of each composite state often possessed extensive privileges, permanently guaranteed by the sovereign, and whenever conditions became difficult, whether through war or weather, regional elites invoked their constitutional guarantees (often termed 'fundamental laws', 'charters' or 'constitutions'), while the central government sought to override them. Such confrontations could and did lead to rebellion. In 1638, when his Scottish subjects refused to accept a new liturgy mandated by Charles I, the king told one of his ministers that 'I would rather die than yield to those impertinent and damnable demands' because 'to yield [is] to be no king in a very short time'. The following year, in Spain, the count-duke of Olivares exhausted his patience with the insistence of the Catalan elite that he must respect their constitutions. He exclaimed: 'By now I am nearly at my wits' end; but I say, and I shall still be saying on my deathbed, that if the Constitutions do not allow this, then the Devil take the Constitutions.' In Catalonia as in Scotland, intransigence soon led to insurrection.[38]

Favourites

Olivares's insensitivity reflected his distinctive status: he was not only Philip IV's chief minister but also his favourite – a courtier who had gained total control of his master's affairs. Favourites abounded across the early modern world, and like minorities and sub-imperialism, their existence made war and rebellion more common. Only the Ottoman empire made the position permanent, in the person of the grand vizier. Elsewhere, favourites often gained a privileged position through the extreme youth of a new ruler: Philip IV and Tsar Alexei came to the throne aged sixteen; Louis XIV of France succeeded aged five (and only started to exercise his powers at twenty-three). In each case, the monarch relied initially upon a much older man, often a member of his household as heir to the throne, to run the government (respectively Olivares, Boris Morozov and Mazarin). In China, the Tianqi emperor ascended to the throne aged fourteen and immediately surrendered his powers to a palace eunuch who had helped to raise him: Wei Zhongxian. But youthful inexperience cannot explain why these rulers continued to rely on their favourites after (often long after) they became adults, or why the institution (although not new) became so much more common during the early seventeenth century than at any other time.

The continued prominence of favourites is all the more surprising in view of the hatred they provoked. The duke of Buckingham, who dominated policy and patronage under both James I of Great Britain and his son Charles, was compared with Sejanus, the tyrannical adviser of the Roman emperor Tiberius – a parallel deeply resented by Charles – and when the duke was murdered in 1628 songs, poems and pamphlets compared the murderer with the biblical David. The fall and suicide of the eunuch Wei the previous year likewise caused rejoicing throughout China, as did the fall of most grand viziers in the Ottoman empire (an event that occurred every four months, on average, in the 1620s). Similar rejoicing would no doubt have greeted the success of any of the numerous assassination plots against

Mazarin or his predecessor, Cardinal Richelieu; or had Tsar Alexei surrendered Morozov to the mob that screamed for his blood in the Kremlin in
1648. According to one of Charles I's more thoughtful English critics, 'The
king's favour is tyrannie, when by that favour a man rules over them in fact,
that can plead neither election nor succession to that power'; and, he
concluded ruefully, if such men were truly deemed necessary, 'a king should
have more than one Favourite, [because] emulation will make them walke
the fairer wayes'. Charles paid no heed.[39]

The rise of the favourites reflected in part the relentless increase in the
administrative burdens that weighed upon monarchs. In the aphorism of
Queen Christina of Sweden: 'If you knew how much princes have to do, you
would be less keen to be one'. Copies and minutes of some 18,000 letters
survive from the office of France's secretary for war between 1636 and 1642,
an average of 2,500 a year; but by 1664 the total had risen above 7,000, and
by 1689 above 10,000. In the Mughal empire, the documents generated by
Shah Jahan and his ministers 'must have numbered in the millions'. In
China, the Shunzhi emperor complained in the 1650s that 'The nation is
vast and affairs of state are extremely complex. I have to endorse all memorials and make decisions by myself without a minute of rest.' His son, the
Kangxi emperor, read and returned fifty memorials on normal days, but
when on campaign the total rose above 400. He later recalled that the paper-
work generated by a rebellion in 1674 forced him to stay up until midnight.[40]

Favourites not only reduced the bureaucratic burden for their masters,
they also simplified the process of decision-making by operating outside
traditional institutional channels. Each sought to monopolize the flow of
both people and information reaching their master and to this end they
promoted their own relatives and clients, excluding all potential rivals.
Richelieu built up a network of *créatures*, literally people he had 'created':
men 'who would be faithful to him and only to him without exception and
without reservation'. The *créatures* worked as a team: whether at court or in
the provinces they exchanged information and did each other favours. They
also took every opportunity to praise Richelieu to the king and made sure
that their advice and proposals coincided with his, since they knew that
their own political survival depended on the cardinal monopolizing the
king's trust. In Spain, as Philip III lay dying in 1621, his favourite the duke
of Uceda came face to face with Olivares, who enjoyed the complete confidence of the heir apparent. 'Now everything is mine', Olivares gloated.
'Everything?' asked the doomed duke. 'Yes, without exception', Olivares
replied, and he immediately set about replacing Uceda's appointees with his
own men.[41]

Reliance on favourites promoted revolts and civil dissension. Rival
courtiers excluded from power might lose patience and rebel, and discontented subjects who hesitated to challenge the Divine Right of Kings found

it easier to justify their disobedience by claiming that the ruler had been deceived by his wicked ministers. The cry 'Long live the king; down with the evil ministers!', common in earlier rebellions, became a constant refrain in the mid-seventeenth century because the monopoly of power by a favourite made it all the more plausible. Ming scholar-officials claimed that the orders issued by Wei Zhongxian lacked imperial sanction; opponents of Philip IV in Portugal and Catalonia claimed that they strove to free the king from the snare or satanic spell cast by Olivares; opponents of Alexei and Charles I both demanded the sacrifice of unpopular ministers who, they claimed, had bewitched their master.

Absolutism and the 'Willingness to Wink'

Circumventing the checks and balances of traditional governments encouraged 'mission creep' on the part of the state. As Sheilagh Ogilvie has astutely observed, the new style of absolutism introduced by monarchs and their favourites 'affected more than taxation and warfare':

> The administrative instruments developed for *these* purposes could also regulate activities previously inaccessible to government, and they could offer redistributive services to a wide range of favoured groups and institutions. Resistance to these *new* forms of redistribution, and competition to control them, were central elements in the 'crisis' of the mid-seventeenth century.[42]

Whereas the traditional bureaucracy of most early modern states contained mechanisms (however rudimentary) by which subjects could present legal protests (however deferentially), the alternative administrations invented by favourites – whether Buckingham, Richelieu, Olivares, Morozov or Wei – brooked no challenge. The imposition of government initiatives by proclamations, often resurrecting or extending a regalian right and enforced by royal judges with instructions to stifle any opposition in the courts, left those affected without legal redress. Everywhere, 'absolute' rulers displayed both inflexibility and ruthlessness in enforcing all government policies, not just those related to war.

David Cressy has explained this phenomenon brilliantly in the context of Stuart England. Whereas the officials of James I 'were inclined to avert their gaze from local difficulties', he has written, those of his son Charles 'went looking for trouble'. England, like most (if not all) early modern states, was rife with conflicts; but until the 1630s 'these conflicts were continually being resolved or mitigated by an overriding insistence on peace'. 'The famed "consensus" of Jacobean England consisted not in mutual agreement on issues but rather a determination to prevent divisive issues from disrupting the body politic. It was a social rather than an ideological consensus, and it worked by winking at the gap between theory and practice.' In the 1630s this 'willingness to wink' disappeared. Instead, Charles and his ministers (especially his bishops) developed a 'remarkable gift for seeing mild irregularity as intransigence, moderate nonconformity as sectarianism, and all disagreement as refractoriness or rebellion'. Eventually, 'in forgetting how to wink they tore the country apart'.[43] By way of example, Cressy cited the insistence that women wear veils when they came to church for the first time after giving birth; the demand that ministers make the sign of the cross at every baptism; and the requirement that communion tables in the centre of the church give way to altars at the east end. Earlier generations had regarded each of these as 'trifling matters' and so, over time, each parish had developed its own ritual to which the congregation became strongly attached. Then, in the 1630s, the central government prescribed all three former practices (along with many others) and excommunicated those who would not conform, measures that affected literally thousands of churchgoers each year. Therefore, when the need to pay for war forced Charles in 1640 to return to the traditional institutional channels and summon Parliament, he faced an avalanche of grievances that paralyzed the transaction of public business.

Some Asian rulers likewise seem to have gone 'looking for trouble' on matters previously regarded as peripheral. Sultan Murad IV forbade both smoking tobacco and drinking coffee throughout the Ottoman empire and had many offenders executed. A seventeenth-century ruler of Borneo forbade his subjects to

> Dress like people from abroad, such as the Hollanders, the people from Keiling, the Biadju, the Makassarese, the Buginese. Let no one follow any of the Malay dressing customs. If foreign dressing-customs are followed this will unavoidably bring misery over the country where this is done ... [There will be] disease, much intrigue and food will become expensive because people dress like those in foreign countries.[44]

Most spectacularly of all, in China, the Qing insisted that all their male subjects shave their forelocks and braid the rest of their hair in a pigtail on pain of death. Initially, the head-shaving edict made sense because it immediately distinguished friend from foe: the long hair of Ming loyalists could be shorn in a few minutes. Shaving the forelocks therefore seemed a perfect test of loyalty; but it also created constant provocation because compliance required constant repetition as each man's hair grew. Nevertheless the Qing refused to back down and instead decreed 'Keep your head, lose your hair; keep your hair, lose your head' (Chapter 5).

Not one of these contentious issues threatened the integrity or security of the state; and not one of them arose from the problems created by the Little Ice Age. With goodwill, statesmanship or just the 'willingness to wink', each of them could have been peacefully resolved; but the exalted rhetoric and claims of those leaders who believed in the Divine Right of Kings or the Mandate of Heaven prevented such an outcome. Instead, they produced crises out of trifles and exacerbated tensions created by more serious problems, thereby increasing the sum of human misery.

Untangling the connections between these distinctive and disruptive aspects of seventeenth-century government, and the increased frequency of war and rebellion, is complicated by the existence of feedback loops. The success of favourites in removing checks and balances made it easier for monarchs to go to war, while monarchs became more dependent on their favourites in times of war because of the need to curtail checks and balances in order to extract more resources. Likewise, interstate war frequently caused intrastate rebellions by driving governments to extract resources from their subjects more aggressively, while intrastate rebellions could turn into interstate wars when alienated subjects secured foreign intervention.

Although these feedback loops varied according to time and place, war proved uniquely capable of uniting opponents of a regime. The imperative to feed Mars led many governments to impose burdens on every social group and all geographical regions, which risked alienating everyone at the same time. To be sure, the heaviest burdens normally fell on the lower ranks of each state, but governments desperate to win a war could also trample on the rights of those with corporate privileges (such as cities, nobles and clergy) in their efforts to extract resources, and on the privileges of certain regions rendered particularly vulnerable by war or the weather.

CHAPTER THREE

'Hunger Is the Greatest Enemy': The Heart of the Crisis

In one of his celebrated *Essays*, published early in the seventeenth century, the English politician and philosopher Francis Bacon warned rulers to ensure that their subjects 'doe not exceed the stock of the kingdome that should maintain them', because a prolonged imbalance between food supply and demand will sooner or later cause hardship, disruption and revolt.[1] Subsequent writers agreed. In 1640 a historian embedded with the army of Philip IV as it passed through the drought-parched fields of Catalonia noted ominously: 'Amid the distress to which human misery reduces us, *there is almost nothing men would not do.*' Eight years later, as southern Italy faced the worst harvest of the century, officials reported 'murmurs among the people that "*it was always better to die by the sword than to die of hunger*"'. In Scotland, during the last famine of the century, an acute observer reminded his compatriots that hunger makes people 'unquiet, rapacious, frantick or desperate. Thus, *where there are many poor, the rich cannot be secure in the possession of what they have.*'[2]

Although the Little Ice Age afflicted almost the entire northern hemisphere, some regions suffered far more than others. This should cause no surprise. Europe west of the Urals covers 4 million square miles, includes climates that range from subarctic to subtropical, and contains hundreds of ethnic, cultural, economic and political divisions: naturally, developments did not take place uniformly in all regions. In Spain, for example, Galicia in the northwest and Valencia in the southeast experienced population decline from about 1615 to the 1640s; but in the centre, although the decline around Toledo also began around 1615, it lasted until the 1670s, and around Segovia, where the decline also ended in the 1670s, it began almost a century before. Ming China covered 1.5 million square miles and, although politically unified, also includes climates that range from subarctic to tropical, and contains hundreds of ethnic, cultural, and economic divisions. Low-lying Shandong in the northeast often experiences both droughts and floods, so that the province rarely generates a surplus – let alone a reserve on which to draw in bad years. Urgent petitions calling for food loans and tax relief therefore emanated from Shandong on an almost annual basis, in the seventeenth century as at other times. By contrast Sichuan province in the west enjoys a mild climate which in most years permits abundant crops of rice, wheat, cotton, sugar, silk and tea – thus reducing vulnerability to the Little Ice Age. Nevertheless, amid the diversity, three broad economic zones are almost always more vulnerable to climate change: marginal farming lands; cities; and macroregions.

Marginal lands proved vulnerable because they produced enough to feed all their inhabitants only during years of optimal harvests; cities, by contrast, proved vulnerable because their prosperity made them strategic targets, which in turn led to the construction of a fortified perimeter that promoted overcrowding, poor hygiene and the spread of diseases inside the walls and, in wartime, exposed the inhabitants to the risk of extensive human and material damage. Finally regional economies consisting of several adjacent towns and their overlapping hinterlands, known as macroregions, proved vulnerable because their well-being depended on the ability to import the food on which their population depended, and to export the specialized goods that they produced. Disruption of either activity, whether at home or abroad, caused almost immediate hardship.

Although the inhabitants of these three economic zones formed a small minority of the global population, they featured disproportionately in the General Crisis. On the one hand, they suffered earlier, longer and more intensely than others, because government policies exacerbated to a unique degree the disruption created by climatic adversity and overpopulation; on the other hand, they harboured a large number of articulate men and women able to publicize their predicament at home and, whenever they could, abroad. Their voices resonated longer and louder than those of others whose experiences may have been more typical or more negative.

Agriculture on the Margin

For most of the sixteenth century, warmer weather permitted the expansion of farming and some of this process took place on lands already close to the limits of viable cultivation. Farmers who cultivated these lands initially reaped spectacular harvests, thanks to the nitrogen and phosphorus that had accumulated in the earth during the centuries when it lay fallow; but once this natural bounty expired, even in good years farming became a high-risk, high-input, low-yield operation that required constant attention to produce even a mediocre crop. In northerly latitudes, as noted in Chapter 1, each fall of 0.5°C in the mean summer temperature decreases the growing season by 10 per cent, doubles the risk of a single harvest failure, and increases the risk of a double failure sixfold. For those farming 300 metres or more above sea level, each fall of 0.5°C in mean summer temperature increases the chance of two consecutive failures a hundredfold.

This cruel calculus applies throughout the northern hemisphere. In Scotland, where the benign climate of the sixteenth century encouraged the cultivation of fields at much higher altitudes and on much poorer soils, the cold and wet summers of the 1640s, which drove down mean temperatures by up to 2°C, brought disaster. In the Lammermuir hills near the English border, three-quarters of the farms were abandoned; and on the Mull of Kintyre, in the west, four-fifths of all townships were abandoned because 'Farmers were not able to plant nor crofters to dig. The corn when it came up did not ripen ... People and cattle died, and Kintyre became almost a desert.'[3] In Sicily some seventy new towns were founded in the sixteenth and early seventeenth century specifically to produce grain for the fast-growing cities of the island. At first farmers harvested up to ten grains for each grain of wheat sown, and more than ten for each grain of barley sown, but the adverse weather of the 1640s drove some yield ratios down to 1:2 – a reduction of 80 per cent, and the lowest recorded in the entire early modern period. Leonforte, one of the new towns, grew from zero to over 2,000 inhabitants between 1610 and 1640, but the drought of 1648, which produced the poorest harvest ever recorded, brought catas- trophe. The town's parish register for that year recorded 426 burials but only 60 births.

Clearing fields for cultivation often involved a pernicious practice that could soon make fertile land marginal: clear-cutting forests. A historian living in Shaanxi province in northwest China recalled that 'flourishing woods' used to cover its hills so that rainfall flowed down in gentle streams, and villagers cut 'canals and ditches which irrigated several thousand [acres] of land'. But as prosperity in the region grew,

> People vied with each other in building houses, and wood was cut from the southern mountains without a year's rest. Presently people took advantage of the barren mountain surface and converted it into farms. Small bushes and seedlings in every square foot of ground were uprooted. The result was that if the heavens send down torrential rain, there is nothing to obstruct the flow of water. In the morning it falls on the southern mountains; in the evening, when it reaches the plains, its angry waves swell in volume and break through the embankments, frequently changing the course of the river.[4]

New and old farms alike ceased to be viable.

The Urban Graveyard Effect

The Little Ice Age forced many farmers on marginal lands to flee to the towns with their families in the hope of finding work, but most of them would be disappointed in part because their flight helped to fuel unsustain- able urban expansion.

The mid-seventeenth century was a 'metropolitan moment': never before had so many people lived in such close proximity. Beijing, the largest city in the world, had more than a million inhabitants, with almost as many in Nanjing. Six more Chinese cities numbered 500,000 or more residents, and a score had 100,000 or more. Mughal India, the most urbanized area in the world after China, included three cities with 400,000 or more inhabitants and nine more with over 100,000. By 1650, 2.5 million Japanese, perhaps 10 per cent of the total population, lived in towns. By contrast, in the Americas, only Mexico and Potosí (the silver-mining centre of Peru) exceeded 100,000 inhabitants, whereas Africa's only metropolis was Cairo, with perhaps 400,000 residents. In Europe, the population of Istanbul, the capital of the Ottoman empire, may have approached 800,000 inhabitants; London, Naples and Paris exceeded 300,000; and ten other European cities numbered 100,000 or more. In the Dutch Republic over 200,000 people lived in ten towns within a 50-mile radius of Amsterdam.

Every one of these metropolitan areas required prodigious quantities of housing, fuel, food and fresh water, as well as schemes to manage traffic, fight fires and keep public spaces clean. Failure to provide these essential services created an 'urban graveyard effect'. As the French demographer Jean Jacquart observed, early modern cities were 'a *mouroir*, a demographic black hole, accounting for disproportionately fewer marriages, fewer births and more deaths'. In London, burials in the seventeenth century were often twice as numerous as baptisms and both maternal and infant mortality was particularly high – nevertheless the city's population almost doubled in the course of the seventeenth century, thanks to massive immigration. As early as 1616, James I predicted with alarm that 'all the country is gotten into London, so as, with time, England will only be London, and the whole country be left waste with everyone living miserably in our houses, and dwelling all in the city'.[5]

According to one of King James's subjects, James Howell, Paris was no better: it 'is alwayes dirty, and 'tis such a dirt, that by perpetual motion is beaten into such black onctious oyl, that wher it sticks no art can wash it off'. In addition, 'besides the stain this dirt leaves, it gives also so a strong scent, that it may be smelt many miles off, if the wind be in one's face as he comes from the fresh air of the countrey. This may be one cause why the plague is alwayes in som corner or other of this vast citie, which may be call'd, as once Scythia was, *Vagina populorum*.' Another contemporary, Xie Zhaozhe, made much the same complaints about Beijing:

> The houses in the capital are so closely crowded together that there is no spare space, and in the markets there is much excrement and filth. People from all directions live together in disorderly confusion, and there are many flies and gnats. Whenever it becomes hot it is almost

intolerable. A little steady rain has only to fall and there is trouble from flooding. Therefore malarial fevers, diarrhoea and epidemics follow each other without stopping.[6]

Of course, city dwellers at virtually all times and in virtually all places have made similar complaints; but in the mid-seventeenth century the problems intensified. By the 1630s the City of London saw population and building densities that 'have probably not been witnessed in Britain either before or since'. Some parishes boasted almost 400 persons per acre (160 per hectare), sometimes with eleven people living in the same room in six-storey houses.[7]

Sir Tony Wrigley demonstrated that the rapid growth of London reflected its progressive emancipation 'from dependence on wood as a source of heat energy by the increasing consumption of coal'. By the late seventeenth century the capital imported almost 500,000 tons of coal a year, but the growing use of fossil fuels damaged the health of city dwellers. In 1656 an 'entertainment' staged in the capital complained that 'the plentiful exercise of your chimneys makes up that canopy of smoke which covers your city'.[8] The sea-coal used by brewers, dyers and other manufacturers contained twice as much sulphur as coal today, and its smoke darkened the air, dirtied clothes and curtains, stunted trees and flowers, blackened buildings and statues, and choked and killed the inhabitants. In an early condemnation of air pollution, published in 1661, John Evelyn claimed that the capital's inhabitants 'breathe nothing but an impure and thick mist accompanied by a fulginous and filthy vapour'. Ladies used ground almonds to clean their complexion, while preachers in churches competed with the constant coughing and spitting of their congregations.[9] The situation was even worse in those Dutch towns where industrial plants burnt peat for brewing, dyeing, soap factories and brick kilns, because (although far cheaper than coal) peat created toxic fumes as well as filthy vapour.

As Christopher Friedrichs has noted: 'Of all the elements, it was not earth, water or air that most persistently threatened the well-being of the early modern city. The most dangerous element was fire' – and the presence of industrial enterprises in the heart of cities increased the risk of fire. So did the use of wood and other flammable material to build cheap and shoddy high-density housing for the influx of immigrants, as well as the practice of cooking on open braziers, using oil lamps and candles, and setting off fireworks during celebrations. 'Oh, this word *fire*!' wrote an English merchant in Java. 'Had it been spoken near me, either in English, Malay, Javanese or Chinese, although I had been sound asleep, yet I should have leaped out of my bed.' He recalled that 'our men many times have sounded a drum at our chamber doors' while he and his colleagues slept, 'and we never heard them; yet presently after, they have but whispered to themselves of *fire* and we all have run out of our chambers'.[10]

Major fires became both more frequent and more destructive in the mid-seventeenth century. A gazetteer of accidental urban fires in England listed over one hundred between 1640 and 1689, at least ten of which consumed over a hundred buildings. In London, when Samuel Pepys emerged from church for the fifteenth consecutive week in 1661 to find suitors pleading for charity because their homes had burned down, he became irritated and 'resolve[d] to give no more to them'.[11] He changed his mind five years later, when fire destroyed St Paul's Cathedral, the Guildhall, the Royal Exchange, 84 churches and 13,000 houses, leaving 80,000 people homeless and causing £8 million of damage. Although Londoners blamed the lord mayor, who had failed to create fire breaks (instead jesting that 'a woman might piss it out'), the true culprit was the climate: after an unusually hot and dry spring, temperatures in the summer of 1666 rose 1.5°C above normal, and a precipitation shortfall of 150 millimetres turned London into a tinderbox.[12]

London was not the only capital city where unusual drought in the mid-seventeenth century facilitated major fires. In Moscow in 1648, after several months without rain, 'within a few hours more than half the city inside the White Wall, and about half the city outside the wall, went up in flames'. A large part of the new Mughal capital Shahjahanabad (now Delhi) burnt down in 1662. Istanbul suffered more (and more devastating) fires in the seventeenth century than in any other period of its history (one in 1660, after a prolonged drought, burned down 28,000 houses and several public buildings) and major blazes regularly devastated Edo, the largest city in Japan. The Meireki fire of 1657, which also broke out after an abnormal drought, destroyed 50,000 homes of merchants and artisans, almost 1,000 noble mansions, over 350 temples and shrines and even the shogun's magnificent new castle, the tallest building in Japan. Perhaps 160,000 people died. No sooner had rebuilding commenced than another major fire 'destroyed an area about 1½ miles in circumference', followed by a third in 1661 and a fourth in 1668 which 'devoured so many houses of nobles and civilians that it is estimated that two-thirds of the city of Edo has been destroyed'.[13]

While all these urban fires were apparently accidental, wars caused many others. During the 1640s, although thirteen English towns experienced accidental fires, soldiers deliberately caused at least eighty more, some of them large (almost 250 houses destroyed in Gloucester in 1643). War also destroyed urban spaces in other ways: constructing or extending fortifications, preparing for a siege, and fire from siege artillery all wrecked buildings. At Exeter, the third-largest city in England, between 1642 and 1646 the defenders deliberately razed all the suburbs, where one-third of the pre-war population had lived, while the enemy's siege guns left 'whole streets converted to ashes'. Although Exeter successfully resisted capture, it did not resume its pre-war size until the eighteenth century.[14]

The prevalence of war meant that every urban space needed walls, but walls rarely provided complete protection to the community within. Although Pavia in northern Italy successfully withstood an eight-week siege in 1655 thanks to its massive walls, the experience still left it ruined: lack of demand destroyed its industries; buying flour before and subsidizing bread prices during the siege bankrupted its treasury; and the besiegers' destruction of all municipal assets beyond the walls seriously impeded economic recovery. Nonetheless, Pavia was lucky: cities captured by force suffered far greater losses. The siege of Mantua in 1629 ended in a sack that reduced its population from 29,000 to 9,000; by 1647 it still had only 15,000 inhabitants and in 1676 only 20,000. The population of Warsaw, the capital of the Polish-Lithuanian Commonwealth, numbered perhaps 30,000 in the 1630s, but this fell below 6,000 after its capture and occupation by Transylvanian and Swedish forces in 1655–57. The victors also destroyed over half its buildings. In China, the Qing conquerors massacred the population and plundered the property of any city that attempted to resist (Chapters 5 and 21).

Floods also contributed to the urban graveyard effect. Because many cities grew up beside rivers and lakes, unusually high precipitation could cause immense flood damage. The worst inundation in the history of Mexico City occurred in 1629, when a combination of heavy rains and inadequate drainage caused the surrounding lakes to rise suddenly, submerging considerable parts of the city for five years. In Europe, the Seine burst its banks and flooded Paris eighteen times in the seventeenth century; and towns in the low-lying province of Holland suffered even more frequently when storms in the North Sea coincided with floods and overwhelmed the dikes.

The reliance of early modern cities on food imported from far away also contributed to the urban graveyard effect. A Chinese magistrate near Shanghai graphically described this danger:

> Our county does not produce rice, but relies for its food upon other areas. When the summer wheat is reaching ripeness and the autumn crops are already rising, the boats of the merchants that come loaded with rice form an unbroken line ... [But] if by chance there were to be an outbreak of hostilities ... such that the city gates did not open for ten days, and the hungry people raised their voices in clamour, how could there fail to be riot and disorder?

His fears turned into reality in 1641–42 when, even without an outbreak of hostilities, global cooling destroyed the rice harvest throughout South China. Perhaps 500,000 people starved to death and public order collapsed.[15]

Palace Cities

'Palace cities', with a large population of otherwise unproductive government officials to feed, were especially vulnerable because the need to
import a high proportion of their food forced them to seek supplies further
afield – and the longer the supply chain, the more susceptible it became to
disruption. In Beijing, every year, 450,000 tons of rice (as well as vast quantities of wheat, millet, beans and other foodstuffs) arrived on huge convoys of
barges travelling up the Grand Canal from the fertile rice paddies of the
Yangzi Valley almost 1,200 miles away. In 1641 drought in Shandong caused
the Grand Canal to dry up, and for some years thereafter fear of bandit
attacks interfered with routine maintenance (dredging, diking, and repairing
the locks), which disrupted the rhythm of the convoys. Since most of the
imported rice went to feed the 300,000 inhabitants of Beijing's Inner City
(the imperial family, bureaucrats, eunuchs, artisans, guards, merchants, and
their households), the failure of the last Ming emperor to feed his own
people no doubt convinced them in 1644 to open the gates of the capital to
a bandit army that promised to restore regular rice supply.

The provisioning of seventeenth-century Istanbul was strikingly similar.
The Ottoman capital imported most of its food – thousands of sheep and
lambs, over 500 cattle and 500 tons of bread daily – because the sultan (like
the Chinese emperor) needed to feed not only his own household but also
the legion of bureaucrats, eunuchs, artisans, guards, merchants, and their
households, as well as students in the colleges and religious schools
(*medreses*) attached to the imperial mosques. Like Beijing, because so much
of the food consumed in Istanbul came from far away – Egypt, the Balkans,
and the lands around the Aegean and Black Seas – natural and human
agency could and did disrupt its supply: in 1620–21, when the Bosporus
froze over; in 1641–43, when unusually weak Nile floods caused an epic
drought in Egypt; and from 1645 to 1658, when enemy fleets repeatedly
prevented ships from entering the Dardanelles. As in Beijing, the failure
of the state to feed its servants led palace personnel to turn on their
ruler, supporting revolts that culminated in regicide in both 1622 and
1648.

In Madrid, another palace city, a special granary was tasked with
supplying the royal court's daily requirement of 30 tons of wheat, and the
magistrates devised a system that obliged every house in each nearby village
to deliver a specified quantity of wheat on a specified day of the week. When
in 1630 this system proved inadequate because of a disastrous harvest, the
magistrates extended it to include over 500 communities within a radius of
60 miles of the capital. In 1647, when 'torrential and persistent rain made
traffic impossible on the roads to Madrid', and 'people cannot get into the
countryside to find dry firewood to heat the ovens, and very few mills still

have a wheel that works, because of the floods', the king was warned that 'if the bread supply fails for a single day, instead of the 100 people who are protesting today, Your Majesty will find the entire population in front of the palace'. To avert this disaster, the Madrid magistrates once again unilaterally extended the grain tribute system, rescinded all exemptions, dismissed all appeals from oppressed villages for relief, and sent agents 120 miles and more from the capital to requisition bread. Thanks to this rapid and radical response, Philip IV never did 'find the entire population in front of the palace'.[16]

Palace cities were not alone in creating a sophisticated supply network vulnerable to disruption. When the burgeoning populations of port-cities around Europe's Atlantic coast (including London, Amsterdam, Antwerp, Lisbon, and Seville) outgrew the capacity of their immediate hinterland to supply their daily bread, they began to import grain from the Polish-Lithuanian Commonwealth, which boasted fertile soil, cheap labour and easy access to water transport. By the early seventeenth century, between 150,000 and 200,000 tons of grain came down the Vistula annually for sale at Danzig, where an average of 1,500 ships loaded and shipped it to Western Europe; but this meant that when war broke out between Poland and Sweden in the 1620s, or when the Danish Sound froze over in 1658, the price of bread soared in the leading port-cities of Atlantic Europe, where the poorer members of its populations starved.

The Macroregions

No settlement in the early modern world was entirely self-sufficient: all of them needed to import at least some items. Even inhabitants of upland villages periodically isolated by winter snows or monsoon rains sometimes had to trudge to the nearest market town to trade handicrafts or surplus agricultural produce for such essentials as salt for food preservation and iron for tools. As population density increased in the sixteenth century, the number of market towns multiplied spectacularly. In China, the number of markets in Zhangzhou prefecture (Fujian) increased from eleven in 1491 to thirty-eight in 1573, and to sixty-five in 1628. In Japan, market towns on the coastal plains stood between 2 and 4 miles apart by the 1630s. By then, English men and women had to travel, on average, only 8 miles to reach the nearest market.

Markets achieved their greatest density around major cities, forming a zone of economic activity that contained the best arable land, the greatest population, the hubs of communication and transport, and the largest capital accumulation. Economists call them macroregions. Ming China contained eight macroregions, each one centred on a river system and separated from the others by natural barriers. The Indian subcontinent also included several, including Gujarat and the Ganges Valley; the Ottoman empire boasted Egypt, the lands around the Aegean, and the Black Sea region; Japan had the Kinai and Kantō plains. Mexico City lay at the centre of the largest macroregion in the Americas. Macroregions of Europe included the Genoa–Turin–Venice–Florence quadrilateral in Italy; the Home Counties in southeast England; the adjacent provinces of Holland, Zealand and Utrecht; and the Île-de-France. Many settlements within these macroregions adopted a high-risk, high-reward economic strategy: they concentrated on producing cash crops that they sold to merchants and manufacturers, and imported much of the food they needed from far afield.

When times were good, macroregions created three major economic opportunities for farmers – but at great risk. First, thanks to rising external demand and stable transport costs, many individual agriculturists made the transition from generalist to specialist, investing in the tools, raw materials and labour required to produce a small range of crops, and sometimes a single crop, for the market. In the duchy of Württemberg (southwest Germany), by 1622 many farmers and even entire communities had ceased to grow grain and instead produced the fine wines for which the area is still famous. This rendered almost everyone dependent on imported grain, and so when the harvest in most of Germany that year produced scarcely half the normal yield, much of Württemberg starved. In eastern China, many farmers also switched to producing cash crops for the market, including sugar, tea, fish, silk and cotton. Initially, these shifts involved little disruption. Rice requires intensive cultivation in March, May and July, whereas cotton requires most attention in April, June and between August and October; so the same labour force could successfully produce both crops. Likewise, fish ponds had long existed in the southern river deltas, with fruit trees planted on the surrounding embankments: the fish (mainly carp) grew fat on the organic matter that fell from the trees, and the muck scooped from the ponds fertilized the trees and the surrounding rice paddy. As growing demand inflated the price of silk, however, farmers began to replace their fruit trees with the mulberries on which silkworms feed, and they also converted rice paddies into fish ponds with mulberries on the embankment.

At first sight this formed a sustainable ecosystem, because almost all the necessary mineral and energy resources were recycled. It was not, however, a closed system: those who concentrated on producing fish and silk could no longer feed themselves and instead depended totally on rice produced miles (sometimes hundreds of miles) away. The same dilemma faced Chinese farmers who abandoned cereals to plant cotton, at first in dry fields or on the ridges between their paddy, and then in their entire holding. Although in some years smallholders could make a fortune from cotton, the crop required twice as much fertilizer as rice and was more vulnerable to floods, droughts and high winds. Since no one stored the annual crop, its market value immediately reflected variations in the climate and the price of fertilizers, as well as in the demand for cotton itself – all of them factors over which farmers had no control. Sooner or later, therefore, either the miracle crop would fail or the market would collapse. In both scenarios, producers would starve.

The second major economic opportunity for farmers created by the emergence of macroregions was land reclamation. In North Holland, the growth of Amsterdam and other adjacent cities encouraged entrepreneurs to reclaim 220,000 acres of lake, estuary and marsh between 1590 and 1640, creating 1,400 large new farms to produce food for the expanding urban population. In China, the sixteenth century saw the completion of over 1,000 new water-control projects, twice as many as the preceding century, and repairs brought many more abandoned projects back into service. In Japan, major land-reclamation projects between 1550 and 1640 more than doubled the area of rice paddy. Nevertheless, hydraulic projects were innately vulnerable. They required constant maintenance: any inundation, however caused, needed immediate remedial action, because the longer the water was left, the harder it was to drain.

The third economic opportunity associated with macroregions was the more intensive cultivation of traditional crops to feed the growing population. The most spectacular increases occurred in southeast China. 'In Guangdong there are fields which get three harvests', wrote a Chinese observer, who added: 'the reason is the warm climate'. 'They obtain three consecutive harvests in one year, two of rice and one of wheat', echoed a visiting European.[17] In both China and Japan, farmers experimented with different types of rice to increase yields – quicker ripening; resistant to cold; resistant to salt (for use near the sea) – until over 150 varieties were in use in Fujian alone, over two-thirds of them found in only one location.

Taken together, these improvements almost doubled rice yields in good years – but only in good years. Drought, cold and any other factor that prevented double-cropping impoverished farmers whose livelihood depended on selling the surplus crops. They also caused shortages or even starvation for their consumers, whether from lack of food or from inability to purchase it – often with long-lasting results. In India, famine and floods in 1627–31 ended the production of both indigo and cotton in Gujarat because, lacking both a market for their goods and food to eat, the weavers fled and never returned. Those who ceased to grow foodstuffs, and instead concentrated on other commercial crops produced for export, such as sugar, tea, indigo or the wide variety of artefacts made from bamboo (writing brushes, rainhats, umbrellas and so on), all shared the same vulnerability: when the supply of staples failed, they lost both their market and their capacity to feed their families. At Luzhou, a once prosperous town in Shanxi province (northwest China), until the disastrous harvest of 1640–42 over 3,000 looms wove imported raw silk, but thereafter:

All the weavers had to take out loans and their debts piled up and ruined them, so by 1644 there were only two or three hundred looms left. Although the weavers worked hard for themselves and for official exactions, toiling day and night with their wives and children, they had to pay all their expenses out of their own pockets and accumulated only debts – how could they go on? Now in 1660 they are thinking of burning their looms and repudiating their debts and, in great sorrow, running away.[18]

The collapse of weaving of course also ruined those who produced the raw silk: sericulture vanished from the neighbouring province of Shaanxi, despite a tradition that went back 2,000 years.

Malevolence and the Macroregions

Besides the catastrophes caused by climate change, the macroregions were also extremely vulnerable to human malevolence. To begin with, agricultural innovations and commercial crops normally required a substantial investment in fixed assets. Sugar production needed roller mills to crush the cane, pans to boil the juice, and trays to dry the crystals; silk manufacture required mulberry trees (each of which took six years to mature), vats for dyeing, and looms for weaving; making cotton fabrics required gins and looms. Such fixed assets could be looted and burned by enemies, just like traditional crops and simple farm equipment – but the damage cost far more and took far longer to repair. Moreover, as with hydraulic projects, what marauders could destroy once they could destroy again. Such action could create a cycle of disorders, because peasants who lost their land faced few alternatives to joining the aggressors, and the floods created excellent redoubts where those aggressors could thrive (see Chapter 5 for some Chinese examples).

A second way in which human agency could cause serious damage to macroregions arose because – although they formed the largest coherent economic units of the seventeenth-century world – they did not form a single market. Instead, in the felicitous image of economic historian Kishimoto Mio, they resembled

numerous shallow ponds connected one to another by channels. Because of their shallowness, the ponds were vulnerable to changes in external economic conditions. For example, too little inflow or too much outflow of money or commodities could easily flood or dry up these ponds and paralyze local economies ... Even if we could calculate the 'total' size of an economy by aggregating the water stock of these ponds, it would not have real meaning in economic history until these ponds are organically integrated into a single economy. More meaningful, perhaps, is to study the flow of money and goods in and out of the 'shallow' local markets from the viewpoint of local inhabitants, listening to their complaints about the destructive effects of external market conditions.[19]

In the seventeenth century such complaints multiplied whenever wars and rebellions closed down markets and trade routes. In 1621 two simultaneous wars – between the Dutch Republic and Spain, and between Sweden and Poland – involved blockades specifically intended to halt the export of Baltic grain: the former because Dutch ships carried most of it, the latter because its profits sustained the Polish war effort. Grain exports through the Danish Sound accordingly plunged from over 200,000 tons in 1618 to 60,000 in 1624 and 1625. Just as the blockaders intended, this fall both ruined Polish farmers and pushed food prices in the Dutch Republic to their highest level of the seventeenth century. Riots broke out in several towns and an alarmed Dutch politician wrote in his journal that 'the plague of God' lay on the land.[20]

A decade later, in East Asia, another blockade crippled those whose economic survival depended on selling Chinese silks in Japan. In the 1630s Shogun Tokugawa Iemitsu first ordered all Japanese residing abroad to come home and forbade all emigration, then prohibited the construction of large ships in the archipelago, and afterwards forbade all trade with the Portuguese. Iemitsu had prepared carefully for the economic impact of these measures upon Japan: he issued new 'frugality and sumptuary laws' designed to reduce the consumption of imported products such as silk, and he also encouraged Dutch, Korean and Chinese merchants to increase their silk imports in order to maintain a steady supply. But he miscalculated: although the Portuguese of Macao lost 'the most lucrative trade that His Majesty [Philip IV] has over here' (just as Iemitsu intended), the Dutch, who expected to gain, also lost because when they imported large quantities of silk, as requested by the shogun, they found that the new frugality and sumptuary laws had decimated demand.[21] For the same reason the Chinese, too, could not sell their cargoes in Japan: the price of raw silk in the Yangzi Valley therefore slumped and its producers starved. Native Japanese importers also suffered because they forfeited the capital previously sent to Macao to buy silks. Many went bankrupt, some fled and a few committed suicide in order to escape their creditors. Everyone involved in the Sino-Japanese silk trade thus experienced serious losses, some of them terminal, because of a political decision over which they had no control and against which they had no defence.

Those living in macroregions were also defenceless against other government initiatives, such as currency manipulation – and during the mid-seventeenth century, an unprecedented number of governments around the world tampered with the currency, both to *make* money when they re-minted existing coins at inflated values and to *save* money when they had to make payments (much as some governments today welcome currency devaluation because it reduces the real cost of their debts and increases the competitive edge of domestic products). The Spanish government took the lead, issuing cheap copper coinage, known as *vellón*, in 1618. Within eight years it had almost completely replaced silver in domestic transactions. Faced by galloping inflation, the government first halted further issues of *vellón* and then halved the face value of all copper coins in circulation. Four more times between 1636 and 1658 the Spanish mints called in existing coins and re-stamped them at a higher value, only to restore the earlier value a few months later in the face of public outcry. In the Ottoman empire, the weight and silver content of the standard silver coin, the *akçe*, dropped in a series of devaluations from 0.7 grams in the 1580s to 0.3 grams in 1640 and it all but disappeared as a medium of exchange. In China, mounting defence spending by the last Ming emperors led first to the issue of large quantities of copper fiat money and then, once copper supplies ran short, to coins adulterated with base metals. The exchange rate of silver to copper coins fell to 1:1,700 in 1638 and to 1:3,000 in 1643. At this point, in desperation, the imperial treasury started to issue paper currency, but since (understandably) no one believed the notes would ever be redeemed that expedient also failed. Worse followed: the Qing refused to accept Ming copper currency as legal tender, and so the exchange rate between silver and copper coins fell to 1:6,000 in 1647. As in Central Europe during the 1620s (page 31), copper coins eventually became worthless.

The importance of silver and gold currency to each geographical area, and to each social group, reflected its reliance on cash as opposed to barter; but trade, especially foreign trade, multiplied the effect of currency variations. To use another insight of Kishimoto Mio:

Sooner or later money creates added income for others through spending. The silver that flowed annually into one regional market in turn created demand in other regional markets through chain-like successions of exchange. For example, producers of raw silk would sell their silk to outside merchants and obtain silver, with which they could buy foods from farmers in the neighbourhood. Those who obtained silver from the sale of food would buy cotton cloth or other miscellaneous commodities with that silver, and so on. If the inflow of silver was to stop for some reason, the silk producers would have no money to buy food, and food producers would also have no money to buy cotton cloth. The decrease of income spreads through a chain reaction.[22]

Kishimoto's model explains why contemporaries paid special attention to the flow of silver and other commodities, because what mattered was not the ability to amass goods but the ability to sell them. In the 1650s, after three decades of war, disease and famine had drastically lowered demand in Jiangnan, a series of warm summers produced bumper rice harvests – but this spelled disaster for farmers. According to one of them, 'This year the price of rice was very low, at a level not seen for several decades. The humblest people in the poorest hamlets all ate fine rice and made cakes, while in my house we ate no midday meal on the last day of the year [traditionally a Chinese feast day].' Or more concisely, in a Chinese aphorism of the day, 'The rich become poor; the poor die.'[23]

'The Haves and the Have-nots'

In the second part of *Don Quixote*, published in Madrid in 1615, Miguel de Cervantes attributed to the phlegmatic squire Sancho Panza a now famous proverb: 'My grandmother used to say that there are only two families in the world: the haves and the have-nots.'[24] When Cervantes wrote, the village of Navalmoral de Toledo, in central Spain, had a population of around 250 families, of which fifty were have-nots who owned no property. Instead they lived in shacks, sometimes without a single piece of furniture – the inventory of their property made when they died recorded no chair, no table, no bed – and survived on what they earned from working for the 'haves'. In addition, twenty widows lived alone in the village without any apparent source of income; and seventeen individuals, described as paupers, lacked even a permanent dwelling place of their own, sleeping in barns or attics in winter and under hedgerows in summer (Fig. 5). In the early decades of the century, whenever a meagre harvest forced up the price of food and required fewer hands to bring it in, these have-nots went hungry, though they rarely starved because their richer neighbours provided alms, while the church used the tithes (a 10 per cent share of the harvest) to provide charity. If dearth continued, however, not only did the number of have-nots increase but the yield of the tithe simultaneously decreased. Thus in 1618, a year of good harvest (and therefore of good tithes), the church of a Spanish village near Navalmoral distributed 12,000 maravedís in alms, but in the 1630s, as harvest yields fell, the sum declined to 2,000 maravedís annually. In 1645, 1647 and 1649, the years that saw the worst harvests of the century, the parish priest wrote sadly in his account-book: 'No charity has been given, since there is nothing to give.'[25]

The have-nots fared little better elsewhere. Even in England, the only European state that boasted an obligatory welfare system (the Poor Law), a failed harvest would double, triple or even quadruple the amount required from the rich to save the poor from starvation so that, in the words of social

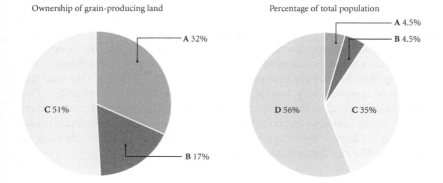

Ownership of grain-producing land

A 32%
C 51%
B 17%

Percentage of total population

A 4.5%
B 4.5%
D 56%
C 35%

Group **A:** 11 families, 4.5% of the population, owned 32% of the land.
Group **B:** 11 families, 4.5% of the population, owned 17% of the land.
Group **C:** 86 families, 35% of the population, owned 51% of the land.
Group **D:** 135 families, 56% of the population, owned no land.

Navalmoral contained 243 families (perhaps 1,100 people)
and 1,000 acres of grain-producing land.

5. The social structure of Navalmoral, Spain, in the early seventeenth century. Of 243 families living in a remote upland village south of Toledo, 11 families owned one-third of the village's land, 22 families owned half of it, and 108 families owned all of it. The rest of the inhabitants were 'have-nots', many of them homeless.

historian Steve Hindle, it is 'difficult to understand how the agricultural labourer and his family got through the year' in the mid-seventeenth century. Paid farm work was no longer 'a living in itself, but simply a vital cash supplement to a subsistence based on the cultivation of cottage gardens and the exploitation of common rights', occasionally augmented by poor relief; and this precarious situation 'rendered imperative the participation of all family members in the production endeavours of the household economy'. In 1657 Edward Barlow, the son of a Lancashire smallholder serving his apprenticeship in a bleaching enterprise, decided at age fifteen that

> I had as good to go seek my fortune abroad as live at home, always in want and working hard for very small gains. Likewise, I had never any great mind for country work, as ploughing and sowing and making of hay and reaping, nor also of winter work, as hedging and ditching and thrashing and dunging amongst cattle, and suchlike drudgery.

So he joined the navy, 'knowing that it could not be much worse whersoever I came'. Many of his contemporaries who stayed at home (in Hindle's lapidary phrase) 'literally worked themselves to death'.[26]

The situation of those living in war zones was even worse. In the Maas Valley of the South Netherlands, the magistrates of St Truiden cancelled their annual fair in 1630 on account of 'these times of war, of shortage of grain, of contagious disease, and of misery'. Four years later the priest in a neighbouring parish recorded in his diary: 'A violent plague struck the village during the months of June and July, taking seventeen victims. Immediately afterwards, war unexpectedly came to us' when three Spanish regiments (that is, troops sent to defend them) 'lodged here. They behaved worse than barbarously: they destroyed everything; they cut trees, completely demolished many houses, and trampled whatever grain they could not steal, not even leaving enough to appease the hunger of the poor farmers. For that reason, we did not collect the tithe this year.' Although none of these individual disasters was unprecedented, they had rarely if ever coincided: throughout the Maasland, in 1634 tithe receipts (which mirrored agricultural production and, as in Spain, provided the principal form of relief for the poor) sank to the lowest level ever recorded between 1620 and 1750.[27]

In China, agricultural production also fell to its lowest levels in the mid-seventeenth century and brought about the collapse of traditional forms of charity. Each county town maintained (at least in theory) a public granary, with smaller additional repositories elsewhere to feed the needy; but by the seventeenth century many lay empty, either through corruption or incompetence. In an attempt to avert disaster, groups of concerned citizens therefore created 'voluntary societies for sharing goodness'. Some distributed aid to impoverished widows, padded winter jackets to the poor and coffins in which to bury unclaimed corpses; others set up soup kitchens and lent money to small businesses in trouble; others still established orphanages, medical dispensaries and schools for the poor. Most voluntary societies, however, provided assistance only to a select few. Some conducted a background check before issuing approved supplicants with a ration card; others helped only those recommended by members or their own relatives. Private charity therefore only scratched the surface of poverty. In 1641, on the eve of the great famine, the founder of a benevolent society in Zhejiang province claimed that although it now helped three to four hundred people (compared with only a few dozen a decade before), he feared that 'the number of persons who are kept alive or given burials is still less than 10 to 20 per cent' of the total poor.[28]

Even this limited charity often ceased during the Ming–Qing transition. In Tancheng county (Shandong), the local elite informed a newly arrived magistrate in 1670 that the area 'has long been destitute and ravaged. For thirty years now fields have lain under flood water or weeds'. Famine, disease and bandits had depopulated the county in the 1630s; the Manchu army ravaged the county and sacked its capital in 1642; relentless rain caused the local rivers to flood, destroying the harvest four times between 1649 and

1659. The county's contribution of forced labour therefore fell from just over 40,000 able-bodied males in the 1630s to under 33,000 in 1643 and to under 10,000 in 1646. By 1670, the new magistrate learned, 'many people held their lives to be of no value, for the area was so wasted and barren, the common people so poor and had suffered so much, that essentially they knew none of the joys of being alive'.[29]

The people of Shandong were not alone in believing that they had faced horrors of unprecedented severity. Others, especially those living on marginal lands, in towns, or in macroregions, wrote similar laments. Enomoto Yazaemon, a Japanese salt merchant who lived near Edo, thought that 'the world was in flames from the time I was fifteen [1638] to the time I was eighteen'. A German cloth merchant lamented that 'there have been so many deaths that the like of it has never been heard in human history'; and a chronicler in Burgundy saw 'everywhere the face of death' when war, plague and harvest failure struck simultaneously. 'We lived from grass taken from the gardens and the fields', he wrote: 'Posterity would never believe it.' A German pastor expressed the same resignation: 'Our descendants will never believe what miseries we have suffered'; while one of his Catholic clerical colleagues asked rhetorically: 'Who could have described so many vile knaves, with all their evil tricks and wicked villainies? ... I would not have had time or opportunity, nor could I have laid hands on enough pens, ink or paper.'[30] The pessimism of these writers becomes explicable only when we look more closely at the scale and the causes of the demographic catastrophe that they had witnessed.

CHAPTER FOUR

Surviving in the Seventeenth Century[1]

The dramatic reduction in food supply in the mid-seventeenth century, whether through human or natural agency, forced many communities to take urgent and extreme measures to reduce their food consumption. The easiest and most effective way to do this was to reduce the number of mouths to feed; and although this process took different forms in different parts of the globe, almost everywhere the population fell steadily from the 1620s until a new equilibrium emerged between supply and demand for basic resources – often not until the 1680s.

The exact scale of demographic contraction is hard to document. In 1654 Abbess Angélique Arnauld of Port-Royal near Paris lamented that 'a third of the world has died', while a generation later, the Chinese emperor asserted that during the transition from the Ming to the Qing dynasties 'over half of China's population perished'. Many surviving statistical data support such claims. Parish registers from Île-de-France, where Port-Royal stood, show that 'almost one-quarter of the population vanished in a single year'. In China, tentative reconstruction of population levels in Tongcheng county in Jiangnan between 1631 and 1645 shows that some areas suffered almost 60 per cent losses.[2] The number of taxable households in western Poland fell by more than 50 per cent between the census of 1629 and that of 1661; further east, tax registers in what is now Belarus showed falls of between 40 and 95 per cent in urban populations between 1648 and 1667 (Chapter 6). In Germany, Pomerania and Mecklenburg in the north, like Hesse and the Palatinate in the centre, apparently lost two-thirds of their population between 1618 and 1648; Württemberg, in the southwest, boasted a population of 450,000 in 1618 but only 100,000 in 1639.

Not all the surviving demographic records reveal the causes of such decline. London's Bills of Mortality, which published weekly totals of burials in each parish of the capital together with the causes of death, included such esoteric afflictions as 'blasted', 'frighted' and 'headmouldshot' as well as different types of homicide ('murthered and shot'), recognizable diseases ('smallpox', 'French pox') and conditions ('childbed'), as well as suicide ('Hang'd and made away themselves'). Taken together, the Bills and other available data nevertheless reveal three distinct mechanisms that reduced the global population during the seventeenth century:

- More deaths through suicide, disease and war;
- Fewer births, either through postponing or preventing marriage, or through abortion, infanticide and child abandonment;
- More migration.

I. Death: 'Never Send to Know for Whom the Bell Tolls'[3]

Suicide

An unprecedented number of people appear to have reacted to adversity in the seventeenth century by killing themselves. In 1637 the preface to a comprehensive English treatise on suicide (326 pages plus index) asserted that 'scarce an age since the beginning of the world has afforded more examples' of people who 'made away with themselves'. 'There are many more self-murderers than the world takes notice of', it announced. 'Yea, the world is full of them.' Greater precision is difficult because, in the words of Machiel Bosman, 'Of all violent crimes, suicide is probably the most poorly documented in the judicial records of early modern Europe'.[4] Nevertheless the surviving evidence suggests that twice as many English men as women took their own lives; that almost one-third of them were under twenty years old; and that about one-fifth were over sixty. Two broad groups of motives stand out. Some felt overwhelmed by a direct threat to their psychological or physical survival, such as fear, bereavement or family strife: several mothers and a few fathers killed themselves after their children died; a nine-year-old boy tried to drown himself because he no longer wanted to live in poverty and misery; a young woman took her own life because she could not marry the man she loved; a twelve-year-old apprentice hanged himself when, after running away from a brutal master, his parents sent him back. A second group of unfortunates killed themselves because they had lost their social standing and could not live with the shame, such as women who became pregnant outside wedlock (especially as the result of incest or rape); and those who had suffered a public humiliation, such as a clerk who shot himself in prison after his arrest for debt. In seventeenth-century Scotland, where more than three times as many men as women 'made away with themselves', just over half were farmers and just over a quarter lived and worked in towns. Since surviving data from other periods show that the number of suicides often rose (or at least became more visible) during times of economic and political crisis, it seems highly likely that 'self-killing' became more common during the mid-seventeenth century.

Suicides in China certainly increased at this time, but not always for the same reasons. Although, as in Europe, motives included melancholy, economic misery and disappointment in love, as well as desperation during a period of turmoil, many sought to humiliate, embarrass or harm someone else. This could be achieved relatively easily because Chinese law insisted that a dead body must remain untouched until the local magistrate came to make a full inquiry – one that often uncovered the provocations, humiliating cruelties, cheating and insults that had forced someone to take their own life. To remove any possible doubt, many desperate people killed themselves at the exact site where the ill-treatment had taken place: the humiliated apprentice in his master's shop; the young childless bride outside the door of her abusive mother-in-law; the members of a starving family in the orchard of a local magistrate who had failed to 'nourish the people' (seen as the primary duty of government officials throughout imperial China).

Such revenge suicides may have increased during the turbulent transition from Ming to Qing rule, as the shortage of resources increased tensions and desperation within each family, but other factors added significantly to the total of women who killed themselves. First, Confucian teaching encouraged virtuous women to commit suicide in two circumstances: those who had been raped or otherwise dishonoured should kill themselves immediately, while virtuous widows should immediately 'follow her husband to the grave' (in the contemporary phrase). The collapse of public order in the mid-seventeenth century dramatically increased the number of women affected by these precepts.

When Qing forces took Yangzhou by storm in 1645 (Chapter 2), many women in the city jumped into wells or hanged themselves while others burned themselves to death in their own homes or slit their throats rather than fall into the hands of soldiers and be raped, enslaved or murdered. Before killing themselves, some young women composed brief autobiographies in prose followed by a few verses expressing anguish at their plight, creating a new genre of literature: *tibishi*, or 'poems inscribed on walls'. The suicide note left by Wei Qinniang, 'a girl from Chicheng' (Zhejiang province), seems typical. Just three months after she got married, soldiers captured her and carried her away from her husband (whose fate she never discovered). Somehow she escaped and 'disfigured my face and covered myself with dirt to obliterate my tracks. During the day I begged by the side of the road and at night I laid low in the blue grass. I swallowed my sobs and wept in secret, fearing that others might find me.' At last Wei found shelter in an abandoned temple where 'I look at my shadow and pity myself: my pretty face has been ruined by dust and wind, and my clothes have all been muddied.' Eventually, 'With no news from home, I recite a few quatrains and in tears write them on the wall. If some compassionate men of virtue would pass them on to my family, it would suffice to make my lonely parents understand.' She then killed herself. Her moving farewell, inscribed on the wall of her temple refuge, is all we know about her.[5] Times of acute disruption also increased suicides among males. Many scholar officials of the Ming killed themselves, especially in the great cities of Jiangnan, in accordance with Confucian teaching: 'the commanders of the troops have to die when they are defeated; the administrators of the state have also to die when the state is in peril'.

A few protested that such sacrifices were needless – Qian Xing, an official who had shifted allegiance from Ming to Qing, argued that 'if everyone dies at the time of national calamity, the whole country will be ruined' – but suicide remained common. When in 1670 a new magistrate took up his post in Tancheng county, Shandong province, he noted that 'every day one would hear that someone had hanged himself from a beam and killed himself. Others, at intervals, cut their throats or threw themselves into the river'. The practice remained common enough in 1688 to occasion an imperial edict that forbade widows to 'treat life lightly' and kill themselves. Instead they must serve their parents-in-law and raise their children.[6]

Suicides rose notably during the mid-seventeenth-century crisis in two other societies: Russia and India. In the former, the practice mainly involved males. From the 1630s onwards, a group of Orthodox Christians became convinced that the end of the world was imminent and immured themselves in hermitages and convents. Some of them, later known as Old Believers, concluded that the tsar was the Antichrist whose mistaken religious innovations had brought the Apocalypse ever nearer; and from the 1660s they defied him. When the government sent troops against them, they killed themselves rather than submit. According to one study, 'the total number of suicides ran into tens of thousands'.[7] In India, by contrast, suicide normally involved women. The Hindu belief that a virtuous woman possessed the power to preserve and prolong her husband's life brought her great status while married, but blame if her husband died. A respectable Hindu widow was expected to expiate her 'guilt' by committing suttee (from *sati*, 'a virtuous woman'), either by casting herself onto her husband's funeral pyre or by being buried alive beside it. Although no official total of suttee survives, a Dutch merchant residing in the Mughal empire in the 1620s claimed that 'in Agra this commonly occurs two or three times a week'.[8] It seems reasonable to suppose that pressure on widows to commit suicide grew in response to higher mortality among males caused by war as well as to increased pressure on resources caused by famine.

Sick to Death

Almost everywhere in the seventeenth-century world, the two most lethal diseases were smallpox and plague. The smallpox virus spreads rapidly and directly between humans by inhalation, and before the introduction of preventive measures it killed around one-third of all those infected, and up to one-half of infected infants and pregnant women. In addition, according to an experienced French midwife, 'almost all women with child that are attacked' by smallpox 'miscarry, and are in great danger of their lives'.[9] Smallpox survivors, after several weeks in agony, often emerged with disfiguring scars, deformed or stunted limbs, or impaired sight. Smallpox spared no one: among the ruling dynasties of Western Europe, it carried off one of Philip IV's brothers in 1641 and his son and heir in 1646; Prince William II of Orange in 1650; two siblings of Charles II of England in 1660 and his niece Mary II in 1694; and Louis XIV's heir in 1711.

Smallpox appears to have become both more deadly and more widespread in the seventeenth century. An English tract of 1665 about 'the tyranny of diseases' devoted a whole section to their apparent 'alteration from their old state and condition'; in particular, it claimed that smallpox was 'very gentle' until 'about forty years ago' and more lethal thereafter.[10] Three considerations support this contention. First, given its highly infectious character, once smallpox entered a community it spread fast; therefore the proliferation of areas of dense settlement, such as cities and macroregions, automatically increased deaths from epidemics. Second, the transportation of African slaves to both Western Europe and America introduced new and apparently more lethal strains of the disease from another continent. Third, communities rarely exposed to smallpox always suffer unusually high mortality at first contact, and as previously isolated areas entered the global economy in the seventeenth century, virtually all their inhabitants succumbed at the same time. This happened not only in America, where the native populations lacked immunity (Chapter 15), but also in East Asia where the Manchus, who had previously lived in small and relatively isolated communities on the steppe, suffered heavy losses when they invaded China and encountered smallpox for the first time. In 1622, shortly after their first incursion into Chinese territory, the Manchu leaders established a Smallpox Investigation Agency to identify and isolate suspected cases; and they later created 'shelters for keeping smallpox at bay', to which those not infected by the disease could escape. Nevertheless in 1649 smallpox killed one of the emperor's uncles as he directed the conquest of south China, and twelve years later it killed the emperor himself.

In the early modern world, epidemics of bubonic plague occurred less frequently than smallpox, but almost all those infected perished in agony. An epidemic could increase deaths in affected urban communities sixfold because, according to Dr Geronimo Gatta, whereas 'in the countryside it is possible to keep a suitable distance between the infected and the healthy', those living in towns had no escape. Gatta knew whereof he spoke: he had just survived a plague epidemic that in 1656 reduced the population of Naples, his native city, from almost 300,000 to perhaps 150,000 in a matter of months.[11] In the neighbouring town of Eboli, almost 1,000 families took Easter Communion in 1656, but scarcely one-fifth knelt at their local altar unscathed one year later; the plague had totally eliminated over eighty families with one or two members, twenty-seven families with three members, fifteen families with four, and fourteen more families with between five and seven. The same epidemic ravaged large parts of the Iberian Peninsula where some cities, including Seville, lost half their population. Some did not recover their pre-plague levels until the nineteenth century.

Many North European cities also suffered catastrophic losses from plague. In 1654 an epidemic struck Moscow with particular ferocity. Within the walls of the Kremlin, 26 people at the Chudov monastery survived but 182 died, while in three convents of nuns 107 survived but 272 perished. In 1663–64 another epidemic afflicted Amsterdam (where it killed 50,000 people), and other Dutch cities, before crossing to London, where it killed well over 100,000 people, with 'between one and two thousand bodies' thrown into graves and plague pits every night during 'the grimmest week for burials in London's long history': 12–19 September 1665.[12]

Plague and smallpox, together with typhus, measles and fever, belong to a cluster of deadly diseases that correlate closely with harvest yields: that is, the number of victims in each epidemic to some extent reflected the food supply. It is therefore not surprising to find that both the frequency and the intensity of these diseases increased amid the famines caused by the Little Ice Age. England's demographic records, which have survived better than those of other European countries, show eight years of high mortality between 1544 and 1666 caused by plague, of which half occurred after 1625. A survey of mortality crises in Italy between the sixteenth and the nineteenth centuries shows that more episodes occurred between 1620 and 1660 than at any other time.

Climatic adversity promoted sickness everywhere. In China, acute drought caused epidemics as well as famines: a study of county gazetteers in Jiangnan revealed over 100 locations affected by disease in the famine year 1641 (see Fig. 7, page 102). Meanwhile at Serres, Macedonia, an Orthodox priest left a particularly vivid account of the same catastrophic combination. The trouble began with the constant rains of summer 1641, which turned into snow during the grape harvest, so that many labourers perished of cold in the fields. The following winter was unusually mild, but snow fell in March and April 1642, and during this climatic anomaly the plague took hold with unsurpassed severity: it afflicted virtually every family in town and country alike, and 'of a hundred who fell ill, only one recovered'.[13]

War, as well as climate, could intensify the effects of epidemics. According to the English merchants in Gujarat, India, the Mughal emperor's campaigns in 1631 prevented 'the supplies of corne to these parts from those others of greatyr plenty; and the raynes hereabout having fallen superfluously', famine ensued. 'To afflict the more', the merchants added, 'not a family hath not been visited with agues, fevers and pestilentiall diseases'. In short, 'Never in the memory of man [has] the like famine and mortality happened.'[14] Northern Italy suffered a similar demographic catastrophe, which began in 1629 when rain destroyed the harvest; then large armies arrived from both Germany and France, where plague already raged, spreading the disease even as they consumed scarce resources and destroyed the fragile infrastructure. Of the 6 million people living in the north Italian plain, perhaps 2 million perished within a year. According to a recent study, 'no other area of Europe came near to the overall losses suffered by the peninsula.'[15]

Like smallpox, plague afflicted infants and pregnant women with espe-cial severity. In Barcelona Miquel Parets, a tanner, recorded in his diary that during the plague epidemic of 1651 'of the poor women who were pregnant at this time ... perhaps two among every hundred survived. If they were in the last days of their term all they could do was to commend themselves to God, as most of them simply gave birth and died, and many of their babies died with them.' Few orphans left at the foundlings' hospital survived, 'because so many were sent there' that not enough wet nurses could be found. As the plague abated, Catalonia suffered droughts for almost a year; and, in addition, the troops of Philip IV arrived to besiege the city (in revolt since 1640). By the time Barcelona surrendered in October 1652, half of those alive the previous year had died.[16]

Climate change also increased the lethality of other diseases. Looking back in the early eighteenth century, the Icelander Mathias Jochumssen sought to explain why the local population had declined over the previous hundred years: although he acknowledged the importance of smallpox and plague epidemics brought by foreigners, he blamed primarily the 'unusually poor standard of living, and rotten food' caused by the adverse climate, which promoted deficiency diseases such as scurvy and reduced fecundity. A correlation of the variations in the weekly death totals recorded in London's Bills of Mortality with the prevailing temperatures supports this interpretation. A fall of 1°C in winter temperatures coincided with a 2 per cent increase in mortality and with a 1 per cent fall in marital fertility. Medical research suggests some reasons: lower temperatures increase deaths through cardiovascular disease, and 'the elderly, the very young, [and] persons with impaired mobility' are 'disproportionately affected because of their limited physiological capacity to adapt'.[17] In addition, the smoke from coal fires (a growing urban problem in the seventeenth century: Chapter 3) carries fine particles that exacerbate cardiac or respiratory diseases such as asthma or bronchitis. Finally, mortality from other common diseases such as mumps, diphtheria and influenza, malaria – all of which kill many people every year – no doubt rose whenever extremes of weather weakened bodily defences and social disruption compromised hygiene.

The Killing Fields

Only wars can eliminate human populations faster than disease, and the increased frequency of hostilities in the mid-seventeenth century sharply increased mortality. Many men died in naval combat. During a battle between the Dutch and English fleets in June 1666, 'the most terrible, obstinate and bloodiest battle that ever was fought on the seas', the Royal Navy suffered over 4,000 men killed, wounded or captured – more than one-fifth of those engaged. Seafaring in the age of sail was a dangerous occupation at the best of times, and some officers 'reckoned there were four deaths by illness or accident for every one killed in action'. Life for sailors ashore could also prove lethal, because neither the government nor the local population could provide adequate support if a fleet stayed in port for long. Just before the 1667 fighting season, the commissioner at an English naval base was 'Sorry to see men really perish for want of wherewithal to get nourishment. One [sailor] yesterday came to me crying to get something to relieve him. I ordered him 10 shillings. He went and got hot drink and something to help him, and so drank it, and died within two hours.'[18]

Battles between armies could also rapidly kill and maim thousands. At Rocroi (France) in 1643 French artillery killed at least 6,000 Spanish veterans in an evening. The following year at Marston Moor (England) the victorious Parliamentarians and their Scottish allies cut down some 4,000 defeated royalists in a little over an hour. Also in 1644, 10,000 men perished at the battle of Freiburg (Germany), and according to the victorious commander 'In the twenty-two years that I have been involved in the carnage of war, there has never been such a bloody encounter.' It is no coincidence that the German term for battle, *Schlacht*, shares its root with 'slaughter'.[19]

The number of casualties rose when circumstances created an asymmetry that pitted the pack against the prey (a common metaphor at the time). Lucy Hutchinson, an eyewitness of the siege of Nottingham Castle in 1644, put it best: 'No one can believe, but those that saw the day, what a strange ebb and flow of courage and cowardice there was in both parties' because, after the catharsis of passing through a killing ground, 'the brave turn cowards, fear unnerves the most mighty, makes the generous base, and great men do those things they blush to think on.'[20] Two factors increased the likelihood of atrocities in mid-seventeenth-century warfare. First, prevailing military conventions (sometimes called the Laws of War) held that rebels who took up arms to oppose their ruler 'ought not to be classed as enemies, the two being quite distinct, and so it is more correct to term the armed contention with rebel subjects execution of legal process, or prosecution, not war'. It followed that in 'a war waged by a prince with rebels . . . all measures allowed in war are available against them, such as killing them as enemies, enslaving them as prisoners, and, much more, confiscating their property as booty'.[21] The worst military massacres often occur during civil wars or during the repression of those regarded by the victors as rebels; and the mid-seventeenth century saw a spate of civil wars and rebellions.

The involvement of protagonists from different religions also raised the risk of atrocities in war. As Blaise Pascal observed in his *Pensées*: 'Men never do so much harm so happily as when they do it through religious conviction' – and the multiple conflicts of the mid-seventeenth century offer abundant examples. The advance of a Cossack army through Ukraine in the Polish-Lithuanian Commonwealth in 1648 was accompanied by the massacre of at least 10,000 Jewish settlers; and six years later, as the Russian tsar prepared to invade the Commonwealth, he gave orders 'to burn alive Poles or Belorussians subsequently captured who would not convert to Orthodoxy' (Chapter 6). During the Thirty Years' War the mere existence of Protestant pastors seems to have provoked Catholic troops to humiliate, torture and kill them, on the grounds that they had 'caused the war'.[22] In 1631 the Protestant city of Magdeburg in Germany refused to admit a Catholic garrison sent by its suzerain, Emperor Ferdinand II, and derisively rejected his demand for surrender. When, after a long blockade, the besiegers launched a successful assault perhaps 20,000 men, women and children perished in the ensuing sack, while fire consumed all but the cathedral and 140 houses.

Few European military leaders in the mid-seventeenth century seem to have felt remorse about massacres carried out in God's name: on the contrary, they often claimed scriptural warrant, urging their soldiers to follow the Old Testament example of Joshua and the Israelites at Jericho and 'utterly destroy all that was in the city, both man and woman, young and old'. In 1645, during the English Civil War, at the fall of Basing House, the heavily fortified mansion of a Catholic peer, Protestant preachers whipped up a religious fervour among the besiegers so that after a successful assault they murdered in cold blood six priests, a former Drury Lane actor (with the sanctimonious comment 'Cursed be he that doth the Lord's Work negligently') and a young woman who tried to stop a soldier from abusing her father – but the newspaper account that conveyed these details reassured readers that the victims 'were most of them Papists' and therefore deserved their fate.[23] Three years later, in Ireland, after English troops carried out a major massacre after taking the city of Drogheda by storm, Oliver Cromwell (their commander) explained that 'in the heat of the action' he had forbidden his troops 'to spare any that were in arms in the town' because 'I am persuaded that this is a righteous judgment of God upon these barbarous [viz. Catholic] wretches, who have imbrued their hands in so much innocent [viz. Protestant] blood'. At least 2,500 soldiers and at least 1,000 civilians (including all Catholic priests) perished.[24]

Nevertheless, death in face-to-face violence rarely formed the largest category of mortality in early modern wars. For example, between 1620 and

1719 some 500,000 Swedish and Finnish troops died in the almost continuous wars waged by their monarchs but only 10 per cent fell in battle (including King Gustavus Adolphus) and only 5 per cent in sieges (including King Charles XII). Of the rest, 10 per cent perished in prison and the remaining 75 per cent succumbed to the normal hardships of war. The stunning scale and impact of premature deaths among the troops emerges from the demographic records of the parish of Bygdeå in northern Sweden. Between 1621 and 1639, Bygdeå provided 230 young men to fight on the continent, where 215 of them died and five more received injuries that left them crippled: only ten remained in service in 1639 – and, since the war still had nine more years to run, the odds of their survival were slim. Military service had become, in effect, a sentence of death for males.

II. Only Women Bleed

The 'Bitter Living' of Women

Wars also had serious demographic consequences for females, especially when they became a primary target. The (Protestant) government in Dublin in 1642 instructed troops fighting Catholic insurgents not to spare any women they encountered, 'being manifestly very deep in the guilt of this rebellion, and, as we are informed, very forward to stir up their husbands, friends and kindred to side therein, and exciting them to cruelty against the English'.[25] Elsewhere, the departure of so many men to fight and perhaps die in the wars feminized many communities. In Württemberg, the number of households headed by single women rose to almost one-third of the total – an unprecedented number; in some villages in Burgundy (France) widows were the *only* surviving householders; in Bygdeå (Sweden), twenty years of compulsory military service had by 1639 more than raised the ratio of men to women from par to 1:3.6, while the number of households headed by women increased sevenfold. Many were war widows: the Swedish central government received thousands of petitions desperately requesting payment of wage arrears, or reimbursement of expenses, due to their deceased soldier-husbands, because the petitioner and her children were starving. A German widow in 1654 spoke eloquently for all these indirect victims of war: '[I am] a poor woman with only a small field or so to my name' who must therefore 'earn a bitter living'. If she (or any other woman living alone) fell afoul of the law, the males who ran the village would sentence her to hard labour; if she failed to behave in a subservient and docile way, they would banish her; and, even if they allowed her to stay, they normally denied her the opportunity to learn or practise a trade (and thus compete with them).[26] A bitter living, indeed.

The lives of urban women in the seventeenth century were seldom sweeter. In many European cities two-thirds of those receiving alms were females, most of them former servants. In a world without domestic appliances or prepared food, Olwen Hufton has noted, 'the first luxury that any family permitted itself was the services of a girl, a maid of all work, to take on the drudgery involved in carrying water – a major time and energy consumer – and coal or wood, going to market or performing laundry service'.[27] But as soon as a recession occurred, that first luxury became the first casualty: employers in hard times had no scruples about getting rid of their maids; and even those who retained their jobs might have to forego wages, and work in return for their keep. The fate of women who worked in industry was no better: unlike male apprentices, who enjoyed some job protection, if their employer fell on hard times, or if they sickened or suffered injury, women would be thrown out of work without any safety net to sustain them.

The bitter living of so many women critically affected population levels because, in the words of Sir Tony Wrigley, 'marriage was the hinge on which the demographic system turned' throughout early modern Europe.[28] Any economic downturn that made it harder for women to marry reduced population growth, both because fewer women married (when bread prices doubled, marriages normally fell by one-fifth) and because the age of brides at first marriage increased. In the sixteenth century, many European women had married at age twenty and gave birth to eight or more children, but in the first half of the seventeenth century the average age of brides rose to twenty-seven or twenty-eight. Few of them gave birth to more than three children. Equally importantly, more women remained celibate: of Englishwomen born around 1566, who might have married in the 1590s, only one in twenty failed to do so; but of those born around 1586, who might have married in the 1610s, almost one in five did not; and of those born around 1606, who might have married in the 1630s, over one in four remained unwed.

These developments reduced population size in the seventeenth century more rapidly than they would do today because of remarkably high mortality among both mothers and children. As the terse French proverb put it, 'A pregnant woman has one foot in the grave', and surviving records suggest that in Europe perhaps 40 women per 1,000 died as a result of childbirth (compared with 0.12 per 1,000 in much of Western Europe today). The death of the mother became even more frequent if the child died in the womb (and since this happens more often at times of scarcity and disease, it would have happened more often during the 1640s). Moreover, as elsewhere in the early modern world, at least one-quarter of all children born died within their first year, and almost one-half died before reaching the age of reproduction. Mothers in the early modern world therefore needed to give birth to *at least* four children simply to maintain a given population level – at least four, because not all of those who survived to adulthood would marry and be fertile, and because periods of adversity trigger important biological and behavioural responses that dramatically reduce fertility:

more spontaneous abortions and fewer conceptions (sometimes because ovulation has ceased); decreased libido; a rise in the age of menarche; a fall in the age of menopause. At the same time, the many diseases associated with famine afflict pregnant women and new mothers more severely, and children conceived in a time of scarce resources tend to be of shorter stature with a reduced life expectancy (Chapter 1). For all these reasons, the birth rate in many if not most communities in the mid-seventeenth century fell below – often far below – the level required to maintain the overall population.

The rising age of brides when they first married, together with the increase in female celibacy, had one further deleterious effect: a rise in illegitimate pregnancies. Mid-seventeenth-century parish registers in Europe reveal that between one-fifth and three-fifths of all births took place only a few months (and sometimes only a few weeks) after the parents' marriage, and that pregnancies among unmarried women also increased. Urban court records throughout Europe are full of testimony from or about young women who came to town looking for work and, when they could not find it, were tricked or coerced into prostitution. One such newcomer to London was reassured by her landlady 'that it is better to do so [become a prostitute] than to steal and be hanged' – a survival strategy unlikely to offer a permanent escape from poverty.[29] Instead, it increased the risk of sexually transmitted diseases and also created a cruel dilemma for women if they became pregnant: they could either abort, abandon or kill their babies at birth, or face the shame and hardship of having to survive by begging (in which case they as well as their babies would probably die).

In most of East Asia, the birth rate also fell in the mid-seventeenth century, but although this decline also reflected a reduction in the reproductive capacity of women, it involved different social pressures. In China, several generations of a family might live together in the same households: one or more parents and one or more of their married sons and their families, forming a single economic, religious, social and demographic unit which could easily number fifty members from three generations. They produced and consumed in common; they performed collectively the appropriate rites for the well-being of family members both alive and dead; they shared the burden of caring for aged and needy relatives; above all, they discussed and determined family size together, because the purpose of marriage was production as well as reproduction. Demographic decisions in many parts of China were seldom individual: they had to be 'negotiated with co-resident kin according to collective goals and constraints.'[30]

Negotiation was important because nearly all Chinese females who reached puberty in the seventeenth century married, often when they were still teenagers, and if a wife failed to produce a son, their husbands might take one or more concubines: a peasant girl purchased for the purpose, a household servant, an entertainer, a prostitute – almost anyone except a woman from their own neighbourhood or 'name'. An active trade in concubines existed in the city of Yangzhou before the sack of 1645, where (according to one contemporary source) 'daughters are as numerous as clouds' and

> At age thirteen or fourteen,
> They are ready.
> Who cares if he is old,
> If he has gold?[31]

To ensure that early and widespread marriage, and the frequent resort to concubines, did not produce more children than each household could support, Chinese families employed one or more of four strategies. First, spouses started to reproduce late: the gap between marriage and becoming a mother for the first time averaged three years in China, compared with eighteen months in Western Europe. Second, wives stopped reproducing early: the mean age of Chinese women when they gave birth to their last child was thirty-three, compared with forty for their West European counterparts, and the average span between the first and last child was eleven years, against fourteen in Europe. Third, Chinese mothers got rid of the children they (or the co-resident kin) did not want. Many families either abandoned for adoption children they could not feed (especially boys, to families who lacked sons of their own) or sold them (especially girls, many of whom became concubines or prostitutes). Fourth, some pregnancies were either avoided or terminated. Early modern Chinese medical literature described many methods of contraception and abortion; Chinese physicians always placed the health of the mother before that of the foetus; and imperial law criminalized neither procuring nor performing an abortion. Parents could and did murder their unwanted offspring – especially girls. The surviving evidence indicates that in the mid-seventeenth-century Chinese families intensified their use of all four strategies for child limitation: starting to reproduce later; stopping reproduction earlier and increasing birth intervals; abandoning more children; and killing off more children at birth.

Infanticide and Abortion

Japanese families also often resorted to infanticide. 'The most horriblest thing of all', wrote Richard Cocks, an English merchant living on the island of Kyushu in 1614, 'is that parents may kill their own children so soon as they are born, if they have not the wherewithal to nourish them.' A Japanese physician living in the same area six decades later confirmed this observation: if a couple had 'five children, they kill two; if they give birth to ten, they kill four or five'.[32] In Japan, as in China, girls usually formed the majority of those killed because, in the words of the eminent sinologist Francesca Bray, 'Infanticide is the most effective way of controlling family size in response to sudden crisis. It is also a fool-proof way of exercising sex selection if all other means fail'. Even the ruling Qing dynasty, for whom detailed data have survived, appears to have killed 10 per cent of all their daughters, many of them at birth, but in the 1640s and 1650s poorer members of the dynasty killed almost twice as many of their daughters as did their richer relatives. Although no other family compiled figures of similar precision, it seems that in Manchuria, the Qing homeland, peasant families killed between one-fifth and one-quarter of their daughters at birth in 'normal' years, and more in years of dearth. In Liaoning, just to the south, the sex ratio among the *last* children born to peasant families stood at 500 boys for every 100 girls.[33] No biological circumstance can account for such gender imbalance: widespread female infanticide is the only possible explanation. At times of economic adversity, poor families in China may have killed as many as half their children at birth.

Infanticide was also widespread in Russia, where early modern households often consisted of three or four families (usually related) living together, and many women married at the age of twelve or thirteen. Children born out of wedlock and children born to the very poor seem to have been killed by their parents far more often than in the West – at least in part because both church and state took a relatively lenient view of the matter. The medieval Russian version of St Basil's Rule forgave women who killed their children 'from simplicity or ignorance, or because of scarcity of necessities'; and although the comprehensive law code issued in 1649 decreed death for a mother who killed a child conceived in adultery, 'If a father or mother kills [their legitimate] son or daughter', the penalty was that 'they shall be imprisoned for a year', and after that 'they shall not be punished'.[34] The law said nothing about foundlings and orphans (a draft decree of 1683 establishing state orphanages remained a dead letter until 1712) – perhaps because Russia's harsh climate reduced their numbers anyway. The available records show that maximum sexual activity among the rural population took place after the harvest, which produced a peak in births just before the next growing season. This in turn produced a peak in infant mortality because the short growing season made it essential for Russian women to work in the fields at seedtime and harvest, which meant that those who were still infants at that point might either receive solid foods too early or else get unsanitary pacifiers that killed them. As the Little Ice Age reduced the length of the growing season still further, the few surviving demographic records show an increase in infant mortality.

Abortion and infanticide were also common in Western Europe. In 1660, according to the dean of the Faculty of Medicine in Paris, the local clergy 'have made a calculation that in the past year, six hundred women have confessed that they had killed and destroyed the fruit of their womb'; and throughout the seventeenth century, infanticide constituted the commonest capital crime tried by the Parlement of Paris, the law court with the largest jurisdiction in early modern Europe. Its cases show that virtually all those accused were women, half of them unmarried and another quarter widows; and that their numbers rose in times of economic hardship. Evidence from rural Tuscany suggests that, as in East Asia, infanticide affected females far more than males. In 1617, 1629, 1635, 1647 and 1658 – all years of dearth – 'the ratio of males to females at baptism was over double'; and during the great famine of 1649–50, although eleven married couples produced nineteen baptized offspring, eighteen of them were boys. Once again, killing female babies at birth is the only possible explanation for this imbalance.[35]

The increasing frequency of abortion and infanticide in seventeenth-century Europe led many states to pass harsh legislation against offenders. An English statute against infanticide in 1624 explicitly targeted unmarried mothers, as even its title made clear: 'Act to Prevent the Murdering of Bastard Children'. Studies of prosecutions brought under the new law show that all victims were illegitimate (mostly born to young domestic servants) and that almost all died on their first day of life (most through strangulation, suffocation, exposure or drowning, though a few had been beaten to death or thrown into a fire). In Germany, the government of Württemberg issued a law in 1658 that enjoined the denunciation of all women suspected of killing their offspring. In almost 130 cases that came to trial over the next forty years, the average age of the mothers (almost all single; almost all sentenced to death) was twenty-five.[36]

Nevertheless, many infanticides in early modern Europe evidently went undetected – despite the diligence of neighbours, many of whom devoted considerable time to watching whether or not single women produced bloodstained laundry once a month. Perhaps the most striking evidence comes from the French city of Rennes, where workmen replacing an old drain in 1721 found about eighty skeletons of infants – and yet no one had been charged with their deaths. The total of murdered infants, in Rennes and elsewhere in Europe, would surely have been higher had not two institutions offered legal opportunities for women of child-bearing age to avoid raising children in times of crisis: nunneries and foundling hospitals.

'Get Thee to a Nunnery'[37]

In Italy, by 1650, perhaps 70,000 females lived in nunneries, most of them in towns. Nunneries housed 8 per cent of the total female population in Bologna, 9 per cent in Ferrara, 11 per cent in Florence and 12 per cent in

Siena. By then there were a similar number of nuns in France, and Spain had 20,000, with many more in Germany, the South Netherlands and Poland. Cities in Russia and the Balkans also included a significant population of females in Orthodox convents, as did those in Catholic Europe's colonies. In every region, the number of nuns increased markedly in the course of the seventeenth century: why? Many women took the veil through a religious vocation; others did so because some disability placed them at a disadvantage in the outside world; a few, no doubt, felt attracted by the lavish lifestyle of certain convents, where servants and slaves made up half the population (and in some nunneries even outnumbered the nuns). A considerable number of young women in the seventeenth century entered convents against their will, however. Elena Cassandra Tarrabotti (1604–52), daughter of a Venetian patriarch, was one of them.

Sent to a nunnery at the age of thirteen by her father, who claimed that he lacked the money for an appropriate dowry, Elena wrote books with titles like *Innocence undone or the father's tyranny* and *The nun's hell* lamenting her lot.[38] Although these works went straight onto the Index of Prohibited Books, the number of daughters from patrician families who took the veil increased as the seventeenth-century crisis advanced. Although patricians justified their 'sacrifice' of a daughter as an act of exemplary piety, they doubtless consoled themselves with the fact that marriage dowries averaged 1,500 ducats, whereas it cost only 400 to place a daughter in even a prestigious convent. Regardless of motive, incarcerating daughters in a cloister proved highly effective in reducing the number of elite women available to marry and reproduce: in 1642, 80 per cent of the daughters of patricians became nuns, and in 1656, 90 per cent.

Many convents also accepted foundlings, and some institutions installed a special wheel to make it easier for mothers to abandon their unwanted offspring: they could place their babies on the wheel outside and then turn for collection within. At the foundling hospital of Milan, where desperate mothers deposited up to 400 babies a year, use of the wheel (*scaffetta*) followed fluctuations in grain prices with sickening regularity: in a year of food scarcity four or five infants might be left on the wheel in a single night. Likewise, the foundling hospital in Madrid took in about 500 children a year and that of Seville almost 300 admissions a year, with notable increases during periods of economic adversity. In cities that lacked a foundling hospital, such as London, up to 1,000 infants a year were 'found on the streets, stalls and dunghills of the capital': their number, too, rose and fell in step with the price of bread.[39]

Parents who abandoned their babies on dunghills probably did not intend them to live, so their action might be seen as infanticide; others, however, left their child in a place where it was sure to be found (such as a *scaffetta*), often with a heart-wrenching note of explanation. By chance, about 150 such notes have survived from the Madrid foundling hospital in the 1620s. All were small – some of them very small, written on a scrap of paper torn from a book or on the back of a document, and in one case on the back of a playing card – and each stated whether the child had been baptized or not, usually together with its name, date of birth and saint's day. A few provided a brief explanation (parents too poor; mother abandoned by her lover; mother died in childbirth; mother unable to produce milk; in one case 'a well-known lady who does not wish to be found out').[40]

Most of those who entered a foundling hospital, whether in Madrid or elsewhere, died there – for the simple reason that almost half of all foundlings were abandoned within their first week, and at that vulnerable age, without the immediate intervention of a wet nurse, they would die. Life expectations therefore fell as admissions rose. In the Hospital of the Innocents in Florence, one-third of the 700 foundlings abandoned in the famine year of 1629 died on their first day; half died within a week; and almost two-thirds died within a month. Leaving a new-born baby on the *scaffetta* may not have been technically infanticide, but the effect was much the same: it provided an additional method, brutal but effective, of rapidly reducing the number of mouths to feed in a crisis.

III. Migration

Millions of people migrated in the course of the seventeenth century in search of safety or better economic opportunities. Many West Europeans travelled to other parts of the continent. Tens of thousands of Russian and Polish families – the majority of them unfree peasants – fled to join the Cossacks living on the fertile 'black earth' lands in the south. Many Russians also went to Siberia (Chapter 6). Between 1600 and 1650, at least 100,000 Scots migrated to the Polish-Lithuanian Commonwealth, and perhaps 40,000 more settled in Ireland. Few of them returned. Many more Europeans went to the Americas. Some, like the Pilgrim Fathers and others who settled in New England, migrated to avoid political and religious policies imposed by governments; others were deported by those same governments, anxious to expel potential troublemakers. As Robert Gray, a London preacher, wrote in a pamphlet entitled *A good speed to Virginia*:

There is nothing more daungerous for the estate of common-wealths then when the people do increase to a greater multitude and number than may iustly parallel with the largenesse of the place and countrey; for hereupon comes oppression, and diuerse kinde of wrongs, mutinies, seditions, commotion, & rebellion, scarcitie, dearth, pouertie, and sundrie sorts of calamities which either breed the conuersion, or euersion, of cities and common-wealths.

Gray noted with approval that 'many Nations' – no doubt he had in mind France, Spain and Portugal – 'perceyuing their people to increase aboue a due and proportionable number, they haue sent their ouerflowing multitudes abroad into other countreyes and prouinces, to the ends they might preserue their owne in greater peace and prosperitie'. Gray urged his compatriots to follow suit, and the Virginia Company duly removed by force the 'super-increasing people from the city' of London to America in order to 'ease the city of many that are ready to starve, and do starve daily in our streets ... for want of food to put into their mouths'.[41]

Chinese families, too, migrated overseas. Thousands, especially from the mountainous southeastern provinces, left for the Philippines and Southeast Asia, either as settlers or to 'service' the European colonies there; and thousands more lived in Batavia (now Jakarta), turning it into a Chinese colonial town under Dutch protection. The establishment of European colonies on Taiwan after 1624 created another opportunity for Chinese co-colonization. Almost immediately, the governor of Fujian province allowed 'several tens of thousands' of those destitute through famine to migrate to the lands around the main Dutch settlement, providing every family with some cash and a cow. By 1683, when Qing forces incorporated the island into the Chinese state for the first time, perhaps 120,000 ethnic Chinese lived on Taiwan alongside the indigenous population.[42]

The seventeenth-century crisis also created a host of fugitives trying to escape death or destitution. Chen Zilong, a scholar-official of the Ming, captured the agonizing fate of such families in his poem 'The little cart':

The little cart jolting and banging through the yellow haze of dusk;
The man pushing behind: the woman pulling in front.
They have left the city and do not know where to go.
'Green, green, those elm-tree leaves: they will cure my hunger,
If only we could find some quiet place and sup upon them together.'
The wind has flattened the yellow mother-wort;
Above it, in the distance, they see the walls of a house.
'There, surely, must be people living who'll give you something to eat.'
They tap at the door, but no one comes; they look in, but the kitchen is empty.
They stand hesitating in the lonely road and their tears fall like rain.[43]

Similar heart-wrenching experiences no doubt took place in the Ottoman empire, which experienced what contemporaries called the 'Great Flight'. In the province of Amasya, Anatolia, a detailed census in 1641–42 recorded a fall of one-third of rural settlements, and four-fifths of the rural taxpayers, since a similar survey undertaken seventy years before. A detailed comparison of the two censuses by Oktay Özel noted that 'bachelors constituted almost half of the adult male population, along with a considerable percentage of landless peasant families' in the earlier survey. These landless unmarried men later migrated to find sustenance: some fled to the cities, others went to study in *medreses*, but the majority joined itinerant bandit armies – and were thereby lost not only to their villages but also to the empire, since most left no offspring, producing a 'diminishing rate of reproduction over generations'.[44]

Thanks to the prevalence of war in the seventeenth century, more men than ever left home to join an army. In India, perhaps one-tenth of the active male population of Hindustan formed part of a sophisticated military labour market because the Mughal emperors, like their Afghan predecessors and their British successors, raised Rajput troops on the northwest frontier and sent them to fight on the eastern and southern borders of their empire. Many of them married and settled there. A similar system of military expatriation characterized the Spanish monarchy. Every year, troops raised in the towns and villages of Castile left to serve overseas, above all in Spanish Italy and the Spanish Netherlands (where, in 1640, Philip IV's troops included over 17,000 Spaniards: a larger concentration than all but a few Spanish towns could boast). Many of these expatriates, like the Rajputs of Mughal India, married and settled abroad. As a Spanish patriarch told his younger brother when he left to fight the Dutch in 1627, 'I don't want you to enjoy the countryside in the Netherlands, but the war. The war must become your home.'[45]

Not all migrants stayed away for ever. Perhaps 6,000 people came to London every year, most to become apprentices to a merchant or artisan in return for instruction in his craft, or domestic servants until they had accumulated enough savings to become financially independent and marry, after which many returned home. Others migrated for a season. Cultivating most staples requires many extra hands at certain predictable, intense, but short periods. In the case of rice, farmers needed to transplant seedlings and add fertilizer as rapidly as possible, and so throughout East and Southeast Asia itinerant labourers followed the rhythm of these activities. In cereal-producing areas, harvesting likewise required many additional hands for a limited period. In Catalonia, for example, every June 'by ancient custom, many reapers [*segadors*] came down from all the mountainous regions and converged on Barcelona' to hire themselves out to farmers with land in the fertile plains of the principality. According to an eyewitness, the *segadors* were 'dissolute and bold men who for the rest of the year lived disordered lives' and 'normally caused disturbances and unrest wherever they went; but the absolute need for their services apparently made it impossible to stop them.'[46] In June 1640, after a prolonged drought had ruined the harvest, the 2,500 restless reapers who arrived in Barcelona almost immediately 'caused disturbances and unrest' that led to the murder of three royal judges and the viceroy, and later to the declaration of the Republic of Catalonia (Chapter 9).

The largest group of migrants in the seventeenth century came from Africa. In East Africa, throughout the seventeenth century, thousands of men and women were captured and taken in caravans across the Sahara desert to Muslim states in the north: some 5,000 slaves arrived in Ottoman Egypt each year, whence they were distributed throughout the empire. In West Africa, far more slaves were delivered to European merchants, who shipped them to the Americas. Around 1640 an official with extensive experience estimated that the slave population of Spanish America stood at about 325,000 and that at least 9,000 new slaves were required each year to maintain this level. In addition, Brazil needed to import over 2,000 slaves annually to replace those who had died; while in 1656 alone 2,000 African slaves arrived in the English colony of Barbados.[47] Thousands more enslaved Africans died en route to the Atlantic coast and in the holding pens where they awaited shipment, or on the voyage itself. All told, during the seventeenth century Europeans enslaved some 2 million Africans, half of them from Angola and most of the rest from the states along the Gold Coast and the Bights of Benin and Biafra (see Fig. 24 on page 351, and Chapter 15).

Negative Compound Interest

The reduction in the human population in the mid-seventeenth century, however caused, had five long-term consequences that created what might be called negative compound interest:

- **Murdering the next generation of mothers.** Widows who killed themselves decimated the current generation of mothers, already depleted by high mortality during and after childbirth (and, in Catholic countries, also by an increase in the number of nuns); but killing or abandoning girls at birth decimated the mothers of the next generation. A female infanticide rate of 10 per cent will reduce population growth in the next generation by at least 20 per cent, and will also increase the number of 'bare sticks' – the common Chinese term for unmarried men. A Fujian gazetteer in the 1650s claimed that almost half of all males remained single because they could not find women to marry.[48]
- **Creating depleted cohorts.** Any short-term crisis that significantly reduced the size of an age cohort, whether through human or natural catastrophe, automatically reduced the ability of that cohort to reproduce itself. A rise of 50 per cent in the prevailing death rate meant that those under fifteen at the time of the crisis would lack the numbers to restore the previous population level. Moreover, by a tragic coincidence (and it seems to be no more than a coincidence) each generation reduced

by a major mortality crisis in the seventeenth century reached marriageable age just as another catastrophe struck. The cohort of Europeans depleted by the plague of 1630–31 had to face the famine of 1661–63, and their children reached the age of reproduction during the crisis of the 1690s. Likewise, the cohort depleted by the crises of 1618–21 and 1647–53 had to face the harvest failures of the mid-1670s. Their children reached the age of reproduction during the Great Winter of 1708–9.

• **Death on the road.** In every age, it is hard to raise a family on the move. Parish registers all over Europe record the frequent death of 'strangers' – men, women and children who perished as they tried to get from one place to another in search of family, work or food. Because each parish kept separate records, their entries rarely allow historians to recreate the losses of migrant families, but qualitative sources sometimes come to the rescue. To take a single example, Peter Hagendorf, a soldier in the Thirty Years' War between 1624 and 1648 who marched almost 15,000 miles around Europe with his regiment, kept a diary in which (among other things) he recorded the births and deaths of his family members, including four children born by his first wife, who died in 1633, and four more born by his second wife. Most of his eight children died as infants (one before baptism, and two more in their first week), leaving only two were alive when the Thirty Years' War ended, one aged five and the other just an infant.[49] If Hagendorf's experience was representative, then active military service was a sentence of death not only for many soldiers, but also for their families; and since perhaps 2 million men fought in the European wars of the mid-seventeenth century, the overall demographic consequences were severe.

• **The hidden costs of migration.** Some villages of northern Portugal, which supplied large numbers of men – both sailors and colonists – for the country's overseas enterprise, boasted fewer than sixty men for every 100 women, and scarcely half the women in the community ever married. Similar gender imbalances characterized other villages with heavy male out-migration, producing two long-term demographic consequences. First, in cereal-producing areas with soils too heavy for women to till by themselves, the departure of too many men (whether to colonize, as in Iberia, or to fight, as in Scandinavia) could reduce food production so much that the community could no longer feed itself and so eventually atrophied. Second, as in Ottoman Anatolia, massive male migration dramatically reduced the number of marriages and therefore the size of the next generation, once again creating a kind of negative compound interest.

In an influential essay *Poverty and famines*, inspired by his experience of the Bengal famine of 1943, Amartya Sen argued that 'Starvation is the characteristic of some people not *having* enough food to eat. It is not the characteristic of there *being* not enough food to eat. While the latter can be a cause of the former, it is but one of many *possible* causes.' That is: famines tend to arise from distribution problems caused by human agency rather than from supply problems caused by nature. Whether or not Sen is correct, similar anthropocentric views have convinced many people faced by food shortages (in the words of Steven L. Kaplan) 'that they were the victims of a terrible conspiracy'. In the early modern world, Kaplan argued, consumers 'uncovered signs that the harvest was not as bad as announced, that unusual and illegal acts were occurring in the grain trade, that the government was not performing as it was supposed to, and so on. As subsistence anxieties deepened ... the conviction grew that the crisis was contrived, that there was a criminal conspiracy afoot against the people, that popular suffering was needless, and that the plotters somehow had to be resisted.'[50]

Many people in the early modern period therefore insisted that government could and must do more to provide their subjects with sufficient food to ensure survival. *Politics drawn from the very words of Holy Scripture*, a treatise written in 1679 for the heir to the French throne by his tutor, Jean-Bénigne Bossuet, stressed that 'The prince must provide for the needs of his people' (the title of one of the book's 'propositions'). Indeed, Bossuet chided:

> The obligation to take care of the people is the basis of all the rights that sovereigns have over their subjects. That is why, in times of great need, the people have a right to appeal to their prince. 'In an extreme famine, the people cried to Pharaoh for food.' The famished people asked for bread from their king, as from a shepherd, or rather as from their father.[51]

This explains why early modern rulers who followed Pharaoh's example and ignored the appeals of their people soon faced protests, and why the greatest population losses occurred in precisely the states that experienced not only famines but also wars, rebellions and revolutions: China, Russia, the Polish-Lithuanian Commonwealth, the Ottoman empire, Germany and its neighbours, the Iberian Peninsula, France, Britain and Ireland.

PART II
ENDURING THE CRISIS

Peace does not become peace in a single day; a crisis does not become a crisis in a single day. Both become what they are through a gradual accumulation.
                                                              Jia Yi, *History of the Han*[1]

Sir John Elliott has pointed out that the 'epidemic of revolutions' in the 1640s 'was not, after all, unprecedented'. In the 1590s, for example, the combination of harvest failure, plague and war reduced Europe's agricultural and industrial production to the lowest levels recorded in three centuries and produced a spate of popular rebellions. Civil war in Anatolia almost brought the Ottoman empire to its knees, and Japanese invasions desolated Korea and destabilized both China and Japan itself. Moreover, the entire northern hemisphere experienced extreme climatic events that caused widespread famine and dislocation. Ottoman chroniclers in Hungary and the Balkans recorded unusually severe winters that froze the Danube solid. Elsewhere droughts and floods ruined the harvests: in Italy, when extreme weather destroyed the 1591 harvest, crowds in Rome mobbed the pope demanding food, magistrates in Naples expelled 2,000 foreign students from the city to reduce food consumption and issued bread ration cards to citizens; and bread prices in Sicily reached their highest level for two centuries. In Scandinavia, people looked back on 1591 as 'the black year in which the grass did not turn green at all'; in 1596 and again in 1597 there was 'so dreadful a hunger that the greater part of the people had to [eat] bread made of bark'.[2]

These striking data from the late sixteenth century remind us that the global cooling of the 1640s, like the 'epidemic of revolutions', 'was not, after all, unprecedented'. They therefore support the scepticism of Niels Steensgaard, who worried that the concept of a seventeenth-century General Crisis had 'become a synonym for what historians in other centuries call "history"'.[3]

This volume respectfully disagrees. It contends, by contrast, that the 1640s saw more rebellions and revolutions than had occurred in any comparable period in world history. Admittedly, most of the new regimes soon failed (the republican experiments in Catalonia, Naples and England collapsed in a matter of days, weeks and years, respectively); but some proved permanent (the Portuguese revolt; the Chinese annexation of Taiwan) or lasted for centuries (the Great Enterprise of the Qing in China; the demise of Spain as a great power; the Protestant ascendancy in Ireland). The seventeenth century also experienced extremes of weather seldom witnessed before and never (so far) since: the only known occasion on which the Bosporus froze over (1620–21); the only time that floods in Mecca destroyed part of the Kaaba (1630); the coldest winter ever recorded in Scandinavia (1641); and so on. This combination of natural and human disasters had profound human consequences. Infanticide and suicide in China rose to unequalled levels; far fewer European women married, and others did so only in their thirties; many Frenchmen, their growth stunted by famine and cold, were of shorter stature than any others on record. These diverse physical indicators all reflect conditions that were both unique and universal – but China, then as now the most populous state on the planet, apparently suffered worst and longest.

CHAPTER FIVE

The Great Enterprise in China, 1618–84[1]

In 1645, depressed by the suicide of the last Ming emperor to rule from Beijing, a gentleman-scholar named Xia Yunyi decided to dictate his memoirs before he killed himself:

> In agony and wrath at the emperor's death, my reason to live has ended ... So what more is there to say? I just fear that, regarding the rise and fall of the state, the advance and retreat of worthy and base men, the origins and ends of bandits and the sources of arms and provisions, those who instruct later generations will miss the realities. Is what I remember worth putting into words? Well, if what I say here has the fortune to survive, later generations will be able to ponder that question.[2]

Later generations did indeed ponder Xia's *Account that will be fortunate to survive*, as well as more than 200 other first-person accounts of the fall of the Ming dynasty, and virtually all laid the blame upon an unprecedented combination of domestic problems and foreign threats. In 2010 the eminent sinologist Timothy Brook reached the same conclusion. 'The fall of the Ming dynasty is many histories', he wrote:

> The history of the expansion of the Manchu empire on the northeast border, the history of the most massive rebellions to wash over China since the fourteenth century, the history of the disintegration of the Ming state, and the history of a major climate episode. Different in the stories they tell, they overlap and together constitute the same history.[3]

The immense size and complexity of the Ming state complicates the task of telling these overlapping histories. By 1600 China included almost 200 million acres of cultivated land, spread over twenty degrees of latitude, with climates ranging from tropical to subarctic. It formed the most ecologically diverse state of its day. In addition, its two major river systems, the Yangzi and Yellow rivers, linked by the Grand Canal, underpinned the most diverse, unified, wealthy and populous economy of the early modern world. The Ming emperors ruled over far more subjects than anyone else. Nevertheless

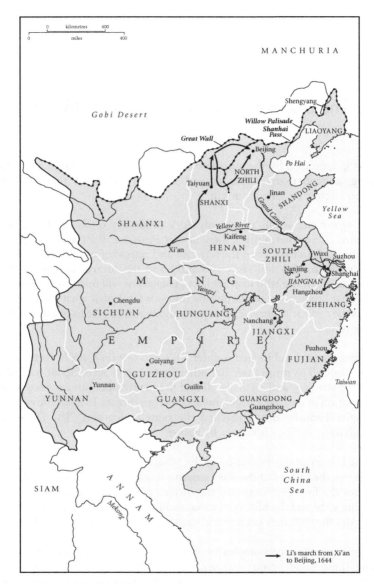

6. Ming China and its neighbours.
Most maps of East Asia use Mercator's projection, which increases the size of Ming China compared with the steppe where the Manchu dynasty achieved dominance in the early seventeenth century. In 1644, the Great Wall failed to protect Beijing (the Ming northern capital) from capture, first by a bandit army from the west and then by the Manchus from the north. Nanjing (the southern capital) fell the following year.

in 1644 the dynasty's northern capital, Beijing, fell twice: first to an army of rebels from the west, and a few weeks later to Manchu invaders from the northeast, who drove out the rebels. The victors then undertook campaigns of conquest that created a state twice as large as Ming China, which endured for more than two centuries. No other political change in the mid-seventeenth century affected so many people, caused so much damage, or created such lasting consequences.

Manchus versus Ming

The late Ming regime suffered from three endemic weaknesses. First, although Chinese subjects revered their emperor as *Tianzi* (Son of Heaven) and assigned him sole power to make authoritative rulings, he could only enforce them in cooperation with a bureaucracy of some 15,000 officials. Although recruited through competitive examinations open to almost all males, the bureaucrats came overwhelmingly from the landowning class, and their personal wealth brought an independence that enabled them to criticize, disobey and even defy an emperor who failed to meet their expectations. Second, conversely, even the senior bureaucrats (the grand secretaries) had limited opportunities to convey their views to the Son of Heaven in person because the later Ming emperors lived isolated in the Forbidden City in Beijing, a walled compound of some 200 acres surrounded by a wide moat where tens of thousands of eunuchs staffed the agencies that managed the imperial household and handled the emperor's correspondence. Third, since the prevailing Confucian ideology aimed to promote harmony and peace, Ming officials tended to favour the least disruptive and least expensive policies. They also formed factions to preserve the status quo and denounced opponents and innovators as traitors.

These weaknesses became prominent in the early seventeenth century, when the Wanli emperor (r. 1573–1620), frustrated by the obstructionism and denunciations of his bureaucrats, left official posts unfilled and refused to sign orders. Instead, he circumvented the civil service by using the imperial eunuchs who answered to the director of the Ceremonial Department (in effect, the emperor's chief of staff and always a eunuch) to carry out his orders. His successor, the Tianqi emperor (r. 1620–27), went further: he made the eunuch Wei Zhongxian his chief minister and used the palace eunuchs, now 80,000 strong, as diplomats, trade and factory superintendents, tax inspectors and government supervisors, even generals and admirals. They formed an alternative administration that reported directly to Wei.

Tianqi's half-brother the Chongzhen emperor (r. 1627–44) initially relied less on eunuchs but proved both stubborn and suspicious in dealing with the bureaucracy. He appointed no fewer than fifty grand secretaries, almost all of whom he removed in response to a remonstrance or denunciation by a jealous colleague. Such inconstancy precluded the formulation and implementation of effective strategies to meet the numerous challenges that faced the empire: instead of discussing how to save it, ministers concentrated on finding someone to blame. The emperor, like his predecessors, therefore came to rely on eunuchs for both civil and military affairs.

As paralysis gripped the Ming government, a small nomadic group living beyond its northeast frontier created a new state. At first their leader, Nurhaci (1559–1626), accepted Ming suzerainty, leading tribute missions to Beijing in person, learning to read Chinese, and studying Chinese history and military practice; but he also united into a single confederation the various tribes that inhabited the steppe of Central Asia. By 1600 Nurhaci commanded at least 15,000 warriors organized into permanent companies of 300 fighting men, normally from the same village and sometimes the same clan, and 'banners' (*gūsa*, the Manchu term for a large military division) commanded by a member of Nurhaci's family. Manchu warriors valued horsemanship and archery; they revered as their ancestors the Mongols who had conquered China four centuries before; and, like the Mongols, they shaved the hair on the front part of their heads and wore the rest in a pigtail behind.

For some time, the new Manchu state prospered from trade (especially from exporting ginseng to China), from war booty (both property and persons) and from agriculture (mostly performed by their slaves); but global cooling changed all that. In China, imperial officials reported in 1615 that they had received a deluge of petitions for disaster relief, and that 'Although the situation differs in each place, all tell of localities gripped by disaster, the people in flight, brigands roaming at will, and the corpses of the famished littering the roads', and the following year an official submitted an *Illustrated handbook of the great starvation of the people of Shandong* (1616). The famine was even worse in Manchuria. When Nurhaci's advisers suggested that invading China might alleviate their problems, he retorted angrily: 'We do not even have enough food to feed ourselves. If we conquer them, how will we feed them?'[4] Nevertheless, the worsening economic situation led him to change his mind: in 1618 he invaded Liaodong, a province north of the Great Wall where subjects of the Ming cultivated wheat and millet during the short but intense growing season. This delivered into Manchu hands a million new subjects – but global cooling soon produced the result that Nurhaci had feared. A chronicle recorded that the year 1620 saw high rice prices, and added 'thereafter there were no more years in which rice was not expensive'.[5]

The Ming failed to exploit the Manchus' problems because an uprising in Shandong province led the emperor to withdraw troops from the northeast frontier. Although they crushed the rebellion before it could spread, their absence allowed Nurhaci to consolidate his control of Liaodong by abolishing the dues and services payable to landlords and confirming all peasants in possession of their land – provided they shaved the front of their heads in the Manchu manner.

The Erosion of Ming Power

Ming China had two capitals. Nanjing (literally 'southern capital'), in the lower Yangzi Valley, never saw the emperor but possessed a full bureaucracy that mirrored the one in Beijing ('northern capital') located at the extremity of China's Central Plain. The strategic vision of the two administrations differed considerably. With the Great Wall barely 30 miles away, ministers in the northern capital normally prioritized defence and the food supply. They needed to monitor developments among the peoples of the steppe, and ensure a constant supply of the essential goods for the capital, the largest city in the world, especially the rice that arrived via the Grand Canal from the fertile paddies in the south, and the coal brought by camel trains from the mines of Shanxi in the west. Ministers in the southern capital, by contrast, needed to maintain peace and productivity in the densely populated lands of Jiangnan, where many farms could no longer feed the families that worked them, either because they were too small or because they produced commercial crops rather than foodstuffs. Even abundant harvests left some people hungry, producing tension between landlords and tenants in the countryside and sometimes disturbances in the towns. Poor harvests almost immediately produced starvation, migration and disorder.

The ministers of both imperial capitals faced three similar problems: a broken fiscal system; a weak military; and ineffective imperial leadership. First and greatest among the fiscal problems was the fact that the Ming never borrowed – a practice that greatly reduced flexibility in times of crisis, since expenditure could be funded only from current revenues, instead of being spread over several years (as in Western Europe). This placed an enormous premium on the efficient collection of the land tax, the Ming government's main source of revenue; but by the 1630s, the distribution of this tax had become extremely uneven. Within each province, counties that had at some point earned imperial disfavour paid extra, whereas others contained tax-exempt lands and therefore paid less. New inequities emerged in the wake of a major tax reform in the sixteenth century, when mounting budget problems led the central government to combine two different obligations – labour services and land tax – into one, known as the Single Whip. Ministers hoped to increase yields by using the detailed household registers of labour services, known as Yellow Books, to allocate tax obligations; but since the gentry enjoyed exemption from labour services, their names did not appear in the Yellow Books and so the Single Whip system allowed them to escape paying the land tax, too. Moreover, local officials determined the annual burden payable by each taxable household, a process that opened the way to corruption and abuse, so that communities were said to dread the annual visit of the tax officials 'as much as if they were to be thrown into boiling water'.[6]

These fiscal discrepancies produced stunning cumulative inequalities. Suzhou prefecture, with 1 per cent of China's cultivated land, paid 10 per cent of the total imperial revenue; Shanghai county paid three times as much as an entire prefecture in Fujian; a single prefecture near Nanjing which had only three counties paid as much as the entire province of Guangdong which had seventy-five counties. An imperial decree that made support of the emperor's relatives the first charge on each regional treasury greatly exacerbated these fiscal inequities: by the 1620s the cost of supporting 100,000 Ming clansmen – most of them with a wife, concubines, numerous progeny and a large household (all tax-exempt) – in some provinces absorbed over one-third of the annual revenues.

As long as China's agrarian economy continued to expand, as it did throughout the sixteenth century, these inequities remained tolerable; and even after the land tax increased from 4 million *taels* in 1618 to 20 million in 1639 (to fund the growing defence expenditure), the underlying tax rate in the richest and most fertile parts of China still lay at or below 20 per cent – a heavy but bearable burden. Moreover 20 million *taels* should have sufficed to support an army of 500,000 men – more than enough to preserve the Ming state – but corruption and weak central control kept much of the yield from reaching the central treasury. By 1644 the government's budget projected receipts of less than 16 million *taels* but expenditure of more than 21 million, with no way to bridge the gap.

The second common problem facing ministers in Nanjing and Beijing was the poor quality of their troops. Ming culture disparaged martial virtues and achievements, which not only discouraged talented members of the elite from serving in the army but also demoralized the troops. Most soldiers did not know how to use their weapons properly, and most officers systematically overstated the size of their armies to the Ministry of Revenue to gain more resources (and understated it to the Ministry of War to avoid being sent into combat). According to a frustrated government inspector, if 100,000 names appeared on the army's lists, only 50,000 soldiers actually served; and, he concluded, 'of those 50,000 men, no more than half are of any use in combat. The Court thus pays for four soldiers but receives the services of only one.' Foreign observers also noted that officers frequently flogged their soldiers, making them 'drop their trousers and lie on the ground, as if they were schoolboys, to receive the blows' – degrading treatment that led many troops to desert and others to mutiny. Over fifty serious military revolts occurred between 1627 and 1644.[7]

The third critical problem facing ministers in both Nanjing and Beijing was the existence of numerous academies where intellectuals discussed current moral and social issues in the light of the teachings of Confucius. Although the goal of Confucianism remained the same for all – the need for each man to perfect himself so that he could serve heaven and human society as a sage ruler or minister – by the early seventeenth century two opposed paths to that goal existed. One required aspiring sages to search for moral principles in the outside world (specifically in the Confucian classics, the words of past sages) and apply them; the other stressed self-examination, and taught that intuition based on each man's 'innate goodness' sufficed to ensure righteous actions. The Donglin (Eastern Grove) Academy in Wuxi, a Jiangnan town, followed the first path: its associates downplayed intuition and introspection in favour of ending corruption and restoring moral rectitude throughout China by applying the ancient virtues enshrined in the classics. They adopted the popular slogan *jingshi jimin*: 'Manage the world's affairs; provide for the people'. Although the Wanli emperor degraded or dismissed ministers who sympathized with Donglin goals, for a brief period after his death they prospered and a branch of the academy opened in Beijing, with a library and a lecture hall where members debated the pressing issues of the day. They used their advantage to attack not only ministers who were corrupt but also those who tolerated corruption in the interests of social harmony. After Nurhaci's occupation of Liaodong, their criticisms focused on 'a man who sucks boils and licks haemorrhoids' whom they blamed for the humiliating defeat: the eunuch Wei Zhongxian, the emperor's childhood friend and now his principal minister.[8]

In 1624 a prominent member of the Donglin Academy submitted to the emperor a memorial that listed crimes committed by Wei against the interests of the state. Almost immediately a host of other officials, many of them also Donglin alumni, submitted similar memorials that called for Wei's dismissal. Their efforts backfired because Wei convinced the emperor that Donglin factionalism formed the true threat to government stability. Warrants went out for the arrest of eleven critics, all Donglin alumni, and imperial agents arrested and brought them to Beijing, where they were carried through the streets in cages like zoo animals, interrogated, tortured, publicly flogged and then murdered. In addition, Wei abolished all private academies as well as compiling a blacklist of military and civilian officials whom he considered sympathetic to the Donglin movement, replacing them with his own protégés (later known as *yandang*, or 'associates of the eunuch').

In 1627, however, the Tianqi emperor suddenly died and his sixteen-year-old half-brother succeeded as the Chongzhen emperor (a reign title that meant 'lofty and auspicious', but later black humour punned on the word *chongzheng*, meaning 'double levy', to create the Double Taxation

Reign). The new ruler declined to protect Wei, who hanged himself, and instead announced an era of renewal. To achieve this he rescinded the order abolishing private academies, exonerated most of the Donglin partisans still in prison, rehabilitated the reputation of those degraded and issued his own blacklist – this time of the *yandang*, most of whom lost their jobs and many of whom were punished. The Chongzhen emperor also urged his senior ministers to end their factional struggle. 'Consider this', he told them at one meeting: 'We have alarms east and west, we have wars north and south, yet [my officials] have no anxiety for the dynasty. All they do is divide into camps, all they talk about is some clique, some Donglin – and of what benefit is that to national affairs?'[9]

The new emperor's pleas failed because, in the perceptive phrase of historian Ying Zhang, in his 'ambition to save the empire in crisis' he 'sought only instant successes and simple solutions'. Each administrative error and policy failure provoked a spate of denunciations and memorials by rivals of the officials deemed responsible. Not knowing whom he could trust to show 'anxiety for the dynasty' or seek 'benefit for national affairs', the Chongzhen emperor changed the heads of the six major departments of state, on average, once every year: a quarter of those dismissed from office he also executed or disgraced.[10]

Factional struggles among China's bureaucrats had two deleterious consequences. First, repeated purges reduced the overall quality of senior bureaucrats. Of the 120 magistrates whose biographies appeared as exemplars for others in the official *Ming history*, composed a generation later, not a single one had served either the Tianqi or Chongzhen emperors. Second, although provincial postings were supposed to be distributed by lot, those who excelled in the civil service examinations went to the more prosperous counties, leaving several poor, remote and problematic areas virtually ungoverned. In 1629, half of the prefectures and counties in impoverished and turbulent Shaanxi province lacked magistrates.

In the words of the official *Ming history*, 'When a dynasty is about to perish, it first destroys its own people of quality. After that, there come floods, drought and banditry.'[11] From the first, the Chongzhen emperor and his dwindling cohort of 'people of quality' had to deal with floods, drought and banditry in the northwest. The adjacent provinces of Shaanxi and Shanxi normally suffer from uncertain rainfall, a short growing season and poor communications; and, in the early seventeenth century, they lacked both adequate granaries to sustain the poor. The brutal punishments for banditry stipulated by the Ming law code (death not only for each bandit but also for all male relatives up to second cousins and for those who concealed or aided them) helped to preserve order for a while; but a prolonged drought that began in Shaanxi province in 1628 (a 'year without a summer') changed the criminal calculus. Starving farmers now abandoned their land and joined the outlaws, enabling the bandit leaders to threaten more prosperous lands to the south. The following year, with Manchuria also afflicted by drought, Manchu forces for the first time broke through the Great Wall and ravaged northern China.

The Chongzhen emperor responded to these developments with two disastrous measures. He withdrew troops from Shaanxi to defend his capital against the Manchus, leaving the rest to fight the bandits without pay and supplies. Many of these abandoned troops, including a soldier named Zhang Xianzhong, joined the bandits. Moreover, to save money, the emperor closed roughly one-third of the courier stations used by foreign envoys, officials and messengers travelling to and from Beijing, and dismissed their employees. This not only reduced the central government's ability to receive news and convey orders in a timely fashion; it also caused great hardship in the northwest, where the courier network had traditionally provided many jobs in an area of high unemployment. Li Zicheng, from a poor Shaanxi family, was one of the couriers thrown out of work. Initially he transferred to the army, but when his unit received no pay he led a mutiny and then (like countless others) joined the bandits. Zhang and Li would both play a prominent role in destroying the Ming state.

Climate change further exacerbated the situation. Some of the weakest monsoons recorded in the last two millennia produced droughts that destroyed the crops in the south and east. A scholar-official lamented that

> Today, people are killed by bandit mobs; tomorrow, they may die at the hands of government troops. No one can tell how many fields have been left fallow by farmers fleeing disaster or in how many fields already-planted crops wither under the sun. No one can count how many families have been broken apart or number the dead lying piled in ditches.

Not surprisingly, many of the starving survivors joined the bandits, who now began to ravage lands as far afield as the Yangzi in the south and Sichuan in the west.[12]

The inequities created by the Ming fiscal system further aided the bandits and hurt agriculture. Even where no bandits threatened, extreme weather adversely affected agriculture. Subtropical Lingnan (the 'lands south of the mountains': Guangxi and Guangdong provinces) saw heavy snowfall in 1633 and 1634, and abnormal cold in 1636: crop yields plunged by between 20 and 50 per cent. Petitions flowed in to the central government from all over the empire, begging for action to end the bandit menace and to relieve the suffering caused by failed harvests, high taxes and bad weather. In 1632 and 1633 government troops made considerable headway against the bandits, gradually pinning them down beside the Yangzi – but in

December 1633 the great river froze over and the rebels escaped across the ice. The cold winter, followed by another drought, led many peasants to join the rebel armies, which soon exceeded 600,000, but the Chongzhen emperor ignored them, concentrating his resources instead on resisting the Manchus.

The Great Enterprise Begins

While Ming China suffered from problems created by the tax system, bandits and the weather, the Manchus improved their military efficiency thanks to some opportune technology transfer. When he declared war on the Ming in 1618, Nurhaci's forces consisted almost entirely of mounted archers whereas his Chinese opponents relied mainly on infantry using firearms. The Manchus' lack of artillery allowed the major towns of Liaodong to hold out – indeed Nurhaci himself received a fatal wound during a siege. The Chongzhen emperor sought to build on this advantage by importing Western artillery, and soon about fifty bronze cannon defended the Great Wall. It was too late. During the 1629 raid into China, Hong Taiji (who, after a savage succession struggle, followed his father Nurhaci as Manchu leader) acquired not only Western guns but also conscripted a Chinese gun crew 'familiar with the new techniques for casting Portuguese artillery'. He also offered huge enlistment incentives to any engineer, and anyone proficient in the art of making and using cannon, who joined him. When the Manchus resumed the war in Liaodong in 1631, they possessed forty Western-style artillery pieces and the crews to work them, and they built palisades and forts to cut off the heavily fortified regional capital. It surrendered within a few weeks. Hong Taiji also incorporated his Chinese volunteers and conscripts alike into the banner system, until by 1642 each of the eight banners of his army had parallel Manchu, Mongol and Han Chinese components.[13]

All this formed part of the Manchus' ambitious programme to challenge the Ming for mastery of China, known as the Great Enterprise (*Da ye*). In 1627 Hong Taiji issued a List of Seven Grievances against both the Ming emperor and the king of Korea: both documents presented the Manchu state as an independent political entity that dealt with its neighbours as a sovereign power. He also established a chancery and six administrative boards, modelled on the Chinese system; devised a new script to communicate his orders in Manchu; and commanded scholars to search historical works for accounts of previous conquests of China by northerners, and of Chinese who switched allegiance to the conquerors at an early stage, as well as examples of bad Chinese rulers whose overthrow had been justified. Hong Taiji also moved to a new capital (present-day Shenyang), named both Shengjing ('Flourishing Capital' in Chinese) and Mukden (from the Manchu word meaning 'to arise'), where his court combined Manchu, Mongol and

Chinese imperial protocol. In 1636 he proclaimed himself the founder of a new multi-ethnic state called *Da Qing* ('Great Qing'), and claimed that the 'mind of Heaven' now guided his actions.

Hong Taiji also decreed how his subjects should look and dress. An edict of 1636 required 'All Han [Chinese] people – be they official or commoner, male or female – [in] their clothes and adornment . . . to conform to Manchu styles . . . Males are not allowed to fashion wide collars and sleeves; females are not allowed to comb up their hair or bind their feet'. Males also had to shave the front part of their head and wear the rest in a pigtail like the Manchus. Two years later, a further decree specified that 'all those who imitate' Ming customs 'in clothes, headgear, hair-bundling and foot-binding are to be severely punished'. Hong Taiji's legislation reconfigured both attire and the body itself to proclaim political allegiance and cultural identity; and it construed failure to conform as treason.[14]

The Little Ice Age Strikes

Just as Nurhaci's decision to invade China in 1618 reflected the global cooling that affected Manchuria more severely than more temperate areas, so the cool and wet weather that ruined several harvests in the 1630s made it imperative for the Manchus to seize as much food as possible from their neighbours. Hong Taiji therefore launched another raid deep into China and also invaded Korea, the most important tributary state of the Ming; but, as in 1618, expansion brought only short-lived relief. Even the new lands under Qing rule no longer sufficed to feed those who lived there, and in 1638 some of Hong Taiji's advisers recommended that he seek a peace treaty with the Ming.

Extreme weather conditions continued to afflict China. Recent climatic reconstructions reveal that the years between 1627 and 1643 saw the 'most severe persistent drought in eastern China' in the past millennium, and that in northern China 1640 (a year of extreme El Niño and volcanic activity: Chapter 1) proved to be the driest year recorded during the last five centuries. According to a magistrate in Henan province, 'there have been eleven months without rain. In the past year people have suffered from floods, locusts and drought. The drought was so bad that people could not plant the wheat, and what little was planted was eaten by locusts . . . The people all have yellow jaws and swollen cheeks; their eyes are like pig's gall.'[15] 1641, a year that also saw extreme El Niño and volcanic activity, brought even worse weather. Jiangnan experienced severe frost and heavy snow, followed by the second driest year recorded during the sixteenth and seventeenth centuries. In July the Grand Canal dried up in Shandong province, cutting off the supply of rice to the imperial capital – something never previously recorded – and smallpox halved the population of some villages.

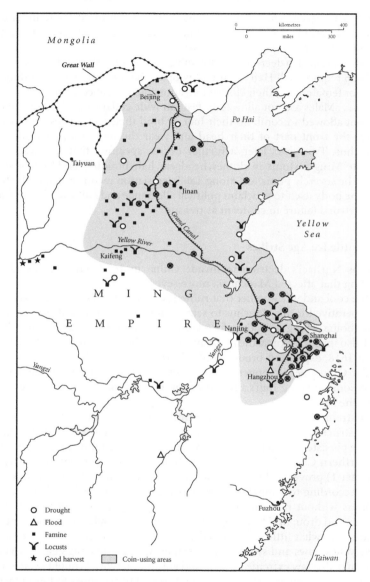

7. Disasters and diseases cripple Ming China, 1641.
Over 100 county gazetteers from the provinces of Henan, Hubei, Shandong and Jiangsu recorded a major natural disaster – drought, flood, famine, locusts – in the year 1641. At the same time, a sudden dearth of silver coins dislocated the economic hub of Ming China.

Confucian doctrine saw famines, flood and droughts as 'heaven-sent disasters' (*tianzai*) and a spate of natural disasters as a cosmic portent that a 'change in the Mandate [of Heaven]' was imminent. The failure of so many harvests in the Chongzhen reign therefore led his ministers to propose drastic remedies. *A complete treatise on agricultural administration*, composed in the 1620s and published in 1637, devoted one-third of the text to famine control, explaining not only how to build and manage granaries but also how to cultivate crops capable of withstanding adverse weather, and which wild plants one could safely eat. By then, however, lack of funds and central direction had left many (if not most) of the public granaries empty and poor harvests made it impossible to replenish them.[16]

Lacking assistance from the central government, individual county magistrates searched for other ways to nourish the people during the worst crisis to afflict East Asia in early modern times. Since the central government devoted all available resources to resisting the Qing, many pinned their hopes on charity to feed the hungry – but this produced too little, too late. When the magistrate of Shanghai county, faced by famine and disease in 1641–42, persuaded some gentry and merchants to 'contribute rice to cook gruel' for communal kitchens, starving people came 'in unbroken streams, leading their elders and bearing their children on their backs. In extreme cases, they fell down dead on the roads before reaching the gruel kitchens; or else, having eaten their fill, they died by the wayside just as they were returning'. Others reported that 'the human price of a peck of rice [barely enough to feed one person for a week] was two children'.[17]

The intensified pressure from Manchus, bandits and climate struck a society already under stress. The fertile soil and benign climate of Jiangnan allowed enterprising farmers to harvest double and even triple rice crops, producing by the 1620s a population density of 1,200 per square mile – one of the highest on the planet. Such overpopulation, coupled with the habit of dividing farms among all sons, had by the 1630s reduced some holdings to half an acre, insufficient to provide enough rice to sustain even a small family. Many farmers survived by raising cash crops such as cotton (which could grow on higher land than wet rice) and tea (which thrived on hillsides), or by switching from rice cultivation to raising silkworms; but this created a new source of vulnerability. Farmers who raised cash crops needed to produce and sell enough both to buy a year's food and to pay their taxes: unfortunately for them, the supply of silver coins suddenly collapsed.

In 1639, probably because the changing Pacific wind system made navigation more hazardous (Chapter 1), two galleons laden with silver sailing from Mexico to Manila foundered with the loss of their entire cargo. Chinese merchants therefore could not sell their silk in the Philippines in exchange for Mexican silver as usual. In normal times, increased trade with Japan might have filled this shortfall because the archipelago (like Mexico) produced silver and craved silk; but in the late 1630s, fearful of foreign influence, the Japanese government restricted all overseas trade, and so little silver left the country (Chapter 3). China's silver imports fell from almost 600 metric tons in 1636–40 to under 250 metric tons in 1641–45, disrupting the trade and production of those who normally used silver, just as the central government increased its tax demands (payable in silver) to unprecedented levels. The farmers of southeast China could scarcely have picked a worse time to rely on cash crops.

The Alienated Intellectuals of Ming China

In the mid-eighteenth century Qin Huitian, a senior government official, opined that 'The demise of the Ming was due to banditry. The rise of banditry was due to famine.'[18] Qin's terse explanation omitted one other critical factor in the demise of the Ming: the role of an alienated academic proletariat whose influence far exceeded their numbers. Throughout China, tens of thousands of male children learned to read, write and memorize a set of classical works on ethics and history, and then took a formal examination in the county town under the personal supervision of the district magistrate, who also graded their papers. Those who passed this examination became eligible to take a more advanced test in the prefectural capital, once again under strict supervision, and all who passed achieved the rank of *shengyuan* (licenciate). Gaining *shengyuan* status brought numerous rewards: a modest stipend; exemption from labour services, physical punishments and certain taxes; the right to wear a special scholar's uniform and to take precedence in all public places over other men, however senior, who lacked the degree. In a famine, they also received government relief before others. All *shengyuan* were expected to prepare for a further round of rigorous examinations, held triennially, which required them to write essays on assigned topics from the classics set by, and supervised by, a team of senior officials who left Beijing in a carefully calibrated sequence so that students could sit the same examination on the same day throughout the state. Those who passed became *juren* (elevated candidates), eligible to take a final set of exams held once every three years in the huge examination halls of Nanjing and Beijing, followed (for the successful) by an additional test administered by the emperor himself. Those who passed gained the coveted status of *jinshi* (advanced scholar): a passport to the highest offices in the state.

This sophisticated system provided a strong administrative backbone for imperial China, and in many ways it served the state well. First, since it was open to most males able to memorize the classical curriculum and write literary Chinese, the central government mobilized its human resources for public service to an extent unequalled in the early modern world. Second, the existence of a common curriculum in a single language and script provided remarkable cultural and linguistic uniformity across the state, despite its immense size and diverse population. Finally, the civil service never completely succumbed to either corruption or absolutism because, however incompetent, capricious or lazy the emperor at the apex of the system, every three years a new cohort of articulate and classically literate men in their prime began their measured progress up China's ladder of success.

Nevertheless, the size of each successful triennial cohort remained very small: seldom more than 500 *juren* candidates and 300 *jinshi*. With such long odds, few candidates passed on their first attempt and some graduated only after many failed attempts, long past their prime. Accordingly, the number of *shengyuan* holders without government jobs soared from perhaps 30,000 in the early fifteenth century to over 500,000 in the early seventeenth – one in every sixty adult males – and below them lurked those who had spent years studying at the county schools but still failed all their examinations.

What happened to this vast academic proletariat? Although those who achieved their *shengyuan* degree were allowed to retain for life the economic and social privileges it brought, the impact of failing subsequent examinations on candidates who had devoted so much of their lives to study (and on the families that had sacrificed so much to support them) could be devastating. Some committed suicide; many suffered a nervous collapse; most went home and either worked while they continued to study (perhaps as secretary to an overworked magistrate or as tutor to the talented son of a rich family), or else vented their frustrations on the local population (a reaction so common that the 'failed candidate' became a stock character in Chinese novels). Some abandoned the examination mill and turned to business or medicine; others still 'ploughed with their ink stand' (in the contemporary phrase), producing biographies, plays, novels, essays, epitaphs and (ironically) manuals on how to pass the civil service examinations.

All these men (and many of their wives and daughters) could read and so become familiar with the unprecedented number of printed works available in urban shops and from rural peddlers. Reading convinced some of these thwarted scholars that they could address China's problems better than their successful competitors now in the bureaucracy – a belief confirmed through the numerous scholarly societies (*wen she*) where educated men debated not only literature, philosophy and history but also practical ways of restoring effective government. In 1629 a scholar from Tiacang, close to both Suzhou and Wuxi, called for these societies to join in an amalgamated federation, which would bring together gentlemen-scholars interested in 'revitalizing and restoring the ancient learning and thus be of some use at a later time. The name of our society, therefore, shall be the Restoration Society [*Fu she*].'[19] The following year, when the society already boasted over 3,000 members, thirty of them took and passed the provincial examination in Nanjing, gaining one-fifth of the total *juren* granted, and the following year twice as many became *jinshi*. Each successful alumnus of the society worked to advance his colleagues within the bureaucracy through patronage and recommendations, and some published collections of model essays couched in a new pragmatic style both to popularize their ideas and also to put pressure on the examiners – for if the authors of such meritorious essays should fail, it would suggest bias and corruption.

Nevertheless, bias and corruption increased. After 1620, the central government, desperate for money, allowed certain localities to sell *shengyuan* degrees; and in 1643, the last full year of Ming rule in Beijing, money even determined the outcome of the metropolitan examination, with the first and second places going to those who paid the highest price. Increasing numbers of examination candidates tried to beat the system by bribing or intimidating the examiners, or by cheating (writing on their clothes or their bodies in invisible ink one of the thousands of successful essays from earlier examinations available in print). Naturally, such practices discouraged other candidates; so did the Chongzhen emperor's habit of punishing any minister who failed to produce instant success. Many talented civil servants therefore either declined promotion or resigned rather than risk disgrace (and possible death) because, as a contemporary proverb put it, 'at any time, the jug could strike against the tiles'. Eventually, many disenchanted scholars threw in their lot with the bandits.

The Rise of the Dashing Prince

The demoralization of the Ming's educated subjects, and most notably of their officials, helped to facilitate the triumph of their enemies. In 1642 Li Zicheng, now known to his followers as the Dashing Prince, captured Kaifeng, once the capital of all China, while later that year a large Qing army broke through the Great Wall, pillaging northern China for several weeks before returning home with enormous booty (including copious quantities of grain to feed their own hungry subjects). In their wake, one town in Shandong reported that 30 per cent of its population had died of starvation, with another 40 per cent so poor that they could only survive by joining the bandits.

This desperate situation produced desperate remedies. The emperor authorized a secret approach to the Manchu leaders to see whether they might negotiate – until a clerk inadvertently published one of the clandestine documents in the government's official gazette. This slip provoked an avalanche of memorials from the bureaucracy condemning such pacifism, leading the emperor to abandon the talks. Meanwhile, some Ming officials agreed to administer their regions in Li's name, and a few accepted an appointment at his 'court', encouraging the Dashing Prince to believe that (like other men from humble backgrounds in Chinese history) he could found his own imperial dynasty.

In 1643, with the aid of disaffected officials, Li Zicheng set up a formal government in Xi'an (once the imperial capital), including a chancery and the traditional six ministries, and then declared himself the founder of a new state, *Da Shun* (meaning 'Great Compliance' [with the Mandate of Heaven]) with a corresponding Shun Era calendar. He also minted his own coins, elevated his leading supporters to noble status, and held civil service examinations to produce his own cadres of administrators. At this stage Li may have thought only in terms of a power-sharing arrangement – either with the Ming or with the Manchus – but early in 1644, perhaps inspired by the lack of effective Ming resistance to the recent Manchu raid, he led his followers in one of the most extraordinary military feats in Chinese history: a 'long march' from Xi'an to Beijing.

Li demanded exemplary behaviour from his troops – 'kill no one, accept no money, rape no one, loot nothing, trade fairly' he ordered – and this tactic encouraged most local commanders and officials in his path to surrender either immediately or after only token resistance. Li pardoned most of them and confirmed them in office. In addition, his officials drafted proclamations that exuded confidence and reassurance: 'Our army is made up of good peasants who have worked the fields for ten generations. We formed this humane and righteous army to rescue the population from destruction.' As soon as he arrived in an area, Li established tribunals that allowed tenants to press claims against their landlords and granted titles to abandoned domain land. To widespread popular acclaim he also arrested, humiliated and executed all Ming clansmen who fell into his hands, while his followers tore down the arches and temples erected by forced labour as memorials to illustrious local dignitaries.[20]

The Chongzhen emperor and his remaining advisers continued to debate extreme remedies: demanding forced loans from all ministers and eunuchs (which raised only 200,000 *taels*); issuing paper money to pay the troops; evacuating the crown prince to Nanjing; making a deal with either Li or the Manchus; recalling the only remaining reliable army, commanded by General Wu Sangui, from the Great Wall to defend the capital. Only the last measure took effect – and even then Wu was still far away when on 23 April 1644, a mere eight weeks after they had left Xi'an, Li and his army stood before the capital of the Ming empire.

Although defended by walls that stretched for over 20 miles, with thirteen huge fortified gates and the largest urban population in the world, Beijing presented a soft target: the garrison had not been paid for five months and food reserves had run low. Some discontented defenders opened one of the outer gates to Li but, since the Forbidden City remained intact, the emperor summoned his ministers and prepared to make a last-ditch stand there. When no one came, he disguised himself as a eunuch and tried to escape, but his own palace guards fired on him and he turned back. So finally, after a reign of seventeen years, he went into the palace garden and hanged himself to avoid being captured, humiliated and executed by his subjects.

Li now controlled most of northern China, including its capital, but he faced three urgent problems. First Zhang Xianzhong, the ex-soldier turned bandit leader (page 99), had conquered Sichuan where he proclaimed himself Great King of the West. Although Zhang did not threaten Beijing, his brutal administration devastated one of the wealthiest provinces of the empire, undermining both the resources and the reputation of the central government. Second, although Li's troops had kept good order on their march to Beijing, they now expected their reward. To avoid looting, the Dashing Prince needed to arrange for taxes to flow into the capital once more. Finally, Li urgently needed to win over the army of Wu Sangui – all that stood between the capital and the Manchu forces beyond the Great Wall.

At first, not knowing whether the crown prince might still be alive somewhere, and anxious to win over Wu and his loyalist troops, the Dashing Prince behaved with great prudence. For example, he sat beside the imperial throne (not on it) to receive the obeisance of the civil and military personnel in the capital. Once again his moderation succeeded: of the 2,000 or more Ming officials in Beijing when it surrendered, fewer than 40 elected to follow the example of their late master by committing suicide – striking testimony to how the Chongzhen emperor had alienated his ministers. Li now ordered the remaining officials in the capital to make substantial contributions to his new treasury, preferably voluntarily but if necessary under duress.

Initially, most officials seem to have regarded their ordeal as a just punishment for failing to serve their late master better (and to commit suicide on learning of his death); but before long Li lost control over his followers, who started to plunder the houses where they lodged and abuse their hosts, and to rape teahouse servants and female entertainers, and finally elite women in their homes. When the Dashing Prince led his troops out of the city on 18 May, only three weeks after their triumphant entry, Beijing rejoiced.

The Tipping Point: China's Battle of Hastings

Li left to deal with Wu Sangui, the Ming general whose troops manned the Shanhai Pass, where the ridge of mountains that carries the Great Wall reaches the sea. Wu had refused to recognize the Shun state, and when a detachment of Li's army attacked him he defeated it, provoking the Dashing

Prince to murder Wu's father and other relatives whom he had captured. Almost immediately, Wu appealed to the Qing for help.

Why did Wu take this fateful decision? Apart from resentment at the death of his relatives, five other considerations played a part. First, commanding scarcely 40,000 soldiers, Wu no doubt feared that the victorious Shun army, numbering perhaps 100,000 men, would overwhelm him if he fought alone (as indeed they almost did). Second, several of his surviving relatives had surrendered to the Manchus in Liaodong, and an alliance with the Qing would ensure their continued good treatment. Third, several Chinese regimes in the past had survived by appealing to a northern neighbour for assistance in time of peril. Fourth, the Manchus had previously shown no interest in permanent conquest: their earlier invasions of China had targeted booty not land. Finally, the Qing ruler Hong Taiji died late in 1643, leaving two quarrelsome brothers as regents for his six-year-old son: Wu probably assumed that this family squabble would fatally undermine Manchu strength and cohesion.

All these considerations were valid, but Wu overlooked the impact of climate change. Long before they received his appeal for help, the Manchus had concluded that the recent famines would force them to invade China or perish: tree-ring series show 1643–44 as two of the coldest years in the entire millennium between 800 and 1800, and the winter monsoons brought little rain, leading to desiccation in Manchuria. The Manchu leaders therefore assembled a Grand Army of 60,000 or more warriors ready to mount another invasion, while their Han Chinese advisers prepared announcements urging their compatriots to support the new invasion: 'The righteous army comes to avenge your ruler-father for you. It is not an enemy of the people. The only ones to be killed now are the [Shun] bandits. Officials who surrender can resume their former posts. People who surrender can resume their former occupations. We will by no means harm you.'[21]

The Grand Army stood poised to break through the Great Wall into Shanxi when Wu's desperate appeal revealed the weakness of the Shun regime. This, Wu informed the Manchu leaders, offered a unique 'opportunity to rip down what is withered and rotten. Certainly there will never be a second chance!' They therefore abandoned the planned invasion of Shanxi and instead led their troops to the Shanhai Pass where Wu, threatened by Li's approaching army, let them through. The Qing regent Dorgon ('badger' in Manchu), in command of the Grand Army, skilfully exploited his advantage. His troops took no part in the struggle between Wu and Li, which sinologist Mark Elliott has called China's battle of Hastings, until the last moment, when they charged the Shun flanks. As a result, Wu's forces bore the brunt of the battle and sustained heavy casualties, leaving the survivors too weak to refuse when Dorgon called upon them to become Qing vassals. If they did this, Dorgon promised, 'your ruler will be avenged. In addition

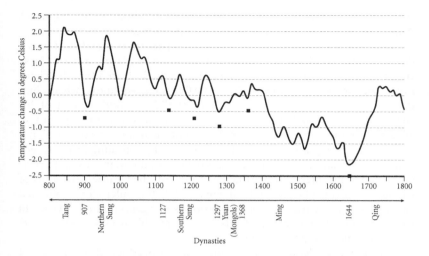

8. East Asian temperatures, 800–1800.
Although temperatures, as measured by tree-ring width, began to fall in the fourteenth century, with a partial recovery in the sixteenth, the lowest temperatures – more than 2°C cooler – occurred in the mid-seventeenth century with the nadir in 1644, the year the Manchus drove the Ming from northern China.

'your posterity will enjoy wealth and nobility as eternal as the mountains and rivers'. Powerless to resist, Wu and his men shaved their heads in the Manchu manner in a dramatic gesture of submission.[22]

Li and his defeated troops fell back on Beijing, which they re-entered on 31 May 1644. The Dashing Prince now had nothing to lose by proclaiming himself emperor, and the hasty enthronement of the first (and last) Shun emperor took place. The following day, realizing that the approach of Wu and the Qing Grand Army made his position untenable, Li set fire to the Forbidden City and ordered his men to retreat.

The population of the capital now prepared to greet Wu Sangui, perhaps accompanied by the missing Ming crown prince; but instead, on 5 June 1644, Dorgon mounted the ceremonial platform they had prepared and told the onlookers: ' "I am the prince regent. The crown prince will arrive in a little while. Will you allow me to be the ruler?" The crowd, astonished and uncomprehending, was only able to lamely answer "yes." ' Since the few remaining palace guards offered no resistance, a detachment of Manchu troops took possession of the smouldering Forbidden City, the hub of the entire Chinese state.[23]

Few people in Beijing realized that a new dynasty had seized power until they read a proclamation issued by Dorgon later that day. It first declared that the 'Great Qing Dynasty' had long sought harmonious relations with the Ming 'hoping for perpetual peace', and in the past had invaded only when their letters were ignored. Meanwhile, bandits had taken control but now the Qing had exacted 'revenge upon the enemy of your ruler-father'. Dorgon took immediate steps designed to win support for the new regime. He announced that the Manchu troops were about to 'unstring their bows' (i.e. stand down), and he reduced the land tax by one-third in all areas that submitted to the Qing. He also visited the body of the late Chongzhen emperor to pay his respects, inviting all former Ming officials to do the same, and he ordered three days of public mourning. This proved an extremely shrewd move, because officials who *now* committed suicide could be seen as expressing loyalty to the Shun regime, and Dorgon capitalized on their dilemma by promising to reinstate and remunerate all former Ming bureaucrats prepared to resume their former office, and an instant one-grade promotion to those who also shaved their heads in Manchu fashion and adopted Manchu dress.[24] Almost all bureaucrats in the capital complied. On 29 October 1644 China's new ruler, the seven-year-old Shunzhi emperor, for whom Dorgon served as regent, entered Beijing. The following day he performed the customary sacrifice to the Supreme Ruler of the Universe at the Temple of Heaven, south of the Forbidden City, and thereby established his claim to be the sole intermediary between Heaven and Earth. A few days later, an official noted in his diary with relief: 'My alarmed spirit has begun to settle. Within the last ten days officials high and low' had resumed work in the capital, so that 'it is just like old times'. He spoke too soon: it would take a generation before all China again enjoyed peace – and, even then, old times would never return.[25]

China Partitioned

Since China had two capitals, the Dashing Prince's capture of Beijing automatically made Nanjing the Ming capital. It possessed many assets. Lying in the lower Yangzi Valley, the richest region of the empire, amid copious supplies of food and a bustling trade, it was the cultural centre of China. Its population was second only to Beijing and it boasted powerful defences. Admittedly the southern capital tended to attract officials with little enthusiasm for war, ill-suited to resist skilful and determined invaders; nevertheless, some weeks before the fall of Beijing, officials in the southern capital made plans for the Ming crown prince to escape and join them and began to mobilize naval and military forces.

The history of China offered a promising precedent – four centuries earlier, the Song dynasty had survived in the south for 150 years after Mongol invaders overran northern China – but the situation in 1644 was very different: crop failures in the 1630s, followed by the worst drought in five centuries (1641–44) had depopulated parts of Jiangnan and left the survivors weak, poor and demoralized. In addition, the sudden scarcity of silver after 1640 (page 104) crippled both trade and tax collection. Amid such weakness, news of the Chongzhen emperor's suicide, and the disappearance of the crown prince, immediately created chaos. Many Ming officials in the south responded to the news of the Chongzhen emperor's death by committing suicide, while several landowning families either set themselves on fire or jumped into wells, and many students either drowned or hanged themselves. Many others considered an interregnum to be a period without laws and acted accordingly. An eyewitness living near Shanghai reported that 'seeing that there was no emperor, the bondsmen made a body of many thousands, and asked their lords for papers of [manumission] because [with the fall of] the Chinese government they were already free. And taking up arms they first turned on the lords in the countryside, killing, robbing and doing a thousand other insults without anyone taking up arms against them.' They also warned magistrates in nearby towns that, unless they received 'the papers of their freedom immediately', they would 'kill all without mercy'.[26]

Not all Ming loyalists gave up. The ministers in Nanjing discussed which member of the imperial family should become the 'caretaker' of the realm until the crown prince reappeared. The prince of Fu, a first cousin of the late emperor, who had supported the hated eunuch Wei's persecution of Donglin, attracted the allegiance of surviving *yandang* officials, but was unacceptable to the self-proclaimed 'righteous ministers'; so although he became first caretaker, and then emperor, bureaucratic factionalism continued to plague the Ming state. Nevertheless, for a moment it seemed that the new ruler in Nanjing might stabilize the situation. He immediately promoted all officials by one grade, scheduled civil service exams, and granted all provinces between the Yangzi and the Yellow rivers tax relief until they recovered from the crisis of the preceding years. He also divided these provinces into four military zones and charged Shi Kefa, one of the few Ming generals to have campaigned successfully against the bandits, with coordinating defensive operations. Shi turned Yangzhou, just north of the Yangzi, into the linchpin of a new defensive system and the springboard for a future offensive to reunite China under its former dynasty.

In Beijing, Dorgon considered his options. When his Chinese advisers urged him to conquer all China, he at first ridiculed the notion – 'Unify it?' he is alleged to have asked: 'We can do no more than gain an inch and hold on to an inch, gain a foot and hold on to a foot' – but he soon discovered that his huge new capital could not survive without rice from Jiangnan, controlled by the Southern Ming, and without coal from Shanxi, controlled by Li Zicheng. Dorgon decided to deal with the second problem first.[27]

Since Li still commanded some 350,000 soldiers whereas the Manchu Grand Army probably now numbered fewer than 100,000 warriors, only half of them Manchus, the regent made several further concessions to win the support of his new Chinese subjects. He barred eunuchs from handling revenues from imperial lands and from participating in court audiences; he also reduced their numbers to about 3,000. He accepted the services of magistrates and ministers who had served both Ming and Shun (although they were later known by the unflattering label *Er chen*, 'ministers who served both dynasties'), and welcomed back many reforming ministers, often associates of the Donglin and other academies, who had lost their posts under the Tianqi and Chongzhen emperors. These officials began to rationalize the legal system (updating the civil code, promulgating new criminal regulations) and to improve revenue collection (forgiving tax arrears in areas that submitted to the Qing and confiscating the vast territories held by Ming clansmen). As Frederic Wakeman, an outstanding historian of the Great Enterprise, observed:

> What is most surprising is how little was actually required in the end to bring the bureaucratic administration up to a reasonable level of efficiency. Adjustments, not replacements; overhaul, not wholesale substitution, were the characteristics of this reform effort. Moreover, the reforms ... were mainly the work of men who had seen service under the Ming, and were now given the opportunities denied them earlier to carry out the kinds of adjustment that would make the system with which they were already familiar work best.[28]

The Qing also courted some important constituencies that the Ming had alienated. Notably, by extolling martial virtues and rewarding military merit, they won over many generals and troops whom the Ming had treated with contempt. Moreover, the ability of the new dynasty to tackle the bandit menace that had made life miserable for so many Chinese won both admiration and support from property owners.

By contrast, two of Dorgon's other initiatives proved highly divisive and unpopular. First, the regent decided to create Tartar towns in a number of strategic cities. In Beijing, Dorgon decreed that only Manchus could live in the Inner City, around the palace, and he forcibly relocated perhaps 300,000 Chinese residents to the Outer City beyond. The Qing took similar steps throughout their territory, eventually creating thirty-four Tartar towns – basically citadels – reserved for Manchu soldiers and their families, while all others had to live outside. This measure naturally infuriated and alienated those compelled to abandon their ancestral homes.

A second innovation proved even more divisive and unpopular because it affected all males, whether living in town or countryside. The Qing had always insisted that their new subjects must make themselves look like Manchus; but realizing that it would take time to acquire new clothes (or adapt old ones), Dorgon at this stage insisted only that all males must shave the front of their heads and wear the rest in a pigtail at the back. He failed to anticipate how much resentment this would cause. Han Chinese culture viewed correct attire and appearance as an essential distinction between civilization and barbarism; and bundling hair into a topknot (*shufu*) formed part of the etiquette that marked the passage of a Han male to adulthood. Dorgon's tonsure decree challenged a fundamental aspect of traditional Chinese culture – and not once but repeatedly, since to keep the front part of the head bald required regular shaving.

Revolts against head-shaving broke out almost immediately in some areas near the capital and, after three weeks, Dorgon prudently backed down. 'Formerly', he explained:

Because there was no way of distinguishing people who had surren-dered, I ordered them to cut their hair in order to separate the yielding from the rebellious. Now I hear that this is directly contrary to the people's wishes, which contradicts my own [desire] to settle the people's minds with civil persuasion. Let each of the ministers and commoners from now on arrange their hair in the old [style], completely according to their convenience.[29]

This compromise convinced most ministers and commoners in northern China to accept the Qing.

Wu Sangui and a part of the Manchu Grand Army now set off in pursuit of Li, who had retreated to Xi'an. He could not hold it: instead, in February 1645, the Dashing Prince took to the hills with a handful of followers, where some months later he met a violent end. Wu, having won control of all his strategic objectives in the northwest (including the vital Shanxi coalfields), now invaded Sichuan to destroy Zhang Xianzhong's Great Western State.

In Beijing, Dorgon bought time by plying the Southern Ming with reas-suring messages that hinted at a negotiated partition of China. 'As for those who have not forgotten the Ming [dynasty] and have supported and enthroned a worthy prince, putting forward their fullest effort with unified hearts to protect the lower Yangzi [against the bandits]', he wrote soothingly to them, 'that is as things should be. We will not stop you. But you should contact us for peaceful, amicable discussions and not rebuff our dynasty.'[30] The Nanjing regime countered with an embassy that thanked the Qing for expelling the bandits from the capital and promised that if they withdrew beyond the Great Wall, they could keep all land beyond it and receive a hand-some annual tribute. This was totally unrealistic: the Manchus had come too far, and needed the resources of the Central Plain too much, simply to go home. Nevertheless Dorgon entertained the embassy from Nanjing with false hopes until, in March 1645, he felt it safe to recall some of his troops from the west and launch a devastating attack on the Southern Ming.

'Keep Your Head, Lose Your Hair; Keep Your Hair, Lose Your Head'

Appalling weather continued to afflict Jiangnan: the winter of 1644–45 saw the weakest monsoon in more than a millennium, prolonging the drought. As the Grand Army advanced southwards from Beijing most cities opened their gates and offered tribute to the conquerors until by May 1645 only Yangzhou held out north of the Yangzi river. The city's makeshift defences proved no match for the heavy artillery brought by the Qing and after it fell the victors sacked it for a week. Only unseasonable rain prevented the fires lit by the looters from consuming the entire city.

This strategic use of terror bore immediate fruit: Nanjing and most other cities in Jiangnan surrendered to avoid the fate of Yangzhou. Nevertheless a few Ming loyalists elsewhere in the south organized resistance, forcing the Manchus once again to find an easy way to distinguish friend from foe – albeit more sensitively than before. Instead of insisting that all Chinese males must shave their heads, a new tonsure decree of June 1645 announced that: 'In all of the places occupied by the Grand Army we will shave the military and not shave the civilians; we will shave the soldiers and not shave the people'. A group of Chinese scholars in the capital unwisely protested against even this compromise, claiming that their traditional system of rites and music required Han Chinese to preserve intact all they had inherited from their parents, so that shaving forelocks was a kind of tonsorial castration. This irritated Dorgon, who insisted that the tonsure decree must be implemented within ten days of its receipt in each locality: disobedience would be 'equivalent to a rebel's defying the Mandate [of Heaven]', and any official who sought 'to retain the Ming institutions and not follow those of this dynasty' would face immediate execution.[31]

This angry overreaction proved a catastrophic error. As a Jesuit eyewitness tartly observed, the Chinese elite now 'grieved and fought more valiantly for their hair and habit than they had done for their kingdom and emperor'.[32] An incident near Nanjing illustrated this point. After the surrender of the southern capital, the new Qing authorities urgently sought information on the resources of the region, dispatching commissioners to secure the tax and population registers of each county. Since the Qing lacked their own reliable cadres, many of these commissioners were former Ming officials: Fang Heng, a young *jinshi* official sent to the town of Jiangyin, was one of them. He arrived still wearing Ming insignia on his robes and was on the point of securing the registers when four Manchu soldiers arrived with the tonsure decree and orders for its immediate implementation. Fang obeyed and prepared a simple Chinese version of Dorgon's proclamation: 'Keep your head, lose your hair; keep your hair, lose your head'. This provoked an armed insurrection that cost the lives of all four Manchu soldiers as well as Fang – but not before he had sent an appeal for reinforcements. Qing troops arrived with twenty-four siege guns, which breached the town walls and enabled a successful assault. Dorgon had ordered his troops to 'fill the city with corpses before you sheathe your swords', and they duly obliged. Tens of thousands of Ming loyalists at Jiangyin kept their hair but lost their heads.[33]

Similar events took place elsewhere as places changed hands (sometimes several times). Yao Tinglin, scion of a gentry family from Shanghai, later recalled how in 1645

> Whenever the Qing army approached every household in the towns and villages would paste pieces of yellow paper on the door with the characters *Da Qing shunmin* ['obedient subjects of the Great Qing'] written on them, but would tear them down as soon as the loyalist rebellion seemed to get the upper hand, only to paste them up again when the Qing troops were supposed to come back.

This explains why Dorgon enforced the tonsure decree so ruthlessly: a man could opportunistically paste up and tear down pieces of paper, but he could not fake his tonsure. By the end of 1645 the Qing had prevailed throughout Jiangnan. In Shanghai, Yao Tinglin recognized that 'nothing would be the same as before: it was to be a new dynasty, people looking differently, new social hierarchies, new rituals, and so forth – in short, "another world, with no restoring of the old order"'.[34]

'China in Tigers' Jaws'

The Qing conquest of Jiangnan marked an important stage in the success of the Great Enterprise because it secured Beijing's food supply. Rice, millet, wheat and beans once again came up the Grand Canal to be stored in the capital's enormous granaries. Although most of these supplies went directly to the Manchu families of the Inner City, other residents benefited because the Qing used the granaries to keep prices low and to run soup kitchens for the poor. They also preserved the traditional examination system for the civil service and tightened up the lax examination standards that had prevailed under the late Ming (page 106), for example by executing candidates found to have cheated.

As one Qing minister wearily noted, however, 'Seizing the empire is easy; ruling it is difficult.'[35] The absence of any effective geographical barrier separating Beijing from the lower Yangzi had greatly facilitated the Qing's progress thus far, but extending control into areas of the south and west loyal to the Southern Ming presented greater challenges, especially since the

Little Ice Age continued to cause hardship and disruption. The winter of 1649–50 seems to have been the coldest on record in both northern and eastern China, and Beijing experienced such a serious drought in 1657 and again in 1660 that the emperor personally conducted prayers for rain. In the south, seventeen counties in Guangdong province reported frost or snow in the 1650s – the highest number in two centuries. The same province also suffered more typhoon landfalls between 1660 and 1680 than at any other time. Moreover, troops raised on the steppe had seldom encountered smallpox and so possessed little immunity when they entered China: the sudden death from the disease of so many Manchu troops and commanders began to undermine campaign plans. The Qing adopted several panic measures to contain the disease (such as banishing from their army any soldier who contracted the disease and ordering that only princes who had survived smallpox should serve as generals) but they failed. Inevitably, not all cases of smallpox were detected before they spread to other soldiers, and some princes eager to acquire honour and booty disregarded the risk of infection, insisted on campaigning and promptly died.

Nevertheless, despite abundant proof that head-shaving united their opponents, the Manchu leaders persevered with the policy. When the principal Ming naval commander, Zheng Chenggong (known as Coxinga in Western sources), seemed willing to defect, the Qing offered him a ducal title and lands, promised that he could keep all his troops under arms, and conceded that all shipping off the coast of Fujian 'shall be subject to your management, inspection and collection of taxes'. Then, just when Coxinga seemed about to accept these terms, the envoys sent by Beijing to handle the final negotiations bluntly stated: 'If you do not shave your head, then you cannot receive the [emperor's] proclamation. If your head is not shaven, then we need not even meet.'[36] Outraged, Coxinga used his warships – some of them armed with Western-style artillery – to dominate all trade in the South China Sea until by 1659 he had gathered sufficient funds and support to mount a campaign up the Yangzi. Thirty-two counties and seven prefectural capitals declared their allegiance before a Qing counter-attack forced Coxinga first to retreat to the coast and eventually to abandon the Chinese mainland, making the island of Taiwan his new base. Lacking a fleet capable of pursuing him there, the Qing now imposed another deeply unpopular measure on their Chinese subjects: in the hope of starving out Taiwan, they ordered all who lived within 20 miles of the mainland's southeast coast to abandon their homes and move inland. They then forbade all seaborne trade, destroyed all buildings and issued orders to kill any human found in the no-go area.

The trade embargo produced what the Chinese called *shu huang*: 'dearth in the midst of plenty'. Famines, war and disorder had already decimated domestic consumption in southeast China: now the government removed all export markets so that, once better weather increased rice yields, supply rapidly outstripped demand. The price of a bushel of rice fell until, according to a local source, 'with such low grain prices, peasant-farmers could not pay taxes or support their families, thus much land was abandoned'.[37]

One cannot attribute this prolonged devastation solely to the Qing. As Lynn Struve noted in her path-breaking study of the Southern Ming: 'The long struggle between the Ming and Qing was not so much a direct clash between two states as a competition to see which side would prevail over, or be defeated by, a third state, so to speak: that of socio-political anarchy'; and 'generally speaking, the Ming lost in this competition more rapidly than the Qing won'.[38] First the Southern Ming court fled south to Fujian, then westward to Yunnan and into Burma where in 1661 Wu Sangui hunted down the last claimant and executed him. After this regicide, the jubilant Qing granted Wu extensive powers in Yunnan, where he settled with his victorious troops and created a prosperous fief. That same year, Coxinga and his followers departed for Taiwan. With that, all organized resistance to the Qing on the Chinese mainland temporarily ceased.

Nevertheless, for several reasons Qing power remained fragile. The death of the Shunzhi emperor in 1661 left a vacuum in which various Manchu factions competed for power, destabilizing the entire government. In addition, fiscal shortfalls caused by the cost of the various Manchu conquests led the central government to decree that all tax arrears must be paid immediately and to threaten that officials would be denied promotion, or even demoted or dismissed, unless they delivered outstanding tax quotas in full. In Jiangnan, a combination of unusually high tax evasion and lingering loyalty to the Ming led the Qing to make good on this threat, imprisoning or executing tax resisters and barring them from holding office. Although these savage measures worked in the short-term – tax evasion by the gentry dropped dramatically – they created a reservoir of discontent: in the words of a Jiangnan gentleman-scholar, the 'laws were like a frost withering the autumn grass'.[39] Meanwhile, global cooling continued to afflict most of China: nine of the fourteen summers between 1666 and 1679 were either cool or exceptionally cool, and a study of Chinese glaciers suggests a late seventeenth-century climate on average more than 1°C colder in the west and more than 2°C colder in the northwest than today.

A further challenge to Qing power arose in 1675 when Wu Sangui submitted a formal petition to the Kangxi emperor requesting, on grounds of old age, that he be allowed to resign his fief of Yunnan, and that his son succeed him. The emperor, now aged nineteen, granted only the first request and took steps to establish central control over the region. Alarmed, two other Chinese generals appointed to rule southeastern provinces as fiefs decided to test Qing resolve by submitting identical petitions.

Most of the imperial council favoured granting the petitions, fearing that rejection would lead to rebellion, and they urged the Kangxi emperor to temporize; but he opted for confrontation and terminated their authority. As the councillors had predicted, Wu Sangui promptly rebelled and (despite having executed the last ruler of the former dynasty) adopted the slogan 'Fight the Qing and restore the Ming' and proclaimed that all men could once again dress and wear their hair in the traditional Chinese fashion. The two other feudatories joined Wu in rebellion (hence the name 'The Revolt of the Three Feudatories'), and so did many other disgruntled Chinese (including some of the Jiangnan gentry imprisoned for tax evasion a decade before, as well as troops from Taiwan commanded by Coxinga's son).

Kangxi continued to underestimate the scale of the challenge he faced, sending a mere 10,000 troops to confront the rebels, who gained control of almost all China south of the Yangzi, as well as of Sichuan and Shaanxi provinces. Wu demanded that the Manchus withdraw beyond the Great Wall, while the Dalai Lama (then as now a major Buddhist leader) offered to broker a deal that would partition China between Wu and the Qing. It took three years and the deployment of over half a million troops to regain all the rebel territories, but once Qing forces had crushed the Three Feudatories, they turned to Taiwan, where in 1683 they forced Coxinga's successor and the last Ming loyalists to surrender. The emperor now authorized the return of local inhabitants to the coastal areas of southeast China from which they had been banned for a generation, and allowed the resumption of maritime trade. After almost seventy years of disruption, destruction and disaster, the Qing had completed their Great Enterprise.

The Cost of Changing the Mandate

The victorious Kangxi emperor now visited the tombs of his ancestors at Shenyang to 'perform the rites of reporting success', and then toured Manchuria. In 1684 he set off to visit Jiangnan: the first emperor to do so in almost three centuries. According to his official diary, his aim was to 'investigate the unknown sufferings of the people'; but he did not investigate too closely. On his 2,000-mile tour, the emperor avoided places punished for their loyalty to the Ming, such as Yangzhou; he slept in the segregated and safe Tartar Towns; and after two months he returned to Beijing.[40]

The imperial party also bypassed the provinces that had suffered the worst devastation, such as Hunan, where a government survey revealed a shortfall in taxpayers of up to 90 per cent. Nevertheless, the emperor must have seen abundant evidence of depopulation and devastation along his route. The total of cultivated land in the empire had fallen from 191 million acres in 1602 to 67 million in 1645, with a partial recovery to only 100 million in 1685, the year after his tour. Human losses were qualitative as well as quantitative. In 1702 historians commissioned by Kangxi to compile the official *Ming history* included the biographies of almost 600 men who killed themselves out of loyalty to the Ming dynasty and of almost 400 women who either 'followed their husbands into death' or committed suicide after being dishonoured. Since these totals only included those who had definitely taken their own lives out of principle, and omitted the rest, the true total must have been far higher: many other members of the elite, both male and female, died because they simply got in the way of soldiers, rebels or bandits, and left no surviving documentary trace. It therefore seems appropriate to speak of a lost generation: in no other part of the seventeenth-century world did such a high proportion of the elite meet a violent end, with the exception of Germany – and Germany was far smaller than China.

The compilers of the *Ming history* omitted many other victims of the transition, notably the millions of Han Chinese who became slaves, either because bannermen took them as booty or because they sold themselves and their families into servitude to escape debts, taxes or starvation. Qing Beijing had a lively slave market, and the dynasty enforced harsh fugitive slave laws. The Dutch diplomatic envoy Johan Nieuhof provided a graphic example of the consequences. As he and his colleagues travelled by boat upriver from Guangzhou towards Nanjing in 1656,

> We saw into what a miserable condition the Chinese were reduced by the last war of the Tartars, who put them upon this slavish labour of towing and rowing their boats, using them worse than beasts at their pleasure, without any exception of persons, either young or old. Often the track'd ways on the riverside are so narrow, uneven and steep, that if they should slip, they would infallibly break their necks, as many times it happens … and if any of them grow faint and weary, there is one that follows, having charge of the boat, who never leaves beating of them, till they go on or die.[41]

Although the Kangxi emperor minimized contact with the vanquished on his southern tour in 1684, he mingled with members of the largest group of beneficiaries from the mid-seventeenth-century crisis: the Manchu warriors and their families. When Nurhaci declared war on the Ming in 1618, he commanded only a few thousand followers, with the grudging support of some Mongol neighbours, and he worried about how to feed them. Twenty-five years later, his son Hong Taiji commanded a powerful confederation of steppe people and ruled over perhaps a million Chinese settlers; but he too worried about how to feed them. In 1684, by contrast, tens of thousands of Manchus and their Mongol allies lived comfortably in the vast empire they had conquered. Each received a salary in silver, a grain subsidy, a housing allowance, arms and ammunition, pensions, loans and land. Most owned slaves.

Eventually, many of the Qing's Chinese subjects joined the ranks of these winners as the new dynasty restored law and order and secured the frontiers, enabling the surviving Han population to prosper and multiply even before the return of a benign climate in the early eighteenth century improved farming conditions. The Qing also solved other problems that had beset their predecessors. The high losses among civil servants, whether through death or resignation, opened up thousands of positions in the bureaucracy and allowed the promotion of many alienated intellectuals whose constant criticisms had weakened the former dynasty. In many rural areas, mortality and migration made available much rich farmland, and allowed surviving farmers to join several small plots together and enjoy greater prosperity. The central government also revitalized the infrastructure of their realm, including roads and granaries, and pioneered the technique of variolation, which dramatically reduced deaths from smallpox (Chapter 21). As Joanna Waley-Cohen has stressed, the Great Enterprise also strengthened 'the empire by uniting its diverse people through the creation of a common basis, one that was founded on loyal pride in imperial achievement and in which all could participate'. When Yao Tinglin completed his *Record of successive years* in Jiangnan in the 1680s, he had no doubt that 'the people of Shanghai were living in a more prosperous and more peaceful world than has ever existed'.[42]

Perhaps Yao's comparison was unfair to the Ming. The Chongzhen emperor had faced problems that probably no ruler could overcome: an inefficient and inequitable tax burden; contempt for martial values at a time when the state faced powerful domestic and foreign enemies; factionalism that paralyzed the bureaucracy; corruption that eventually entered even the examination system; and, above all, the Little Ice Age. As Timothy Brook wrote, 'No emperor of the Yuan or Ming faced climatic conditions as abnormal and severe as Chongzhen'. Indeed, he added, 'the greatest puzzle might well be to figure out how the Ming remained standing for as long as it did'.[43] The invaders, for their part, twice placed their Great Enterprise at risk, and thereby prolonged the suffering of their subjects: once in 1645 with Dorgon's tonsure decree and again in 1673 with Kangxi's decision to remove the Three Feudatories. Eventually, however, Qing military prowess prevailed and the new dynasty went on to conquer extensive territories in Inner Asia, doubling the area controlled from Beijing to more than 4 million square miles (larger than all of Europe west of the Urals and far larger than China today). Westward expansion might have continued further, had the Qing not encountered the outposts of an even larger state expanding eastward: Romanov Russia.

CHAPTER SIX

The 'Great Shaking': Russia and the Polish-Lithuanian Commonwealth, 1618-86[1]

The Humiliation of Russia

In September 1618 troops commanded by the Polish crown prince Władysław Vasa stormed Moscow. Although the assault failed, Tsar Michael Romanov (r. 1613-45) agreed to the Truce of Deulino, by which he relinquished all Russian lands conquered over the previous decade by the Polish-Lithuanian Commonwealth, which now became the largest state in Europe, twice the size of France. The tsar had little choice: twenty years of famines, rebellions, civil wars, and invasions by both Sweden and Poland had reduced Russia's population by perhaps one-quarter. In some areas more than half the villages and even entire towns had been abandoned. The intervening period soon became known in Russian history as *smuta*: 'The Great Trouble'. A generation later, in 1648, angry Muscovites broke into the Kremlin, sacked the apartments of the tsar's leading ministers and murdered two of them, triggering rebellions elsewhere in the empire. Astonishingly, Michael's son Alexei Romanov (r. 1645-76) not only survived this crisis but went on to vanquish the Polish-Lithuanian Commonwealth, regain all the lands surrendered at Deulino and extend the bounds of Russia until it covered 6 million square miles, making it the largest state in the world.

Since messages between the tsar and these new acquisitions might take two years to arrive, Russia's rulers faced an important dilemma: whether to leave local affairs in local hands, and risk political disintegration, or to maintain central control and sacrifice both efficiency and potentially constructive local initiatives. Admittedly, certain environmental features reduced the dilemma somewhat. Although the major rivers of Siberia – Ob, Yenisey and Lena – run north–south, their tributaries form an almost continuous east–west waterway from the Urals to Lake Baikal. Likewise the broad rivers that run from Muscovy south toward the Black and Caspian Seas – Dnieper, Donets, Don and Volga – allow communications by boat in the summer and on the ice in winter. These natural corridors permitted not only mass migration, the transmission of orders, tribute and trade, but also dramatic military raids: Cossack adventurers captured Sinop in Anatolia in 1614 and Azov near the Crimea in 1641.

The principal strategic challenges to the Russian state nevertheless lay elsewhere. Smolensk, forward bastion of the Polish-Lithuanian Commonwealth, stood only 250 miles west of Moscow; Narva, a Baltic outpost of the

Swedish state, stood 500 miles to the northwest. Each of these three states adhered to a different branch of Christianity: Russia was solidly Orthodox, Poland-Lithuania was predominantly Catholic, and Sweden was overwhelmingly Lutheran.

Russia's two western neighbours, both ruled by branches of the House of Vasa, also covered vast areas. The Swedish state stretched from the southern shore of the Baltic to the North Cape, a distance of over 600 miles, and the Polish-Lithuanian Commonwealth (*Rzeczpopolita*) stretched from the Baltic almost to the Black Sea, also a distance of over 600 miles. Yet although the dimensions of all three states dwarfed the size of their West European neighbours, their populations remained far inferior. Russia and the Commonwealth scarcely boasted 11 million inhabitants each, while the Swedish crown included only about 2 million. By way of comparison, both France and the Holy Roman Empire contained about 20 million inhabitants. These figures disguise striking differences in population density. Whereas parts of Western Europe boasted twenty-two inhabitants per square mile, Poland averaged eight, Lithuania six and Muscovy not even one. Moreover, according to the first reasonably complete Russian census, in 1678, almost 70 per cent of the tsar's subjects lived in the lands north of Moscow while only 1 per cent inhabited the vast expanse of Siberia. The rest lived in the steppe south of Moscow, especially on a zone of *chernozem* ('black earth', so called because of its colour) some 200 miles wide that ran from the Black Sea into Siberia: 270 million acres of soil so fertile that, according to a Western traveller, the 'grass grew so tall that it reached the horses' stomachs' and, wherever farmers chose to sow grain, 'everyone is confident of a rich annual harvest'.[2]

Throughout the seventeenth century three groups vied for control of the black earth. Initially, Muslim Tartar raiders based on the Crimea and owing nominal obedience to the Ottoman sultan predominated, but gradually Russian farmers spread southwards, first along the banks of the Don and then along the Dnieper, where they faced subjects of the Polish-Lithuanian Commonwealth settling in Ukraine.[3] Paradoxically, the colonization of their southern frontier destabilized both states. In Russia, the cost of defending the borderlands required heavy taxes, while the southward flight of so many peasant families to the new lands led northern landlords to demand draconian measures from the tsar to prevent the haemorrhage of their servile labour force. In the Polish-Lithuanian Commonwealth, the southwards migration not only of serfs but also of Jews trying to escape the restrictions imposed on them by various towns increased the burdens on those who remained.

The fact that the Russian, Polish and Swedish monarchies were all composite states also caused instability. The tsar ruled over numerous distinct ethno-linguistic groups, including Muslims along the middle and lower Volga and nomads in Siberia, some of whom preserved considerable political autonomy and even some of their own institutions after incorporation. Likewise, although the kingdom of Sweden boasted remarkable religious and administrative uniformity, its overseas territories – Finland, Estonia and (later) parts of Germany – retained great political autonomy, as well as their own separate representative assemblies, languages and legal systems. Each component of the Polish-Lithuanian Commonwealth retained its own legal systems, treasury, army, and local Estates (Sejmiki). In addition, the Commonwealth boasted great ethnic diversity: large communities of Tartars, Scots and Armenians lived among the Poles, Germans, Lithuanians and Ruthenians.

The constitution of the Commonwealth deepened this diversity. At the death of each monarch, representatives from every region assembled in a federal Diet (the Sejm) to negotiate concessions from the various claimants before electing one of them king. Thereafter, representatives from the Sejmiki met together for six weeks at least once every two years, with emergency sessions when necessary, and at the end of each Diet, a plenary session debated all the legislation recommended for enactment. At this stage, the veto of just one representative on just one issue required the king to dissolve the Diet without passing any legislation (not even measures already agreed upon). Although both foreign contemporaries and most subsequent historians castigated the *liberum veto* as a weakness that doomed the Commonwealth to decline, they exaggerated: no one used it until 1652 – and even in that year a new Diet convened four months later and passed all pending legislation. In fact, the *liberum veto* safeguarded regional rights (which is why attempts to replace unanimity with some form of majority rule always failed); and the Sejm offered the earliest example in world history of a federal parliament that bound together a multi-national and multi-ethnic state.

The Commonwealth's principal weakness lay in its religious pluralism. Catholicism predominated in Poland, but in both Lithuania and Ukraine it competed with a powerful Orthodox Church and, after 1596, with a distinct Uniate Church created specifically to reconcile an important group of Orthodox Christians with Rome. In addition, the Commonwealth was home to numerous other religious groups. Each landlord had the right to determine the faith of his subjects, and major cities had the right to grant toleration to whomever they pleased. The city of Lwów (Lviv) contained thirty Catholic churches (and fifteen monasteries), fifteen Orthodox churches (and three monasteries), three Armenian churches (one of them a cathedral) and three synagogues. The Roman Catholic hierarchy, strongly supported by the crown, used a wide range of economic, social and political inducements to exploit this diversity and win converts. Their success can be measured by the fact that although the federal Diet in 1570 included fifty-nine non-Catholic lay senators, that of 1630 included only six. Over the same period the number of Protestant communities in Poland fell from over 500 to scarcely 250, and large numbers of Orthodox clerics and laity joined either the Catholic or the Uniate Church.

Nevertheless, for the first half of the seventeenth century the Commonwealth held its own against its neighbours, largely because it embraced Western military technology. King Sigismund III (r. 1587–1632) created special musketeer infantry formations, standardized artillery calibres and added field guns – all modelled on Western prototypes. Although progress remained slow, mainly because the Polish nobles resisted any measure that might enhance the power of the monarchy (such as hiring foreign mercenaries, arming serfs, and fortifying royal cities), the military effectiveness of the Commonwealth's troops steadily improved.

The humiliating Great Trouble convinced Tsar Michael Romanov that he must imitate the military methods of the Commonwealth. He therefore welcomed foreign military advisers to train and command so-called New Formation Regiments equipped with Western weapons; and when, on the death of Sigismund III in 1632, the Sejm bargained with Crown Prince Władysław before electing him, the tsar launched an invasion to recapture the lands sacrificed at Deulino. The Russians swiftly captured twenty towns and laid siege to Smolensk, but Smolensk held out until Władysław, now elected king of Poland, arrived with a relief army and besieged the besiegers. In February 1634 the Russian commanders surrendered, and a few weeks later Michael reluctantly concluded the Eternal Peace of Polianovka, by which he not only confirmed all Poland's territorial gains at Deulino but also promised to dissolve his New Formation Regiments and to pay a huge war indemnity. The cost of this new defeat created tensions and problems that would shake the Romanov state to its foundations.

The 'Imaginary Little World' of the Muscovites

Understanding the Russian crisis of the mid-seventeenth century is complicated by the paucity of sources. Only the archives of the Siberian chancellery have survived relatively intact for the early modern period. Only Sweden maintained a permanent diplomatic representative in Moscow, and on some events his dispatches form the only surviving source. Most Russians refused to talk to foreigners because they were either too fearful or too absorbed in what one of them dismissed as 'upholding their imaginary little world'.[4] Luckily for historians, the natural archive of the period fills some of the gaps in the human archive. Climatic reconstructions reveal notably cooler and drier conditions in the 1640s in Ukraine; poor harvests in the south in 1642; drought and a plague of locusts in 1645 and 1646; and early frosts and poor harvests in the south in 1647 and 1648. When the government carried out a land survey in 1645–46, the commissioners found that many communities had shrunk significantly in both size and wealth since the previous survey two decades earlier. One of the few detailed studies of this census data (for Karelia, the area between Lake Ladoga and the White Sea) reveals that from 1628 to 1646 the overall number of households declined by one-quarter, and gentry households by almost one-half, while the proportion of landless peasants doubled.

Tsar Michael nevertheless increased the tax burden on this dwindling population. Besides the indemnity payable to Poland, the tsar embarked on costly measures to improve the defences of his dominions. He stockpiled foreign munitions, recruited more Western officers to train his New Formation Regiments and ordered the construction of almost fifty fortified new towns linked by wooden and earthen ramparts that ran for 800 miles from the Dnieper to the Volga. The western half of Russia's 'Great Wall' became known as the Belgorod Line, after the largest new town in the region, and the eastern half as the Simbirsk Line, after the town on the Volga where it terminated (Fig. 9).

These lines, built between 1636 and 1654, represented one of the greatest military engineering feats of the seventeenth century – but the cost proved crippling. Apart from the capital outlay required for construction, each of the new towns and fortresses required a permanent garrison. In addition, each summer, the government mobilized an army of servitors (*deti boiarskie*: gentry who held lands in return for military service) and sent them south to guard against Tartar invasion. Beyond the fortifications, the central government also supported thousands of Cossacks (perhaps from *kazac*, the Turkic word for 'free man'): fiercely Orthodox warriors who patrolled the steppe to disrupt any attack by the Muslim Tartars or their Ottoman overlords.

Despite the general uncertainty and the high costs of defence, Russia continued to expand southward. Whereas the growing season in Karelia begins in April and ends in September, around Belgorod it begins a month earlier and ends a month later. This important discrepancy enhanced the attraction of the *chernozem* farmland now protected by the Belgorod and Simbirsk Lines. Admittedly, cutting down forests to create fields gave rise to hotter summers, colder winters and a greater risk of extreme events such as droughts in the region, but *chernozem* continued to produce bountiful crops in all except years of exceptional climatic adversity (such as 1647 and 1648). Furthermore, in the 1630s the tsar exempted the pioneers from paying taxes for a decade or more. Not surprisingly, this combination of advantages attracted a substantial migration of peasants from the north. Some joined the Cossacks beyond the Belgorod and Simbirsk Lines, seeking Tartar booty, and others fell victim to Tartar raiders and begged to be ransomed; but most settled and prospered on the hundreds of thousands of acres of virgin farmland now available behind the lines.

9. The Russian empire and the Polish-Lithuanian Commonwealth.
Imperial Russia had to defend itself against three enemies: Sweden and the
Polish-Lithuanian Commonwealth in the west, and the Crimean Tartars in the
south. Against the latter, the Romanov tsars constructed the fortified Belgorod and
Simbirsk Lines that made use of natural features such as forests and rivers. In 1670,
the lines also halted a Cossack invasion. The Polish-Lithuanian Commonwealth
had no such defences, and in 1648 a Cossack invasion fatally weakened the state,
opening the way to invasions by both its Russian and Swedish neighbours. The
Commonwealth briefly ceased to exist.

This mass migration created a crisis for the servitors of the northern regions, on whom the tsars traditionally depended to defend the empire. To finance their military service, most servitors claimed that they required the labour and services exacted from at least twenty unfree inhabitants of their lands, but by the 1630s, at least in the Moscow region, they owned on average only six peasant households. Servitors therefore repeatedly petitioned the central government to compel fugitive serfs to return and to abolish the time limit for their recovery, but although in 1636 the tsar decreed that all former serfs in the south must return, he limited the measure to those who had fled *before* 1613 and he exempted those who had fled to Siberia.

The Tsar and his 'Slaves'

The rapid territorial growth of imperial Russia led to an equally rapid expansion of the central bureaucracy. Tsar Michael created forty-four new departments of state in Moscow (known as *prikazy*: chancelleries), and his son Alexei (r. 1645–76) created thirty more. Both the volume of business handled by each *prikaz* and the staff employed to handle it grew exponentially. For example, the military chancellery employed fifty-five men in the 1620s but over a hundred in the 1660s, and issued orders that ranged from major troop movements down to how many cubic metres of earth should be moved to create ramparts and how many timber beams of a specified dimension should be used to construct the main gate of a frontier town. Nevertheless, even as imperial Russia became more hierarchical, one exception remained. In the admirable formula of Valerie Kivelson: 'All interactions with the state, whether legal disputes in court or communications between provincial and central offices, were framed as humble petitions addressed directly to the tsar.' 'Humble' seems an understatement: the phrase for 'petition' in Russian, *bit' chelom*, literally means 'to beat one's brow [to the ground]', and all petitioners addressing the tsar called themselves 'Your slave'. Even nobles used demeaning diminutives of their names ('I, little Ivan, Your slave, beg you . . .').[5] Since petitions to the tsar formed the only legal channel by which subjects might request redress of their grievances, and since government intervention in daily life (and, consequently, opportunities for abuse) steadily increased, both grievances and petitions multiplied. To cope with them, Michael Romanov created a special petitions chancellery and one of its officials always travelled with him in order to direct each petition to the appropriate department of state.

On Michael's death in 1645 his sixteen-year-old son and successor Alexei faced so many petitions, many of them from groups of servitors protesting about the shortage of serf labour, that the Swedish envoy in Moscow believed they presaged a general uprising; but the new regime pacified the petitioners by reducing the military service obligation of the gentry by one day and promising to undertake a new census of all Russia as a preliminary step towards abolishing all restrictions on reclaiming fugitive serfs. Inspectors from the central government duly visited each administrative region to ascertain the number of taxable households (and whether they paid less or more now than before), the amount of land under cultivation (as well as land abandoned) and the economic activities of each village (including estimates of crop yields). They also compiled lists of all serfs who had fled during the previous ten years. But no edict abolishing the time limit for recovering fugitive serfs materialized – not least because Alexei's chief adviser (and brother-in-law) Boris Morozov had welcomed so many runaway serfs to his own extensive domains and concealed them from the government inspectors.

Morozov acted much like royal favourites in Western Europe. First he drove from office those who had advised the previous tsar and then, in order to fund defence spending, he made some highly unpopular fiscal decisions. He instructed another of the tsar's brothers-in-law, treasury minister Peter Trakhaniotov, to freeze the wages earned by the *streltsy* – the corps of musketeers recruited from the gentry, on whom the government relied to preserve law and order in the capital – creating a group of discontented subjects who were both organized and armed. He also approved a proposal from Nazarii Chistyi, a prosperous Moscow merchant, to introduce indirect imposts modelled on Dutch prototypes. In 1646 imperial decrees imposed a stamp tax on the paper required for all official transactions and created state monopolies for the sale of tobacco and salt. The salt monopoly, in particular, provoked widespread resentment because it increased prices fortyfold. Faced with such brutal inflation, consumers purchased far less salt (causing sales and tax revenues to fall dramatically) and protested. Late in 1647 the tsar brought some 10,000 musketeers into the capital to preserve order, but when popular unrest continued he reluctantly revoked the salt monopoly.

Other anomalies in the Russian tax system also caused unrest. Morozov entrusted control over the commerce of the entire capital to another of his relatives, Levontii Pleshcheev, who extorted bribes of unprecedented size in return for rendering justice. Moreover, in an attempt to make good the shortfall caused by the revocation of the salt monopoly, the treasury began the ruthless collection of tax arrears from the previous two years. A recent study estimates that these measures together tripled the tax burden for 1648, but in doing so Morozov had managed to alienate almost simultaneously town elites, servitors and ordinary taxpayers.[6]

In April 1648, Morozov mobilized the servitors, ready to march south and parry an anticipated Tartar raid, but when no incursion occurred he disbanded them without pay. The Moscow contingent was still in the capital, alienated and angry, when the following month some of the tsar's subjects decided to present their grievances to him in the accustomed manner: through a formal petition. These servitors no doubt shared the outrage of the crowd as they watched the imperial entourage intercept the petition so that the tsar would not see it. Instead, he and his family left the city on a pilgrimage to a nearby monastery.

'The Whole World is Shaking'

According to Patrick Gordon, a Scottish immigrant who served in the tsar's army for almost four decades, the Russians were 'morose, avaricious, niggard, deceitful, false; insolent and tirrannous, where they have command; and, being under command, submissive and even slavish, sloven and base, and yet overweening and valuing themselves above all other nations'.[7] The failure of Tsar Alexei to receive his subjects' petition stirred up these passions, and in his absence crowds gathered in and around the churches of the capital to vent their grievances. They decided to try again to make him listen and on 11 June 1648 a large group emerged from the city to greet their ruler as he returned from his pilgrimage, bearing not only the customary bread and salt but also copies of a petition denouncing the corruption of Levontii Pleshcheev. Once again, the imperial entourage intercepted the documents and Alexei's guards used their whips to drive the crowd back. At one point they opened fire, but a few of the enraged protestors followed the tsar into the Kremlin with their formal supplication; Morozov had them imprisoned. Half an hour later, as the tsarina returned from the monastery in her carriage, the crowd tried to present their grievances to her, but once again the guards drove them back. At this, 'the entire crowd, totally exasperated, threw stones and used cudgels against the guards'.[8]

Both parties had transgressed important boundaries: the tsar had unexpectedly and inexplicably rejected a petition, the only legitimate avenue for subjects to bring their problems to his attention, and a group of subjects had forced their way into the Kremlin, a sacred space for both church and state. A far worse transgression occurred the next day during a religious festival, when the royal family left the Kremlin to worship. Crowds surrounded the tsar as he returned from church, begging that he accept their supplication. Although the document used the customary self-abasing style of petitions to the tsar ('Your slaves and poor subjects'), it also contained some notable innovations. It claimed to speak in the name of the 'persons of all estates and the whole people' (not just of Moscow or a particular group), and it lamented the intolerable 'bribes, presents and gifts' demanded by bureaucrats in return for any official act (the tsar could easily identify the offenders, the document claimed, because they 'build themselves houses that are not commensurate with their stations'). The supplicants boldly reminded the tsar of 'the story in your tsarist palace [the Kremlin] of the Greek king in Constantinople, Justinian', who averted divine punishment by issuing laws 'in all his land to abstain from all injustice and the oppression of the poor', and continued:

> Now Your Tsarist Majesty can do exactly the same, if you wish to avoid the punishment of God which now threatens your kingdom, and let the unjust judges be rooted out, get rid of the incompetent, punish all bribery and injustice, the impeding of justice, and all unfairness, stave off and prevent the many innocent tears which fall, protect the lowly and the weak from violence and injustice, for then God Almighty will surely show mercy towards us and turn away from punishment.

In case Alexei missed the point, the suppliants reminded him that his father Michael had been 'designated and chosen by God and the whole people as lord' to rule 'in the time when the Muscovite territory was nearly completely consumed by evil persons'; and warned that currently 'the whole people in the entire Muscovite territory and its adjoining provinces are incited to revolt by such injustice, thereby causing great chaos in your Tsarist capital Moscow and in numerous other places . . . Nothing good will follow from postponement and delay', they concluded ominously: rather 'this can also bring your Tsarist Majesty great trouble and tribulation'.[9]

Once again the imperial entourage tore this petition 'into tiny pieces and flung them in the suppliants' faces' and used their whips on the crowd, which pursued the tsar towards the Kremlin. This time, when Morozov ordered the musketeers to close the gates to the complex, they refused and so several thousand demonstrators gathered in front of the tsar's apartment and demanded immediate satisfaction of their complaints, which now included the execution of the hated Pleshceev. When Morozov appeared on the palace balcony to reason with the crowd, they shouted back 'Yes, and we must have you too!' Morozov's servants urged the musketeers in the Kremlin square to fire on the crowd; but the troops (no doubt remembering Morozov's refusal to pay their arrears) replied 'that they would not fight for the nobles against the common people'. They also told the tsar to his face that 'They would not make enemies of the people for the sake of the traitor and tyrant Pleshceev'.[10]

Together, troops and rioters broke into Morozov's luxurious apartments in the Kremlin, smashing his fine furniture with axes and crushing his jewels to dust; but 'they suffered not the least thing to be carried away, crying aloud *To naasi kroof*, that is to say, "This is our blood"'. As the terrified tsar pleaded for moderation, the crowd murdered three of his household officials and forced their way into the residence of Nazarii Chistyi, architect of the hated salt monopoly, and killed him with their axes and clubs, exclaiming 'Traitor, this is for the salt!' The protestors then attacked the residences of

Pleshcheev, Trakhaniotov and other ministers of Alexei who had given offence, and torched some seventy residences of nobles and merchants. Then they 'ran' (the word used by all the sources) back inside the Kremlin and demanded that the tsar surrender Pleshcheev, Trakhaniotov and Morozov. Alexei immediately surrendered Pleshcheev, but asked for two days to consider the fate of his other ministers. Perhaps surprisingly, the demonstrators agreed – but no sooner had they dispersed than fires broke out in five places across Moscow. Thanks to the prolonged drought, the fire spread rapidly and, according to the terrified Swedish ambassador, 'within a few hours more than half the city within the White Wall, and about half of the city outside the wall, went up in flames'. Some calculated that 50,000 homes and 2,000 people perished in the flames. On 15 June crowds re-entered the Kremlin and demanded the immediate surrender of Trakhaniotov and Morozov.[11]

In the words of an irate citizen: 'the whole world is shaking ... There is great shaking and the people are troubled'. The fate of Romanov rule now hung in the balance. Some protestors declared 'His Imperial Majesty to be a traitor as long as he refused to banish Morozov from his court and capital'. A few even asserted that 'the tsar is young and stupid' and, though attributing his obstinacy to wicked advisers, added 'The devil stole his mind' – an accusation of Satanism that challenged the tsar's claim to be the champion of Orthodoxy and, by the same token, his right to rule.[12]

Realizing the seriousness of his situation Alexei now delivered Trakhaniotov to the crowd. He also promised, kissing a golden crucifix held by the patriarch, that Morozov would become a monk and retire to some distant monastery, and would never again hold government office. Because of the evident religious solemnity, the crowd believed him and once again dispersed, while the tsar distributed a substantial cash payment to all musketeers to regain their loyalty. He also ordered his Western officers to 'train 20,000 soldiers near the Swedish border', but long before these loyal troops were ready, a new confrontation arose.[13]

Perhaps imagining that the crisis had passed, Morozov recommended that Alexei should not pay the servitors still in Moscow a promised bonus, and on 20 June a new petition in the name of the gentry, merchants and 'people of all ranks' again insisted that Morozov must go. It also demanded that the tsar summon the Zemskii Sobor (the Assembly of the Land). In desperation, Alexei recalled the ministers of his father whom Morozov had displaced in 1645, and sent the hated chief minister into exile under heavy guard. Over the next eight weeks more than seventy petitions flowed in from discontented groups and the tsar approved all of them: to appease native merchants, he abolished the trading privileges of the English; to content ordinary Muscovites, he promised money to rebuild structures damaged by the fire; to pacify the servitors, he paid their promised bonus and also returned all fugitive serfs found on Morozov's domains. Alexei also reluctantly agreed to convene the Zemskii Sobor.

These sweeping concessions reflected far more than the troubles in Moscow. A combination of drought and locusts ruined the harvests of 1647 and 1648 throughout Russia, even on *chernozem* lands, creating not only widespread food shortages but also heightened anxiety, and (as the supplication of 12 June had predicted) revolts also broke out in numerous other places, from Kozlov and Kursk in the south, and Novgorod and Pskov in the west, to Tobolsk and Tomsk in Siberia.[14]

The Great Compromise of 1649

Although the 'great shaking' affected large areas of Russia, the fate of the Romanov state was decided in Moscow. According to one of tsar's advisers, it was 'the rebellion of the common people' of his capital, backed by the local gentry and the musketeers, which led Alexei to summon representatives of his subjects 'not willingly, but out of fear'.[15] First, an elite group of clerics, nobles, servitors, merchants and representatives of the main towns assembled in the Kremlin in July 1648 and requested that the tsar should 'order to be written up on all sorts of judicial matters a law code and statute book [*Ulozhennaia kniga*], so that henceforth all matters would be done and decided according to that statute book'. They also wanted the tsar to convene the Zemskii Sobor to ratify the new laws.[16] The tsar duly established a five-man commission to draft a new law code, who had completed their task by October when some 600 representatives arrived in Moscow to attend a Zemskii Sobor.

For a moment it seemed that the assembly (and therefore the new law code) was doomed because, having bought the musketeers' compliance with another substantial gift, Alexei summoned Morozov back to Moscow where he resumed his post as chief minister. Fearing that this blatant breach of the tsar's solemn promise would produce another wave of popular violence, some leading Muscovites fled the capital and others moved into the residence of the Swedish envoy, defended by his staff and some of the tsar's musketeers. Perhaps this reaction unnerved Morozov, or perhaps he realized that a comprehensive law code might benefit the central government, for he allowed the assembly to continue its deliberations. Many of the provisions of the new law code (*Sobornie ulozhenie*), which contained almost 1,000 articles, closed off the avenues that had led to the troubles. Article 1 ('Blasphemers and church troublemakers') ensured the sanctity of churches and church services (precluding any further attempt to intercept the tsar on religious holidays), and Article 3 ('The sovereign's palace court') did the same for the sovereign's palace (making any forced entry to the Kremlin treason). Article 10 ('The judicial process') decreed that, in future, subjects must submit their petitions through the local governor to the appropriate chancellery, not to the tsar: anyone who tried to circumvent this procedure would be imprisoned and flogged. Article 19 ('Townsmen') required all citizens to register in the town where they resided on publication of the law, and forbade them and their descendants ever to leave it.

The *Ulozhenie* also made numerous concessions demanded in past petitions. It abolished tax exemptions enjoyed by most privileged groups, allowed the pursuit and recapture of all former citizens who had fled in order to escape paying taxes, and settled the matter of fugitive serfs. According to Article 11 ('Judicial process for peasants'), no peasant could legally leave the domains of his lord; no time limit existed for reclaiming fugitives; all those who had previously fled could now be reclaimed; and peasants lost ownership of their personal property (the law deemed their goods to be possessions of their lord). Moreover, these measures applied to the serf's family: 'a peasant who married a fugitive, or the child of a fugitive, was transferred with his spouse when the latter was reclaimed by the fugitive's rightful lord'. Almost immediately government commissioners began to track down and return fugitive serfs. Although in some areas (notably the southern frontier and Siberia) fugitives could still live in relative safety, the *Ulozhenie* deprived perhaps half the rural population of their freedom of movement. In future, noblemen could henceforth buy and sell serfs (and their families), move them around, trade them and even use them as wagers in card games.[17]

The government had resisted restricting peasant migration for as long as possible because of its importance for the expansion of Siberia and the southern frontier; but Alexei's dependence on his servitors both to preserve domestic law and order and to fight foreign enemies eventually led him to give way. The trauma of the 'great shaking' must have made the sacrifice of the serfs seem a cheap price to pay for the return of political stability – especially since the Zemskii Sobor showed no interest in demanding more concessions (at a time when Alexei could hardly have refused to make them). The compromise of 1649 brought gains to both sides: landholders gained total control over their serfs for the next two centuries; the restoration of domestic harmony allowed the tsar to rearm. By the end of the year, the Swedish ambassador reported that foreign officers were drilling Russian soldiers in Moscow 'almost daily, because they must become capable of training the others who are to be enlisted'. They would allow Alexei to annexe large parts of the Polish-Lithuanian Commonwealth.[18]

The Ukrainian Revolt

The Eternal Peace of Polianovka bound only the two principal signatories, and so technically it lapsed with the death of Michael Romanov in 1645; but Władysław IV made clear his desire to renew the peace, and immediately sent an embassy to Moscow to negotiate not only this but also a common strategy for an attack on the troublesome vassals of the Ottoman sultan: the Crimean Tartars. Anticipating Moscow's agreement, Władysław set out to persuade the Cossacks living along the Don and Dnieper rivers in Ukraine to spearhead the venture.

Until 1569 Ukraine formed part of the Grand Duchy of Lithuania, a state with a weak government and a strong Orthodox Church; but in that year Lithuania joined Poland in a closer union, and the region passed under Polish control. This transfer produced three destabilizing consequences. First, the crown strove to impose Polish officials, laws, troops and the Catholic faith on the newly incorporated lands. Second, the underpopulated Ukraine attracted many immigrants from many areas: Poland, Lithuania, Russia; the Balkans and even further afield. Some settled on the black earth farmlands (just as fertile as those in Russia: page 126), while others went to the towns – both to new settlements founded by the nobles and to established cities like Kiev – where they formed a reservoir of malcontents. The third destabilizing consequence arose because the crown granted huge domains in Ukraine to a few great nobles, on the grounds that they required extensive resources in order to coordinate frontier defence. By 1640 about one-tenth of the landholders controlled two-thirds of the population. To maximize the yield of their vast new domains, these nobles appointed aggressive estate managers, and tasked them both with collecting tolls, taxes and rents more efficiently and with exporting as much as possible of the crops raised on the rich soil. Many of the managers came from the expanding Jewish population of Ukraine, which rose from perhaps 3,000 in 1569 to at least 45,000, living in almost 100 communities, by 1648.

The relative ease of river access (for both settlers and exporters) meant that these developments occurred primarily along the Dnieper and its tributaries, challenging the independence of Cossacks there who lived from a combination of fishing, hunting, farming and raids to secure Tartar booty (both human and material). Nevertheless, since the crown still needed Cossacks to defend its southern border it maintained a register of veterans, each of whom received an annual stipend. The Cossacks elected their own officers and a commander, the hetman, who in wartime led them on campaign; but in most years, the Cossack register included scarcely one-tenth of the available warriors. Only they were entitled to a stipend, leaving a disgruntled and heavily armed population living along the Dnieper 'below the rapids' (*Zaporizhia*), for whom raiding offered the only chance of preserving the lifestyle they had adopted.

In 1630 the Cossacks rebelled, appealing to 'both clergy and laymen of the Greek [Orthodox] religion' because their faith 'was being taken away, and asking them to stand up for the faith'.[19] Alarmed by this alliance of Cossacks and clergy, the government ended the unrest by increasing the number of registered Cossacks from 6,000 to 8,000. Nevertheless, the influx of Polish settlers to the Dnieper Valley continued, as did the demands of the new noble landowners. In 1635 the federal Diet provocatively reduced the number of registered Cossacks and called for a fort to be built at Kodak on the lower Dnieper, garrisoned by units of the regular army. These measures provoked a new revolt by the Cossacks, who sacked Kodak and murdered its garrison. Although payment of wage arrears to the registered men allowed the capture and execution of the leading rebels, according to Adam Kysil (a Ukrainian nobleman appointed by King Władysław as a commissioner to pacify the revolt) the Cossack problem remained 'a boil perennially on the verge of bursting'; and indeed, almost immediately, another rebellion broke out. Once again, the Cossacks claimed that they sought to defend the Orthodox faith as well as the civilian community, and to that end (according to a chronicler) they 'treated the Poles with contempt, killed the Germans like flies, burned towns and slaughtered the Jews like chickens. Some burned monks in Roman Catholic churches.'[20]

Such brutality alienated many supporters, allowing Polish troops to force the rebels to surrender. They reluctantly swore an oath, drawn up by the secretary of the Zaporozhian Cossacks, Bohdan Khmelnytsky, that they would henceforth obey the crown in all matters. They agreed to another reduction in the number of registered Cossacks, undertook not to attack the Tartars (or the Ottomans) without express royal permission, promised to take orders from a Cossack commissioner appointed by the crown (instead of from their elected hetman) and accepted that the commissioner would appoint the colonels and captains formerly elected by their men. Naturally, Władysław appointed Polish magnates as commissioners; equally naturally, the commissioners appointed their Polish supporters to serve as colonels and captains. The newcomers soon began to exact much the same taxes and services from the Cossacks as from their peasants, and they punished non-compliance with confiscation of property and even outlawry.

Kysil, who negotiated the treaty that ended the revolt, urged Władysław not to press his advantage too hard, but the king continued to grant huge estates in Ukraine to leading Polish nobles, who continued to increase the burdens on the peasantry. He also stationed Polish troops in the leading cities where, in the absence of punctual pay, they extracted food, lodging and other goods at gunpoint. Rabbi Nathan Hannover, an eyewitness, described the cumulative impact of these changes in his *Abyss of despair*, an account of the massacre of the Ukrainian Jews in 1648–49. Władysław, he observed, had

> raised the status of the Catholic dukes and princes above those of the [Ukrainians], so that most of the latter abandoned their Greek Orthodox faith and embraced Catholicism [or joined the Uniate Church]. And the masses that followed the Greek Orthodox Church

became gradually impoverished. They were looked upon as low and inferior beings and became the slaves and the handmaids of the Polish people and of the Jews ... Their lives were made bitter by hard labour in mortar and bricks, and in all manner of heavy taxes, and some [lords] even resorted to cruelty and torture with the intent of persuading them to accept Catholicism.[21]

Although anti-Semitic exaggeration has distorted the record (for example claiming that Jews had acquired the lease of churches and only allowed Christian services to take place in them in return for hefty payments), Jewish estate managers often did gain extensive powers over the rural population, including the exclusive right to distil and sell vodka: this meant that they operated the only taverns in the region, where they could charge whatever prices they liked, and they called in the army to destroy illegal stills. Unsurprisingly, these measures alienated the local population from their Jewish neighbours.

Rabbi Hannover did not mention another factor that helped to precipitate the rebellion that would cost half of Ukraine's Jews their lives and property: adverse weather. The failure of the 1637 revolt triggered a mass migration of Cossacks to the lower Dnieper which, even at the best of times, suffered from extreme humidity and heat in summers and intense cold in winters – but, as in other parts of the northern hemisphere, these were not the best of times. The diary of Marcin Goliński of Kraków recorded the deteriorating situation, with an exceptionally cold summer in 1641 (during which the sparse grain harvest ripened late and the wine was sour); spring frosts in 1642 and 1643 that blighted all crops; and heavy snow and frosts in the early months of 1646 that gave way to daily rains, making the roads impassable. Further south, a French resident noted that even in normal years, the Ukrainian winter 'has as much power and force to destroy anything as fire has to consume', and that plagues of locusts in 1645 and 1646, followed by the cruel winter of 1646–47, destroyed the harvests and made it impossible for Cossack communities along the lower Dnieper to feed themselves.[22]

Amidst this climate-induced adversity, Władysław's plan to launch a massive Cossack raid against the Tartars possessed considerable appeal: war would increase the number of registered Cossacks drawing salaries to 12,000, offering them a providential escape from overpopulation and hunger. But the federal Diet refused to approve the necessary funds for the venture, while the Cossack commissioner and his subordinates behaved outrageously towards individual Cossacks. This included Bohdan Khmelnytsky, secretary of the Cossack Host: when he was away in Warsaw in 1646 to receive the king's instructions for the projected Ottoman campaign, a Polish officer seized some of Khmelnytsky's property and publicly flogged one of his sons so brutally that he subsequently died; another officer allegedly abducted Bohdan's betrothed and married her; and the following year, Polish troops ravaged his domains. Khmelnytsky fled to the unregistered Cossacks of the lower Dnieper, but found little solace there: autumn and winter 1647 brought rains that destroyed crops and caused widespread floods; spring 1648 was exceptionally hot and dry, and locusts destroyed the harvest. An inscription carved into the cathedral wall of Old Sambor, not far from Lviv, said it all: 'There was great hunger throughout the Christian world.'[23]

Khmelnytsky chose this moment to declare that he possessed letters signed by Władysław authorizing the Cossack Host to mobilize against their oppressors. Although the documents were almost certainly forged (nobody ever claimed to have seen the originals), many believed Khmelnytsky's claim that he held some sort of commission from the king – thanks in part to his charismatic character, which impressed virtually everyone he met. Born to a minor noble family in Ukraine around 1595 and educated at a Jesuit college, he acquired extensive military experience and boasted personal contacts not only in Istanbul (where he had spent two years as a captive) but also in Warsaw (where he met the king and his leading ministers). Although at first Khmelnytsky commanded barely 250 followers, within a few weeks several thousand Cossacks, including registered men whose pay had fallen into arrears, rallied to his cause and elected him their hetman. He also requested and received assistance from the sultan's vassal, the Tartar khan of the Crimea, where (according to a chronicle), 'last year [1647] there was no harvest, and now the cattle, sheep and cows are dying.'[24]

In May 1648, Khmelnytsky led his Cossack followers and their Tartar allies towards Kiev where they ambushed a Polish army, capturing or killing the hated Cossack commissioner and almost all the Commonwealth's regular troops. A few weeks later King Władysław died, creating an inter-regnum, and all over the region peasants rose up against their lords while Orthodox clerics called for vengeance against the Catholics. According to one of the Polish nobles who fled, 'Every peasant has either killed his lord or driven him out with just the shirt on his back, his life and his children.' Most abandoned their armouries and arsenals, leaving the Cossacks to appro-priate the weapons within.[25]

Although Kysil persuaded Khmelnytsky to halt his army and send back his Tartar allies, a Cossack leader nicknamed Crook-nose continued to march northwards, encouraging the native population to turn on the Jews. Rabbi Hannover left a chilling description of the massacres that started in June 1648 at the town of Nemyriv, where the local population helped Cossacks to enter the citadel in which the Jewish residents had taken refuge. As the raiders cut down the men, many women jumped off the walls and drowned themselves. In all, Hannover estimated that 6,000 died in two days. July saw similar events further north: 2,000 Poles and 12,000 Jews killed at Polonne when the Ukrainians persuaded servants within the town to open the gates ('Why fight us to protect the nobles?'), with more at Zasław (the rabbi's home town) and Ostrog (where the Ukrainians also looted and wrecked Catholic monasteries).[26]

It is hard to establish the exact scale of the killings. Early Jewish estimates range from 80,000 to 670,000, but more recent calculations suggest 10,000 killed in the violence itself, with at least 8,000 more fleeing to Jewish communities elsewhere (from Amsterdam to Egypt), and perhaps 3,000 sold to the Tartars as slaves. An unknown number survived only because they converted to Orthodoxy. In all it seems likely that the Jewish population in Ukraine fell by half in the summer of 1648. In the opinion of a Jewish chronicler in Kraków writing later in the century: 'from the time of the Destruction of the Temple there has not been such a cruel slaughter in the community of the Lord'.[27]

Three unrelated developments now affected the outcome of the revolt in Ukraine. First, although Khmelnytsky had written a letter to the tsar requesting that he take the entire Cossack Host under his protection, the letter arrived just as Moscow erupted in rebellion: Alexei could not help. Second, the murder of the Ottoman sultan Ibrahim ruled out aid from the Crimean Tartars, because the khan expected a summons to restore order in Istanbul (Chapter 7). Finally, while the Polish nobility debated the election of a new king, they appointed three joint commanders for the regular army – and even then created a supervising committee of senators to keep them in check. This divided command made it easy for Khmelnytsky to rout the Poles again in September 1648 and lead his men towards Warsaw (see Figure 9). Eventually the Sejm elected Władysław's brother John Casimir as their king and also authorized a truce with the Cossacks. Khmelnytsky was agreeable (he had supported the candidacy of John Casimir), and he led his men back to Kiev.

Since the great nobles had all fled, and the lesser ones who stayed lacked any corporate identity, by default the Cossack Host became the focus and mouthpiece for the entire community; and the flight of the Catholics and most other religious minorities left the Orthodox clergy of Kiev in control of the pulpits. They hailed Khmelnytsky (despite his Catholic upbringing) as 'Moses, saviour, redeemer, and liberator of the Ruthenian nation from the slavery of the [Poles], he who was given by God, hence called Bohdan [which literally means 'God-given']'. Whenever Khmelnytsky attended church he received pride of place, and the metropolitan bishop compared him with Constantine the Great, the founder of Orthodox piety, and hailed him as 'prince of Rus' (thus the descendant of the ancient rulers of Kiev).[28]

This concerted clerical campaign to turn Khmelnytsky into a national hero affected his style of leadership. He put it best in a speech early in 1649 to the negotiators sent by John Casimir. 'True enough, I am a wretched little man, but God has granted that I am the sole ruler and autocrat of Rus', he began. 'The time was ripe to negotiate with me when I was being sought and hounded ... on the Dnieper', or 'when I was on the march to Kiev'; but now 'I shall fight to free the whole Ruthenian nation from Polish bondage. At first I fought for my own damages and injustice – now I shall fight for our Orthodox faith.'[29] Khmelnytsky therefore submitted an ultimatum to the crown that included many ecclesiastical demands: apart from a general amnesty, and maintaining the number of registered Cossacks entitled to a state salary at 12,000, they required the admission of Orthodox prelates to the federal Diet, the restitution of all former Orthodox churches taken over by the Catholics and the expulsion of all Jesuits and Jews.

These extreme demands reflected in part the landmark winter of 1648–49 and the prospect of a further harvest failure: lacking crops and wages from the central government, war alone offered the Cossacks the prospect of sustenance for the coming year. In addition, if the nobles and Catholic clergy who had fled should ever return, they would surely exact a terrible revenge upon the former rebels – amnesty or no amnesty. Khmelnytsky therefore wrote a series of letters begging assistance from Tsar Alexei, the Don Cossacks and the Tartars. Although the tsar, still facing urban rebellions, again refused to aid the Cossacks openly, he allowed them to import bread and other foodstuffs duty-free, which saved many from starvation; and he invited prominent churchmen from Kiev to come to Moscow to forge a union of all Orthodox churches (under Muscovite aegis, naturally). Khmelnytsky had more success with his overture to the Tartar khan, who arrived in person with a huge following in August 1649. Together they ambushed another Polish army, this time led by King John Casimir in person, near Zboriv on the Strypa river. The resulting Zboriv Agreement granted almost everything in Khmelnytsky's ultimatum: a general amnesty, the expulsion of all Jesuits and Jews and toleration of Orthodox worship. In addition, the new king promised that no Cossack would be tried by non-Cossacks; no royal troops would be stationed in Ukraine; some Orthodox bishops could sit in the Diet; no Jewish settlement would be allowed in the region; and the number of registered Cossacks entitled to government pay would rise to 40,000 – a huge increase that in effect made Khmelnytsky (now confirmed as hetman) the head of a new autonomous unit within the composite Polish-Lithuanian Commonwealth.

The Zboriv Agreement solved nothing. The Tartars forced Khmelnytsky to allow them to enslave numerous Christian families as the price of their withdrawal to the Crimea, which caused general outrage. The Diet refused to ratify the king's concession that Orthodox bishops could join its deliberations and instead voted money to raise a new army, which in 1651 defeated the Cossacks and their Tartar allies and occupied Kiev. Khmelnytsky reluctantly signed a treaty that conceded some territory to the Commonwealth and reduced the number of Cossacks on the register. At the same time, however, the hetman renewed his plea for Tsar Alexei to send military assistance.

The Tipping Point: The Commonwealth Dismembered

This time the tsar was more sympathetic. According to Peter Loofeldt, a Swedish diplomat in Russia, Morozov (once again at Alexei's side) welcomed the chance 'to seek war with Poland over the disputed borders' as a means of diverting the 'increased hatred and resistance' directed towards him. But, Loofeldt continued, when Alexei asked the Zemskii Sobor to discuss assisting the Cossacks, 'The secular lords protested strongly against war, saying that "it is indeed easy to pull the sword from the scabbard, but not so easy to put it back when you want, and that the outcome of war was uncertain".'[30] Only an unrelated religious development undermined this consensus and allowed Morozov to get his way.

In 1652 Alexei appointed the zealous monk Nikon as patriarch of Moscow and encouraged him to undertake a comprehensive campaign of church reform. To improve the behaviour of the laity Nikon forbade smoking, swearing, working on Sundays and pagan practices (such as celebrating the winter solstice and staging carnivals during Lent), and he limited liquor sales (one store in each town and one bottle per customer at a time, with no liquor sales at all on Sundays, holidays or during Lent). To raise clerical standards, Nikon censured drunken priests and required holy offices be chanted audibly. To demonstrate that the Russian Orthodox Church was the one true descendant of the Church of the Apostles, Nikon introduced liturgical practices from Greek Orthodox communities, collected and ritually defaced all icons painted in the Western style, and sponsored Old Testament imagery suggesting that Moscow had become the New Jerusalem.

Morozov soon persuaded Nikon to see the war to save the Orthodox Cossacks as part of this reform programme. The Patriarch duly 'declared the undertaking to be a holy enterprise and the tsar as the protector and saviour of all the persecuted brethren of the old Greek religion, comparing him with the kings David, Josiah, and Constantine the Great, who wanted to defend' their faith. 'This had quite an impression upon ordinary people', wrote Loofeldt, 'making them much more willing to become involved in attacking Poland again. So preparations for the war now began in earnest.'[31]

The apparent dysfunction of the Commonwealth encouraged Russian belligerence. The first use of a *liberum veto* in the Sejm of 1652, which brought about the dissolution of the Diet before it had voted the taxes necessary for a new campaign, encouraged several cities in Lithuania to defy the central government and attempt to ally with Khmelnytsky. The tsar now sent two diplomatic delegations westward: one went to Warsaw to demand the return of Smolensk and other territories ceded at Polianovka, which (as Alexei had anticipated) the Sejm rejected; the other went to Ukraine with an offer of Russian protection, provided that Khmelnytsky ended his alliance with the Tartars, which (again as Alexei had anticipated) the hetman accepted. In October 1653 the Zemskii Sobor voted unanimously to declare war on the Commonwealth and to place Ukraine under Russian protection.

The long distance and dangerous conditions meant that the tsar's envoys only reached Khmelnytsky at his camp in Pereiaslav in January 1654 – but their timing could scarcely have been better. Commonwealth forces had defeated Khmelnytsky's army the previous year, convincing him that he could not succeed alone. He therefore asked his followers to choose one of four allies: the sultan, the Crimean khan, the Polish king or the tsar. Unanimously they chose their co-religionist Alexei, and the tsar's officials spent the next few days administering oaths of loyalty. They also invested Khmelnytsky as hetman in the tsar's name; promised that Russian troops would arrive to strengthen defences against the Poles; and agreed that the number of registered Cossacks should be 60,000. In addition, they confirmed all the concessions made by John Casimir at Zboriv. Cossack envoys accompanied the tsar's officials back to Moscow, where in March they formally accepted the terms of what became known as the Union of Pereiaslav.

The continuing unrest in Lithuania convinced Morozov and Alexei that they should launch a pre-emptive invasion of the Grand Duchy, and in May 1654 the tsar led an enormous army, numbering perhaps 100,000 troops and including all the New Formation Regiments equipped with the latest Western weapons, towards Smolensk. In July they re-entered the siegeworks abandoned twenty years earlier. After three months, the city surrendered to the tsar. Alexei also captured Vilnius, the capital of Lithuania, and proclaimed himself its grand duke.

The tsar's rapid and complete success alarmed Sweden. Realizing that if he remained neutral, Russia might conquer the entire Commonwealth, King Charles X Gustav (r. 1654–60) and his council debated whether they should help the Poles resist another Russian onslaught, or attack themselves and secure some territory before the Commonwealth collapsed. They decided on the latter, and in July 1655, just as Alexei entered Vilnius, Charles invaded Poland. He entered Warsaw in September and Kraków in October. When John Casimir fled the country, most of his magnates accepted Swedish authority. It was the most complete and rapid state breakdown ever seen in early modern Europe.

Ruin and 'the Deluge'

Sweden's spectacular military success rested on slender foundations. Charles X commanded only 36,000 troops – far too few to hold down the vast tracts of the Commonwealth that now lay at his feet – and he lacked a clear plan for exploiting his sudden triumph. In May 1656 Alexei declared war and laid siege to the heavily fortified Swedish outpost of Riga, but he could not mobilize sufficient resources to prevail in two wars at the same time. In order to concentrate on his new war with Sweden, the tsar concluded an armistice with the exiled John Casimir, who swiftly regained Warsaw.

The Russian volte-face infuriated Khmelnytsky, who still regarded the Commonwealth as the Cossacks' greatest enemy. He therefore refused to follow the orders received from Moscow and break with Sweden: instead he tried to build an anti-Polish coalition (going so far as to propose an alliance to Oliver Cromwell), but many Cossacks defied Khmelnytsky and continued to obey the tsar. The hetman's death in 1657 did not end this division: most Cossacks on the west bank of the Dnieper continued to favour Poland, whereas those on the east bank mostly favoured Russia. These developments plunged the region into a period of bloody anarchy that Ukrainian historians eloquently call *Ruina*, and Polish writers, with equal eloquence, call *Potop*: 'the deluge'.

Three factors prolonged the conflict. First, geography confined hostilities to largely predictable locations. In particular, the impenetrable forests forced both troops and supplies to use the rivers, where they could be more easily intercepted; and the huge marshlands provided refuge to defeated units, allowing them to regroup for attacks on enemy lines of communication and reducing the chances of a knock-out blow. Second, the Russian army pursued a policy of deliberate brutality that eventually proved counter-productive. Alexei had instructed his generals 'to present the inhabitants of Belorussian towns with written surrender requests, but if they spurned them, to burn alive Poles or Belorussians subsequently captured who would not convert to Orthodoxy'; and when troops under the tsar's personal command took Vilnius, the capital of Lithuania, they started fires that raged for seventeen days and killed perhaps 8,000 people.[32] Throughout the annexed areas, Russian troops persecuted Catholics, Jews and members of the Uniate Church. Third, natural disasters intensified the devastation caused by the war. Plague epidemics struck Poland twice in the 1650s and the Little Ice Age produced more long cold winters and disastrous harvests – including the landmark winter of 1657–58 when (according to a courtier) in central Poland 'no one can remember such a long winter'. The thaw only began in April.[33]

Eventually, the protagonists became too exhausted to continue fighting: Alexei concluded an armistice with Sweden at the end of 1658, and a peace three years later that sacrificed all his gains. It is hard to disagree with the lapidary verdict of Brian Davies: 'Nothing had been gained from the war'.[34] The Commonwealth, too, concluded peace with Sweden in 1660 but for another six years continued its campaigns against the tsar and his Ukrainian allies.

During 'the deluge', the Polish-Lithuanian Commonwealth encountered all four Horsemen of the Apocalypse – pestilence, war, famine and death – with catastrophic consequences. Thousands perished in sieges; tens of thousands became the victims of persecution, like the Jews of Ukraine, slaughtered by Cossacks who saw them as oppressors, or those of Mogilev, slaughtered by Russian troops who feared they might side with an advancing Polish army. In all, the population of the Commonwealth fell by at least one-third.

The population of Russia also plummeted through a lethal combination of climate change, war and plague. Tree-ring, pollen and peat-bed data show that the springs, autumns and winters between 1650 and 1680 were the coldest recorded in Russia during the past 500 years, and crops repeatedly either failed or produced little food. An epidemic of bubonic plague in 1654–57 also caused widespread depopulation. An official survey found that the epidemic had killed four-fifths of the monks and three-quarters of the nuns living in convents within the Kremlin, as well as half the officials in the Foreign Ministry and nine-tenths of those in the Revenue Ministry. Alexei's wars also created a fiscal disaster. According to Richard Hellie, an eminent historian of imperial Russia, 'the real cost of the Muscovite military establishment in the mid-1650s must have been at least 3 million rubles a year', so that 'above one-eighth of Muscovy's productive resources went just to pay for the army'. Building the Belgorod and Simbirsk Lines absorbed perhaps as much again.[35]

The tsar therefore introduced desperate expedients to balance his budget, including currency debasement. The imperial mint began to issue copper coins as well as silver, but in 1662, lacking funds to defeat Poland, the tsar decreed that taxes could only be paid in silver. The exchange rate between the two currencies now shot up from parity to 1:15, and producers refused to accept copper coins for their goods, including grain, creating a famine in Moscow. As in 1648, thousands of inhabitants and soldiers gathered to petition the tsar to hand over the ministers whom they blamed for their plight, while others looted and burned offenders' houses. This time the tsar's foreign troops restored order. After executing some 400 protestors out of hand and ordering hundreds more to be branded on the cheek (for future identification) and imprisoned, the tsar sent thousands of offenders and their families into permanent exile in Siberia and the lower Volga. Yet they did not suffer in vain: the following year, the tsar recalled all copper money and prices quickly fell to their earlier levels.

John Casimir threw away the golden opportunity presented by these disorders by overreaching (just as Alexei had done by attacking Sweden). He demanded that the Sejm elect his successor during the own lifetime, but opponents of this innovation repeatedly used the *liberum veto*, forcing the king to dissolve the Diets four times in 1665 and 1666. This starved the

government of resources for the war against Russia, provoked an aristocratic rebellion that created new devastation, and encouraged the Cossack
leaders to seek Ottoman support once more. With Russia and the
Commonwealth both paralyzed, representatives of John Casimir signed a
truce at Andrusovo (a village near Smolensk) in February 1667 that brought
the Thirteen Years' War to an end, ceding all Alexei's conquests (Smolensk
and the lands along the upper Dnieper) and transferring to the tsar Kiev
and all Ukrainian lands east of the Dnieper. But Russia, too, still faced many
problems.

Russia's Religious Schism

Before Alexei left Moscow on campaign in 1654 he conferred the title 'great
sovereign' on Nikon, which allowed the patriarch to sign decrees in the tsar's
name during his absence. Nikon now oversaw the printing of three separate
editions of a new liturgy that incorporated numerous innovations, and at a
Church Council in 1656 he enjoined the exclusive use of the new ritual and
denounced as heretics all who objected. Although the changes were small,
they could not be ignored. For example, requiring priests and congregations
to make the sign of the cross with three fingers instead of two, as Nikon
insisted, affected the most common visible symbol in Orthodox worship.

The Old Believers (as Nikon's critics came to be known) argued that the
true Christian faith was timeless and not subject to change: to add or
subtract anything was to destroy the truth. They also argued that the Russian
Orthodox Church alone had (in the words of Avraamii, a 'holy fool' who
became one of the earliest and most articulate Old Believers) preserved the
'true Orthodox faith, handed down by the holy Apostles, confirmed at the
seven ecumenical councils and sealed with the blood of the holy martyrs'.
Heresy, he added, 'comes in by a single letter of the alphabet'.[36] It followed
that, if Russia alone had preserved the true faith, every divergent belief or
practice must be heretical. Unlike Nikon, the Old Believers despised the
numerous Greek Orthodox prelates who visited Moscow to raise money for
their beleaguered churches, because they all lived under Ottoman rule – a
clear sign of divine disfavour. They also dismissed the new liturgies
produced by Moscow printers because they were based on texts produced
by Latin Christian printers in Venice.

In Russia, as elsewhere in Europe, many lay people saw strict observance
of traditional religious practices as their pathway to salvation – especially
after a plague epidemic in 1654–57 intensified devotion to local icons and
liturgical practices that had apparently saved a family or a community.
Many people later recalled that 'the plague had led them to change their way
of life', and all over Russia ordinary people began to participate in commem-
orative religious processions, worship at votive chapels dedicated to plague

victims, construct special churches in a single day and discover new relics – all to exalt their devotion to Mother Russia. They also fasted, self-flagellated and prayed more than ever before.[37]

Imposing religious uniformity in such a climate throughout the largest state in the world presented unique challenges. It was easy for discontented clergy to turn local insecurities into intransigence – and imperial Russia had no shortage of discontented clergy. Many, including Old Believers, lived in the small religious communities and hermitages that abounded in remote areas; others were priests, monks and nuns who wandered the country, and who were often defrocked for resisting authority. In addition, every parish had its own service book (usually manuscript), and it took time for the new printed texts to reach outlying areas; moreover, since individual copies were expensive, many parishes remained without one. Nevertheless, two more church councils in 1666–67, presided over by Alexei in person and attended by all Russian and many foreign bishops, repeated the injunction to use only the new liturgy and anathematized as 'heretics and recalcitrants' all those who refused to conform (anathemas that remained in force until the 1970s). Most critics of the reform now recanted, but a handful remained obdurate. Some, like Avraamii, died at the stake in Moscow as heretics; others went into exile in the far north (where the tsar's servants also put many to death); but, as long as they lived, these and other Old Believers tirelessly gathered and copied ancient church texts and wrote martyrologies of the fallen.

Stenka Razin

Moscow's religious intransigence contributed to another major uprising in 1670 by the Cossacks of the Don and Volga valleys. Over the previous quarter-century, thanks to peasant migration from the north, the number of Don Cossacks had tripled to perhaps 25,000, which severely strained the scarce resources of the region. The situation remained tolerable in the settlements along the lower Don and its tributaries, where an elite of long-established Cossack families, known as householders, monopolized the best hunting, fishing and grazing, pocketed subsidies from Moscow and monop-olized political power; but conditions deteriorated for the newcomers, significantly known as the 'naked ones' (*golytba* or *golutvennye*). They lacked land, property and subsidies, surviving only if householders gave them work. In the words of a government report, 'runaway peasants have come from neighbouring areas with their wives and children, and as a result there is now great hunger on the Don'.[38]

The Truce of Andrusovo in 1667 hurt the Cossacks. It deprived them of a fruitful source of booty and also reduced pressure on the tsar, no longer dependent on their services, to send grain and munitions punctually. Later that year, one of the disgruntled householders, Stepan ('Stenka') Razin therefore decided to lead an expeditionary force of naked ones on a major raid. Like Bohdan Khmelnytsky in Ukraine, Razin was familiar with the corridors of power long before he became a rebel: he had visited Moscow and also headed an embassy from the tsar to neighbouring steppe rulers. Now, despite the tsar's express prohibition, Razin and his men sailed down the Volga to the Caspian, where they took advantage of a series of natural disasters that weakened Iran (Chapter 13) to institute a reign of terror in 1668 and 1669.

Razin's remarkable success in securing wealth during a famine year won over many more naked ones, and in spring 1670 he decided to lead his forces back to the Volga and advance on Moscow. His declared aim was to move 'against the Sovereign's enemies and betrayers, and to remove from the Muscovite state the traitor boyars' as well as tsar's councillors, provincial governors and town magistrates, 'and to give freedom to the common people'.[39] Razin's followers, who now numbered 7,000, determined otherwise: having advanced up the Volga as far as Tsaritsyn, which they captured, they voted to move south again and seize control of all Russian settlements down to the Caspian. This they accomplished thanks to two strokes of luck: the Crimean Tartars, who normally attacked Russian outposts along the lower Volga each summer, campaigned elsewhere, relieving Razin of a potential enemy; and Alexei had deported many of those involved in the Moscow revolt of 1662 to the lower Volga, and these exiles, eager for revenge, betrayed their towns and joined the rebels.

Razin's position remained vulnerable, because the economy of the entire lower Volga depended on the goods and grain sent by the tsar, so he persuaded his followers that they must campaign against Moscow in order to secure the supplies without which they would starve. As he moved north in summer 1670, he won the support of others alienated by Moscow's oppressive policies – fugitive serfs, political exiles, unpaid soldiers, Muslims dispossessed by Christians – as well as of oppressed peasants anxious to take revenge on their brutal masters and of townsfolk impoverished because the tsar imposed a trade embargo on all areas in revolt. Many women joined the movement and disseminated propaganda; a few, including Razin's mother, commanded rebel detachments. Many clerics alienated by Nikon's reforms provided spiritual support and drafted 'seditious letters' (as the government termed them) inviting those along Razin's line of march to help him 'eliminate the traitors and the bloodsuckers of the peasant communes', and proclaiming that he was on his way to Moscow 'to establish the Cossack way there, so that all men will be equal'. The clerics cleverly tailored each message to its audience – for example, writing to Muslim communities in their own language and claiming 'this is our watchword: *For God and the Prophet*'. Razin even claimed that he had received letters of support from Nikon and that he represented the crown prince. By September 1670, the revolt that originated with a subsistence crisis on the lower Don affected a swathe of territory along the Volga that stretched for 800 miles and even subjects of the tsar in Siberia and Karelia received copies of Razin's 'seditious letters'.[40]

The revolt failed because of the lines created expressly to protect the capital against attacks from the south (page 126). The tsar sent reinforcements, bonus pay and extra ammunition to garrisons along the fortified lines, which foiled the attempts of Razin and some 20,000 men to secure Simbirsk, at the eastern end of the line. After a month, a counter-attack drove him back with heavy losses. He himself suffered serious head and leg wounds, impairing his aura of invincibility, and he retreated to the fortified base on the Don from which he had started. Meanwhile his brother, with another army, failed to take Voronezh, a key fortress on the Belgorod Line, and also fell back. When the tsar's New Formation Regiments approached, the Cossack householder elite decided that it would be prudent to arrest Razin, which they did in April 1671, and to send him to Moscow where Alexei had him tortured and executed.

Nevertheless, when Alexei died in his bed a few days later, leaving two minor sons to succeed him, the troubles resumed. Abroad, war broke out with the Turks in 1677 and lasted for four years; at home, opposition resumed in the smaller monasteries, which still resented efforts to put Nikon's liturgy into effect. In May 1682 a dispute over which of Alexei's sons should succeed led a group of musketeers, resentful of the growing prominence and higher pay of the New Formation Regiments, to storm the Kremlin. They received support from Old Believers in the capital and, the following month, submitted a petition calling on the Regent Sofia, Alexei's oldest child, to restore the traditional liturgy. Sofia agreed to a debate on the validity of the Nikonian reforms, but when an Old Believer spokesman suggested that not only the patriarch but also her father had been a heretic, Sofia stormed out of the meeting and fled, leaving the capital under the control of the mutineers until she could raise an army capable of defeating them. She had her revenge two years later when, having regained Moscow without firing a shot, Sofia issued legislation against the Old Believers: all who failed to attend their local parish church must be questioned; all suspected of heresy must be tortured; all who refused to recant must be burned at the stake; and anyone who sheltered dissenters must be severely punished. Henceforth, Old Believers flourished only on the periphery of the state. Nevertheless, although religious dissent never again threatened the integrity of the Romanov state, by 1900 one Russian in six was an Old Believer, and the faith today numbers millions of adherents.

The New Order

According to Frank Sysyn, 'modern national relations' in Eastern Europe 'begin with the Khmelnytsky revolt' of 1648. Even a flurry of provincial rebellions a generation later failed to change the new political configuration in which, in the words of Brian Davies, 'the circle of serious contenders for hegemony had been narrowed to the Ottoman Empire and Muscovy'.[41] The vast Polish-Lithuanian Commonwealth never recovered from the mid-century crisis: its population had declined by at least one-third; in 1668, one year after signing the Truce of Andrusovo, John Casimir abdicated and emigrated; and in 1686 the Eternal Peace of Moscow perpetuated the transfer of all territory ceded to Russia.

War and rebellion also ruined Ukraine. The measure of self-rule secured by Khmelnytsky between 1648 and 1654 did not survive; and even in 1700 the former Cossack state contained scarcely a million inhabitants – one-third of its former size – and those survivors remained dependent on support from the tsar. Although they retained some internal autonomy for a while longer, they could never again boast that 'the tsar rules in Moscow, but the Cossacks rule on the Don'.

Conversely, the Romanov regime emerged strengthened from the mid-century crisis. Even the *Ulozhenie* (law code) of 1649, which stemmed directly from the 'great shaking' of the previous year, reinforced the power of the tsars. Admittedly the code gave most landholders what they most wanted – total control over their serfs – but it also established the obligations of servitors and musketeers to the tsar (and the punishments for those who fell short) and it extended the fiscal responsibilities of townspeople. The code also offered equality before the law to Russians and non-Russians alike – one no longer needed to convert to Orthodoxy, or even to speak Russian, in order to plead in a Russian court – and it standardized court protocol, forensic procedures and fees. A single document thus laid out the entire Russian social and legal system with the tsar at the apex, and it proved a bestseller: two editions of 1,200 copies issued by the Government Printing Shop sold out almost immediately, and regional administrators as well as many individuals soon possessed their own copy.

Alexei's wars also brought significant material gains, adding perhaps 120,000 square miles (an area about the size of modern Italy or Nevada) and over a million new subjects to the Romanov state, which not only wrested important concessions from the Polish-Lithuanian Commonwealth at Andrusovo and again at the Eternal Peace of Moscow, but also forced both the Ottoman sultan in 1681 and the Chinese emperor in 1689 to recognize the tsar as their equal. Russia became for the first time a great power, a position it has never lost.

CHAPTER SEVEN

The 'Ottoman Tragedy', 1618-83[1]

'The Greatest Empire That Is, or Perhaps That Ever Was'

At the dawn of the seventeenth century the Ottoman empire overawed European visitors. A Venetian diplomat marvelled that it had acquired territory that 'included 8,000 miles of the circuit of the world' and 'a great part' of Asia, Africa and Europe. An English traveller considered it 'the greatest empire that is, or perhaps that ever was'. Such awe was justified: the sultan ruled over 20 million subjects and a million square miles. Although Istanbul lay more than 700 miles from Vienna, and 1,000 miles from Baghdad, thanks to the empire's efficient logistical infrastructure imperial couriers carried orders from the capital to officials in Hungary and Mesopotamia in fourteen days or less; and an army leaving the Bosporus in spring could normally reach either the Tisza or the Tigris in ten weeks.[2]

Nevertheless, between 1620 and 1690 the empire saw two regicides, three depositions, and significant loss of territory in both Europe and Asia. Until recently, the nature of this decline remained obscure because, in the words of a recent article, 'The seventeenth century has been the black hole of Ottoman history'; but careful study of the available human and natural archives reveals that the lands around the eastern Mediterranean suffered more from both the Little Ice Age and the General Crisis than almost any other part of the northern hemisphere.[3]

Effective Ottoman government depended on a complex balance of forces. Every sultan claimed absolute and indivisible jurisdiction in all matters not explicitly covered by existing Islamic law, and issued decrees on fiscal, penal and administrative matters that every subject must obey. Although some sultans occasionally acted as supreme judge and heard cases in person, they normally ceded executive authority to a single minister, the grand vizier, and their council heard and decided the thousands of petitions that flowed in from subjects every year. In the seventeenth century most sultans lived in the imperial palace in Istanbul and seldom strayed far from its grounds. This arrangement conferred immense power on those who controlled access to the imperial apartments, and particularly to the palace harem where the sultan's concubines lived under the supervision of several hundred eunuchs. Since no seventeenth-century sultan married, each concubine who bore a son intrigued ceaselessly to ensure that her offspring succeeded and, later, to influence his policies. The sultan's mother was the most powerful woman – and often the most powerful person – in the empire.

The Ottomans divided their subjects into two categories: the *reaya* ('subjects who pay taxes') and the *askeri* (literally 'of the military', those who served the state). Among the latter, a select group of soldiers and government officials known as the 'sultan's slaves' (*kullar*), exercised great power. Until the 1630s, the sultans recruited their slaves by conscripting boys from Christian communities under their control in the Balkans and Anatolia, a practice known as *devşirme* ('gathering' in Turkish). Once the boys reached Istanbul, they began rigorous training to turn them into obedient, skilled and Ottomanized converts to Islam, and afterwards they either joined the Janissaries (literally 'new troops', infantry equipped with firearms) or became palace officials (though they too received a military training – appropriately enough in a state that regarded war as its principal activity). The *devşirme* system enhanced the power of the Ottoman dynasty in three distinct ways. First, only the sultan's slaves could legally own and use firearms. Second, since each regional centre boasted a similar provincial cadre – governor, provincial council, treasurers, garrison commandant and presiding judge – the imperial government could rotate its 'slaves' easily from one post to another because, wherever they went, they encountered familiar administrative systems, procedures and expectations. Finally, the young converts who prayed, ate, slept and trained together developed a remarkable cohesion and loyalty; and, since they could never leave the sultan's service, risking loss of life and property if they refused or disobeyed an order, they normally formed a principal pillar of the state. Nevertheless the system possessed one obvious weakness: without constant supervision, the sultan's slaves might usurp their master's power and dictate policy.

Muslims by birth dominated only two professions in the Ottoman empire: the heavy cavalry and the clergy. In the sixteenth century ethnic Turks served as cavalry troopers (*sipahis*: from the Persian word meaning 'army'), maintaining themselves and their followers with a fief (*tımar*) granted by the sultan. By 1600, however, few fiefs produced enough to support a trooper and his retainers and their number had fallen to around 8,000. The central treasury therefore began to pay salaries to the troopers who formed part of the permanent garrison of Istanbul and various provincial capitals, and their numbers rose to 20,000 by 1650. This increase placed an intolerable strain on the central treasury, which sometimes could afford to pay either the Janissaries or the cavalry, but not both, leading to rivalry and sometimes pitched battles between them.

The clergy, or *ulema* ('knowledgeable people' in Arabic), were all Sunni Muslims. They not only provided public worship and religious education but also administered pious foundations and served as judges. At their head stood the chief mufti (*şeyhülislam*), appointed by the sultan and paid a state salary like the rest of the clergy, who received a stream of requests from the central government to certify (usually in the form of a written opinion, or *fatwā*) that a proposed action or edict conformed to Islamic law. On occasion, the chief mufti withheld such certification, producing a political crisis that might result in his deposition or, in extreme circumstances, his murder. One seventeenth-century chief mufti survived for just half a day. The sultans also founded *medreses* (literally 'places of study') and paid preceptors to provide basic education in Arabic grammar and syntax, in logic and in rhetoric, as a prelude to instruction in theology and law.

The size of the clergy tripled between 1550 and 1622, reflecting a rapid expansion in the number both of *medreses* and mosques: each student could find a position as preceptor, preacher or judge upon graduation, together with a state salary. When the empire ceased to expand in the seventeenth century, graduates might have to wait years for the chance to pass the examination, without which they could not obtain the licence required to teach and preach. Even those who gained their licence often remained at the bottom of the hierarchy because a select group of families (known as *mevali*) virtually monopolized the elite positions. Most of the chief muftis and senior judges appointed between 1550 and 1650 were related; indeed almost half came from just eleven families. This concentration of power in so few hands naturally created widespread frustration among the other clerics who found their careers blocked.

Frustration also mounted among some Muslims who did not belong to the regular clergy but nevertheless claimed supernatural powers. Some wandered from one community to another, surviving on alms from the faithful; others lived in solitary ascetic devotion in one place; and others still practised as healers – like 'Slimey' Hüseyn in the Istanbul hippodrome, who claimed that his snot healed those on whom it landed. Many more men and women, known as *sufis*, believed that the path to God lay through experience rather than scholarship and therefore performed their devotions publicly, sometimes accompanied by music and dance. The majority belonged to one of Islam's religious orders, each one headed by a sheikh, and either lived in or supported one of the order's lodges.

Although the Ottoman empire lacked a tradition of collective action by the clergy, extreme climatic events in the mid-seventeenth century, as well as the multiplicity of political and economic problems, provided charismatic preachers of all persuasions with convincing evidence of divine discontent and the need for rapid and radical change. Many began to take their message directly to the faithful in passionate sermons delivered in mosques during Friday services, attended (at least in theory) by all males throughout the empire. On several occasions, their preaching imperilled the stability of the Ottoman state itself.

Climate and Depopulation

Climate change did not affect all parts of the empire with equal force. The coastal plains that surrounded the Mediterranean, which formed the heart of the empire, coped best with the Little Ice Age because farmers from Greece to Morocco normally enjoyed sufficient sun and rain to produce cereals, vegetables, tobacco and even cotton without irrigation. By contrast, farmers on the hills and plateaux overlooking the sea who produced cereals and a few vegetables still needed to irrigate their crops. In this zone, even small climate changes could produce major problems. The situation was worse further inland, where farmers could produce crops only if they invested in extensive irrigation systems (Fig. 10). Here, even a short drought or unseasonable frost could ruin the entire harvest. In several regions of Anatolia, the number of rural taxpayers fell by three-quarters between 1576 and 1642, and almost half of all villages disappeared, while throughout Anatolia heavy spring precipitation in both 1640 and 1641 and droughts later in the decade destroyed many harvests and no doubt caused even more depopulation.

Balkan farmers also suffered intensely during the Little Ice Age. Surviving tax registers show that the population of Talanda (central Greece) and of

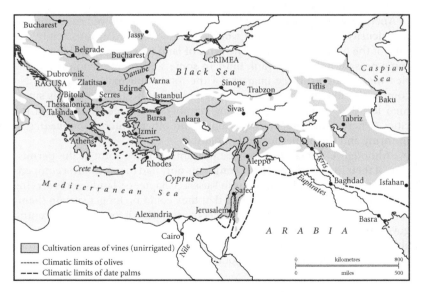

10. The climatic zones of the Ottoman empire.
The staple crops of the eastern Mediterranean – olives, vines and date palms – can survive even where little rain falls in summer; but a shift in rainfall patterns, as occurred in the mid-seventeenth century, can cause long-term damage. The more frequent droughts and colder weather also destroyed grain and citrus fruit.

Zlatitsa (Bulgaria) fell between 1580 and 1642 by almost 50 per cent. Around Bitola (Macedonia) in 1641 one-quarter of all taxpaying households were abandoned. To the east, in Serres, farmers found an abundance of grapes when they began to harvest them in September 1641, but then came 'so much rain and snow that many workers died through the great cold'.[4] It was much the same in other parts of the empire. In Palestine, repeated droughts ruined many settlements, including the Jewish religious centre of Safed, where a traveller in the 1660s noted that the population had fallen from 70,000 to 9,000, and the number of textile workshops had declined from 3,000 to 40. Egypt, too, experienced drought when the Nile reached its lowest level of the century in 1641–43, and again in 1650, because El Niño episodes produced low summer rainfall in the Ethiopian highlands and the Sudd swamps of Sudan, which began the river's annual water cycle. Because 'Egypt is a desert with a river running through it', a poor Nile flood drastically reduced the crop yields of the entire province – which in turn reduced the food available to supply Istanbul, the sultan's armies, and the holy cities of Arabia – and also forced charitable institutions throughout the empire to close their doors (and their soup kitchens), increasing local misery.[5]

As usual, overpopulated areas, where supply barely satisfied demand even in good years, felt the effects of the Little Ice Age most acutely. In parts of Anatolia, a benign climate in the sixteenth century allowed rural population density to reach levels that would 'never be reached again, even by the turn of the twentieth century'. The price of land rose steeply and the size of some peasant holdings shrank until, in some settlements, landless unmarried men made up three-quarters of the total adult male population. No community like this can long survive, and from the 1590s onwards single men left their villages in increasing numbers for three destinations: the cities (where some sought employment and others entered the *medreses*), the army and the numerous outlaw bands.[6]

Although Ottoman law forbade peasants to leave without the permission of their lord, until the 1630s emigrants only had to pay modest compensation if they did so. Thereafter, as in Russia (Chapter 6), landowners serving in the imperial cavalry complained that they could no longer sustain themselves while fighting for the sultan; but despite edicts from the central government demanding the forcible return of fugitives, nothing seemed able to halt the exodus. The central problem of the Ottoman economy now became a chronic shortage of labour.

'The Ottoman Tragedy'[7]

Against this backdrop of rural crisis, a series of political upheavals called by one contemporary historian 'the Ottoman tragedy' rocked the empire from 1617 to 1623. It began with an unprecedented dynastic crisis. All male members of the Ottoman dynasty normally lived in sealed apartments in the imperial palace, appropriately known as 'the cage', until one of them became sultan and executed all the others. In 1595 Sultan Mehmed III had followed tradition and executed all nineteen of his brothers, as well as pregnant slaves in the harem, and he later executed the crown prince on suspicion of treason, so that at his death in 1603 only two male members of the Ottoman dynasty survived: his sons Ahmed (aged thirteen, who became the new sultan) and Mustafa (aged four). Prudence dictated that Mustafa should be allow to live (even if, some speculated, he was 'nurtured like an innocent little sheep who must soon go to the butchers') and he was still alive when Ahmed died in 1617.[8] His survival created unprecedented confusion: should the next sultan be Mustafa, now aged eighteen, or Ahmed's oldest son Osman, aged fourteen? Initially, supporters of the former prevailed but he behaved so erratically that, after three months, a household faction imprisoned him – the first sultan to be deposed by a palace coup – and engineered the proclamation of Osman in his stead.

Osman, too, ruled erratically. His preceptors had instilled in him a determination to follow the Prophet Mohammed's injunction to 'enjoin right and forbid wrong', and the new sultan soon forbade the cultivation and use of tobacco (on the grounds that it wasted money, induced laziness and imitated a habit introduced by infidels). He also punished senior religious leaders who had supported Mustafa's succession, especially the elite clerical families, by abolishing their salaries during periods of unemployment and in retirement, as well as their right to appoint a successor (usually one of their relatives). Osman thus made the most powerful Muslim clerics his bitter enemies – an especially unwise move, given the severity of the winter of 1620–21. For forty days ice floes closed the Bosporus and prevented grain from reaching Istanbul. The vast city began to starve, but Osman exacerbated the shortages by requisitioning supplies for a campaign against Poland, which had attacked one of his vassals in the Balkans. He left the capital in May 1621 but severe cold and rain (this was an El Niño year), combined with unexpectedly tenacious Polish resistance, forced him to conclude a humiliating truce. When the sultan and his demoralized troops returned to Istanbul in January 1622, they found it in the grip of famine and high prices.

The poor performance of Osman's troops in Poland convinced him of the need to replace the elite Janissaries and *sipahis*, who formed the core of both the field army and the garrison of Istanbul, with troops from Anatolia, Syria and Egypt. At first the sultan stated that he would undertake a pilgrimage to Mecca, but then he ordered the principal institutions of government, and the imperial treasury, to cross the Bosporus to Asia. Once again the sultan's timing proved unfortunate. Nile floods had ruined the harvest in Egypt in 1621, and then came a drought that reduced the food supply even further. According to one chronicler, the sultan's troops protested that they 'could not go into a desert without water, and without doubt their animals will perish too'; according to another, they asked rhetorically 'After the Polish campaign, what fool among the soldiers would go now?'[9] On 18 May 1622, the appointed day for the sultan's departure, the Janissaries of the city garrison demanded that Osman remain in the capital and hand over to them those who had advised him to move. They returned the following day, this time accompanied by the chief mufti and other members of the religious elite. When the grand vizier came out to negotiate, they murdered him and burnt down his residence, and then alienated officials opened the gates of the Topkapı Palace complex and the mutineers swarmed in. One group found Mustafa (who had languished in the cage since Osman's accession) and took him to the Janissaries' mosque, where they proclaimed him sultan again, while another found Osman and dragged him through the crowded streets of the capital in a cart, subjecting him to public insults, and cast him in prison where he was strangled – the first regicide in Ottoman history.

Several provincial governors in Anatolia, including those whom Osman expected to provide the troops for his new army, refused to recognize the coup and turned on the Janissaries and *sipahis* in local garrisons. The capital too remained in uproar as food prices rose to the highest levels recorded in the entire seventeenth century. In desperation, the government paid new bribes to retain the loyalty of the Janissaries and, when this created an unacceptable budget deficit, reduced the silver content of the currency to its lowest level of the century. The next few months saw five grand viziers come and go (some of them murdered). Shops and markets stayed closed, food ran short and plague spread until in January 1623, the coalition of mutineers and senior clerics who had engineered Mustafa's restoration deposed him again. The chief mufti declared that a madman could not be sultan and proclaimed Osman's oldest surviving brother, the eleven-year-old Murad, as his successor – the fourth sultan in six years.

Several groups now vied for control of imperial resources and power: Murad IV's Greek mother, Kösem Sultan; the senior palace officials, especially the eunuchs; the Istanbul garrison; and the senior clerics. Thanks to their intrigues, the average tenure of a grand vizier in the 1620s fell to four months, and rebellions and roaming bandit gangs in Anatolia deprived the central treasury of revenues. The unpaid garrison of Baghdad mutinied and betrayed the city and most of southern Iraq to Iranian troops.

Murad IV's Personal Rule

The nadir of Murad's rule occurred in 1630–31, which coincided with another El Niño episode. Over 3 metres of rainfall flooded Mecca, a city that normally sees virtually no precipitation, destroying two walls of the Kaaba.

Extreme weather also disrupted operations by the Ottoman army fighting in Mesopotamia: in January 1630, wrote one chronicler, 'the Tigris and Euphrates all overflowed and floods covered the whole Baghdad plateau', and another compared the ceaseless rains with 'the times of Noah'. The following August, by contrast, the waters of the Tigris sank so low that no ships could use it, leaving the army 'desperately in need of munitions and provisions'.[10] The Nile, too, fell far below the level needed to irrigate the fields, causing a major famine; and drinking water ran short in Istanbul. At the same time, rural insurrections and bandit activity peaked: in 1630 and 1631, the sultan's council issued over 150 orders in response to complaints about brigands and peasant revolts, almost seventy arising from abuse by provincial officials, and over fifty more condemning local elites who colluded with brigands.[11]

In 1632 the chaos in Istanbul reached such a pitch that the leading protestors demanded an emergency meeting with the sultan in person to secure the redress of their grievances. Murad complied, and reluctantly delivered to the mob his grand vizier, chief mufti and several other ministers. All were murdered, and for several weeks insurgents held the city to ransom, threatening to burn down the house of anyone who refused to pay them. Eventually, the exasperation of the citizens of the capital allowed the sultan to eliminate many of those involved in the disorders, purge corrupt judges and crack down on bribery. Perhaps 20,000 perished in the disorders – but Murad, now aged twenty, at last held the initiative and began eight years of personal rule. Every year, he ordered the execution of hundreds of officials and subjects for the failure to maintain local roads, indiscipline on campaign, selling adulterated bread and a host of other infractions. In addition, suspecting that his critics hatched plots in coffeehouses and taverns, Murad ordered all of them to close.

The crisis of authority in the empire favoured the reform programme of the empire's most prominent Muslim preacher, Kadizade Mehmed (1582–1635), the son of an Anatolian judge (and thus a member of the clerical elite), who spent time in both a *medrese* and a Sufi lodge before coming to Istanbul in 1622, the year of the regicide. He soon deduced that the chaos around him stemmed from a failure to adhere strictly to the dictates of Islamic law, and decided to take his message directly to the faithful through sermons in which he repeatedly quoted the *hadith* (prophetic tradition), 'Every innovation is heresy; every heresy is error; every error leads to hell'. He predicted that God would continue to punish the entire empire until everyone returned to the beliefs and practices of the Prophet Mohammed, and criticized innovations associated with Sufis, condemning their habit of singing, playing, chanting and dancing while reciting the name of God (on the grounds that the Qur'an expressly forbade entertainment). He also condemned new social habits, such as shaking hands or bowing before superiors, allowing women to prophesy, and wearing or using anything except traditional Muslim garments and artefacts.

Sufi leaders reacted to Kadizade's criticisms swiftly and in kind. Those who served as preachers used their sermons to mobilize support among both *medrese* graduates and their congregations for their innovations; but the tide turned in 1633 when, following another prolonged drought, a great fire in Istanbul destroyed at least 20,000 shops and houses, the barracks of the Janissaries and the state archives. Kadizade blamed the disaster on religious 'innovations' and warned of more catastrophes to come unless they ceased. After one particularly vivid sermon, his followers (known as Kadizadelis) sacked the taverns of the capital. Since Murad did nothing to stop the disorders, the Kadizadelis advanced one step further: citing the Qur'anic injunction to 'enjoin right and forbid wrong', they called on the faithful not only to amend their own lives but to seek out and punish sinners. Individual Sufi sheikhs were denounced and beaten, their lodges vandalized and their adherents given the choice of either reaffirming their faith or being put to death.

Since Murad saw coffee shops and taverns as potential centres of sedition he supported the Kadizadelis' efforts, killing those who gathered to consume coffee, alcohol and tobacco because in doing so he believed he was also killing his political critics; but his intervention kept everyone on edge because they never knew when some small miscalculation or oversight might lead their cruel and capricious sultan to take their lives too. No doubt Istanbul breathed a collective sigh of relief when in 1640 Murad died after a short illness – during which he characteristically threatened to kill his physicians unless he recovered, and tried to have his sole surviving brother Ibrahim strangled (he had already killed all three of their siblings). Instead, Ibrahim became sultan.

The Mad Sultan

The new ruler was twenty-five: the first sultan in a generation to reach the throne as an adult and also, as the sole surviving male member of the entire Ottoman dynasty, the first to rule unchallenged by rivals. Nevertheless, he had spent his life confined in the cage in the Topkapı Palace, reading the Qur'an, practising calligraphy and living in constant fear of sharing the violent fate of his other brothers. Just like Murad, he therefore came to the throne lacking political experience. For the first four years Kara Mustafa Pasha, Murad's last grand vizier, managed public affairs competently. Abroad, he promoted peaceful relations with both Iran and the Austrian Habsburgs; and he recovered Azov from the Cossack adventurers who had seized it (Chapter 6). At home, Kara Mustafa stabilized the coinage (albeit after another sharp devaluation), initiated a new land survey in an attempt to create a more equitable tax base, reduced the garrison in the capital and banned the Kadizadelis from delivering inflammatory sermons. Although a provincial governor defied him and led an army within sight of Istanbul, the capital remained loyal and the rebellion crumbled.

This loyalty seems surprising because the summer months of 1640, 1641 and 1642 all saw ceaseless rain as well as plague, while a drought in Egypt reduced the supplies of several staples normally consumed by the palace. Kara Mustafa nevertheless found alternative sources to feed the capital, and in 1643 his reforms had created a small surplus in the Ottoman treasury; but he failed in one crucial respect: he could not cure Ibrahim's numerous medical complaints. The sultan suffered from perpetual headaches and repeated attacks of physical exhaustion; he also worried that he, the last surviving male of the dynasty, might be impotent. In 1642, since the doctors provided by Kara Mustafa failed to provide remedies, Ibrahim turned to charlatans recommended by his mother, Kösem Sultan. One of these, Cinci Hoca, seems to have cured at least Ibrahim's impotence, because in the next six years he sired several children, including four future sultans; but Cinci Hoca used the imperial favour thus earned to build a faction against Kara Mustafa, and early in 1644 engineered his downfall.

The next twelve years would see the rise and fall of twenty-three treasurers, eighteen grand viziers, twelve chief muftis and countless provincial governors. Since each official tried to get rich and to enrich his followers as quickly as possible, the 60,000 government officeholders of 1640 had grown to 100,000 by 1648. The number of troops paid by the Ottoman treasury also increased, from 60,000 to 85,000, due to the outbreak of war with the Venetian Republic.

Late in 1644 some galleys of the Knights of Malta seized a convoy carrying pilgrims from Istanbul to Mecca. Some died in the struggle, including the former chief eunuch of Ibrahim's harem, and the victorious galleys steered to the Venetian island of Crete, where authorities allowed them to put ashore some of their spoil and captives and take on supplies. This breach of neutrality infuriated Ibrahim, and he ordered immediate retaliation. In June 1645 Janissaries stormed ashore and gained control of most of Crete – but the Venetians counter-attacked by blockading the Dardanelles, which cut off not only relief for the Ottoman garrisons on Crete but also the supply of grain to Istanbul. The war lasted until 1669, costing the lives of some 130,000 Ottoman troops and absorbing around three-quarters of the imperial budget. It also coincided with more episodes of extreme weather.

Floods in 1646 and drought in 1647 destroyed the harvest surpluses on which Istanbul depended, creating another food shortage. The Ottoman capital normally consumed some 500 tons of bread each day, of which about half went to feed the palace employees, the capital's garrison and the students in the *medreses*, so that the sultan's immediate entourage were among the first to feel (and resent) any shortfall. Late in 1647 a new grand vizier persuaded his master to retreat to his private quarters in the Topkapı Palace where he could be shielded from adverse news; but Ibrahim's behaviour in seclusion became ever more eccentric. He developed an extravagant taste for luxury items, especially furs, and became irrationally impatient for the goods he craved: on occasion he made the shops of the capital open at midnight while his men requisitioned items for him and his concubines.

Funding the war as well as the sultan's exotic tastes proved a challenge. Since the continued predations of pirates and bandits kept many rural areas depopulated, the Ottoman treasury moved away from dependence on taxes on agricultural production, long the mainstay of the treasury, in favour of personal taxes – especially the *avariz* (levies paid in either cash or kind, such as chickens for the imperial kitchens or repairs on road and bridges) and the *cizye* (a poll tax on non-Muslims). Each year, the treasury held a public auction at which it sold the right to collect these personal taxes to the highest bidder, receiving payment in cash and in advance. When even this did not suffice to fund Ibrahim's war with Venice, his ministers searched for new ways to raise money: they imposed excise duties on more goods, demanded that even senior clerics pay taxes, sold still more public offices (even judgeships) and withheld pay from the Istanbul garrison. Resentment in the capital mounted, intensified by the continued Venetian blockade of the Dardanelles, which cut off its principal source of food, and by a plague of locusts that destroyed crops in Moldavia, another area that supplied staples to Istanbul. As usual the resulting dearth affected acutely those who depended for their daily bread on the sultan: the officials and garrison of the palace.

Then in June 1648 major tremors rocked Istanbul. According to Kâtib Çelebi, 'an earthquake like this has not been seen in our times. According to some experienced and enlightened ones, when an earthquake happens during daytime in June, blood is shed in the heart of the empire.' Four minarets at the Hagia Sophia collapsed and the mosque built by Ibrahim's father Ahmed sustained severe damage during Friday prayers, killing several thousand worshippers. The earthquake also demolished the city's main aqueduct so that drinking water ran short just as the summer heat began. The Kadizadeli preachers blamed these natural disasters on the failure to follow the teachings of the Prophet, and a Venetian observer in the city reported that 'the wise made diverse predictions of unrest in the city in the near future'. 'The wise' were right.[12]

A Second Regicide

In August 1648, at an audience with the sultan, the chief mufti demanded the dismissal of the grand vizier, Ahmed Pasha. Ibrahim started to shout abuse and attacked bystanders with his stick, but a group of disaffected

Janissaries strangled Ahmed and threw his body into the street where the crowd swiftly dismembered it. They also wrote a letter to Ibrahim, demanding that he get rid of his concubines and his furs, pay the arrears of his troops and restore all the goods unjustly confiscated from his subjects. Ibrahim promptly tore their letter up, whereupon the chief mufti issued a *fatwā* ordering the sultan's deposition on the grounds that he was incapable of ruling the empire and protecting the Muslim faith.

Guards now persuaded Ibrahim that, for his own protection, he should retire to an inner apartment of the palace – and then locked the door behind him. Shortly afterwards Janissaries accompanied by the chief mufti entered the palace, and finding Ibrahim already confined they sought out his oldest son, Prince Mehmed, aged seven, and hailed him as the new sultan. The next day Ibrahim escaped from confinement (probably freed by one of his concubines) and, with a sword in his hand, searched the palace intending to kill Mehmed, but guards overpowered him and locked him in the cage where he had lived before his accession. Balthasar de Monconys, a French visitor in the capital, claimed that 'such a peaceful revolt has never been seen: the whole process took only forty hours, and affected only the sultan, his chief vizier and one judge'.[13] Nevertheless, Ibrahim continued to scream and rage in his sealed room, gaining some sympathy from members of his entourage and thus raising the possibility of a restoration. To avoid such a development, on 18 August 1648 the chief mufti issued another *fatwā* that legitimized the second regicide in Ottoman history.

Up to this point, the capital had remained calm but now rioting broke out, led by *medrese* students and junior palace officials. Both groups gathered in the Hippodrome to make a formal protest against the chaos; but Janissaries surrounded and butchered them. Led by Kösem Sultan (Mehmed's grandmother as well as Ibrahim's mother), the Janissaries retained power in the capital until the summer of 1651, when continuing food shortages, heavy taxation, currency devaluation and military defeats provoked a new round of protests. The city's tradesmen, who claimed that they had received a dozen demands for new taxes that year alone, shut their shops and demanded that the chief mufti go to the palace to demand reforms. The frightened young sultan agreed to abolish all taxes imposed since the reign of Suleiman the Lawgiver a century before, but Kösem Sultan once more called on the Janissaries to restore order. This time she faced a more formidable adversary within the palace: Turhan, Mehmed's mother, supported by a faction among the eunuchs. In a desperate bid to retain power, Kösem decided to kill the young sultan and replace him with one of his brothers – one with a more docile mother – but Turhan's supporters acted first and had Kösem strangled and dragged naked from the palace.

As long as Venetian forces maintained their blockade of Istanbul and held out on Crete, palace factions in Istanbul conducted a lethal struggle for power. A stream of ministers attempted new policy initiatives, only to lose their heads when they failed to produce instant success: in 1655 and 1656, seven grand viziers came and went, some within a matter of weeks. One lasted a single day. In March 1656 an attempt to balance the budget through savage currency devaluation provoked mutiny because the city garrison found that no shopkeepers would accept payment in the virtually worthless new coins. The Janissaries once again marched on the palace and demanded that the sultan surrender thirty of his 'wicked advisers'. Mehmed reluctantly complied, and they were hanged from a tree in the main square of the capital – the fourth regime change in eight years.

With no effective government to oppose them, the Venetians routed the Ottoman fleet and occupied the Aegean islands of Tenedos and Lemnos, which virtually isolated Istanbul from the Mediterranean. Many inhabitants feared that the Venetians would next launch a direct attack, sold their property and left the capital; provincial governors refused to send money to the central administration. 'Public talk ran very hard' against the sultan, a foreign visitor noted, 'so that upon the least unlucky turn and new disgrace in their public affairs, he stood in great hazard of a revolution'.[14] The future of the Ottoman state seemed bleak indeed.

The Return of Stability

The Kadizadeli preachers, as usual, blamed all disasters on religious innovation. They had lost some influence while Kösem Sultan dominated government policy, because she favoured Sufis, but after her murder in 1651 the Kadizadelis secured new legislation against smoking and drinking. Five years later they laid plans to demolish all Sufi lodges, and all minarets except one on each mosque 'to ensure that Istanbul would reflect the Prophet's Medina' – but the onslaught never took place, because on 15 September 1656 Mehmed IV appointed Köprülü Mehmed Pasha as grand vizier.[15]

A former *devşirme* boy from Albania, now aged almost eighty, Köprülü had previously held only minor government offices, but he possessed a deep understanding of the inner workings of the Ottoman state. Before accepting office as grand vizier, he conducted intense negotiations with Turhan Sultan, Mehmed's mother, and the chief mufti to authorize a pre-emptive strike against the Kadizadelis. When a few days later they refused to call off their planned attack on the Sufi lodges, Köprülü had their leaders arrested and exiled. Köprülü also mobilized religious support for an offensive against Venice. He ordered all the palace pages named Mehmed (after the Prophet) to recite the opening verse of the Qur'an every day until the end of the campaign; and he commanded 101 men to recite the entire Qur'an 1,001 times in the principal mosques of the capital.

Confident that these measures would produce a miracle, Köprülü led the Ottoman fleet in person to fight the Venetians, and he was not disappointed: the Venetian garrisons surrendered Lemnos and Tenedos, at last breaking the blockade of the capital and restoring the vital supply of food from Egypt. Next, Köprülü led an army into the Balkans, apparently intending to attack Venetian possessions on the Adriatic coast, but a revolt in Anatolia called him back. The rebel leader, Abaza Hasan, governor of Aleppo (the third city of the empire), had won widespread support from other regional governors in Anatolia and Syria, and preachers began to hail him as the Messiah who would restore the Islamic community to purity. He and his supporters demanded that the sultan depose his grand vizier.

This presented Mehmed IV with a major challenge. Although he realized that refusal to sacrifice Köprülü would probably unleash a civil war, the sultan persuaded the chief mufti to issue a *fatwā* condemning the rebels: 'Since they committed an act of oppression against the sultan, their blood can be shed lawfully: those who cause Muslim armies to abandon their fight with infidels by perpetrating sedition are worse than infidels.' Multiple copies of the *fatwā* went out, together with a call for all adult males to fight against the rebels.[16] The Little Ice Age helped rescue Köprülü: the landmark winter of 1657–58 followed by a failed harvest in Anatolia made it impossible for the rebels to feed their troops, and support for the insurrection dissipated. Early in 1659 Abaza Hasan and his lieutenants surrendered against a promise of clemency – only to be executed. Köprülü Mehmed Pasha now sent a trusted lieutenant to round up renegades, end unjustified tax exemptions and confiscate all illegally held firearms.

The Little Ice Age also produced the greatest setback of Köprülü's five-year tenure as grand vizier. In 1659–60 the lands around the Aegean and Black Seas experienced the worst drought in a millennium: no snow fell over the winter and no rain fell during the spring. The combination of drought and extreme heat turned the wooden buildings of Istanbul into a tinderbox, and in July 1660 a fire of unparalleled ferocity destroyed two-thirds of the capital, causing the minarets on the mosques to burn like candles. The chief casualty was the area where most of the capital's Jewish and Christian populations lived: seven synagogues and at least twenty-five churches burned to the ground.

Köprülü Mehmed Pasha nevertheless survived even these disasters, and the following year he died in his bed and passed on his office of grand vizier to his son Köprülü Fazıl Ahmed: a peaceful transition with few parallels in seventeenth-century Ottoman history (and the first time a son succeeded his father in the office). Almost immediately the new grand vizier, who had previously been a preceptor in a *medrese*, invited a charismatic Kadizadeli preacher named Vani Mehmed Efendi to join him in Istanbul. Following the example of Kadizade Mehmed a generation before, Vani attributed the

Great Fire to the failure to follow the religious practices of the first Muslims. Once again, the sultan paid heed and forbade the consumption of tobacco, coffee and alcohol; condemned musical performances, public singing, dancing and chanting; prohibited any unsupervised meetings between unmarried people of the opposite sex; and insisted on the strict enforcement of the Sharia. He also destroyed popular Sufi tombs and exiled or executed Sufi leaders. Kadizadeli preachers claimed that the disproportionate destruction of property belonging to Jews in the Great Fire of Istanbul was a sign of divine displeasure and called for legislation to prevent them from returning to the area. Köprülü Fazıl Ahmed obliged by confiscating all the lands where synagogues had stood before the fire and put them up for auction: since he forbade non-Muslims to make a bid, the area became Islamized. To symbolize the change, and to proclaim her own growing authority, Turhan Sultan sponsored the completion of the huge New Mosque (Yeni Cami) in the area. When it opened in 1665, Vani Mehmed Efendi became its first preacher and his sermons there continued both to assert the supremacy of Muslim traditions and to criticize the Jews.

The Messianic Moment of Shabbatai Zvi

These changes profoundly unsettled and alarmed Jews throughout the Ottoman empire. Some were already on edge because the Jewish calendar had just begun a new century (5400: 1640 of the Christian calendar) and a new century often occasioned a resurgence of Jewish Messianism. In addition, the years before 5400/1640 had witnessed a weakening of the authority of both rabbis and traditional texts (Torah and Talmud) among some Jewish groups, in favour of new sources of authority. One of these was Kabbalah (literally 'something received'), a strain of Judaism that flourished in Safed, reeling from economic decline (page 154), exalting mysticism and revelation as well as tradition and Torah, and revering prophets and healers as well as rabbis. In 1650 Manasseh ben Israel published *Esperança de Israel* (*Hope of Israel*), soon translated into other languages, predicting that the redemption of the Jews and the end of the world fast approached.

Four disasters heightened anxiety within the global Jewish community. First, in 1645, Portuguese settlers attacked the settlements created by the Dutch in Brazil. Whereas the Dutch colonial regime had actively encouraged Jewish settlement, the Portuguese destroyed Jewish property and killed (or delivered to the Inquisition) all the Jews they could find (Chapter 15). Second, that same year, the outbreak of war between Venice and the Ottoman empire ended the lucrative commerce that had sustained Jewish communities living in both states. Third, in 1647 Philip IV declared state bankruptcy in Castile and confiscated the capital of all loans made to his government over the previous two decades, most of them by Jewish bankers from Portugal. All were ruined. Fourth, in Ukraine the Cossacks massacred thousands of Jews (Chapter 6).

Harrowing reports of all these catastrophes reached Izmir (Smyrna), a prosperous port-city whose Sephardic community included a religious student called Shabbatai Zvi. One day in 1648, while walking in solitary meditation outside the city, he 'heard the voice of God speaking to him: "Thou art the saviour of Israel, the Messiah, the son of David, the anointed of the God of Jacob, and thou art destined to redeem Israel, to gather it from the four corners of the Earth to Jerusalem."' From that moment on, Shabbatai later told his disciples, 'he was clothed with the Holy Spirit' and felt empowered to behave in extravagant ways. To prove his point, he demolished a synagogue door with an axe on Shabbat, arranged a wedding ceremony in which he married the Torah, and pronounced repeatedly the forbidden Tetragrammaton. Such ostentatious flouting of Jewish laws led the rabbis of Izmir first to condemn him as a fool and then in 1651 to banish him.[17] Shabbatai travelled widely, living in various cities of the Ottoman empire until his outrageous behaviour caused his expulsion, until in May 1665 a young kabbalist called Nathan of Gaza proclaimed that Shabbatai was the true Messiah.

Much had happened since Shabbatai first made the same claim two decades earlier. He himself had married a refugee from the Ukrainian massacres and so became personally aware of the catastrophe; a group of Portuguese Jewish exiles in Izmir published a new edition of Manasseh ben Israel's *Hope of Israel*; and the destruction and displacement of the Jewish community in Istanbul after the Great Fire of 1660 raised anxieties among their co-religionists throughout the Ottoman empire. In addition, many Christians calculated from a passage in the Book of Revelation that the world would end in 1666 and predicted that, immediately beforehand, a charismatic leader would unite all the Jews of the world, wrest Palestine from Muslim control and then convert to Christianity.

By 1665, therefore, many Jews and Christians were predisposed to accept Nathan of Gaza's claim – circulated by sermons, letters, and a remarkable series of forged documents – that the long-awaited Messiah had come. Acclamation began in Safed, the centre of Kabbalistic study, where twenty prophets (both male and female) proclaimed Shabbatai's messianic status. Letters exchanged between Jewish scholars and study groups spread the news rapidly and before long prophets in Aleppo, Izmir, Edirne, Thessalonica and Istanbul had proclaimed Shabbatai to be the Messiah. By spring 1666, rabbis in Morocco, Tunisia and Libya had also had become staunch adherents, and reports in newspapers and pamphlets kindled enthusiastic support for the movement in the Jewish communities of Europe. In Istanbul, the entire Jewish community eagerly discussed 'the imminent establishment of the kingdom of Israel, and the fall of the Crescent and of all the royal crowns in Christendom'; and when Shabbatai arrived many underwent 'transports of joy such as one can never understand unless one has seen it'.[18]

The grand vizier imprisoned Shabbatai as soon as he arrived, but since the capital's Jewish community continued to fast and pray instead of working and paying taxes in September 1666 the sultan's council presented Shabbatai with two alternatives: they would execute him (and, according to some sources, all his followers) unless he either proved immediately by some miracle that he was the Messiah, or else converted to Islam. Shabbatai chose the latter course, living as a Muslim pensioner of the sultan until he died a decade later. So ended 'the most important messianic movement in Judaism since the destruction of the Second Temple'.[19]

The Tipping Point

Köprülü Fazıl Ahmed's fifteen-year tenure as grand vizier, one of the longest on record, saw not only the return of stability to Istanbul but also notable territorial gains. He led an invasion of Hungary and captured a number of fortresses before securing an advantageous truce with the Austrian Habsburgs; he also joined the soldiers in the trenches in Crete until in 1669 they forced the last Venetian garrisons to surrender. The whole island now became an integral part of the Ottoman empire, and remained so until 1898. Fazıl Ahmed followed up these victories with three campaigns against Poland, one of them led by the sultan in person, forcing the Commonwealth to cede parts of Ukraine. He even managed to balance the state budget. In 1675, to celebrate all these successes, Mehmed IV and his grand vizier held 'a scrupulously orchestrated exhibition of dynastic splendour and munificence' that lasted fifteen days.[20]

Few observers of this pageant could have guessed that twelve years later mutinous troops would force the sultan to abdicate. The process began that same year, 1675, one of the two 'years without a summer' in the seventeenth century, which initiated a period of intensely cold winters and unusually dry springs. A major eruption of Mount Etna in 1682 seems to have reduced crop yields all over the eastern Mediterranean and produced a particularly cold winter and wet spring. Fazıl Ahmed did not have to face these challenges, since he died in 1676, succeeded as grand vizier by his brother-in-law Merzifonlu Kara Mustafa, who almost immediately invaded Ukraine to exploit his predecessor's gains. Having done so, he made an advantageous settlement in 1681 – the first formal treaty between the sultan and the tsar. Two years later, however, the Ottomans rejected the Austrian Habsburgs' offer to renew the truce between the two empires, despite the refusal of the chief mufti to authorize a declaration of war. Instead, encouraged by Vani Mehmed Efendi, still the most influential preacher in the empire, Merzifonlu Kara Mustafa prepared a campaign to capture Vienna, the Habsburg capital.

The campaign went badly from the start. Christian forces broke the Ottoman blockade of Vienna, killing large numbers of the besiegers, and the Venetians captured several Ottoman outposts along the Adriatic coast. Mehmed IV executed Merzifonlu Kara Mustafa and exiled Vani, but he lacked the resources to keep his army in Hungary supplied: in September 1687 it mutinied and marched on Istanbul, forcing Mehmed to abdicate (the fifth forcible removal of a sultan in sixty years). In 1699, the Ottomans signed peace treaties that ceded most of Hungary to the Habsburgs and parts of Greece to the Venetians, marking the first major territorial retreat of the empire in almost three centuries.

The speed with which the gains and stability of the Köprülü era evaporated requires explanation. Vienna, to be sure, lay at the outer limit of the Ottoman state's effective reach: even if the sultan's troops had captured the Habsburg capital in 1683, they would probably have lost it to a spirited Christian counter-attack soon afterwards. Moreover, although the Köprülüs helped the Ottoman state to recover from bankruptcy, they failed to accumulate a substantial reserve in the central treasury. Finally, the Little Ice Age seems to have struck the Ottoman heartland with particular force. Most of the eastern Mediterranean suffered drought and plague in the 1640s, the 1650s and again in the 1670s; the winter of 1684-85 was the wettest recorded during the past five centuries; the winters of the later 1680s were at least 3°C cooler than today.

Nevertheless, both climatic adversity and human ineptitude in the Ottoman empire had been far worse during the 1640s and 1650s – yet no territory was lost to the Christian powers until 1683 (on the contrary, the Turks acquired Crete). The relatively late tipping point seems to have stemmed from a significant change in the military balance. For much of the seventeenth century, the principal western states deployed most of their resources in fighting each other: the Habsburgs fought in Germany (1618–48); Spain fought the Dutch Republic and France (1621–59); Poland fought the Cossacks, Sweden and Russia (1621–29, 1632–34, 1648–67). In the course of these domestic struggles, the armed forces of the major western states made important technological advances that the Ottomans could replicate only with difficulty, if at all. At sea, European sailing warships could fire a broadside capable of destroying an oar-driven galley, but the Ottomans proved incapable of constructing galleons to match them. On land, the Europeans constructed fortresses of enormous sophistication, and at the same time developed siege techniques capable of taking all but the strongest fortress: very few of the fortified places captured from the Turks after 1683 in Hungary and the Adriatic ever returned to Ottoman control. European armies also deployed massed musketry salvoes and artillery barrages to far greater effect in battle. The Ottoman empire could still imitate successfully – its Janissary corps rapidly perfected musketry salvoes – but it seemed incapable of innovating. Its decline was therefore relative rather than absolute: although the Ottomans eventually recovered from the seventeenth-century crisis, their European rivals recovered more quickly and more completely.

CHAPTER EIGHT

Bloodlands: Germany and its Neighbours, 1618–88[1]

The Long Shadow of the Thirty Years' War

In the early seventeenth century, some 20 million people lived within the Holy Roman Empire of the German Nation (which included almost all of modern Germany and Austria, as well as Slovenia, the Czech Republic, and parts of western Poland and eastern France). Although its population was much the same as that of France, public authority in the empire was divided among some 1,300 territorial rulers. At the apex stood the seven Electors (*Kurfürsten*) who met periodically to choose each new emperor: the archbishops of Mainz, Cologne and Trier (all of them ruling small states in the Rhineland); the Electors of Saxony and Brandenburg and the king of Bohemia (all of them ruling large territories further east); and the Elector Palatine (who governed territories both on the lower Rhine and on the border of Bohemia). Collectively, the seven Electors governed almost one-fifth of the empire's population, and at meetings of the imperial Diet (*Reichstag*) they formed the most prestigious of its three colleges. Spiritual princes who ruled fifty fiefs, together with lay colleagues who ruled another thirty-three fiefs, formed the Diet's second college; while fifty Imperial Free Cities, most of them in the south and west of Germany, formed the third. About a thousand lesser rulers, both lay and secular, most of them also in the south and west, governed the rest of Germany; but they lacked direct representation in the Diet. The political geography of the empire therefore possessed striking disparities. Whereas just four states, all Protestant (Brandenburg, Saxony, Pomerania and Mecklenburg), dominated the sparsely populated north and northeast of Germany, the more populous south and west contained a multitude of smaller political entities, some Protestant but most Catholic.

Throughout the empire, rulers large and small strove to reinforce their independence. They created schools and universities to train clergy and teachers who would faithfully follow the state religion; they created tariff barriers designed to protect local production as well as to raise revenues; and they spent heavily on defence, fortifying their towns and raising militia units, and in many cases joining a confessional alliance (the Protestant Union from 1608, the Catholic League from 1609). All of them sought financial independence. As Maximilian, ruler of Bavaria from 1598 to 1648, wrote at the beginning of his reign: 'I believe that we princes only gain respect, both from spiritual and secular powers, according to "reason of state"' – adding, by way of explanation: 'only those who have a lot of land or a lot of money get that respect'.[2] Over the next two decades Maximilian doubled his tax revenues and used the proceeds to build state-of-the-art fortifications, fund the Catholic League and create a war chest of 4 million thalers.

Despite these impressive achievements, Maximilian was not an independent prince: like all other German rulers he owed obedience to the Holy Roman Emperor, an elective office held since the mid-fifteenth century by one of the Habsburg archdukes of Austria. The Habsburgs also held the elective crowns of Hungary and Bohemia; but there, unlike Germany, the representative assemblies (Estates) had secured religious toleration. In 1609 the Estates of Bohemia extorted from King-Emperor Rudolf II the Letter of Majesty, an edict that guaranteed full religious toleration throughout the kingdom, and they created a standing committee, known as the Defensors, to ensure that the concessions went into effect. Rudolf and his successor, Matthias, fought back by replacing Protestant officials with Catholics, often either foreigners or recently ennobled townsmen: both groups became dedicated supporters of the crown and, in return, the crown showered rewards and offices on them. Matters came to a head in the winter of 1617–18 when Matthias, supported by his cousin and heir-designate Ferdinand, ordered his regents in Prague to prohibit the use of Catholic endowments to pay Protestant ministers, deny civic office to all non-Catholics and forbid Protestant worship in all towns built on Catholic lands. The Defensors determined that these measures contravened the Letter of Majesty and summoned the Bohemian Estates to assemble in Prague in May 1618. Meanwhile Protestants excluded from power produced polemics that criticized the Catholic-dominated court.

The Prague Spring

Unusually cold and wet conditions ruined the harvests in 1617 and 1618 throughout Central Europe, so that economic tension was already high by the time the Bohemian Estates convened. When the regency council declared the meeting of the Estates illegal, in May 1618, some of its members invaded the council chamber and threw three ministers out of a high window in an event that became known as the Defenestration of Prague. The Estates then created a provisional government and began to raise an army in preparation for the inevitable Habsburg backlash.

Religious tensions also ran high in Germany, fuelled by Protestant celebrations in 1617 of the centenary of Martin Luther's successful defiance of the papacy. Hans Heberle, a village shoemaker living near Ulm, a city with both Catholic and Protestant inhabitants, later wrote in his diary that 'This Jubilee became one of the causes of the war' because it set German

Protestants and Catholics at each other's throats. The Bohemian Estates exploited these tensions and appealed to the German Protestant Union for military and financial assistance; but although the Union raised an army of 11,000 men 'to protect liberty and law' and 'to maintain our religion like true patriots', it refused to deploy them outside Germany.[3] Then in March 1619 Matthias died, creating a double succession dilemma, one as king of Bohemia and the other as Holy Roman Emperor, because both were elective offices. On 25 August 1619, although they had previously recognized Ferdinand as king-designate, the Estates of Bohemia elected as their new ruler Elector Frederick of the Palatinate, the director of the Protestant Union as well as son-in-law of James I of England and nephew of Maurice of Nassau, the dominant figure in the Dutch Republic.

Had this momentous event occurred only slightly earlier, it would have affected the choice of the next Holy Roman Emperor, because the king of Bohemia was one of the seven Electors. Instead, on 28 August 1619, the Electors (or their representatives) in Frankfurt-on-Main, unanimously chose Ferdinand as the next Holy Roman Emperor. Neither he nor his colleagues knew that, 300 miles away in Prague, he had just been deposed as king of Bohemia.

Frederick of the Palatinate, aged twenty-three, now held the fate of Europe in his hands. If he accepted the crown offered by the Bohemian Estates, an observer in Frankfurt wrote presciently, 'Let everyone prepare at once for a war lasting twenty, thirty or forty years', for 'the Spaniards and the House of Austria will hazard everything to recover Bohemia . . . rather than allow their dynasty to lose control of it so disgracefully.' By contrast, if the Bohemian cause was 'neglected and by consequence suppressed' a British diplomat mused, 'this business of Bohemia is like to put all Christendom in confusion' and the German Protestants 'are like to bear the burden of a victorious army'. Frederick saw the matter in providential terms: he had received 'a divine calling that I must not disobey. My only end is to serve God and his Church', and in November 1619, eighteen months after the Defenestration, he travelled to Prague for his coronation, while his supporters laid siege to Vienna, Ferdinand's capital.[4]

This marked the zenith of Frederick's fortunes. The following month, disputes among his commanders sabotaged the siege of Vienna; Duke Maximilian of Bavaria promised to help Ferdinand defeat his rebellious subjects; and in return Ferdinand promised to reimburse all of Maximilian's military expenses and grant him any of Frederick's possessions that he could take (including, in the event of total victory, Frederick's Electoral title). The Catholic League authorized Maximilian to raise 25,000 men, and the following summer the new army invaded Bohemia.

Frederick's troops fell back until in November 1620 they made a desperate stand on White Mountain (Bílá Hora), just outside Prague. The engagement lasted only two hours but, as an English diplomat on the spot noted, 'The loss of cannon, the baggage, reputation, is the Imperials' victory who, as it seems, hold Bohemia now by conquest'.[5] Frederick, whom Catholic polemicists cruelly dubbed the 'Winter King', because his reign had lasted barely one year, promptly fled from Prague. Catholic troops invaded and occupied his German territories.

Ferdinand immediately pressed home his advantage. He created a judicial commission that tried his opponents, executing almost thirty and condemning almost 600 to lose their lands – which he promptly sold in order to pay off his troops. It was, in the words of Peter Wilson, 'the largest transfer of property in Europe before the seizures during the Communist takeover after 1945'.[6] These changes, though brutal, affected relatively few; another initiative affected almost everyone. Although his forces had won their victory relatively swiftly, the cost of the campaign far exceeded Ferdinand's available funds. He therefore allowed a consortium of his creditors to mint debased currency, which he used it to pay his debts. Neighbouring states soon followed this example and set up mints expressly to produce cheap coins, causing runaway inflation known as the *Kipper- und Wipperzeit* (the see-saw era) in much of Central Europe. The wages of urban workers fell by half between 1619 and 1621, and by half again by 1623, provoking protests in several central European towns. By the time Ferdinand decreed a return to the old coin values in 1624, many families had lost almost all their savings.

The effects of devaluation were felt far beyond Germany, because it reduced the cost of manufactures there so drastically that no one bought imported goods. The number of English woollen cloths exported to the continent from London fell from over 100,000 in 1618 to 75,000 in 1622, and even those brought no profit because of the devaluation. At the same time, cheap foreign imports flooded the English market, reducing domestic demand for home manufactures. Germany's other trading partners also suffered. Spanish Lombardy, whose economy depended heavily on German demand, also experienced a dramatic setback: between 1618 and 1622 unemployment in the capital rose to 50 per cent and the trade of Cremona (a centre of cloth production) fell by 90 per cent. In Spanish Naples, where the poor harvests of 1619, 1620 and 1621 coincided with orders from Madrid to send money to assist Ferdinand, urban and rural property values fell by about half, and the economy of the kingdom languished. In the biblical phrase used by a Neapolitan economist of the day, 'Seven lean years began in 1623.'[7]

The Crisis of the Dutch Republic

Meanwhile, civil war almost engulfed the Dutch Republic. Neither history nor tradition linked the seven provinces that had rebelled against Habsburg Spain in the sixteenth century and formed an independent federal state.

Friesland and Gelderland had spent much of the fifteenth century fighting Holland and Zealand, with Groningen, Overijssel and Utrecht as both prize and battleground. All preserved their local laws and liberties intact, and they even spoke different languages: Dutch in the western provinces, Frisian in Friesland, and either Oosters or Low Saxon in the eastern provinces. The religious complexity of the republic was even greater. According to one foreign observer, one-third were Calvinists, another third Catholics, and the rest Anabaptists, Lutherans, Jews, members of some other sect, or atheists.

The House of Orange-Nassau provided a degree of cohesion. Count Maurice of Nassau (prince of Orange after 1618) served as captain-general of the confederate army and governor (stadtholder) of five provinces, and a cousin governed the other two. Certain institutions also served the entire republic – the Council of State (for military matters), the admiralty boards and the audit office – but executive power rested with the individual assemblies (the *Staten* or states) of the seven United Provinces, each of which sent delegates to the States-General. This central body remained surprisingly small, seldom more than twelve deputies (and often only four or five), because most decisions (including those on war, peace and taxes) required ratification by the seven provincial assemblies; and securing the consent of each provincial assembly required much time and patience, because even there the delegates could settle little without reference to their principals. In all, perhaps 2,000 magistrates (known as regents) in fifty-seven towns, plus scores of noblemen, participated directly in the republic's decision-making process. It was not democratic by modern standards, but in virtually no other major state in the early modern world did so many people help to shape national policy.

With so many protagonists in decision-making, friction frequently developed – particularly over contentious issues such as finance, religion and foreign policy. Holland, for example, normally favoured peace with Spain because war was expensive and Holland paid more in taxes than all the other provinces combined. Its southern neighbour Zealand, by contrast, normally opposed peace with Spain because the province prospered from piracy at Spain's expense and from the revenues generated by tolls and passports to trade with the enemy. The Calvinist clergy, too, vociferously favoured war with Catholic Spain, using not only the pulpits but also their contacts with individual regents (often their close relatives) to sabotage peace initiatives. Most of those who had fled Catholic persecution in Flanders and Brabant likewise opposed any settlement with those who had driven them into exile.

The discussions of 1607–9 over whether or not to make peace with Spain polarized these diverse groups. On one side stood the regents of Holland and Utrecht, led by Jan van Oldenbarnevelt, the chief executive of Holland, who wanted peace abroad and a measure of religious toleration at home. They faced a coalition that comprised Zealand and other areas that profited from the war; most Calvinist ministers; and most southern immigrants, supported by Maurice of Nassau (who stood to lose much of his influence and patronage once the republic demobilized its armed forces). Although the views of Holland prevailed – in April 1609 the republic concluded a twelve-year truce with Spain – domestic divisions soon deepened because of a controversy between two Calvinist theologians at Leiden University. Francis Gomarus, a refugee from Flanders, preached that everybody's spiritual destiny had been determined from the beginning of time; Jacob Arminius, a Hollander, argued that an individual's life choices affected their chance of salvation. A vicious controversy on these issues of salvation gathered momentum and the States of Holland, alarmed by attacks on Arminians, passed a resolution authorizing each town to raise special militia units to preserve law and order whenever necessary, and ordered them to obey the orders issued by the town that raised them. Maurice, who sided openly with the Gomarists, condemned these measures and toured the inland provinces to remove Arminians from office.

Oldenbarnevelt and his supporters retaliated by raising more militia units, whereupon Maurice used regular troops to disarm and disband them, and to arrest Oldenbarnevelt. Maurice also purged the town councils, replacing experienced magistrates with novices who lacked the ability to monitor the stadtholder's policies effectively, and also removed all Arminians from schools and universities. In 1619, a National Synod of the Calvinist Church condemned the Arminians as heretics and 'perturbers of the peace' in both church and state, and deprived some 200 of their ministers of their livings; and Maurice persuaded the States-General to sentence Oldenbarnevelt to death. He then encouraged and funded the efforts of his nephew Frederick of the Palatinate to recover his lands, and engineered a renewal of the war with Spain in 1621, when the Twelve Years' Truce expired.

Enter Denmark

Frederick's renewed defiance prompted Emperor Ferdinand II (r. 1619–37) to honour his promise (page 171) to transfer the Electoral title to Maximilian. This decision outraged public opinion inside and outside Germany because it was unconstitutional – the Golden Bull of 1356, universally regarded as the fundamental and immutable law of the Holy Roman Empire, ordained that the Electorate should remain in the Palatine house in perpetuity – and Frederick exploited his advantage to forge alliances with France, England, Sweden and Denmark, as well as the Dutch Republic. All promised to fight until Frederick had recovered his forfeited lands and titles. They also promised to support Christian IV of Denmark when he mobilized 20,000 soldiers to achieve this goal.

Christian commanded prodigious resources. His dominions stretched from the North Cape to Holstein in Germany and from Greenland to Oland in the Baltic, and included both sides of the Danish Sound. Since every ship entering or leaving the Baltic needed to pass beneath the guns of Elsinore Castle, the Danish king derived a huge income from the Sound Tolls; and by 1625 his total assets amounted to almost 1.5 million thalers. Christian was a devout Lutheran who firmly believed in the existence of a Catholic conspiracy that aimed to extirpate Protestantism throughout the empire – a fear confirmed by the relentless campaign of Catholicization by Ferdinand and his allies after the battle of White Mountain, expelling Protestant ministers, prohibiting public Protestant worship and reclaiming secularized church lands. Christian proclaimed himself Defender of German Liberties, and in spring 1626 he led his army across the Weser and marched towards the Palatinate.

This development alarmed the Catholic commander, Count Tilly, who warned that 'The Danes hold great advantages: they will act first and overwhelm us.'[8] Ferdinand therefore ordered Albrecht von Wallenstein, a military entrepreneur who had grown rich from buying up confiscated lands in his native Bohemia, to raise and maintain an army of 24,000 men. Wallenstein managed to send reinforcements to Tilly just before he confronted the Danish army at the battle of Lutter-am-Barenberg, where Christian lost half his soldiers and all his field artillery. For the next two years Catholic troops occupied Jutland as well as northern Germany and forced the inhabitants to finance them through 'contributions': payments in cash and commodities made directly to local troops. Paradoxically, this system required a steady increase in the size of his army to collect the contributions and Wallenstein therefore raised more troops, which in turn necessitated more contributions, until by 1628 he commanded (and had to support) no fewer than 130,000 men. Meanwhile, the supplies available dwindled due to a combination of cold winters, late springs and wet summers, culminating the 'year without a summer', 1628, during which neither grain nor grapes ever ripened in many areas.

Many sought scapegoats for the combination of extreme weather and oppressive soldiery: an unparalleled spate of witchcraft trials occurred, some involving the execution of hundreds of suspects at a time, while others blamed the Jews. The most serious unrest occurred in Upper Austria, which rashly joined the Bohemians in revolt but fell to the army of the Catholic League. After White Mountain, the emperor granted the territory to Maximilian as a pledge against the repayment of his war expenses and allowed him to use the duchy's taxes both to pay interest on this debt and to sustain the army of occupation. Taxes in the region rose fourteenfold, all payable in silver at a time when the *Kipper- und Wipperzeit* had destroyed the savings of most taxpayers. The cowed population might have tolerated these burdens had the government not also imposed unpopular religious measures: it expelled all Protestant pastors and schoolteachers, allowed Catholic creditors to foreclose on Protestants in order to force the sale of their property, and decreed that all residents of the duchy must attend Catholic worship or leave by Easter 1626. A rebel army led by a prosperous Protestant farmer, Stephen Fadinger, routed the governor and his troops and chased them back to Linz (the duchy's capital), which they besieged. They also begged Christian of Denmark to send aid, but his defeat at Lutter prevented this; then a bullet killed Fadinger in the trenches around Linz, and a general assault on its defences failed. Eventually 12,000 imperial troops restored order and executed scores of rebels. The government again ordered Protestants either to convert or leave.

Despite this victory, resentment of the exactions levied by Wallenstein's army grew and a group of Catholic rulers called on the emperor to reduce its size, end all recruiting, replace 'contributions' with taxes raised and administered by civilians, and appoint a special commissioner to audit the general's accounts. Although Ferdinand rejected these demands, he attempted to placate his Catholic allies with religious innovations. He informed them that, after nine years of war, he wished to reconfigure the religious contours of the empire, and in particular to repossess church lands secularized by Protestant rulers. These steps, he claimed, would be 'the great gain and fruit of the war', and in this (he promised his fellow Catholics) he did not 'intend, now or in the future, to have to bear the responsibility before posterity of having neglected or failed to exploit even the least opportunity'.[9]

'The Root of All Evils'

As long as Christian of Denmark remained in arms, the emperor deemed the opportunity too risky; but once peace negotiations got underway he prepared a document known as the Edict of Restitution that required Protestant rulers to return to the Catholic Church all lands secularized since the Peace of Augsburg of 1555, which had brought Germany's religious wars to an end. He had 500 copies of the edict secretly printed and distributed, with instructions for simultaneous publication on 28 March 1629. Once Christian made peace, Tilly and Wallenstein deployed their troops to enforce the edict, making no distinction between previously loyal and rebellious Protestant states, and the secularized lands of six bishoprics and 100 convents were soon back in clerical hands, with 400 more convents scheduled for restitution. Naturally such a drastic measure caused uproar among Protestants everywhere. On the title page of one copy of the edict – one printed in a *Catholic* stronghold – a contemporary hand added the words *Radix omnium malorum*: 'The root of all evils'.[10]

International affairs now distracted the emperor from further enforcement of the Edict of Restitution. The Dutch captured a treasure fleet sailing to Spain from America, compromising the Spanish Habsburgs' ability to fund their troops in northern Italy, where they sought to crush the defiant duke of Mantua, and in the Netherlands, where they faced the largest Dutch army ever assembled to that date. Ferdinand therefore instructed Wallenstein to send one expeditionary force to the Netherlands and another to help the king of Poland withstand an invasion by King Gustavus Adolphus of Sweden. When a French army crossed the Alps into northern Italy (Chapter 10), Ferdinand diverted Wallenstein's expeditionary force from the Netherlands to Italy, while his troops helped the Poles to inflict a stinging defeat on Gustavus and French diplomats brokered a truce that freed the battle-hardened Swedish army to invade Germany.

These developments compelled Wallenstein to recruit yet more troops, until by spring 1630 he commanded 151,000 men, spread out over Germany and northern Italy, and the increased demands for contributions led the German Catholic rulers to insist that Ferdinand dismiss his expensive general. The emperor reluctantly agreed to meet the Electoral College at Regensburg to resolve this and other contentious issues.

Since his election as emperor, Ferdinand had removed several rulers who opposed him and confiscated their lands, created a huge imperial army and issued the Edict of Restitution – all without convening an imperial Diet. Representatives of German rulers and foreign powers alike therefore converged on Regensburg in the summer of 1630, anxious to pressure the Electors to restore the *status quo ante* and, especially, to get rid of Wallenstein. Since the cost of his army far exceeded the available resources, the imperial commander himself made no effort to stay on, and in August 1630 Ferdinand duly replaced him with Count Tilly, commander of the army of the Catholic League, who reduced the imperial forces by two-thirds. The Electors capitalized on this success by extracting a promise from the emperor that, in future, 'no new war will be declared other than by the advice of the Electors', but he refused to modify the Edict of Restitution.[11] Ferdinand thus made two fatal errors at Regensburg. By sacrificing Wallenstein, he lost the one man who might have defeated all his enemies; by retaining the edict intact, he convinced the Lutherans of North Germany that they would soon experience its terms themselves.

The Tipping Point: The Rape of Germany

Gustavus Adolphus arrived on German soil at the head of a powerful army just as the Electoral meeting commenced. He immediately issued a manifesto in five languages that rehearsed his fear that the Habsburgs aimed to dominate the Baltic. Only at the end did the manifesto mention, briefly, the maintenance of German liberties as a motive for invasion. It said nothing about saving the Protestant cause. Gustavus therefore attracted little support: his sole German ally when he landed was the small port-city of Stralsund, and only the dispossessed and those under direct threat of imperial occupation (such as the city of Magdeburg) initially declared for him. The outlook for the Swedish expeditionary force looked bleak until October 1630, when Louis XIII concluded an alliance with Gustavus that promised to provide 1 million livres annually for five years to finance a war for 'the safeguarding of the Baltic and Oceanic Seas, the liberty of commerce, and the restitution of the oppressed states of the Roman Empire'. Gustavus promptly raised more troops and occupied the duchies of Mecklenburg and Pomerania, turning the eastern Baltic into a Swedish lake.[12]

Count Tilly and his troops concentrated on subduing the Protestant city of Magdeburg, Sweden's only ally in central Germany, until in May 1631 they stormed and sacked it. The following month Habsburg and French negotiators, assisted by a young diplomat in papal service named Giulio Mazzarini (known as Jules Mazarin in France), signed a treaty that brought peace to northern Italy and freed all imperial troops there to return to Germany. Once again it seemed as if Ferdinand would be able to expel the Swedes single-handed, but Tilly rashly decided to confront Gustavus before reinforcements arrived from Italy. At Breitenfeld, near Leipzig, the superior discipline and firepower of Gustavus's troops put an end to the Catholic tide of victory: Tilly lost 20,000 men, his field artillery and his treasury.

Rather like Lutter five years before, the battle of Breitenfeld transformed the military balance. Meeting with no effective resistance, Gustavus sent one army to Bohemia, allowing the return of the exiles who had fled in 1620, and led another through Germany's Catholic heartland to the Rhine. His principal Swedish minister, Axel Oxenstierna, created a government-general to administer all the occupied territories and organized contributions to maintain the 120,000 victorious soldiers deployed across northern Germany.

Ferdinand now saw no choice but to recall Wallenstein because, as one of his councillors put it: 'Now we shout "Help, Help", but nobody is listening.'[13] But before the general could recruit a new imperial army, Tilly rashly attacked the nearest Swedish positions. Gustavus won another convincing victory and, accompanied by a jubilant Frederick of the Palatinate, systematically plundered Bavaria. Maximilian pleaded with Wallenstein to send help, but instead the new general retreated with his army to the Alte Veste, a huge fortified camp near Nuremberg that Gustavus vainly besieged for two months before withdrawing northwards. Then, on 16 November 1632, the king launched a surprise attack and at the battle of Lützen, near Leipzig, Wallenstein lost so many soldiers that he decided to retreat, but the Swedes, too, suffered heavy casualties – including Gustavus himself.

For the next sixteen years rival armies moved to and fro across the empire, seeking to gain a decisive strategic advantage, but instead leaving death, devastation and insecurity in their wake. Oxenstierna, who after Gustavus's death became both head of the Swedish regency council and director of the Protestant war effort in Germany, tried to negotiate a favourable settlement with Wallenstein until in February 1634 Ferdinand had his general murdered (on the grounds that he was supposed to wage war, not make peace). The following September, Oxenstierna's forces suffered a stinging defeat at Nördlingen and withdrew to the Baltic, as Ferdinand exploited his advantage to conclude the Peace of Prague with some of his German enemies in May 1635. Once again the emperor overreached, declaring war on France. Oxenstierna renewed the alliance with France, promising (among other things) that neither side would make peace without the other. The alliance – and therefore the war – would last for thirteen more years.

The battle of Breitenfeld in 1631 saved the Protestant cause, but in doing so it transformed both the scope and the impact of the war. Previously, most Germans had seen hostilities as both exceptional and temporary: in the words of a miller's daughter near Nuremberg, until the summer of 1632 'we had indeed heard of the war, but we had not thought that it would reach us [here]'. Many rural areas saw public order collapse as the secular and religious elite fled whenever their side suffered a defeat, and many local burial registers record parishioners who died of hunger, sometimes after eating 'such things as heretofore [even] the pigs didn't eat'. Abnormal weather added to the misery. In Bavaria, an abbot living in the Alpine foothills filled his diary in these years with reports of the misery caused by the 'coldest winter', the 'raging storms', the 'wintery spring' and the 'stormy summer'; and, in 1642, by 'a flood that was worse than any in human memory', by 'hailstones weighing up to a pound', and by a hoarfrost that covered the fields in mid-June and ruined the harvest. He also noted the 'multiplication' of wild boars and wolves, 'making the roads and paths of all places unsafe'.[14]

Peace Breaks Out in Germany

Peace nevertheless proved elusive, in part because leaders who had gone to war for the sake of religion found it hard to moderate their position. In 1634–35, Emperor Ferdinand started the process by convening a committee of twenty-four theologians whom he asked whether he could make religious concessions to achieve peace without falling into mortal sin. One group of theologians argued that 'God, who up to this point has rescued our most pious emperor from so many dangers, in this extremity will also show us the way either to continue the war or to obtain a better peace'; but most approved limited religious concessions to Protestants as the lesser of two evils. Ferdinand therefore signed the Peace of Prague. When in 1640 the Diet of Regensburg debated whether Protestants might retain secularized church lands, even the hard-headed Maximilian of Bavaria ordered his advisers to consult theologians about their scruples against making peace with Protestants; and once again most theologians favoured modest concessions, provided they improved the chances of peace.[15]

The agreements reached at Prague and Regensburg settled most of the German disputes, and they formed part of the final peace settlement; so why did the Thirty Years' War drag on until 1648? Although most German protagonists sought the redress of a specific *past* injustice, first Sweden and then France had invaded Germany primarily to prevent a potential *future* injustice – namely a Habsburg victory that might threaten their national security. This meant that the 'two crowns' (as contemporaries called France and Sweden) could not be bought off with the transfer of one or even several tracts of land (although both did indeed make territorial demands): instead they refused to sign any agreement until they had created structures to guarantee its implementation.

European history offered no precedent or guidance for achieving such aims and so, despite prayers, pamphlets, broadsheets, medals and plays calling on the two crowns to make peace, they continued to make war. Sweden (and, to a lesser extent, France) wanted to create a balance of power both between the emperor and the states of the empire, and between the Catholics and Protestants of Germany. The two crowns insisted that only the imperial Diet, not the emperor, could legally declare war, and that all territorial rulers with seats in the Diet should have the right both to arm themselves and to make alliances.

The two crowns hoped that these measures would make any future war in Germany virtually impossible because, in the words of a Swedish diplomat,

> The first rule of politics is that the security of all depends on maintaining the equilibrium between each individual state. When one begins to become more powerful and formidable, the others throw themselves onto the scale [*Waagschale*] by means of alliances and federations in order to offset it and maintain a balance.[16]

They hoped to impose their vision on the diplomats, representing almost 200 European rulers (150 of them German), who late in 1643 began to arrive in Westphalia: those of the Protestant states gathered in Osnabrück, those representing Catholic states assembled 30 miles away in Münster. Negotiations ceased for a few months, when Sweden launched a surprise attack on Denmark (page 182), but in December 1644 the four major protagonists (France and Sweden on the one side, Spain and the emperor on the other) exchanged documents that set out their peace terms. They refused to suspend hostilities while they negotiated, however, which meant that the demands of each side waxed and waned with the changing fortunes of war – as Mazarin well knew: 'We have always said that we would increase our demands to reflect how events improved in our affairs', he reminded his diplomats at the peace conference, and so his mutable bargaining position reflected 'how much the situation has changed in our favour recently'.[17]

The 1645 campaign gave France and its allies a decisive advantage. A Swedish army invaded Bohemia and routed a Habsburg force under the personal command of Emperor Ferdinand III (r. 1637–57), and French troops destroyed the field army of the Catholic League. Ferdinand now faced the inevitable and drew up in his own hand a secret instruction for his chief negotiator at Westphalia that authorized a layered sequence of humiliating concessions on all the major issues. In religious matters, Ferdinand hoped to turn the clock back to 1630, the high water mark for the Catholic cause following the Edict of Restitution; but, if that proved impossible, he would settle for 1624; and 'in extremo casu' (when writing his final concessions, the emperor invariably switched from German to Latin) he would accept 1618, the optimal date for the Protestant cause. In political matters, Ferdinand would allow the Palatine and Bavarian branches of the Wittelsbach family to hold a seat in the Electoral College alternately, but 'ad extremum' he would create an additional (eighth) Electorate so that each branch would enjoy permanent representation. Sweden could keep eastern Pomerania plus, 'if it cannot be avoided', the archbishopric of Bremen and parts of Mecklenburg for the lifetime of the present ruler – and 'in ultimo necessitatis gradu' in perpetuity. 'In extremo casu', France could annexe the Habsburg lands in Alsace, and 'in desperatissimo casu' Breisach too.[18] Since further defeats placed Ferdinand in desperatissimo casu, his negotiators eventually made all these concessions, and in September 1646 they signed a preliminary treaty with France – but the terms would only come into effect when Sweden also made peace.

The Congress of Westphalia now turned serious attention to thorny religious issues, such as a normative date for the religious settlement. Like Ferdinand, the German Catholics (supported by France) wanted to turn the clock back to 1630, while the Protestants (supported by Sweden) pressed for 1618: in the end the peace congress annulled the Edict of Restitution and settled on a normative date of 1624. Those who had fled their homes to avoid religious persecution since then received the right to return (thus granting, for the first time in European history, legal protection to religious refugees). The Congress also determined that, in the future, any religious change required an 'amicable composition' between Catholics and Protestants, instead of a simple majority vote – a remarkable compromise in such a devout age.

Even after settling the crucial religious issues, the war dragged on for six more months while the French tried to extract a promise from the emperor that he would never again assist the Spanish Habsburgs, and the representatives of the Swedish army tried to extract 30 million thalers to pay its wage arrears. The Little Ice Age indirectly facilitated a compromise on the latter. The summer of 1647 was exceptionally cold: according to a Spanish diplomat in Münster, July was 'like November' and in August 'the weather is so cold that it could be the end of October'. The winter that followed proved unusually long and hard – in March 1648, a Bavarian nun recorded that 'there came such a great cold spell that everyone might have frozen' – followed by an exceptionally wet summer.[19] The bedraggled Swedish troops eventually settled for 5 million thalers, payable in stages, and on 24 October, the plenipotentiaries of the major states signed the complex peace instruments. The Thirty Years' War was over.

Multiple copies of the peace agreement rolled off the printing presses and some provisions went into immediate effect: Protestants returned to the cities and territories from which they had been banned (provided Protestant worship had existed there in 1624, the agreed normative date); church lands reclaimed under the Edict of Restitution changed hands again. Specific amnesties also took immediate effect. The son of Frederick of the Palatinate assumed a seat in the Electoral College, now expanded to eight Electors, and those deprived of lands and goods for supporting France or Sweden (though not those condemned for rebellion) received them back. Those who had sought refuge abroad now returned, and as the war-weary German governments collected enough money to pay their arrears, foreign troops began a phased withdrawal. In October 1650 Prince Karl Gustav and the rest of the Swedish high command, their work done, embarked at Wismar and sailed home. They found their fatherland on the brink of revolution.

Denmark and Sweden on the Edge

The extreme weather that afflicted most of Germany throughout the 1640s also ruined harvests in Scandinavia, causing bread prices to climb far beyond the reach of families already weakened by two decades of war. A remarkable combination of adverse social, dynastic and constitutional circumstances then almost brought both the Swedish and Danish monarchies to their knees between 1648 and 1651. Denmark suffered more. Although Christian IV did not intervene directly in Germany after his defeat in 1629, he exploited the continental involvement of his rival Sweden by meddling in its internal affairs until 1643, when Swedish troops suddenly invaded Jutland and occupied all Danish territories east of the Sound, and the Swedish navy routed the Danes. These hammer blows forced Christian to accept a humiliating peace that ceded to Sweden several Danish territories and, more significantly, virtual exemption from the Sound Tolls. Although the king retained some prestige as a sort of national patriarch – few Danes could recall any other monarch, since he had reigned for sixty years – henceforth he deferred to the nobles on the Council of the Realm. When he died in February 1648, Christian left a constitutional crisis because the Estates had not yet recognized a successor. Even though his oldest surviving son, Crown Prince Frederick, was the only viable candidate, the Council of the Realm (which according to tradition acted as the executive during an interregnum) delayed his election until he agreed to a coronation charter that forbade the monarch to involve the kingdom in foreign wars.

The new king faced a difficult task. The recent Swedish occupation had caused widespread damage to farms and a sharp drop in agricultural production; now disastrous harvests almost doubled the price of bread, and most areas also suffered from plague. These natural disasters came on top of sharp economic setbacks. The end of the war in Germany had led to a sudden drop in foreign demand for Danish agricultural produce, and to the return of many demobilized soldiers in need of domestic employment. In addition, the Council of the Realm imposed heavy taxes to liquidate the debts of previous wars – taxes from which its members, by virtue of their noble status, were exempt. This combination drove many smaller land-owners into debt as they raised capital to repair the damage of the war years and pay their taxes at a time when their profits plunged, creating a dangerous divide between the great nobles who ran the government and everyone else.

The Little Ice Age also contributed to a similar crisis in the Swedish monarchy. A prolonged period of cold weather had reduced crop yields and trade; the harvest of 1650 'was the worst Sweden had known for fifty years or was to know for near fifty more'; and in March the Stockholm bakers fought each other at the city gates to secure scarce flour.[20] As in Denmark, the harvest failure coincided with unprecedented fiscal pressure to liquidate the debts created by Sweden's continental war. Although no foreign troops had crossed the frontiers of the kingdom and caused damage, the constant demand for taxes and recruits created widespread hardship.

Queen Christina, Gustavus Adolphus's daughter, who came of age in 1644, did nothing to solve these problems. Not only did she spend vast sums on herself (court expenses soared from 3 per cent of the state budget in 1644 to 20 per cent in 1653), she also alienated so many crown lands that her revenues fell by one-third. Once her officers and soldiers returned from Germany, demanding their wage arrears and rewards for their services, 'donations were given faster than the land registers could record them [and] sometimes were given twice over'. Christina also doubled the number of noble families in Sweden within a decade, creating a new title almost every month. Such prodigality caused bitter divisions among the queen's subjects, as did the continuing political dominance of Oxenstierna and his aristocratic allies who by 1648 held twenty of the twenty-five seats on the council.[21]

The opposition drew strength from the fact that Christina lacked an heir. After she made clear in 1649 that she did not intend to marry, she worked towards securing the succession for her cousin Charles Gustav, for which she required the approval of the Diet (*Riksdag*), which included not only noble, clerical and urban chambers but also one for the peasants. The queen summoned delegates to assemble in Stockholm in July 1650 despite the probability that they would seize the opportunity to air their numerous grievances.

The Diet began with a concerted attack by the urban chamber on the number and privileges of the nobles. 'Do they want to introduce into Sweden the same servitude for men born free that prevails in Poland?' the delegates asked indignantly. They also complained that the queen preferred to appoint nobles to all the best positions in church and state, depriving others of valuable career opportunities – a development that caused especial frustration because Sweden now boasted more university graduates than ever before.[22]

The grievances of the peasants were more vehement and more radical. They complained not only of excessive demands from their lords but also about the alienation of crown lands, which delivered peasants to noble control, decimating crown revenues (because the noble lands paid little of no tax to the state), and also reduced the size of the peasant chamber in the Diet (because only peasants on crown lands could take part). These issues proved a rallying cry for all three non-noble Estates because, in the words of Archbishop Lennaeus of Uppsala:

When the nobility have all the peasants subject to themselves, then the Estate of Peasants will no longer have a voice at the Diet; and when the Estate of Peasants goes under, [the Estates of] Burghers and Clergy may easily go under too ... and since the Estate of Nobles has all the land in the kingdom under its control, where is the crown's power? For he who owns the land is the ruler of the land.[23]

The three Estates held joint meetings, forged common resolutions, claimed that the will of the majority of the four Estates should prevail, and refused to discuss the queen's proposals before she had redressed their grievances. Led by the burgomaster and town secretary of Stockholm (both lawyers), and the historiographer royal, they cloaked their demands with appeals to the Fundamental Laws of the kingdom and published them in a joint supplication drawn up in October 1650. The document included demands 'that all without distinction shall enjoy equality before the law' and 'that all private prisons and torture … may be abolished'. No crown lands should be alienated in future, and those already alienated should be recovered whenever possible. The supplication even condemned Sweden's foreign policy: 'What have we gained beyond the seas, if we lose our liberty at home?'[24]

The supplication circulated widely, yet the opposition failed to achieve any of its goals. Sweden possessed no plausible alternative leader except Charles Gustav, and he had nothing to gain from overthrowing his cousin. The major nobles likewise had nothing to gain from overthrowing Christina, and in any case plenty of contemporary examples showed the dangers of rebellion. The Swedish council regularly received and discussed the latest news about uprisings in other states – especially in England. According to one councillor, just as the troubles 'there in England originated with impatient priests, so it is also occurring here. It sets the worst possible example and does much harm'.[25] Queen Christina skilfully exploited her critics' divisions. She won the nobles' goodwill by promising not to revoke the grants of crown land she had made to them. She divided the other Estates by offering limited concessions to each of them: the clergy received some of the privileges they asked for (such as a guarantee that the crown would favour only orthodox Lutheran theology); the leading townsmen were promised open access to some crown offices (albeit mostly in remote areas); and the peasants won some limitations on the labour services they could be required to perform for their lords. In October 1650 the deputies recognized Charles Gustav as heir presumptive and dispersed.

Although a political victory for the queen, the Diet proved a fiscal failure: only the general recovery of alienated crown land (a process known in Sweden as a *reduktion*) could have solved the financial crisis facing the monarchy and without it, despite imposing new indirect taxes, Christina could not pay the wages of her soldiers, sailors and household servants. She also lacked any resources for relief when the harvest of 1652 failed all over Scandinavia. Shouts of 'Death to the nobles' and 'Devil take the bailiffs' soon rang out and, in one area, the peasants elected a 'king', with councillors, and drew up a list of nobles whom they intended to murder; but after royal troops had repressed the insurgency, the peasant king was broken on a wheel and his councillors (one of them a priest) were hanged.

An ambassador who travelled through the areas affected by the uprising remarked on the overall poverty of the population, the neglect of the roads and the dead animals in the fields; and one might wonder why these dire conditions failed to provoke broader unrest? The central government believed that the answer lay in its military system. 'The only means to keep the peasant under discipline is conscription', according to one councillor: that is, the constant forced migration of Swedish and Finnish young men to fight on the continent removed both potential leaders and marginal (and therefore dangerous) elements from the population.[26] Of the 25,000 Swedish and Finnish soldiers sent to Germany in 1630 and 1631, more than half died within two years and many more sustained serious injuries (Chapter 4). As young men began to realize that military service was virtually a sentence of death, recruiting efforts faltered. Over the course of the Thirty Years' War, Finland supplied some 25,000 young men to fight on the continent – equivalent to perhaps one-quarter of its total adult males – but although six conscription drives in Viborg province during the 1630s produced some 4,000 men, eight drives in the 1640s produced fewer than 3,500, and eight more in the 1650s produced fewer than 2,500. Those who managed to avoid the draft included deserters (some hid in the forests before the first muster, others fled during the march to the coast), the injured (some of them clearly with a self-inflicted wound), and the sick – including the unusual if not unique claim of Jakob Göransson who, when conscripted in 1630, asserted that 'every month he has a period like a woman and during that time he lies as if he were dead'.[27] Yet whether they served, deserted or menstruated, no conscript could take part in peasant insurgency. The war that ruined so much of Germany probably provided a safety valve for both Finland and Sweden and thus paved the way for the abdication of Christina and the peaceful succession in 1654 of her cousin as King Charles X Gustav.

The Second Crisis of the Dutch Republic

As soon as Dutch Republic concluded peace with Spain in 1648, some inhabitants looked back on the war years with nostalgia. 'War, which has made all other lands and countries poor, made you rich', wrote one pamphleteer in 1650: 'Your country used to overflow with silver and gold; the peace [with Spain] makes you poor.'[28] At first sight, such claims seem ridiculous. By the 1640s, almost 90 per cent of the republic's total expenditure went on defence, creating a huge tax burden, mostly in the form of indirect taxes: in the city of Leiden, excise duties accounted for 60 per cent of the price of beer and 25 per cent of the price of bread. Still, revenues fell far short of the republic's military and naval spending: between 1618 and 1649, the debt of the States of Holland soared from under 5 million to almost 150 million. The war also harmed the republic's economy in other ways. Villages near the frontiers paid protection money to enemy garrisons or else risked being ravaged; merchants who shipped goods abroad risked having them confiscated; privateers in Spanish service not only caused serious direct losses – in 1642 alone, they captured 138 Dutch ships – but also forced up freight and insurance rates.

Many in the republic, led by the States of Holland (which, thanks to its critical role in financing the war, had regained some of the power lost in 1618: page 174), therefore favoured a settlement with Philip IV. In 1635

France had declared war on Spain and, in concert with the Dutch, launched an immediate assault on Philip's possessions in the Netherlands. Although French forces made little progress before 1640, thereafter its forces made some major gains (see chapters 9 and 10 below). Each victory caused alarm in the republic. 'France, enlarged by possession of the Spanish Netherlands, will be a dangerous neighbour for our country', declared the States of Holland; it would be 'Hannibal at the gates' echoed a pamphleteer.[29] Popular opinion shifted towards concluding peace before Spanish power collapsed totally. Nevertheless, hammering out a settlement acceptable both to the stadtholder, now Maurice's brother Frederick Henry, and to all seven provinces, proved difficult. Zealand held out (mainly because its privateers prospered from the war) but after prolonged haggling, in January 1646 the delegates of the other six provinces left for Münster in Westphalia, headquarters of the Spanish delegation at the German peace congress.

From the first, Philip IV's negotiators assured the Dutch that their master was now ready to concede full sovereignty to the republic. They also slyly leaked a French proposal to marry the young Louis XIV to a Spanish princess, with the Netherlands as her dowry. This duplicity, as well as the spectre of the French just across the border, encouraged the Dutch to counter with a list of seventy-one conditions, almost all of which Philip IV accepted. The States of Holland therefore recommended that the States-General upgrade the talks from a truce to a full peace. Again, Zealand dissented and again it was overruled: in November 1646, by six votes to one, the Dutch agreed to work for peace, and two months later the two parties signed a provisional agreement that ended both the fighting and the economic sanctions.

Almost immediately, a run of bad harvests drove up food prices and created popular pressure for tax reductions that only peace could bring. At the same time the ceasefire on both land and sea caused a surge in Dutch trade, further fuelling pressure for a permanent settlement that would perpetuate such prosperity. In May 1648, at Münster, the delegates of Spain and of six Dutch provinces solemnly swore to uphold a permanent peace. The longest revolt in European history was over.

Supporters of the settlement predicted glittering prosperity, universal harmony and even a new Golden Age as soon as the war ended. Some improvements duly occurred – freight and insurance rates declined even further; trade with Spain, Spanish Italy and Spanish America soared – but the benefits accrued mostly to the merchants of Holland. Many of the republic's other citizens felt worse off in 1648–50 than before. They certainly suffered the same appalling weather as other parts of the world: in some areas it rained every day between April and November 1648, so that the hay and grain rotted in the fields, and a local bard wrote a poem entitled 'The rainy weather of the year 1648'. Then came six months of frost and snow during which the canals froze over, stopping all barge traffic. Many complained of 'the winter that lasted six months'. The summer of 1649 was also unusually wet, and the summer of 1650 unusually cold. Between 1648 and 1651, grain prices in the republic stood at their highest level for a century.[30]

Paradoxically, peace with Spain intensified the impact of poor harvests in frontier regions. As soon as the fighting ceased, the central government reduced the garrisons in the fortified towns in the east and south, and since the Dutch state normally paid its troops in full and on time, the dramatic reduction in military consumers put many local suppliers (such as tailors, saddle-makers, boot-makers and innkeepers) out of business. Zealand also suffered because the province had invested heavily in creating a colony in northwest Brazil, and its leaders acquiesced to peace with Spain only in return for promises from their neighbours that they would help to fight Brazilian settlers still loyal to Portugal (Chapter 15). An expeditionary force duly set out, but the Portuguese settlers routed it.

In 1650 the Spanish ambassador in The Hague warned his government that 'the common people certainly do not like the peace, blaming it for all the shortages they suffer, especially that of grain, without remembering the sterility of past years'.[31] The ambassador was unaware that William II had already decided to exploit the general discontent to his own advantage. The previous autumn he confided to his cousin William Frederick that, unless the States of Holland ceased to insist on further reductions in the armed forces, he intended to eliminate its leaders, above all the magistrates of Amsterdam. Unbowed, the States of Holland ordered several more units on their payroll to disband, but the prince immediately instructed the units' commanders to disregard these orders. He also complained to the States-General – where Holland had but one vote among seven – about the province's usurpation of his power. As in 1618, the States-General authorized the prince to visit every town in Holland and remove every magistrate who had opposed him.

Amsterdam now offered to negotiate an agreement over troop reductions, but when the prince refused, it composed a statement that reminded everyone that the stadtholder was the servant of the state and not its master. Outraged by such defiance, on 30 July 1650 the prince arrested and imprisoned his leading critics in the States of Holland, believing that his cousin William Frederick and 12,000 soldiers had just forced their way into Amsterdam – but contingency disrupted his plan. The previous night, a large part of the troops destined for the operation lost their way in a thunderstorm and arrived at the rendezvous thoroughly soaked. While they dried out, a postal courier bound for Amsterdam rode by and (since the troops had no orders to detain passers-by) warned the city of its peril. Thanks to this extraordinary chance, by the time William Frederick and his men finally arrived, the magistrates had armed the citizens, closed all the gates and flooded the moat around the city.

William did not give up. He joined his troops outside the walls of Amsterdam, and after a few days the city meekly agreed to surrender its outspoken magistrates and acknowledged the stadtholder's sole right to issue orders to the army, determine troop levels and decide foreign policy. Next, with his domestic base secure, the stadtholder issued an ultimatum to Philip IV: unless Spain immediately opened peace talks with France, the Dutch Republic would declare war again – but suddenly the prince took to his bed, stricken by smallpox, and on 6 November 1650 he died. Since he lacked an acknowledged heir, both his newfound powers and his threat of a new war with Spain died with him. The States of Holland immediately freed their imprisoned colleagues and invited representatives from all the provinces to join them in a special session of the States-General, armed with full authority to fill the unprecedented constitutional vacuum.

Even before the Great Assembly met in January 1651, the States of Holland took several revolutionary steps. Most notably, they resolved not to appoint a stadtholder for the province and instead made clear that all troops took their orders from them; and they authorized the patrician elite of each town to choose their own magistrates and name their representatives to the States. The Great Assembly confirmed both these initiatives and introduced some more. It appointed a Holland nobleman to command the republic's army and reduced the autonomy of the military: henceforth courts martial would try only military offences (such as desertion and disobedience), leaving soldiers accused of all other crimes to be tried by civil courts. In religious affairs, the States granted freedom of worship to Catholics and Jews; and, although they agreed to tolerate non-Calvinist churches only where they already existed, this effectively guaranteed toleration for all who desired it. The assembly also sensibly decided to ignore pleas to prosecute those who had promoted William II's agenda, which restored the domestic harmony imperilled by the controversies of the previous four years.

This constitutional revolution received widespread praise – the philosopher Baruch Spinoza termed it 'the System of True Liberty' – and it brought unprecedented prosperity to most parts of the republic for a generation. Nevertheless, there were two linked weaknesses: the very prosperity of the Dutch provoked envy and attacks by its neighbours; and, without a stadtholder to coordinate military and naval operations, the republic had difficulty in holding its own. When Britain attacked in 1652, the Dutch navy lost almost every battle; and although it managed to close off both the Baltic and the Mediterranean to British shipping, which led to peace in 1654, that same year the last Dutch outposts in Brazil surrendered to Portugal. The republic did rather better when Britain attacked again in 1664, but it forfeited New Netherland, the last Dutch outpost in North America. Then, in 1672, Britain attacked a third time, this time in alliance with France.

Dutch forces crumbled before the invasion of Louis XIV at the head of 130,000 troops, and one month later the States-General reluctantly appointed William III, posthumous son of the late prince of Orange and now aged twenty-two, commander of its army and navy. It was almost too late: having captured every Dutch town in his path with scarcely a struggle, in June 1672 Louis entered Utrecht in triumph. As rioting rocked the principal towns of Holland, the remaining delegates in the States-General voted to surrender on ignominious terms. Luckily for them, Louis rejected their offer and demanded more, allowing Prince William to restore order and focus the energy of everyone on resisting the French – but his efforts might still have failed without a sudden change in the weather. Extreme drought in the spring of 1672 had facilitated the French invasion by reducing the level of the Rhine and other rivers so that Louis's cavalry could wade across and create a bridgehead, allowing his engineers to build bridges for the infantry. The drought also prevented the Dutch from using their ultimate defensive strategy: opening the dikes to create a water barrier between Holland and the French. Until mid-July, the water rose painfully slowly, but then heavy and prolonged rain rendered all routes into Holland impassable, and at the end of the month Louis left Utrecht and returned home. Under the firm hand of William III, Dutch naval and military organization steadily improved until first England (1674) and then France (1678) made peace.

The Danes 'Forge their Own Chains'

After the humiliating concessions extracted in return for his coronation in 1648, Frederick III of Denmark worked hard to reach a better relationship with his subjects, and in spring 1657 he persuaded both the Council of the Realm and the Diet to authorize a declaration of war against Sweden, since Charles X appeared to be mired in a simultaneous war against Poland and Russia (Chapter 6). The war almost cost Frederick his kingdom. Charles left Poland at once to deal with his new enemy, marched across Germany and occupied much of Jutland. The Swedes had planned an amphibious attack on Copenhagen, but the onset of one of the coldest winters of the Little Ice Age in mid-December suggested another possibility. According to an English diplomat at the scene:

> The extraordinary violent frost was by this time increased to such a degree, that the Little Belt which divides Jutland from the isle of Funen was so intensely frozen, as suggested to the Swedish king an enterprise

(full of hazard, but not disagreeable to a fearless mind edged with ambition) of marching over the ice into Funen with horse, foot and cannon.

The astonished Danish defenders 'made large cuts in the ice' but with the extreme cold they soon 'congealed again'. The Swedes therefore stormed ashore and swept all before them because 'Funen [and] the other Danish isles are all open and unfortified, and have no defensible places'.[32] For the first time in European history, meteorologists now decided military strategy: they persuaded Charles to disregard the misgivings of his senior officers and follow the itinerary where they indicated that the ice had frozen hard enough to allow some 8,000 Swedish veterans with their artillery to cross from Funen to Zeeland and approach Copenhagen. Although the Danish capital lacked the strength to resist a siege, Charles (ignorant of its weakness) granted an immediate ceasefire in return for Frederick's promise to surrender almost half his kingdom and to send military and financial assistance to Charles's campaign in Poland.

The following year, when it became clear that Frederick had no intention of sending the promised assistance, Charles determined to reduce Denmark 'to the position of a province of Sweden': its nobility would be exiled, its recalcitrant bishops replaced with docile Swedes and its university moved to Gothenburg (on the Swedish side of the Sound).[33] But before he could attempt these ambitious goals, the redoubtable Charles X suddenly died, leaving a four-year-old son under a regency council. A few months later the two governments concluded a treaty that obliged the Swedes to restore a few of the lands they had captured but advanced their frontier permanently to the Sound. Never again would Denmark be able to control shipping entering and leaving the Baltic.

Frederick III nevertheless faced other serious problems. The adverse weather (of which the frozen Baltic was merely the most memorable extreme event) had drastically reduced harvests, and this shortfall, coinciding with a plague epidemic and enemy occupation, devastated many areas of his kingdom. The total population of Jutland and the home islands fell by about a fifth between 1643 and 1660; some parishes claimed that three-quarters of their farms lay abandoned; and almost half of the Danish clergy died between 1659 and 1662. In addition, the war had created huge debts. To address these issues, Frederick III ordered the representatives of the nobles, the towns and the clergy to assemble in Copenhagen in September 1660.

All members of the Danish Diet agreed that the kingdom's fiscal crisis could be solved only by reducing spending and raising taxes – but here, consensus ended. Representatives of the burghers and clergy insisted on creating new excise taxes to be paid by everyone without exception, but the nobles insisted on exemption. After four weeks of haggling, on 14 October 1660 a group of outraged clerics and citizens proposed extensive reforms to the prevailing political system, including the abolition of the elective character of the monarchy in favour of the hereditary principle – a move that would involve revocation of Frederick's 1648 coronation charter and thus of the privileged position enjoyed by the nobility. The authors of this revolutionary proposal probably meant it merely as a tactical manoeuvre to scare the nobles into agreeing to pay their share of the new taxes, but the king swiftly exploited his unexpected opportunity.

One week later, Frederick doubled the guards on the capital's ramparts and closed its gates, ordering all ships to stand off so that no one in the city could leave, and on 23 October 1660 delegates of all three Estates gathered in the royal palace and offered Frederick full hereditary rights. The king graciously accepted, ending over a century of aristocratic dominance, and appointed a constitutional commission to propose the necessary changes to perpetuate his new status; but after a couple of days, the commissioners obsequiously declared that Frederick himself should formulate a new constitution. The Diet agreed, asking only that the king should not dismember the kingdom or change its faith, and that he should respect the ancient privileges (without specifying which). On 28 October, Frederick received the unconditional homage of his people. The Danish Diet would not meet again for two centuries.

The king and his advisers now remodelled the central government (ironically imitating the system of the hated Swedes) with administrative colleges, a supreme court to replace the judicial functions previously exercised by the Council of the Realm and an army raised through conscription. Frederick also sold almost half the crown lands to pay off his war debts and introduced new taxes, direct as well as indirect, that the nobles had to pay along with everyone else. In 1665 the new 'absolute and hereditary monarchy' received its definitive form in the Royal Law (*Kongelov*), a constitution that remained in force until 1849 (making it the longest-lasting constitution in modern European history). The king received 'supreme power and authority to make laws and ordinances according to his own good will, to expound, to alter, to add and take from, indeed simply to abrogate laws previously made by himself or by his forefathers and also to exempt what and whom he pleases from the general authority of the law'. Other clauses gave the crown 'the supreme power and authority to appoint and dismiss all officials, high and low'; 'supreme power over the clergy, from the highest to the lowest'; and sole 'control over the armed forces and the raising of arms, the right to wage war, to conclude and dissolve alliances with whom and when he sees fit, and to impose duties and other levies'.[34]

Frederick III had achieved in Denmark by popular consent what many of his fellow monarchs failed to win by force. Thirty years later Robert

Molesworth, sometime British ambassador in Copenhagen, still found it hard to believe that in just four days an entire kingdom had 'changed from an estate little differing from aristocracy to as absolute a monarchy as any is at present in the world'. Molesworth cruelly continued: 'To the [Danish] people remained the glory of having forged their own chains, and the advantages of obeying without reserve'. They were now 'all as absolute slaves as the negroes are in Barbados; but with this difference, that their fare is not so good'.[35]

'The All-Destructive Fury of the Thirty Years' War': A Myth?[36]

Of the tens of millions of Germans who lived through the Thirty Years' War, fewer than 250 eyewitness accounts have survived; moreover, only nine authors were female. The geographical distribution was also atypical: two-thirds of the authors (including all the women) lived within a quadrilateral bounded by Münster, Magdeburg, Basel and Munich. Most eyewitnesses covered only part of the period: many began to record events only after the war affected them; others stopped before the war ended, and several left gaps in their narratives – usually because it became too dangerous to write. The motive for writing also varied. None of the authors seems to have envisaged publication: the majority kept a private record to remind their families or their descendants of the horrors that they had experienced, and a few wrote explicitly just 'for myself'.[37]

Despite these disparities, and although none of these authors knew what the others had written, they described many of the same experiences. Three-quarters of the civilians stated that they had been plundered by troops – some of them repeatedly. Over half reported having to flee their homes at least once; almost half reported the murder by soldiers of individuals whom they had known personally; about one-fifth described being assaulted themselves. The most striking omission is rape: only three of the printed accounts written by civilians mentioned violence committed against women they knew. Even in 1789, the publisher of a war memoir that mentioned a rape omitted the victim's name because her family still lived in the area.[38]

Although human history is full of people who claim that they experienced misfortunes unparalleled in other ages, subsequent research has corroborated the extreme claims of those who lived through the Thirty Years' War. Of 800 surviving German parish registers for the period 1632–37, all except five recorded a significant mortality crisis. Large numbers of soldiers also perished: the records of the Swedish army reveal that 2 million soldiers, most of them Germans, died as a direct result of the war. Hans Medick and Benjamin Marschke estimate that the war killed at least 5 million people.[39] Moreover deaths, whether through violence, starvation or disease, formed only one of three variables that affected all early modern populations. In addition, during the Thirty Years' War:

- Births fell because brides postponed their marriage and conceived fewer children, either through abstinence or infertility
- Migration soared as civilians left their homes either to find security or sustenance elsewhere, or to join an army.

The combination of these variables magnified the demographic decline in each community and restrained its rate of its recovery. Losses through adult mortality were, paradoxically, the easiest to make good because death might leave a vacant farm or firm that would provide economic opportunities that allowed the next generation to multiply. Migrants were more difficult to replace, because either they took their fortune with them or else they had none to take: only a community with assets would attract migrants from elsewhere to replace those who had left. A shortfall in births – especially female births – proved the hardest to replace in demographic terms because, apart from creating a missing generation (children who would have been born but for the trauma of war), a generation that numbered fewer mothers would itself produce fewer children, so that the deficit perpetuated itself.

Given the political fragmentation of early modern Germany, all national aggregations are tentative, and trying to identify regional demographic variations is even more hazardous. Nevertheless a careful study of local records led John Theibault to suggest that the overall figures conceal major disparities – that, in effect, most losses occurred in a relatively small area: 'More than a quarter of the population of the empire may have lived in areas that had no losses or lost less than 10 per cent of their population, while only about a tenth lived in areas that lost more than half their population.'[40] Specifically, in the southern half of Germany, the Rhine Palatinate (a battleground almost from the moment when Frederick accepted the Bohemian crown) and Württemberg (savagely contested between Catholics and Protestants from 1631 until the war's end) lost at least one-half of their pre-war population. Bavaria, Franconia and Hessen seem to have lost between one-third and one-half. In the northern half of Germany, Mecklenburg and Pomerania (both of them occupied by Sweden) lost at least one-half of their population; while Saxony and Brandenburg (also a battleground almost continuously from 1631 to 1648) lost between one-third and one-half. Many strategically important regions suffered staggering material as well as demographic losses: two-thirds of the buildings in the once prosperous and densely populated countryside around Magdeburg and Halberstadt were destroyed between 1625 and 1647; the debts of the city of Nuremberg rose from under 2 million gulden in 1618 to over 7 million by 1648; and so on.

11. The depopulation of Germany during the Thirty Years' War, 1618–48. The political fragmentation of Germany means that the demographic impact of the Thirty Years' War can only be reconstructed from regional data. These reveal a few areas, notably in the northwest (and in Austria and Switzerland in the south), that lost one-tenth or less of their pre-war population; while the Bohemian lands, where the war began, lost up to one-third. By contrast, some Protestant areas in the northeast and southwest lost over half of their people, while parts of the Catholic south lost between one-third and one-half.

The war destroyed culture as well as people. The booksellers of Germany, the birthplace of the printing industry, brought out 1,780 titles in 1613 but only 350 in 1635; and the international book fair at Frankfurt collapsed. Many of the thriving urban music societies of Germany closed their doors – the Musikkranzlein at Worms and Nuremberg, the Convivia Musica at Görlitz, the musical colleges at Frankfurt and Mühlhausen – and princes reduced their musical patronage. Even popular music declined: of over 600 songs from the period 1618–49 that make some reference, direct or indirect, to the Thirty Years' War, scarcely 100 appeared after 1634.

Many intellectuals fled to avoid the war. The largest single group came from the lands ruled by Ferdinand II. The polymath Jan Amos Comenius, who supported the Bohemian revolt, left his native Moravia after White Mountain and sought refuge in Poland, where he started work on an 'encyclopaedia of universal knowledge', which he believed could solve the world's problems. He later moved to Holland, England and finally Sweden, with the intention of founding a special college where colleagues could work on his project. The poet Martin Opitz, from Silesia, also fled to Poland after White Mountain and ended his days there as a refugee. The astronomer and mathematician Johan Kepler had to flee twice: first from Graz in 1600, when Ferdinand expelled all Protestants in Styria, and then from Linz in 1626 when the brutal suppression of the Upper Austrian peasants' revolt (page 176) made him fear for his life. In 1675 the artist and art historian Joachim von Sandrart published his *German academy of architecture, sculpture and painting*, in which he provided numerous biographies of artists whom 'bloodthirsty Mars' had forced to flee or (like Sandrart himself) compelled to 'give up laborious copperplate engraving and take up painting in its stead', because a canvas was more portable in an emergency. Others lost their work through theft or malice. Thomas Robisheaux has argued that the war made its impact not just through 'the harsh and inhuman conduct of the soldiery, but from the way all social, political, and religious order vanished and so contributed to the wild disorder and confusion at every level of society'.[41]

In the conclusion to her classic study of the Thirty Years' War, first published in 1938, Dame Veronica Wedgwood stated sadly: 'The war solved no problem. Its effects, both immediate and indirect, were either negative or disastrous. It is the outstanding example in European history of meaningless conflict.'[42] The evidence presented in this chapter lends some support to her verdict: not only Germany but also parts of Italy, the Dutch Republic, Sweden and Denmark all suffered effects that were 'either negative or disastrous'.

Nevertheless, the Peace of Westphalia brought eventual benefits not only to Germany but also to at least some of its neighbours. Pressure from France and Sweden eventually created a political balance within the Holy Roman Empire by enhancing the power of the Diet (which after 1648 alone possessed the authority to declare war) and of territorial rulers, while reducing that of the Habsburgs. The new principle of 'amicable composition' reduced the risk of another religious war: for almost a century, no German state declared war on another – and none would ever again wage war for religion. Moreover, once the states of Central Europe ceased to be ravaged

by religious and civil wars, foreign powers lacked a plausible excuse to intervene in their disputes – a development that promoted international stability.
The Congress of Westphalia also provided a new model of conflict resolution. Similar international peace conferences would terminate the major
wars between European states – the Pyrenees (1659), Breda (1667), Nijmegen
(1678), Rijswijk (1697) and Utrecht (1713) – paving the way for the Concert
of Europe that would successfully maintain peace among the great powers
after 1815.

CHAPTER NINE

The Agony of the Iberian Peninsula, 1618-89[1]

'The Target at Which the Whole World Wants to Shoot its Arrows'

At his accession in 1621, aged sixteen, Philip IV governed an empire on which (as his spin doctors put it) 'the sun never set': Spain and Portugal; Sardinia, Lombardy, Naples and Sicily; the southern Netherlands; as well as the Spanish and Portuguese colonies in the Americas, the Philippines, Asia and Africa. Nevertheless, this global extent brought weakness as well as strength. A letter written in 1600 to Don Balthasar de Zúñiga, a senior diplomat, underlined the strategic dilemma:

> We are gradually becoming the target at which the whole world wants to shoot its arrows; and we know that no empire, however great, has been able to sustain many wars in different areas for long. If we can think only of defending ourselves, and never manage to contrive a great offensive blow against one of our enemies, so that when that is over we can turn to the others.

This assessment proved prophetic. During the next two decades, the Dutch Republic forced Spain to recognize its de facto independence and seized some Iberian outposts in Asia and Africa, and several Italian states successfully broke free of Spanish influence. In 1619 Zúñiga, now Spain's chief minister, lamented that 'when matters reach a certain stage, every decision taken will be for the worst, not through lack of good advice, but because the situation is so desperate that it is not capable of remedy'.[2]

Zúñiga died in 1622, ceding his position as chief minister to his nephew, already the king's *privado* or favourite: Don Gaspar de Guzmán, count of Olivares and later duke of San Lúcar (hence his clumsy title 'the count-duke') – a man who, at least initially, rejected such pessimism. 'I do not consider it useful to indulge in a constant, despairing recital of the state of affairs', he chided a critic. 'I know it, and lament it, without letting it weaken my determination or diminish my concern; for the extent of my obligation is such as to make me resolve to die clinging to my oar till not a splinter is left.' At Christmas 1624, Olivares presented his master with a comprehensive reform programme; but almost immediately, Britain declared war and attacked Cadiz, and the duke of Savoy laid siege to Genoa, Spain's most important ally in the Mediterranean.

Olivares neglected reform while he concentrated on these strategic challenges, and on recovering from the Dutch both Bahía in Brazil and Breda in the Netherlands. Success in all these endeavours led him to crow to a colleague that 'God is Spanish and favours our nation these days'; but the need to react simultaneously to attacks in so many different areas convinced him of the importance of an integrated imperial defence strategy. A few days after the relief of Cadiz, he unveiled a grand proposal called the Union of Arms.[3]

The scheme aimed to create a rapid reaction force of 140,000 men drawn from the monarchy's various component parts: if any of them came under enemy attack, a portion of the force would immediately rally to its rescue. Olivares expected that the Union would not only share the costs of imperial defence, but also 'familiarize' – the word used in government circles – 'the natives of the different kingdoms with each other so they forget the isolation in which they have hitherto lived'. In January 1626, the king and his favourite set out from Madrid to sell the Union of Arms to the Corts (representative assemblies) of Aragon, Catalonia and Valencia. Afterwards, they intended to move to Lisbon and there prepare an invasion of Ireland in retaliation for Charles I's attack on Cadiz.[4]

The Union of Arms had little chance of success, because Olivares used unreliable data to fix the obligations of each part of the monarchy. He calculated that Catalonia's population numbered 1,000,000, so that it should provide 16,000 paid soldiers; but subsequent research suggests that Philip had only 500,000 Catalan subjects. Moreover, by unilaterally imposing the demands of the central government on regional authorities accustomed to autonomy, the Union provided a common focus for previously separate grievances. In Aragon, where the king and his minister called for a permanent standing army of 3,333 soldiers, with a further 10,000 as a strategic reserve, tenacious opposition forced them to accept just 2,000 for fifteen years. The Valencian Corts granted only one-quarter of the crown's request. The Catalan Corts refused to vote anything.

Undeterred, in July 1626 Philip signed orders that put the Union of Arms into effect in Portugal, Italy and the Netherlands, and commanded the Council of the Indies to apply it to America and the Philippines; but the scheme provoked universal opposition. The treasury of Castile therefore continued to bear the brunt of defending the monarchy, forcing the king to issue a Decree of Bankruptcy in February 1627 that froze the capital of existing loans, most of them from Genoese bankers, and suspended all interest payments. Olivares had already secured an undertaking from Portuguese bankers, almost all of them New Christians (as people of Jewish descent were termed) to lend over 1 million ducats, and just as the government hoped, the emergence of these rivals led the Genoese bankers to accept low-interest bonds in repayment of their old debts, and also to provide new loans. On the very day the government finalized these generous arrangements, news arrived at court that the duke of Mantua had died, leaving a disputed succession: Philip and Olivares regarded the coincidence as providential and decided that, notwithstanding the numerous wars already afoot, they could afford to intervene in order to prevent a French candidate from gaining Mantua. They would soon regret their choice.

The Portuguese bankers expected repayment of their massive loan from the silver bullion scheduled to arrive in Spain from America, but in September 1628 a Dutch fleet ambushed the entire treasure fleet and captured its cargo. Even a year after the disaster, Philip admitted that 'whenever I speak about it my blood boils in my veins, not for the loss of money, because I pay no attention to that, but for the reputation that we Spaniards lost'.[5] Without the anticipated silver, Philip's armies abroad received no funds for several months, and in 1629 the Dutch recaptured the heavily fortified city of 's-Hertogenbosch and almost 200 surrounding villages, and Louis XIII led a large army across the Alps to support the French claimant to Mantua. Olivares predicted (with uncanny accuracy) that France had just started a war that would last for thirty years.[6]

Such setbacks might have disposed some rulers towards peace, but Philip gushingly reassured his councillors that 'none of these losses that I have suffered and continue to suffer have afflicted or discouraged me, because God Our Lord has given me a heart that has room for many troubles and misfortunes without becoming overcome or fatigued'. All his wars must therefore continue, and in October 1629 he announced his intention to travel first to Italy and then to the Netherlands to take personal command of his troops. Realizing that his great adventure would involve enormous costs, he invited each of his ministers to suggest ways of funding it.[7]

Characteristically, the king appointed a committee of theologians to evaluate the proposals received. Equally characteristically, the committee rejected every suggestion that called for reduced spending (that the king should stay in Spain; that he should make peace in Italy in order to have more money to fight in the Netherlands; that he should make 'the best peace treaties possible, postponing until a better occasion the royal intentions of Your Majesty in the hope that God will return to fight for His cause'). Instead the committee approved proposals for three new taxes: a stamp duty on all official documents (*papel sellado*); a national salt monopoly (*estanco de la sal*); and the retention of part of the first year's salary of every newly appointed office-holder, both secular and ecclesiastical (*media anata*). It also made the radical suggestion that the new taxes should be imposed universally throughout the empire, and not just in Castile.[8] Olivares welcomed the theologians' report, and ordered ministers to draw up plans to introduce each new tax, to be paid by everyone, even the clergy, throughout Castile and Portugal. Although some ministers warned about the dangers inherent in such innovations, the king pressed ahead.

Open Opposition Begins

The new taxes appealed to Olivares in part because they were regalian rights, which could be imposed and changed at will; but since the precedents for most of them lay deep in the past (many had not been levied for decades if not centuries), government apologists ransacked historical works for justifications while opponents of the taxes searched for counter-precedents that restricted or precluded each royal initiative. In Catalonia, lawyers published historical accounts of the Fundamental Laws or Constitutions of the principality that no ruler could violate; clerics published tracts defending the duty to preach in Catalan (not Castilian, as the central government insisted); and trade officials wrote discourses in favour of protecting economic goods produced in Catalonia against imports (especially from Castile). All three groups of Catalan writers fostered a sense of 'us versus them', and a distrust of everything that emanated from Madrid.

Olivares's opponents had no difficulty in recruiting university-trained historians and lawyers to ferret out precedents, because Spain, like other European countries, possessed a surfeit of them. By 1620 some 20,000 students attended university in Castile, representing more than one-fifth of their age cohort. The majority studied law and what today would be called liberal arts. Only a small proportion of these students entered government service: many others devoted their learning to researching and writing critiques of government policy.

The government's critics gained strength when extreme weather caused a series of catastrophes throughout the Spanish Monarchy. In 1626–27 the worst floods ever recorded inundated Seville; in 1629 disastrous floods left much of Mexico City under water for the next five years; in 1628–31 Spanish Lombardy suffered from a drought-induced famine and plague, which killed about one-quarter of the population; in 1630–31, according to a contemporary chronicler, Lisbon 'lacked everything, especially grain' because of drought.[9] Castile also suffered from the climatic downturn. Madrid sent officials as far as Andalusia and Old Castile to requisition additional grain and the capital's granaries distributed twice as many bushels of wheat in 1630–31 as in any other year. The rural population was less fortunate. To take a single example: at the village of Hoyuelos, near Segovia, the tithe yield (a fixed percentage of the harvest) fell from nineteen bushels of wheat in 1629 to two in 1630 and to only one in 1631. Some towns and villages lost half of their population and the king's ministers warned that, owing to 'the shortage of grain, Your Majesty's vassals find themselves in great need and incapable of serving you as they would wish'.[10]

Olivares ignored these warnings and instead began a massive recruiting drive throughout Castile to create armies numerous enough to win all his foreign wars, expecting the salt monopoly to fund everything; but, as with the Union of Arms, the lack of accurate demographic data doomed the venture. Ministers tried to estimate the total size of Castile's population, and thus project salt consumption, on the basis of the number of religious indulgences distributed in recent years; but this failed to take account of the death and migration of consumers, which sharply reduced the demand for salt. Once he realized the scale of the shortfall, Olivares demanded that every household declare under oath its anticipated salt consumption for the coming year, which would cost 69 reales per bushel – a figure that included no less than 58 reales of tax. Consumers who underestimated their consumption could acquire more salt later at a higher price, but they could not take less.

Such heavy reliance on regalian rights provoked widespread opposition. Some Castilian taxpayers resisted the salt monopoly passively, either declaring that they would consume no salt at all or giving improbably low estimates; others organized protests. Philip now convened the Cortes of Castile and called on the delegates to 'give the last drop of blood in your veins, if necessary, to uphold, defend and preserve Christianity'. He provided them with details on each recent campaign and its cost, and warned that Spain now faced 'the greatest, the most urgent and desperate situation that has arisen or could arise'. The assembly responded that the extreme weather, failed harvests and high mortality precluded raising any new taxes – unless the king agreed to abolish the salt monopoly. Reluctantly, he did so.[11]

Olivares and his master therefore decided to make another personal attempt to persuade Catalonia to participate in the Union of Arms but Catalonia, like the rest of the Iberian Peninsula, had suffered a series of natural disasters. During a prolonged drought in winter 1627–28 the clergy of Barcelona led no fewer than thirty-four processions to pray for rain. Their prayers were answered with storms that washed away another harvest. Then in 1630 a new drought caused food prices to rise sharply while trade and industry slumped. Barcelona introduced bread rationing and, although the authorities foiled a plan by the *segadors* to storm the customs house, starving citizens attacked the city's granaries, pulled half-baked loaves from the ovens, and devoured them. Coming on top of such hardships, the demand in 1632 for new taxes to fund the Union of Arms rallied virtually all Catalans around their Constitutions and so, as in 1626, even the presence of the king and his chief minister failed to persuade the Corts to vote any new taxes. Once again they departed empty-handed.

Open unrest had already broken out among the Basque population of the northern lordship of Vizcaya. Although technically part of the kingdom of Castile, and so subject to the new salt tax, Vizcaya (like Catalonia) boasted powerful privileges (*fueros*) and a local representative assembly (the Junta General). In 1631, during a meeting of the Junta, 'some women from the coastal areas' denounced local officials who tried to impose the salt tax as 'traitors' whom 'it would be better to kill'. Local officials prudently suspended both the session and the salt tax. One year later, when the magistrates of Bilbao tried again to collect the new salt duty, the crowd sacked the houses of the tax collectors. News of these events appalled the king: 'I saw these papers and this *consulta* with sadness', he scribbled on the dossier forwarded by his council, 'to see in Spain something unheard of for centuries'.[12] Nevertheless he trod warily, dispatching mediators rather than troops; and when mediation failed, he merely imposed an economic boycott on Vizcaya until 1634, when he abolished the salt monopoly and issued a general pardon.

The defiance of Catalonia and Vizcaya, coupled with the success of Gustavus Adolphus and his Protestant allies in Germany, deeply depressed Olivares. 'You get to the mountaintop', he wrote despondently to a colleague in autumn 1632, 'and then everything falls, everything goes wrong. We never see a comforting letter; not a dispatch arrives that does not tell us that everything is lost because we had failed to provide the money'. Realizing that his problems arose in part from trying to fight on all fronts all the time, Olivares tried to maintain the peace in at least some areas. For example, he instructed the viceroy of India that he must 'always maintain peace with the Mughal' emperor 'since he is our very close neighbour and his power [encircles] our territories. If he is offended, he may break off with us to the great damage' of Portuguese India, 'which is not in a condition to resist so great an enemy'.[13] Olivares saw no need for such restraint in European affairs: instead, he convinced his master that sending a large army overland from Italy to the Netherlands would at a stroke drive the Protestants out of southern Germany and induce the Dutch to make peace. At first the gamble worked amazingly well. In the summer of 1634 Spanish troops crossed the Alps and joined the imperial army in Germany, where together they routed the Swedes at the battle of Nördlingen: 12,000 Protestants perished and 4,000 more, including the Swedish commander, fell prisoner. The triumphant Spaniards then marched on to Brussels while the imperialists reoccupied almost all of southern Germany.

Olivares could be forgiven for hailing Nördlingen as 'the greatest victory of our times', but he nevertheless remained in thrall to a domino theory: that failure to defend any imperial interest would imperil the rest. He therefore advocated a pre-emptive strike on France, with eleven coordinated assaults from Spain, Italy, Germany, the Low Countries, the Atlantic and the Mediterranean, and convinced himself that 'there is no possibility that the blow can misfire'.[14] Of course it misfired: no state could make large-scale preparations on so many fronts and escape detection. Louis XIII therefore prepared his own pre-emptive strike. He secured a Dutch commitment to invade the Netherlands; he persuaded Sweden not to make a separate peace with the Habsburgs in Germany; and in May 1635 he sent a herald to declare war on Philip IV. At first Olivares rejoiced, since he could now present his own strike as a response to unprovoked French aggression. Fatefully, he selected Catalonia as the principal theatre of operations, envisaging that Philip would go there in person and invade France with 40,000 men.

In the event, another disastrous harvest in Catalonia left both the army and the civilian population starving, and a French invasion of the Low Countries diverted the Spanish troops tasked with advancing on Paris. So although Olivares had mobilized (and paid for) two fleets and 150,000 soldiers, they achieved nothing. Undaunted, he prepared another knock-out blow for 1636 – and this time he came close to achieving his goal. An imperial army invaded eastern France, and troops from the Netherlands captured Corbie and came within sight of Paris.

This success convinced Olivares that one more assault would crush France, and to fund it he again resorted to manipulating regalian rights: he recalled all copper coins and re-stamped them at three times their face value, and he introduced stamp duty (following the recommendation of the committee of theologians). Once again, the Little Ice Age thwarted his efforts. Dearth afflicted Andalusia, and flash floods destroyed half the houses in Valladolid, a former capital of Castile. Olivares saw no alternative to reducing the cost of the cheapest stamped paper required for simple transactions from 10 maravedíes to 4, 'because poor people were not transacting any business because of the tax'.[15]

So in 1637, Olivares turned once more to Catalonia for help, ordering 6,000 men to mobilize for the defence of their fatherland, citing a 'constitution' known as *Princeps Namque*; but the Catalan authorities objected that the measure could be invoked only when the sovereign resided in the principality. Since Philip remained in Madrid, they refused to obey. When a royal army eventually invaded France, it contained troops from various parts of the monarchy but not a single Catalan – and few Portuguese, because a serious rebellion had just broken out there.

The Portuguese Emergency

At the heart of the unrest lay a dispute over who should pay for the defence of Portugal's overseas empire. Already in 1624, when the Lisbon authorities begged the king to send funds to recover Bahia in Brazil from the Dutch, ministers in Madrid complained that 'the Portuguese are by their nature grumblers and spongers'.[16] Although on this occasion Olivares relented, organizing a massive and successful Luso-Spanish relief expedition, the Dutch capture of Pernambuco five years later led to a new dispute: Portugal begged for central funds to defend their empire but Madrid insisted that the Portuguese must pay more themselves.

Union with Spain had never been popular with some Portuguese, but until 1620 many members of the elite studied at Spanish universities, served in the Spanish army and administration, married Castilian spouses and wrote in Castilian. Thereafter, the loss of overseas trade and territory to the English and the Dutch caused widespread resentment. Olivares himself contributed to the economic crisis by terminating the profitable overland trade between Portuguese Brazil and Spanish Peru, and by encouraging the Inquisition to round up Portuguese merchants in search of any Jewish practices (over half the Portuguese merchant class was of Jewish descent, and many languished in prison while inquisitors carried out their lengthy enquiries). These developments reduced the taxes and profits of colonial trade, which provided two-thirds of the total revenues of the Portuguese crown as well as the livelihood of most of the kingdom's merchant elite.

As in Castile, Olivares intended to increase revenues by manipulating regalian rights, for example by imposing the *media anata* and a salt monopoly, and he asked the Portuguese Cortes to approve them. When they refused he forced the vicereine, the king's cousin Margaret, to introduce other fiscal innovations to pay for imperial defence. On his orders she withheld one-quarter of all payments due in pensions and bond interest; extracted forced donations from the towns; required nobles to raise troops for the crown; increased excise duties by 25 per cent; and imposed a tax on meat and wine known as the *real d'água*, previously levied by towns (not the crown) as an emergency measure. She also prepared to impose a tax on capital and rents and ordered officials throughout Portugal to register the possessions of every household, however poor. In August 1637 at Évora, the third largest city in the kingdom, a crowd gathered outside the house of the mayor and instructed him to stop the compilation of the new registers. When he refused, the crowd stormed his house and made a bonfire of his furniture before torching the houses of those compiling the registers and those who had collected new taxes (such as the *real d'água*). Posters appeared in the streets signed by 'Manuelinho', formally the town's simpleton but now hailed as 'secretary of the young people, ministers of divine justice', threatening the 'tyrant Pharaoh' (King Philip) and his agents with death because they had imposed new taxes without popular consent. Within a few weeks, some sixty places in southern Portugal had joined the rebellion. Some would defy Madrid for six months (Fig. 12).

The count-duke feared that if Philip gave the rebels of Évora what they wanted, 'not only the rest of Portugal but also all His Majesty's realms in Europe, in America and in India would want the same – and with very good reason, because they would risk nothing in doing so since they would know that one miserable town, just by rebelling, had obliged its king to agree to terms favourable to them'. Nevertheless, recognizing the need to settle the matter before either the French or the Dutch intervened, Olivares declared himself ready to offer a pardon based on 'the most generous models; and to that end we are studying what we did in Vizcaya'. The king, too, favoured

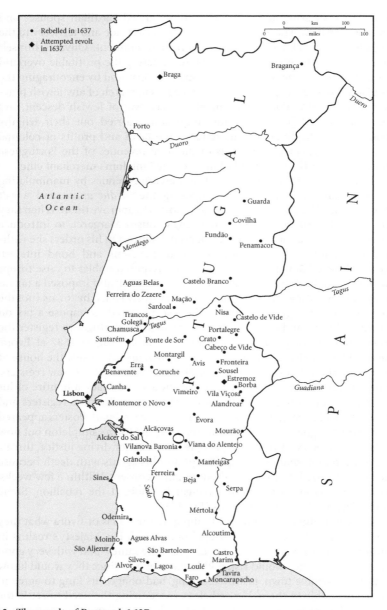

- Rebelled in 1637
- Attempted revolt in 1637

Bragança

Braga

Porto

Duero

Duero

Atlantic
Ocean

Guarda

Covilhã

Fundão

Penamacor

Mondego

Aguas Belas

Castelo Branco

Ferreira do Zezere

Mação

Sardoal

Trancos

Nisa

Castelo de Vide

Golegã

Chamusca

Tagus

Portalegre

Santarém

Ponte de Sor

Crato

Cabeço de Vide

Montargil

Avis

Fronteira

Erra

Coruche

Sousel

Estremoz

Benavente

Borba

Canha

Vimeiro

Vila Viçosa

Lisbon

Alandroar

Montemor o Novo

Guadiana

Évora

Alcáçovas

Mourão

Alcácer do Sal

Vilanova Baronia

Viana do Alentejo

Grândola

Manteigas

Ferreira

Sines

Beja

Serpa

Sado

Mértola

Odemira

Alcoutim

Moinho

Agues Alvas

São Aljezur

São Bartolomeu

Castro Marim

Silves

Alvor

Lagoa

Loulé

Faro

Tavira

Moncarapacho

P O R T U G A L

S P A I N

Tagus

12. The revolt of Portugal, 1637.
An epidemic of revolts against Spanish rule affected almost all the southern half of the kingdom, in some areas lasting six months.

clemency. Although he gave orders to prepare 10,000 troops to invade if necessary, recognizing that although 'these rebels may be prodigal sons, they are still sons', he suspended collection of the *real d'água*.[17] This combination of stick and carrot worked well: by the end of 1637 all the towns in revolt had submitted and the king issued a generous pardon – which was indeed modelled on that for Vizcaya – that condemned only five rebels to death and another seventy to the galleys.

Olivares at Bay

The pacification of Portugal temporarily salvaged Olivares's reputation, as did the repulse of a French attack on the port of Fuenterrabía in 1638; but elsewhere Spain's enemies triumphed. In Germany, French forces captured Breisach on the Rhine, cutting communications between Lombardy and the Netherlands; in West Africa, the Dutch took São Jorge da Minha, the oldest Portuguese colony in the tropics; in India, they blockaded Goa, the capital of Portuguese India; in Brazil, they attacked the viceregal capital, Salvador, with a large fleet, and the citizens only just managed to repel them. In 1639 Olivares therefore sought to regain the initiative with two characteristically dangerous and expensive gambles. First, he appointed the relatively inexperienced count of La Torre as governor and captain-general on sea and land of the State of Brazil, entrusted him with a fleet of forty-six ships carrying 5,000 men, and instructed him to bring the Dutch to battle. Second, he decided to launch another invasion of France from Catalonia with the express intention of forcing the Catalans to become 'directly involved, as up to now they seem not to have been involved, with the common welfare of the Monarchy'. He therefore sent troops raised in Castile, Italy and the South Netherlands to the principality, expecting the Catalans to feed, lodge and pay them.[18]

Olivares overlooked not only the strength of Catalan antipathy but also the geographical obstacles to mounting major military operations in the principality. In the north the frontier with France ran through high mountains and arid plateaux, in the west a barren wasteland separated Catalonia from Aragon, and in the south the Ebro Delta prevented easy communication with Valencia: each geographical barrier made it hard to launch a Habsburg invasion of France. Even within the principality, steep hills and deep river gorges made interior communications a labyrinth. In addition, the unpredictable climate limited agricultural yields in many upland areas, leaving little surplus to feed an army.

Contingency created a further political obstacle to Olivares's decision to turn Catalonia into the principal theatre of operations. Once every three years, a small boy stood beside a silver urn containing 524 slips of paper, each one bearing the name of an eligible member of the Corts, and drew out slips until he had the names of two clerics, two nobles and two burgesses. These six men immediately began a three-year term as the Diputació, or Standing Committee of the Corts, whose main task was to ensure that their ruler respected and obeyed the Constitutions of the principality. In 1638 the small boy drew from the silver urn a slip of paper bearing the name of Pau Claris, a canon of Urgell Cathedral and a trained lawyer, who became the senior clerical diputat, followed by another bearing the name of Claris's cousin Francesc de Tamarit, who became the senior noble diputat.

Unfortunately for Madrid, both of the new diputats possessed a passionate and uncompromising devotion to their native land and its Constitutions, seeing the Catalans as God's chosen people and condemning every political innovation by Madrid as tyranny. Since Olivares's decision to bring the war to Catalonia was bound to produce innovations, the stage was set for political confrontation. It began in earnest when the French captured the frontier fortress of Salces in 1639 and the judges in Barcelona received orders to ignore their oath to observe the Constitutions whenever they conflicted with the needs of the army. '[If] the Constitutions do not allow this', Olivares informed the viceroy, Don Luis de Queralt, count of Santa Coloma, 'then the Devil take the Constitutions'. Neither the viceroy nor the judges dared remind the all-powerful favourite that this attitude would surely drive the new diputats to protest; instead, in the words of a perceptive observer, they preferred to 'write, consult, doubt and obey'. In less than a year, almost all of them would pay for this complicity with their lives.[19]

The Revolt of the Catalans

Early in 1640, thanks to relentless pressure by the judges and naked black-mail by Santa Coloma (who promised a patent of nobility to all landholders who spent thirty days with the army), an army composed largely of Catalans recaptured Salces – but the victory impressed neither the royal commanders in Catalonia, preoccupied by the need to lodge and feed their victorious troops, nor Olivares, who now issued orders to raise 6,000 new troops in the principality for service in Italy. Anticipating trouble, Santa Coloma forbade any lawyer to take up complaints lodged by peasants against soldiers; and when the diputats protested at this further innovation, Olivares ordered a magistrate, Miquel Joan Monrodón, to arrest Tamarit and commanded the church authorities to prosecute Claris. An eyewitness underscored the dangers inherent in such arbitrary policies: 'in truth the greatest grief of the downtrodden is removing their ability to ask for redress', and (he continued ominously) 'Amid the distress to which human misery reduces us, there is almost nothing men would not do'.[20]

Extreme weather added to the distress. Following a meagre harvest the previous year, no rain fell on the fields of Catalonia in spring 1640, producing a drought so intense that the authorities declared a special holiday to allow the entire population to make a pilgrimage to a local shrine to pray for water – one of only four such occasions recorded in the past five centuries. Since still no rain fell, unless the villagers could exclude the troops, they faced starvation; but when Santa Coloma de Farners, a hamlet about 60 miles northeast of Barcelona, refused orders to quarter an approaching regiment of Castilians, the viceroy sent the same magistrate who had arrested Tamarit to intimidate them. Even loyalists felt that the viceroy had made a serious error in choosing Miquel Joan Monrodón – a man 'irritable by nature, hasty, arrogant and proud' – to perform a task that required 'guile rather than force'. They were soon proved right. On 30 April 1640, Monrodón ordered the arrest of any civilian found carrying a firearm. The villagers responded by chasing him and his officials into the local inn, which they then set on fire, shouting 'Now you'll pay for putting Tamarit in prison.' Monrodón and most of those inside died in the flames.[21]

The villagers of Santa Coloma de Farners now rang the church bells to summon aid, and hundreds of armed men soon stood ready to protect the community. Their resistance forced a regiment travelling behind the Castilians to make a sudden detour and they took their revenge on the next village on their route, Ruidarenes, where they burnt down several houses and the church. Meanwhile, outraged by the murder of Monrodón, the viceroy ordered his troops to march back to Santa Coloma de Farners and burn it to the ground. This they accomplished on 14 May, leaving the church and most homes in ashes. For good measure, they also destroyed the rest of Ruidarenes.

That same day, the local bishop excommunicated the troops for sacrilege, and shouts of 'Long live the king and death to the traitors' soon rang out as the inhabitants of some fifty villages attacked any soldiers still billeted in their area, and plundered the property of royal officials and loyal inhabitants. A little rain now fell, saving the harvest, but everyone feared what might happen when the *segadors* entered Barcelona for their annual hiring fair on 7 June 1640, the Feast of Corpus Christi – everyone except Santa Coloma, who on 6 June obeyed an order from Madrid to send the squadron of galleys that normally defended Barcelona, manned by most of the city's garrison, to fight the French. The preconditions for disaster were now all in place.

That night hundreds of *segadors*, some armed and others carrying firewood, entered the city under cover of darkness. On the morning of the Corpus Christi holiday they circulated among the crowds amid occasional shouts of 'Long live Catalonia' and 'Death to the Castilians' until someone sighted a servant of Monrodón, who fled into his house with the *segadors* in hot pursuit. When a shot from within killed one of them, his comrades used their firewood to burn down the door, broke into the house and sacked it. The crowds now sought out the hated judges, who fled from one convent to

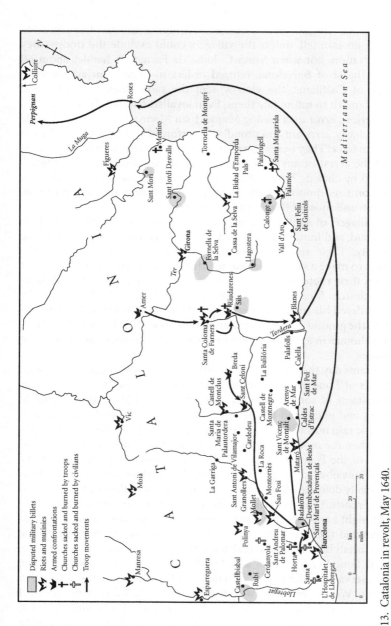

13. Catalonia in revolt, May 1640.
Even before the 'Corpus de Sang' in Barcelona on 7 June 1640, a large part of Catalonia was in arms, with both royal troops and outraged peasants involved in confrontations that in some cases culminated in the burning down of churches.

another seeking sanctuary as the *segadors* burnt their possessions and papers, smashed the windows and walls of their homes, and even chopped down the trees in their gardens. When some priests tried to intervene, they angrily replied that the judges 'had watched the Castilians burn churches and sacraments', and although 'they could have stopped it, they had done nothing; it was thus reasonable that they should pay for it'.[22]

When the crowd – by now some 3,000 strong – came under fire from the house of a government minister, they chased its occupants to the royal shipyards, where the viceroy had taken refuge. With no galleys and a depleted garrison, the *segadors* soon forced an entry, seized the weapons stored in the royal arsenal, and fanned out to search for those who had fired on them. The viceroy tried to escape along the beach, but two of the pursuers intercepted him and stabbed him to death.

This news left both Philip and Olivares stunned, but they knew that the Catalan cause was far from united. Many nobles remained loyal to the king; a few bishops joined the rebels, most did not; and although most monks and friars defied Madrid, most nuns did not. In August 1640, the king issued a declaration that accused the Catalans of treason and instructed the marquis of Los Vélez, from an eminent Catalan family, to mobilize an army to restore royal control. He also announced that he would hold another Catalan Corts and restore 'in the principality the free exercise of justice, which has been violated and made impossible by certain wicked and seditious persons'. Almost immediately loyalists in the southern town of Tortosa staged a counter-revolution, seizing the leading rebels and hanging fifteen of them. They also stressed the danger of anarchy posed by the uprising – but their cause suffered a setback with the publication of some of Olivares's intemperate letters to the late viceroy, expressing his contempt for the Constitutions and suggesting that Madrid really did intend to destroy Catalan autonomy.

In September the Diputació summoned some 250 clerics, nobles and burgesses to Barcelona, where the assembly, known as the Junta de Braços, drew up a formal refutation of Olivares's charge of treason, asked for support from Valencia, Aragon and Mallorca, and took steps to mobilize resistance. Olivares's policies had alienated every major social group in Catalonia. The nobles resented the appointment of non-Catalans to lucrative military commands and the repeated demands for military service without pay or reward; the clergy resented the appointment of non-Catalans to plum posts and the imposition of heavy taxes; and the town oligarchs resented the ceaseless pressure from Madrid to provide loans that they knew would never be repaid. The Diputació now administered an oath of loyalty – over 300 citizens took the oath on the first day; over 1,000 had done so within a week – and began to raise both troops and money. They also summoned a special committee of theologians to debate the legality of their policies (and published its favourable findings); and they commissioned pamphlets depicting themselves as lawful subjects enforcing a contractual relationship with their ruler, and as the avenging hand of God against the royal troops who had profaned His churches.

Claris and his colleagues sent copies of these publications to Madrid, Naples, Valencia, Zaragoza and other outposts of the monarchy, as well as to Paris and Rome, where they soon won support. As soon as news of the revolt arrived in the Dutch Republic, the States-General set up a special committee to organize assistance, recognizing their 'common interest' with all other rebels against Philip IV – but only France actually sent help, and by the end of 1640 it numbered only 3,000 foot and 800 horse. A public bond issue in Barcelona raised over 1 million ducats, with purchasers from among all sections of the population, but that sufficed to pay only 8,000 soldiers. Even given the strategic advantages conferred by geography, such a small force seemed unlikely to protect the principality against the royal army of 23,000 infantry, 3,100 cavalry and 24 artillery pieces that Los Vélez reviewed on 7 December 1640.

At first the campaign went very well for the king. Cambrils, the first major stronghold in the path of his army, resisted for three days but then surrendered. Los Vélez argued that 'neither oath nor word oblige the king when dealing with his vassals', and therefore hanged several Catalan officers as his troops massacred about 500 other local defenders in cold blood.[23] This example of strategic terror encouraged other towns to surrender, and the royal army continued its advance on Barcelona. The revolt of the Catalans seemed doomed. Then news arrived that revolution had broken out in Lisbon.

The Revolt of Portugal

The Portuguese rebellion – the only one in mid-seventeenth century Europe to achieve permanent success – originated in 1634, when three Portuguese gentlemen (*fidalgos*) visited the Royal Armoury in Madrid and viewed with disgust the collection of trophies secured when Spanish forces had invaded their country to enforce a union with Castile in 1580. There and then they took an oath to restore Portugal's independence, but by 1637 only five *fidalgos* had joined the plot: they therefore failed to exploit the revolt in Évora, and many other places, that year. Duke John of Bragança, Portugal's premier aristocrat, also stood aloof. He possessed enviable fiscal, ecclesiastical and military resources and, unique among Europe's aristocracy, he could create nobles himself; but he therefore avoided any entanglement that could put his pre-eminence at risk. He lent full-hearted support to the central government against the rebels in 1637; and when the following year Cardinal Richelieu offered to send French troops if he rebelled (and threatened to support a rival claimant to the Portuguese throne if he did not), the duke again did nothing.

Two of Olivares's henchmen – Miguel de Vasconcelos, the principal minister in Lisbon, and his brother-in-law Diogo Soares, secretary for Portuguese affairs in Madrid – established a stranglehold on the administration of Portugal in the course of the 1630s. They controlled all patronage, packed the administration with their relatives and clients, and sold offices and titles of nobility to the highest bidder. In addition, despite virulent anti-Semitic feeling in the kingdom, they protected the New Christian bankers on whose loans Philip's armies and navies depended. Despite widespread hostility, Vasconcelos and Soares managed to marginalize all their rivals, but they remained dangerously isolated. As Vasconcelos warned Soares in September 1640, 'anything might provoke a great conflagration ... I tell you the truth, dear colleague: I don't know what to do.'[24]

Vasconcelos's misgivings were well founded. Few Portuguese now saw any benefit in the union with Castile; nor did vassals living overseas. When a new viceroy of India arrived at Goa early in 1640 he found that the 'constant presence of Dutch naval squadrons over the past four years' and the 'lack of relief from Portugal' had paralyzed trade and led to the loss of many outposts in Ceylon.[25] In Brazil, the leading colonists wrote urgent letters in 1640 to remind Philip, their 'father, king and lord', that they lived 'in a faraway place, persecuted for so many years with the losses of war, with the robberies and cruelties of our enemies'. The indigenous population helped the Dutch, showing them routes to the interior, killing planters and burning sugar mills; the need to fortify their borders deprived them of the labour of their slaves on the plantations; and the slaves themselves 'seem on the brink of rebellion'. When the colonists heard about 'the events in Catalonia' most of them despaired, because their needs would now inevitably move further down the list of imperial priorities, causing further 'delays in sending us relief'.[26]

Despite these deep-seated problems, Philip IV lost Portugal because of another miscalculation by Olivares. In November 1640, the count-duke ordered Bragança to raise a regiment of his vassals and lead it to the Catalan front in person. The duke, fearful that he would never be allowed to return, pleaded that the responsibilities of his estate precluded his departure; but Olivares 'replied that this was unacceptable'. British ambassador Arthur Hopton saw the danger clearly: if another rebellion materialized in Portugal, it was 'almost certaine that the duke of Braganza shall ingage himselfe herein'. And indeed, on 1 December 1640, it was Bragança's agent in Lisbon who gave the order for the discontented *fidalgos* (who still numbered only forty) and their followers to storm the viceregal palace.[27]

Even with the duke's support, the conspirators ran grave risks. The Portuguese capital was the largest city in the Iberian Peninsula, with perhaps 170,000 inhabitants, and it normally boasted a powerful Spanish garrison – but here too, Olivares's earlier policies worked against him. He had withdrawn to the Catalan front all but two of the companies guarding the viceregal palace, and they proved unable to stop intruders from hunting down Vasconcelos whom they murdered. Shortly afterwards, one of the *fidalgos* appeared on the palace balcony and shouted 'Long live King John IV' – meaning Bragança – and the crowd immediately took up the cry. To avoid further bloodshed, the vicereine ordered the outnumbered Castilian troops to surrender and a few days later Bragança arrived in the capital to receive oaths of fealty from his new subjects.

The Tipping Point

For a few weeks, Olivares pinned his hopes of recovering Portugal on a quick victory in Catalonia, where Los Vélez and the royal army continued to advance, meting out exemplary punishment to all towns taken by force. Yet these examples failed to intimidate the people of Barcelona. When news of the massacre at Cambrils arrived, crowds hunted down and killed any Castilians they could find; and when they heard of the fall of Tarragona, they murdered three royal judges and some other loyalists, disfigured them with repeated blows and shots, and then hanged them from the gallows in the city square. Shaken by the rising disorder, Claris persuaded the Junta de Braços to proclaim the Catalan Republic on 16 January 1641; but when, five days later, Los Vélez's troops captured and sacked nearby Martorell, the Junta gave up its brief bid for independence and recognized Louis XIII as the new ruler of the principality. In return, the French king agreed that all officers of justice, clerics and military governors would henceforth be Catalan; that the Constitutions would prevail; that several unpopular taxes would cease; and that the Inquisition would continue its work with full powers (a special concern of the Junta). He even graciously agreed that the Catalans could choose another lord if he did not treat them well.

Bernard de Duplessis-Besançon, Richelieu's personal representative, now took charge of military affairs within Barcelona, as the city's clergy (reinforced by those who had retreated from surrounding villages rather than surrender) organized round-the-clock services and processions to beg the city's patron saints for protection against the besiegers. According to one royalist, 'the efforts of the most committed friars in their writings and pulpits' upheld the cause as much as 'the *segadors* in the streets'.[28] They needed all their persuasive powers when on 24 January 1641 a herald from Los Vélez arrived and threatened that unless the city surrendered at once, it too would be sacked.

As Duplessis-Besançon remarked in his memoirs, 'In war, the least circumstance, difficult to assess [at the time], often produces major effects', and indeed, on the same day that Los Vélez issued his summons, a ship arrived in Barcelona bearing two envoys from John IV of Portugal with an offer of alliance.[29] This re-energized the Catalans and thousands of men and women flocked to join the city's defenders, turning the tide: the royal assault failed and many royal soldiers (especially the Portuguese) deserted as they retreated. Los Vélez soon lacked sufficient men to mount a formal siege and fell back, pursued by French troops.

Olivares found temporary consolation in the sudden death of Pau Claris, probably poisoned by royal agents, and began to contemplate diverting resources from the Catalan to the Portuguese front. As he pointed out to Philip, 'unless action is taken immediately against Portugal, there will be no chance of recovering that kingdom for many years, since each day's delay will make the enterprise more difficult'; but, once again, circumstances beyond his control thwarted him. The count-duke had planned to use the silver aboard the 1640 fleet from America as security for loans to fund the 1641 campaign, but none arrived: although the viceroy of Mexico had amassed over 750,000 ducats to send to Spain, he decided to hold the entire shipment back 'because of so many rumours of enemies' ready to ambush the fleet. With no silver to offer, Spain found few bankers willing to lend.[30]

As Olivares had feared, Spanish inaction allowed John IV to consolidate his control over Portugal. The Cortes acclaimed him as their new king and, after he abolished the hated regalian rights newly imposed by Madrid, they voted new taxes (including a property tax – the very imposition that had provoked the revolt of Évora three years before). They also presented a list of grievances for redress, including improved administration of justice, punishment of corrupt public officials and better government control over bankers who collected the taxes assigned to repay their loans. Other grievances revealed deep social divisions. The nobles and representatives of the towns in the Cortes begged their new king to limit the number of clerics (Portugal boasted 30,000 seculars and 25,000 regulars), reduce the jurisdiction of the ecclesiastical courts and levy more taxes on the church. The towns called for the closure of all universities and colleges for five years (save only for Coimbra), because they produced more clerics and lawyers than the kingdom needed. The towns and the clergy wanted strong measures against New Christians (none should be allowed to become doctors, lawyers, priests or public officials). The Cortes thus targeted individual groups and sought to return to 'normal times' because 'it is not wise to innovate': they showed little interest in national issues or collective interests.[31]

Olivares tried to exploit these social divisions by supporting a conspiracy hatched by a group of Portuguese prelates, nobles and New Christian bankers to assassinate John IV and restore Madrid's control, but they failed. The new king executed most of the noble conspirators, suspended the clerics and appropriated the revenues of their sees, and allowed the Inquisition to persecute New Christian merchants. He also sponsored sermons and printed propaganda against Castile, and his agents concluded treaties with France, Sweden, Britain and, most important, the Dutch Republic, which in August 1641 sent a fleet to defend Lisbon. Bragança would not become a Winter King like Frederick of the Palatinate (Chapter 8).

The Fall of Olivares

Olivares enjoyed no respite. News now arrived of sedition in another previously loyal part of the monarchy, the kingdom of Aragon, and suspicion fell principally on the viceroy: the duke of Nochera, a Neapolitan. In 1639 Nochera had warned about the discontent in Aragon caused by recruiting, taxation and loss of trade with France; the following year he sent a defeatist letter urging the king to conciliate the Catalans 'or they will be like Hydra, where instead of one you find seven'; now he offered his services as intermediary between the two sides. Olivares interpreted this as treason and ordered the duke's arrest. He died in prison.[32]

Some nobles in Castile also became restless. Although many of them privately disapproved of the expensive yet unsuccessful policies pursued by Olivares, they made no headway because they lacked a constitutional forum for expressing their grievances (aristocrats no longer attended the Cortes). Early in 1641, two grandees took matters into their own hands. Olivares had ordered the duke of Medina Sidonia, together with his relative and neighbour the marquis of Ayamonte, to initiate secret talks to see whether John IV (married to the duke's sister) might be reconciled with Madrid: instead, they sought Portuguese support to 'turn Andalusia into a republic' with Medina Sidonia at its head. King John offered to send a flotilla from Lisbon to Cadiz, where it would burn the Spanish warships in the harbour and seize the treasure fleet expected from America. Evidence of the plot reached Madrid only a few days before the Portuguese ships arrived off Cadiz, and this time Olivares reacted both promptly and forcefully. When Medina Sidonia ignored a direct summons to come to court, an envoy arrived with a vial of poison and orders either to bring the duke to Madrid or 'to send him to meet his Maker'. Realizing that the game was up, Medina Sidonia complied and threw himself (literally) at the king's feet, unchivalrously blaming everything on Ayamonte.[33]

This, the first aristocratic plot in Castile for 150 years, took place amid another spate of climatic disasters. In spring 1641 a prolonged drought threatened the harvests of Castile; in August 1642 a tornado struck the city of Burgos with such force that it destroyed the nave of the cathedral; the years 1640–43 saw the highest precipitation ever recorded in Andalusia; and in January 1642 the Guadalquivir burst its banks, flooding Seville.

The government added to the misery through desperate measures to find money for the wars with Catalonia and Portugal – especially through another devaluation of the currency of Castile, calling in all copper coins and re-stamping them at triple their value. As Hopton reported:

Concerning the state of this kingdom, I could never have imagined to have seen it as it is now, for the people begin to fail, and those that remain, by a continuance of bad success, and by their heavy burdens, are quite out of heart ... The greatest mischief of all is that the king of Spain knows little of all this, and the count-duke is so wilful as he will break rather than bend.[34]

King Philip was not quite as ignorant about the public affairs of his monarchy as Hopton supposed, however. He clearly comprehended the seriousness of the first memorandum he received with news of the Portuguese rebellion, because his rescript is stained with tears; and in spring 1642 he finally emulated his brother-in-law Louis XIII and joined his troops. It did little good: he failed to retake Lleida (Lérida) but the French captured Perpignan, the second city of Catalonia, allowing them to advance to the Ebro.

Philip returned disconsolately to Madrid to find that almost all his nobles had boycotted his court: only one aristocrat joined him in the royal chapel on Christmas Day – the son of the murdered count of Santa Coloma. Criticisms of the count-duke became more public, with special venom reserved for the lavish Buen Retiro palace, which Olivares had built for his master on the outskirts of Madrid. Many saw the palace as a symbol of his misguided policies: even its name, 'Retreat', served as a stick with which to beat the minister as his armies retreated on all fronts. In January 1643 Philip gave his favourite and chief minister of almost twenty-two years permission to retire.

The fall of Olivares led to a long-overdue reappraisal of the monarchy's strategic priorities. Two weeks after the count-duke left court, the king and his Council of State together reviewed imperial priorities and concluded that less money should go to Germany, Italy and the Netherlands until all enemy forces had been expelled from Spanish soil, and that the war in Catalonia must take precedence over recovering Portugal. Philip therefore ordered his commanders in all theatres except Catalonia to assume a defensive posture. He also opened peace talks with both France and the Dutch, and instructed Don Francisco de Melo, governor-general of the Spanish Netherlands, to forward conciliatory personal letters to his sister Anne of Austria, now the regent of France, with orders to 'deploy every seemly and feasible means to secure a treaty'.[35]

Melo was a career diplomat who had gained his exalted office by default: he was the senior Spanish minister in Brussels in 1641 when smallpox claimed his superior – but he reassured the king that his relative lack of military experience did not matter, because nowadays 'a mere doctor of philosophy' could lead an army to victory. The following year he vindicated this arrogant view: leaving a skeleton force to hold the Dutch at bay he defeated a French army in battle and captured five French towns. Melo's success seems to have weakened Spain's desire for peace. When the Council of State met early in 1643 in the presence of the king to discuss whether to open peace talks with France, the count of Oñate (another career diplomat) argued that although he favoured negotiations at some future time, he opposed them at this juncture: 'We should leave a little time for Time', he quipped.[36] Melo therefore invaded France again, laying siege to the heavily fortified town of Rocroi, but this time a relief force arrived promptly, drove off the cavalry and then attacked the Spanish infantry until they either died or surrendered.

Although the battle of Rocroi did not destroy the empire on which the sun never set, it transformed its strategic vision. The king, now thirty-eight years old, sought consolation in spiritualism, summoning men and women renowned for their prophetic powers from all his dominions to give advice on what he should do next, culminating in a 'summit' of prophets in October 1643. For the next twenty-two years, Philip wrote a holograph letter once every two weeks to one of the spirit mediums, Sor María de Ágreda, begging her to erect a barrier of prayer against his enemies that might take the place of the human resources he lacked. Philip's ministers also hoped that God 'would give Your Majesty's armed forces the results that reflect the justice of your cause' – but in view of 'the variety of events and accidents that normally occur in wars', they recommended that Spain concentrate its resources on fighting the Dutch and the Catalans, requiring Italy to fund the other wars of the Monarchy, as well as paying for its own defence.[37]

Even so, the cost of continuing Philip's wars exceeded what his Spanish subjects could bear. Don Juan Chumacero, the minister responsible for law and order in Castile, had already warned the king that his vassals 'cannot withstand the burden of their taxes, so that everything may collapse at the same time'. In particular, he feared that the towns will 'shake off their yoke at the same time, through frustration' at the endless demands of the tax collectors, especially when 'the crops are generally poor' because 'storms have destroyed a large part of them, and what has been harvested is of poor quality'. Don Luis de Haro, Olivares's successor as chief minister, felt equally pessimistic when early in 1646 he went to Cadiz to get the Atlantic fleet to sea. 'Everything comes down to difficulties and more difficulties' because of the extreme weather, he complained. 'Three months of snow and rain, and the harshest weather ever seen by Man' had created 'difficulties or rather impossibilities', and he contemplated suicide: 'Sire, I do not know how to deal with this, unless it is to drown myself.'[38]

The following winter brought continual rain. According to a Madrid news-letter, 'In Spain, and (they say) in all Europe, the era of Noah's flood came again with a vengeance, because the rains that fell were so heavy and so continuous, and the rivers rose so excessively, that commerce and communication ceased between the cities, towns and villages. Many lives were at risk; many buildings collapsed.' When the harvest failed again in 1647 (an El Niño year), Chumacero despaired: 'There is no shortage of people who blame to Your Majesty, saying that he does nothing, and that the council is at fault – as if we had any control over the weather!' Other ministers reinforced Chumacero's point: 'Hunger is the greatest enemy', they warned the king, 'and in many states the shortage of bread has provoked unrest that ended in sedition'. They knew whereof they spoke: adverse weather had just produced 'the most important urban uprising in Castile since the revolt of the Comuneros' over a century before.[39]

The Green Banner Revolts

Just when the harvest of 1647 seemed safe, all over Andalusia 'the weather turned very cold, even worse than the coldest January day'. Freak frosts killed the ears of grain and produced the worst harvest of the century. It also left little seed corn for the next year: according to a chronicler, 'the peasants did not sow one third of what they should have sowed'. In March 1648 the senior magistrate of Granada reported that he had never seen so many chil-dren begging in the street and noted that the foundlings' hospital was full and could scarcely feed those already there. A loaf of bread cost triple its usual price. A group of men armed with swords and clubs now marched on the city hall shouting 'Long live the king and down with the bad govern-ment!' Rumours circulated that 'the people want to elect a king and declare a state of rebellion', but instead they elected a respected gentleman, as the new chief magistrate: the pious (and aptly named) Don Luis de Paz ('Peace'). The new leader did his best to bring to market all the grain stored in the city and assure bread at a reasonable price until the next harvest; but for several weeks Granada defied the central government.[40]

In Madrid, the king's ministers proclaimed that disorder should always be punished without mercy – but opined that the extreme circumstances of 1648 made moderation advisable. The troubles had arisen from 'need and hunger', and many vassals lived 'on the edge of desperation'; therefore, the ministers reasoned, 'because of the unfavourable disposition of the weather, we need to give way and dissimulate in order to avoid greater setbacks'. The king agreed: later that month he paraphrased these views in one of his breast-beating letters to Sor María de Ágreda. He favoured clemency, he told her, 'because it is not possible to squeeze my vassals more, as much because of what they would suffer as because of the risk we would run of suffering more misfortunes'.[41]

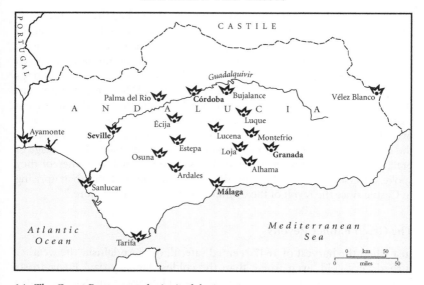

14. The Green Banner revolts in Andalusia, 1647–52.
The extent of these urban revolts has been seriously underestimated by historians: they probably involved more people and places than the revolt of Catalonia. All the towns and cities named on the map experienced rebellions in these years.

The king settled two other contentious issues that same year – his diplomats at Westphalia signed treaties that ended the long-running wars in both the Netherlands and Germany (Chapter 8) – but he stopped short of concluding a general peace. Although his chief negotiator at Münster observed pointedly, 'I leave it to the superior intelligence and prudence of Your Majesty to consider if this is the right moment' to settle with France or 'to remain at war while all Europe makes peace', the outbreak of the Fronde revolt led Philip to miss his chance (Chapter 10) – but the continuation of extreme weather prevented him from exploiting to the full his rival's weakness.[42] In October 1648 the city council of Cadiz lamented that 'the grape harvest of this year is ruined' and the following month reported that 'for several days the butchers have had no meat'. The following spring, in Seville, 'The darkness, wind and rain on Maundy Thursday' were the worst ever known in the city, one chronicler recorded; 'it was as cold as January', wrote another. A devastating plague epidemic carried off half the city's population.[43]

In spring 1652 high food prices triggered a new wave of insurgency, starting in Córdoba, where several hundred men from a poor parish took to the streets shouting 'Long live the king and death to the evil government!' They found huge quantities of grain hidden in the houses of the rich, and to avoid further disorders, the clerical and lay leaders of the city set up an interim government charged with supplying cheap bread. Desperate to avoid any confrontation that might compromise his ability to fund his wars, the king ordered the immediate dispatch of 6,000 bushels of grain from Madrid to Córdoba; and, as in the case of Granada in 1648, he accepted that the protestors had not 'intended to forget their obedience to me' but had rather acted through 'the anguish caused by hunger, their lack of foresight in not laying by the wheat required for their sustenance, and the exploitation of many people who sold wheat at excessive prices'. Philip therefore pardoned them all.[44]

Such leniency created a dangerous precedent: everyone could see that collective violence had succeeded in Córdoba, because the city turned into an oasis of plenty and starving people from all over Andalusia arrived to consume the grain graciously provided by the king. In Seville, rioters now attacked the houses of those suspected of hoarding grain as others entered the city's arsenal and distributed armour, weapons and even artillery. Others still broke into the prisons and freed the inmates. The magistrates made haste to restore all copper coins to their former value, abolished recent royal taxes (including excise duties on food and the hated stamp tax) and proclaimed – falsely – that the king had issued a general pardon. These measures pacified the situation until, a month later, a large force of city merchants and gentlemen suddenly attacked the headquarters of the rebels. Lacking the expertise to use their artillery, they surrendered. With that, the cycle of revolts known as the Green Banner (*el Pendón Verde*), which affected some twenty Andalusian towns and involved as many people as the revolt of the Catalans, ended.

The Spanish Phoenix?

Against all the odds, the year 1652 turned out well for Philip IV. His troops recaptured Dunkirk, the principal port of the South Netherlands; Casale, reputedly the strongest fortress in Italy; and Barcelona, whose surrender led almost all of Catalonia to submit. Although French forces, assisted by a few Catalans, retained the northern areas, after twelve years of savage and continuous war Madrid again controlled most of the principality. These successes seem to have given the king a false sense of security and he rejected a French peace offer that included an end to all assistance to Portugal, because he refused to abandon the prince of Condé, Louis XIV's cousin, who had defected to Spanish service. The war therefore continued and Spain lost more ground.

Castile experienced exceptional precipitation throughout the 1650s, reducing the yield of one harvest after another, and Philip's advisers became alarmed. They warned him in January 1659 that peace with France, whatever the price, 'is absolutely essential for the conservation of the Monarchy of Your Majesty' – adding that 'experience has shown that the more we delay peace talks, the more we lose'. This time the king declared his willingness both to sacrifice Condé (on the grounds that 'when one places in the balance the conservation of the Monarchy, the importance of this necessity cannot be compared with that of the prince of Condé') and to marry his daughter María Teresa to Louis XIV (although in making this concession he compared himself with Abraham sacrificing Isaac). Later that year, the Peace of the Pyrenees ended twenty-five years of continuous war.[45]

Philip was now free at last to concentrate on the reconquest of Portugal, but almost two decades of independence had allowed the Bragança regime to consolidate its position both at home and abroad. Portugal controlled Angola and Brazil, generating both trade and tax revenues, while French and British troops arrived to defend the frontiers. Nevertheless, in 1663, after signing yet another decree of bankruptcy, Philip launched a powerful invasion that captured Évora. Its fall provoked an attempted coup in Lisbon, but an Anglo-Portuguese army mounted a successful counter-attack. The councillors of Philip IV summed up the futility of the situation perfectly. 'A truce with Portugal is the only way to ensure that we will not lose everything and to repair the desperate state in which we find ourselves', they lamented – only to continue lamely that 'considering that the army is already on campaign, and that it would not be right to sacrifice all the treasure it has cost, or to despair of some happy outcome, it seems we should wait and see what happens'.[46] Only the king's death in 1665 opened the path to peace.

Of the vast Lusitanian empire that Philip IV had inherited, heralds proclaimed his sickly four-year-old son 'Charles I of Portugal' in only two tiny outposts: Ceuta and Tangier. After two more inconclusive campaigns, the regents accepted English mediation, and in 1668 they signed a peace that recognized Portugal as an independent kingdom with the same borders it had enjoyed before the union with Spain. They restored all the property confiscated from Spaniards who had sided with Bragança, and even the duke of Medina Sidonia received the lands confiscated from his father after his conspiracy to create an independent Andalusia. It was the most humiliating treaty ever signed by the Spanish Habsburgs – and still the monarchy did not enjoy peace.

In 1667, Louis XIV declared war on Spain on the grounds that the terms of the Peace of the Pyrenees had not been fulfilled, and his troops seized Lille and other towns in the Spanish Netherlands, which became a permanent part of France when the parties made peace. Further aggression by Louis secured more territorial gains, in part because harvest failures in 1665–68 and 1677–83, plus another plague epidemic in 1676–85 and yet more harvest failures in 1685–88, prevented Spain's demographic and economic recovery. In 1687 the abuses that normally arose from billeting troops during a time of dearth provoked another rebellion in Catalonia, and thousands of peasants marched on Barcelona. Just as in 1640, some ministers in Madrid advocated firm repression, but others had learned their lesson: recognizing that 'force is a very dangerous remedy when the civilians are more powerful than the army' they hesitated to do anything rash lest 'all Catalonia be lost in a few hours, as we saw in the days of the count of Santa Coloma'. Nevertheless, France again intervened, conquering much of the principality (including Barcelona) and also invading the Spanish Netherlands (destroying most of Brussels in a ferocious bombardment). It is hard to dissent from the verdict of the Venetian ambassador in 1695: 'The whole of the present reign has been an uninterrupted series of calamities.'[47]

Counting the Cost

What had Spain gained by fighting so many wars during the seventeenth century? In material terms, nothing: it acquired no new territory and instead lost the Portuguese empire, Jamaica, parts of the Netherlands, and northern Catalonia. Everywhere, Philip IV had far fewer subjects at his death than at his accession. In Naples, the revolt of 1647–48 cost the lives of at least 6,000 people and the plague of 1656 tens of thousands more (Chapters 4 and 14). A recent study concluded that the war against the Catalans 'between 1641 and 1644 constituted the worst demographic catastrophe for many of the Aragonese frontier communities'.[48] The Castilian population living near the Portuguese border also fell dramatically because the king's troops exacted so much money, food and other local resources. The burden on smaller communities often led their inhabitants to flee because they could not feed both themselves and the troops billeted upon them. Baptisms in Extremadura fell by more than a quarter.

Nevertheless, most of Castile lay far from any theatre of operations and so avoided direct devastation by troops: the steady demographic decline apparent in surviving parish registers therefore stemmed from other causes. It is possible to identify three potential culprits. First, extreme weather, starting with the drought of 1630–31, ended the viability of many settlements on marginal land, and subsequent climatic anomalies periodically caused sterility and dearth elsewhere. Second, every year saw the emigration of thousands of people from Castile. Some went involuntarily, as conscripts to fight for the king overseas or as prisoners captured by North African pirates. Many more took ship for America because, in the words of the French ambassador in 1681, 'they cannot live in Spain'. Inevitably, this reduced the number of subjects at home.[49]

The third culprit for the depopulation of Castile was taxation. According to calculations by I. A. A. Thompson, the crown's taxes absorbed approximately '8 percent of national income in the 1580s and 12 percent in the

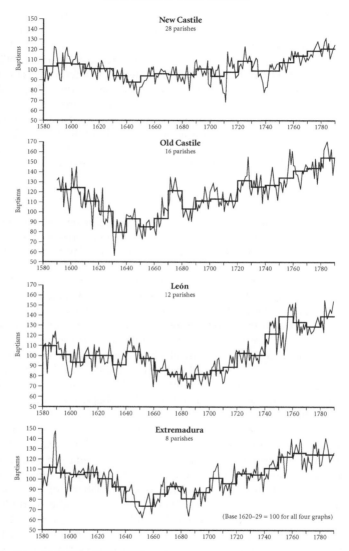

15. Baptisms in Castile, 1580–1790.
Baptismal registers from 64 Spanish parishes all showed a fall during the seventeenth century, but with different rhythms. Old Castile suffered its sharpest fall during the drought of 1630–1, New Castile and Extremadura during the plague years of 1649–50, and León during the famine of 1662. Extremadura also suffered losses because of the war with Portugal between 1640 and 1668. Other regions adversely affected by Philip IV's wars – Aragon, the Basque provinces, Navarre and Valencia – are not represented.

1660s' – an increase of 50 per cent – and, on top of this rising burden, taxpayers also had to satisfy the competing demands of the church, of land-lords (for the rural population) and of the towns (for the rest).[50] The system of making each community pay a fixed quota for each tax, whatever the number of taxpayers, hit smaller communities particularly hard in times of falling population, because almost inevitably a point came when the remaining taxpayers could no longer afford the collective assessment and so abandoned the settlements. As Alberto Marcos Martín forcefully stated:

> In the last analysis, we should judge a fiscal system (any fiscal system, past or present) not by what it assesses but by what it collects (and by which methods and processes it chooses to accomplish this task), and by what it does, what it creates, and how it spends what it has amassed in taxes and contributions.[51]

By this yardstick, the record of the Spanish Habsburgs is abysmal. Their borrowing, mainly from foreign bankers, created a sovereign debt far beyond Spain's capacity to service: over 112 million ducats by 1623, over 131 million by 1638, almost 182 million by 1667, and almost 223 million by 1687. In addition, public-sector borrowing undercut local manufactures and encouraged a rentier mentality among those sectors of the population with the potential to be entrepreneurs. At the same time, the need to raise and create taxes to repay lenders led to onerous fiscal expedients with high social and economic costs. Most tax revenues were remitted abroad, to fund armies and navies fighting to achieve international goals that mattered to the dynasty but not to most Spaniards. Between 1618 and 1648, the govern-ment exported to pay its armies at least 150 million ducats – a sum almost exactly equal to the increase in Castile's public debt.

Timothy Brook has pointed out that, in China, no previous emperor had 'faced climatic conditions as abnormal and severe as Chongzhen', and one could make much the same claim for his contemporary, Philip IV. During his reign, Spain suffered extreme weather without parallel in other periods, particularly in 1630–32 and 1640–43. Nevertheless, more than any other seventeenth-century ruler, Philip intensified the impact of climate adversity by disastrous policy choices. In the words of Arthur Hopton in 1634: 'It is no wonder that many of their designs fail in the execution, for though this great vessel [the Spanish monarchy] contains much water, yet it has so many leaks it is always dry' – in other words, in trying to do too much, the crown achieved nothing.[52]

The Spanish Habsburgs seem to have convinced themselves that foreign wars, however expensive and inconclusive, offered the best way to defend Spain itself. As Philip once observed: 'With as many kingdoms and lord-ships as have been linked to this crown it is impossible to be without war in some area, either to defend what we have acquired or to divert our enemies.'[53] Nevertheless, although the king spent every day of his forty-four-year reign at war – against the Dutch (1621–48), the French (1635–59), Britain (1625–30 and 1654–59) and in the Iberian peninsula (1640–68), as well as in Germany and Italy – he could surely have avoided (or ended more swiftly) some conflicts, and thereby reduced the fiscal pressure that crushed his subjects and provoked so many of them to rebel.

In 1650 an English statesman in Madrid marvelled at the capacity of Spain's leaders for self-deception. They were, he wrote, 'a wretched, miserable, proud, senseless people and as far from the wise men I expected as can be imagined; and if some miracle do not preserve them, this crown must be speedily destroyed'. A generation later, one of those leaders made the same point: 'I fear deeply for Italy; I am very worried about Catalonia; and I never forget about America, where the French already have too many colonies. We cannot govern by miracles for ever.'[54] The miracles ceased when Carlos II died childless in 1700, and a savage succession war resulted in the partition of the Spanish monarchy, with the larger part falling to a grandson of Louis XIV, whose descendants rule Spain to this day.

CHAPTER TEN

France in Crisis, 1618–88[1]

La Grande Nation?

'Both geographically and socially', wrote Lloyd Moote in 1971, the Fronde revolt in France (1648–53) 'was the most widespread of all the rebellions in mid-seventeenth-century Europe'.[2] Its extent should not cause surprise, because France was the largest state in Western Europe, covering almost 250,000 square miles. More striking is the extensive social support for the revolt: almost all the leading nobles defied the crown at some point, as did judges and civil servants, cardinals and curates, lawyers and doctors, industrial workers and field hands. About 1 million French men and women died, either directly or indirectly, because of the Fronde.

Like its neighbours, Spain and Britain, France was a composite monarchy, the product of territorial unification during earlier centuries. Seven provinces on the periphery (Brittany, Burgundy, Dauphiné, Guyenne, Languedoc, Normandy and Provence) retained considerable autonomy, guaranteed by their own fiscal institutions, sovereign law courts, and representative assemblies (the *États*: these seven provinces were therefore known as *Pays d'États*, 'provinces with Estates'). The central government in Paris directly controlled the remaining two-thirds of the country (known as the *Pays d'élections* after the tax officials, the *élus*, who apportioned tax quotas). Traditionally, legislation and taxes for the kingdom had been voted by another representative assembly, the *États généraux* (Estates-General), but after 1614 it met only once. As a result, Louis XIII (r. 1610–43), like his brothers-in-law Philip IV and Charles I, augmented his revenues whenever possible by manipulating existing taxes and enforcing regalian rights, and he relied on his senior judges (those sitting in his kingdom's ten Parlements, or sovereign law courts) to enforce them.

Louis XIII ruled the most populous state in Europe, with perhaps 20 million inhabitants, and French farmers normally produced an abundance of wheat and barley as well as some new crops, more resistant to climatic adversity, such as maize, buckwheat, beans, tomatoes and potatoes. Louis's father, Henry IV, placed French finances on a sound basis: he stabilized the public debt by repudiating foreign obligations and unilaterally reducing the interest payable on remaining loans; and he raised revenue by introducing indirect taxes, including the paulette, a nine-year agreement

16. Seventeenth-century France.
The kingdom comprised both *Pays d'élections*, governed directly from Paris, and *Pays d'États*, most of them on the periphery, where the central government ruled through provincial institutions. The crown suppressed the Estates of Dauphiné in 1628, and those of Normandy in 1666, and gradually increased its authority over the rest.

(renewable) that allowed holders of government office to pass on their posts to anyone they chose in return for annual payments to the treasury. The paulette soon became indispensable to the government, because it yielded up to 10 per cent of total revenues, but it gave rise to a serious disadvantage. It granted all members of the state bureaucracy, including the judges, almost complete job security as well as the right to sell or bequeath public offices as they pleased. The crown could no longer control them.

Seventeenth-century France was also weakened by the coexistence of two religious communities that had spent the later sixteenth century fighting each other. In 1598 Henry IV issued the Edict of Nantes, which guaranteed France's Protestants (often known as Huguenots, who made up about 10 per cent of the kingdom's population), the right to think, speak, write and worship as they wished within their own homes. They could also worship publicly in specified areas where they were numerous, and meet at regular intervals in assemblies to discuss both religious and political issues. The edict also legalized Catholic worship and assemblies everywhere, and the French Catholic Church rapidly gained in strength: diocesan seminaries to train the secular clergy rose from eight in 1614 to perhaps seventy in 1660, and the total number of convents in France doubled between 1600 and 1650.

Almost as soon as he began to exercise effective power Louis XIII determined to eliminate the Huguenots as a prelude to embarking on costly foreign adventures. He claimed that the Edict of Nantes did not apply to Navarre (of which he was also king), and when the Huguenot National Assembly protested he invaded, unleashing a religious war. This ruled out an assertive French foreign policy, and a succession of cold winters and summers caused dearth and disruption at home until in 1624 Louis made Armand-Jean Duplessis, Cardinal Richelieu, his chief minister. The cardinal continued to prioritize domestic over foreign policy, largely because his patron, the king's mother, Marie, supported a group of courtiers (known as the *dévots*) whose chief political goal was the destruction of Huguenot independence. He and the king in person laid siege to La Rochelle, despite its state-of-the-art fortifications, and did nothing while Spain mobilized troops in northern Italy to prevent a French claimant from acquiring the duchy of Mantua (Chapter 9). Luckily for Richelieu, Spain had not prevailed by the time La Rochelle surrendered in October 1628. He and Louis immediately led a powerful army across the Alps – but their attempt to dominate northern Italy failed in part because extreme climatic events produced another economic crisis.

France Goes to War

A sequence of unusually wet winters and summers either reduced or destroyed the crops in much of France, culminating in a famine that coincided with a plague epidemic in 1630–31. The catastrophe also reduced the demand for industrial goods and paralyzed trade, making it far harder to mobilize the necessary human and material resources to win the war. Richelieu also helped to undermine his foreign policy by provoking a new domestic crisis. Convinced that the provincial Estates were shirking their fiscal responsibilities, the cardinal sent special commissioners into Burgundy, Dauphiné, Languedoc and Provence with orders to raise taxes. In addition he threatened to suspend the paulette when its nine-year term

ended unless the senior judges provided full support to these commissioners. This was a dangerous gambit. The central government depended
absolutely on the services of its 25,000 civil servants and especially on the
200 judges of the Parlements, who decided cases referred to them by lower
courts and reviewed the edicts issued by the king. No royal legislation could
be enforced until after the local Parlement registered it, and both before and
after registration the judges had the power to modify government edicts in
the light of legal appeals concerning exemptions and exceptions.

The king could enforce registration by making a personal appearance in
Parlement, a process known as a *lit* ['bed'] *de justice*; but such action could
cause embarrassing confrontations. At a *lit de justice* in 1629 at the Parlement
of Paris, the largest and most powerful tribunal in the kingdom, Louis
forced his judges to register some contentious tax edicts, but they reminded
him that 'Great in the law though he is, the king will not wish to overturn
the basic laws of the kingdom ... Our power is great too.' As harvest failure
and plague caused widespread hardship, judges in other tribunals emulated
the defiance of their Paris colleagues.

Alarmed by this domestic unrest and infuriated by the failure of France's
foreign wars, the *dévots* determined to get rid of Richelieu. They argued
that unless France withdrew from all foreign commitments and allowed
time for retrenchment and reform, the monarchy would collapse, and on
10 November 1630, Marie publicly humiliated Richelieu and banished him
from her presence. The next day, the cardinal returned to tender his resignation, as protocol demanded, to find Marie locked in conversation with her
son. She immediately launched into a tirade about the cardinal's wickedness,
telling Louis that he must choose which of them he wished to retain as his
advisor. Richelieu left disconsolately, and his enemies flocked to congratulate
Marie on her victory (and to claim the offices that the cardinal and his
numerous relatives and clients would soon vacate.) Louis, however, spent the
rest of the day pondering the decision required by his mother and eventually
summoned Richelieu. Together, they planned how to govern without Marie.

After the 'Day of Dupes' (as contemporaries called it) Marie fled abroad,
never to return, and Richelieu executed, imprisoned or banished those who
had rashly revealed their opposition. Nothing now restrained his resolve to
check Habsburg expansion, whatever the domestic cost, and he promised
large subsidies to both Sweden and the Dutch Republic. This support
required tax increases, however, which produced more protests: in 1631,
another famine year, six more urban revolts rocked France; and 1632 was
little better. Louis made some concessions – he reluctantly renewed the
paulette for nine years – but in 1634 he reached the conclusion that, given
the size of the subsidies he had already paid to the Swedes and the Dutch, it
would cost only 1 million livres more each year to declare outright war.
Therefore, the king informed Richelieu, 'I believe it is better for us to attack

them now than to wait for them to attack us'. In May 1635 a French herald delivered a declaration of war on Spain.[3]

War and Insurgency

Contrary to Louis XIII's prediction, outright war increased state expenditure by a great deal more than 1 million livres. Apart from an immediate rise in taxes, the government had to feed and lodge troops at a time when the war depressed trade and industry in many areas, causing widespread economic hardship. Fourteen urban revolts occurred in 1635, the year war broke out, and although a reasonable harvest helped to preserve order in the countryside, a wet winter followed by an unusually stormy spring produced widespread opposition to recruiters and tax collectors in 1636. Many prepared formal complaints to the crown. Shaken by an invasion from the Spanish Netherlands that captured Corbie and threatened Paris (Chapter 9), the king graciously agreed to overlook the seditious nature of these assemblies and forgave tax arrears; but prosecuting the war soon led to new demands for both men and money, provoking in 1637 one of the largest popular uprisings in French history: the Croquants of Périgord, in southeastern France.

The trouble began when Louis ordered royal judges to sequester grain for the troops assembled to attack Spain, revoked forgiveness of past taxes, and increased the principal tax on land, the taille, by about one-third. The speed of the hostile reaction stunned everyone. Men from a group of forest villages formed an army that included many of the veterans who lived among them, led by the lord of La Mothe la Forêt, who called on each parish to produce twenty recruits and 5 livres a day for their support. When he had drilled and trained the veterans, La Mothe led some 8,000 men to Bergerac, a largely Protestant town, which welcomed them. Several more members of the elite rallied to the Croquant cause and started to compose sophisticated manifestos. Many have survived, filled with complaints about the 'insupportable, illegitimate and excessive [taxes] unknown to our forefathers', exacted by financiers who 'consume the poor labourers down to the bone', coupled with laments about the loss of trade and industry caused by the war. They petitioned the king (who, they claimed, had been deceived by his evil counsellors) to restore justice and liberty by reducing their tax burden to its pre-war level.[4]

The government managed to restore order in Périgord only by recalling troops from the Spanish front. Afterwards, in an attempt to discourage others from rebellion, judges arrived with instructions to seek out 'those with something to lose' and to administer 'exemplary punishments' that would 'horrify the rest of the rebels'. A dozen leaders, including several gentlemen, were degraded and publicly executed; but many others (including La Mothe) faded back into the backwoods from which they had come – in many cases becoming bandit leaders.[5]

Open revolts like that of the Croquants formed only the 'public transcript' (to use the language of James Scott) of French popular resistance.
Many other communities, especially in remote areas, defied the government
in ways that have left few archival traces: people delayed tax payments as
long as possible and fled whenever recruiters approached. Elsewhere, the
mere threat of violence led local authorities to make concessions that
averted open revolts. Although such tactical retreats preserved public order
in the localities, they left the central government short of taxes – taxes
whose yield it had already alienated to the *partisans* (bankers who had
loaned money under a contract, or *parti*, that assigned repayment, with
interest, from a specified source of income).

Much of France's war expenditure came from the taille, and as usual
with early modern taxes, the burden did not fall evenly. Certain areas (such
as enclaves ruled by foreign princes) and social groups (including university
professors and judges as well as nobles, clergy and the tax collectors themselves) did not pay the taille at all. As fiscal needs rose, the government
naturally tried to revoke exemptions, but although this strategy produced a
short-term advantage, in the form of increased tax yields, it brought a long-
term danger because those newly subjected to fiscal pressure often favoured
and sometimes fomented tax strikes and tax revolts by others. In addition,
some provinces paid far more taille than others, and within provinces some
areas enjoyed exemption – which meant that the rest had to pay more. Each
round of tax increases accentuated these disparities. The tax burden on
Lower Normandy, for example, quintupled between 1630 and 1636, and in
1639 the chief treasury minister warned Richelieu that the region 'pays
almost a quarter of the taxes of the entire kingdom'.[6] Small wonder that
Lower Normandy saw the greatest of the seventeenth-century tax revolts:
the *Nu-pieds* (the 'Bare Feet').

The *Nu-pieds* revolt took its name from the 10,000 workers who walked
barefoot on the sand as they carried wood to heat the saltpans on the
beaches of Lower Normandy, and then carried salt from the pans to nearby
towns. In July 1639 they murdered an official who they believed had brought
orders to extend to Normandy the *gabelle* (the salt monopoly that existed in
other French provinces, requiring each householder to agree in advance to
purchase a set quantity of salt at inflated prices). In other violent demonstrations, angry crowds in Lower Normandy murdered about 100 people –
most of them tax collectors – and maimed or threatened many others;
they also burnt down tax offices and the houses of those suspected of
enriching themselves from taxes. Sometimes the perpetrators numbered a
few score, at other times a few hundred, and eventually they coalesced to
form the 'Army of Suffering' with perhaps 5,000 members. As with the
Croquants two years earlier, a minor nobleman (calling himself 'General
Nu-pieds') took charge and used local veterans to command and drill his

followers. General Nu-pieds also composed and circulated manifestos that denounced the new salt tax and called for a return to the charter of 1315 that formed the basis of Normandy's incorporation into the kingdom of France.

As with the Croquants, Richelieu responded to the Nu-pieds by sending in troops, who in November 1639 defeated the Army of Suffering in a pitched battle – albeit only after several hours of hard fighting, thanks to the skill of veterans among the rebels – and executed the captured leaders and imprisoned the rest. Judges later toured the duchy to impose exemplary punishments 'to rule out any recurrence of similar disorders in the future'.[7]

The domestic unrest undermined France's war effort. For a moment, in 1640, the revolts of Catalonia and Portugal against Philip IV held out the promise of a speedy victory (Chapter 9), but these hopes were quickly dashed when the count of Soissons, a disaffected cousin of the king exiled for his part in an earlier plot, issued a manifesto that promised 'to restore everything to its former place: re-establishing the laws that have been overthrown; renewing the immunities, rights and privileges of the provinces, towns and personages that have been violated'. The conspiracy might have achieved these goals had Soissons not foolishly lifted the visor of his helmet with a loaded pistol and pulled the trigger as he did so. Another plot led by Louis's favourite courtier, who aimed to get rid of Richelieu as a prelude to making peace with Spain and retrenching at home, came within an ace of success.

The tide turned in 1642, when Louis's troops forced the surrender of Perpignan and advanced to the Ebro river. Although Richelieu died at the end of the year, and Louis followed him to the grave in May 1643, shortly afterwards French forces defeated an invasion from the Spanish Netherlands at Rocroi. This victory gave a spectacular boost to Anne of Austria, regent for her four-year-old son Louis XIV, advised by Cardinal Jules Mazarin; so although another disastrous harvest in 1643 caused the price of bread to reach its highest level in half a century, the war continued.

For a while, Mazarin's gamble worked. With better harvests, popular revolts abated and tax revenues increased, permitting new conquests at the Habsburgs' expense in Flanders, Germany, Italy and Catalonia. Still, the cardinal's colleagues feared for the future, and as soon as the 1646 campaign ended they advocated negotiating a settlement. Mazarin expressed some sympathy, even extolling 'the art of quitting when one is ahead, because then one keeps what one has won', but he failed to follow his own admirable advice.[8] By 1647 the crown had not only spent in advance all its revenues for the year, but also anticipated those for 1648, 1649 and part of 1650. Funds to keep the army fighting could now come only through creating more new taxes and mortgaging their future yields to a banker in return for immediate advance payments in cash; but each edict creating a new tax required the approval of the Parlements, which might either refuse to register the edict (preventing its collection) or modify it (reducing its application, and therefore its yield, in some way).

To avoid this outcome, Mazarin silenced outspoken judges and other critics by issuing *lettres de cachet* (which placed the recipient under house arrest), and he took the young king to a *lit de justice* at the recalcitrant Parlement of Paris and obliged its judges to register controversial new tax edicts. Some judges denounced such conduct as tyranny and thereby emboldened some of those affected by the new taxes to withhold payment; but Mazarin paid no attention, and in November 1647 sent another set of edicts creating new taxes for registration. Three proved especially contentious: an excise duty on foodstuffs coming into Paris; a tax on lands alienated from the royal domain; and the creation of several new public offices for sale to the highest bidder. The cardinal made clear that the paulette would not be renewed until all these measures had been registered.

Mazarin had chosen a dangerous moment for confrontation. As a Parisian diarist observed, these new taxes came at a time when the government had imprisoned 'people of all social backgrounds, not after due process before a court of law, but simply on a warrant from the royal council and a list [of names] signed by the minister of finance': in the year 1646 alone, some 25,000 people went to jail for failing to pay their taxes. The sparse harvest of 1647 left both the capital and the court short of food, yet the cardinal refused to heed his predecessor's warning about Paris: 'One must never awake this great beast', Richelieu had written. 'It should be left asleep.'[9]

Paris was the largest city in Christendom in the mid-seventeenth century, with 20,000 houses and over 400,000 inhabitants, but until 1647 it enjoyed exemption from most taxes. The new edicts threatened to change this, and just after the new year hundreds of Parisians gathered outside the Palace of Justice, where the Parlement debated the new edicts, chanting 'Naples, Naples' – a pointed reminder that another capital city had recently rebelled in protest against an unpopular tax (Chapter 14). Someone in the crowd struck one of the judges as he emerged, and when guards attempt to make an arrest, the women in the crowd counter-attacked and drove them away. Two days later, when the queen regent went to hear Mass in Notre Dame, several hundred women 'shouted at her and demanded justice'.[10] That night Anne and Mazarin deployed troops around the capital – but, in response, the city's militia companies assembled and ostentatiously tested their firearms as a sign that they would fight if attacked. Instead, on 15 January 1648 the young king was brought to another *lit de justice* at the Parlement of Paris to force registration of the unpopular tax edicts.

During the debate, Pierre Broussel, a senior judge, made a bold statement in defence of the Parlement's right to reject 'royal actions contrary to the well-being of the state and God's commandments, as these edicts are, not only because they contain clauses prejudicial to the well-being of the state but because they were presented in contravention of the customs and protocol of the assembly, which must always enjoy its powers freely'. Like opponents of government innovations elsewhere, Broussel sought to ground the discussion of individual grievances on general principles, citing in support Scripture, history and institutional custom. Another senior judge, Omer Talon, reminded the young king: 'Sire, you are our sovereign lord. Your Majesty's power comes from above and, after God, you are responsible for your actions to no one except your own conscience. But your glory requires that we should be free men and not slaves.' Warming to his theme Talon reviewed the hardships created for both townsmen and civil servants by the new taxes that the Parlement had just been forced to register and then, turning to Anne, he boldly admonished her:

> Tonight, in the solitude of your oratory, think of the sorrow, bitterness and consternation of all the servants of the state who today see their goods confiscated, even though they have committed no crime. And add to that thought, Madam, the desperation of the countryside, where the hope of peace, the honour of battles won, and the glory of provinces conquered cannot feed those who lack bread.[11]

The Revolt of the Judges

The speeches of Broussel and Talon immediately appeared in print and encouraged several groups directly affected by recent taxes to petition Parlement to modify the edicts they had just been forced to register. The most surprising petitioners were the *maîtres de requêtes*: lawyers who worked for the royal council. On appointment each *maître* gained noble status, which exempted him and his family from paying most taxes for ever, so aspiring lawyers were willing to pay 150,000 livres, plus the annual paulette, in order to enjoy such lucrative positions. They therefore objected vehemently to a tax edict forcibly registered on 15 January 1648 that created twelve new similar positions, because it would reduce the resale value of existing offices. The following month, the maîtres went on strike and asked the Parlement to reject the edict. The judges not only agreed to investigate their grievance: they also authorized an examination of the other edicts just registered.

Opposition by the Paris judges stimulated defiance by colleagues elsewhere. The Parlement of Brittany arrested and imprisoned the officials sent from Paris bringing similar tax edicts for registration, and the Parlement of Toulouse sentenced to hard labour any excise collector who began to collect the new duties. The order to double the number of judges in the Parlement of Provence excited such passionate opposition that the first man to purchase one of the new offices was stabbed to death, and posters went up warning other prospective purchasers to expect the same. Everywhere tax payments ceased.

Mazarin chose to ignore these consequences of his policies because (not without reason) he felt supremely confident that the revolt of Sicily and Naples would lead Philip IV to use force to regain control, which would benefit France: troops sent to Naples would come from Catalonia, allowing French forces there to make progress, and a Spanish attack on Naples would 'kindle rather than put out the fire' there and also prevent Spain from defending its positions in northern Italy.[12] Mazarin also rejected the beleaguered king of Spain's offer to cede all France's gains in the Low Countries permanently, and those in Lombardy and Catalonia for thirty years, in return for peace. Had Mazarin accepted these terms – far better than France would ever receive – he could have immediately diverted his troops in the Netherlands to Germany and extracted better terms at the peace congress in Westphalia; but instead, in the hope of gaining yet more, he poured all available resources into new campaigns in Catalonia and Lombardy.

The decision to continue the war with another Catholic monarch alienated not only the *dévots*, but also another religious group who became known as Jansenists. In 1640 a huge Latin treatise entitled *Augustinus* had appeared from the pen of Cornelius Jansen, a theologian at the university of Leuven in the Spanish Netherlands. It argued at great length that humans had fallen so far from their original innocence and perfection that only the most rigorous and sincere devotion could merit salvation: conventional piety would not suffice. Resenting the numerous editions and translations of earlier anti-French polemics by Jansen, Richelieu secured papal condemnation of the *Augustinus* and had it banned. Nevertheless, shortly after the cardinal's death Antoine Arnauld published *Frequent communion*, an eloquent tract in French that popularized the main ideas of Jansen's virtually unreadable Latin folio tomes. In particular, it condemned the practice of taking frequent communion, advocated by the Jesuit Order (among others), as a way of appeasing God: instead, Arnauld argued, the laity should take the sacrament only when they had purged all impiety from their hearts and minds. Following Richelieu's example, Anne sent a copy of *Frequent communion* to Rome and requested a papal condemnation. She also planned to send its author there for trial on charges of heresy – but whereas Jansen was a subject of Philip IV, Arnauld was the son of a French judge and scion of a prominent Parisian family. The judges of the Parlement of Paris argued not only that France had plenty of theologians competent to determine the orthodoxy of Arnauld and his work, but that sending a French subject to Rome would open the door to papal intervention in the affairs of the French church. In the words of Orest Ranum, Anne's decision gave Arnauld and 'the Jansenist cause more support in Parisian society than he or his predecessors had ever hoped for. For the first time, radical members of the

Parlement argued that the Queen Mother was a foreigner subverting French laws'.[13]

The opposition of the judges and Jansenists, together with another cool wet spring that spoiled the crops, led Mazarin to change his mind about peace: in May 1648 he indicated to Madrid a willingness to open talks on the basis of the generous terms offered by Spain five months before. By then, however, Philip had made peace with the Dutch and repressed the revolts of both Naples and Sicily: he therefore determined to increase military pressure on France. 'All we need now', mused a Spanish diplomat, 'is some moderate victory in the Netherlands in order to start some commotion in France that will open up a highway to an honourable peace.'[14] A few weeks later, a far greater commotion than even the most optimistic Spaniard could have imagined paralyzed the French capital.

The Tipping Point: The Barricades of Paris

In April 1648, Mazarin attempted to drive a wedge between his opponents. He offered the judges of the Parlement of Paris (and only them) the opportunity to renew the paulette on normal terms. Other officials could renew only if they agreed to forego their salaries for four years; and the *maîtres de requêtes*, still on strike, lost their right to renew altogether. This blatant attempt to divide and rule backfired disastrously. The judges of the Parlement expressed open support for all their colleagues and invited the other three sovereign courts that met in Paris – the Cour des Aides and the Chambre des Comptes, which handled tax appeals and audits, and the Grand Conseil, which heard ecclesiastical disputes – to send delegates to meet with them in a special room in the Palace of Justice known as the Chambre Saint-Louis. This encouraged other discontented groups of civil servants to defy the regent, starting with the officials who administered direct taxes in the provinces: the *trésoriers de France*. By mid-May 1648 the entire fiscal machinery of the French state had ceased to function.

Realizing the dangers that now faced them, Anne and Mazarin put pressure on the judges by withdrawing the offer to renew the paulette, prohibiting any further meetings in the Chambre Saint-Louis and imprisoning some *trésoriers* in the Bastille – but these provocations fuelled further defiance. Starting on 30 June 1648, fourteen judges from the Parlement and six from each of the three other central courts met daily in the Chambre Saint-Louis to discuss the various grievances of the kingdom. Mazarin initially rejected their demands, but after learning that Spanish troops were preparing another invasion he agreed to renew the paulette for nine years for all officials, reduce the taille rate for the current year, forgive all tax arrears from before 1647 and free all those imprisoned for non-payment of taxes.

Mazarin felt confident that these measures would end the crisis. In July 1648 he enthused that, 'for fear of something worse', a consortium of creditors had 'promised to provide a fixed sum to prolong the war as long as the obstinacy of the Spaniards makes it necessary. We had spent all [the revenues] of this and the next two years; now we have found a way of spending them a second time.' Then, to the cardinal's horror, the Parlement of Paris voted to investigate the accounts of the consortium of creditors for evidence of fraud and excessive profits. All lending to the government immediately ceased, and Mazarin reluctantly instructed his negotiators at Münster to end the war in Germany on the best terms available. 'It is almost a miracle', he observed, 'that amid so many self-made obstacles we can keep our affairs going, and even make them prosper; but prudence dictates that we should not place all our trust in this miracle continuing any longer.' The diplomats must 'make peace at the earliest opportunity.'[15]

Ironically, just one week after Mazarin signed this letter, the prince of Condé routed the Spanish invasion – but the cardinal immediately threw away his advantage. Calculating that a *Te Deum* for the victory celebrated in Notre Dame Cathedral would provide an opportunity to arrest the leading judges, including Pierre Broussel, the queen regent personally invited all of them to attend. Some of her targets escaped, including Broussel, who lived close to Notre Dame, but a detachment of guards arrived in hot pursuit and abducted him from his home.

The arrest of Broussel, who was a militia captain as well as a judge, caused general outrage and dense crowds began to rampage through the streets smashing doors and windows, and shouting 'Long live the king! Free the prisoner!' To prevent looting, the militia companies mobilized and deployed the heavy chains that most Paris streets still maintained for emergency use, behind which they erected over 1,200 barricades to repulse any attack. The militiamen declared that they would not lay down their arms until the regent freed Broussel, while at the royal palace the guards made clear that they would not fire on their compatriots. Gangs of young Parisians now took to the streets with slings, known as *frondes*, which they used to smash the windows of the opulent Palais Mazarin. That 'weapon of the weak', reminiscent of David's overthrow of Goliath, gave its name to the revolt that would last for five years: the Fronde.

Although Broussel's release and triumphant return home the following day calmed tempers and brought down the barricades, a 'Great Fear' continued to grip the French capital as householders worried that their property might be damaged if either the 'populace' or the troops attacked. The judges, led by Broussel, took advantage of the uncertainty to continue their scrutiny of recent tax edicts, rejecting some outright and asking Anne to modify others. They also did nothing to prevent a torrent of literary attacks, called *Mazarinades*, on her chief minister.

The Fronde

Madame de Motteville, a perceptive member of the regent's household, lamented that 'the people, in the hope of saving themselves from dues and taxes, dreamed only of tumults and changes' and she suspected that the anti-royalist revolts in England as well as in France formed part of an 'evil constellation that menaced the well-being of kings'[16] Anne of Austria evidently shared this pessimistic view, because she issued a proclamation that accused the judges of plotting with Spain to seize her sons, hinting that they aimed to establish a republic, and she commanded all four sovereign courts to leave the city immediately for four separate destinations. She also ordered the prince of Condé and his victorious army to blockade the capital.

The regent's maladroit actions united her opponents. Since she had lumped all the judges together as traitors, even those who had remained loyal until this point now subscribed to a declaration that Mazarin was 'an enemy of the king and the state'. They also set up a wartime administration with committees to handle military, financial, diplomatic and other business, and made common cause with other opposition groups in Paris. City magistrates loaned them 1 million livres; the Paris clergy (many of them Jansenists), led by Paul de Gondi, archbishop-designate and later cardinal de Retz, also lent their support with advice, sermons and pamphlets; many nobles flocked to Paris to join the judges, bringing their troops with them. In January 1649 the royal family fled from their capital.

Initially, most noble adherents of the Fronde acted out of principle. Richelieu had imprisoned them, sequestered their property and demolished their castles not only for plotting but also for lesser offences such as duelling. Now they sought to dismantle the machinery of prerogative rule that had humiliated them. They also favoured opening peace talks with Spain, believing that continuing the war would ruin France. Although the nobles lacked an established forum for discussing grievances – they could meet legally only in the Estates-General (none had been called since 1614) or in the provincial Estates (and by 1649 only those on the periphery of the kingdom still met regularly) – at first this scarcely mattered because it proved so easy to circulate their views in pamphlets. Paris boasted over 350 print shops, many of which had prospered from producing the multiple copies of government edicts required for distribution throughout the kingdom. When the collapse of royal authority ended this lucrative source of business, printers compensated by publishing cheap political diatribes. The total number of pamphlets published from 1649 to 1653 far exceeded those produced during the rest of the century – a statistic made more remarkable by the fact that the winter of 1648–49 lasted almost six months, with intense cold followed by a rapid thaw and copious rains that caused the Seine to burst its banks, inundating the city hall and surrounding houses.

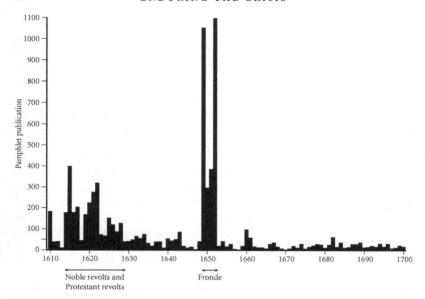

17. Pamphlet publication in seventeenth-century France.
The Bibliothèque Nationale de France, in Paris, boasts a remarkable collection of pamphlets, whose contours reflect the political stability of the kingdom. The assassination of Henry IV in 1610 led to the publication of over 100 items, and the subsequent rebellions of Protestants and nobles sometimes produced over 400 items in a year; but these totals were dwarfed by the *Mazarinades* (tracts attacking the chief minister Cardinal Mazarin), which exceeded 1,000 in 1649 and again in 1652.

Floods south of Paris, which normally provided most of the capital's bread, combined with Condé's blockade, caused the price of a loaf of bread to leap from 9 to 18 sols in February 1649. Since an unskilled labourer at this time earned 12 sols a day (assuming he could find work), this increase spelled starvation. Those who lived outside Paris had to contend not only with snow, floods and famine but also with Condé's troops. As the regent's entourage travelled by coach near the capital in February 1649 it passed 'through several villages where we noted a frightening desolation. They had been abandoned by their inhabitants. The houses were burnt and torn down, the churches pillaged.'[17] Some took refuge in a nearby convent, like Port-Royal-des-Champs southwest of the capital, where Abbess Angélique Arnauld (sister of Antoine) did her best to provide food and protection. Within the capital, starving crowds surrounded the Palace of Justice and shouted 'Give us bread or give us peace.' They welcomed an agent of Philip IV who arrived to invite the Parlement to mediate a peace between France and Spain. The Fronde seemed about to succeed – but, later that day, news arrived that in London Charles I had been tried, condemned and beheaded by his subjects.

The regicide produced a wave of incredulity, indignation and revulsion throughout France, and to distance themselves from it, the Paris judges offered their condolences and a pension to Charles's widow (and Louis XIV's aunt), Henrietta Maria, and opened peace talks with the regent. In March 1649, in return for Anne's promise to confirm all her previous concessions and grant an amnesty to her opponents, the Parlement of Paris annulled its edicts against Mazarin.

Yet the Fronde continued. Many nobles, whose paramount goal had been to remove the cardinal from power, persevered with their defiance, and crowds of Parisians gathered outside the Palace of Justice shouting 'No peace! No Mazarin!' and, more alarmingly 'Republic!' At this point, Philip IV again offered to open peace talks with France, but (in view of the weakness caused by the Fronde) he now demanded not only the withdrawal of all French troops from Catalonia and Lorraine but also an end of all aid to Portugal. When Mazarin angrily rejected these demands, in exasperation, the prince of Condé joined the Fronde.

No surviving evidence suggests that Condé wanted to displace his royal cousin on the throne, or dismember France by carving out a state of his own: he probably aimed to replace Anne as regent. He seemed close to achieving this goal when, in January 1650, Mazarin acknowledged that he was Condé's 'very humble servant' who would further the prince's interests in all matters. But the prince's triumph lasted just two days: Anne's guards suddenly arrested and imprisoned him, together with his two principal allies.

Condé's supporters and relatives in Normandy, Burgundy and Guyenne promptly renounced their obedience to the central government – despite desperate material conditions. According to a contemporary history, 1651 began with 'a deluge of rain that seemed to presage the misfortunes that later afflicted the poor kingdom' of France. 'War had led to excesses; taxes had ruined the population; famine had sent many to their graves; and despair had led to uprisings.'[18] Virtually all French troops at home were either mutinous, because they lacked pay and food, or commanded by nobles in open defiance of the crown. Even areas that saw no fighting suffered: in Provence, the combination of plague and the highest grain prices of the century provoked almost seventy popular revolts; in Picardy, local clergy visiting their parishioners found families too weak even to answer the door because they had not eaten for several days.

In February 1651 Anne released Condé; Mazarin, fearing that the former prisoner would seek revenge, fled to Germany. He left Paris in chaos: the endless rain caused the Seine to burst its banks again, flooding many houses; the hundreds of aristocrats who had gathered to demand Condé's release now created a forum, the Assembly of Nobles, where they formulated grievances against the central government (non-payment of interest on bonds; lack of employment and salaries; repeated violations of their traditional immunity from taxation, recruiting and billeting). Unusually, the Assembly admitted no distinction of rank, so that its printed pronouncements went out in the names of princes, peers, barons and ordinary gentlemen alike, all of whom took turns to preside at the daily meetings. The Assembly insisted that Anne convene the Estates-General, and in order to get the nobles to go home, she eventually promised to convene one on 8 September 1651.

It seems astonishing that the nobles believed her. Although several groups optimistically drew up lists of grievances (*cahiers de doléances*) the regent did not even bother to issue writs to summon the deputies: the Estates-General would not meet again until 1789. Realizing that he had been duped, Condé fled Paris and signed a formal alliance with Philip IV. Soon afterwards, Anne and Louis also fled the capital again to join Mazarin, who had re-entered France at the head of a powerful contingent of German troops.

The court's departure from Paris provoked another flood of *Mazarinades* (no fewer than 1,100 appeared in the course of 1652, sometimes ten on the same day). This time the Jansenists entered the fray, led by Robert Arnauld, brother of Antoine and Angélique, with a pamphlet entitled *The naked truth*, which offered a penetrating analysis of the origins of the Fronde that laid equal blame on Mazarin and Condé – and for good measure compared them with Cromwell, the 'usurper and tyrant of England' and 'the Mohammed of this century'.[19] Condé returned to Paris briefly in April 1652, but when his supporters seized the city hall and murdered his leading opponents, popular opinion turned against him. The prince fled to Bordeaux, where a coalition of craftsmen, lawyers and merchants had set up a city government, known as the Ormée, which requested assistance from Cromwell. With the help of some English advisers, the city's leaders drew up a republican constitution as shouts of 'No kings! No princes!' echoed through the streets.

Mazarin now prepared a counter-attack. He persuaded a group of financiers to provide him with enough loans to win over some of the judges by paying their debts and granting them pensions, while Anne conferred lands and titles on potential supporters: seventeen noble families became 'peers of the realm (*duc et pair*)', a relatively new and extremely rare honour. In October 1652, Louis XIV re-entered his capital and immediately held a *lit de justice* that annulled all legislation enacted by the Parlement over the previous four years. He also forbade the judges in future from meddling in 'affairs of state and the management of finance'; prohibited any proceedings 'against those whom the king has appointed to govern' (such as Mazarin);

abolished the Chamber of Justice established to investigate profiteering among the crown's bankers; and exiled his most obdurate critics.[20]

Nevertheless, Louis still faced many domestic enemies. Bordeaux remained in full revolt, supported by Spanish and British troops, while Condé and several regiments commanded by his clients and noble allies fought in the Spanish Netherlands for Philip IV. Above all, the Paris clergy openly opposed Mazarin, who in December 1652 arrested cardinal de Retz. From prison, Retz begged his noble relatives and clerical colleagues to organize an uprising that would secure his release, and a clerical Fronde began among the Paris clergy, almost all Jansenists who already resented the government's persecution of those who revered the *Augustinus*. In May 1653, Mazarin persuaded the pope to issue a bull that categorically condemned Jansenism, naming five propositions allegedly contained in Jansen's writings as heretical; but this tactic served to intensify the opposition of the Paris priests – some fifty in number – who now assembled regularly to discuss the affairs of the archdiocese.

The clerical Fronde drew strength from Mazarin's inability to crush his other opponents. Although royal troops forced the surrender of Bordeaux, ending the Ormée, Condé took command of Spain's armies in the Netherlands and in 1654 seemed poised to march on Paris. At the same time Retz escaped from prison, intending to rally his supporters once more. Hundreds of people converged on Notre Dame, where the cathedral chapter – in defiance of the government – sang a *Te Deum* to celebrate their hero's escape, but chance now saved Mazarin: as Retz fled, he fell off his horse and dislocated his shoulder. He was confined to his bed.

Although these developments largely restored domestic peace, France suffered extensive damage from the landmark winter of 1657–58. Snowmelt and heavy rain caused many rivers burst their banks, including the Seine which flooded Paris for the third time in a decade; and since farmers could not sow their crops, the subsequent harvest was very poor. Even Mazarin had to recognize that, since he lacked the resources to mount another campaign, it was time to practice 'the art of quitting when one is ahead' and in spring 1659 he accepted Philip IV's peace overtures and gained for France northern Catalonia and parts of the southern Netherlands (Chapter 9).

The litany of war, climatic adversity and political crisis eventually demoralized even the most resilient. 'If one ever had to believe in the Last Judgment', wrote one Parisian in 1652, 'I believe it is right now.' The following year, another claimed that 'two-thirds of the inhabitants of the villages around Paris are dead of illness, want and misery'. Abbess Arnauld feared that the general desolation 'must signify the end of the world'; while in 1655 the king's uncle Gaston of Orléans declared that 'the Monarchy was finished: the kingdom could not survive in its present state. In all the Monarchies that had collapsed, decline began with movements similar to the ones he

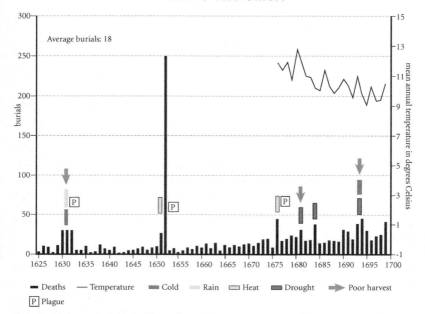

18. War, climate and mortality in Île-de-France during the seventeenth century. The records of Créteil, a village 7 miles southeast of Paris, reveal a close link between climate and mortality throughout the seventeenth century: burials rose with the cold and rain of 1631 and 1693–4, and the heatwave of 1676. The worst demographic crisis of the village – indeed, of all Bourbon France – occurred in 1652 during the Fronde revolt, when a combination of floods, a catastrophic harvest, and military operations increased burials fourteenfold. Annual temperatures fell sharply in the 1690s.

discerned [now]; and he launched into a long list of comparisons to prove his statement from past examples.'[21] Some historians have gone even further than the duke, suggesting that the plight of the French monarchy during the Fronde was greater than in 1789, and that the territorial integrity of the kingdom might not have survived a Frondeur victory.[22]

The Sun King

Nevertheless, the French monarchy did survive and Louis XIV became the most powerful ruler in Western Europe. In part, this outcome arose from contingency. Abroad, France benefited from miscalculations by its enemies. The intransigence of Philip IV between 1648 and 1655, when he held the advantage, saved Mazarin from having to trade land for peace, and Spain's eventual decision to make extensive territorial concessions appeared to vindicate the policies that the cardinal had pursued so recklessly. At home, Mazarin's most dangerous rivals either overplayed their hand (like Condé) or else lost the initiative at a critical moment (like Retz). Moreover, his opponents lacked a single agenda: judges, princes, nobles, clerics, urban patricians and popular leaders all stood for something different and often devoted more energy to destroying their erstwhile allies than to wresting concessions from the crown. They also lacked a unifying ideology: Protestantism had become a spent force, and Jansenism still looked to Rome rather than to Paris for inspiration. The nobility, too, was deeply divided: the hereditary ('sword') aristocracy despised the newer 'robe' nobility, made up of judges and other royal officials who had purchased their exalted status. Had the Estates-General met in 1651, the former would probably have sat with the urban delegates in the Third Estate (as they had done in 1614) rather than with the latter in the Second Estate (as they would do in 1789).

Such fundamental divisions eventually enabled the crown to present itself as the only alternative to anarchy. Mazarin had already shown the way, because peace with Spain enabled him to eliminate several groups of his domestic opponents. He terminated the assemblies of both the Paris clergy and the *trésoriers de France* (whose defiance had started the Fronde), and he dissolved an association of devout Catholics known as the Company of the Holy Sacrament. The legislation that ended the Company's legal existence also sought to outlaw meetings of all religious 'confraternities, congregations and communities' in France, on the grounds that 'under the veil of piety and devotion' they might foment 'cabals and intrigues'.[23]

After Mazarin's death in 1661, Louis XIV eliminated any individual or corporation that might offer an alternative focus of loyalty, starting with the late cardinal's wealthy and corrupt finance minister, Nicholas Fouquet, whom Louis arrested and charged with treason. He then set up a commission of inquiry to investigate all who had grown rich from lending money to the crown in wartime, leading to the conviction of almost 250 individuals for defrauding the treasury and fines that totalled 125 million livres – almost twice France's annual budget.

The commission of inquiry was one of several royal initiatives that (ironically) addressed grievances voiced by the Frondeurs. Others included Louis's decision to create fewer new offices, end the punitive taxation of office-holders and renew the paulette automatically. In addition Fouquet's successor, Jean-Baptiste Colbert, paid both loan interest and official salaries punctually and in full; redeemed many costly long-term bonds; and reduced reliance on direct taxes like the taille (whereas indirect imposts had produced less than 25 per cent of the royal budget in the 1640s, they brought in 50 per cent by the 1670s). Although Colbert's fiscal reforms only raised the crown's notional revenue from 83 million livres in 1661 to 95 million in

1667, they more than doubled the amount that actually reached the central treasury: from 31 million to 63 million livres.

Louis also excluded from the central government three groups that he believed had grown too powerful during the regency: his relatives (even Anne of Austria lost her seat on the council and he never consulted his brother on matters of state); his nobles (in the course of his long reign, he appointed only two to the rank of minister of state); and his prelates (there would be no more Richelieus or Mazarins). Instead, Louis entrusted major tasks to men of relatively humble origins (like Colbert, whose father was a bankrupt provincial draper) with whom he met either separately or in council, reserving all the key decisions for himself. In the provinces, Louis appointed permanent officials with wide-ranging functions (reflected in their full titles 'Intendants of justice, police and finance'). In the *Pays d'État*, a series of edicts eliminated the power of the Parlements to challenge royal edicts before registering them, and Louis held no *lits de justice* after 1675. The local Estates likewise lost their right to present grievances before voting taxes. The language of politics in France had shifted from negotiation and compromise to obedience and subordination.

Louis tackled with similar vigour another inherited domestic challenge: the Huguenots. He issued a stream of edicts that restricted the religious and personal liberty of France's 750,000 Protestants; he quartered troops on the households of those who refused to convert; and in 1685 he revoked the Edict of Nantes (page 229). Afterwards, the king ordered the destruction of all Protestant churches, proscribed all Protestant worship, closed all Protestant schools and ordered all Protestant ministers either to convert or to leave the kingdom within two weeks. At least 200,000 Protestants followed their pastors into foreign exile, but the rest remained and, at least outwardly, conformed.

Louis also used force to impose his will on the cities of France, assailing their physical as well as their political strength. He ordered the demolition of almost all fortifications in the interior – including those of Paris, replacing its walls with the tree-lined Grands Boulevards that dominate the right bank of the city to this day. Instead, Louis's military architects created a ring of artillery fortresses around the periphery of France, christened *le pré carré* ('the duelling field'), designed to prevent any enemy from reaching the heart of the kingdom. The king also steadily increased the size of France's standing army. In the 1650s, through desertion and demoralization, the effective strength of France's armies probably did not exceed 150,000 men, and after the Peace of the Pyrenees the total fell to around 55,000; but during the 1670s, France's armies exceeded 180,000 men in peacetime, rising above 250,000 in wartime – the largest and most expensive defence establishment in Europe. Soldiers became more numerous in France than clerics, and military buildings (barracks as well as fortresses) dwarfed even the largest cathedrals and monasteries.

Plus ça change?

Like Richelieu and Mazarin before him, Louis XIV viewed Christendom as a hierarchy in which some states naturally played a greater role than others: he therefore believed that an enduring balance of power in Europe required France to be pre-eminent. Every one of the peace conferences of his reign – the Pyrenees in 1659, Breda in 1667, Nijmegen in 1678, Rijswijk in 1697 and Utrecht in 1713 – convened to end a war that France had begun. Also like Richelieu and Mazarin, Louis fought his wars at a time when climatic adversity reduced available resources. Tree-ring data from western France show a long period of cooler and drier weather in the later seventeenth century, with several years of extreme adversity. The year 1672, when the Dutch War began, saw the worst harvest in a decade (thanks to drought followed by downpours), and those of the two succeeding years were scarcely better. Then came 1675, a 'year without a summer', almost certainly reflecting the powerful eruption of two volcanoes in Southeast Asia. In July, in Paris, Madame de Sévigné complained to her daughter that 'we have the fires lit, just like you, which is very remarkable.' Her daughter lived in Provence, where a diarist lamented that 'the seasons were so disorderly that all the crops were incredibly late' and predicted that they would not be harvested until late October, 'something never seen here before'. He was right: the grape harvest of 1675 ripened later throughout France than in any other year since records began.[24]

Admittedly, Louis took some steps to ameliorate the adverse effects of the Little Ice Age. During the famine of 1661–62 he bought grain in Aquitaine, Brittany and the Baltic and brought it to the capital – something neither Richelieu nor Mazarin had attempted – and this became standard operating procedure in times of dearth for the rest of the Old Regime and beyond. Elsewhere, however, the king showed little concern and so each period of climatic adversity produced not only famine but also a spike in popular rebellions. At least 300 revolts broke out in 1661–62, a few of them suppressed only after troops arrived; and 1675 saw popular protests in western France from Bordeaux to Nantes. Those in Lower Brittany reached the scale of earlier uprisings. Louis had introduced several new indirect taxes to fund the Dutch War – notably a duty on all products made of tin, a state monopoly on tobacco, and a stamp duty on all legal documents – and obliged the Parlement to register them. As before, this produced cries of 'Long live the king – without excise taxes', and several Breton cities saw assaults on the offices that issued stamped paper, as peasants stormed and sacked the mansions and castles of their lords. By September 1675 the movement had a leader, the lawyer Sébastien Le Balp, a collective name (*les Bonnets rouges*: 'the Red Caps'), and several printed lists of grievances; but it collapsed when a royalist managed to murder Le Balp, and regular troops entered the duchy to carry out systematic reprisals. Louis signed a general amnesty in February 1676 that excluded more than 150 people (including fourteen priests) in order to concentrate on the next campaign against the Dutch.

The footprint of the year without a summer thus resembled that following climatic disasters in the first half of the seventeenth century; but there were also differences. The revolts of 1675 neither inspired sequels not did they attract aristocratic involvement. Louis XIV would face no more Frondes. He might, by contrast, have observed a grim consequence: of the thousands of men who enlisted in his armies, those born in 1675 and 1676 were the shortest in stature, with an average height of 159 centimetres (page 21). Moreover, among that disadvantaged group, those born in the west of France, the areas that had rebelled, were the shortest cohort of Frenchmen *ever* recorded, thus vindicating the claims made by some of Louis's subjects, from Madame de Sévigné to Sébastien Le Balp, that the conditions they faced were 'something never seen here before'. Even the Sun King was no match for the Little Ice Age.

CHAPTER ELEVEN

The Stuart Monarchy: The Path to Civil War, 1603-42[1]

'Great Britain': A Problematic Inheritance

The civil wars that racked seventeenth-century Britain would have been impossible without the creation of a new composite state in 1603, when James Stuart of Scotland succeeded his childless cousin, Elizabeth Tudor, as ruler of England, Wales, Ireland and the Channel Isles. It was an unequal union from the first. England's population in 1603 exceeded 4 million, whereas Ireland's was less than 2 million, and Scotland's well under 1 million. London, with 200,000 inhabitants in 1603 (and perhaps 300,000 in 1640) had no equal in the Stuart monarchy.

England and Scotland had spent much of the previous four centuries at war, leaving a reservoir of mutual hatred and suspicion, and the economic, social and political differences of the two kingdoms were exceeded only by their incompatible religious establishments and doctrines, each enforced through a panoply of laws and courts. Although both were officially Protestant, in Scotland bishops appointed by the king contended with regional assemblies (known as presbyteries) of parish ministers who followed the theology of John Calvin; whereas in England the monarch, who was also the head of the church, appointed all bishops and upheld a Protestant theology hostile to both Catholics and Calvinists (normally known as Presbyterians). England's Catholics inspired disproportionate popular hatred and fear, because although they numbered under 5 per cent of the total population, they included many prominent adherents (including the spouses of both James and his son Charles and many of their courtiers) as well as some extremists (such as the group led by Guy Fawkes who in 1605 attempted to blow up the royal family and both Houses of Parliament in the Gunpowder Plot).

Ireland, too, proved a troubled inheritance. In 1603, after a bitter nine-year struggle, English forces managed to suppress a major Catholic rebellion, and soon afterwards James confiscated the lands of many former rebels and granted them to settlers from Britain. By 1640 some 70,000 English and Welsh, and perhaps 30,000 Scots, had settled in Ulster (the northern province of the island), mostly in new towns and in plantations (lands confiscated from the native Irish and granted to groups of British immigrants). These newcomers joined either the Protestant Church of Ireland (closely modelled on the Church of England, with bishops appointed by the crown) or one of a growing number of Presbyterian communities; but everywhere they remained heavily outnumbered by Catholics obedient to bishops and abbots appointed by Rome.

To overcome such diversity, and to guard 'against all civill and intestine rebellion', James strove to foster a common loyalty among his subjects. He assumed the title 'king of Great Britain' and stated that 'his wish above all things was at his death to leave one worship to God, one kingdom entirely governed, [and] one uniformity in laws' throughout his realms. In Ireland, his agents completed the work of their predecessors in imposing English law and administrative practices until, by 1612, according to an official addicted to metaphors, 'the clock of civil government [in Ireland] is now well set, and all the wheels thereof do move in order; the strings of this Irish harp, which the civil magistrate doth finger, are all in tune'. That tune became steadily more Protestant.[2]

James also strove to Anglicize his native Scotland. In secular affairs he worked through the Privy Council, a body of nobles and officials sitting in Edinburgh whose proclamations had the force of law. After 1612, a standing committee of the Scottish Parliament, known as the Lords of the Articles, prepared legislation for ratification by the full assembly and enabled James to control the parliamentary agenda. Sometimes arguments arose over the imposition of taxes – which tripled from 1606 to 1621 – but every increase eventually received approval. The only area of policy where James encountered spirited resistance was religion. His efforts to impose English liturgical practices when he returned to Scotland in person in 1617 produced bitter opposition from the General Assembly of the Church.

Shortly after James's brief visit, foreign events began to undermine his authority in other areas. Elector Frederick of the Palatinate, husband of James's daughter Elizabeth, accepted the crown of Bohemia and provoked a counter-attack by both the Austrian and Spanish Habsburgs, who confiscated his hereditary lands (Chapter 8). James decided to assist his son-in-law both by sending him money and (later and very grudgingly) troops, and also by seeking to marry his son and heir Charles to a Spanish princess, with the restoration of the Palatinate to Frederick as (in effect) her dowry. When marriage negotiations languished, in 1623 Prince Charles travelled to Madrid together with the duke of Buckingham, his father's favourite, to advance his suit in person. Meanwhile, to facilitate the 'Spanish Match', James relaxed the anti-Catholic penal laws in England – a move guaranteed to alarm and alienate his Protestant subjects at any time, but especially during an economic depression.

The El Niño autumn of 1621 brought rains that ruined the harvest throughout Britain. Even in areas that normally exported grain, the magistrates feared that 'this time of so extraordinary a want both of corn and of work' would 'breed in those of their condition a dangerous desperation'.[3] Amid such tensions, and with tracts, poems and sermons railing against the Spanish Match, foreign ambassadors predicted rebellion should Prince Charles bring back a Spanish bride; but the marriage negotiations failed and the prince's return as a bachelor precipitated ecstatic rejoicing. Charles and Buckingham became national heroes, and they exploited their popularity by persuading James to declare war against Spain, primarily to put pressure on Philip IV to restore Frederick of the Palatinate's confiscated estates. To pay for the war, James summoned Parliament, where Charles and Buckingham persuaded the House of Commons first to impeach a minister who opposed the war and then to vote £300,000 in new taxes to fight it.

Although this assembly earned the epithet '*Felix Parliamentum*, the happy Parliament', it created serious problems for Charles. Resort to impeachment (common in the fifteenth century but little used since then) set a dangerous precedent that could be used against any royal official who displeased Parliament. Moreover, as James himself had observed many years before, 'a wise king will not make warre against another, without he first make provision of money' – but in 1624, although the war with Spain was predicted to cost £1 million a year, the English treasury had no adequate provision of money. This shortage forced Charles to reconvene Parliament as soon as he succeeded his father in March 1625 to request more funds to prosecute the war. The acrimonious debates that marked this and the three subsequent sessions destroyed both the king's popularity and the national unity created by the failure of the Spanish Match. In the words of Richard Cust, the king's most perceptive biographer, 'Charles's honeymoon with the English people was over'. How did it end so fast?[4]

'The Crisis of Parliaments'

One major reason for popular disenchantment in the later 1620s lay beyond Charles's control. No sooner had he secured a declaration of war on Spain than another run of poor harvests drove up food prices, which drastically reduced the demand for manufactured goods, and plague broke out in London, killing some 40,000 people and halting trade. On the Isle of Wight, normally a grain-exporting region, a landowner noted in 1627 that 'the coldness of the summer and the great fall of rain in August and September' had ruined the harvest, and 'the winter of 1629 was one of the wettest that ever I knew'.[5] Charles could hardly have chosen a worse time to wage war.

The king's policy decisions exacerbated the situation. In the hope of creating an anti-Habsburg alliance to support his Palatine policy, he proposed marriage to Henrietta Maria of France – but her brother Louis XIII demanded a measure of toleration for English Catholics before he would consent. Accordingly, Charles suspended the penal laws again, provoking predictable anti-Catholic agitation. In addition, he and Buckingham (still the royal favourite and chief minister) decided to spend the funds voted by Parliament for the new war on two ambitious ventures: an army to raise the Spanish siege of the Dutch city of Breda and a fleet to capture the Spanish port-city of Cadiz. By the time Charles met his second Parliament in 1626, both operations had failed (Chapter 9) and these setbacks, combined with famine, plague and the failure to persecute Catholics produced widespread exasperation.

The English Parliament was a volatile body. With almost 150 peers and bishops in the House of Lords and over 500 members of Parliament (MPs) in the House of Commons, it formed the largest representative assembly in the early modern world. Securing approval for the crown's proposals and avoiding any attempt to exploit the crown's fiscal needs to extract concessions, therefore required careful management not only of general elections and of the two Houses, but also of the capital city in time of Parliament. No seventeenth-century government succeeded in these vital tasks, creating a fundamental and recurring instability at the heart of the state.

Charles summoned and dissolved Parliament three times between 1626 and 1629. In each instance, the House of Commons began by blaming the crown for disastrous policies and demanded redress for its grievances before it would vote money to fund the king's wars; Charles took offence at its demands, tried to bully it into submission by arresting recalcitrant members; and when those efforts failed, he dissolved it. Instead the king raised money for his war by demanding loans from leading subjects (and imprisoning those who refused) and by enforcing regalian rights; but his decision in 1627 to declare war on France as well created a critical need for parliamentary funding.

Much was at stake when Parliament assembled the following year. As one MP put it, 'This is the crisis of Parliaments. By this we shall know if Parliaments will live or die', adding that 'our lives, our fortunes, our religion, depend on the resolution of this assembly' because 'if the king draws one way, and the people another, we must all sink.' Some MPs nevertheless refused to vote further taxes until Charles had redressed what they saw as the abuses of previous years, invoking principles and precedents drawn from a careful study of history and the classics, until after several weeks of wide-ranging debate, MPs incorporated their various grievances into a single document, known as the Petition of Right. This document not only required the crown to cease billeting soldiers and sailors in private homes and to stop subjecting civilians to martial law, but also forbade the imposition of taxes without parliamentary consent and the imprisonment of any subject without showing due cause – two prerogatives that Charles regarded as an integral part of monarchical power. MPs also begged Charles to 'take a further view of the present state of your realme' and to dismiss Buckingham, whom they blamed for 'the miserable disasters and ill successe that hath accompanied all your late designes and actions'.[6] Once again the king responded with an immediate dissolution of Parliament.

By the time Charles made peace with both France and Spain in 1630, his debts totalled four times his annual revenues. In addition, both agrarian malaise and bad weather continued. The year 1629 brought 'so wonderful and great a flood as had not been seen of forty years'; 1630 saw widespread harvest failure; the summer of 1632 was 'the coldest that any then living ever knew'; spring 1633 was 'wet, cold and windy' and the following autumn 'a marvellous ill seed-time'; summer 1634 brought drought; and the following winter saw such 'very intense cold' that the Thames froze over. Then came two more summer droughts, that of 1636 so 'excessive' that 'everyone declares that there is no memory of such a misfortune in England, whose usually damp climate is so changed that the trees and the land are despoiled of their verdure, as if it were a most severe winter'.[7]

Nevertheless, Charles managed to govern without Parliament – in part because England's foreign trade expanded: since war ravaged Europe while Britain enjoyed peace, many merchants shipped their wares through English ports, paying customs duties. By 1639 the king had discharged his debts and raised his annual revenues to £900,000, almost twice as much as a decade before, thanks to increased customs duties and the ruthless enforcement of regalian rights, such as fines for encroaching on royal forests, fees from those eligible to become knights and, most lucrative of all, 'ship money', a levy to pay for the Royal Navy. Legal doubts surrounded all these extra-parliamentary income streams, but when a few courageous subjects challenged their legality Charles brought a test case to the law courts and in each case the judges ruled in his favour. The king's courts, formerly the arbiters of his subjects' lives, had become partisan protagonists. When in 1637 the judges upheld the legality of Ship Money, some landholders argued that their king was now 'more absolute than eyther France or the great duke of Tuscany'.[8]

Charles also strove to become more absolute in religious matters. Although he allowed the laws against Catholics to lapse, he vigorously persecuted Presbyterians. Throughout the 1630s his bishops enforced conformity to the doctrine and liturgy of the Church of England by summoning resisters – and there were thousands of them – to their courts and, when convicted, handing down harsh punishments. Those who remained unbowed came before the king's courts, especially the Courts of High Commission and Star Chamber, where they received even harsher punishments.

Although these legal victories by the king caused some to grumble, many of his subjects would probably have agreed with the claim of Edward Hyde, a prominent royalist and author of the influential *History of the rebellion and civil wars in England*, that in the 1630s England 'enjoyed the greatest calm and the fullest measure of felicity that any people in any age for so long time together have been blessed with'.[9]

The Scottish Revolution

Charles managed to offend his Scottish subjects from his accession in 1625, when as part of his mobilization plans for the war with Spain he resolved to create among his kingdoms 'a strict union and obligation each to [the] other for their mutual defence': a sort of Union of Arms inspired by Spain's ill-fated scheme (Chapter 9). To obtain funds for the Scottish component of the Union army, the new king announced an Act of Revocation, a device traditionally used by Scottish monarchs at their accession to reclaim lands usurped from their immediate predecessor; yet despite numerous precedents, the manner in which Charles presented his initiative provoked widespread opposition. Just like the Edict of Restitution promulgated in Germany that same year, the Revocation subsequently appeared even to loyalists as the 'root of all evils' (page 176). Looking back in the 1640s, the historian Sir James Balfour saw in it 'the groundstone of all the mischief that followed after, both to this king's government and family', and believed that it 'laid open a way to rebellion'.[10]

The resentment generated by the Revocation had scarcely abated before Charles took steps to ensure that just as his monarchy 'has but one Lord and one faith, so it has but one heart and one mouth . . . in the churches that are under the protection of one sovereign prince'. The king sought to end the 'diversity, nay deformity' of worship he had observed when he visited Scotland in 1633 because 'no set or public form of prayer was used, but preachers or readers or ignorant schoolmasters prayed in the church' extemporaneously. He charged the energetic but inflexible William Laud, archbishop of Canterbury, with devising a remedy.[11]

Laud prepared a Code of Canons for Scotland that prohibited extempore prayer and other ancient liturgical customs, which Charles published by virtue of 'our prerogative royal, and supreme authority in causes ecclesiastical' – apparently forgetting that the Church of Scotland recognized no such supreme authority. The king also ordered every church to buy and use a Prayer Book and, when reminded that no Scottish Prayer Book existed, he instructed Laud to prepare one that contained set prayers and responses based on (albeit not identical to) English practice. Then, using 'Our royall authority, as king of Scotland', Charles enjoined the universal and exclusive use of the new Prayer Book, with effect from Sunday, 23 July 1637. Clerics who failed to acquire a copy and to use it on that date would be declared rebels and outlaws.[12]

The king had chosen a dangerous moment to innovate because Scotland, too, suffered economic and climatic adversity. The 1630s was the second coldest decade in the past 800 years, and in June 1637 the Privy Council in Edinburgh issued emergency legislation to deal with a plague epidemic, an acute shortage of coins and a universal 'scarcity of victuals' because of the

poor harvest. Small wonder, then, that imposing the new Prayer Book, derisively known as Laud's Liturgy, unleashed a revolution – especially since its
opponents were already well prepared.

In April 1637 a group of ministers led by Alexander Henderson (a man
of obscure origins whose abilities as a preacher and organizer would soon
catapult him to international prominence) met secretly in Edinburgh with
the 'matrons of the kirk' (the wives of prominent laymen) and warned them
that the king aimed to abolish Scotland's traditional forms of worship, and
thus imperil their chances of salvation. A strange accident provided irrefutable confirmation of Henderson's claim: once the government's printer had
corrected the proof sheets of Laud's Liturgy, he discarded them – but, since
good paper was valuable, the sheets were promptly used by 'the shops of
Edinburgh to cover spice and tobacco' and so became public knowledge,
convincing everyone that 'the life of the Gospel' would be 'stolen away by
enforcing on the kirk a dead service book'. The matrons of the kirk therefore
ordered their maidservants to protest whenever it was first used.[13]

The maidservants obliged. No sooner had the dean of Edinburgh begun
to use Laud's Liturgy at the morning service in St Giles Cathedral on Sunday,
23 July than the young women sitting on their folding stools at the front
'with clapping of their hands, cursings and outcries, raised such a barbarous
hubbub in that sacred place that not any one could either hear or be heard'.
The protestors then 'threw the stools they sate on at the preacher' and then
'did rive [rip] all the service bouk[s] a peisses'. The dean, together with the
judges and magistrates at the service, ran for their lives; and when they tried
to use the new Prayer Book in the afternoon, the crowd stoned them.[14]

Charles responded by commanding the Scottish Privy Council to punish
all 'authors or actors' of these disturbances, and to enforce use of the new
Prayer Book forthwith. The council duly summoned the leading members of
the Edinburgh clergy – but instead of obeying the monarch, it determined that
'the service books cannot be orderly used in the kirks', and authorized ministers to continue to preach in the traditional form. It also set free those imprisoned for riotous conduct.[15] Henderson and his colleagues used their reprieve
to draw up a 'supplication' against religious innovations to be presented to the
king in the name of the godly nobles, burgesses and ministers of Scotland.
Charles regarded this protest as sedition and ordered the committee to
disperse; but instead Henderson, ably assisted by Archibald Johnston of
Wariston, a determined and devout Edinburgh lawyer, drafted a document
that they called the National Covenant. Although it claimed to safeguard
'the true worship of God, the majesty of our king, the peace of the kingdom,
for the common happiness of ourselves, and the posterity', its content was
profoundly subversive since it condemned all innovations in ecclesiastical and
secular government made since the Union of 1603. Moreover, it obliged every
Scottish householder to take a solemn and public oath that they would 'to the

uttermost of our powers, with our means and lives' defend 'the foresaid true religion, liberties and laws of the kingdom *against all sorts of persons whatsoever*' – a formula that could be used to justify rebellion.[16]

On the third Sunday in March 1638, as arranged by Henderson and his colleagues, in each Scottish parish the congregation rose to its feet and, with right hands raised, repeated in unison the oath to uphold the Covenant 'against all sorts of persons whatsoever'. They then signed their names, after which (Wariston reported) 'such a yell' came from the throats of the assembled crowd 'as the like was never seen or heard' – and indeed Scotland, perhaps the world, had never seen such an exercise in popular democracy. After more sermonizing, a messenger set forth for London bearing both the Covenant and a list of eight demands (composed by Wariston) 'containing the least of our necessary desires to settle this church and kingdom in peace' to present to the king. Wariston then 'prayed the Lord to preserve us from that great sin of retiring one single inch in this cause of God out of diffidence and worldly fears'. For him, at least, there would be no surrender and no compromise.[17]

Charles, too, showed intransigence. He informed the marquess of Hamilton, whom he selected to serve as his personal representative in Scotland, that since nothing 'can reduce that people to their obedience but only force', he 'would rather die than yield to those impertinent and imperious demands' – although for the time being he cynically authorized Hamilton to make empty concessions to the Covenanters. 'Flatter them with what hopes you please', he wrote, 'your chief end being now to win time ... untill I be ready to suppress them'. In October 1638, since it was too late for military intervention, Charles even gave his consent for the General Assembly of the Scottish Church to convene for the first time in twenty years, and hundreds of ministers and pious laymen (many of them heavily armed) attended its opening session. 'It is more than probable that these people have somewhat else in their thoughts than religion', warned Hamilton, who presided over the assembly in the king's name: religion, he feared, 'must serve for a cloak to rebellion'. He, too, now saw no way for the king to regain control except through a full-scale invasion.[18]

Wariston and his colleagues studied carefully the theories of political resistance advanced by continental writers and, thus enlightened, they spread their views through pamphlets and preaching. They also cultivated links with known opponents of Charles's policies in England (in the hope of stirring up opposition that might divert Charles from attacking Scotland); and persuaded Chancellor Axel Oxenstierna, who equated Scotland's cause with Sweden's war of independence over a century before, to send them almost thirty heavy guns, 4,000 muskets and 4,000 suits of armour, and to allow over 300 Scottish officers in Swedish service to return home, including General Alexander Leslie, a veteran with thirty years' experience of continental warfare.

The Tipping Point

This gave the Scots a critical military advantage in opposing the king, who had decided to lead 20,000 men to the Scottish border in person, while an Irish army invaded the southwest and Hamilton with the Royal Navy block-aded the east coast and landed troops to assist opponents of the Covenant in the northeast. It was a promising strategy, but extreme weather delayed the mobilization of Charles's English army. Spring 1639 saw a 'most grievous tempest of wind, thunder, lightning and rain', then a ten-week drought followed by 'the greatest wind that ever I heard blow', and finally 'abound-ance of raine [which] made foule travelling'. 'I feare', a royalist commander fretted, that if the cold 'continues, it will kill our men'.[19] Nevertheless, at the end of May 1639, Charles reached the river Tweed, the border between the two kingdoms, where he and his 20,000 soldiers fortified a camp facing a Scottish army entrenched just across the river.

The king's inexperience allowed Leslie to trick him. According to an English officer, 'the great bruite [rumour] of the ennemye's strength, and their able commanders, did beget a distrust in most, and a murmure in others'. The English did not discover until later that Leslie had drawn up his troops expressly 'to beguile men's view' and prevent them from realizing that they faced scarcely 12,000 half-starved Scots, many of them poorly armed. Therefore, instead of leading his far larger army in an attack, on 18 June 1639 Charles signed a ceasefire and opened negotiations with his rebellious Scottish subjects.[20]

Just as he had insisted on leading his army in person despite his lack of military experience, Charles now insisted on conducting the negotiations in person, despite his lack of diplomatic experience. The Scots insisted that their sovereign ratify the acts of the last General Assembly of the Church (which meant abolishing all bishops), summon a new Parliament, and return for trial and punishment the 'incendiaries' (as they termed the king's supporters who had fled to England. The Pacification of Berwick eventually granted all these demands and required the king to demobilize his army and lift the naval blockade.

Charles had just made, in John Adamson's words, 'the greatest single mistake of his life'.[21] By failing to exploit his military superiority, he forfeited his best (if not his only) chance of victory over his Scottish rebels; and by retreating from his earlier stated position ('I would rather die than yield to these impertinent and damnable demands') he discredited not only himself but also his leading advisers. As Hamilton predicted, the Covenanters 'will give no credit' to anything he said, 'but will still hope and believe that all their desires will be given way to'.[22] Equally damaging, the Scottish crisis generated numerous pamphlets in England that initiated debate on issues that had previously been taboo: liturgy and church government, the limits of authority and obedience, even justifications for resistance. Worst of all, Charles was now bankrupt: the campaign itself, which had cost £1 million, drained the English treasury of its reserves; and the king's craven concessions emboldened many to withhold payment of ship money and other regalian rights (and led many royal officials to leave offenders alone, fearing another change of the royal mind). Tax revenues therefore declined sharply. To solve these problems Charles turned to Thomas Wentworth, lord deputy of Ireland.

Charles had sought to fund his Union of Arms scheme in Ireland by offering Catholics there concessions, known as the Graces, in exchange for new taxes to pay for the island's defence. Among other things, he promised to relax the requirement that all holders of public office must recognize the king as Supreme Governor of the Church of Ireland (something no Catholic could do), and also to guarantee the titles of all families that had held their lands for sixty years (which would virtually end the further creation of plantations). As soon as he had made peace with France and Spain in 1630, however, Charles reneged on these promises and instead ordered strict application of anti-Catholic laws, commanded the dissolution of all Catholic convents, and required all magistrates either to take the Oath of Supremacy (recognizing Charles as Supreme Governor of this Church) or be dismissed. The king entrusted enforcement of these measures to a group of militant Protestant landholders, who carried out their work with efficiency and enthusiasm – until Charles reversed course yet again in 1632, when he named Thomas Wentworth, an Englishman with extensive administrative experience but no Irish connections, to serve as lord deputy.

Wentworth realized that the royal innovations would alienate most segments of Irish society, but he relied on their deep-seated mutual hatred to prevent cooperation. In this he succeeded: by 1639 none dared defy him and the Irish treasury had accumulated a small surplus. He therefore left Dublin for London full of confidence and, supported by Laud and Hamilton, he persuaded Charles to reconvene the English Parliament and demand funds for a new invasion of Scotland. Wentworth, now raised to the peerage as earl of Strafford, returned to Dublin and persuaded the Irish Parliament to authorize taxes sufficient to raise an army of 8,000 foot and 1,000 horse for the same purpose, and mobilization began at once. As the new troops converged on Ulster, the Irish province nearest to Scotland, the new earl returned to London just in time to take his seat in the first English Parliament to meet in eleven years.

England on Edge

The 500,000 Englishmen entitled to vote in elections seized the unexpected opportunity of a new Parliament to protest against the controversial policies of the previous decade, rejecting candidates who had collected regalian rights or enforced Laud's liturgical innovations. Instead, their representatives took with them to Westminster long lists of grievances which they debated at length, ignoring the government's pleas for new taxes. Charles tolerated this irritating behaviour until he learned on the morning of 5 May 1640 that the House of Commons planned to debate a motion urging 'reconciliation with . . . his subjects in Scotland'. Since this would destroy the entire moral foundation of his Scottish policy, the king dissolved the assembly, soon to be known as the Short Parliament.[23] Angry crowds soon roamed the streets, and a group of about 500 surrounded Lambeth Palace, Laud's official residence as archbishop of Canterbury, because they blamed him for the king's decision to dissolve the assembly. It was the first major episode of mob violence in the Stuart capital; it would by no means be the last.

Laud was not in Lambeth Palace because, as soon as Charles returned from dissolving Parliament, he summoned his archbishop to a war committee convened to discuss whether, despite the lack of parliamentary funds, 'the Scotts are to bee reduced or noe?' According to notes taken at the meeting by Sir Henry Vane, the secretary of state, some councillors favoured a compromise – 'If noe more mony than what proposed, howe then to make an offensive war?' one asked – but Strafford dismissed this concern because a 'defensive warr' would involve 'losse of honor and reputacon'. He then argued that since 'the quiett of England will hold out longe', the king should 'goe on with a vigorous warr, as you first designed'. He also stressed that 'You have an army in Ireland, which you may imploy here to reduce this kingdome' (the ambiguity of 'here' and 'this kingdome' would come back to haunt him). 'Scotland shall not hold out five monthes', Strafford predicted: 'One sumer well imployed will doe it.' The earl also argued that the king was now 'absolved from all rules of government, beinge reduced to extreame necessitie', and 'everythinge is to be done as power might admitt' – that is: since Parliament had refused its support, the king could and should fund his army via regalian rights and forced loans from the leading London merchants.[24]

Failure to raise sufficient funds by these means left Charles dangerously exposed. He had ordered 35,000 English troops to mobilize at York for service against Scotland and could hardly abandon his war plans without 'losse of honor and reputacon'. His decision to postpone the rendezvous until mid-August multiplied the risks because an outbreak of plague prevented some recruits from reaching York, and thanks to the El Niño episode of that year much of England experienced a very late spring with 'an abundance of rains and cold winds'. Even Strafford, appointed to lead the king's army, arrived late: by 24 August 1640 he had only reached Huntingdon, where he found 'the waters mightily risen and the ways as foul as Christmas'.[25]

In Dublin, with Strafford absent, news of the dissolution of the Short Parliament led to a hiatus in collecting the taxes already voted. This delayed mobilizing the army designed to invade Scotland from Ulster, so that although infantry regiments had assembled by June, the cavalry had not. In addition, they lacked both weapons and transport to Scotland. The troops therefore remained in Counties Antrim and Down, consuming local supplies already much reduced by a run of bad harvests.

The Scottish Parliament reassembled and, in the words of Sir James Balfour, took advantage of these setbacks to overturn 'not only the ancient state government, but fettered [the] monarchy with chains'. It passed a Triennial Act, requiring an assembly to convene at least once every three years, with or without a royal summons, excluded all bishops from the assembly and created an elaborate structure of standing committees to govern Scotland whenever Parliament was not in session. The leading Covenanters also received a secret letter signed by seven English peers promising, if the Scots invaded, to 'unite themselves into a considerable body, and to draw up a Remonstrance to be presented to the king' that would contain the grievances of both Scotland and England, to which they would 'require' (not request) redress.[26] The treasonable promise of collusion by the seven peers persuaded the Scots to make a pre-emptive strike. On 20 August 1640, General Leslie, who had spent the previous year training his soldiers, led 18,000 men across the Tweed. Scotland and England were at war again.

Charles left London that same day for York. There he 'spake with the lords, colonels and gentlemen', urging them to march against the Scots; but instead of mobilizing, on 28 August they sent Charles a 'humble petition' protesting that the previous year's campaign had cost the county over £100,000, so that 'for the future, the burden is so heavy that we neither can, nor are able, to bear it'.[27] In the meantime, Leslie led his army across the Tyne, routed the small English force facing him at the battle of Newburn and captured Newcastle, before sweeping south to capture Durham as well. The king, leading his army from York towards Durham in person, panicked and fled.

Many in England openly rejoiced at their king's defeat. When news of Newburn arrived in London, church bells rang out in celebration and Laud warned his fellow councillors that 'We are at the wall', and that the only way to organize effective resistance to the Scots was by summoning a council of peers 'or the calling of a Parliament'.[28] The dissident English peers honoured their promise to their Scottish colleagues and sent to the king a petition asserting that 'By occasion of this war, your revenue is much wasted, your subjects burdened with ... military charges ... and your whole kingdom become full of fear and discontents'. To solve these problems, the peers proposed a simple and immediate solution: peace with the Scots. At greater length, they also complained about 'the sundry innovations in matters of religion'; 'the great increase of Popery, and employing of Popish recusants';

the rumours 'of bringing in Irish and foreign [i.e. Catholic] forces'; 'the urging of Ship Money' and imposition of sundry taxes on 'the commodities and manufactures of the kingdom'; and 'the long intermission of Parliaments'. To address these issues, the peers demanded that Charles

> summon a Parliament within some short and convenient time, whereby the causes of these and other great grievances which your people lie under may be taken away, and the authors and counsellors of them may be there brought to such legal trial and condign punishment as the nature of the several offences shall require.[29]

On 5 September 1640, Charles summoned all the English peers to meet him in York.

Foul weather continued to impede the transaction of public affairs – 'we have had so great rains these two days, and the waters are so out', Vane complained, 'that there is scarce means to pass anywhere upon the roads' – but in late September just over seventy peers had joined their king. In his opening speech to the assembled nobles, Charles announced that 'I desire nothing more than to bee rightly understood of my people, and to that end I have of myself resolved to call a Parliament.' It would assemble in Westminster on 3 November 1640.[30] The king next asked for advice on what to do about the Scots and expressed the hope that his nobles would fund a campaign to avenge Newburn. They refused, and instead insisted on the appointment of sixteen peace commissioners with full powers to conclude an armistice with the Scots, their number to include all of the seven peers who had urged the Scots to enter England – men who could hardly deny the invaders' demands, given that the Scots possessed their treasonable letter.

The king had surrendered a significant part of his prerogative – the power to make war and peace – and the peace commissioners used their newfound authority to conclude the Treaty of Ripon, which made crucial concessions to the Scots. It left them in control of the Tyneside collieries, on which London depended both for its manufactures and for heating – a measure that gave the Scots critical leverage because, until Parliament authorized the taxes necessary to buy them off, the capital would starve and freeze. It stipulated that negotiations for a final settlement would take place at Westminster, which gave the dissident English peers and their allies in Parliament and in Scotland a unique opportunity to reshape the entire political structure of Charles I's monarchy. It also promised the Scots the enormous sum of £850 a day to maintain their soldiers in England until a permanent settlement could be arranged – a requirement that prevented Charles from dissolving the new Parliament until it had voted sufficient funds to disband both of the armies in England and also Strafford's army in Ulster.

In his opening speech to Parliament on 3 November 1640, Charles could not refrain from reproaching his audience. Had the previous assembly believed him, he chided, 'I sincerely think that things had not fallen out as now we see', and he called for an immediate vote of funds for a campaign to expel the invaders. His belligerent stance did not lack supporters: a number of MPs felt either bound to obey the king, right or wrong, or insulted by the Scots' success ('if the Scots should be too refractory, [let us] bring them by force of true English courage to reason'; 'We should get them out by fair means or foul').[31]

This emboldened Charles to summon Strafford (still commander of the royal armies in both England and Ireland and still the strongest advocate of resuming the war against the Scots) back to London, where he immediately started to strengthen the Tower's defences. Rumours spread that he and the king were preparing charges of treason against those whom they now knew had corresponded with the Scots, and on 11 November the MPs therefore made a pre-emptive strike, accusing the earl himself of high treason. He was immediately arrested, halting not only plans for a third campaign against the Scots but also the transaction of most public business.

Furious at this turn of events, the king decided both to dissolve Parliament and to summon officers serving with his army in Yorkshire to rescue the earl from prison; but the news leaked, and thousands of Londoners rushed to Westminster to form a human shield while the Commons finalized a Bill of Attainder against Strafford. It passed by 204 votes to 59, revealing how far Charles's actions had shifted opinion against the earl and his master. On 1 May 1641 the king paid another visit to Parliament to explain that he did not believe Strafford to be guilty of high treason and to express the hope that the two Houses would vote for the lesser charge of misdemeanours. His audience remained unmoved. As the earl of Essex (whose own father had been executed for treason) put it, if Strafford lived, Charles would restore him to his former offices 'as soon as the Parliament should be ended', and then he would seek his revenge. By contrast, said Essex, shaking his head as he uttered his most memorable phrase: 'Stone-dead hath no fellow'.[32]

Perhaps realizing that his speech to Parliament had missed its mark, Charles initiated another plot to seize control of the Tower and free his faithful minister, but again news leaked out and by nightfall on 2 May 1641 a crowd of about 1,000 had gathered near the Tower to make sure that the plotters stayed outside while Strafford stayed inside. The following day, a crowd of some 15,000 people assembled at Westminster, both to protect Parliament and to protest the king's attempted coup. 'Surely', the craftsman Nehemiah Wallington wrote in his diary, 'I never did see so many together in all my life. And when they did see any Lords coming they all cried out with one voice "Justice! Justice!"' One of the lords claimed that someone in the crowd warned him that 'if they had not justice tomorrow, they would either take [i.e. lynch] the king or my Lord Strafford' – the first recorded suggestion of regicide as a solution to England's political problems. John Pym, one of the most effective speakers in the House of Commons, convinced his colleagues that they faced a 'popish conspiracy' that aimed 'to subvert and overthrow this kingdome'. He also exploited the general panic to rush through a document entitled the Protestation, with authorization to print enough copies for every parish in the kingdom.[33]

Like the Scottish National Covenant (which probably served as its model), the Protestation required a universal public pledge to defend the established church against its enemies and to seek punishment for all who had endeavoured 'to subvert the fundamental laws of England and Ireland, and to introduce the exercise of an arbitrary and tyrannical government'. Again as in Scotland, the clergy read the document aloud to their congregations from the pulpit before subscribing their names; and then they called upon 'all masters of families, their sons and men-servants' to 'subscribe his name, with his own hand or mark' in a special register. Tens of thousands, including apprentices and servants, duly swore the oath and affixed their signatures or marks, and then flocked 'in troops to the Parliament house with the Protestation on the top of their swords', while the city militia swaggered through the streets with the 'Protestation fastened to their pikes or hats'. Over 3,000 of the Protestation returns survive today, bearing the names of some 37,000 individuals. 'Never before', wrote David Cressy, 'had so many subjects been invited to act as citizens, regardless of rank.'[34]

On 7 May the House of Lords approved Strafford's Bill of Attainder and a delegation from both Houses set out to present it to Charles, along with another bill that forbade the dissolution of Parliament without its own consent. A crowd estimated at 12,000 escorted the delegation to Whitehall Palace, which they then blockaded for thirty-six hours, shouting slogans, until Charles – having begged his bishops and ministers to find a way to save Strafford, and having wept at the council table when they could not – at last signed both documents. The crowd now swelled to perhaps 200,000, and gathered around the scaffold on Tower Hill to gloat over the earl's execution. Afterwards those who had come from out of town to see the spectacle 'rode in triumph back, waving their hatts, and with all expressions of joy, thro' every town they went, crying "His head is off, his head is off! . . . and breaking the windowes of those persons who would not solemnize this festival with a bonfire'.[35]

Parliament had already sat continuously for longer than any previous assembly, but having gained the upper hand it now made the most of its advantage. The day after Charles signed Strafford's death warrant, Parliament approved a draft peace treaty with the Scots; and the day after the earl's execution, it voted the funds required to demobilize the English and Scottish armies in the north. Until those funds arrived, however, in Vane's anguished words 'We are here still in the labyrinth and cannot get out.' Soon after it assembled, Parliament had urged 'persons in every county of the kingdom' to exploit 'the present opportunity, by giving true information' to document cases of royal misrule during the previous decade: almost all of England's forty counties obliged, presenting petitions that bore over half a million signatures.[36]

England had seen nothing like this exercise in direct democracy, and what became known as the Long Parliament used the petitions to prepare a stream of legislation designed to destroy both the agents and the apparatus that had enabled Charles to govern England for eleven years without their participation. The two Houses had brought impeachment proceedings against over fifty senior ministers, including Archbishop Laud, twelve bishops, and almost half the judges, many of whom languished in the Tower pending trial; and had submitted bills for the king's signature that declared most of his regalian rights illegal and abolished both Star Chamber and High Commission. Charles grudgingly assented to all these measures and also approved a peace treaty that ended the 'late troubles' with his Scottish subjects, one of whose clauses forbade the king to make 'war with foreigners without consent of both Parliaments'. On 11 August Charles left London to oversee the demobilization of the Scots and English armies.[37]

If the king's opponents had been able to stop there, they might have retained the upper hand. Instead, the composite nature of the Stuart monarchy – which had previously been such an asset to them – turned into an asset to the king. The Scots now insisted on immediate compliance with their demand for 'one confession of faith, one form of catechism . . . and one form of church government in all the churches of His Majesty's dominions' – modelled, of course, on Scotland's Presbyterian system. It could not be done. Ireland's Catholic majority was invincibly opposed, and few of Charles's English opponents were Presbyterians.[38] Moreover, unlike the Scots, England lacked a single rallying point: instead of the National Covenant, there were two rival documents, the Protestation and the Prayer Book. Choosing between them would divide communities and even families throughout the kingdom, and Charles decided to build up a royalist faction in both Scotland and Ireland to exploit these divisions.

The Irish Revolution

The Irish political elite were deeply troubled by these developments in Charles I's other kingdoms. One protagonist in the Irish rebellion later recalled hearing 'that the Scotts hadd peticioned the Parliament howse of England that there should not bee a Papist left alive either in all England, Ireland or Scotland', so that the only effective response was 'to ryse vpp in armes and take all the stronghouldes and forts into their handes'.[39] Ireland's Catholic leaders were therefore receptive to hints from Charles that if they organized military and financial support for him against his enemies in Britain, he would confirm the Graces (page 258). Working principally through the earl of Antrim, with whom he enjoyed a close relationship, the king apparently approved a plan 'that the castle of Dublin should be surprised and seized' and an army of 20,000 men mobilized to be 'employed against the [Irish] Parliament' and then 'against the Parliament of England if occasion should be for so doing'. In other words, Charles envisaged starting a civil war first in Ireland and then in England.[40]

Antrim did his best, but again the climate intervened. Adverse weather ruined the harvest in Ireland in 1641, as it had done in 1639 and 1640, causing widespread food shortages. The province of Ulster suffered worst because the presence of the troops raised by Strafford for the invasion of Scotland soon consumed all available resources. Land rents fell by half, creating great tensions between natives and newcomers. In this highly charged atmosphere, in August 1641 the king gave his consent for the Irish Parliament to debate the Graces; but when news of this concession arrived in Dublin, the Lords Justices (a commission of Protestants entrusted by Charles with governing Ireland after Strafford's fall) immediately dissolved the assembly.

This surprise move, which blocked the road to constitutional reform for the foreseeable future, outraged Catholic members of the Irish Parliament. Seeing that the Scots' armed insurrection had secured concessions, first from the king and then from the English Parliament, they concluded that only military strength could now overthrow 'the tyrannicall governement that was over them' and decided to 'imitat Scotland, who gott a privilege by that course'.[41] Over the next few weeks, one group of conspirators under Connor, Lord Maguire, made plans to take Dublin Castle, while another led by Sir Phelim O'Neill, a prominent landowner and a justice of the peace, would seize all the fortresses in Ulster garrisoned by Protestants. They agreed to act simultaneously on 23 October, a market day in Dublin, which would make the arrival of the conspirators in the capital the night before less conspicuous. Maguire intended to equip his followers with the weapons stored in Dublin Castle and then force the English government to grant religious and political freedom, but his plot was betrayed by one of the conspirators on the night of 22/23 October. The Lords Justices immediately arrested Maguire and sent urgent messages to warn Protestant officials in the provinces. They arrived too late. Just as the Dublin plot was discovered, O'Neill and his allies used a variety of ruses to capture the major fortresses in Ulster, and insurgents elsewhere immediately declared their opposition to British rule. Well over one hundred members of the Irish Parliament eventually joined the rebellion.

As with the Scots Covenanters, the initial aims of the Irish confederates were conservative: they did not seek the return of forfeited lands, only an end to further plantations; they did not demand independence from England, only an end to London's power to alter the status quo; and they did not strive to overthrow Protestantism, only to end the persecution of Catholics. In the absence of Maguire, however, groups of Catholics exploited the temporary collapse of public authority to settle scores with local Protestants.

Although several confrontations were intensely personal – some attackers stabbed, hanged, burned or drowned neighbours whom they had known for years (Chapter 17) – most Catholics did not intend to kill their victims but rather to humiliate and expel them by stripping them and turning them out to fend for themselves. The Little Ice Age often rendered these activities lethal, however. Extremely cold weather prevailed in October 1641, just as the rebellion began, heralding 'a more bitter winter than was of some years before or since seen in Ireland', with severe snow and frost afflicting the whole island. Therefore the 'stripping of soe many thousands, men, women and children of all sorts and ages, in such season of the year would have infallibly killed them' as they tried to flee.[42] The surviving accounts of those affected by the uprising record more deaths from 'snow and frost' and 'extreme cold' than directly from violence, indicating that the Little Ice Age at least doubled – and may have more than doubled – the number of Protestants who met an unnatural death in autumn 1641.

The most harrowing and heart-wrenching (and for English readers, the most inflammatory) accounts involved the suffering of women and children. A Protestant sailor recorded how, shortly after the uprising began, he and 'his wife and five smalle children' were 'stript of all their clothes' by their Catholic neighbours. That night, 'flying away for safftie naked in the frost, one poore daughter of his, seeing him and her mother greeve for their generall misery, in way of comforting said she was not cold, nor would crye', but immediately afterwards 'she died by that cold and want. And the first night this deponent and his wife, creepeing for shelter into a poor [shack], were glad to ly upon their children, to keep in them heate and save them alive.'[43]

Sectarian passion also increased the death toll. Some of the Catholic clergy, especially in Ulster, presented the rising as a Crusade, a chance to regain Ireland for the True Faith, and encouraged Catholic gangs to round up Protestant settlers (Scots as well as English) and either stab them to death, burn them alive in their houses or drive them into icy water where they perished. As soon as they could, Protestants responded in kind, ordering the troops 'sent into the enemies [Catholic] quarters to spare neither man, woman nor child'.[44]

How many died in the violence? Few paused to count the corpses at the time, and some who did found the task overwhelming. The most vivid

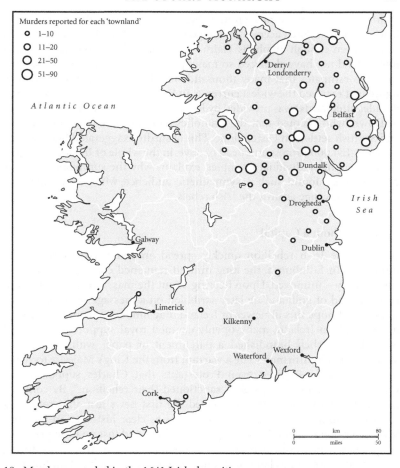

Murders reported for each 'townland'
- 1–10
- 11–20
- 21–50
- 51–90

Atlantic Ocean

Derry/
Londonderry

Belfast

Dundalk

*Irish
Sea*

Drogheda

Galway

Dublin

Limerick

Kilkenny

Waterford   Wexford

Cork

19. Murders recorded in the 1641 Irish depositions.
The circles show the number of murders reported for each 'townland' in the
depositions taken from over 3,000 survivors (almost all of them Protestants) by a
team of judges in the immediate aftermath of the Irish uprising in October 1641.
About one-fifth of those interviewed reported deaths. Almost all occurred in areas
(notably Ulster) where Catholic natives outnumbered Protestant newcomers by
two or three to one. In all, well over 4,000 perished.

memory of Anthony Stephens, a farm hand, was seeing about 140 people
at Coleraine buried 'in one deepe holle or pitt, and layd soe thick and
closse together as he may well compare it *to the makeing or packing up of
herrings*' – a peculiarly disturbing image. In all, Stephens was 'perswaded
there died noe fewer within three months after the begining of the Rebellion

within the said towne of Colraine then seven or eight thowsand of the Brittish nation'.[45] Such accounts outraged the many British Protestants who read it, even though Stephens' claim was impossible – Coleraine, a small town, could not have sheltered so many people. What mattered at the time were the estimates that many thousands had been cruelly murdered. The figure that received the widest currency in Britain at the time (and for many decades afterwards) was provided by Reverend Robert Maxwell, archdeacon of Down, who claimed that the Catholics had massacred 154,000 English and Scottish settlers in Ulster alone. This absurdly exaggerated figure (there were not 154,000 Protestants, dead or alive, in the whole of Ireland), coupled with horrifying individual examples, explains why the survivors and their co-religionists found such a sympathetic audience when they called for immediate revenge against the Irish rebels.[46]

A King Without a Capital

News of the Irish rebellion quickly spread around Charles's composite monarchy. In Edinburgh, the king himself remained curiously – to some, suspiciously – unmoved. Upon hearing about the massacre, he went out to play a round of golf; and he later scribbled on a message from one of his ministers, 'I hope this ill newes of Ireland may hinder some of theas follies in England.' In Ireland, many openly claimed royal support. O'Neill and other Ulster rebels brandished 'a parchment or paper with a great seal affixed which he affirmed to be a warrant from the King's Majestie for what he did', which convinced even Protestants that Charles supported the Catholics and had perhaps even sanctioned their rebellion.[47] By contrast, the English Parliament, which received the first news just after returning from its summer recess, saw the massacres as clear justification for their fears of a general Catholic uprising against them and lost no time in organizing countermeasures. It resolved to mobilize Scottish help in restoring Protestant control in Ireland, and solicited loans from leading Londoners to raise and arm troops for an immediate counter-attack. But who would control these soldiers?

John Pym, now so prominent in parliamentary business that he was known as King Pym, feared that Charles might use any troops raised for Ireland against his English opponents, and so he compiled a remonstrance with 204 individual points, asserting that without redress of outstanding grievances 'we cannot give His Majesty such supplies for support of his own estate, nor such assistance to the Protestant party beyond the sea [in Ireland], as is desired'. The 204 points included not only the demands for religious uniformity made by the Scots but also many constitutional novelties based on Charles's concessions to the Scots, such as the requirement he appoint only officials approved by Parliament.[48]

Like the Petition of Right in 1628 (page 252), the Remonstrance of 1641 situated individual actions by Charles within the overall framework of a Catholic conspiracy to subvert the 'fundamental laws' and religion of England and Ireland. Not all MPs accepted this, and after fourteen hours of bitter debate, it passed the Commons by only 159 votes to 148; but Pym made sure that copies were available for purchase the following day, 24 November 1641. On the 25th Charles entered London, escorted by over 1,000 soldiers from the recently disbanded northern army.

For the next six weeks rioting rocked the capital. Gangs of unemployed young men roamed the streets of London shouting 'down with the bishops, hang up the popish lords'. The king responded in kind: he ordered the lord mayor to 'kill and slay such of them as shall persist in their tumultuous and seditious ways and disorders'; he commanded his courtiers to start wearing swords; and he built a barracks just outside Whitehall Palace to accommodate the soldiers he had brought with him from Yorkshire. Clashes between anti-royalist gangs and Charles's guards steadily increased until, following the worst frosts in living memory, early in January 1642 some 200 Londoners armed with staves and swords marched through the cold to Whitehall shouting anti-Catholic slogans. One threw a 'clot of ice' at the soldiers guarding the palace gates, who promptly gave chase and injured several civilians.[49]

On 3 January 1642 the Commons asked the London magistrates to call out the city's Trained Bands (its militia) to protect them, but Charles forbade this move. Instead he presented to the House of Lords articles of impeach-ment against one peer and five MPs, ordered his agents to seal and search their residences and sent a messenger to Parliament to demand their imme-diate arrest. The king's printer published and distributed the articles of impeachment against them. The House of Commons responded by ordering the unsealing of the residences, refusing to deliver the five MPs and calling for the printer of the 'scandalous publication' to be punished.

This triple slap in the royal face, combined with unseasonable floods that prevented about 200 MPs from returning to the capital after the Christmas recess, encouraged Charles to undertake a coup d'état. According to one source, it was his wife Henrietta Maria who triggered this disastrous course of action: 'Go, you coward', she allegedly yelled at him, 'and pull those rogues out by the ears, or never see my face more.'[50] Unfortunately for her plan, one of the queen's confidantes overheard this exchange and sent a messenger with 'timely notice' of Charles's intentions to the House of Commons. Heavy rains had turned the streets of London into a quagmire and so on the afternoon of 4 January the messenger got from Whitehall to Westminster faster than Charles and his troops. As his soldiers brandished their weapons at the door to the Commons chamber, Charles entered and 'commanded the Speaker to come out of his chair, and sat down in it himself, asking divers times whether these traitors were there'. When no one replied, he carefully scrutinized the faces in the chamber before uttering his most famous words, 'All my birds are flown'; after which he rose and returned empty-handed to Whitehall.

Charles's flagrant breach of parliamentary privilege caused the Commons to cease their deliberations, and as he returned through the streets to Whitehall, the king found that shopkeepers had pulled down their shutters and stood menacingly at their doors bearing arms. Worse, 'the rude multitude followed him, crying again "Privileges of Parliament! Privileges of Parliament!"' and clutching the Protestation. Charles experienced 'the worst day in London', according to one eyewitness, 'that ever he had'.[51]

Charles's attempted coup suggested that he was prepared to use violence against his English subjects, just as he had done against the Scots. On the night of 5/6 January 1642 rumours flew 'that there were horse and foot coming against the city. So that the gates were shut and the [port]cullices let down and the chains put across the corners of our streets, and every man ready in his arms'. In open defiance of the king, the London magistrates at last called out the Trained Bands, which escorted the five MPs denounced by the king as traitors back to Westminster in triumph. Since they now heavily outnumbered the 'cavaliers' (as Charles's swordsmen were known), on 10 January the king fled with his family to Windsor Castle. Henrietta Maria fully realized the significance of this move: her husband, she told an ambassador, 'was now worse than a duke of Venice'.[52]

Charles I: A Problematic King

The newly created 'kingdom of Great Britain' was a composite state, and therefore had a lower political boiling point than other polities (Chapter 3). Composite states required particularly sensitive handling when a ruler embarked on war, as Charles I did between 1625 and 1630 and again in 1639 and 1640. One can reasonably object that no sovereign could have foreseen the extreme weather that would complicate military operations, but Charles could hardly plead ignorance of the fact that *any* war would force him to raise new taxes, and that this would inevitably cause both a clash with the House of Commons and popular resentment; yet he decided to press ahead.

Charles likewise seemed oblivious to the disruptive consequences of changing traditional forms of worship at a time of economic crisis and spiritual uncertainty. Tens of thousands of Charles's subjects became involved in the political process mainly if not solely because they believed the king's policies imperilled their salvation. First in Scotland and then in England, ordinary citizens subscribed their name to public documents – the National Covenant and the Protestation, respectively – that they hoped would preserve their ancient faith, even though doing so set them on a collision course with their sovereign.

Once again, Charles could hardly plead ignorance. As Kevin Sharpe noted, Charles worked hard at being king and exhibited an 'obsession with ordering'. He regularly presided at Privy Council meetings; he read and annotated incoming correspondence and scrutinized the credentials of candidates for state offices; and in matters of religion, he commanded while his bishops executed. His personal intervention in the crafting and promulgation of both the Canons and the Prayer Book for Scotland did not stand alone. Charles also demanded from Laud an annual account of his ecclesiastical province, which he read and returned with a barrage of schoolmasterly comments ('This must be remedied one way or other; concerning which I expect a particular account of you'); demands for further information ('I desire to know the certainty of this'); and promises to back up his archbishop's decisions with the full force of the law ('Informe mee of the particulars, and I shall command the judges to make them abjure').[53]

The obsessive personality is not rare among rulers, and in Charles's case it might have stemmed from his childhood, when he was overshadowed until age twelve by his charismatic brother Henry, whose death was deeply mourned. Henry set a standard that Charles could never match – not least because of the latter's diminutive stature and a life-long stammer. It is less easy to explain two other character defects that complicated relations between the king and his subjects: inconstancy and irresolution. James I had once assured the English Parliament that 'I will not say anything which I will not promise, nor promise anything which I will not sweare; what I sweare I will signe, and what I signe, I shall with God's grace ever performe.' Charles was different: although he frequently and ostentatiously gave his 'word as a king', he often later reneged. His policy towards the Scots in 1638–39 oscillated from implacable obstinacy ('I would rather die than yield to those impertinent and damnable demands') to abject capitulation (at the Pacification of Berwick), with the result that his subjects gave 'no credit' to anything he said. Likewise, in 1641 he first promised Strafford that he would not suffer for his loyalty and then signed his death warrant; and the following year, after many vehement denials, he signed into law a bill depriving bishops of their right to vote in the House of Lords. Such retreats, in the words of Lord Clarendon, 'exceedingly weakened the king's party' strategically as well as tactically, because many of his supporters 'never after retained any confidence that he would deny what was importunately asked'.[54]

Nevertheless, opposing the king in the expectation that he would eventually give way was a high-risk strategy. Charles had often displayed both intolerance and vindictiveness. Admittedly, as Sharpe noted, the king executed not a single subject for treason or crimes of state (a striking contrast with both his fellow monarchs and the republican regime that followed), but he imprisoned and banished those who criticized him.

Moreover, in 1628 he urged his judges to torture John Felton, Buckingham's assassin, and twelve years later he wrote out in his own hand the warrant authorizing the torture of a man suspected of leading the attack on Lambeth Palace after the dissolution of the Short Parliament. In 1639 and again in 1640 he led an army to suppress his Scottish subjects, and two years later he would surely have executed the five MPs he accused of treason had Parliament passed their bills of attainder (they were, after all, guilty as charged).[55]

For many of his opponents, Charles's actions not only reeked of political arbitrariness: they also raised fears of a Popish Plot. Every English parish church was supposed to have on public display a copy of John Foxe's *Book of Martyrs*, filled with graphic examples of how in the past Catholics had tortured and killed English Protestants. A few political leaders in 1640 had witnessed Spain's attempt to invade England in 1588, some could remember the Gunpowder Plot in 1605, and almost all recalled the Spanish Match in 1623. Fear of a Catholic takeover therefore formed a permanent part of opposition rhetoric in Stuart England. Its resurgence in 1641–42, in the wake of the traumatic news from Ireland served up almost daily in lurid pamphlets, was entirely predictable.

These circumstances explain Essex's insistence during the trial of Strafford that 'stone-dead hath no fellow' – even though forcing the king to commit judicial murder significantly increased the risk of civil war. As Charles later wrote, 'The failing to one friend has indeed gone very near me; wherefore I am resolved that no consideration shall ever make me do the like' again. He refused to trust or keep faith with those responsible; and henceforth he made promises that he had no intention of keeping, because 'I have set up my rest upon the justice of my cause, being resolved that no extremity or misfortune shall make me yield, for I will either be a glorious king or a patient martyr.' In doing so, he plunged all his kingdoms into the most turbulent and destructive decades they would ever experience.[56]

Politics, it is often said, is the art of the possible – but what exactly was possible in early Stuart Britain? Wariston explicitly ruled out 'retiring one single inch in this cause' and he was not alone. In June 1638 Hamilton summed up Charles's Scottish dilemma with remarkable perspicacity: 'How far your Majestie in your greatt wisdome will think itt [fit] to *wink at ther madnesis*, I dare not nor presume to advise', but 'I dare assure you, till sume part of their madness hes left them, *that they will sooner loose ther lives than leive the Covenantt, or part frome ther demands.*'[57] Once Charles had decided to impose a Prayer Book, come what may, nothing short of full independence for the Scottish church would have satisfied Wariston and his associates. Likewise in Ireland, after the sudden dissolution of the Dublin Parliament in the summer of 1641, only implementation of the Graces would have satisfied Maguire and his fellow conspirators.

As early as 1638 Wariston became convinced that a peaceful resolution of the tension that developed in the Stuart monarchy in the 1630s could only be achieved through 'the Lord's removal of Charles', either from disease (he contracted smallpox in 1632, but in a mild form) or some accident such as a fatal fall from his horse (which would kill his grandson, William III). Given the temperament of the protagonists, once subjects like Wariston, Maguire and Essex decided to oppose a monarch like Charles I, civil war became the most likely if not the inevitable outcome.

CHAPTER TWELVE

Britain and Ireland from Civil War to Revolution, 1642–89

In 'the cruel and unnatural wars fought in recent years',

> much innocent blood of the free people of this nation hath been spilt, many families have been undone, the publick treasure wasted and exhausted, trade obstructed and miserably decayed, vast expence and damage to the nation incurred, and many parts of this land spoiled, some of them even to desolation.

This bleak assessment formed part of the indictment read out in January 1649 by John Cook, solicitor-general for the Commonwealth of England, at the trial of Charles I for war crimes.[1] The king's execution ten days later brought about Britain's only experience (thus far) of republican government, its first written constitution, the first effective political union between all parts of the Atlantic archipelago and the foundation of the first British empire – but, as Cook noted, these achievements came at a high cost. The civil wars killed about 250,000 men and women in England, Scotland and Wales, or 4 per cent of the total population; several hundred thousand men and women were maimed or rendered homeless; tens of thousands more were taken prisoner and enslaved by the conquerors. In addition, poor harvests produced in Scotland a famine of which 'the lyke had never beine seine in this kingdome heretofor, since it was a natione'; and 'so great a dearth of corn as Ireland has not seen in our memory, and so cruel a famine, which has already killed thousands of the poorer sort'.

> More people died during the course of the 1640s and 1650s [in Ireland] than in the rebellion of 1798 or in the civil wars of the twentieth century. Proportionally, the conflict resulted in a greater demographic catastrophe than the potato famine of the 1840s, with the population loss estimated at over 20 per cent.[2]

The civil wars also caused unprecedented material damage. In England and Wales, the total cost of property damage exceeded £2 million, with over 11,000 houses, 200 country houses, thirty churches and half a dozen castles destroyed (and many more seriously damaged). Moreover, to pay and deploy its armies, the central government in London extracted over £30 million in taxes and fines from the population, while the soldiers extracted many millions more directly. It is impossible to calculate a global cost for the war, but the experience of Cheshire between 1642 and 1646 offers an eloquent example: its inhabitants paid at least £100,000 in taxes and a further £120,000 in goods and services requisitioned directly by the soldiers. When sequestration, plunder and wanton destruction are included, the civil war cost Cheshire at least £400,000. By contrast, the county's annual Ship Money assessment, which had featured so prominently in provoking the constitutional crisis of 1640, had been just £2,750.

Such precise calculations of material damage remain unavailable for Charles I's other dominions, but documents from the 1650s reveal many abandoned farms in the Borders and the Western Isles of Scotland, and in parts of Ulster in Ireland. In addition, the rioting in Edinburgh in 1637 began a train of events that resulted in the demise of Scotland as an independent nation for almost a decade; and the Irish troubles that began in 1641 opened social and cultural wounds that remain unhealed to this day.

The Uncivil Wars

A 'Great Fear' swept England in the winter of 1641–42, comparable in intensity to the one that gripped France in 1789; but instead of reflecting fear of famine it arose from the perceived 'danger from the papists and other ill-affected persons' who stood 'ready to act the parts of those savage bloodsuckers in Ireland if they be not speedily prevented'.[3] The mention of Ireland was significant because news of the massacres that followed the uprising on 23 October 1641, and rumours that the king himself had sanctioned the revolt, seemed to authenticate the long-standing fear of a similar atrocity in England. More than a quarter of all pamphlets published in London in the next few months carried news of Ireland, including the *Remonstrance of diverse remarkable passages concerning the church and kingdom of Ireland* drawn up by Dr Henry Jones, one of the judges who collected sworn depositions from survivors of the uprising (page 267). Having presented the depositions of over 600 victims to Parliament, Jones included a lurid selection of them in his *Remonstrance*. This information had an immediate and dramatic impact: the House of Commons, shamed and shaken, offered 2.5 million acres of Irish land, to be confiscated from the rebels, as security to those who would 'adventure' funds to raise troops to restore Protestant control, and to authorize an emergency tax of £400,000, to be assessed and collected by Parliament, not by the king.

In March 1642, Charles grudgingly assented to the Impressment Bill, which allowed Parliament to use duress to raise soldiers for the defence of Ireland, and then departed for northern England. In his absence, Parliament continued to usurp the crown's executive functions – appointing men whom it trusted to take charge of the Tower of London, the arsenals of Hull and Portsmouth, and the Royal Navy – and in June presented the Nineteen Propositions, a document that sought to curtail yet more of the crown's executive powers: privy councillors, ministers, judges and even the tutors for the king's children could henceforth take office only after parliamentary approval; there must be no more Catholic queens; Parliament must approve in advance the marriage of any member of the royal family; the king must accept Parliament's right to raise soldiers when it chose.

Charles, now at York, warned his subjects that if he accepted the Nineteen Propositions Parliament would surely 'destroy all rights and properties, all distinctions of families and merit', leading to another 'Jack Cade or a Wat Tyler' (the leaders of popular revolts in 1450 and 1381, respectively), so that 'we shall have nothing left for us but to look on' until 'this splendid and excellently distinguished form of government end in a dark, equal chaos of confusion'. He therefore rejected the document.[4] The king also reminded his subjects that 'I am constant for the doctrine and discipline of the Church of England as it was established by Queen Elizabeth and my father, and [am] resolved (by the grace of God) to live and die in the maintenance of it.' He used a portable printing press to publicize that he alone could enact the measures 'whereby the good and quiet people of our kingdom' – what today would be called the silent majority – 'may be secured, and the wicked and licentious may be suppressed'.[5]

This time, Charles's political instincts proved sound. The proliferation of radical religious groups, especially in London, alarmed many ordinary English men and women; so did Parliament's failure to end the widespread economic dislocation and to prevent gangs of unemployed workers in Essex and Suffolk from sacking the houses of local Catholics and Protestant royalists. Members of the elite who dreaded either a collapse of public order or a religious free-for-all now made their way to York where they joined courtiers terrified that they might share the fate of Strafford and Catholics alarmed by Parliament's blood-curdling rhetoric. By July 1642 barely one-third of the House of Commons, which had once contained over 500 members, turned up to vote and the House of Lords mustered only thirty peers (one-quarter of the total). Many of the rest had rallied to the king.

The migration of royalists (as they would soon be called) encouraged those who remained at Westminster to take more radical steps. Pym and his supporters lashed out at anyone who spoke out against them: scores of 'scandalous ministers' and discontented laymen joined half the bench of bishops and half the judges in prison. They also created a Committee of Safety, which levied forced loans on pain of confiscation (something expressly prohibited by the Petition of Right: page 252); and, conscious that a successful English challenge to Charles still required Scottish involvement, they pushed through a religious programme expressly aimed to win the support of the Covenanters – thereby further alienating those English men and women who preferred the Book of Common Prayer.

Meanwhile rumours that Parliament planned to send troops to extirpate them led Ireland's Catholic leaders to take steps to defend themselves. They created a formal 'confederation', with its own General Assembly and Supreme Council headquartered in Kilkenny, which for the next seven years governed much of Ireland and raised its own army and navy to repel any attack from England. They need not have worried. In August 1642, Parliament resolved to use the funds raised for the invasion of Ireland to raise instead an army of 10,000 volunteers for its own defence, and it named the earl of Essex as lord general. Charles responded with a proclamation 'for the suppressing of the present rebellion under the command of Robert, earl of Essex', and signed commissions ordering his supporters to raise troops – an unequivocal declaration of war.

The king now embarked on his third campaign in four years and advanced southward at the head of an army of 14,000 volunteers. Essex led those who had rallied to the defence of Parliament northward until on 23 October 1642, the first anniversary of the Irish rebellion, they met the royalist army in pitched battle at Edgehill – the first fought on English soil for well over a century. Both sides suffered serious losses, and Essex withdrew northwards so that nothing stood between Charles and London; but instead of exploiting his advantage the king marched to Oxford, which he fortified and made his 'temporary capital'. The royalists now held Wales, the west of England and most of the Midlands and the north, while Parliament controlled the southeast, a few ports elsewhere and the navy. To win the war, Charles needed only to take London whereas his opponents could declare victory only after they had forced the king to surrender and secured control of the whole country.

Scotland's leaders decided to exploit the leverage provided by this unequal equation, offering to send Parliament military assistance on condition that every Englishman over the age of eighteen swore to accept the Covenant and that the English Parliament took an oath to impose Presbyterianism in both Ireland and England. Although this programme enjoyed limited appeal in England and virtually none in Ireland, a year of desultory fighting convinced the parliamentary leaders that they had no choice other than to accept the divisive demands of the Scottish Covenanters. In January 1644 another Scottish army entered England and, together with Parliament's forces, defeated the king's field army on Marston Moor, near York, in the greatest battle ever fought on English soil. The royalists lost over 5,000 soldiers and control of northern England.

Yet Marston Moor did not end the war. Charles persuaded the Irish Catholic Confederates to send an expeditionary force to Scotland, compelling the Scots to recall most of their forces from England to defend their homeland; he lured Essex and the main parliamentary army deep into Cornwall, where in September 1644 he forced their surrender; and the following month, he almost defeated the earl of Manchester, commander of the rest of Parliament's field army. These signal failures outraged many MPs including Oliver Cromwell, one of the architects of victory at Marston Moor, and he proposed (in effect) that Parliament dismiss both Essex and Manchester and instead create a new national army. Sir Thomas Fairfax became lord general of the New Model Army, to number 22,000 men, with Cromwell as his second-in-command, and in June 1645 they brought the king to battle at Naseby. This time, Parliament's victory proved decisive, for they not only inflicted heavy casualties but also captured Charles's coach, containing his personal archive.

'The king's letters taken at the late fight at Naseby' proved as important (a London newspaper claimed) 'as all the wealthe and souldiours that we tooke'. John Wallis, a young mathematician, deciphered the secret correspondence between Charles and his wife, and the House of Commons published a selection, accompanied by a commentary, in a pamphlet entitled *The king's cabinet opened*. A royalist writer later admitted that this publication had 'uncloath'd the king'.[6] How could a pamphlet with fewer than fifty pages achieve so much?

Three recent developments had created an unprecedented public interest in England in political developments both at home and abroad. First, thanks to the proliferation of schools in England and Wales, by 1640 a significant part of the population could read about the affairs of the kingdom directly. Second, thanks to the collapse of royal censorship, interested readers had a multitude of publications about politics at their disposal: over 2,000 printed works appeared in England in 1641, more than in any previous year, and the number doubled in 1642 – an annual total unequalled for almost a century. Moreover, although only one newspaper appeared in England in 1639, over sixty came out in 1642, each issue filled with foreign and domestic news so that readers could now 'feel how the pulse of the king and kingdom beats' on a regular basis. Third, many of these readers had developed (in the phrase of Noah Millstone) 'a way of seeing grounded in suspicion, the prevalence of deceit and the conviction that things were not as they seemed', specifically where governments were concerned. According to a troubled Suffolk clergyman, everyone seemed 'disposed to speake the worst of state businesses, & to nourish discontente, as if there were a false carriage in all these things'.[7] In short, England in the 1640s boasted the most animated public sphere in the early modern world, 'where claims and counterclaims could be asserted and negotiated, and where the range of princely and imperial power could be questioned and contested'.[8]

Few pamphlets or newspaper articles packed the punch of *The king's cabinet opened*, however. Charles's letters complained about the stultifying

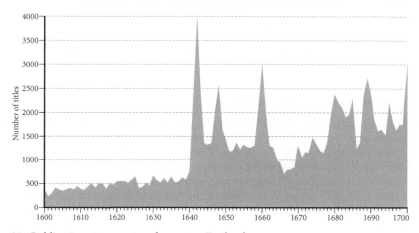

20. Publications in seventeenth-century England.
As in France a few years later, the abolition of government censorship triggered an
unprecedented surge in printed works (as reflected in the titles of all known
published works). The negotiations between Parliament, the army and the king in
1647–48, and the debate in 1659–60 over the best form of government for the
British state, likewise produced surges of printed works; but the totals of 1641–42
would not be equalled for almost a century.

dullness of his entourage in Oxford, ridiculed the peers and MPs who had
rallied to him, and made fun of the ambition of his ministers. In addition,
the king clearly transacted 'nothing great or small' without his wife's 'privity
and consent', even though (in the sanctimonious language of the editors)
'she be of the weaker sexe, borne an alien, bred up in a contrary religion'.
Even more compromising were the king's attempts to solicit foreign military assistance; his promises to repeal the penal laws against English
Catholics in England 'as soon as God shall inable me to do it'; and, worst of
all, his promise to grant full toleration to the Irish Catholics provided 'they
engage themselves in my assistance against my rebels of England and
Scotland'. The last pages of the pamphlet contained a devastating selection
of Charles's public statements that he flatly contradicted in his private
letters. As the editors observed: 'The king will declare nothing in favour of
his Parliament so long as he can finde a party to maintaine him in this
opposition; nor performe anything which he hath declared, so long as he
can finde a sufficient party to excuse him from it'. *The king's cabinet opened*
robbed the king not only of his 'cloathes' but also of his credibility.[9]

After Naseby, Charles gradually lost his other assets. The remaining
royalist garrisons in England surrendered; the Scots crushed the Irish
invaders in battle; factional rivalry prevented the Confederation of Kilkenny

from sending the king any relief from Ireland. As Fairfax's army prepared to besiege Oxford, in May 1646 Charles therefore fled the city and surrendered to a detachment of Scottish troops, creating a power vacuum in England and Wales.

The New Model Army Takes Charge

Without the guiding hand of John Pym, who died in 1643, the Westminster Parliament became bitterly divided between a Presbyterian majority committed to imposing the Solemn League and Covenant throughout the three kingdoms and a minority, known as Independents, who opposed it. The former found much support in London and the surrounding counties, which perhaps concealed from them the fact that the rest of the country detested Presbyterianism – as did many officers and men in the New Model Army.

Early in 1647, after the surrender of the last royalist outposts in England, the Presbyterian caucus at Westminster voted to send to Ireland about half the New Model Army, who would receive six weeks of their wage arrears, and to disband the rest with nothing. This was outrageous – Parliament owed at least £1 million in unpaid wages to its soldiers, who had fought with great valour and outstanding success – and Fairfax refused to discuss any demobilization plan with the commissioners sent by Parliament until they had clarified questions on full payment of arrears, indemnity for past actions and other matters. Pamphlets on these issues circulated among the rank and file, including one stating that no soldier should go to Ireland until all had received their pay arrears in full.

Irritated by the unexpected (though foreseeable) resistance to its plans, Presbyterian MPs declared the authors of these pamphlets to be 'enemies of the state and disturbers of the public peace'. This accusation outraged the entire army, and at the urging of Henry Ireton, deputy commander of the cavalry (and Cromwell's son-in-law), several regiments elected representatives, popularly known as Agitators, who met with their officers. Together they resolved that they 'would sooner die than disband without the utmost farthing of their arrears'.[10] Amazingly, Parliament failed to perceive the imminent peril they faced and peremptorily ordered Fairfax to disperse his troops as a prelude to disbanding them. Instead, with the full support of his senior officers, Fairfax summoned all units to assemble near the town of Newmarket on 4 June 1647.

Once again, contingency intervened. Parliament had recently paid the Scots £400,000 in return for a promise to withdraw their troops from England and surrender of the king. While the Presbyterian leaders debated the terms of the new constitutional settlement to be imposed on him, Charles remained under guard at his palace of Holdenby, in the Midlands, until on 3 June George Joyce, formerly a tailor and now a cornet (the most junior commissioned rank in the cavalry) arrived at Holdenby with 500 troopers and seized the king. Fearful for his fate, Charles persuaded Joyce to take him to a royal hunting lodge near Newmarket – the place where (unknown to him) Fairfax had summoned his troops to assemble the following day.

Possession of the king's person greatly strengthened Fairfax's hand, and Parliament now offered his troops full payment of arrears and an act of indemnity for all wartime actions, hoping that these improved terms would destroy the army's unity – and, indeed, many officers and men accepted the deal and departed for Ireland, but most of them were Presbyterians sympathetic to Parliament. The Independents who replaced them approved a radical pamphlet prepared by Ireton in association with the Agitators, entitled *The solemn engagement of the army*, which pledged that they would remain under arms until Parliament had not only satisfied their material grievances but also granted certain legal protections to both the soldiers and 'other freeborn people of England, to whom the consequence of our case does equally extend'.[11]

The troops assembled at Newmarket also approved the creation of a General Council (comprising two officers and two Agitators from each regiment, as well as Fairfax and his senior officers) which promptly resolved to march on London, taking with them the king. They also took with them a printing press, which produced multiple copies of a radical declaration (again mostly drafted by Ireton) stating that 'We shall, before disbanding, proceed in our own and the kingdom's behalf to propound and plead for some provision for our and the kingdom's satisfaction and future security', by virtue of the fact that 'We were not a mere mercenary army, hired to serve any arbitrary power of a state, but called forth and conjured by the several Declarations of Parliament to the defence of our own and the people's just rights and liberties'. The document drew explicit parallels with the Scots, the Dutch, the Portuguese 'and others' who had secured their goals through armed resistance to arbitrary power, and demanded that Parliament do five things:

- Provide full accounts of the money levied and spent during the war;
- Issue a 'General Act of Oblivion' in order to remove 'the seeds of future wars or feuds';
- Grant freedom of worship for all law-abiding subjects;
- Conduct a drastic electoral redistribution that would 'render the Parliament a more equal representation of the whole' kingdom; and then
- Dissolve itself so that the king could summon a new assembly according to the new franchise.[12]

For the first time in English history, a powerful group of subjects demanded both freedom of conscience and electoral reform – issues that would dominate political debate in the entire Western world for the next two centuries. It is easy to forget that they were first articulated and discussed by the West's first national army.

The 'Young Statesmen'

In July 1647 the army's General Council discussed a yet more radical document, entitled *The heads of the proposals*, which detailed the army's terms for reaching a settlement with the king and also proposed a new English constitution, with a Parliament elected every two years according to a franchise apportioned according to the distribution of taxes; the abolition of regalian rights and imprisonment for debt; and the right of those accused in criminal trials not to incriminate themselves. One cavalry trooper expressed a sense of awe as he and his colleagues debated 'the settling of a kingdom', because 'we are, most of us, but young statesmen'; nevertheless, after ten days of animated debate, the council formulated a set of proposals that the senior officers shared with the king. Having read them, Charles enquired what they would do if Parliament rejected their suggestions. After an awkward silence, Colonel Thomas Rainborough blurted out: 'If they will not agree, we will make them'.[13]

The king had identified a crucial weakness: London now boasted powerful new walls and could easily withstand a siege. Nevertheless, the capital was restless; the recent harvest had proved poor, driving up food prices; and the tax burden to support an army that refused to disband grew. Hardship and discontent increased, trade and industry atrophied, and the kingdom became 'a melancholy, dejected, sad place'.[14] These tensions created new divisions in Parliament between the Independents who wanted to appease the army and the Presbyterians who did not, and popular demonstrations multiplied outside the Palace of Westminster. The Speakers of both Houses, followed by over sixty MPs and peers, now fled to the army and appealed for its protection, and a week later sympathetic units of the capital's garrison opened one of the city gates, which allowed Fairfax and his soldiers to enter London without firing a shot. They wore laurel leaves as they escorted the parliamentary refugees back to Westminster in triumph and installed the king (under guard) at Hampton Court Palace. They then withdrew to Putney, strategically located on the Thames between Parliament and the king, where the army's General Council resumed its discussions on how best to arrange the affairs of the kingdom.

The composition of the army had changed since it left Newmarket because many London radicals, known as 'Levellers', took many of the places vacated by Presbyterians. Some secured election as Agitators. Unlike most of the radical groups that flourished in the 1640s, the Levellers had no religious programme but instead demanded wide-ranging social and political reforms. In October 1647, at Putney, the General Council debated *An agreement of the people for a firm and present peace upon grounds of common right*, a concise and eloquent pamphlet approved by the Agitators arguing that sovereignty should pass to a single-chamber assembly, chosen every two years by an electorate 'proportioned according to the number of the inhabitants'. The primary function of this body would be to secure certain 'native rights' for all Englishmen: freedom from religious compulsion and from conscription; a universal indemnity 'for anything said or done in reference to the late public differences'; laws that applied to all citizens equally; and a written constitution. The document concluded by calling on the army to impose this revolutionary programme by force if necessary.[15]

The Putney Debates involved some fifty officers and men, including some of the new Leveller Agitators, and hot exchanges on the franchise (and its limits) continued until the weary stenographers laid down their pens. We therefore do not know for certain the outcome, but the majority apparently agreed that all Englishmen, 'if they be not servants or beggars, ought to have voices in electing those that shall represent them in Parliament'. Although the senior officers imposed a news blackout, Agitators circulated printed copies of the *Agreement* and invited soldiers to sign it in anticipation of another march on London, this time to dissolve the Long Parliament by force. Despite some reservations, Fairfax presented to Parliament a remonstrance in which he pledged to 'live and die with the army in the lawful prosecution of these things following; first, for the soldiery', full pay arrears, an amnesty, and 'provision for maimed soldiers, and the widows and orphans of men slain in the service'; and

> Secondly, for the kingdom, a period to be set for this Parliament (to end so soon as may be with safety), and Provision therewith to be made for future Parliaments; for the certainty of their meeting, sitting and ending; and for the freedom and equality of elections thereto, to render the House of Commons (as near as may be) an equal representative of the people that are to elect.[16]

The Second Civil War

The king had followed the Putney Debates closely and, fearing that he would never recover his authority, in November 1647 he fled from Hampton Court. He was soon recaptured and, this time, the army (who saw his flight as breaking his parole) locked him up and began to discuss depriving their royal prisoner of his office. Fully aware of this development, Charles decided to accept the propositions clandestinely proffered by commissioners from the Scots, his erstwhile captors, who had concluded that their English allies were not capable of making good on all the concessions they had made. In December 1647 the king signed a secret Engagement to establish Presbyterianism in England for a trial period of three years and allow a committee of theologians nominated by both sides to determine a permanent religious settlement. In return, the Scots 'Engagers' promised to send an army 'into England for preservation and establishment of religion, for defence of His Majesty's person and authority, and restoring him to his government'.[17] The Scots had thus declared war on the Westminster Parliament and the New Model Army.

For a time, economic hardship worked in favour of the king and the Engagers. Throughout England, 'bad weather ruined the harvests of corn and hay for five years from the autumn of 1646 onwards, and every succeeding year until the harvest of 1651 exacerbated the problems left by the previous one'. In London, according to James Howell, 'a famine doth insensibly creep upon us, and the Mint is starved for want of bullion. Trade, which was ever the sinew of this island, doth visibly decay and the insurance of ships is risen from two to ten in the hundred.' Howell continued: ' 'Tis true we have had many such black days in England in former ages, but those paralleled to the present are as the shadow of a mountain compared to the eclipse of the moon.' The Leveller John Wildman agreed. In January 1648 he warned the House of Commons that 'trading was decayed and the price of food so excessive that it would rend any pittiful heart to heare and see the cryes and teares of the poore, who protest they are almost readie to famish'. Wildman claimed that clothiers who 'set at work formerly 100 did not now set at work above a dozen'. He predicted that 'a sudden confusion would follow, if a speedie settlement were not procured'.[18]

The Second Civil War now began, but Cromwell soon defeated Charles's Scottish allies, and Fairfax crushed the royalists of southeast England, after which the victors converged on London, angry and embittered. On 20 November 1648 they demanded that Parliament execute 'capital punishment upon the principal author and some prime instruments of our late wars': namely the king and his leading supporters. The following morning troops surrounded the palace of Westminster and either excluded or arrested all MPs thought likely to vote against putting the king on trial. This reduced the House of Commons to a Rump (as it became derogatorily known) of scarcely 150 MPs, which obligingly created a High Court of Justice consisting of 135 judges, including officers and MPs (many of whom refused to serve), to try the king.

Proceedings began on 20 January 1649, when Charles was brought into court under guard and Solicitor-General John Cook read out the charges against him: 'out of a wicked design to erect and uphold in himself an unlimited and tyrannical power to rule according to his will', the king had tried 'to overthrow the rights and liberties of the people', and to that end had 'traitorously and maliciously levyed war against the present Parliament and the people therein represented'. After considering the evidence of eyewitnesses concerning twelve specific atrocities (what today would be called war crimes) in which the king had taken part, on 27 January 1649 fifty-nine members of the High Court of Justice 'adjudge[d] that he, the said Charles Stuart, as a tyrant, traitor, murderer, and publick enemy, shall be put to death, by the severing of his head from his body'. Three days later, a masked executioner cut off his head.[19]

Creating the British Republic

The Rump, now the supreme executive and legislative authority in England, at first did not know what to do with its unlimited power. It took three weeks for them to remove the name 'king' from all legal documents and to vest the monarch's executive functions in a Council of State; six weeks to abolish the House of Lords; and almost four months to declare 'the people of England and of all the dominions and territories thereunto belonging' to be a 'Commonwealth'. Meanwhile, a volume entitled *Eikon basilike* (*The king's image*), in which Charles had 'set down the private reflections of my conscience, and my most impartial thoughts, touching the chief passages ... in my late troubles', circulated widely both at home and abroad.[20]

Eikon basilike rallied royalist opinion just as the enormous sums consumed by the army and navy eroded support for the Rump. Three permanent taxes established by Parliament in 1643 proved particularly burdensome: customs duties, whose yield was applied directly to the navy; a tax on wealth and income called the Assessment, ironically allocated among towns and shires on the same basis as the hated Ship Money, to pay the army; and excise duties (initially levied only on alcohol and certain goods considered non-essentials, but later also on staples) used to pay the army and also discharge public debt. Collection proved particularly burdensome in 1649 because of the extreme weather and failed harvest: even in London, normally the best-supplied region of England, the price of flour reached a level unequalled for another half century, and the Bills of Mortality showed a marked excess of burials over baptisms. Domestic opponents of the new regime multiplied. In April 1649 some Levellers accused the Rump of tyranny, and their imprisonment provoked a demonstration by hundreds of women and a protest signed by over 10,000 people. New radical groups also criticized the Rump: the Fifth Monarchists wanted to set up a regime ruled by 'saints' to prepare for the imminent Second Coming; the Diggers proclaimed that all property should be held in common; the Ranters believed that they had discovered an internal divinity that freed them from conventional morality; and the Quakers admitted no distinction in social rank between men and women, or between rich and poor.

Foreign critics of the new republic also multiplied. Immediately after the regicide, the Scots Parliament provocatively proclaimed its allegiance to 'Charles the Second, *king of Great Britain and Ireland*' and, even more provocatively, declared that before he could exercise his royal powers, he must promise to uphold 'the security of religioun, the union betwix the kingdoms, and good and peace of [all] his kingdoms according to the Solemn League and Covenant'.[21] This was, in effect, a new declaration of war on England. In Ireland, the Catholic Confederates also recognized Charles II as their legitimate king and paid 18,000 soldiers to fight for his cause: by July 1649 only Dublin and Londonderry lay beyond their control. In addition, royalists controlled the Scilly and Channel Islands, Virginia proclaimed its allegiance to Charles II and outlawed those who denied that he was the rightful king of England. Several other colonial governors soon followed suit. Only Rhode Island formally recognized the Commonwealth: colonists elsewhere regarded the regicide as 'a very solemn and strange act' and declined to commit themselves before they had evidence that the new regime enjoyed divine approval.[22] In Europe, virtually no government recognized the Commonwealth; the tsar of Russia expelled all English merchants; and royalist exiles murdered the diplomats sent by the republic to Spain and Holland.

Faced by such hostility, the Rump debated the terms of an oath of allegiance to be taken by its officials. A proposal to include language approving Parliament's trial of Charles I was defeated by thirty-six votes to nineteen (an indication both of the small size of the Rump and of doubts concerning the regicide). Eventually all state employees and members of the armed forces as well as all clerics, teachers and students at universities and schools swore that 'I do declare and promise that I will be true and faithful to the Commonwealth of England as it is now established, without a king or House of Lords' – but even so, one member of the Rump lamented that 'all the world was and would be their enemies' and that 'the whole kingdom would rise and cut their throats on the first good occasion'; another committed suicide on the first anniversary of the regicide; and a third died a month later of depression.[23]

Ireland presented the most immediate problem for the Commonwealth, and in August 1649 Cromwell sailed for Dublin with 12,000 New Model Army veterans and fifty-six siege guns. Their brutal sack of Drogheda and Wexford persuaded most of the remaining rebel strongholds to surrender, and within a year the government in London controlled Ireland more effectively than it had ever done. Cromwell returned in triumph to London in June 1650 – but almost immediately news arrived that Charles II planned to return to Scotland, invade England and impose the Covenant on all his hereditary dominions. The Rump therefore decided to launch a pre-emptive strike, once again led by Cromwell.

The Scots suffered from several disadvantages. Since 1636, the kingdom had experienced the worst sustained drought and cold in a millennium, culminating (according to historian Sir James Balfour) in 1649 with a cereal harvest of 'small bulke', so that the prices of foodstuffs 'of all sortes were heigher than ever heirtofoe aney[one] living could remember'. He also recorded a panic reaction common to many countries during the Little Ice Age: claiming that 'the sin of witchcraft daily increases in this land', and fearing divine punishment if it continued, the Scots Parliament issued some 500 witchcraft commissions in 1649–50, resulting in more executions for sorcery than at any other time in the country's history.[24] The Covenanter leadership also attempted to avert divine punishment by purging its army of all 'malignant, profane, scandalous persons', but this fatally weakened its strength when in July 1650 Cromwell did what Charles had failed to do a decade before: cross the Tweed with a large army of veterans, an artil-lery train and a large fleet to cover his right flank. At the battle of Dunbar on 3 September some 3,000 Scottish soldiers died and 10,000 more fell prisoner.

For many people at home and abroad the crushing victory at Dunbar confirmed the republic's legitimacy. In Massachusetts the Reverend John Cotton hailed it as a convincing sign that God approved of the new regime, led his parishioners in a special day of thanksgiving and wrote a personal letter of congratulation to Cromwell. In Paris, Thomas Hobbes put the finishing touches to the first masterpiece of political philosophy in the English language: *Leviathan, or the matter, form, and power of a common-wealth, ecclesiastical and civil.* The book, 'occasioned by the disorders of the present time', argued that 'If a monarch subdued by war render himself subject to the victor, his subjects are delivered from their former obligation, and become obliged to the victor' – a perfect justification for the Rump.[25]

Naturally, Charles II disagreed. Refusing to accept that God would uphold a regicidal regime against an anointed king, he assembled in Scotland a force of royalists and Covenanters whom he led southward in summer 1651, covering 330 miles in three weeks to Worcester, where they hoped to receive reinforcements from the surviving English royalists. Instead, on 3 September 1651, the anniversary of Dunbar, Cromwell once more prevailed: another 3,000 Scots fell on the field and 10,000 more became prisoners. Relatively few escaped, like Charles II, to fight another day. To Cromwell, this 'most remarkable, seasonable and signal victory' was 'for aught I know, a crowning mercy' to the young republic.[26] The Rump organized a victory parade in London, which included some 4,000 Scots prisoners who promptly went into penal servitude – some to drain the Fens and mine Tyneside coal, the rest to labour in the American colonies – and declared that 3 September should forever be celebrated as a day of thanksgiving.

Creating the First British Empire

As soon as Cromwell had triumphed at Dunbar, he told the Rump that the time had come to export England's revolution: 'You shall shine forth to other nations, who shall emulate the glory of such a pattern, and through the power of God turn into the like', he told them.[27] The assembly, by now meeting almost every weekday to exercise its comprehensive executive and legislative functions, accepted the charge with enthusiasm, starting with North America.

Although most of London's merchant elite had supported the king during the Civil Wars, most of those who traded with the American colonies supported Parliament: they repeatedly loaned money, and eight of them sat in judgement on Charles I. In return for their material and moral support, the colonial merchants demanded protection for their trade against royalist privateers, and the Rump obliged in 1651 with the Navigation Act, which stipulated that all imported goods should be carried either on English vessels or ships from the country of origin. The following year a fleet left England to enforce Commonwealth control over the Caribbean, and captured twenty-seven Dutch ships trading with the prosperous royalist outpost of Barbados. The Dutch regarded this as a declaration of war, but after eighteen months of bitter naval conflict made peace and promised to respect the Navigation Act. The republic's warships also secured Barbados and forced the royalist governor of Virginia to surrender. The Rump also allowed the colonial merchants to ship far more African slaves to the American colonies, whose population may have quadrupled during the 1650s (Chapter 15).

The Commonwealth also brought Scotland to heel. A month after the 'crowning mercy' at Worcester in September 1651, the Rump issued a declaration that called for political union between Scotland and England predicated on religious toleration; pardon for virtually everyone not still in arms; abolition of all existing legal jurisdictions in favour of the English system of justices of the peace; and destruction of all insignia of royalty. Representatives from each Scottish shire and town reluctantly accepted the incorporation of the two kingdoms and agreed to send representatives to a new Union Parliament. 'What had begun back in 1637 as a rebellion to prevent Scotland's reduction to the status of an English province', David Scott ironically observed, 'had ended in precisely that fate.'[28]

These measures were mild compared with the Rump's treatment of Ireland. Once again the colonial merchants took the lead. They had lent large sums to uphold the Protestant cause in Ireland; now they demanded the confiscated lands that Parliament had offered as collateral security. In August 1652 the Rump passed a comprehensive Act for the Settling of Ireland, which required everyone to accept the authority of the Commonwealth; condemned all participants in the 1641 uprising to lose their lives and property; deprived all landowners (Protestant or Catholic) who had 'born[e] arms against the Parliament of England or their forces' of two-thirds of their estates; and confiscated between one-fifth and one-third of the lands of any Irish Catholic who could not demonstrate 'constant good affection to the interest of the Commonwealth of England' between 1641 and 1650. This meant that all Irish landowners were deemed guilty unless they could prove themselves innocent, and for two years a special High Court of Justice heard evidence against those condemned by the act. The depositions taken down after 1641 by Henry Jones and his colleagues, which named those who had harmed or robbed the Protestant settlers, now came into their own. Organized by county, and ominously entitled Books of Discrimination, the evidence served to deprive hundreds of Catholics of their lives and tens of thousands of their property. In 1640 Catholics owned about 60 per cent and Protestants 40 per cent of the island's cultivable land, but by 1660 the Protestants owned 80 per cent and the Catholics only 20 per cent. The redistribution of Irish land represented one of the most dramatic and permanent consequences of the seventeenth-century crisis.[29]

The Rump had thus created the first British empire – and yet it lasted fewer than five years because it failed to take one final step dear to the New Model Army, whose troops had largely created that empire. It refused to arrange elections for a new Parliament according to a franchise based on personal assets, not just property, with additional representatives for Scotland and Ireland.

The Road to Restoration

In April 1653 Cromwell lost patience. At his direction, in the middle of one of the Rump's debates on constitutional reform, a detachment of musketeers marched in, removed the Speaker by force from his seat and cleared the chamber. A new Council of State (numbering, with suitable biblical symbolism, twelve besides the lord general) replaced the Rump as a caretaker executive and invited 140 representatives, nominated by godly communities in Scotland, Ireland, England and Wales to assemble in London and frame a constitution for the new state. After four months of fruitless discussion Cromwell again sent in troops to clear the chamber, and instead adopted a written constitution drawn up by his allies called the Instrument of Government, which entrusted supreme legislative authority in 'England, Scotland and Ireland, and the Dominions' to a lord protector, advised by a council. Every three years the protector must convene a

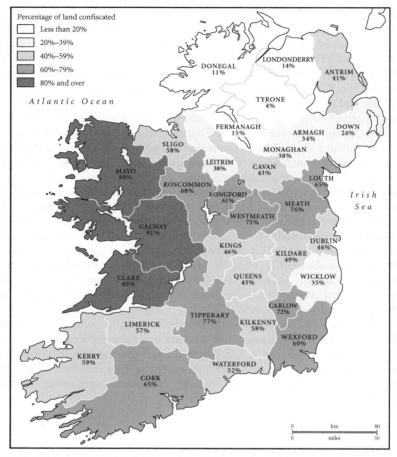

21. The redistribution of confiscated Irish land, 1653–60.
The overall pattern of redistribution, which reduced the proportion of lands owned by Catholics from almost half to less than one-quarter, conceals even more dramatic regional shifts. A detailed reconstruction for three Ulster counties showed that only 5 of 58 Gaelic Catholic landowners in 1641 retained their lands 20 years later. Most of the rest had been forcibly resettled in the far west.

Parliament comprising thirty members from Scotland and thirty from Ireland, together with 400 representatives from England, Wales and the Channel Islands, elected from new constituencies created according to their respective tax obligations in which every man with assets worth £200 or more could vote (unless they had fought against Parliament). Cromwell, who became Lord Protector of England, Scotland and Ireland for life, summoned the first Parliament elected by the new franchise to assemble on 3 September 1654, the national holiday. The Instrument of Government thus created a new political framework that balanced monarchy (protector), aristocracy (council) and democracy (Parliament). It also, for the first time in Western history, guaranteed freedom of public worship to almost all Christians. For the only time in its history, Britain had a written constitution.[30]

An attempt to carve out a British empire in Latin America fatally undermined the Protectorate. The same colonial merchants who had invested in the subjugation of Ireland, Barbados and Virginia now offered to fund a 'Western Design' aimed at 'promoting the glory of God, and enlarging the bounds of Christ's kingdom' by creating a British empire in the Caribbean. Late in 1654, a fleet carrying 9,000 soldiers and sailors left England for the Caribbean, but they captured only Jamaica. Although the island later became a powerful base for future operations, most of the expeditionary force died there. Soon afterwards its commander sailed for home.

Cromwell reacted to the failure of the Western Design by suppressing all but a handful of newspapers (for fear of the political consequences of widespread press criticism); dividing England into twelve military districts, each under a major-general (for fear that the fiasco might encourage royalists to rebel); and decreeing a day of public fast and humiliation because 'the Lord has been pleased in a wonderful manner to humble and rebuke us, in that expedition to the West Indies'. Nevertheless, the Lord continued to humble and rebuke the regime. In London, burials exceeded baptisms in 1652, 1654, 1656 and 1658; and the winter of 1657–58 seemed the worst that any 'man alive had known in England: the crow's feet were frozen to their prey; islands of ice enclosed both fish and fowl frozen'.[31] In the depths of this landmark winter Cromwell convened another Parliament to advise him, but after a month of wrangling once again he dissolved it – the third time he had so acted. He planned to summon another, but died on the national holiday and the anniversary of his two greatest victories, 3 September 1658. His eldest son, Richard, automatically became lord protector, despite the fact that he lacked any military or executive experience (he had served on the council for only nine months), and he immediately convened a new Parliament. Soon after the session started, an MP launched into a bitter denunciation of the Protectorate: 'In five years we have had greater mal-administration than in five hundred years before. . . . The people care not what Government they live under, so as they may plough and go to market.'[32] But though critics proved adept at pulling down what they did not like, they could not agree on what should take its place.

The fundamental obstacle to creating a stable republic lay in Oliver Cromwell's practice of employing men from a wide spectrum of political opinions, from former royalists to ardent Presbyterians, in order to create a broadly based polity – because it created a regime cemented only by loyalty to the protector. Once his death removed that bond, the regime was doomed. In April 1659 the army leaders demanded that Richard Cromwell dissolve the new Parliament, since in three months it had achieved so little; having done so, he resigned as lord protector. After two weeks of anarchy, the army leaders reconvened the Rump, which resumed its discussions on how to create a permanent form of republican government. It, too, made little progress until frustrated army units surrounded Westminster and dissolved the assembly yet again.

This vacuum of power allowed General George Monck, commander of the English garrisons in Scotland, to assemble an army and march on London to restore the Rump, while army officers in Ireland loyal to the Rump seized Dublin Castle and gradually gained control of the other garrison towns of Ireland (just as the conspirators of 1641 had hoped to do). The senior officers in London reacted by restoring the Rump, which (although barely fifty strong) resumed its executive and legislative functions. It faced two pressing problems: the accumulated wages and arrears of the republic's armed forces, which totalled at least £2 million (equivalent to a whole year's revenue); and George Monck, who reached London with his troops on 3 February 1660.

Monck now held the fate of the entire Anglo-Atlantic world in his hands. He approached the surviving members of the Long Parliament, excluded because they had opposed the trial of Charles I (page 284), and secured a promise from each one that, if readmitted, they would authorize writs for new parliamentary elections and then dissolve themselves. Once they had agreed to these terms, on 21 February Monck's musketeers escorted them back to the Commons Chamber and stood guard until they kept their promise. Monck, for his part, agreed to accept whatever constitutional arrangement the new assembly might approve. He, too, kept his promise.

The first general election held since 1640 was contested by an unprecedented number of candidates and produced, perhaps surprisingly, an assembly dominated by royalists. It convened on 25 April 1660 and immediately authorized all peers currently in England to come to Westminster and form a House of Lords in the traditional manner. Some members of Parliament, including Monck (an MP as well as commander-in-chief of all army units in Britain) hoped to impose conditions on Charles II in return for his restoration, but the king-in-exile pre-empted this with the Declaration of Breda (named after the Dutch city where he had taken up residence), which made four key concessions. Charles promised a free and general pardon to all who pledged loyalty to him, except those whom Parliament should exclude; a resolution by Parliament of all disputed titles to property; religious toleration for all who lived in peace, unless Parliament decided otherwise; and a promise to accept whatever measures Parliament took to pay the arrears of Monck's soldiers.

Despite its apparent magnanimity, the Declaration cleverly made Parliament responsible for all the difficult and unpopular outstanding decisions: whom to punish, whom to tolerate, whom to reward, and whom to tax. On 1 May, after hearing the Declaration read out, the peers at Westminster formally resolved 'that according to the ancient and fundamental laws of this kingdom, the government is, and ought to be, by King, Lords and Commons', and a week later, both chambers declared that Charles II had been England's lawful king since the death of his father and invited him to return.[33] Charles landed at Dover on 25 May and issued writs for a new Parliament according to the traditional franchise. Once again, it produced a House of Commons dominated by royalists, later known as the Cavalier Parliament.

The 'Happy Restoration'

Republican insignia were the first to go: coats-of-arms and flags were removed from all public places, warships renamed, monuments smashed, coins recalled and melted down. January 30, the anniversary of the regicide, replaced 3 September as a national holiday – and on the first one, in 1661, the corpses of Cromwell and other republican leaders were disinterred and strung up on the common gallows at Tyburn. By then, the courts had tried and punished perhaps 100 of their colleagues, restored the confiscated estates of some 800 royalist landowners (summarily depriving those who had usurped them during the interregnum), and allowed a further 3,000 royalist families to buy back their sequestered property.

Charles II thus honoured the first two promises contained in the Declaration of Breda – a general pardon and a resolution of property disputes – but he failed to persuade Parliament to approve the religious toleration he favoured. Instead, the assembly enacted measures to punish Catholics and all who refused to 'conform to the liturgy of the Church of England as it is now by law established'. A Test Act required all preachers, town magistrates and teachers at schools and universities, and later all peers and members of Parliament, to demonstrate their orthodoxy. Those who refused immediately lost their posts, and fines awaited all who failed to attend their parish church each Sunday, all who attended a non-conformist meeting of five or more, and all officials who failed to enforce the law. Persistent non-conformists went to prison – including John Bunyan, who composed *The pilgrim's progress* (the most popular work of prose fiction in English written in the seventeenth century) during his twelve years in jail.

The Cavalier Parliament also restored to Charles II most of the powers exercised by his father. Although both the prerogative courts (Star Chamber and High Commission) and the regalian rights (like Ship Money) that had caused such controversy in the 1630s disappeared for good, the king gained sole control of the kingdom's armed forces (the dispute that had precipitated the Civil War) as well as the right to convene and dismiss Parliament and the right to appoint and remove judges at will. He also inherited the three principal taxes that had formed the mainstay of the republican regime – customs, excise and the assessment – and, as the English economy burgeoned, the first two revenue streams steadily increased, making the crown largely independent of Parliament in peacetime. From 1679 to 1684 only 7 per cent of the crown's income came from parliamentary taxes and from 1684 to 1688 the figure fell below 1 per cent. A return to personal rule by the monarch seemed a distinct possibility.

The Restoration in Scotland and Ireland took a different form because both kingdoms had recognized Charles as their monarch (in 1649) and because the concessions contained in the Declaration of Breda applied only to England. In Scotland, Charles restored the separate Parliament and judicial courts abolished by Cromwell; voided all legislation passed by his father's opponents; and (like his father) handpicked his privy councillors, who both implemented the policies decreed in London and named 'the Articles' (the committee that shepherded through Parliament all legislation desired by the king.) A few leading Covenanters (including Wariston) went to the scaffold for their role in bringing down the monarchy. In religious matters, Charles did not tamper with the traditional liturgy, the immediate cause of the revolution in 1637, but required all ministers to accept the authority of bishops (and deprived of their livings the 300 who refused). He also gained the right to maintain a considerable standing army, ready to stifle any opposition. In Ireland, Charles also restored the Parliament and legal system abolished by Cromwell – but otherwise did very little. Although after the Restoration a Court of Claims heard petitions from those who had lost their lands, it reinstated few royalists. By contrast, almost 8,000 veterans of Cromwell's army and many civilian supporters of the republic received confirmation of their titles to lands confiscated from the rebels; and every year, on 23 October, Protestant clerics delivered sermons to remind their congregations that, despite the massacres carried out on that day in 1641, the rebellion had failed and its perpetrators had received due punishment.

The king changed even less in England's colonies. As in Britain and Ireland, orders went out to restore the established church, but Charles instructed his officials in America 'not to suffer any man to be molested or disquieted in the exercise of his religion' and he showed little enthusiasm for hunting down even the three regicides who fled there (all eventually died of natural causes in Connecticut). He also continued the Commonwealth's policy of populating the colonies with African slaves, British criminals, political opponents and paupers; he reissued the Navigation Act that protected the colonists from foreign competition; and he retained and developed Jamaica, the republic's sole American conquest from Spain.

Nevertheless, the Restoration monarchy at first seemed precarious. In England poor harvests in 1661–62 caused a marked rise in deaths and fall in marriages and conceptions; 1665 witnessed an epidemic of bubonic plague that killed perhaps a quarter of London's population; in 1666 a great fire destroyed most of the capital's historic centre; and in 1667 the Dutch, on whom Charles had rashly declared war, launched a surprise attack up the Thames, coming almost within sight of London to destroy several ships of the Royal Navy at anchor.

The next decade proved more peaceful, but in 1679 and 1680 part of England's political elite tried repeatedly to push through parliamentary legislation to exclude from the succession Charles's brother and heir James, an open and devout Catholic whom many suspected of wanting to turn Britain into an absolute monarchy like 'France, where the subjects are at the disposal of the king for life and limb, and to invade other nations' property for the luxury of the Court; and little men of low fortunes are the ministers of state.'[34] Yet despite three general elections in two years (a record still unmatched in British history), Charles managed to avoid signing an Exclusion Bill into law. He also managed to rule without Parliament for the rest of his reign. Only the reckless behaviour of James, who became king when Charles died without legitimate offspring in February 1685, frittered away all these gains.

The Glorious Revolution

At first, the new monarch seemed secure on his throne. A few months after his accession Charles's illegitimate Protestant son, the duke of Monmouth, invaded, but he attracted little support and James's troops swiftly defeated him (Chapter 17). Encouraged by this success, James reduced the size of the electorate to the English House of Commons: he pressured the nobles and gentry in each county to agree in advance on who should represent them, and thus avoid a contested election; and he revoked the charters of parliamentary boroughs, reissuing them in a form that allowed the crown to appoint – and remove – the officials who chose its members of Parliament.

James had an ulterior motive in modifying the franchise in this way: he was determined to repeal the Test Acts that restricted government posts to members of the Church of England, and to this end he also engaged in a practice known as 'closeting', in which he personally interviewed members of England's political elite to ensure that, when next summoned to Parliament they would favour repeal the Test Acts. He dismissed those who refused and replaced them with those more compliant – often either Catholics or Dissenters (Protestants who did not belong to the Church of

England). 'Not since the Norman Conquest', wrote Sir John Plumb, 'had the crown developed so sustained an attack on the established political power of the aristocracy and major gentry'.[35]

Because this process took longer than James had anticipated, instead of summoning Parliament he used his prerogative powers in April 1688 to issue a Declaration of Indulgence. In it, the king suspended 'all manner of penal laws in matters ecclesiastical' and decreed that all his subjects might 'meet and serve God after their own way or manner'. Catholics, Quakers, Jews and all other religious groups would now enjoy the same freedoms as Anglicans. Moreover – and here lay the fatal flaw – James required all members of the Anglican clergy to read out his declaration from their pulpits.[36]

Just before the day appointed for the public reading, seven bishops (including the archbishop of Canterbury) presented a petition to the king stating that they could not comply because no English king had the unilateral power to suspend a parliamentary statute (namely the Test Acts). Only Parliament could do that. James responded by charging the seven bishops with treason, but on 10 July 1688 a jury found them not guilty. They emerged to a tumultuous welcome.

That same day, seven peers signed a letter inviting James's son-in-law, Prince William of Orange, to invade England with sufficient force to defeat James's army and navy. Just as their predecessors in 1640 had invited a Scottish invasion and promised widespread support (Chapter 11), the seven peers assured William that 'there are nineteen parts of twenty of the people throughout the kingdom who are desirous of a change' of regime, and urged him 'to venture upon the attempt' before the end of the year.[37] By a combination of superb logistics and good luck, on 5 November (15 November by the Gregorian calendar), 500 Dutch ships arrived off the Devon coast and the prince began to disembark his 23,000 veteran troops with a powerful artillery train; but contrary to the assurances of the seven peers, scarcely any noblemen provided active support as the prince marched on London. In the event it scarcely mattered, because the steady stream of defectors fleeing James's court undermined the king's will to resist and just before Christmas a detachment of Dutch troops discreetly escorted him from his London palace to a ship that took him to France, creating an interregnum for the second time in a generation.

The political situation in London during the winter of 1688–89 was similar to that during the winter of 1641–42: the anointed king had fled, and although most of his opponents agreed that they did not want him back, they remained divided on what they wanted in his stead. But whereas John Pym could rely only on the London Trained Bands, William of Orange commanded veteran troops who garrisoned all key points in and around London: they remained there until those who had invited the prince over, and their allies, agreed to make him their sovereign. In February 1689 William and his wife Mary, James's elder daughter, accepted an invitation from a convention of peers and commoners to become joint sovereigns in England. In return – indeed in the same document – the new monarchs promised to redress specified grievances (such as recognition that 'suspending the laws, or the execution of laws, by regal authority, without consent of Parliament, is illegal') and guarantee basic liberties.[38] They also approved several critical pieces of legislation, notably the Mutiny Act (which left the crown in charge of the kingdom's armed forces, but only for six months at a time, which meant that Parliament must meet frequently in order to renew it), and the Toleration Act (which, although it did not mention Catholics – nor indeed the word 'toleration' – abolished the mechanisms that had allowed bishops and judges to enforce conformity). In addition, instead of allowing the crown to levy customs and excise, Parliament retained control of all public revenues and from this allotted William enough to defray the expenses of his household, officials, judges and diplomats. Parliament had regained almost all the powers acquired in 1641 – and this time it would keep them.

After the Revolution

The wisdom of this minimalist Glorious Revolution (as it became known) appears most clearly in comparison with events elsewhere in the Stuart monarchy. In Scotland, the insistence of the Presbyterian majority on excluding from power all who did not share their views strengthened the Jacobites (as the supporters of James II and his descendants were known), and although a Jacobite uprising in 1690 failed, the victors lacked the strength to resist another forced union with England in 1707. The Scottish Parliament would not meet again until the late twentieth century. In Ireland, the Jacobites (with French aid) triumphed until in 1690 William invaded in person at the head of his foreign troops and won the battle of the Boyne, consolidating the Protestant Supremacy established by Cromwell so that it would last for more than two centuries (three centuries in Ulster).

The return of extremely cold weather and a succession of disastrous harvests soon jeopardized Revolution Settlement. In London, John Evelyn reported in May 1698 that such unseasonably cold weather 'had not ben known by any, almost, alive', with 'all tree fruits ruined and threatning the rest with famine'.[39] Scotland suffered far more: in upland regions, cold and wet summers caused the harvest to fail every year between 1688 and 1698, a year when the Scottish government lamented the onset of 'a perfeit famine, which is more sensible than ever was known in this Nation'. The population of the northern kingdom fell by one-tenth in the course of the decade, with losses up to one-third in upland communities. Even in the 1780s, a survey of Scottish parishes recorded several areas abandoned at 'the end of the last century, when that part of the country was almost depopulated by seven years of famine: and now they lie neglected, along with many thousand acres, in like situation, in different parts'.[40]

The political legacy of the seventeenth-century revolutions also endured. In a debate on the repeal of the American Stamp Act in 1766, an English MP asked 'Shall we stay until some *Oliver* rises amongst them?'; and nine years later an *Essay upon government adopted by the Americans, wherein the lawfulness of revolutions are demonstrated*, published in Philadelphia, juxta-posed a section on 'the late civil war' in the 1640s, which dwelled upon 'the barbarous murder of King Charles I' by 'a few particular persons', with one on 'The Revolution justified' of 1688. If a monarch should fail to protect his subjects, 'they may refuse him their obedience', the anonymous author conceded (following Thomas Hobbes),

> but this does not give them any power over his person: [if] such prin-ciples and such practices upon such pretences were to be allowed, they would make the right of princes and the peace of society the most precarious thing that can be; and lays [rulers] open to the insults of every *Massinello*, who has but impudence enough to charge the government with popery or tyranny . . . and cunning enough to time it with some popular discontent.[41]

The *Essay* therefore expressed the hope that America might experience a bloodless revolt by 'the whole society', like the Glorious Revolution of 1688, and avoid the violence of the 1640s.

To Thomas Babington Macaulay, writing in 1848 as 'all around us the world is convulsed by the agonies of great nations', the events of 1688–89 in England, 'of all revolutions the least violent, has been of all revolutions the most beneficent'. It brought to an end the chronic instability that had char-acterized the composite monarchy of the Stuarts and ushered in a state dedicated to the vigorous promotion of economic development, broad reli-gious tolerance and free competition among political interests, characteris-tics that still define liberal democracies today. The Glorious Revolution therefore remained, as Macaulay proudly proclaimed, 'our last revolution'. It provided for England, albeit at a very high cost, a complete and permanent escape from both the General Crisis and the Little Ice Age.[42]

PART III
SURVIVING THE CRISIS

In 1623 an Italian preacher, Secondo Lancellotti, set out to refute those who complained about the unprecedented harshness of the world. His best-selling book *Nowadays, or how the world is not worse or more calamitous than it used to be,* identified forty-nine 'fallacies' held by contemporaries whom Lancellotti called *hoggidiani* – 'whiners' – and then listed examples in each of the forty-nine categories to prove them wrong. 'Princes nowadays are *not* more avaricious or indifferent towards their subjects than they used to be', he chided, and 'human life nowadays is *not* shorter, so that men do *not* live for less time now than they have done for thousands of years'. Lancellotti devoted his last chapters to natural phenomena. He reviewed recent accounts of famines, fires and plague epidemics as well as natural phenomena (such as earthquakes, floods and cold weather) 'nowadays', and argued that such catastrophes in the past had been far worse. According to Lancellotti, life had never been so good – but proving his case took over 700 pages.[1]

Nowadays sold so well that Lancellotti wrote a sequel (asserting that science and literature, too, were 'not worse than before'), but his vision was deeply flawed. To claim that seventeenth-century princes were no more 'avaricious or indifferent towards their subjects' than their predecessors (in itself hardly a ringing endorsement) obscured the fact that, in many cases, their misguided policies caused immense harm. The data in Part I of this book reveal that human life was indeed shorter than in the past, and that famines, fires and epidemics as well natural phenomena (not only earth-quakes, floods and cold weather but also volcanic eruptions and El Niño episodes) all increased markedly.

Not surprisingly, as the seventeenth century advanced, the ranks of 'whiners' swelled and their assessments became ever more pessimistic. 'The worst news keeps coming in from everywhere', the Spanish intellectual Francisco de Quevedo lamented to a friend in 1645. 'I cannot be sure whether things are breaking up or have finally broken up. God knows!' A few years later, the Spanish Jesuit, Baltasar Gracián, published *El criticón*

(*The fault-finder*), a vast allegorical novel that divided human life into four seasons, each divided into chapters that Gracián characterized as a crisis of some sort. Each of the thirty-eight crises he described presented a bitter and desolate survey of the human condition. In Paris, Thomas Hobbes complained about the 'continual fear and danger of violent death' in which he and his contemporaries lived.[2]

Nevertheless, Lancellotti had a point. Some of the pessimists lived longer than their 'whining' might have predicted: although Gracián was only fifty-eight when he died, Quevedo died aged sixty-five and Hobbes died aged ninety-one. Moreover, all three died in their beds of natural causes. Although unparalleled hardships befell many of their contemporaries who lived in composite states, urban areas, marginal lands, and macroregions, those in some other regions largely escaped. Although government oppression at a time of climatic catastrophe provoked major rebellions in two of the Italian states ruled by Spain – Sicily and Naples – strategic concessions brought peace in a matter of months. Mughal India, Safavid Iran and Tokugawa Japan all experienced extreme weather events and some rebellions in the seventeenth century, but they avoided the fatal synergy between human and natural factors that elsewhere turned crisis into catastrophe. Parts of sub-Saharan Africa and the Americas appear to have remained largely unscathed by both the Little Ice Age and the General Crisis. The experience of these regions supported Lancellotti's vision: life was *not* worse there, or 'more calamitous than it used to be'.

The reasons for this circumstance differed – a century of civil war in Japan; planned parenthood in Iran; elimination by Europeans of indigenous peoples in both Africa and the Americas – but the result was the same: the Little Ice Age struck societies in which the demand for food did not already exceed supply. This seems to have mitigated disaster and, in the case of Japan, it also promoted a strong recovery.

CHAPTER THIRTEEN

The Mughals and their Neighbours[1]

'The Most Potent Monarchs on Earth'

The Mughals ruled an area half the size of Europe and a population of perhaps 100 million (the same as the whole of Europe and second only to China). Most of their subjects lived in a fertile crescent running from the mouth of the Indus river through the well-watered and densely populated lands of the Punjab, along the rich Ganges Valley to the Bay of Bengal. Farmers there cultivated many crops, sometimes securing two harvests each year. Although primarily an agrarian state, the empire included three cities with 400,000 or more inhabitants and nine others with over 100,000 people; and Shah Jahan, who ruled from 1627 to 1658, created the only capital city in the world built entirely during the mid-seventeenth century. In both the cities and the countryside, craftsmen manufactured a vast range of high-quality goods for export.

Although the emperors ruled substantial areas in this heartland directly, from the 1570s onwards they granted the rest to prominent supporters (known as *mansabdars*, literally 'men who hold rank') in return for serving in the imperial army with a specified number of troops. Shah Jahan maintained over 400 *mansabdars*, but not all received the same amount of land: by far the largest share went to his four sons, who between them ruled almost 10 per cent of the empire. Each grant of territory was known as a *jagir*, literally 'holding place' because the emperors regularly rotated their *mansabdars* from one *jagir* to another. As Stephen Dale has pointed out, 'This system required accurate land-revenue estimates, which in turn necessitated land surveys in order to make assignments that generated sufficient funds to support the number of troops commanded by each officer. These features necessarily generated an enormous financial bureaucracy.' Controlling such a vast and diverse empire turned the Mughal emperors into workaholics. Whether in their capital, on progress or on campaign, every day they publicly bestowed titles and promotions, received petitions, heard claimants and dispensed justice: a chronicler claimed that in the 1660s the emperor 'appears two or three times every day in his court of audience . . . to dispense justice to complainants'.[2]

The Mughals constantly presented their rule as divinely sanctioned. Shah Jahan's father Jahangir (r. 1605–27) enrolled and initiated religious

disciples and sought advice from eminent Muslim holy men. His son
Aurangzeb (r. 1658–1707) could recite the entire Qur'an from memory and
enjoined his judges to uphold the Sharia. The Taj Mahal, built by Shah Jahan
between 1631 and 1653, emphasized the central role of Islam in his government: lengthy inscriptions from the Qur'an explained how each part of the
structure and the surrounding gardens replicate Paradise, with the dome
representing the throne of God. The emperor ordered that his own tomb
be placed immediately below that dome, together with an epitaph that
described him as *Rizwan*, the gatekeeper of Paradise.

Nevertheless, the Mughals did not place their trust in God alone. Even
the names they chose breathed absolute power: *Jahangir* means 'conqueror
of the world'; *Shah Jahan* means 'king of the world'; Aurangzeb took the
regnal name *Alamgir* ('world conqueror') at his accession. John Ovington
noted that each emperor constantly employed 'a numerous army to awe
his infinite multitude of people and keep them in absolute subjection',
and that he did so in person. According to Aurangzeb, 'An emperor should
never allow himself to be fond of ease and inclined to retirement', and he
warned his successors that they should 'always be moving about as much as
possible':

It is bad for both emperors and water to remain at the same place;
The water grows putrid and the king's power slips out of his control.
In touring lie the honour, ease and splendour of kings:
The desire of comfort and happiness makes them untrustworthy.[3]

True to his own advice Aurangzeb, like his predecessors, spent over one-third
of his reign on the move – although, also like them, he seldom strayed further
than 800 miles from Delhi. Three factors limited their radius: the Mughal
emperors moved slowly (never more than 10 miles a day and often less)
because it took time to receive in person (and overawe) the major vassals
along their route; they normally returned to the capital before the monsoon,
which halted travel throughout Hindustan between July and October; and
moving too far or too long from any given area might encourage rebellion.
Nevertheless, rebellions occurred and the Mughals often ended them by
negotiation and compromise rather than by force, not least because India
boasted a vast military labour market that favoured the seller: the emperor
was merely the largest, never the only, ruler recruiting troops each year.
According to calculations by Mughal ministers, some 4 million men in
northern India possessed military equipment and training, and the emperor
needed to raise enough of them not only to execute his own designs, but also
to prevent any of his rivals from creating an army capable of mounting a
challenge. They therefore maintained an army of 200,000 cavalry and 40,000
infantry, supported by elephants, horses, camels and oxen.

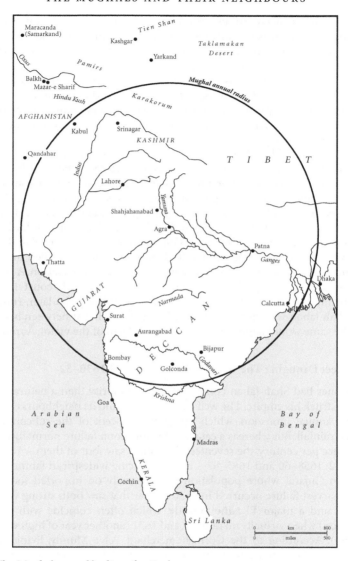

22. The Mughal annual 'radius of action'.

Although the Mughal emperors went on progress every year, they almost always returned to their capital ahead of the annual monsoon. Since this limited the time for travel to nine months, and since the court travelled at an average speed of about 5 miles a day, the emperor's effective 'radius of action' was around 800 miles. Significantly, this included Kabul (which the Mughals managed to retain) but not Qandahar or Balkh (where repeated campaigns failed).

Since such concentrations of humans and animals could not live off the country, the emperors not only took with them on campaign enough money to pay their troops punctually and in cash, but also prepared strategic reserves of specie along the line of march, and arranged for bankers travelling with the army to transfer revenues from outlying territories to their camp. This practice allowed troops to buy their food from merchants who set up bazaars every time the army halted. The Mughal military system had no peer in the seventeenth century, and it allowed the emperors to rule their territories effectively.

Nevertheless, like other dynasties with Central Asian origins, the Mughal regime suffered from a serious weakness: the system of tanistry, by which each ruler emerged only after a process of ruthless competition, sometimes between siblings but also between father and sons (Chapter 2). When rumours spread in 1622 that Emperor Jahangir had fallen ill, his eldest son, Shah Jahan, demanded sole command of the imperial armies and, when he failed to receive it, led an army from Gujarat in the west to Bengal in the east to create a coalition strong enough to depose his father. He failed, and only the surrender of his own sons as hostages secured a reconciliation. A succession struggle ensued when Jahangir died in 1627, with court factions supporting the claims of various rival princes; thirty years later, rumours that Shah Jahan was mortally ill would trigger a civil war between his sons; and the same would happen in 1707 after the death of the victor, Aurangzeb.

'A Perfect Drought': The Great Indian Famine of 1630–32

No sooner had Shah Jahan consolidated his authority than a natural catastrophe struck his empire. The well-being of India and its neighbours depends on the annual monsoon, which brings 90 per cent of the subcontinent's annual rainfall. But whereas a catastrophic monsoon failure normally occurs only once per century, the seventeenth century saw four of them – 1613–15, 1630–32, 1658–60 and 1685–87 – each producing widespread famine, especially in Gujarat whose population relied heavily on imported food. The worst harvest failure occurred in 1630 (a year that saw both strong volcanic activity and a major El Niño episode, which often coincide with a weak monsoon) when virtually no rain fell, and 1631 (another year of high volcanic activity). According to the Cornish merchant Peter Mundy, living in the Gujarati port of Surat, 'All the highways were so full of dead bodies that we could hardly pass from them without treading on or going over some.' Then, in 1632, the region suffered 'such inundations as have not been known or heard of in those parts', unleashing water-borne epidemics, such as malaria and dengue fever, with the result that 'Not a family' escaped 'agues, fevers and pestilential diseases ... so that the times here are so miserable that never in the memory of man [has] any the like famine and mortality happened.'[4]

Although no census and (as yet) no natural archive exist to quantify the scale of the disaster, Mundy believed that 'The famine itself swept away more than a million of the common or poorer sort; after which the mortality succeeding did as much more amongst rich and poor'. Foreigners did not escape: of the twenty-one Englishmen living in the East India Company's factory at Surat in 1630, seventeen died in the following three years. In Goa, 500 miles to the south, the viceroy of Portuguese India estimated that famine and plague had killed 4 million Mughal subjects, and (he claimed), although a hardened soldier, he could hardly bring himself to recount the suffering he had seen.[5]

As usual during such climatically induced catastrophes, as food prices rose the purchasing power of most families fell, and since buying staples now absorbed almost all their income, the demand for manufactured goods also fell. Both manufacture and trade therefore atrophied, producing unemployment among artisans and transportation workers; and neither the production of cotton or indigo in the region ever recovered their previous levels. After the famine, what 'was in a manner the garden of the world is now turned into a wilderness, having few or no men left to manure their ground, nor to labour in any profession'. Mundy predicted that the region 'will hardly recover its former estate in fifteen, nay in twenty years'.[6]

Mundy's forecast proved to be overly pessimistic only because Shah Jahan used some of the vast resources at his disposal to respond to the disaster with a series of vigorous countermeasures. He established soup kitchens and almshouses 'for the benefit of the poor and destitute. Every day sufficient soup and bread was prepared to satisfy the wants of the hungry'. At this time the emperor resided in the capital of one of the Hindu states of the Deccan he had recently conquered, and he ordered that, as long as he remained there, '5,000 rupees should be distributed among the deserving poor every Monday ... the day of the emperor's accession to the throne. On twenty Mondays, one *lach* [100,000] of rupees was given away in charity'. Also, since 'want and dearness of grain had caused great distress in many other countries', Shah Jahan forgave taxes worth 7 million rupees, 'amounting to one-eleventh part of the whole revenue' in order to 'restore the country to its former flourishing condition and the people to affluence and contentment'.[7]

Shah Jahan also sponsored a number of other initiatives to promote economic recovery after the monsoons resumed their normal rhythm: he donated ploughs to the poor as he visited each region, 'so that the forests might be cleared and land cultivated' in order 'to populate the country', and he and his principal courtiers founded hundreds of market towns. The emperor also took steps to increase exports. He sponsored the cultivation of cotton, sugar cane, silk, tobacco and indigo in Bengal, to attract European merchants; and he ordered the construction of great ships in Gujarat, which allowed him to participate in the lucrative carrying trade between Mughal India and the Persian Gulf and Red Sea. Thanks to this combination of tax breaks and stimulus spending, within a decade Gujarat (as well as other regions) once again reported a revenue surplus.

Shah Jahan also spent prodigiously on other things: 10 million rupees on his Peacock Throne; 7 million on the Taj Mahal; 7 million more on building a new capital, Shahjahanabad (now Delhi); and almost 1 million on creating the Shalimar Gardens at Lahore. He also waged war almost constantly – against the Hindu states of the Deccan, against the Portuguese in Bengal and against the Sikhs. Nevertheless, despite all this expenditure, after two decades on the throne Shah Jahan had accumulated a reserve of 95 million rupees. Three reasons explain this unique fiscal outcome. First, the productivity of both rural and urban workers, coupled with the fertility of many areas under Mughal rule, yielded high tax revenues. Second Shah Jahan's wars often turned a profit, both in booty and in the massive number of captives sold into slavery in Central Asia. Third, the emperor limited his expenditure on foreign-policy goals. Although he campaigned in the Deccan, he never seems to have contemplated an attack on the Tamil lands further south or on Sri Lanka; and although he expanded into Bengal and founded Jahangirnaga (now Dhaka), a Mughal city that by 1640 numbered over 200,000 inhabitants, he made no serious plans to advance into Burma. As Sanjay Subrahmanyam observed: 'We may see the Mughal empire as an unfinished project, in a terri-torial sense, but also as one that had a proper sense of its limits.'[8] The sole area of Shah Jahan's imprudence was Afghanistan.

Afghanistan: The Perpetual Battleground

The Mughals prided themselves on their direct descent from both Chinggis Khan and Timur (Tamerlane), and their official chronicles frequently noted plans to recapture Central Asia, their ancestral home; but only Shah Jahan attempted to achieve this goal. In 1639, his troops seized Qandahar in southern Afghanistan from its Safavid defenders, and the emperor and a massive entourage crossed the Khyber Pass (3,000 feet above sea level) for the first time. They spent the summer at Kabul as they prepared to conquer the lands beyond, but the Uzbeks, who occupied the Mughal homeland, mounted such an effective defence that the emperor beat an ignominious retreat. In 1646 a succession dispute between two Uzbek rulers, one of whom appealed for Mughal assistance, led Shah Jahan back to Kabul, whence his troops at last crossed the Hindu Kush and occupied the fertile alluvial plain around the great trading city of Balkh.

This was a remarkable achievement: the Mughal expeditionary force was probably the largest ever to enter the region before the Soviet invasion of 1979. Nevertheless, even at the best of times, Afghanistan lacked both the crop surpluses and infrastructure required to feed such a large slow-moving army – and the Little Ice Age was far from the best of times. Unusually heavy snowfalls in the 1640s reduced both the campaign season and the quantity of supplies that could be brought over the passes from India, while a shorter growing season throughout the Himalayan region reduced local crop yields. The Mughal troops therefore depended for their survival on what they could forage from the countryside, but the scorched-earth policy pursued by the Uzbeks made foraging impossible. So although the gold, jewels, silk and brocade of the imperial army, as well as their weapons, horses and war elephants, initially impressed the population of Balkh, the winter of 1646–47 brought such intense cold that the Mughal garrisons 'burned themselves in the fires they lit for warmth, and no one left their house for fear of being frozen'.[9]

Nevertheless, in spring 1647 the emperor returned to Kabul for the third time and his son Aurangzeb led another army across the Hindu Kush. The prince used his war elephants as well as his new troops to defeat Uzbek attacks, but three months later he admitted defeat and abandoned Balkh. It took the survivors a month to regain Kabul, and they arrived just in time to accompany the emperor back across the Khyber Pass. This encouraged the shah of Iran to demand the return of Qandahar – and when no satisfactory answer came, in 1649 he besieged and recaptured the city. Three subsequent Mughal attempts to regain Qandahar all failed.

Shah Jahan's efforts had advanced the frontier of his empire scarcely 30 miles beyond Kabul, yet it cost the lives of perhaps 40,000 soldiers and over 35 million rupees. The war also caused extensive damage to Afghanistan but the fragmentary nature of the surviving sources makes it difficult to assess the exact impact. The records of the shrine at Mazar-e Sharif ('tomb of the holy one'), close to Balkh, shed a rare beam of light. The shrine had prospered in the earlier seventeenth century, as local Afghan rulers granted lands and founded *medreses*, until its activities involved over 3,000 people either as producers or consumers; but many local merchants and farmers fled when Mughal troops entered the valley, so that in their first year the army of occupation collected only half of the previous year's revenues, and in their second year only half of that. Grain prices skyrocketed, and fire-wood (essential for survival at such a high altitude) could not be found. Only in 1668 did the administrator of the shrine dare to leave Mazar-e Sharif to secure a charter from the nearest ruler that restored the lands and revenues lost 'because of the turmoil in the world'.[10]

The Crisis of Mughal India

Shah Jahan's favourite son, Dara Shikoh, resided with him at court and boasted fiefs (*jagirs*) almost equal to the combined fiefs of his three younger brothers, who resided in distant provinces as the emperor's representatives, and from time to time commanded major campaigns on his behalf. Shah Jahan intended this arrangement to avoid any rebellion similar to the one that he and his brothers had mounted against their father, Jahangir, but he reckoned without the ambition, ability and charisma of Aurangzeb. In 1636, the young prince began to govern the newly conquered regions of the Deccan for his father, and over the next eight years he built up a retinue of loyal followers whom he rewarded with lands acquired in frontier warfare. He constructed a new city, which he named Aurangabad, and both there and in his conquests he made lavish grants to individual mosques and *medreses*, forged ties with religious leaders and encouraged conversion to Islam by new subjects.

Shah Jahan next entrusted the reconquest of Central Asia to Aurangzeb and, although his efforts failed, the prince increased his experience of command. He also became convinced that his brother Dara Shikoh had held back supplies and reinforcements for his campaigns, and in 1652 Aurangzeb met with his two other siblings to 'make plans for the preservation of their life and honour and the management of their affairs' in case the emperor suddenly died. Aurangzeb then returned to the Deccan where he built up a cadre of talented and devoted military officers in preparation for the inevitable succession struggle, and with their assistance he led successful campaigns against the empire's southern neighbours, securing substantial treasure and tribute. Aurangzeb thus possessed immense political, military and financial resources when, late in 1657, rumours spread that his father 'was unable to attend to business' through illness and that Dara Shikoh had taken 'the opportunity of seizing the reins of power'.[11] The princes governing Gujarat and Bengal at once took control of local treasuries, issued their own coinage and had Friday prayers read in their name.

The ailing emperor responded by charging Dara Shikoh with restoring order, but he foolishly divided his forces and sent armies simultaneously against both usurpers. Aurangzeb now joined his rebellious sibling in Gujarat, while his troops secured Agra and, with it, the imperial treasury and the person of Shah Jahan. The emperor offered to partition the empire among his sons but, confident that he could triumph alone, Aurangzeb declared himself emperor and campaigned against each of his brothers in turn, executing them and any of their sons who defied him. He kept his father in prison until he died.

Unfortunately for the new emperor's subjects, as a Mughal chronicler observed, these 'disturbances and the movement of large armies for two years in different parts of the empire', coincided with another failure of the monsoons. In 1659 southeast India saw 'so great a famine' that, according to resident English merchants, 'the people [are] dying daily for want of food'. In Gujarat 'the famine and plague' became 'so great' that (as in 1630–32) they 'swept away [the] most part of the people'. The monsoon failed again in 1660, and English merchants in Madras lamented the 'great dearth' in the southeast 'now these eighteen months'. Their colleagues in Gujarat believed that 'never famine raged worse in any place, the living being hardly able to bury the dead', and (according to a Mughal chronicler) 'the scarcity of rains' caused famine in the Ganges plain.[12]

Aurangzeb set up ten soup kitchens in Agra, with more in surrounding communities, and he obliged leading noblemen to do the same. He also issued an edict abolishing taxes. It did not bring relief. In 1662 a major fire ravaged the capital, Shahjahanabad; in Gujarat the English merchants predicted that extreme drought would 'utterly dispeople all these parts' because 'there are more than 500 families of weavers that are already fled, and the rest will certainly follow, if the famine should increase'; in Bengal, famine, drought and disease also caused a sharp increase in the price of staples; and, in the south, Kerala experienced three years of drought.[13] A French physician residing in Delhi after the famine noted that 'of the vast tracts of country constituting the empire of Hindustan, many are little more than sand, or barren mountains, badly cultivated, and thinly peopled; and even a considerable portion of the good land remains untilled from want of labourers'. A new revenue survey compiled for Aurangzeb's treasury showed that receipts from the ten core provinces of the empire had fallen 20 per cent below pre-war levels. The civil war also allowed Shivaji, a Maratha nobleman, to build up a powerful Hindu state that defied the Mughals: Shivaji sacked Surat, their principal port, in 1664 and again in 1670.[14]

Although Shireen Moosvi, an economic historian, has suggested that the crisis of 1658–70 formed the watershed between Mughal expansion and decline, the empire continued to expand for four decades more.[15] In the 1670s Aurangzeb captured, tried and executed the charismatic Sikh leader Tegh Bahadur, and thus halted a movement that had won both Hindu and Muslim converts; and in the 1680s he defeated a Rajput revolt led by one of his own sons, conquered the Deccan kingdoms of Bijapur and Golconda, and captured and executed the leader of the Maratha confederacy. After these spectacular successes, however, Aurangzeb spent the rest of his reign campaigning against the Marathas in the Deccan, despite monsoon failures in 1686 and 1687 (both years of strong El Niño activity), which intensified the deleterious effects of his campaigns.

Despite his prodigal spending on war, when Aurangzeb died in 1707 his treasury contained 240 million rupees – far more than the total inherited from his father and almost certainly more than the resources of any other ruler in the world. Nevertheless, the Mughals' Rajput, Sikh and Maratha neighbours stood poised to exploit the inevitable succession war that erupted immediately after the emperor's death. Even the wealthiest state in the world could not overcome the weaknesses caused by tanistry.

Southeast Asia: Turning Plenty into Poverty

Several rulers in the Indonesian archipelago in the early seventeenth century also boasted vast wealth – and for much the same reasons as the Mughal emperors: the tax yield from unusually fertile land and profitable commerce. Nevertheless, also like Mughal India, their lands suffered from the unusual frequency of both El Niño and volcanic events in the mid-seventeenth century. Most of the archipelago normally receives 100 millimetres (4 inches) or more of rainfall every month (there is no dry season), but tree-rings from central Java between 1643 and 1671 reveal the longest period of drought ever recorded. Not a single year during this period received the normal amount of rain, and 1664 was the driest year recorded in the last five centuries. Surviving written sources from Java, the most densely populated island, show famines in six years and poor rice harvests in ten more between 1633 and 1665. Those from the outer islands of the archipelago show famines in two years, and poor rice harvests in ten more, with universal shortfalls in four years: 1633, 1657, 1660 and 1664.

Three natural advantages mitigated these effects of the Little Ice Age. First, since the archipelago straddles the equator, changes in solar energy there have far less impact than in more northerly latitudes. Even during the great drought, abundant arable land remained available. Second, the tropical climate and fertile soil of the archipelago allowed a wide variety of edible crops to flourish, some of them resistant to adverse weather. Third, the densely populated coastal areas managed to avoid some consequences of the drought by importing rice from unaffected areas. Dutch East India Company ships normally brought rice from India to relieve famines, because crop failures in the archipelago and in the subcontinent rarely coincided. For all these reasons, although 20 million people inhabited the Indonesian islands in the mid-seventeenth century, few starved. The situation changed when, despite such natural abundance, human agency managed to create a crisis.

The eagerness of several groups of merchants – Portuguese, Dutch, English, Muslim, Chinese and Japanese – to acquire individual spices inexorably led each island to concentrate on raising a single cash crop (nutmeg on the Banda Islands, cloves on Amboina, and so on), abandoning all others. Until the 1630s, local growers prospered and, thanks to export taxes, so did their rulers; but this state of affairs changed rapidly as political and military events closed off one market after another. In 1638, the Japanese government prohibited all long-distance trade (Chapter 3); in 1641 the Dutch East India Company captured Melaka from Portugal, which gave them a stranglehold over all ships sailing through the Straits; and Chinese maritime trade contracted after the fall of the Ming in 1644. These developments conferred enormous commercial advantages on the Dutch, who exploited their control of the sea lanes to exclude competitors and increase the price of the food they imported to Indonesia. They could always blockade the ports of producers dependent on imported foodstuffs, creating artificial famines that compelled the victims to sell spices cheap.

The falling price of spices and the rising cost of food staples had profound consequences for the peoples of the archipelago. Some destroyed their pepper vines and reintroduced rice cultivation, others stopped producing spices for reasons of security. One Muslim prince forbade his subjects to plant pepper 'so that they did not thereby get involved in war, whether with the [Dutch] or with other potentates' – but achieving this outcome was easier said than done because the Dutch usually attacked in alliance with a rival 'potentate'.[16] In southern Sulawesi, the sultans of the prosperous port-city of Makassar overran the neighbouring state of Bone and then defied Dutch pressure to grant them monopoly trading privileges, instead welcoming all traders. They prudently invested some of the resulting revenues in constructing a large fortress (Sombaopu) around the royal palace which, on the seaward side, boasted walls 14 feet thick with four bastions equipped with heavy guns donated by Danish, English and Portuguese traders, all of whom maintained factories in the city. After the fall of Melaka in 1641, between 2,000 and 3,000 Portuguese transferred their activities to Makassar, and the sultanate enjoyed unprecedented prosperity until Arung Palakka, the exiled prince of Bone, sought Dutch help. In 1667, at Palakka's request, Dutch warships bombarded the new fortifications of Makassar until they had secured important trading concessions. Two years later, claiming that the sultan had breached the agreement, the allies returned and took the city by siege. Arung Palakka now became the new sultan of Makassar, and the Dutch created a citadel from which they controlled all seaborne commerce in southern Sulawesi.

The war between Makassar and the Dutch came at a high cost for all protagonists: it left much of Makassar in ruins and, although the Dutch eventually gained from the elimination of a competitor, several decades passed before profits from trade equalled the cost of the two naval expeditions.

The Enigma of Iran

The surviving evidence suggests that Safavid Iran, too, should have undergone a major crisis during the mid-seventeenth century. Confederations of pastoral nomads had long dominated Iran's history, providing all of its ruling dynasties, including the Safavids, and also the shah's elite troops: the *Qizilbash* ('red heads', named after their red turbans). Around 1500 AD the first Safavid shah, Ismail, the charismatic head of a Sufi brotherhood, had claimed to be the Messiah and called upon all who shared his Shi'ite faith to support him. His project succeeded only in Iran, leaving his country – with sizable Jewish, Hindu, Sunni Muslim and Armenian Christian minorities – surrounded by the Sunni Ottomans, Mughals and Uzbeks. By 1600 the shahs ruled scarcely 10 million subjects (compared with 100 million Mughal and 20 million Ottoman subjects), leaving their state extremely vulnerable whenever their mighty neighbours decided to declare war.

At the core of the Safavid state lay the Iranian plateau, a vast but arid interior basin with its large settlements mostly located around the edge, whereas the centre is sparsely populated. The French jeweller Jean Chardin, who lived in Iran for much of the 1660s and 1670s, attributed the survival of the Safavid state, despite such natural disadvantages, to the innovations of Shah Abbas (r. 1588–1629). In politics, Abbas abandoned tanistry: to avoid the risk of sons challenging and deposing their fathers (as he himself had done), he killed or blinded his own sons and created a harem where his grandchildren would live until one of them succeeded. This brutal practice, repeated at each accession, put an end to succession struggles (although, as in the Ottoman empire, it also produced inexperienced rulers). He counterbalanced the power of the *Qizilbash* by creating a force of slave soldiers (*ghulam*) whom he armed with muskets and artillery. Abbas also undertook economic reforms: he improved roads and created a corps of highway police to protect their users; constructed bridges and caravanserais; and fostered the cultivation of cotton, rice and silk. He also brought Armenian merchants to his new capital, Isfahan, where they oversaw the production and sale of silk thread, which soon became both Iran's most lucrative export and the shah's principal source of revenue. Ironically, Shah Abbas used the silver produced by selling silk to the Ottoman empire to fund a war that wrested Iraq and most of the Caucasus from Ottoman control. He also captured both the Persian Gulf ports of Gombroon (which he renamed Bandar Abbas after himself) and Hormuz from the Portuguese, opening further markets for Iranian silk. In all, Abbas virtually doubled the size of the Safavid state, and Isfahan grew into a metropolis of perhaps 500,000 inhabitants.

As soon as Abbas died in 1629, his oldest grandson, Safi, murdered or blinded all his male relatives in the hope of avoiding civil war. He nonetheless faced a serious uprising against high taxes in the silk-producing region of Gilan led by a discontented *Qizilbash* who claimed to be the Messiah; the new shah prevailed only after executing him and 2,000 of his followers. Three years later, after thwarting an attempt to poison him, Safi massacred all the surviving female descendants of Abbas and also all clerics at his court save one, who served as the shah's minister of religion, education and justice. The eunuchs and *ghulam* of the palace ran the rest of the government, but they failed to conserve Abbas's gains. In the north, Cossack raiders terrorized the lands around the Caspian Sea; in the east, Uzbek chieftains launched repeated raids; and, in the west, the Ottomans recaptured Iraq, leading Safi in 1639 to agree to the Peace of Zuhab, which returned all of Mesopotamia to Ottoman control (Chapter 7).

Although the Safavids would never again rule Iraq, their brief period of dominance produced long-lasting consequences. The Iranians had exalted their new Shi'ite subjects and oppressed the Sunni majority, and so when Safi relinquished the region, he abandoned a large Shi'ite population. They would suffer political exclusion and oppression until the collapse of Ottoman control until 1916, and again under Saddam Hussein in the later twentieth century. As in Ireland (Chapter 12) the roots of an insurgency that has lasted until today originated in the sectarian struggles of the mid-seventeenth century.

Iran itself enjoyed a peace dividend after the Peace of Zuhab, which reopened the Silk Road to the Mediterranean. The shah and his Armenian merchants prospered, and when Safi died three years later, his ten-year-old son Abbas II (r. 1642–66) succeeded peacefully. In 1649, at the request of the Afghans, he captured Qandahar from the Mughals and retained it despite three desperate sieges (page 307), commemorating his success in magnificent historical wall paintings that can still be admired in the gardens of the Chihil Sutun ('forty columns') palace in Isfahan – one of several splendid architectural complexes built by Abbas II to adorn his capital. Stephen Dale has suggested that Chihil Sutun may represent the shah's 'own perception of his reign as a kind of Safavid golden age', but soon after its completion a run of extreme climatic events caused widespread suffering.[17]

Iran regularly experiences droughts, high winds, violent hailstorms and earthquakes; but the second half of the seventeenth century saw far more of these natural disasters than usual. For six months in 1663 the northwest of the country received neither rain nor snow, so that 'wells dried up and crops withered', and in 1665–66 a poor harvest led to bankruptcies among the merchant community. Plague and famine raged for another three years. In addition, repeated devaluations of the currency created economic instability; the Cossacks of Stenka Razin raided lands around the Caspian Sea (Chapter 6); and prolonged rains ('the worst in living memory') destroyed 2,000 houses in the wine-producing area of Shiraz. When Jean Chardin returned to Iran in 1676, he believed that the wealth of the country had diminished by 50 per cent since his first visit a decade earlier.[18]

Two factors helped the Safavid state survive both economic instability and political ineptitude. First, its rulers rebuffed numerous invitations from Russia, Venice and others to join their hostilities against the Ottomans. Rather like the Mughal empire, we may see the Safavid state as 'one that had a proper sense of its limits'. Second, and no less important, Chardin attributed the stable population of Iran during the Little Ice Age to the 'great many women who make themselves abort and take remedies against growing pregnant.'[19]

Although the Qur'an is silent on reproductive matters, several *hadith* enjoined birth control, and Islamic law allows abortions within the first three months of pregnancy (albeit officially only to preserve the life of the mother and not as a way to control population size). By the seventeenth century, those able to read medical treatises in Arabic could find descriptions of almost 200 techniques of contraception and abortion, and pharmacies stocked many of the medications mentioned therein. Some of these treatises explicitly allowed contraception (including early abortions) 'when times are bad' or 'to attempt to escape having too many dependants, or to escape having any dependants at all'. Iran therefore avoided the overpopulation that plagued states elsewhere.[20]

Nevertheless, the Safavid state remained fragile and survived in part because it lacked external challengers: early in the eighteenth century the shahs proved powerless to halt attacks by prevent the Ottomans and the Russians, both of whom made territorial gains. Later their Afghan neighbours rebelled and murdered the last male members of the dynasty, before defeating the Mughals and conquering Hindustan in 1739.

Since both the Safavid and Mughal states eventually collapsed into anarchy, some might infer that they merely deferred catastrophe rather than averting it; but this is unfair. Like other parts of south and southeast Asia, although their territories suffered natural disasters (notably droughts) in the mid-seventeenth century, they escaped political catastrophe, and for most of their subjects life was indeed 'not worse or more calamitous than it used to be'.

CHAPTER FOURTEEN

Red Flag over Italy[1]

Sicily in Revolt

Spain began to impose heavier taxes on its Italian subjects in 1619, when Philip III demanded 3 million ducats to pay for the troops he sent to assist Emperor Ferdinand II defeat his enemies in Germany and Bohemia (Chapter 8). Despite their unequal size and wealth, the king asked for 1 million each from Lombardy, Naples and Sicily – but Sicily presented a particularly tempting fiscal target. The island's fertile soil normally produced yields of seven to ten grains of wheat and nine to eleven grains of barley for each grain sown, the highest recorded in seventeenth-century Europe. Thanks to a benign climate, the population of the island doubled from 600,000 in 1500 to 1,200,000 in 1623. Three-quarters of them lived in towns, including 130,000 in Palermo and perhaps 120,000 in Messina, the two principal cities.

Nevertheless, there were two Sicilies. The west and centre of the island, including Palermo, the administrative capital, produced and exported primarily grain; while the east, including Messina, the commercial capital, produced and exported primarily silk. Despite their prosperity, both parts of the island were economically vulnerable. Messina and its hinterland produced virtually no grain, so its prosperity depended on the ability to import bread and export textiles. When war broke out between Venice and the Ottoman empire in 1645 (Chapter 7), silk exports from Messina fell by one-quarter, throwing thousands out of work. A recession could strike western Sicily with equal severity and suddenness because drought could render barren the marginal lands around the new towns founded in the sixteenth century on the central uplands (Chapter 3). During the 1640s, yield ratios on some farms fell to 1:3 – the lowest recorded on the island during the entire early modern period – which drastically reduced the grain available to feed Palermo.

The decision by Philip IV and his council in 1643 to make Italy contribute more for Spain's foreign wars (Chapter 9) almost doubled the tax burden on his Sicilian subjects over the next four years. The government of the island, headed by a viceroy, met these demands with many of the same fiscal strategies adopted elsewhere during the mid-seventeenth-century crisis: it turned to regalian rights (such as the *media anata* and stamp duty);

it bullied the island's parliament into voting new taxes and sold the right to collect them to bankers in return for cash advances; and it imposed excise duties on almost all commodities in common use. When these measures fell short of the king's demands, viceroys also alienated crown lands and rights (although this reduced revenue in the long term), sold public offices (although this undermined the loyalty and integrity of the civil service) and issued bonds (although the interest payable greatly increased expenditure). The fiscal pressure forced major towns to take similar measures: they sold lands, imposed new taxes and borrowed money at interest (by the 1640s, the public debt of Palermo stood at over 4 million ducats). Both central and local governments therefore lacked resources when a series of natural disasters struck the island.

Starting in September 1645, rain fell almost continuously on Sicily for a year, ruining the winter crops and drastically reducing the yield of the summer harvest. In addition, floods washed away houses and bridges, gales destroyed the olive trees, and an eruption of Mount Etna caused widespread damage. Nevertheless, Philip IV commanded his viceroy to send 300,000 bushels of grain to Spain and ignored all objections that the island's own food shortages made this impossible. He also forbade Messina's magistrates to subsidize grain prices (as they had been doing) to mitigate the impact of rising food prices on the poor, and in August 1646, with wheat prices higher than ever previously recorded, bakers implemented the only alternative to raising bread prices: they reduced the size of a standard loaf.

A crowd of women and children now took to the streets of Messina, brandishing the smaller loaves on the ends of pikes and shouting 'Long live the king and down with the evil government', and rioters killed one of the magistrates, torched the houses of two others and stoned the homes of several nobles. The viceroy, the same marquis of Los Vélez who had performed so dismally in the campaign against the Catalans in 1640–41 (Chapter 9), now reacted swiftly and effectively: he deployed the soldiers aboard the galleys stationed at Messina to arrest the rioters; sent grain from the government's strategic reserve; and travelled to the city in person to restore order.

Although these measures pacified Messina, they did nothing to address the crisis caused by a severe drought that ruined the harvest in the rest of Sicily. A serious epidemic broke out in Palermo, killing hundreds each week as bread prices reached their highest level in three centuries. The local clergy organized processions to pray for rain and beg forgiveness for the sins that, they claimed, had brought God's punishment on the land. Then a miracle occurred: it rained for two days and the crops began to grow again. Just as popular anxiety subsided, 'a Sirocco wind blew day and night, so fierce that it dried the breath in your throat and killed off both grain and fruit crops' – but on 19 May 1647 a second miracle occurred: a ship arrived in Palermo harbour carrying several tons of grain. Unfortunately the ship also brought letters from the king repeating the threat that unless city magistrates ended the bread subsidy they would have to pay for it personally. The bakers therefore received orders to reduce the size of the standard loaf 'in order to align the price with the cost'.[2]

The prospect of mass starvation, the outpouring of religious zeal and then two apparent miracles of deliverance created emotional overload in Palermo. The following day about 200 people, many of them boys and women, gathered outside the city hall and shouted 'Long live the king and down with the evil government', 'Big loaves, no excise' and, more simply, 'Bread, bread'. Their cries attracted a much larger crowd, and some began to throw stones at the windows and set fire to the doors. Others broke into the principal prison and freed over 1,000 inmates.[3]

The convicts transformed the situation. One of them, Antonino la Pelosa, incited the crowd to storm the treasury and burn the tax documents within. In a desperate attempt to restore order, Los Vélez abolished excise duty (*gabella*) on five basic foodstuffs and fixed the prices at which each should be sold; and he restored the bread subsidy so bakers could produce larger loaves for the same price. More surprisingly, he also deposed the magistrates who had decreed the reduction; granted citizens the right to elect two magistrates themselves; and formally pardoned not only all demonstrators but also those freed from prison. The violence subsided until the magistrates, perhaps still fearful that the king would hold them responsible for any food subsidy, insisted that all items except bread should be sold at cost price. La Pelosa and his supporters denounced this move as a breach of faith and burned down the houses of officials and merchants involved in tax collection. Famished citizens in other Sicilian towns drew the obvious lesson from the ease with which Palermo had won concessions, and crowds led by women and boys took to the streets shouting 'Long live the king of Spain, down with the excise'. Almost everywhere the magistrates complied, and if they refused, the crowds broke open the prisons and burnt down the houses of the rich until they received the concessions they demanded.

Although the viceroy captured and executed La Pelosa, his concessions created a fiscal crisis in Palermo, because without the excise duties the city lacked money to pay its creditors. On 1 July 1647 Los Vélez met with the newly elected magistrates and the guild leaders, and they agreed to impose taxes that targeted the rich: duty would henceforth be paid on every window, door and balcony of the city's houses; on every pound of tobacco; and on every horse-drawn carriage. They also decreed that there would be no exemptions for either nobles or clerics. This represented the first progressive fiscal system of early modern Europe, and it might have produced a lasting settlement had not news arrived of a revolution in Naples.

Red Flag over Naples

The kingdoms of Sicily and Naples were adjacent, enjoyed numerous cultural and commercial ties, obeyed a common master, and each closely monitored events across the straits. Reports of the Palermo revolt spread swiftly to the city of Naples, where posters criticizing the Spanish government soon appeared. There was much to criticize. With perhaps 350,000 inhabitants, Naples was the largest city in the Spanish monarchy and one of the largest in Europe, and many observers contrasted the ostentation of the city's patrician elite with the destitution of the rest of the population, many of whom eked out a living in 'high-rise' apartment buildings in the city centre, in shacks around the periphery, or in the streets. The Neapolitans called them *lazzari* ('Lazaruses') because their ability to arise from their beds and walk seemed miraculous – but also considered them 'prone to rebellions, to revolutions [*rivoluzioni*], to break laws, customs, obedience to superiors' and 'capable of reducing everything to disorder with every tiny movement'.[4]

To minimize the risk of disorder, viceroys exploited rivalries among different social groups. The patricians belonged to one of six groups of families (known as *seggi* or 'seats'), and a representative from each served on the city council, together with a candidate chosen by the viceroy from a list submitted by the non-patrician householders (misleadingly known as the *eletto del popolo*, 'man elected by the people') who served for a six-month term, which the viceroy could terminate or prolong at will. Despite his title, the holder reflected the views of the government. In 1620 a major political crisis ensued when the lawyer Giulio Genoino, then serving as *eletto del popolo*, proposed a programme of constitutional reform that included equal representation in the city's government of patricians and *popolo*. One of his manifestos justified the new division on the basis of their respective numbers – '300,000 against 1,000' – but although the viceroy endorsed the plan, Madrid vetoed it and condemned Genoino to prison.[5]

Genoino and other members of the city's intellectual elite had consulted legal and historical sources, which convinced them that the Spanish government had tampered with their 'ancient constitution'. They found, for example, that 'Naples was born a free Republic, divided into Senate and People'; that it had paid no taxes until the time of the Normans; and that it had 'elected a doge pleasing to the [Byzantine] emperor or to another friendly prince who offered protection'. The historian Camillo Tutini published an erudite but controversial book which argued that, since Roman times, *popolo* and patricians had shared power in Naples equally, with the implied argument that they should do so again.[6] By contrast, Neapolitan reformers kept silent about another innovation: the capital's exemption from all taxes imposed on the rest of the state. Representatives from the rest of the kingdom met every two or three years in a Parliament dominated by the nobility and voted taxes that, for the most part, fell on their vassals. It could scarcely be otherwise, because by 1640 only ten towns besides Naples remained under direct royal control: the crown had sold the rest, along with most of its domain lands, to nobles. As in other countries, it had also sold titles of nobility – the 161 peers of 1613 rose to 271 in 1631 and to 341 in 1640 – and to raise cash it also sold to nobles (both new and old) public offices, regalian rights and jurisdiction over their vassals (including the right of final appeal). Once they had acquired these comprehensive legal rights, the nobles used armies of retainers to compel their tenants to sell produce to them far below market price, to exact numerous (and sometimes new) feudal services and to suppress any opposition by force. Such excesses generated anger and protests throughout the kingdom.

The mounting fiscal burden imposed by Madrid exacerbated these dangerous tensions. Each year from 1637 to 1644 over 1 million ducats left the kingdom to pay for Philip IV's wars – one viceroy protested that Naples contributed more than the Americas to imperial defence – but in 1641, with the revolts of Catalonia and Portugal to suppress, and war with France on multiple fronts, the government's demands reached an unprecedented 9 million ducats and 14,500 troops. In response, the viceroys tripled taxes by using much the same combination of expedients as other regional governments of the Spanish monarchy. They also resorted to loans from local bankers, which dramatically increased public sector borrowing, until by 1647 the total public debt approached 150 million ducats – not far short of the kingdom's gross domestic product. The viceroys also squeezed the eight public banks of Naples to surrender their cash deposits, which they then exported, issuing paper currency in lieu of the absent specie (probably the first true paper money in Europe). By 1647 all eight banks were effectively bankrupt, while private financiers demanded ever-higher interest on each loan raised for the wars in Spain and Lombardy: 8 per cent in the 1630s, 40 per cent in 1641, 55 per cent in 1642, and 70 per cent in 1643.

These fiscal developments affected the entire population of the kingdom. Tens of thousands of Neapolitans subscribed to the loans – either directly or through bankers – attracted by the high interest secured on the yield of future taxes, and so they faced ruin when those taxes failed to materialize. Moreover, as in Castile, depopulation reduced tax revenues. A census of the kingdom (excluding the capital) in 1595 revealed some 550,000 households, but half a century later another census recorded scarcely 500,000, so that the unremitting fiscal pressure from Madrid fell upon a shrinking tax base, forcing the viceroys to search for new sources of revenue. In 1642 the parliament of Naples agreed to raise an unprecedented 11 million ducats, financed largely by a hearth tax to be paid by every household in the kingdom with exemption only for those in the capital – on condition that it would not be asked to vote new taxes for a decade. This agreement meant that only the magistrates of the capital could raise new taxes, and they duly imposed duties on certain imports, notably tobacco and fruit, until (according to a city chronicler) 'there was no item of food that did not carry as much tax as its actual cost'. He underestimated: by 1647 taxes on some items tripled their sale price.[7]

Although the fruit excise targeted the rich, because most fruit went to well-to-do households, its imposition had caused riots in the past and it now symbolized unjust and oppressive taxes. Nevertheless, when the *seggi* offered to advance 1 million ducats against the yield of a new fruit excise the viceroy accepted. Before long, 'placards went up throughout Naples, inciting the people to "make a revolution" like that of Palermo'; and unknown persons blew up the excise office, located in the Piazza del Mercato, the city's central marketplace, surrounded by the shacks and high-rise projects that housed the poor.[8] In June 1647, as he passed through the piazza on his way to hear Mass at the church of Santa Maria del Carmine, which housed an image of the Virgin widely reputed to perform miracles, the viceroy received numerous protests against the fruit tax. Don Rodrigo Ponce de León, duke of Arcos, had been viceroy of Naples since 1646, and before that viceroy of Valencia: he should have known better than to promise to suspend a levy whose yield had already been promised to creditors, because it required him to find immediately an alternative source of revenue. Arcos proposed a tax on those who owned carriages (as Los Vélez had done in Sicily), but when the *seggi* (who owned most of the carriages) vociferously objected, the viceroy crumbled and re-imposed the fruit excise.

At this point, a French fleet appeared off Naples, and the viceroy immediately sent part of the city's galley squadron, manned by part of the garrison, to drive them away. Coincidentally, an urgent command arrived from the king to send troops to Genoa, and so Arcos dispatched another detachment of the garrison aboard the rest of the galleys, leaving Naples virtually undefended. Some loyal Neapolitans marvelled at these decisions, taken when the viceroy was well 'aware of the ill-will of the Neapolitans on account of the new tax on fruit, and knew equally well that for the same reason – excessive taxes – the people of Palermo and of almost every other place in Sicily had rebelled'. They feared that Arcos had committed political suicide. They were right.[9]

Since popular violence often occurred on Sundays and holidays, when the streets and the taverns were full, Archbishop Ascanio Filomarino of Naples decided to cancel the customary celebrations honouring the Virgin Mary on 16 July, because they involved a ritual battle in the Piazza del Mercato between two teams of young men from the area, dressed as 'Moors' and 'Christians' and armed with sticks. But he did not cancel the rehearsals which took place in the piazza every Sunday morning.

On the morning of Sunday, 7 July, a dispute broke out between stall-holders in the Piazza del Mercato and local producers over who should pay the fruit excise, and when the *eletto del popolo* arrived to try and restore order (and secure payment of the tax), the women in the marketplace 'began to shout "Long live the king and death to the evil government"'. Suddenly a man dressed in white overalls and wearing a red cap leaped onto a fruit stall and shouted 'No excise! No excise!' and threw first fruit and then stones at the *eletto*.[10]

The demagogue was a 27-year-old fishmonger named Tommaso Aniello, known locally as Masaniello, born and raised in one of the lanes adjoining the Piazza del Mercato, and leader of the team of 'Moors' during the festival. He had already drilled his *ragazzi* (boys), dressed in red and black, to a high level of cohesion. Now, 'in the twinkling of an eye, thousands and thousands of common people' came into the square and under Masaniello's lead they seized weapons stored in the tower of the Carmine church and 'unfurled the red flag on the tower of the church as a sign of war'. When refugees from Sicily among the crowd 'called them cowards because they were satisfied with only one thing and incited them to demand everything, as had happened in Palermo', Masaniello led the crowd through the streets towards the viceroy's palace.[11] Meanwhile, others forced open the prisons and freed the inmates. Around midday, 30,000 insurgents gathered before the viceroy's palace and demanded the immediate abolition of all excise duties. Fearful of the fiscal consequences of such a sweeping concession (estimated at 5 million ducats in lost revenue), Arcos agreed to abolish only some taxes. This angered the crowd, which surged forward, and the viceroy's guards (who had orders not to fire) fell back. Arcos was lucky to escape the fate of Viceroy Santa Coloma in Barcelona, seven years before.

Masaniello had chosen his moment well. After departure of the galleys and most of the garrison Arcos had only 1,200 soldiers left to preserve order in one of Europe's largest cities, whereas reinforcements for the protesters armed with 'ploughshares, pitchforks and shovels' poured in from the surrounding countryside. The following day, two important figures emerged at Masaniello's side: Giulio Genoino and his nephew Francesco Arpaja, both veterans of the attempt to change the city's constitution in 1620. Together with Masaniello, they compiled a list of houses to be sacked and burnt each night. Archbishop Filomarino, acting as mediator, persuaded Arcos to abolish all excise duties and to issue a general pardon, in order to restore at least a temporary peace, but once again the viceroy erred: his pardon unwisely characterized the protesters as rebels. Such disrespect provoked a new wave of violence, and the houses of several more ministers and tax collectors went up in flames.

On 9 July 1647 Genoino and his associates drew up a list of twenty-two *capitoli* (articles) that demanded not only a comprehensive pardon but also a host of specific concessions which, they claimed, had been granted to the city in previous charters – including an end to all excise duties on food, the equalization of the tax burden between the capital and the provinces, and the selection of the *eletto del popolo* by the popular assembly (just as Genoino and his supporters had proposed three decades before.) As the list was read out in the crowded Carmine church, a member of the audience shouted out an objection: the Spanish government had reneged on promises made to its opponents in the Netherlands, Catalonia and Portugal, so what could be done to secure compliance this time? In response, Arcos not only granted all the rebels' demands but also promised that the insurgents could bear arms until confirmation of his concessions arrived from Madrid – a dangerous step, since Masaniello's well-drilled militia now numbered 10,000, and he deployed them skilfully to prevent the royal galleys from re-entering the harbour when they returned from chasing off the French.

The revolt of Naples might have ended at this point had a group of heavily armed horsemen not ridden into the Piazza del Mercato and tried to assassinate Masaniello. They missed their mark, and when the crowd overpowered and tortured them, the ringleaders revealed that they had the viceroy's approval for their attempt, that they had planted barrels of gunpowder around the square (which they intended to detonate in order to kill as many protestors as possible), and that they had poisoned the city's water supply (in order to kill the rest). News of the plot radicalized the rebels, some of whom now called for an independent republic. In desperation, Arcos appointed Genoino as head of the treasury and Arpaja as *eletto del popolo*. He also promised them that the rest of the kingdom would pay a new hearth tax to replace the city's excise duties and that a new census would be undertaken to establish a more equitable tax base. The viceroy asked for only one thing in return: the elimination of Masaniello.

Many insurgents were now prepared to sacrifice their leader. Genoino and the lawyers despised him, those whose property had been torched wanted revenge and many felt alienated by his increasingly erratic personal behaviour. On 16 July four conspirators (each of whom later received a handsome reward from Arcos) murdered Masaniello, and the crowd mutilated his body. Many expected this event to end 'the great revolution of the people', but once again unforeseen and unrelated developments transformed the situation.[12]

One of the viceroy's guards celebrated Masaniello's murder by riding through the streets shouting that the viceroy would again reduce the size of loaves of bread. Several bakers anticipated this order, whereupon angry consumers seized their pikes and, spearing the small loaves, marched on the viceregal palace in protest. They also recovered Masaniello's mutilated body, and a procession of 40,000 men and women followed his coffin through the streets 'saying the rosary and the litany, to which they added "St Masaniello, pray for us"'. Archbishop Filomarino himself conducted the exequies.[13]

News also arrived that other parts of the kingdom had followed the example of the capital, starting with Salerno, where on 10 July a crowd of peasants and citizens demanded the abolition of all excise duties. When the excise collectors refused, the crowd torched their houses. By the end of the month, in over 100 towns 'the people took cruel revenge for the ill treatment they had received from their lords' until, according to one observer, 'There was not a village left that has not experienced revolution, with arson, murders and robbery'.[14] Then news arrived of a fresh uprising in Palermo.

Red Flag over Sicily

It took only four days for tidings of Masaniello's revolt to reach Palermo, where it immediately upset the delicate balance of forces created by Viceroy Los Vélez in the wake of the May protests. Giuseppe d'Alesi, an artisan who had witnessed the dramatic events in Naples, returned to Palermo bearing a copy of the twenty-two articles conceded by Arcos and at once began to make plans with colleagues to secure similar concessions. On the next religious festival – 15 August 1647, Assumption Day – the conspirators rode through the streets shouting 'Death to the evil government, out with the Spaniards'. The viceroy and his outnumbered Spanish guards fled.

City crowds now hailed Alesi as their chief (*capopopolo*) and burnt over forty buildings belonging to nobles and merchants before negotiating an agreement with the viceroy acceptable to all parties. The resulting Forty-nine Articles included concessions to the common people (abolition of excise duties throughout the kingdom), the guilds (who would henceforth appoint three of Palermo's six city magistrates as well as many subordinate officers) and the lawyers (who secured a promise that the legal system would be reformed). In addition, Los Vélez swore that henceforth only native Sicilians would hold secular and ecclesiastical posts and pensions.

The viceroy also fomented opposition to Alesi until on 22 August a combination of disaffected guildsmen and vengeful nobles murdered the *capopopolo* and his closest supporters. Then, just like Arcos in Naples, Los Vélez erred: he arrested guild leaders, whereupon thousands of armed citizens waving red flags took to the streets until the viceroy surrendered control of both the fortifications and the government of the city to the guilds. The new masters tried to provide cheap food, but since the sparse harvest caused bread prices to rise, this could only be done by subsidizing the bakers, at a cost of over 1,200 ducats a day. By the end of October the city's deficit stood at almost 150,000 ducats.

The guildsmen of Palermo hoped to make common cause with protesters elsewhere in the island, but given the fierce particularism of Sicilian towns, and the refusal of Messina to join the cause, this remained a chimera. Los Vélez now issued a proclamation ordering all who had fled the capital to return within three weeks or face confiscation of their goods: eventually the return of the fugitives, mostly royalists, gave him numerical superiority over the insurgents and he forbade anyone to carry arms without a permit. Then the stunning news arrived that a powerful Spanish fleet had tried to take Naples by storm. And failed.

The Empire Strikes Back

News of the disorders in Palermo reached Madrid on 16 June 1647. As the councillors of Philip IV immediately observed, the conflict with France had turned the entire Mediterranean into a war zone, which meant that royal commands took weeks and even months to arrive. This delay made it 'impossible to provide remedies from here in time, because of our distance from the place where they are needed, so that before the dispatches carrying orders can arrive there, the danger may have ceased or increased'. Concessions, therefore, could not be avoided. At first, news of the revolt of Naples failed to change the councillors' prudent stance, 'because the state of affairs over there changes from one moment to the next, and what seems appropriate today might not be so tomorrow'; nevertheless, Philip IV took several steps to free up resources.[15] In January 1647 he declared himself ready 'to give in on every point that might lead to the conclusion of a settlement' in the Netherlands and signed a ceasefire with the Dutch Republic (Chapter 8); and eight months later he informed Sor María de Ágreda that 'In these stormy times, it is better to utilize deceit and tolerance rather than force' where rebels were concerned. But this time he deceived his confidante, because he had just ordered his navy to set sail for Naples.[16]

Poor planning and bad luck had often thwarted the plans of Philip IV, but this time promised to be different. His illegitimate son Don John of Austria had already set sail from Cadiz with Spain's main battle fleet to fight the French, but since they had only reached Minorca when news of the revolt of Naples arrived, Philip ordered his son to sail there instead, in the hope of ending the troubles speedily.

The vast city had become a soft target. After the murder of Masaniello, various groups disrupted public order: on a single day, students, silk workers and the musicians of the viceroy's palace took to the streets to protest their situation. Fearing for his safety, Genoino fled to Arcos seeking protection but instead the viceroy promptly arrested him and sent him to Sardinia with secret instructions that he should be murdered on arrival. By contrast, Arcos accepted from Arpaja, still *eletto del popolo*, a new constitutional programme (known as the Fifty-eight Articles) which confirmed a division of power in the city between nobles and the people; exiled those whose houses had been torched; reserved all offices in the kingdom for natives; and replaced the royal judges with twelve new ones, all of them local lawyers who had proven their ability to defy the Spanish authorities. Not without reason, local chroniclers called these developments 'a new revolution'.[17]

Such was the situation when, on 1 October 1647, Spain's main battle fleet carrying 9,000 troops arrived off Naples. Instead of using his advantage to negotiate from strength, Don John of Austria unwisely accepted Arcos's advice to launch an immediate attack. The fleet bombarded the city in preparation for a massed assault until artillery fire from the city, directed by Gennaro Annese, an armourer who lived in the Piazza del Mercato, forced it to withdraw from the harbour. Don John agreed to a truce, but refused to ratify the Fifty-eight Articles, and on 25 October 1647 Annese 'read out in a loud voice' a letter from the French ambassador in Rome offering on behalf of Cardinal Mazarin to send a fleet and money to support the 'Most Serene Republic of Naples', which would 'henceforth live under the protection of the king of France'. Spanish power in Italy – and therefore perhaps the future of Spain as a great power – now hung in the balance.[18]

The Republic of Naples

Annese did not lie when he announced that the new Republic of Naples enjoyed French support. Although Mazarin did not 'believe that the [republican] project could succeed' and was reluctant to accord it official recognition, Duke Henry of Guise, who happened to be in Rome when the troubles began, claimed to have a letter of commitment signed by Louis XIV, and he travelled to Naples to provide assistance.[19] The new regime displayed remarkable self-confidence, issuing more than 250 printed edicts over the winter of 1647–48 that commanded or forbade a wide variety of acts in the name of the republic; aspiring to convene a States-General (on the Dutch model) for the twelve provinces of the kingdom; and urging all Neapolitans to defy the Spaniards 'in defence of the liberty of this kingdom, which involves at the same time working for the longed-for liberty of Italy'. It also disseminated publications about other successful revolts against Spain (for example, by the Dutch and the Catalans); commissioned an official history of its achievements; and sponsored a series of paintings of the main events of the revolution from a talented young artist, Micco Spadaro.[20]

Nevertheless, ensuring an adequate food supply remained the principal problem. The grain harvests of 1647 and 1648 were extremely poor, and a blockade of the city by the nobles cut off most supplies, leading Annese to decree in December 1647 that bakers must reduce the size of the standard loaf from 40 to 24 ounces, and must supply the militia with loaves before anyone else, 'so that we can continue the present war'. Both measures naturally provoked widespread hostility; and when French warships at last arrived off Naples, the duke of Guise exploited the discontent to proclaim himself *Duce* (leader) of a new Royal Republic of Naples. He also published a new constitution and issued coinage bearing his own image.[21]

These developments completely unhinged Philip IV. In October 1647 he signed a new decree of bankruptcy, confiscating the capital of his creditors and regaining the sources of revenue assigned to them; four months later he reluctantly offered to conclude a permanent peace with the Dutch that acknowledged their full independence and allowed them to keep all their conquests in Asia and America. The king also offered humiliatingly favourable terms to the French, but Mazarin unwisely rejected them, confident that he could do better by sabotaging the Hispano-Dutch peace talks and exploiting the revolts of Naples and Sicily. He miscalculated on both counts: the Dutch accepted Philip's terms and, shortly afterwards, the Royal Republic of Naples began to fall apart. The fleet sent by Mazarin to protect Naples failed to defeat its Spanish adversaries and withdrew; and although the *Duce* continued to issue grandiloquent proclamations to his new subjects – for example, ordering public banks to reopen, 'giving our word as a prince' that all deposits would be repaid – few obeyed his orders.[22]

The Tipping Point

According to Vincenzo d'Andrea, a lawyer who had advised Masaniello and Genoino and now served as Annese's principal counsellor, at this point the Neapolitans had five choices: they could declare for Spain, for France, for Guise, for the nobles, or for the former republican leaders. One of these choices disappeared in March 1648, when letters arrived in the city reporting 'a revolution of the people of Paris' in which the crowds chanted 'Long live Naples, we don't want any more taxes or wars' (Chapter 11). Far from feeling flattered, the Neapolitans realized that France could not now help them. D'Andrea therefore opened secret talks with Don John (whom Philip had appointed viceroy in place of the hated Arcos) and with his principal political advisor, the count of Oñate. Both men agreed to confirm most points in the Fifty-eight Articles: a general pardon; abolition of stamp duty, the *media anata* and all excise duties on foodstuffs; appointment only of native Neapolitans to most offices; confirmation of those appointed by the republic; and equality of votes between the nobles and the 'people' in the city's government. In addition, they promised compensation for the damage caused by the bombardment the previous October and agreed that citizens could retain their weapons until the king solemnly ratified all his concessions. On 6 April 1648 Annese opened the gates of Naples and Don John and his Spanish troops marched in.

Despite the general pardon, Don John arrested Annese and a number of others, and had them executed for collaborating with the French. He also summoned the *capopopoli* of all cities to come to Naples at once 'on pain of being deemed rebels': all were imprisoned and some executed. Nevertheless,

Philip IV deemed it prudent to confirm the other concessions granted by his son and Oñate, and he put the duke of Arcos on trial as soon as he returned to Spain. D'Andrea became a trusted adviser to Oñate, who even persuaded the king to confirm in office the judges appointed by the republic.[23]

As soon as news of Don John's failed attack on Naples reached Palermo the guilds there, fearful of similar aggression, demanded restoration of their traditional privileges. Dispirited by this new setback, and by the prolonged rain that reawakened fears of famine, Los Vélez sickened and died – but the disappearance of such an unpopular and unsuccessful minister actually benefited Spain. Cardinal Teodoro Trivulzio, the primate of Sicily who boasted extensive diplomatic and military experience, happened to be in Naples and Don John immediately appointed him acting viceroy of Sicily. Trivulzio arrived in Palermo just two weeks after the death of his predecessor and deftly exploited the divisions among the insurgents to assert his authority.

Paradoxically, the extreme weather helped him: gales blew down walls and trees while the rains caused floods on a scale unequalled in living memory; farms on marginal lands harvested virtually nothing and so sent no grain to the cities that had become dependent on them; an unusually cold winter filled both the hospitals and the church shelters for the homeless. It was easy for Trivulzio to represent all these misfortunes as evidence that God disapproved of the rebellion. The demoralized guilds evacuated the fortifications, and Trivulzio demanded that everyone surrender their weapons and ordered the *capopopoli* of other cities to relinquish power. The collection of the new excise taxes began again on 1 September 1648, just before Don John arrived in Palermo with his galleys and troops.

Once again, the prince prudently made major concessions. Collection of the restored taxes became the responsibility of a specially elected body with wide-ranging powers; imposts fell mainly upon luxury goods (tobacco, bottled wine and carriages) and less on foodstuffs; and there were now no exceptions. When collection of the new excises commenced, Trivulzio paid his share first and formally waived all rights to exemption for himself and all other clerics. The king, for his part, took care to appoint only Sicilians to vacant positions on the island. When a lucrative church living fell vacant in July 1648, he wrote 'Although I could appoint a foreigner to this abbacy without prejudice to Sicily, to please the kingdom I want [the appointee] to be a native'.[24] It was a far cry from the imperious commands of earlier years.

A Final 'Epidemic of Uprisings'

Despite these concessions, in 1672 Messina staged 'the most important domestic conflict faced by the Spanish monarchy in the second half of the seventeenth century'.[25] The city had received major rewards from the crown in return for its loyalty during the uprising in 1647–48: a monopoly of all silk exports from the island, exemption from several taxes, and the promise that every viceroy would spend as long in the eastern city as he did in Palermo, bringing with him all the institutions of the central government. Needless to say, Palermo bitterly opposed this concession (the presence of the central government brought profit as well as prestige) and few viceroys complied, leading the senate of Messina to send a stream of petitions and envoys to Madrid to demand compliance. When in 1669 the viceroy (who obdurately remaining in Palermo) imposed a new tax from Messina, widespread protests broke out.

In Madrid, the ministers of the infant King Carlos II resolved that 'the affairs of Messina have reached such a state that no other remedy will be effective other than using force' – even though some voiced concern that this might drive the Senate 'to despair so that they commit the ultimate crime: to surrender to the enemies of this crown the port that is the key of the two kingdoms of Naples and Sicily'. The viceroy responded by appointing as *straticò* (the chief magistrate and garrison commander) Don Luis del Hoyo, an official 'versed in all the political arts, having served in the "school" of the count of Oñate during the time of the revolution of Naples'.[26] Shortly afterwards, excessive heat and drought destroyed the harvest throughout the island: grain prices skyrocketed, forcing magistrates to introduce rationing and subsidize the price of bread – in spite of which the streets of many cities with beggars and country roads with bandits.

The existence of factions in most towns exacerbated the tensions, especially in Messina, where the Senate accused the *straticò* of exploiting the dearth to undermine the city's independence. Hoyo took his revenge in March 1672 when he encouraged a crowd of protestors to assault the Senate building and open all the city's gaols, releasing some 500 prisoners into the streets. After an uneasy two-week truce, the senators and their supporters, under cover of the religious festivities on Ash Wednesday, organized another street protest – but the *straticò* persuaded the crowds to burn down the houses of the leading magistrates. The victorious faction, known as *Merli*, outlawed the Senate's supporters, known as *Malvizzi* (factional names derived from two varieties of Sicilian sparrow), so that when the viceroy arrived in May 1672, 'the city was virtually deserted, the nobles and principal citizens having fled'.[27] He and Hoyo now set out to destroy Messina's autonomy, revoking many of its privileges and imposing new taxes.

Then orders arrived from Madrid to send all available resources to fight in a new war against France. Like the duke of Arcos in Naples a generation before, the viceroy complied – even withdrawing the galleys that normally patrolled the Straits of Messina – and thus leaving Messina virtually defenceless when on 7 July 1674, the anniversary of Masaniello's rebellion, the *Malvizzi* staged another protest. This time they succeeded, and although they professed loyalty to the king, and displayed his image prominently on the facade of the Senate building, they arrested and executed over fifty *Merli*, exposing their mutilated bodies to public contempt.

The viceroy of neighbouring Naples urged Madrid to overlook this provocative behaviour, and to show 'mildness, kindness and dissimulation' to the rebels because 'the same influences and grievances' existed elsewhere; and, he continued, 'as we saw in the years 1646 and 1647, having begun in Palermo, the epidemic of uprisings immediately attacked Naples'. His words went unheeded. Instead government forces blockaded Messina, hoping to force its submission, and the *Malvizzi* responded by 'committing the ultimate crime' identified five years before: they invited Louis XIV to take their city under his protection.[28]

At first Louis declined the honour, thinking that any revolt that lacked aristocratic backing would soon fail, but in September 1674 he changed his mind and announced that 'it is in my interest not to allow a fire that has spontaneously broken out to be extinguished'. A French fleet brought troops and food to Messina, and for the next three years Sicily became a minor theatre of operations in Louis XIV's Dutch War (Chapters 8 and 10). Nevertheless, Louis had little interest in liberating Sicily from Spanish rule. When financial pressure compelled him 'to deploy my forces only in the places where they are absolutely necessary', he abandoned the island to its fate. French warships evacuated the last of the king's troops and several hundred *Malvizzi* in March 1678.[29]

The Spanish authorities now exacted their revenge, abolishing Messina's tax exemptions and its leading institutions (including the Senate, the mint, the university and even the office of *straticò*). In addition, the victors shipped the city's historical archive and its principal art collections to Spain; razed the palace in which its Senate had met (replacing it with a statue that showed Carlos II killing the hydra of rebellion); and erected a vast citadel to preclude any further unrest. On top of all this, another drought struck Sicily – grain prices in 1679 again reached famine levels – which compounded the misery caused by new taxes and the abandonment of many farms as a result of the war. Messina's population in the 1681 census stood at 62,279, barely half its pre-war total.

As Luis Ribot observed, the rebellion of Messina 'put to the test the entire Spanish system in the kingdom of Sicily, laying bare both its weaknesses and its strengths'.[30] The *Malvizzi* had received support from neighbouring communities in eastern Sicily and also from sympathizers across the straits – several barons of Calabria, eager not only to make an exorbitant profit but also to make trouble for their suzerain, sent supplies and reinforcements to Messina – but the rebels and their French allies made little further progress. No community welcomed them spontaneously and no anti-Spanish uprising occurred elsewhere on the island.

Given that bitterly opposed factions like the *Merli* and *Malvizzi* existed in every town, and given the four-year duration of the conflict, the absence of a general uprising is notable: those who hated Spanish rule had ample opportunity to act, yet few did so. By contrast, pro-Spanish disturbances took place in some areas under French control while several areas under government control organized demonstrations of loyalty. In part, this outcome reflects the ancient hostility between the leading cities of Sicily. Messina had rebelled alone in 1646 and it had stood aloof when Palermo and other cities rebelled the following year. It also reflected a tradition of hostility towards the French, starting with the massacre known as the Sicilian Vespers in 1282, which ended French rule over the island.

Nevertheless, Spain did not prevail solely by default. Despite the later xenophobic propaganda of the *Risorgimento*, Spanish rule was accepted in the seventeenth century by most of its vassals overseas. In the Netherlands, for example, many citizens of Lille lamented their transfer from Spanish to French rule after Louis XIV captured the town in 1667. They regarded the ensuing settlement as 'a peace without joy, because it left us under the king of France', and for several decades they continued to celebrate the births and marriages of the Spanish royal family, toasted the health of Carlos II, and protested at any criticism of him.[31] Had France deprived Spain of its Italian territories, too, similar loyalty to the Habsburgs would probably have persisted.

Furthermore, concessions made by Madrid turned important groups of its vassals into 'stakeholders' in the regime. In Sicily, in the wake of the rebellion of 1647–48 the Spanish government curtailed the political powers of the nobles, abolished tax exemptions for the privileged and entrusted collection of the new taxes (which fell upon luxury goods rather than everyday items) to a new elected body. In addition, henceforth the treasury paid interest on its bonds at 4 per cent for Palermitans, 3.5 per cent for other Sicilians, and 3 per cent for foreigners, and it created a special fund to amortize public debt in an orderly fashion. The new system proved so successful that it remained in force, with only minor alterations, until the revolution of 1860. Spain pursued similar policies in Naples after Masaniello's revolt. This flexibility, and the desire to placate the elite, help to explain why, even though catastrophes caused by both human and natural agents continued, the revolt of Messina remained the only serious uprising faced by the Spanish Habsburgs in Italy after 1648. Unfurling the red flag had ceased to be an attractive option.

CHAPTER FIFTEEN

The Americas, Africa and Australia[1]

Most of the surviving human and natural archives from the mid-seventeenth century relate to Europe and Asia. Despite the immense size of the other three inhabited continents (16 million square miles for North and South America, almost 12 million for Africa, and 3 million for Australia), the human archive is sparse, because few indigenous populations left written or pictorial records that can be precisely dated. Even where an abundant natural archive (above all tree rings) reveals the footprint of the Little Ice Age, its human impact remains obscure. Thus many Europeans in North America realized that the indigenous population was declining rapidly – in New Mexico, 'where three Pecos had lived in 1622, only two lived in 1641 and only one in 1694'; in New England 'by the 1640s the number of Iroquois (and of their Indian neighbours) had probably already been halved' – but none suggested the probable causes.[2] In Australia, only archaeology and the natural archive provide reliable testimony but (as elsewhere) much of it lacks chronological precision. Historians can reconstruct the experience of humans living in these areas in the seventeenth century only when literate residents or travellers from other regions – most of them Europeans – compiled written sources that have survived.

The Americas

Substantial records exist for the European colonies in North and South America: the French settlements along the St Lawrence river and the Great Lakes; the English colonies of New England, the Chesapeake and the Caribbean; the Spanish viceroyalties of Mexico and Peru; and coastal Brazil. Despite the distance that separated these colonies, and their environmental differences, their histories in the seventeenth century shared six striking similarities:

- From Newfoundland to Patagonia, the Americas experienced notably colder winters and cooler summers in the middle decades. 1675, a 'year without a summer', remains the second coldest recorded in North America during the last six centuries.

- Areas normally affected by El Niño suffered more because the phenomenon occurred twice as frequently in the mid-seventeenth century: there were more rain and floods along the Pacific coast and throughout the Caribbean, more droughts in the Pacific Northwest and more cold winters in the Atlantic Northeast.
- Both seismic and volcanic activity along the continent's Pacific shore increased.
- Almost all surviving harvest records show dearth in the 1640s and 1650s.
- In the words of John McNeill, 'From Canada to Chile, the Americas in the seventeenth century served as a playing field for the ambitions of several European statesmen and countless independent warrior-entrepreneurs.'[3] In several regions, wars of unusual ferocity intensified the impact of climatic adversity: the Pequot War and King Philip's War in New England; the Beaver Wars in New France; and the Dutch-Portuguese struggle in Brazil.
- All the indigenous peoples who came into contact with Europeans, whether directly or indirectly, suffered losses – sometimes catastrophic losses. In New England and New France (but only there), this decline was partially offset by a dramatic growth in the number of settlers, both through strong immigration and because many colonists appear to have lived longer.

Anglo-America

According to Benjamin Franklin, writing in 1751:

> Marriages in America are more general, and more generally early, than in Europe. And if it is reckoned there, that there is but one marriage per annum among one hundred persons, perhaps we may here reckon two; and if in Europe they have but four births to a marriage (many of their marriages being late), we may here reckon eight, of which if one half grow up, and our marriages are made (reckoning one with another) at twenty years of age, our people must at least be doubled every twenty years.[4]

Franklin's demographic speculations were certainly correct for New England. As early as 1634, John Winthrop commented on the unusually low mortality among the settlers around Massachusetts Bay, and a few years later a group of men from New England who briefly returned to their native land thanked God for 'blessing us generally with health and strength ... more than ever in our native land; many that were tender and sickly here [in England] are stronger and heartier there.' And they all knew why: 'God has so prospered the climate to us that our bodies are hailer, and children there born stronger' and lived longer.[5]

Church records confirm these claims. Most of the female settlers in seventeenth-century New England married before they turned twenty-four, and half of all settlers seem to have survived to age seventy. Completed marriages (ones where both parents survived to bring up their children) produced on average six children, and most of them (unlike those born in Europe) survived to childbearing age. Thanks to this remarkable fecundity, and to continued immigration, the settler population of New England increased from about 14,000 in 1640 to over 90,000 in 1700 – a sixfold increase in two generations.

The experience of British colonists elsewhere in North America was different. Almost from its foundation in 1607 Jamestown, Virginia, experienced (in the words of its first governor) 'a worlde of miseries', because drought caused its early settlers 'to feele the sharpe pricke of hunger', and in desperation 'many of our men this starveinge tyme did runn away unto the salvages'. The 'salvages' could offer little help because the years 1607–12 saw the longest drought registered in the Tidewater region in eight centuries, and it affected natives as well as newcomers.[6] Demographic growth therefore remained slow: although at least 6,000 English men, women and children had come to Virginia from England since 1607, by 1624 the colony still numbered only 1,200.

Several other circumstances contributed to this slow growth. First, although in 1618 the Virginia Company recruited and sent out far more colonists than before, it failed to send sufficient provisions to feed them – and the newcomers arrived just as a new drought reduced the local crops. Many died. In 1622 drought again forced natives and newcomers to compete for the scarce food, culminating in a massacre that cost the lives of almost 350 English men, women and children; and then (a sorrowful settler reported) came 'a generall sicknes, insomuch as wee have lost I believe few lesse than 500, and not manie of the rest that have not knockt at the doores of death'.[7] Not until the 1680s, when the white population of Virginia numbered 60,000, did it become self-sustaining.

That milestone took even longer to reach in England's Caribbean colonies, thanks largely to tropical diseases – especially two mosquito-borne viruses that thrived in the wetter conditions created in the region by increased El Niño activity: malaria and yellow fever. Although more than 200,000 Europeans came to Barbados, Jamaica and the Leeward Islands in the course of the seventeenth century, even in the 1690s these islands' combined white population stood at only 40,000.

Mortality among the indigenous inhabitants of Anglo-America was also high. Archaeologists have found around Massachusetts Bay several Native American mass graves from the early seventeenth century that lack the customary grave goods, suggesting unusually rapid mortality, probably due to smallpox; while Thomas Morton, who arrived in 1622, found piles of 'bones and skulls upon the severall places of their habitations'. The copious evidence of sudden death 'made such a spectacle' that 'it seemed to mee a new-found Golgotha'.[8]

Morton's chilling image reflects the crucial difference between European and Native American aetiology. Although the native peoples of the Americas suffered from a variety of illnesses before they came into contact with Europeans, to many early colonists they seemed robust, healthy and 'unusually free from any apparent physical defects and deformities'.[9] In part, this reflected the absence of diseases that produced stunting and disfigurations, such as smallpox and measles; but this created a virgin population with no immunity when the Europeans arrived. Widespread vulnerability, combined with the probability that several Old World diseases (notably smallpox and yellow fever) became more virulent in the seventeenth century (page 71), explains not only why Morton encountered a 'new-found Golgotha' in Massachusetts, but also why the same uncommonly high mortality among the indigenous population of New England persisted. In Massachusetts, Winthrop commented on this phenomenon in 1634, musing that 'If God were not pleased with our inheriting these parts, why did he drive out the natives before us? And why does He still make room for us, by diminishing them as we increase?' Nevertheless, some early settlers of New England still felt overwhelmed by what Roger Williams of Providence colony called in 1637 the 'ocean of troubles and trialls wherin we saile'. Williams had in mind not only extreme weather and dearth, but also the Pequot War.[10]

For a time, European diseases spared the Pequot, whose members occupied some 2,000 square miles of southern New England and by 1630, thanks to their numbers and their strategic location, controlled almost all the overland trade of the English colonies. Then in 1633 a devastating smallpox epidemic apparently eliminated 80 per cent of their number. This loss destabilized the entire region and the following year John Winthrop noted that the Pequot 'were now in war with the Narragansett whom till this year they had kept under, and likewise with the Dutch', so that 'by these occasions they could not trade safely anywhere'. In 1635, a hurricane so powerful 'as none now living in these parts, either English or Indian, had seen the like' struck New England: it destroyed houses, 'blew down many hundred thousands of trees', and flattened 'all the corn to the ground, which never rose more'. A winter of extreme frost and snow followed, ruining another harvest. In this precarious situation, the stream of settlers arriving from England created an intense competition between natives and newcomers for maize, on which both groups relied. As Katherine Grandjean has noted: 'Corn was central to waging the Pequot war'.[11]

A sortie by colonists from Fort Saybrook (Connecticut) to seize Indian corn in August 1636 provoked a Pequot counter-attack, following which the colonists 'pursued through the wildernesse, slew and took prisoners' and enslaved all the Pequot they could find. Two years later, representatives of the victorious English and their allies gathered at the Dutch trading post in Hartford to divide up both the vanquished and their assets. The Treaty of Hartford forbade the surviving Pequot to use their name and native language, or 'to live in the country that was formerly theirs but is now the English's by conquest'. It also incorporated the first Fugitive Slave Law in North American history: any former Pequot who escaped must be returned to his or her original captor. The gender ratio among the Pequot after 1640 sank to one male for every twenty females, and the tribe continued to decline until by the beginning of the twentieth century it numbered just sixty-six people.[12]

Maize seized from the Pequot during the war helped the colonists in New England to survive another 'very hard winter' in 1638–39 when 'the snow lay from November 4th to March 23rd, half a yard deep', and a drought the following summer; but then came the second-coldest winter in a century, in 1641–42, when Massachusetts Bay 'was frozen over, so much and so long, as the like, by the Indians' relation, had not been so these forty years'. After this, drought and cold significantly reduced the harvests throughout New England – just as the outbreak of the English Civil War (according to Winthrop) 'caused all men to stay in England in expectation of a new world, so as few coming to us, all foreign commodities grew scarce, and our own of no price'.[13]

The situation of the colonists remained perilous throughout the decade. New England supported Parliament, but Virginia and other southern colonies remained loyal to King Charles. Barbados, by far the most prosperous English settlement in the New World, saw neutrality as the best way to survive: its freeholders swore 'not to receive any alteration of government, until God shall be so merciful unto us as to unite the king and Parliament', and vigorously pursued a policy of free trade.[14] Initially the regicide changed little – only Rhode Island immediately recognized the English Commonwealth as the legitimate government – but in 1652 the young republic dispatched one fleet to enforce its authority in the Caribbean and another to subdue Virginia. It also created a Council of Trade to promote overseas commerce and passed a Navigation Act that restricted all trade in the Anglo-Atlantic to English merchantmen (Chapter 12). In addition it sent to America thousands of defeated British and Irish opponents who toiled alongside tens of thousands of slaves imported from Africa; and because New England enjoyed 'peace and freedome from enemies, when almost all the world is on a fire' (in the words of a group of boastful colonists in 1643), the migration of freeborn English men and women also rose. These various migrations increased the population of the Anglo-American colonies to 200,000 by 1660.[15]

Sir William Berkeley, governor of Virginia, realized that by 1675 this influx had created an unsustainable situation. 'All English planters on the main[land] covet more land than they are safely able to hold', he observed, with 'the Indians complaining that strangers had left them no land to support and preserve their wives and children from famine.'[16] His analysis contained much truth. The English planters coveted more land not only because immigration and the natural increase praised by Benjamin Franklin had swelled their numbers, but also because they cultivated cash crops like tobacco, which heightened their need for arable land; and their livestock also required space – a lot of space: their cattle trampled down crops as they grazed while their hogs rooted up clams and corn stores, and although the natives begged the newcomers to fence in their pastures, the newcomers retorted that the natives should fence in their crops. In addition, the cooler climate of the mid-seventeenth century reduced the total yield of crops, increasing anxiety in both communities about their long-term prospects for survival; while the virtual extinction of the fur-bearing animals of New England, especially beavers, through excessive hunting meant that natives had fewer trade goods to offer settlers, and so whenever the colonists' courts fined Native Americans for some transgression, land was often the only asset they had left with which to pay. This combination of pressures precipitated King Philip's War, which almost ended the existence of New England.

In 1671 the Court of Massachusetts Bay imposed on Metacom, chief (*sachem*) of the Wampanoag, known to the English as King Philip, a fine of £100 – a large sum that he could pay only by surrendering some of his land to the colonists. Outraged, Philip forged alliances with his neighbours, acquired guns and constructed forts until June 1675, a 'year without a summer' when unusually cool weather threatened the crops of natives and newcomers alike, when he launched a coordinated attack on New England.

Philip commanded far more fighting men than the colonists, but two critical weaknesses undermined his cause: the tensions inherent in any alliance, and the lack among the tribes of any tradition of following a single leader – indeed, several of the traditional enemies of the Wampanoag fought for the colonists. So although (to quote Berkeley again) Philip 'made the New England men desert about 100 miles of ground they had divers years seated and built towns on', he and his allies suffered some serious defeats.[17] Above all, the unusually cold winter of 1675–76 froze the swamps that normally protected an important Narragansett fort, allowing a colonial army to march across the ice to the fort and slay all within it, as the Mohawk (members of the Iroquois Confederation) assaulted Philip's winter encampment, forcing him and his followers to disperse. For a while, another poor growing season combined with hostilities to produce dearth in New England, but the colonists received supplies by sea and from each other, whereas Philip's supporters starved. In addition, colonists conducted joint operations with their Native American allies until in August 1676 they cornered Philip and killed him.

King Philip's War dragged on for a few more months, but Indian power east of the Connecticut river was broken, never to return. Perhaps 3,000 Native Americans died through combat, disease and hunger; 2,000 more fled; and another 1,000 were sent in servitude to Bermuda. The colonists, too, suffered significant losses: Philip's forces destroyed seventeen of the ninety settler towns in New England before the war, and inflicted damage on over fifty more. Some of the smaller outposts perished for ever, and 'the work of a generation would be required to restore the frontier districts laid waste by the conflict'. It was, in proportion to population,

> the costliest in lives of any American war. Out of a total population of some thirty thousand, one in every sixteen men of military age was killed or died as a result of war; and many men, women, and children were killed, carried into captivity, or died of starvation or exposure as a result of the Indian raids ... Losses on this scale among the mature male population posed a real threat to the colony's continued prosperity, perhaps even its survival.

The war also disrupted trade, inflicted material losses of at least £150,000, and 'eliminated so much of the capital invested in colonization by the two founding generations that per-capita income did not achieve 1675 levels again until 1775'.[18]

Unrest also threatened the prosperity of Virginia. In July 1676, some Chesapeake planters demanded that Governor Berkeley sanction attacks on their Indian neighbours, and when he refused malcontents led by the well-connected Nathaniel Bacon issued a declaration in the name of 'the Commons of Virginia' that commanded 'in His Majesty's name' the arrest of the governor and his supporters 'as traitors to the king', and in September they captured and sacked Jamestown, the colonial capital. In the event, the unhealthy environment of the Tidewater nipped rebellion in the bud – dysentery carried off Bacon and many of the other rebellious planters newly arrived from England – but several commentators linked these troubles with those in Ireland in 1641 ('The tyranny of the natives exceeds that of the rebellion in Ireland, if possible'). The government in London therefore dispatched a fleet of fourteen warships with 1,300 regular troops, and orders first to pacify Virginia and then to proceed to Boston and restore order in New England.[19]

James, duke of York, the future James II, played a leading role in this initiative. He was already proprietor of the North American territories acquired in 1674 from the Dutch, and his governor, Sir Edmund Andros, played a leading role in saving New England by mobilizing the Iroquois against King Philip. In 1677 Andros sealed a lasting agreement, known as the Covenant Chain, with the entire Iroquois Confederation, which brought peace to the Anglo-Indian frontier from Maine to the Carolinas – but colonial expansion westward ceased for almost a century. James now took other steps to restrain the colonies. As in England (see Chapter 12 above), he revoked charters and other royal concessions, and in 1686 created the Dominion of New England, with Andros as his first governor general.

Three years later Andros was in Boston, consolidating his authority, when news arrived of the Glorious Revolution in England. He was immediately seized and imprisoned, and the New England colonies promptly restored the form of government laid down in their confiscated charters. They later recognized William and Mary as their sovereigns, but their capacity for independent action worried the new monarchs: although they dismantled the dominion created by their predecessor, the new colonial charters they issued gave the crown far more authority.

New France and the Beaver Wars

In several respects, the history of the lands along the St Lawrence and around the Great Lakes in the seventeenth century resembled that of neighbouring New England. Above all, the European population increased very rapidly. Almost 70 per cent of all women who arrived in the colony from France married before they turned twenty, one-half of the married settler families produced four or more children and one-quarter produced ten or more. After 1650, some parishes in Québec registered three or four births for every death, and the number of French settlers increased from 800 in 1660 to 15,000 by 1699. Nevertheless, as in New England, even this spectacular increase failed to offset the decline of the Native American population.

In part, the decrease was deliberate. Several indigenous nations were matrilineal and their womenfolk sometimes used herbs for abortion (and perhaps also for contraception, since they rarely seem to have had two pregnancies in less than two years). These practices mitigated the intensity of the mid-seventeenth-century crisis by easing demand for limited or falling food supplies, but they offered no protection against European diseases. According to Adriaen van der Donck, a Dutchman who lived in the Hudson Valley in the 1640s, the Native Americans 'affirm that, before the arrival of the Christians, and before the small pox broke out among them, they were ten times as numerous as they now are; and that their population had been melted down by this disease, whereof nine-tenths of them have died'.[20]

Many Native Americans also died in internecine wars, many caused by the insatiable demand of Europeans for the thick pelts of the North American beaver, which made warm, rainproof hats and coats. In the early seventeenth century the twelve matrilineal clans that formed the Wendat people (called Huron by the French), who farmed and hunted in the woodlands north of Lake Ontario, began to exchange beaver pelts for European metal artefacts, above all tools; by contrast the Mohawk, members of the Iroquois Confederation which controlled the woodlands south of the lake, exchanged them primarily for brandy and firearms. Before long, the Iroquois not only possessed 'pieces, powder and shot' but became 'far more active in that employment' – firing guns accurately – 'than any of the English, by reason of their swiftness of foot and nimbleness of body, being so quick-sighted, and by continual exercise'.[21] For a time the Wendat continued to prosper, since metal tools increased the yield of their maize crops, and their numbers rose to perhaps 25,000; but contact with Europeans also exposed Wendat traders to measles and smallpox in a form to which they apparently possessed no resistance. After a savage epidemic in 1639–40 the Wendat numbered scarcely 12,000, depriving them of the strength to resist Iroquois attacks – especially since the acquisition of European firearms and ammunition, and the drunken rages induced by consuming European brandy, gave Iroquois warfare a new ferocity.

The 'design' of the Iroquois, according to the Jesuit missionary Isaac Jogues in 1643, 'is to take, if they can, all the Hurons; and, having put to death the most considerable ones and a good part of the others, to make of them both but one people and only one land'. Jogues failed to note that both Iroquois and Wendat fought primarily to secure captives to replace their losses from disease, with the women (especially the widows) separating the captives who would live and breed from those to be tortured and killed. The Wendat persuaded the French settlers and missionaries living among them to provide firearms and teach them how to build stone forts to defend their villages more effectively, but an all-out Iroquois assault in spring 1649 forced them to take refuge on Gahoendoe island in Lake Huron. Unfortunately for them, intense drought killed almost all the maize they planted, so that (in the words of Kathryn Magee Labelle), 'Instead of a place of refuge, [the island] became a death trap.' An excavation in 1987 of a fort and the adjacent Wendat village on the island revealed a grave filled with the tiny skeletons of malnourished children, victims of Europeans' obsession with firearms and fur hats, as well as of the Little Ice Age.[22]

The experience of most other indigenous populations in North America during the seventeenth century remains a mystery. European visitors to the central Mississippi Valley in the mid-sixteenth century encountered impressive hierarchical chiefdoms, but the French explorers who visited the same areas in the 1670s reported widespread desolation. Since the area had limited lands capable of feeding large settlements from maize cultivation, Patricia Galloway, a leading historian of the Choctaw of the Mississippi Valley, has speculated that 'a point came when some of the river valleys could no longer support their population, or when populations right at the edge of need experienced a run of bad years or even climate change, and the people had to disperse or starve'. Tree-ring series from the region reveal intense droughts in the mid-seventeenth century, and these probably doomed any remaining settlements, but greater precision seems impossible without a more extensive human archive. Galloway has dubbed this period the 'Black Hole' of Native American history beyond the Appalachians.[23]

Despite all the setbacks, however, in 1700 Native Americans still far outnumbered Europeans in North America; and thanks to their growing proficiency with both firearms and horses, they controlled most of its territory. The native inhabitants of the Mississippi and Arkansas river valleys may have been fewer than before but they still occupied their ancestral lands, from which they excluded all European settlers, negotiating with them (if at all) from a position of great strength. To the north lay 'the Middle Ground' (a term coined by Richard White), where Native American peoples avoided dependence on any single group of Europeans and, if they negotiated, imposed their own diplomatic protocols, such as the Covenant Chain. Given the ravages of climate change, lethal diseases and savage wars, such resilience was a remarkable achievement.[24]

'This Land of Brazil'

Although European colonists in the Americas sometimes went to war against each other in the seventeenth century as part of broader conflicts – the English against the Dutch in the 1660s, and against the French in the 1690s – they seldom fought for long, or systematically destroyed European property. Brazil formed a significant exception. By 1630, perhaps 60,000 people inhabited the entire viceroyalty, roughly half of them Portuguese and the rest African slaves and Native Americans; and the captaincies of Bahia in the centre, with its capital at Salvador, and of Pernambuco in the north, with its capital at Olinda, each boasted around 12,000 settlers, many of them living on sugar plantations where *engenhos* (mills) made sugar from cane to sell in Europe. Then two catastrophes occurred: sailors aboard two fleets from Lisbon introduced yellow fever, which swiftly decimated the unprotected indigenous population of the colony; and a Dutch expeditionary force captured Pernambuco, destroyed Olinda and gradually expanded its control southwards down the coastal plain. In 1641 the Portuguese settlers celebrated when they heard of the Portuguese Revolution (Chapter 9), but any hope that shifting their allegiance from Philip IV to John IV would improve their lot soon faded. In 1641–42 smallpox 'raged so violently among the Indians that entire villages were almost totally extinguished. The survivors retreated into the forests since they no longer dared to remain in their homes'.[25]

By contrast, Dutch Brazil prospered. By 1644, some 15,000 settlers inhabited the fertile coastal plain that stretched from the Amazon Delta to the São Francisco river, almost half of them inhabiting the handsome new capital Mauritsstad (now the heart of Recife), built with bricks and tiles shipped out from Holland and boasting a fine palace for the governor, numerous churches and the first Jewish synagogue in the New World. When a Dutch expeditionary force captured Portuguese Angola, the principal source of slaves required to produce sugar, the economic future of the new colony seemed secure; but everything changed in 1645. An uprising by the Portuguese settlers of Brazil drove the Dutch back into a few coastal fortresses and for the next nine years, civil war raged between the two groups of colonists and their Native American allies. In 1648 Dutch warships entered the Bay of All Saints, the heart of the Portuguese colony, and torched the *engenhos* around its shores, and in retaliation the Portuguese burned so many *engenhos* in Pernambuco that the province lost for ever its position as the colony's leading exporter of sugar. In all, at least 20,000 Dutch settlers perished in Brazil, and at the fall of Mauritsstad in 1654 the survivors lost almost all their assets.

'Panic in the Indies'

In the rest of Latin America, news of the revolt of Portugal ignited what Stuart Schwartz has called 'panic in the Indies'. According to Don Juan de Palafox y Mendoza, bishop of Puebla and inspector-general of the viceroyalty of New Spain in July 1641, 'the whole monarchy trembled and shook, since Portugal, Catalonia, the East Indies, the Azores and Brazil had rebelled'. Rumours spread that rebels had murdered the Spanish garrison in Salvador and that the settlers had made common cause with the Dutch to destroy Spanish power. Both stories proved false, but when an agent from Lisbon arrived in Cartagena and attempted to seize the treasure galleons assembled in the port, 'apprehension and hysteria' gripped the capital, whose 6,000 Portuguese residents were (Palafox claimed) armed to the teeth and likely to enjoy the support of their numerous African slaves in any rebellion.[26]

The adverse climate, probably related to unusually frequent volcanic eruptions and El Niño events, made the situation even more volatile. The Valley of Mexico normally receives copious rain from May to August, but in 1639 it suffered the first of five years of drought, during which the price of maize quintupled, leaving granaries without food and citizens without water. Blame for these disasters gradually shifted to the viceroy, the marquis of Escalona, who had married the duke of Bragança's sister (which might tempt him to transfer New Spain to Portuguese hands) and had Jewish ancestors (which, some thought, led him to favour local New Christians). In June 1641 Palafox wrote a secret letter to the king suggesting that Escalona would be better employed elsewhere, preferably in Europe. Philip took no chances: he signed three distinct secret orders, authorizing Palafox to use whichever one he judged appropriate. One invited Escalona to return to Spain, where the king claimed to need his advice; a second censured the duke's conduct and ordered him to transfer his authority to Palafox; the third authorized the bishop to have Escalona killed. In each scenario, Palafox would become both archbishop of Mexico and interim viceroy. In June 1642, as soon as he received the package containing these remarkable letters, Palafox's agents arrested Escalona and confined him in a convent outside the town under guard until he could be shipped back to Spain.

Plots, both real and imagined, also rocked other parts of Spanish America. When the viceroy of Peru first received news of the Lisbon revolt, he did not believe it but on receiving confirmation from Buenos Aires, together with a plea for help, he sent troops across the Andes to secure the settlement. In addition, conscious that he was '500 leagues from the Caribbean and 800 leagues from Buenos Aires', the viceroy fretted that Portuguese residents of Lima might rebel, perhaps winning over some of the city's African slaves, and so he rounded up and disarmed some 500 of the Portuguese, forcibly moving the younger males to camps inland.[27]

These harsh measures made Spanish America safe again from political danger, but offered no protection from natural disasters. Earthquakes destroyed churches, houses and fortifications in Santiago de Chile (1647), Concepción de Chile (1657) and Lima and Callao, Peru (1687). In Argentina, increased rainfall caused the Paraná river to burst its banks in 1621–22, 1651–52 and 1656–58 (all of them El Niño years), inundating the regional capital, Santa Fé: despite its splendid public buildings, the colonists eventually abandoned the city in favour of a site on higher ground. New Spain, by contrast, suffered repeated droughts. Between 1641 and 1668 officials in the capital authorized eight public processions of Our Lady of Guadalupe, an image reputed to possess a miraculous ability to produce rain. In 1642 the northern town of Monterrey experienced 'such a shortage' that inhabitants 'sold rotten maize that could no longer be eaten – something neither heard nor seen [before] in New Spain, even in times of greater famine.'[28]

According to a report in 1638 from the missions of Sinaloa and Sonora (modern New Mexico), 'of the 300,000 Indians who had been baptized by the Jesuits, only one-third were still alive', and soon afterwards prolonged drought forced Christian communities on the Salinas plains to use their own urine both to irrigate their crops and to make bricks for the lavish mission churches, while Apache bands (also suffering from the extreme drought) launched repeated attacks. By 1678, every mission had been abandoned. The Rio Grande basin also suffered from extreme weather: 'increased rainfall and perhaps the lowest temperatures of the Little Ice Age' in the 1640s and 'possibly the worst drought in the Spanish Colonial period' in the 1660s, combined with repeated raids by Apaches. According to Daniel T. Reff, the population 'declined by over 80 per cent between 1608 and 1680' and half of the Christian pueblos were abandoned.[29]

In the Caribbean, the heavy rains associated with the numerous El Niño episodes created optimal breeding and feeding conditions for the vector of both malaria and yellow fever: the mosquito. The first yellow fever pandemic in the New World began in 1647, a year that saw a 'strong' El Niño. In Barbados, attracted both by the sugar and by population densities above 200 per square mile, the mosquito-spread disease killed one of every seven Europeans on the island as well as decimating other communities in the Caribbean and Yucatán. A local chronicle reported that in 1648 'there was bloody vomit and we began to die' – a clear reference to yellow fever, to which neither Europeans nor Native Americans possessed inherited immunity – followed by 'such a hard and extraordinary drought that it rendered the land sterile and produced such intense heat' that wildfires raged throughout Yucatán. 'Almost half the Indians perished with the mortality caused by the plague, famine and smallpox which since the year 1648 until the present one of 1656, in which I am writing this down, have so exhausted this land.'[30]

Two demographic strategies intensified the impact on women of these natural disasters. According to Maria Sybilla Merian, a botanist who toured the Caribbean in the later seventeenth century, both African and Native American women frequently aborted their offspring to spare them a life of servitude and humiliation. Merian provided a detailed description of the peacock flower (*Poinciana pulcherrima*), one of over a dozen abortifacients in use in the colonial West Indies, asserting that 'The black slaves from Guinea and Angola have demanded to be well treated, threatening to refuse to have children. In fact, they sometimes take their own lives because they are treated so badly.'[31] Many European and *mestiza* women also had no children because they entered convents. The number of nuns (which, as in Europe, included many women claustrated against their will in times of economic hardship) and their servants reached astonishing proportions: by 1700, 20 per cent of the total female population of Lima lived in convents, thereby reducing the number of potential mothers.

Seventeenth-century Latin America also experienced some popular revolts, both by groups of settlers and by indigenous peoples. The most serious settler rebellions took place in the Altiplano of modern Bolivia and Peru, highly vulnerable to colder and drier weather and also prone to violent rivalries between the rich silver-mine-owning families, many of them of Basque origin, and the more numerous poor settlers from other parts of Spain, many of them descended from the first conquerors. This rivalry erupted into an armed struggle in the vast mining complex of Potosí, which ended only with the execution of forty ringleaders in 1625, and the 'mountain of water' that submerged much of the town the following year when the reservoir of Caricari broke (page 5), drastically reducing the city's silver output and ruining most mine owners. Three decades later Andalusian prospectors discovered new silver deposits at Laicacota and founded a town that soon became the fourth largest in the region, with many Basque inhabitants. Almost immediately, the region experienced a series of disasters: the maize and coca harvests failed from 1659 to 1662, epidemic diseases struck in 1660–61 and then came an outbreak of violence that pitted Basque and native mineworkers (both Indian and *mestizo*) against the Andalusian mine owners. Government forces managed to restore order but in 1665 Basque insurgents, supported by sympathetic magistrates, seized control of Laicacota and its rich mines. The following year the Andalusian mine owners, aided by a militia of Indians and *mestizos*, counter-attacked, chased out the magistrates and sacked the town to shouts of 'death to the Basques'.

No sooner had news of this open defiance arrived in Lima than rumours circulated that the indigenous inhabitants of the capital intended to flood the city while other groups throughout the viceroyalty rose up, killed all the Spaniards and restored Inca rule. The new viceroy, the count of Lemos, acted promptly and executed the alleged ringleaders, but then Basque fugitives from Laicacota convinced him that 'the kingdom had come within an ace of a great disaster' and persuaded him to lead a punitive expedition into the Altiplano to restore royal authority (and their control). Lemos, the first viceroy to visit the area for a century, executed over sixty insurgents, imprisoned many more and razed the rebel strongholds; but when the government in Madrid investigated the troubles at Laicacota, generating 25,000 pages of testimony, it concluded that Lemos had overreacted. In reality there had been no risk of a 'great disaster', only manipulation of government power by factions: claims that Spanish rule 'tottered on the verge of insurrection' had no substance.[32]

The Pueblo revolt in present-day New Mexico a decade later was very different – although once again climate played a crucial role. The prolonged drought and cold in the northern Rio Grande basin, intensified by increased volcanic activity and incessant Apache raids, led to the abandonment of many settlements until a group of Pueblo speculated that the unusually heavy snows over the winter of 1679–80 would produce floods sufficient to prevent supplies and reinforcements from reaching the Spanish settlers the following year, leaving them uniquely vulnerable. They calculated correctly: in August 1680 a coordinated uprising expelled the Spaniards from Santa Fé, the regional capital, after which the victors set about eradicating all traces of Christian worship. Communities abandoned mission villages, some for settlements on the high plateaux that could be better defended, others to join the Apache and Navajo, and craftsmen began to produce traditional artefacts once again. Even though continuing drought weakened the Pueblo and allowed the Spaniards to return to Santa Fé in 1692, they abandoned many of their exploitive practices and concluded a formal peace with the Pueblo, whom they began to treat as allies to be courted rather than conquered. As Andrew Knaut has noted, the revolt achieved a unique feat: 'a complete setback of European expansion in the New World'.[33]

American Exceptionalism?

Compared with the political and social upheavals elsewhere, including other parts of the American continent, the Spanish colonies seem to have largely escaped the Global Crisis, despite experiencing severe climatic adversity. As Peruvian historian Margarita Suárez has observed, 'Rather than a general crisis, the seventeenth century produced a crisis in Spain's ability to extract economic benefits from its colonies' – that is, the crown lacked the power to exacerbate the impact of natural disasters via costly and detrimental policies.[34] Instead, local elites encouraged the cultivation of specialty crops for export, often to other colonies: by the 1650s Peruvian sugar, conserves, wines and vinegars were shipped to Chile, Panama and even Mexico, where they displaced Spanish competition. Most of these goods came from relatively modest units of production that proved hard for the government to monitor and tax, even had local officials proved vigilant – but they did not. Reports of administrative corruption, inefficiency and extortion reached such a pitch that when in 1663 the crown appointed special inspectors to investigate treasury officials in Peru, they uncovered over 1 million pesos in unpaid back-taxes. Although the viceroy had some corrupt officials executed, revenues remitted to Spain continued to fall: the average sum leaving Lima every year for Seville dropped from 1.5 million pesos in the 1630s to below 150,000 pesos in the 1680s.

These developments meant that more wealth remained at home, as reflected in the output of the leading mints of the viceroyalty, which produced twice as many coins in the 1680s as in the 1620s, and in the magnificent public buildings (especially churches) constructed in virtually every colonial town during the later seventeenth century. The colonial elites of Spanish America had become stakeholders in the imperial system, much like their colleagues in Spanish Italy (Chapter 14). There, too, rebellion ceased to be an attractive option.

Africa

Sub-Saharan Africa forms the largest area of the globe for which evidence of a seventeenth-century crisis is either ambiguous or absent. We know that both extreme climatic events and important structural changes occurred there, but it is extremely difficult to link them.

The Cape of Good Hope, at Africa's southern tip, offers a good example of the limitations of the available sources. In 1652 Dutch settlers established a small fort to guard the stores to refit and replenish ships sailing between Southeast Asia and the Netherlands, and soon afterwards began to establish farms beyond the walls of their fort. The local Khoekhoen complained to Governor Jan van Riebeeck that the Dutch 'were living upon their land, and they perceived that we were rapidly building more and more as if we intended never to leave'; and they 'declared boldly that this was not our land but theirs'. Since their words made no impression, the Khoekhoen opposed the colonists' expansion by force, but after about a year of indecisive fighting they resumed negotiations. Their leaders once again 'strongly insisted' that the Dutch 'had been appropriating more and more of their land, which had been theirs all these centuries, and on which they had been accustomed to letting their cattle graze etc. They asked if they would be allowed to do such a thing supposing they went to Holland'. Concerned that the Khoekhoen were getting the better of the argument, 'they had to be told that they had now lost the land as the result of the war and had no alternative but to admit that it was no longer theirs', van Riebeeck recorded dryly in his journal. 'And we intended to keep it.'[35] The Khoekhoen eventually agreed that they would graze their herds only on lands unoccupied by Dutch farmers, but before long the influx of colonists and the ravages of European diseases (above all smallpox) among the native population tipped the balance permanently in favour of the newcomers. A century later, Cape Colony was the largest Dutch outpost overseas.

The shifting balance of power at the Cape in the mid-seventeenth century thus had effects that were both important and long-lasting – but can it be linked to the Little Ice Age? Proxy data from tree-rings, stalagmites, pollen and lake sediments all suggest that 'the climate of the interior of South Africa was around 1°C cooler in the Little Ice Age', and that the seventeenth century saw a steady rise in precipitation, probably because of a southward shift in the Intertropical Convergence Zone (ITCZ), the band of storm clouds that circle the Earth near to the equator, where rain-bearing winds from the South Atlantic meet the dry trade winds that prevail further north.[36] The Dutch newcomers thus enjoyed numerous advantages: they had chosen the sole concentration of temperate land in a largely tropical continent, an area populated by herders and gatherers who (unlike other indigenous peoples in the region) lacked iron, and above all iron weapons, and who could therefore be either expelled or enslaved with relative ease; and they arrived there at a time of plentiful precipitation. This, coupled with their superior efficiency and technology, is why the Dutch prospered.

The southward shift of the ITCZ after 1630, probably because of cooler temperatures in northerly latitudes associated with the Little Ice Age, may have brought higher precipitation to South Africa but it caused desiccation, droughts and famines elsewhere. The flood waters of the Nile in Egypt fell to some of the lowest levels ever recorded in 1641–43, 1650 and 1694–95, suggesting severe drought in the Ethiopian highlands where the Nile rises. Records from modern Senegambia mention famine fifteen times during the seventeenth century, with especially intense events in 1639–42, 1666–68, 1674–76 and 1681. Visiting Gorée Island (just off modern Dakar) during the last of these great famines, the French trader Jean Barbot reported that 'several thousand persons have lost their lives and a greater number their liberty' and that the survivors 'looked like perfect skeletons, especially the poor slaves'. Barbot also stated (presumably on the basis of local information) that 'there were even worse famines in 1641 and 1642'. In the interior, Timbuktu and the region around the Niger Bend also suffered an intense famine at this time, with '1639 and 1643 virtually without rain, and low rainfall in the intervening years', and after two years of drought they experienced famine again in 1669–70; Lake Chad fell to the lowest level ever recorded in the 1680s.[37]

The climate of West Africa is determined primarily by the rainfall regime. *Sahara* means 'wilderness' or 'desert' in Arabic, and even the Sahel, the semi-arid belt south of the Sahara that stretches from the Atlantic Ocean to the Red Sea, receives only 100 millimetres (4 inches) of rain in most years – a meagre precipitation sufficient to sustain only small nomadic groups that herd camels, sheep and goats on hardy seasonal vegetation. The lands just south of the Sahel receive between 100 and 400 millimetres (4 to 16 inches) of rain per year, which make it possible to breed and raise cattle, sheep and goats. Somewhat further south, annual precipitation of between 400 and 600 millimetres (16 to 24 inches) allows the cultivation of millet, the hardiest of cereal crops. Further south still, annual rainfall exceeds 600 mm and produces the rolling tropical grassland known as savannas, which stretch from latitude 15° North to latitude 30° South. Farmers there can grow sorghum and other rain-dependent crops.

It is important to note the vulnerability and incompatibility of these farming strategies. Small variations in rainfall can produce major consequences. If the rains fail in an area that normally receives 400 millimetres of precipitation, all agriculture ceases and its cultivators must either become pastoralists or migrate – but those who remain will soon encounter nomads forced by the same desiccation to migrate south in search of grazing for their herds and flocks. Nomads and herders therefore compete directly for land that has become marginal, and to this end may deploy either force or the threat of force.

In the savannas, reduced rainfall could also produce drought, war and regime change. Since rainmaking formed an essential skill for successful rulers in West Africa, a prolonged drought might call this ability into question and lead to a challenge from someone, such as a religious leader who claimed supernatural powers (in Muslim areas often a Sufi sheikh: *marabout*

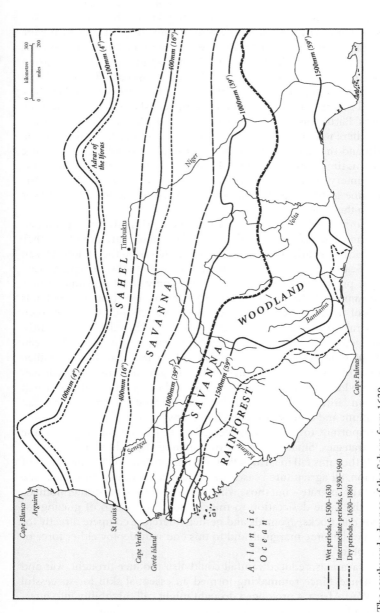

23. The southward advance of the Sahara from 1630.
Rainfall patterns in West Africa range from an annual average of 4 inches in the Sahel on the fringe of the Sahara desert to almost 60 inches in the rainforests along the coast, with several critical thresholds that determine what types of agricultural endeavour are viable. After about 1630, each of these thresholds moved southwards, forcing farmers to migrate.

in many West African languages). A ruler whose rainmaking skills faltered during a drought might deflect challenges through a successful war, using surprise and mobility to expel or subjugate neighbouring populations and provide his followers with access to adequate water and land. This no doubt explains the migration of nomads from what would soon be called the Empty Quarter of the Sahel to prey upon the farmers immediately to the south, which in turn led those displaced to attack their own southern neighbours, and so on, in a tragic domino effect.

Louis Moheau de Chambonneau, a French trader in Senegal, attributed the famine of 1674–76 largely to a civil war caused by a charismatic Muslim *marabout*, aged about thirty, who attracted many adherents in the various Wolof kingdoms along the Senegal river by 'preaching penitence entirely naked, disdaining clothes, with his head entirely shaved. He spoke only of the law of God and of welfare and freedom'. Chambonneau called the sheikh 'Tourbenan' (from the Arab and Wolof word *tuub*, meaning a convert to Islam), while Arabic sources called him 'Nasir al-Din' (and claimed he came from Mauritania), but everyone agreed that 'the war of the *marabouts*' targeted those whom the sheikh deemed insufficiently Islamized. Chambonneau reported that in 1674 the new French base at St Louis, at the mouth of the Senegal river, already suffered a 'scarcity of foodstuffs because of these wars', but the situation deteriorated when one of the Wolof rulers counter-attacked and 'for the entire year 1676 did nothing but kill, take captives, pillage and burn the countryside ... destroying the millet harvest and cutting it down while still green, forcing the local population to eat boiled grass'. When Chambonneau sailed up the river to trade, 'whole families offered themselves to me as prisoners, provided they were fed, having reached the extremity of killing each other in order to steal some food'.[38]

Worse followed. Tourbenan's version of Islam enjoined, among other things, a shift from agriculture to herding, and he demanded that his disciples cease sowing crops. This shift left them totally unprepared when severe drought returned in the 1680s. Desperate for food, Tourbenan's converts killed their animals to stay alive, but having done so, they either fled or starved. The number of slaves known to have been transported by sea from the states of Senegambia reflects these changes, almost doubling from over 8,000 in the 1670s, the decade of the war of the *marabouts*, to over 14,000 in the 1680s. Even a century later, visitors found no farmers in the region.

Similar instability, also caused by a combination of human and natural causes, prevailed further south. After visiting the Bights of Benin and Biafra in 1678–79 and 1681–82, Jean Barbot asserted that local states were 'ruined by the continual wars which have caused continual famines'. According to historian John Thornton, 'probably more than half of the people of Atlantic Africa lived in polities that measured around fifty kilometres across and had only a few thousand inhabitants'. These polities did not wage wars for land, but for people: since African legal systems did not regard land as private property, 'ownership of slaves in Africa was virtually equivalent to owning land in western Europe or China'.[39] Until the mid-seventeenth century, when the ITCZ migrated southwards and the climate deteriorated, most conflicts for slaves in West Africa remained small-scale and involved elite warriors who fought with javelins and clubs; but thereafter rulers created armies of slaves and mercenaries, equipped with muskets, who fought over far larger areas and took far more captives. This change in turn triggered an arms race in which rulers eager to acquire firearms for their defence traded them for slaves, feeding a dramatic expansion in the transatlantic slave trade as European demand for slaves to work their American sugar plantations escalated.

Forced Migration: The African Slave Trade

African slavery existed long before the arrival of Europeans, and took two forms. Men and women captured during wars or raids usually became slaves; so did those found guilty of serious crimes, if they were fined a sum larger than they could pay: unless their kin-group helped them, offenders were sold as slaves (the sale price paid their fine). 'The distinction between these two processes was crucial', Robert Harms reminds us: slaves condemned for a crime 'were unlikely to run away because they had no place to go' and so often remained in the area; whereas a captured slave 'had quickly to be taken from the point of capture so that he [or she] could not return home'. The demand of both European and Arab traders for cheap labour increased both forms of slavery, but especially the latter. According to a missionary who interviewed slaves in the Caribbean about their background, 'most of them were captured in open warfare'.[40]

Four distinct slave trades existed in early modern Africa: north, across the Sahara and up the Red Sea; east, across the Indian Ocean; and (by far the largest) west, across the Atlantic. The fate of perhaps 2 million African men, women and children forcibly deported to the Americas during the seventeenth century has recently become better known thanks to the remarkable database constructed from almost 35,000 documented slave voyages (perhaps four-fifths of the total) by David Eltis and his colleagues. Almost one-half of the slaves came from West Central Africa (the 500 miles of coast on either side of the Congo estuary); almost one-quarter came from the Bights of Benin and Biafra (the coastal areas on either side of the Niger Delta); and almost one-tenth from Senegambia. Until 1641, ships from Spain and Portugal accounted for 97 per cent of the slave trade organized in Europe, but the separation of the two crowns and the ensuing war between them (Chapter 9) opened the way for French, Dutch and British ships to deport slaves to work on their rapidly expanding sugar plantations in the

	Senegambia	Sierra Leone	Windward Coast	Gold Coast	Bight of Benin	Bight of Biafra	West Central Africa	Southeast Africa	All regions combined
1601–10	16,251	0	0	0	583	507	74,532	51	91,923
1611–20	18,700	0	0	211	4,880	1,154	124,795	62	149,807
1621–30	8,185	0	0	0	4,854	3,884	165,502	821	183,246
1631–40	7,650	0	0	0	3,546	1,855	105,679	0	118,730
1641–50	28,563	521	0	4,250	5,307	28,749	56,014	0	123,404
1601–50	*79,353*	*521*	*0*	*4,461*	*19,170*	*36,149*	*526,522*	*934*	*667,110*
1651–60	30,836	989	707	1,790	15,681	27,102	79,325	0	156,429
1661–70	21,126	151	0	22,233	30,918	40,300	110,210	4,399	229,337
1671–80	11,821	0	0	32,083	40,843	47,434	78,819	11,776	222,775
1681–90	23,493	1,539	0	19,218	117,956	21,996	65,570	10,931	260,703
1691–1700	23,797	2,268	0	33,630	163,246	40,777	74,694	2,770	341,182
1651–1700	*111,073*	*4,947*	*707*	*108,954*	*368,644*	*177,609*	*408,618*	*29,876*	*1,210,426*
1601–1700	190,426	5,468	707	113,415	387,814	213,758	935,140	30,810	1,877,536

24. Slaves captured and shipped from Africa, 1600–1700.
The Europeans exported twice as many slaves in the second half of the seventeenth century as in the first. Of this total, one-third came from West Central Africa, one-tenth from Senegambia and Sierra Leone, and over half from the Gold Coast and the Bights of Benin and Biafra. The table does *not* include the overland deportation of slaves living in sub-Saharan Africa to the Ottoman empire.

Caribbean. Of 160,000 slaves who disembarked in the Americas during the 1680s, over 141,000 went to the Caribbean.

The Little Ice Age played its part in increasing the slave trade, not only via the southward shift of the ITCZ but also via an increase in epidemics and wars in West Central Africa. This area supplied the largest number of slaves for the transatlantic trade – of an estimated 12,569 slaves deported to the Americas in 1639, all but 285 came from West Central Africa – and research by James Fenske and Namrata Kala suggests that agricultural production in this region was particularly susceptible to climate change. This in turn affected the slave trade: 'exports declined when local temperatures were warmer than normal' while colder years, like those of the Little Ice Age, saw 'increased slave exports'. Droughts became more frequent in the mid-seventeenth century, producing (as Joseph C. Miller demonstrated) 'violence, demographic dispersal, and emigration' – and more slaves for the Americas.[41] The competition for slaves, combined with the harsher climate, also caused political fragmentation. Most notably, the kingdom of Kongo, one of the few large states of Atlantic Africa with perhaps 500,000 subjects, succumbed. Its kings had long maintained an ambiguous relationship with the Portuguese, converting to Catholicism but resisting persistent Portuguese efforts to impose economic and political control until 1665, when a severe episode of drought, locusts and disease disrupted life in the interior and led to a succession dispute. A Portuguese force from Angola invaded and routed the main army of Kongo at the battle of Ambuila (Mbwila), killing the king and ending the independent existence of his realm. Henceforth, regional chiefs maintained themselves by fighting wars to secure slaves, whom they sold to European traders in return for guns and ammunition, prolonging the vicious cycle. The decennial total of slaves exported from West Central Africa leaped from almost 60,000 in the 1650s to almost 100,000 in the 1660s, and exceeded 120,000 in the 1670s. Right up to abolition of the slave trade, the region continued to export more slaves than any other part of the continent.

Did this high level of forced migration perhaps mitigate the impact of the Little Ice Age on the humans who remained in sub-Saharan Africa? After all, migration from a community normally reduces its food requirements, and certain areas experienced significant out-migration – especially when we include not only deportees who reached America, but also those who succumbed to disease as their captors took them from drier inland areas to the coast, where they lacked immunity to new disease environments, and those who perished in the overcrowded and unsanitary coastal holding pens or on the voyage. Malnutrition, ill treatment and despair all took their toll: many slaving ships' logs record the cause of death of their precious cargo, and they included 'stubbornness' and 'lethargy' as well as 'dysentery', 'scurvy' and 'a violent blow to the head'. Not for nothing did the

Portuguese call the vessels that carried slaves from Africa to America *tumbeiros:* coffin ships.

If we total the forced migrations from East, Southeast and West Africa, seventeenth-century European and Arab traders between them apparently deported some 30,000 slaves annually from a continent with perhaps 100 million inhabitants. At first sight this may make the overall impact seem marginal (except, of course, for the victims), but the slaves did not come from all parts of the continent. On the contrary, most seem to have lived within 150 miles of coast – and, within that restricted area, often in specific regions. Although we currently lack data on the exact origin of seventeenth-century slaves, those from the eighteenth century show a remarkable degree of geographic concentration. For example, the profile of slaves who embarked on vessels leaving the Bight of Biafra reveals that most originated in the same small region: the Cameroon Highlands. In addition, over half the slaves deported to the Caribbean from the Bight of Biafra in the seventeenth century were female and one-tenth were children, and over one-third of those deported from West Central Africa were female and almost one-fifth were children. Deporting women and children intensified the impact of emigration by removing (in effect) the next generation.[42]

It nevertheless seems unlikely that even the areas of Africa most involved in the slave trade lost more than 10 per cent of their population through forced migration, whereas a drought or a drought-induced epidemic could wipe out three times as many people. So to return to the earlier question: did forced migration mitigate the disruptive impact of the Little Ice Age on those humans who remained in sub-Saharan Africa? The answer may well be affirmative. If we add famine and disease mortality to the high level of forced migration in coastal regions, the synergy of human and natural agency in sub-Saharan Africa may have removed one-third of the total population, as it did in much of Asia and Europe, and these disasters may have assisted those who remained to survive even prolonged climatic adversity.

Australia

Australia, the driest inhabited continent, covers 5 per cent of the world's land mass and yet it has (and has probably always had) one of the lowest population densities in the world. The reason for this disparity lies in a combination of climate and isolation. Only the southeastern and southwestern corners of Australia boast a temperate climate and fertile soils, but since they lie furthest away from the other continents, until the late eighteenth century they remained almost entirely isolated from the rest of the world, both demographically and economically. Deserts and semi-arid lands, known as the Outback, cover more than two-thirds of Australia, and the annual rainfall in some locations there can vary from under 100 millimetres to over 900 millimetres.

As in sub-Saharan Africa, the critical variable is rainfall: even in the temperate zone, drought and the threat of drought are constant concerns.

Australia, in the words of the pioneering historian of climate Richard Grove, has 'above all other places a claim on the epithet, "the el Niño continent"'.[43] This means that the droughts that afflict China, Southeast Asia, Indonesia and India in El Niño years also afflict Australia, and that the doubled frequency of El Niño events in the mid-seventeenth century (page 14) would have struck Australia with unusual force. In particular, the continent presumably experienced the same major drought registered in nearby Indonesia from 1643 to 1671 (page 310). Reconstructed tree-ring sequences from Tasmania appear to confirm this surmise, showing a succession of poor growing seasons in the mid- and late seventeenth century, decades that saw the 'most prolonged cool period in the past 700 years'.[44]

How much would these climatic events have affected the population of Australia in the seventeenth century? To survive the extreme climate of the Outback even in normal times, native Australians devised distinctive adaptive strategies. In the Western Desert the annual life cycle began with a wet season from December to February, when huge thunderstorms unleash torrential rains, which provide abundant water but do not immediately produce food. Therefore, at least in the mid-twentieth century, family groups moved around, foraging until seeds, root crops and fruit began to appear in March. Then they settled down, living by waterholes on the plains and harvesting crops for about two months (albeit often amid periodic drizzle and temperatures that fall to 6°C at night). This 'cold time' ended in August as temperatures rose rapidly and the landscape gradually dried out. Men now set fire to vegetation on the plains, both to trap game and to improve the yield of seeds and tubers the following year, while women prepared and stored the vegetables that would sustain the group through the rest of the year. Eventually in the 'hot time' or 'hungry time', as temperatures rose to 50°C, the waterholes on the plains dried up, forcing the families to retreat to rock holes for the rest of the year. There they reduced their daily activities and tried to make their food and water last as long as possible, but since drought and heat stress limited foraging, the average calorific daily intake might fall as low as 800 per person – less than half of what is necessary to sustain even someone of small stature. Sometimes the weak drank blood drained from the stronger members of the group in order to survive until the rains returned and allowed them to leave their rock holes and spread out on the plains once more.

Although the environment of the Western Desert is harsh, its human inhabitants used up to 120 native plants to satisfy their needs. Of these, seventy yield edible parts and over forty produce seeds, which can be husked, winnowed, ground and turned into a paste to be either baked in the campfire or eaten raw. In addition, tubers and bulbs were pulled up and roasted, and large game (such as kangaroo), once caught, were gutted and grilled, and small game (birds, lizards and snakes) were baked.

Since both the seasons and the availability of each resource followed the same pattern each year, the Aboriginal populations of the Western Desert survived thanks to an intimate environmental expertise, combined with knowledge of when each source of nutrition would become available as the seasons changed, and the use of fire to trap game and stimulate future crops. The one unpredictable variable was the duration of the 'hungry time', which determined whether or not there would be sufficient water and food to sustain the group from one annual cycle to the next, and thus how many would die (either from thirst or starvation) and how many would be born (because famine amenorrhoea would prevent conception).

Scott Cane, whose research on Aboriginal subsistence strategies in the Western Desert is summarized above, noted one other important feature of life in the 'hungry time': 'If the rains fail to come, tensions run high and fights are common'. All Aboriginal groups carried weapons – some of them offensive: spears (often used with spear-throwers which produced a velocity of 100 miles per hour and accuracy up to 50 yards), boomerangs and clubs (sometimes with sharp shells attached to the head). Clearly these weapons were used against people, as well as game, because warfare features in many Aboriginal oral traditions, and the first British settlers in 1788 'found Aborigines with wounds that could only have [been] caused by fighting with other Aborigines'.[45] It seems probable that wars between rival groups – like fights between group members – became more common in years when the 'hungry time' dragged on and reduced essential resources, as it did in the mid-seventeenth century. It also seems probable that a fatal synergy between natural and human factors prevailed in Australia, and that its population shrank when the reduced resources available failed to satisfy its minimum demands.

Cane believed that his study of the Aborigines of the Western Desert presented 'the last reliable data on hunter-gatherer subsistence economies from arid environments anywhere in the world', because the way of life he described ended in the 1950s when the hunter-gatherers 'moved from the Desert onto cattle stations, missions and government settlements, scattered around the desert fringe'.[46] In the seventeenth century, however, hundreds of thousands (if not millions) of hunter-gatherers populated the various arid environments of the planet. Although the absence of a relevant human archive precludes certainty, it seems likely that in large parts of Africa and the Americas, in Central Asia and in the far north of Europe, those nomadic populations evolved coping strategies similar to those of the Aboriginal population of Australia, and that those strategies proved only partially adequate whenever the annual 'hungry time' grew longer. If so, the only large area that registered rapid and sustained population growth in the seventeenth century, apart from New France and New England, was Japan.

CHAPTER SIXTEEN

Getting It Right: Early Tokugawa Japan[1]

The Pax Tokugawa

Whereas seventeenth-century Europe suffered wars, revolution and economic collapse, seventeenth-century Japan experienced rapid demographic, agricultural and urban growth, political stability and no wars.

Tokugawa Japan: The First Century

Year	Population size (in millions)	Urban population (in millions)	Arable land (in million hectares)	Harvest yield (in million bushels)
1600	12	0.75	2	100
1650	17	2.5	2.4	115
1700	30	4	8	150

The figures above are striking not just because of the dramatic population growth – some parts of the archipelago experienced a fourfold increase during a century when most of the world experienced a sharp demographic decline – but because of the simultaneous increases in urban population, total land cultivated and harvest yields. Over 7,000 new villages sprang up, many of them on lands brought under the plough for the first time thanks to complex hydraulic engineering projects (153 in 1601–50 and 227 in 1651–1700). Average rice production per village rose from around 2,000 bushels in 1645 to over 2,300 bushels in 1700.[2]

Japan also experienced 'urbanization without precedent in history'. Between 1600 and 1650 the number of people living in towns and cities tripled, and between 1651 and 1700 it almost doubled again, with most of the urbanites dwelling in over 100 'castle towns'. Kanazawa, for example, with some 5,000 inhabitants in 1583 when it became the headquarters of the largest domain in western Japan, had 70,000 by 1618 and perhaps 100,000 by 1667. Edo (as Tokyo was then known) grew from little more than a fishing village in 1590, when it became the headquarters of the Tokugawa domain, to a metropolis of perhaps a million people a century later.[3]

These unique achievements did not stem from a benign environment: on the contrary, the Japanese archipelago has always been extremely vulnerable to climate change. Its northern areas are subject both to the Chishima Current, which brings Arctic water southwards, and the Yamase Effect, which produces cool air for considerable periods of the summer. Both climatic events can cause crop failures. Furthermore, most of Japan consists of mountains thrown up by the collision of the Earth's tectonic plates, which has three adverse consequences. First, the archipelago – like many other parts of the Pacific Rim – has an unusual number of active volcanoes, and their eruptions could and did reduce local temperatures. Second, the majority of the Japanese population lived (and still lives) on the coastal plains of three islands – Honshu, Shikoku and Kyushu – and the abrupt gradients that rise from these coastal plains make it harder to bring new land under cultivation. Finally, pressure to house and heat the rapidly growing population led to clear-cutting the tree cover (that is, removing all grades of wood rather than only certain trees) on those steep slopes, which caused serious soil erosion and magnified the risk of frost, flood and drought.

According to Conrad Totman, an eminent environmental historian of Japan, the combination of aggressive cultivation of recently deforested land and the prevalence of clear-cutting on steeply sloping ground 'crowded the biological boundaries of crop viability', made the transition from relative abundance to ecological overload in marginal lands exceptionally abrupt and 'increased the portion of total food production chronically at risk of failure'.[4] The archipelago therefore could not escape the effects of the Little Ice Age. During the landmark winter of 1641–42 the first winter snow fell on Edo six weeks earlier than usual, and according to the memoirs of Enomoto Yazaemon, a merchant living nearby, 'on New Year's Day, pots and pans full of water froze and seemed likely to burst; and one foot of frost covered the fields. Thereafter I observed seven snowfalls until the spring.' The prolonged cold weather caused the price of rice to rise from 20 silver monme in 1633 to 60 in 1637–38, and to 80 in 1642.[5] Even in Osaka, the 'kitchen of Japan' where in normal years merchants, lords and officials maintained huge stockpiles of food, rice became so scarce in July 1642 that 'the common man cannot maintain himself, his wife and his children, so that many people died of hunger'. Crowds congregated before the house of the city governor 'begging His Excellency to provide them with some means by which they could stay alive' until 'the aforesaid governor distributed rice from various storehouses, and from the granary in Osaka Castle, to the destitute at a low price. This ended the disturbances.'[6]

Disturbances at Shimabara, in the southern island of Kyushu, proved harder to end because they arose from the imposition of oppressive taxes at a time of climatic adversity. According to a Dutch merchant living nearby, when the lord of Shimabara demanded 'taxes and demands for so much rice that they could not be met', his agents tied up those unable to pay and dressed them in 'clothes made of straw', which they then set on fire. They also humiliated 'their wives by stringing them up with their legs entirely bare'. Outraged by such atrocities, weary of being asked 'to pay far more in taxes than they are able to do', and unable to 'subsist on roots and vegetables', villagers resolved 'to die one single death instead of the many slow deaths to which they were subject': so in December 1637 they rebelled. Their defiance encouraged peasants of the neighbouring Amakusa islands, also long abused by their lord, to kill his magistrates and the soldiers sent to restore order, after which they joined the rebels of Shimabara. European missionaries in the region had converted many Japanese to Christianity, including Amakusa Shirō, a sixteen-year-old boy who claimed to be the reincarnation of Christ; and many of them joined the revolt. So did 200 discontented samurai (warriors), who offered invaluable military advice. Some 25,000 insurgents, marching 'under banners bearing the sign of the cross', now burned down the headquarters of the Shimabara domain, and collected food and weapons before retiring to the neighbouring castle of Hara, on a promontory surrounded by the sea. Amakusa Shirō 'preached and celebrated Mass twice a week', confidently proclaiming that 'judgment day is at hand for all Japan' and that 'all Japan will be Christian'. It took an army of over 100,000 men sent by the central government three months to take Hara Castle by storm, after which they slaughtered everyone within – including Amakusa Shirō.[7]

The unrest at Osaka, Amakusa and Shimabara proved a turning point in Japanese history. The first four decades of the seventeenth century had seen some forty major rural revolts (*hōki*) and 200 lesser uprisings (*hyakushō ikki*), as well as almost eighty feuds fought between the major landholders; but during the next eighty years both revolts and feuds virtually ceased. The food riots in Osaka also had no sequel: most Japanese towns remained peaceful for a century or more. Most remarkable of all, the Shimabara campaign proved to be the last major military action in the archipelago for two centuries. Seventeenth-century Japan thus presents a curious contrast with the rest of the world: although initially its experience did not differ markedly from that of other countries, since it suffered from both the Little Ice Age and the General Crisis, in the 1640s it broke free. How?

The Industrious Revolution

The distinguished Japanese historian Hayami Akira has identified two paths of escape from the tyranny of subsistence agriculture. The first path is capital-intensive and labour-saving: investing money to make agricultural production more efficient, thereby creating a pool of cheap labour to fulfil factory demands and so facilitate an Industrial Revolution. The second path is the exact reverse: a labour-intensive and capital-saving strategy, which Hayami christened 'the Industrious Revolution', in which peasants escaped from subsistence farming by investing more time and energy, rather than more money, in their endeavours. Although improved farm tools and techniques played their part in Japan's Industrious Revolution, output improved primarily because families rationalized production and worked both harder and longer. 'Self-exploitation', Hayami argued, is the principal explanation for why the amount of cultivated land quadrupled and agricultural output doubled in Tokugawa Japan between 1600 and 1868.[8]

Japanese families adopted four additional prudential strategies to ensure that demand for basic resources would not outrun supply. First, many people worked away from their homes and families for prolonged periods, and in some villages up to one-third of all adolescents left to work either in a neighbouring community or in a town. Hayami's research revealed that 'the lower the social stratum, the more people work away from home, and the higher their age upon return to the village to marry'. On average, the daughters of poor Japanese families married five years later than their wealthier sisters – a delay that significantly reduced the number of children they could bear.[9] Second, those women who stayed home worked long hours in the field, which no doubt both reduced fertility and increased infant mortality (Chapter 4). Third, in the absence of milk from animals (for few Japanese farmers raised livestock), mothers breast-fed their children intensively, often exclusively, until the age of three or four, a practice that normally suppresses ovulation. Fourth, as in China, when families faced economic hardship in spite of these prudential strategies, they regularly resorted to abortion and infanticide.

Until the early seventeenth century the Japanese term *mabiki*, 'thinning out', was used only for crops, but thereafter it became the metaphor of choice for infanticide – and infanticide became more common. 'As we observe from the recent population registration', the lord of Sendai (north of Edo) noted in 1677,

> The number of people is increasing greatly, and we estimate that within ten or fifteen years, there will be grain shortages. If people multiply wantonly like the brood of birds and beasts, there will be mass starvation. Even if adults can be saved, infanticide will be impossible to avoid.

Yet, as Fabian Drixler has pointed out, poverty was neither a necessary nor a sufficient cause of infanticides. Some occurred because the unfortunate child was born at a time, on a day or in a year deemed inauspicious, others because one parent had reached an age deemed unlucky (thirty-three for women, forty-two for men), or simply because they were twins. The Industrious Revolution also contributed, because 'many couples were unwilling to divert labour, especially the mother's, from production to child-rearing'.[10] The same considerations also led to a rise in contraception and abortion. Popular medical manuals published in the 1680s and 1690s included charms and spells to avoid and end pregnancy, and one, entitled *A guide to women's good fortune* (*Fujin kotobukigusa*, 1692), explained how to abort a foetus.

Besides these four negative strategies for survival in time of hardship, Japanese villagers implemented certain positive policies designed to promote collective survival. Above all, the average community was divided into many holdings of different sizes: one or two large ones, rather more middle-sized ones and a majority of small or very small ones. Although this distribution pattern was true of villages throughout the early modern world (see Fig. 5 above), in Japan many farmers included in their households both servants and sub-tenants, while most villagers with no land were attached to one of the landholding households. Documents often referred to the house-hold head as *oyakata* ('one who takes the role of parent') and to servants and sub-tenants as *kokata* (or 'child': in Japanese, the term 'orphan' does not mean 'without parents' but 'without family'). Each village was a cluster of mutually dependent households, not a collection of autonomous farming units. Ideally, the *oyakata* furnished the capital goods needed periodically by the smaller households, while the *kokata* provided the labour required at certain crucial periods by the larger farms (above all transplanting rice seedlings, a task that had to be effected in each field within a matter of hours). Communities also cooperated to perform collective functions that exceeded the resources of individual households, such as building or re-thatching a dwelling, repairing the dikes, or dredging the irrigation chan-nels. Above all, during food shortages, the *oyakata* were expected to feed their *kokata* (whether servants or sub-tenants) and not leave them to starve.

All these strategies helped to mitigate the impact of the Little Ice Age, but two other factors played a greater role. On the one hand, although Japan had enjoyed the same benign climate in the sixteenth century as the rest of the northern hemisphere, a century of civil war (known as the 'Warring States Era': *sengoku jidai*), which ended in 1615, ensured that most of the archipelago was underpopulated rather than overpopulated. On the other hand, the Warring States Era left a favourable political legacy: the ceaseless power struggle eventually eliminated alternative foci of power until only one remained: the Tokugawa dynasty. In 1614 Richard Cocks termed Tokugawa rule 'The greatest and powerfullest tyranny that ever was heard of in the world', and over the next two centuries the dynasty coordinated responses that neutralized some of the worst effects of the Little Ice Age and created conditions favourable to rapid economic and demographic growth.[11]

'The Greatest and Powerfullest Tyranny That Ever Was Heard of in the World'

Although Japan had been an empire throughout its history, by the sixteenth century the emperor exercised no executive authority. Instead political and military power in the archipelago was divided among *daimyō* ('great names') until, in the last decades of the century, the warlord Toyotomi Hideyoshi eliminated or neutralized all his rivals and imposed a series of measures that promoted social and economic stability. He commanded farmers throughout Japan to surrender all their 'swords, bows, spears, muskets, or any other form of weapon' and instead 'engage solely in cultivation'; and he decreed that samurai could no longer be farmers and farmers could no longer be samurai.[12] Although a few samurai gave up their weapons and remained in their community as farmers, most relocated with their households to the new towns that grew up around the headquarters of the local *daimyō* where they became salaried armed retainers. These measures both separated the samurai from their rural power base and demilitarized the countryside.

Hideyoshi also commissioned a vast cadastral survey. Inspectors toured the archipelago to measure all land parcels, to identify their purpose (rice paddy, dry field, residential lot) and their quality (from 'superior' to 'very inferior'), and to estimate their potential productive yield according to a standard measurement: the *koku* (approximately five bushels in the case of rice, the commonest but not the only commodity assessed). Land was therefore measured in the number of *koku* it could produce: the *kokudaka*, which became the basic unit of taxation. Hideyoshi permitted no exceptions or exemptions to the survey: if they encountered resistance, his officials must 'pursue a lord to his castle and put him to the sword along with all of his vassals' and 'kill all the recalcitrant peasants in a whole district'. He thus created a far more complete inventory of the productive capacity of his country than any other ruler of the early modern world.[13]

Hideyoshi's death without an adult heir in 1598 reopened the civil wars, but two years later the *daimyō* Tokugawa Ieyasu (d. 1616) defeated a coalition of opponents and in 1603 secured from the emperor the title 'shogun' (in full, *Sei-i taishōgun*: 'Great generalissimo who overcomes the barbarians'). Ieyasu and his immediate family now controlled the major towns and about one-quarter of Japan's arable land; other relatives and long-term allies held the rest as fiefs.

Ieyasu levied no direct taxes on the *daimyō*: instead he requested donations for specific purposes (such as building materials and labour to expand and fortify his headquarters at Edo) and invited each of his allies to spend prolonged periods in his new capital, where he could keep them under surveillance. He also continued Hideyoshi's practice of accumulating information to enhance his power. His cartographers used the cadastral survey to produce a uniform national map that showed all the provinces and towns, as well as all the sea routes and harbours, the roads and post-stations, and the travelling distances by land and sea between the principal centres; and he invested in a comprehensive communications infrastructure centred on a network of trunk roads. Each road had numerous checkpoints, where travellers had to show their papers, and post-stations with fresh horses, porters, supplies and lodgings, linked by relays of professional runners who travelled in pairs (one carrying documents or small parcels and the other with a lantern so that they could travel by night and day). This provided the Tokugawa with an unparalleled knowledge of public affairs and unrivalled ability to react to an emergency on the periphery of the state.

As soon as he had destroyed the remaining strongholds of his opponents in 1615, Tokugawa Ieyasu issued a plethora of regulations ranging from the petty (court nobles must micturate only in urinals) to the drastic: henceforth each *daimyō* could retain only one castle and must destroy all the rest. Ieyasu died the following year, but his son Hidetada (who ruled until 1623) and grandson Iemitsu (r. 1623–51) inherited his title of shogun and further consolidated and expanded the power of the central administration. They periodically dispatched inspectors to assess each fief's defensive disposition, legal system, economic means and general morale. An unfavourable report, or provoking peasant rebellion through heavy-handed policies could result in forfeiture of a fief. From 1615 to 1651 the three Tokugawa shoguns confiscated the fiefs of ninety-five *daimyō*, who thus lost all means of support – as did their retainers, who became *rōnin* (masterless samurai), and they transferred a further 250 fiefs from one *daimyō* to another. As Harold Bolitho has observed: 'Never in the history of Japan had so much violence been done to local autonomy.'[14]

Iemitsu extended his control over *daimyō* in other ways. He forbade them to build big ships, levy tolls or engage in disputes among themselves; required them to maintain roads, bridges and post-stations in their domains; and commanded them to extirpate Christianity and to decide all legal cases according to the laws of Edo. He also transformed their visits to Edo into a closely managed system known as 'alternate attendance' (*sankin kōtai*: the first word meant 'reporting for audience', the second 'to rotate').[15] Each *daimyō* now had to reside in the shogun's capital for twelve months out of every twenty-four, and keep his principal consort and heir there permanently (in effect as hostages). Iemitsu appointed different months of the year for each *daimyō* to 'report for audience', both to prevent possible plotting among them and to avoid depleting any area of all its local leaders at the same time. In addition, at all checkpoints on the roads leading to and from Edo his guards searched for weapons coming in, women going out (since a lord who removed his wife might be plotting treason) and anyone who lacked written permission to travel. He expected each *daimyō* to travel to Edo with an entourage of samurai and servants appropriate to his rank, and by 1700 the various aristocratic compounds in the capital housed at least 250,000 people.

Iemitsu also issued codes to regulate the behaviour of other groups of subjects. His regulations for villagers on Tokugawa lands codified appropriate behaviour in annoyingly comprehensive detail, with special regard for the type of clothes and ornaments permitted for each social group. His edicts for town dwellers regulated the production and distribution as well as the consumption of goods, urging craftsmen, artisans, artists and architects to increase their productivity through hard work. This comprehensive legislation formed the foundation of the Industrious Revolution.

In the 1630s Iemitsu also banned all foreign trade, and residence abroad, by any Japanese (except for a compound near Busan in Korea and another at Okinawa in the Ryukyu Islands, where a few Japanese merchants handled trade with the Asian mainland). For a while, Iemitsu tolerated the presence of Portuguese merchants, although in 1636 he confined them to Dejima, an artificial island in Nagasaki bay connected to the mainland by a single bridge and dependent on the Japanese authorities for everything, even drinking water. Three years later, blaming missionaries for the Catholic overtones of the Shimabara rebellion, the shogun expelled all Portuguese from Japan; and when an embassy with over fifty members returned in 1640 to plead for the restoration of free trade, he had them all executed. The following year Iemitsu confined all Dutch merchants to the vacant island of Dejima, and for almost two centuries they remained the only Europeans allowed to visit and trade with the archipelago.

The savage treatment of the Portuguese formed part of a coordinated campaign to control the religious beliefs of Iemitsu's subjects. In 1638 he required everyone living on Tokugawa domain lands to present proof to the local magistrate that they belonged to a Buddhist temple; in 1665 his successor extended the same requirement to *daimyō* lands; and from 1671 the proof had to be presented annually. The magistrates compelled anyone suspected of Christian beliefs to trample on images of the Virgin Mary to demonstrate publicly their indifference. Those who refused, and any missionaries captured, were tortured and executed. Tokugawa apologists propagated an alternative faith by sacralising the new dynasty: its founder, Ieyasu, became *shinkun* ('divine ruler') and by 1624 Japan boasted forty shrines dedicated to him. Many more followed, including the stunning architectural complex constructed in 1634–36 by Iemitsu at Nikkō, 80 miles north of Edo, which covers over a square mile and contains more than 500 paintings and 5,000 sculptures.

The shoguns also sponsored tracts that mixed texts drawn from Buddhist, Confucian and Shinto sources to explain how the dynasty had acquired the Mandate of Heaven, and how Japan's warrior code (*Bushidō*) formed the ideal instrument to preserve it. Most of those who wrote such tracts were either warriors or the sons of warriors, and they stressed absolute obedience to authority as the supreme virtue for subjects, exalted military norms in peace as well as in war and compared the primary task of civil leaders with that of generals: directing and coordinating the movement of great masses of people. Suzuki Shōsan, a samurai who became a monk believed, like Thomas Hobbes in England (page 287), that subjects owed obedience to any ruler who provided them with peace and justice.[16]

Coping with the Kan'ei Famine

Tokugawa Japan enjoyed several important advantages compared with other states confronting the Little Ice Age. At the local level, the *oyakata/kokata* system provided a safety net for many of the most vulnerable people; and the *kokudaka* system created granaries that could be opened in case of famine. The separation of *daimyō* and samurai from their hereditary lands, together with the 'sword hunts', made resistance more difficult to organize; while the stream of edicts regulating behaviour both accustomed the central government to take the initiative in social and economic matters and predisposed its subjects to obey. Nevertheless, climatic adversity placed Japan under severe stress. During the severe winter of 1641–42 (according to the memoirs of Enomoto Yazaemon) 'the corpses of those who had starved to death filled the streets'; Edo 'was full of beggars clad only in straw'; and 'from 50,000 to 100,000 people starved to death in Japan'. One upland village informed the shogun in 1642 that the famine had eliminated one-third of its population: 147 householders had starved to death, ninety-two had been forced to sell all their land, and thirty-eight more had fled.[17]

To cope with the Kan'ei famine (named after the Japanese era in which it occurred), Iemitsu convened a series of emergency meetings with officials from regions around the capital to discuss appropriate measures. He set up food kitchens and shelters for the starving, and instructed all *daimyō* and city magistrates to do the same; he authorized magistrates to release rice held in the government's granaries both to the starving and to farmers who lacked seed grain; and he ordered *daimyō* residing in Edo as part of the *sankin kōtai* system to return to their domains and organize famine relief. Most striking of all, he forbade *daimyō* to impose labour services on their peasants without government permission, and he drastically reduced his own demands on his subjects.[18]

When, despite all these prudent measures, food prices continued to rise, Iemitsu ordered farmers to plant only staple crops (for example, no tobacco and other cash crops could be planted as long as the famine lasted), prohibited the use of rice to make *sake*, and decreed that the peasants of each village would be held collectively responsible for paying its tax quota, so better-off farmers must help the rest. He also authorized confiscation of any abandoned smallholding to the common good and required villages to maintain roads and bridges so that foodstuffs could reach famine areas swiftly. His officials erected signs all over the country reminding farmers to tend their fields, bring their crops to market and be frugal. In 1642, when certain granary officials and rice merchants were denounced for withholding rice reserves in the hope of getting a better price, Iemitsu had eight of them executed, required four others to commit suicide, and exiled many more after confiscating their property. The hoarding ceased. He also decreed that 'because people are suffering extreme poverty of the poor harvests, *daimyō* must be careful to avoid measures that would make their situation even worse'; and when the lord of Aizu nevertheless provoked a peasant uprising, Iemitsu immediately confiscated his fief.[19]

Iemitsu's comprehensive and proactive response to the famine seems to have worked. Although Japan, like other areas in the northern hemisphere, continued to suffer periodic climate adversity, after the 1640s surviving records no longer mention people dying in the streets. Moreover, the number of revolts by vassals against their *daimyō* dropped from seventeen between 1631 and 1640 to nine between 1641 and 1650. The shogun did not rest on his laurels: to prevent any recurrence of disorders and subsistence crises, Iemitsu undertook more cadastral surveys and issued a stream of further edicts in 1648–49, later known as the Kei'an Laws.

Some decrees micromanaged everyday behaviour. The shogun instructed villagers to arise early each morning to cut grass and pull weeds, to cultivate fields all day and to spend their evenings making ropes and sacks. They should eat only barley and millet, except on a few specified holidays, leaving any rice they produced to pay their taxes; they must drink neither *sake* nor tea; they should plant trees around their house to supply firewood; and their toilets should have ample storage for human waste to provide fertilizer for their crops. Other of the Kei'an Laws aimed at reducing conspicuous consumption. Townsmen must not build three-storey houses, use gold in their homes, ride in palanquins or wear wool capes, and their servants must not wear silk; only headmen could wear cotton rain capes and use umbrellas (everyone else must use straw capes and hats). Not even underwear escaped regulation ('The loincloths of sumo wrestlers should not be made of silk'). *Daimyō*, for their part, must not commission elaborate woodcarvings, metal ornaments, lacquered mouldings or lattice work in their dwellings, and they could only serve modest meals, accompanied by a small (prescribed) amount of sake. Iemitsu left no doubt concerning the rationale for all this: at a time of general crisis, he stated, it was imperative to conserve resources. 'Unless you are generally frugal, you will not be able to govern the country. If the superiors indulge increasingly in luxury, the land tax and corvées of their subordinates will increase and they will be in distress.' The shogun also sponsored public works that increased food production (especially canals, land reclamation and irrigation schemes: the annual rate of construction doubled after the 1640s) and set up a system of emergency loans that were immediately available to *daimyō* in the wake of a natural disaster (whether fire, flood, earthquake or volcanic eruption) and repayable in easy stages.[20]

The shogun's example spread to *daimyō* lands. After the peasant rebellion at Aizu in 1642, Iemitsu transferred the fief to his own half-brother, Hoshina Masayuki, who immediately imitated the shogun's policies. He initiated a new survey of the rice-producing capacity of each village, eliminating land made barren by floods or landslides. He granted tax relief to villages whenever their crops failed, and he reduced the overall tax rate 'in order to help those whose need is greatest and to prevent peasants who might otherwise default from being forced to become indentured servants'. He also established funding agencies that made loans (some of them interest-free) to villagers in distress or to outsiders wishing to settle in the fief, and sponsored land-reclamation schemes that substantially increased the areas under cultivation. Thanks to these measures, between 1643 and 1700 the population of Aizu – a fief with some 200 villages – rose by 24 per cent, and although the tax yield rose by 12 per cent, the average tax per capita fell by 11 per cent. Other *daimyō* strove to emulate Tokugawa 'policies even when they were not, strictly speaking, required to do so' and always took care to articulate policies that fell within the broad framework established by the shogun.[21]

Pursuit of a risk-averse foreign policy formed the last of Iemitsu's critical efforts to preserve Japan from crisis. Not only did he strictly limit all contact with foreigners, confining to islands in Nagasaki bay first the Portuguese, then the Dutch, and lastly also the Chinese, he also forbade all except occasional Korean and Ryukyuan embassies to enter the country. More important, he scrupulously avoided foreign intervention. Admittedly, Iemitsu offered asylum to Ming refugees, but he went no further: in 1646 and again in 1650 he rejected requests for military assistance from Ming loyalists who opposed the Qing (Chapter 5). Likewise, in 1637 and in 1643 he declined Dutch invitations to launch a joint attack on Spanish Manila. The importance of avoiding foreign entanglements cannot be overestimated. Whereas Europe knew only four years of peace during the seventeenth century, and China knew none, Tokugawa Japan knew only four years of war (and none after 1638). By avoiding war, which drained the revenues of most other early modern states, the shoguns managed to keep tax rates relatively low but still accumulate resources with which to respond effectively in case of a natural disaster.

The Tipping Point: Onwards and Upwards

In 1651 Iemitsu died after a long illness, leaving a ten-year-old son to succeed him, guided by a council of regency – but since two of the senior regents had been the shogun's lovers protocol required that they immediately commit suicide. This created a vacuum of power for which some enemies of the Tokugawa had long prepared.

Iemitsu's success in avoiding war both at home and abroad had deprived the samurai of their *raison d'être*, and many of them either taught or studied at schools and academies. In the words of Mary Elizabeth Berry: 'For peacetime soldiers, stripped of battleground activity and notoriously underemployed by the shogunal and *daimyō* bureaucracies, learning became both a rationale for privilege and an opportunity for work – as doctors, political advisors, tutors, teachers, and authors'.[22] But not all samurai managed to make this transition. Each time the shogun confiscated a fief he added to the thousands of *rōnin*, or masterless samurai, who harboured bitter resentment towards the regime. Reports of Iemitsu's prolonged final illness encouraged several of them to lay plans to seize power as soon as he died.

Yui Shōsetsu, who taught at a military academy in the capital, led one group of samurai conspirators that aimed to capture and blow up the Edo arsenal (whose deputy commander they had suborned), set fires in twenty places around the capital, seize the great castle built by Iemitsu as his headquarters and kill the remaining regents during the ensuing chaos. They might have succeeded but one of the leading conspirators was ill at the moment of Iemitsu's death and Yui decided to wait until he recovered. During this interval, another leader developed a fever, and in his delirium betrayed details of the plot. The government therefore managed to nip the conspiracy in the bud, crucifying or beheading more than thirty rebels. Tokugawa power remained intact for the next two centuries.

There was much more to Tokugawa success in surmounting the crisis of 1651 than contingency and a few executions, however. The arbitrary policies pursued by the first three Tokugawa shoguns had destroyed or weakened their opponents so effectively that no viable alternative focus for loyalty remained. The emperor, the major temples and most of the *daimyō* had all incurred heavy debts (often through providing the donations demanded by the shoguns for their building projects, and in maintaining the lavish mansions in Edo required by the alternative attendance system). They therefore lacked the resources to exploit the vacuum of power caused by Iemitsu's death. In addition, the limit of one castle per fief left the *daimyō* at a severe disadvantage in challenging the central government, which garrisoned dozens of strongholds strategically located throughout the country – including the great castle in Edo that, according to a Dutch envoy, 'can be compared to one of the largest walled cities in Europe' and contained enough weapons to equip 100,000 soldiers.[23]

In any case, the military effectiveness of Japanese warriors of 1651 was not what it once had been. Many of the samurai who attended on their lord in Edo had only a tenuous link with his fief (some men, born in Edo, had never met their colleagues in the fief), and whether or not they lived in Edo, none were combat ready. The siege of Shimabara in 1637–38 marked the only military operation most of them could remember, and even then only samurai from a few fiefs had seen action; the rest lacked combat experience. Moreover, since the Tokugawa stored huge quantities of weapons in its arsenals, and closely monitored (and reduced) the production of guns, any armed confrontation between Tokugawa and *daimyō* forces risked becoming a bloodbath.

On the positive side Tokugawa rule, though arbitrary, had brought tangible benefits to almost all social groups. The *daimyō* gained because the shoguns protected lesser lords against their larger neighbours: for 250 years, no *daimyō* attacked the lands of another (a major contrast with the sixteenth century). Instead, those with a grievance sought redress in Edo. Towns prospered because the samurai and other retainers who relocated from the villages increased the demand for both food and artefacts. Merchants appreciated the creation of beacons, lighthouses and rescue facilities, which made seaborne trade safer, while improved roads and bridges facilitated land commerce. All these developments increased the demand for manufactured goods: a manual published in 1637 tabulated over 1,800 'notable products' for sale in Japan.[24]

Tokugawa rule also brought peace and prosperity to the peasantry. Politically, the shoguns promoted mechanisms for defusing contentious events: although rebellion inexorably brought repression, protestors often achieved at least some of their goals (albeit normally posthumously). Economically, the state's fiscal demands declined. As Hayami Akira observed: 'Taxes in Tokugawa Japan were based on the principle of establishing a fixed level of production, and levying taxes on that' – that is, most communities continued to pay taxes on the basis of the surveys carried out in the 1590s, which excluded the yield of new or improved arable lands. Thus a tax rate of 50 per cent on a village with a registered production 1,000 *koku* of rice in the 1590s required the payment of 500 *koku* in tax, but if by 1650 production had increased to 2,000 or more *koku* the village would still pay only 500 *koku* in tax. (This would be like taxing US farmers today according to the yield of their fields in, say, 1945.) In addition, neither the shogun nor the *daimyō* taxed income from the non-agricultural production of their peasants (cotton fabric, silk thread, paper, soy sauce, and so on) and they levied no regular taxes on incomes, inheritances or commerce.[25]

These measures not only favoured the Industrious Revolution but also promoted economic growth. Whereas past generations of peasants had worked only for sustenance and to pay their taxes, increased market demand and the prospect of retaining their profits encouraged greater production;

and since low tax rates by the state also benefited landlords, they too promoted the Industrious Revolution. Some imported new strains of rice and improved existing ones, allowing farmers to select the seed best suited to local conditions; others distributed iron-tipped farming tools among their peasants and promoted technological improvements in civil engineering (above all irrigation and water supply). All peasants gained from the requirement that samurai migrate from their villages to the lord's castle, because whereas resident samurai could determine the assets and income of each peasant household by personal inspection, surveyors sent by absentee samurai could more easily be deceived.

Despite all these benefits, the Tokugawa regime might have crumbled in 1651 had it failed to address the principal grievances that motivated the conspirators. Perhaps shaken by the arrival of news of Charles I's execution by his subjects (Chapter 12), the regents drastically reduced demands for *daimyō* donations to Tokugawa building projects; and henceforth they seldom interfered in how the *daimyō* ran their domains, allowing them to issue their own currency as well as their own legal codes. They also did their best to solve the problem of samurai unemployment. Instead of confiscating fiefs and creating more *rōnin*, the regents provided more salaried positions for samurai – for example, employing 1,200 of them as an elite fire brigade in Edo after the Meireki fire, and paying others to employ their pens in creating an ideology of unconditional obedience. Herman Ooms has noted the extraordinary durability of the tracts written by samurai like Suzuki Shōsan (page 364). In the 1930s, 'when an even sharper delineation of nationhood was needed, one that could mobilize the Japanese to the highest degree', the state deployed the absolutist writings of the seventeenth-century Tokugawa apologists. They created the only ideology 'Japan has ever had. Social and political values in present-day Japan maintain the structure they received in the seventeenth century'.[26]

Japan in Print

Just as the Tokugawa knew what they liked, they also knew what they did not like. In 1630 Iemitsu issued an edict banning thirty-two books in Chinese (most of them translations of European works), and also set up a censor's office in the Confucian Academy in Nagasaki to examine and report on all foreign books arriving in the city, the principal entrepôt for trade between Japan and the outside world. On the basis of the censors' reports, magistrates burned and banned any condemned work, and obliterated or removed from other works any page that contained any reference to Christianity. In addition, the shogun commanded booksellers throughout Japan to bring in for scrutiny any Japanese work that mentioned foreign religions. Those found with Christian literature faced draconian penalties.

The Tokugawa censors showed less vigilance in non-religious matters. On paper, the government proscribed a wide range of writings: unauthorized items about the dynasty itself, its advisors and its policies; works that contained pornography; anything about 'strange events that have happened recently' (including adultery, vendettas and major fires as well as foreign news) or 'that deals with new and curious matters'. An ordinance of 1686 not only forbade the publication of 'such outrageous materials as reckless songs and rumours about recent events' but also ordered the arrest of 'those who sell these items on street corners'. Nevertheless, practice rarely matched theory. From the 1680s (if not before) street vendors sold broadsheets containing news of current non-political events thanks to a subterfuge: although their activity breached the ban on mentioning current events in publications, they wore masks. The government therefore tolerated them. Authors circumvented censorship by producing fictionalized accounts: Japan's best-known playwright, Chikamatsu Monzaemon, even published and staged a play about the 1637–38 Shimabara uprising by ostensibly setting it in the twelfth century. Except where Christianity was concerned, Tokugawa censorship was far from the thought control exercised by the Qing, the Romanovs, the popes and other European rulers.[27]

The Tokugawa nevertheless fully appreciated the power of print to promote their political and social agenda. Starting in 1643, just after the *sankin kōtai* system became mandatory, Edo printers began to publish personnel rosters giving the name, rank, age, crest, income and address of each *daimyō*; their families and their retainers; their schedule of attendance on the shogun; the distance travelled from each fief; the gifts presented; and so on. Starting in 1659, other printed rosters listed the principal officeholders of the Tokugawa, both in the capital and in the provinces, together with their duties, address, stipend, deputies, tenure in office and previous appointments. Tens of thousands of copies flowed from the presses every year, each one duly updated.[28]

The spate of official publications transformed literacy rates in Japan. According to Eiko Ikegami, in 1600 'most Japanese with good reading ability – including the ability to read Chinese characters – were either upperclass townspeople or farmers who did not have to perform manual labour'; but now every one of the archipelago's 70,000 villages required at least some literate males to read out and copy each text into a special register before passing it on to the next village according to a fixed schedule (the headman of the last village on the schedule certified that the original had completed its required circulation). The same was true of the trade guilds and city wards: all required men with literate skills. The print trade was also transformed. Before 1590 the total output of Japanese printers, most of them attached to Buddhist monasteries, did not exceed 500 titles, almost all of them Buddhist religious texts; but by 1625 another 500 titles had appeared and in 1666 the first *List of*

Japanese and Chinese books in print included over 2,500 titles. The pace continued to quicken, with almost 4,000 titles in the 1670 list and over 7,000 in the 1692 list.[29]

This rapid expansion reduced but did not eliminate the preponderance of religious texts, which made up almost half of the total titles even in the 1690s – but the other half displayed an amazing intellectual range. Books of *haikai no renga* ('playful linked verse') poetry grew fastest: from 133 titles in 1670 to 676 in 1692. Bashō Matsuo was only one exponent among many masters of *haikai*, and he could undertake the famous journeys during which he composed his verses only because other *haikai* enthusiasts all over Japan welcomed and entertained him. These practitioners also paid to enter poetry competitions, some of which offered substantial prize money. In 1694, the year of Bashō's death, a Kyoto booklet announced the results of a recent competition that had attracted over 10,000 entries.[30]

Although some entrants no doubt learned to compose poetry from Bashō or another master in person, any aspiring poet could consult printed reference books that taught the rules of composition. By the 1690s, Japanese readers could find manuals that described almost every aesthetic pursuit and hobby: how to arrange flowers, perform the tea ceremony, play *shamisen* and write letters. They could also consult travel guides, or illustrated works containing the different patterns of the kimonos and other garments worn by the actors and courtesans of the big cities. Fifty-five titles in the 1685 *List of Japanese and Chinese books in print*, and 119 titles in the 1692 list, described the 'amorous arts'. Despite the shoguns' ban on pornography, some of these works contained striking woodblock illustrations (often hand-coloured) representing sexual acts – both heterosexual and homosexual – with the protagonists' genitals prominently displayed.

The prolific writer (and former samurai) Asai Ryōi popularized these and other delights in a book entitled *Tales of the floating world*, published in 1661, four years after the Meireki fire destroyed three-quarters of Edo. He wrote in the name of those who

> live only for the moment, turning our full attention to the pleasures of the moon, the snow, the cherry blossoms and the maple leaves; singing songs, drinking wine, diverting ourselves in just floating, floating; caring not a whit for the pauperism staring us in the face, refusing to be disheartened, like a gourd floating along with the river current: this is what we call the *floating world*.[31]

The 'floating' took place in two principal locations, both of them new in the seventeenth century. The first was the theatre district in each major city, where actors (and until 1627 actresses) staged *kabuki* (literally 'not straight') plays, an art form that emerged from a combination of classical Noh drama with Shinto dances and popular pantomimes. The second was the pleasure-quarter (*akusho*: literally 'bad place') licensed by the magistrates of almost every major town. Edo's pleasure-quarter was by far the largest, in part because the *sankin kōtai* system brought so many single adult males to the capital: by the 1680s, several tourist guides to Edo included the names, ranks and residences (plus, in some cases, the fees, physical appearances and specialties) of over 1,000 courtesans.

Getting It Right?

Tokugawa Japan thus did not differ significantly from the rest of the world in exposure either to the Little Ice Age or to other natural disasters (such as urban conflagrations): both caused widespread devastation. The principal distinctions lay elsewhere. First, where other regions became overpopulated, thanks largely to the expansion of agriculture into marginal areas, Japan began the seventeenth century underpopulated, thanks largely to the Warring States Era. Second, Tokugawa rulers made policy choices that mitigated rather than exacerbated the effects of adverse climate. The risk-averse foreign policies of the shoguns spared their subjects not only from the damage caused by armies on the rampage but also from paying higher taxes. Moreover, from 1615 onwards, the Pax Tokugawa gradually turned the warrior elite into urban consumers living on stipends and dependent upon a commercial economy that delivered cheap labour, goods and services. It was therefore in their interest to promote initiatives (such as land-reclamation schemes and improved crop strains) that maintained or increased that supply of cheap labour, goods and services.

The Pax Tokugawa nevertheless came at a cost. In political terms, the shoguns deprived their subjects of many freedoms: none could engage in foreign trade, travel abroad, embrace a religion prohibited by the regime or handle certain forms of literature. Vassals could no longer organize collective protests against abusive lords, or petition higher authorities for redress (both became a capital offence): their only legal recourse was collective migration to a neighbouring domain, a procedure known as *chōsan ikki* ('organized flight'). In economic terms, although the Industrious Revolution significantly increased output, it required not only relentless and ruthless self-exploitation by producers. Farmers (in the lapidary verdict of Conrad Totman) 'became trapped in an inflexible, high-risk, high-input, low-yield operation that could be sustained only by the most attentive husbandry', and many of the upland and northern areas brought under cultivation 'crowded the biological boundaries of crop viability (for reasons of both climatic marginality and soil character) and exacerbated the danger of crop failure'.[32] In addition, avoiding war led to the neglect of military innovation, so that when in 1863 navies arrived from states that had invested heavily in military technology, the Tokugawa regime was powerless to resist them and had to accept humiliating trading agreements. Rebellions soon overthrew the shogunate.

By then, however, the Tokugawa system had brought peace to all Japan for more than two centuries – an unparalleled achievement for such a large population – and protected the archipelago from the famines that afflicted much of the northern hemisphere in the 1690s, the climax of the Little Ice Age when average temperatures fell 1.5°C below those of the later twentieth century. The first major food crisis after the Kan'ei famine of 1641–42 did not occur until 1732 – a respite of almost a century: another unparalleled achievement. For most subjects of the Tokugawa, after the 1640s the Global Crisis was something that happened to other people.

PART IV
CONFRONTING THE CRISIS

Many seventeenth-century writers attributed the violent disorders they saw around them to innate defects of human nature. According to Thomas Hobbes in 1641, 'man's natural state, before they came together into society, was war; and not simply war, but the war of every man against every man'. Blaise Pascal put the same point more concisely a decade later: 'All men by their nature hate one another'. Many of their contemporaries saw life as a zero-sum game in which assets could only be redistributed, not created – or, in the aphorism of Francis Bacon: 'whatsoever is some where gotten is some where lost'.[1]

In every community, this zero-sum mentality led to intense, destabilizing, endless competition in order to protect assets against encroachment, in which 'the tools of victory were skill, ingenuity, hard work and perseverance'. Townsfolk, living in more complex entities with greater contrasts between rich and poor, also sought strength in numbers. They joined guilds for economic concerns, confraternities for religious and social issues and factions for politics, each one capable of generating programmes and slogans, as well as mass protests. In 1699, the Scottish philanthropist and politician Sir Robert Sibbald observed that protests became more violent whenever resources ran short, because 'those that are of a firy and active temperament' turned 'unquiet, rapacious, frantick or desperate'. Many contemporary sources provide supporting evidence. In the northern Italian city of Bologna, whose excellent records have been closely studied, homicides quadrupled in the seventeenth century largely because of the vendettas waged between extended families (relatives, clients, servants), which regularly left streets and public places, even churches, full of dead and mutilated bodies.[2]

Lu Kun, a gifted bureaucrat of late Ming China, offered a comprehensive analysis of the principal threats to law and order. 'From of old', he informed his emperor, 'there have been four kinds of people who like to rebel':

First are those who have no means of support, no food or clothing, whose families are in difficulties, and who consider rebelling in hopes of delaying their demise. Second are people who do not know how to behave, who have high spirits and violent natures, who violate the laws to make life easier for themselves, who are fond of jade and silk and sons and daughters but cannot get them legitimately, and who think that if there is a rebellion they can steal what they want. Third are the people of heterodox beliefs ... whose teachers preach and attract crowds, and who will respond to and join up with anyone who calls them. Fourth there are the people without self-control, who turn petty rifts into major fights, who think only of being strong, who hope only for a change, and who take no pleasure in the existing peace in the world.[3]

The categories identified by Lu Kun applied not only to China but also to other areas in the seventeenth century. Chapter 17 examines the motives and protocols of those who rebelled because they 'have no means of support, no food or clothing' due to economic adversity (especially dearth, unemployment, high taxation and state oppression), as well as the fellow travellers with 'high spirits and violent natures' who 'think that if there is a rebellion they can steal what they want'. Chapter 18 considers protests by 'people of heterodox beliefs' (that is, critics of the government) 'who hope only for a change' – above all the nobles, clerics and intellectuals who created ideologies to underpin political as well as economic grievances, and advanced alternative solutions for the problems of the day. Chapter 19 focuses on how these and other groups of protestors managed to 'preach and attract crowds', using all available media to spread the word about their grievances and their strategies of redress, in the hope that attracting wider domestic and foreign support would not only gain greater concessions but also avoid repression by the authorities. In the mid-seventeenth century, the combined actions of these 'unquiet, rapacious, frantick or desperate' people would bring almost half the states of the northern hemisphere to their knees.

CHAPTER SEVENTEEN

'Those Who Have No Means of Support': The Parameters of Popular Resistance[1]

Public and Hidden Transcripts

Collective resistance was perhaps the most common human reaction to the seventeenth-century crisis. As a disgruntled English landowner observed, 'The meaner sort of people [are] always apt to rebel and mutiny on the least occasion' – and, indeed, the total number of food riots in England rose from twelve between 1600 and 1620 to thirty-six between 1621 and 1631, with fourteen more in 1647–49. In Germany and Switzerland, more than half of all major peasant revolts recorded in the seventeenth century took place between 1626 and 1650; in France, popular revolts peaked in the middle decades of the century.[2]

The Frequency of French Popular Revolts, 1590–1715

Date	Aquitaine (southwest France)		Provence (southeast France)	
	Number	Annual average	Number	Annual average
1590–1634	47	1	108	2.4
1635–1660	282	11.3	156	6.3
1661–1715	130	2.7	110	2

Records from other states reveal a similar peak of revolts in the mid-seventeenth century. In Russia, a wave of urban rebellions in 1648–49 shook the central government to its foundations; in Japan at least forty rural uprisings took place between 1590 and 1642, a total unmatched for two centuries; in China the number of armed uprisings rose from under ten in the 1610s to over seventy in the 1620s and over eighty in the 1630s, affecting 160 counties and involving more than a million people.

Nevertheless, official documents record only a small fraction of the occasions when the standard of rebellion was unfurled: most revolts left scarcely a trace in the surviving archives, especially if the protestors defied the authorities with words rather than deeds. As John Walter observed, 'grumbling was the easiest and probably the first weapon of the weak' in the seventeenth century, and it took place mainly in unregulated spaces such as alehouses or coffee houses; and he speculated that 'dearth years doubtless saw increased grumbling'.[3] No doubt years of armed conflict, religious

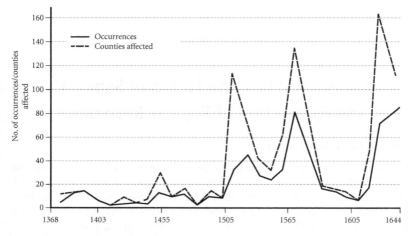

25. Collective violence in Ming China, 1368–1644.
James Tong found 630 cases of collective violence (rebellion and banditry) recorded in the surviving Gazetteers of Ming China. Of these, four-fifths occurred in the second half of the dynastic era, affecting almost all of the 1,000 counties included in his survey and reaching a peak between 1620 and 1644.

innovation and political tension produced the same phenomenon. Yet even if historians could compile a comprehensive list of all the opprobrious words, insulting gestures and grumbles, it would still fall far short of a complete inventory of popular resistance during the seventeenth century because many protestors strove to prevent their resistance from leaving a trace in the public record.

Anthropologist James C. Scott has suggested that the poor normally adopt a risk-averse strategy when dealing with the elite and with the state, waging 'defensive campaigns of attrition' that included 'foot-dragging, dissimulation, desertion, false compliance, pilfering, feigned ignorance, slander, arson, sabotage'. These everyday acts of disobedience, Scott noted, 'require little or no co-ordination or planning; they make use of implicit understandings and informal networks'. They therefore did not make head-lines but instead left what he termed a hidden transcript preserved, if at all, in narrative accounts and oral traditions.[4]

Even when material conditions were at their worst, and mass revolts occurred with unparalleled frequency, the 'defensive campaigns of attrition' far outnumbered them, because the risk-averse strategies discerned by Scott reflected the central concern of the poor in all periods: survival. To make sure they had enough food to survive, the poor applied a simple calculation: 'How much do I have left?' Only when the calculation dictated a shift from clandestine to open resistance did collective protest move from the hidden to the public transcript.

Articulating Grievances

In the seventeenth century, three scenarios provoked open resistance with great frequency: a failed harvest; the arrival of troops requiring food and lodging; and a tax increase. Each scenario had its own tempo. Because of the prevailing peccatogenic outlook (Chapter 1), most disasters in the early modern world initially led those afflicted to seek someone to blame. This process often began with introspection – a community tried to expiate its own collective sins through acts of penitence (processions, rogations, pilgrimages, self-flagellation) – but if this did not work, attention shifted to local individuals or groups whose conduct might have offended God. Some blamed their neighbours, denouncing them as witches or Jews in an attempt to eliminate them through the legalized violence of the courts. Others suspected their superiors of selfishly creating an artificial shortage, and took the law into their own hands: irate crowds might threaten grain merchants and bakers who failed to provide sufficient food at an affordable price, or carters and bargees who transported grain, and force them to distribute their precious stock either free or at an artificially low price.

If these strategies failed, hungry crowds might blame local authorities. Sometimes they demanded that magistrates impose affordable prices at the market, purchase supplies elsewhere or conduct an inventory of all grain reserves; at other times they attacked the town granary, the mansions of anyone suspected of hoarding grain, and (in Christian countries) the abbeys and church barns that stored the yield of the tithes. Communities might also try to prevent outsiders from exporting food from the area, whether merchants hoping to maximize profits or agents buying up grain on behalf of a city or an army – although this often escalated the level of collective violence. Normally, intimidating the local people who handled grain required only threats and the confiscation of their goods, and influencing local authorities might only require throwing a few stones; but preventing the export of grain could involve beating or even killing the perceived perpetrators.

Billeting was the second major precipitant of popular revolt. Throughout Europe, soldiers quartered in a community demanded free light and heat, clean bed linen, and three meals a day from their hosts. In Spain, according to a contemporary source, 'The soldier consumes in just one week what the farmer expected would feed himself and all his children for a month', which helps to explain how the pressure of billeting troops played such a large role in generating popular revolts in Catalonia in 1640 and in Ulster the following year.[5]

Taxes, naturally, also produced an abundance of revolts – especially in wartime. In seventeenth-century Aquitaine, for example, more than half of all known uprisings against tax collectors took place in the war years from 1635 to 1659. The unpopularity of increasing existing taxes explains why early modern governments often resorted to alternative fiscal strategies in wartime – either taxing items previously exempt or taxing categories of subjects previously exempt – but this too was a high-risk strategy. The decisions to impose a salt monopoly in Vizcaya in 1631 and in Normandy in 1639, a new tax on property in Portugal in 1637 and in Paris in 1648, and a new excise duty on fruit in Naples in 1647, all triggered major revolts (Chapters 9, 10 and 14). In both Portugal and Castile, the attempt to make the clergy pay stamp duty provoked determined resistance; in France, the crown's erosion of tax exemptions enjoyed by the aristocracy and the civil service eventually turned both social groups into rebels. The resort of Charles I and his ministers to regalian rights such as Ship Money imposed heavy burdens on social groups previously exempt. In the county of Essex, where scarcely 3,000 households normally paid the taxes voted by Parliament, over 14,000 households faced Ship Money assessments – indeed, the king's commissioners wrote '£0-0s-0d' beside a few names, showing that they had visited and assessed even the poorest residents in the realm. No wonder the new tax turned so many of Charles's subjects against him.

Provoking opposition was doubly dangerous for governments in times of hardship, because, as an agent of the French government perceptively noted in 1643, 'Once armed and in rebellion, the people will use its weapons against everyone who asks them for money'. A generation later, a Qing magistrate made the same point: 'Incidents that begin as expedients to get food for empty stomachs often end up as organized rebellion', because a threat to one of life's necessities – food, jobs, welfare, traditional rights – could provoke an entire community to unite and rebel.[6]

Deterrents to Collective Violence

Given the existence of multiple grievances, and their correlation with adversity, it is surprising that revolts did not occur even more often. In his study of peasant societies, James Scott discerned four factors that normally deter even desperate people from open resistance. First, the need to earn a daily wage formed a powerful restraint on rebellion: a family that did not work – whether because on strike, in rebellion, or unemployed – might not eat. Second, vertical links of kinship, friendship, faction, patronage and ritual in each community created ties between the dominant and the dominated that discouraged violent action. Third, and paradoxically, any economic development within the community that increased social divisions also militated against collective action. A shift towards producing crops (especially industrial crops) for export normally created groups of prosperous cultivators who, as long they could sell their goods, remained largely insulated from the frustrations and sufferings of those still tied to subsist-ence farming, which significantly reduced the likelihood of unified resist-ance. Fourth, in most farming communities of the early modern world, the poor depended for their survival on deference and subordination: better-off neighbours were more likely to provide relief in time of need to those who had always shown respect and obedience, whereas neglect or surliness might lead to denial of charity and even expulsion from the community. However much the poor may have resented their subordination and humiliation, their circumstances compelled them to conform: they might try to negotiate the terms of subordination, but they rarely dared to challenge it.[7]

For all these reasons, despite desperate economic circumstances and apparently intolerable provocation, the poor normally ensured that their protests did not break any laws. They took care not to steal (often ostenta-tiously burning the property of their victims and beating anyone seen looting) and they rarely carried prohibited weapons. In England, at least, those who destroyed property did so two by two, because the law stated that a riot (severely punished) began only when three or more people became involved. In both England and China, where even some judges had doubts whether laws remained in force during an interregnum, resentful subjects might await the interval between the death of one monarch and the procla-mation of the next to seek revenge for a perceived wrong.

'Women Can Do No Wrong'

Except for a few privileged groups, those who tried to form associations larger than the family to achieve social and economic goals ran grave risks, and yet in early modern Europe women often led collective protests. The depositions made after the Irish uprising of 1641 'are full of references to the women being "more forward than the men", or "more fierce and cruel than the men". Women wield weapons, . . . serve in armies and provide leadership roles . . . They act as gaolers, lookouts and guides, as well as spies and "intel-ligencers". A study of French popular revolts found that 'the most constant element was the presence of women': they outnumbered men in more than half of all known rural food riots and in more than three-quarters of all known urban food riots. Some protests consisted entirely of women.[8] Protestors throughout the Dutch Republic chanted the slogan 'Women can do no wrong!' When the magistrates of Haarlem invited bids for collecting a new excise duty in 1628, a group of women attacked the first man to make a bid, shouting: 'Let's sound the drum and send our husbands home; then we'll get the bastard and beat him up because we cannot be punished for fighting.' English magistrates agreed. 'If a number of women (or children under the age of discretion) do flocke together for their own cause, this is none assembly punishable', wrote the author of a standard handbook for magistrates in 1619; and that same year the Court of Star Chamber, asserted (in a case that involved breaking down the fences around enclosed fields) that women were 'not subject to the lawes of the realme as men are, but might . . . offend without drede or punishment of lawe'.[9]

In part, this double standard reflected the realization that a woman and her family might cross the threshold between survival and starvation in a matter of days if not hours. An English pamphlet published during food riots in Essex in 1629 pitied the local weavers who 'cannot live unless they bee paied every night, many hundreds of them havinge no bedds to lye in, nor foode; but from hand to mouth mainteyne themselves, their wives and children'. In France, half a century later, 'you could hear women in the market-place cry out that they would rather slit their children's throats than watch them die of hunger'. In such circumstances, survival might require resistance and even revolt. Nevertheless European women lost their immunity if resistance got out of hand. During the famine of 1629 in Essex, Ann Carter, a butcher's wife, led a large crowd of women to prevent the export of grain from the region and forced the would-be exporters to pour their grain into their bonnets and aprons. Her conduct on this occasion reflected the accepted protocols of early modern protest, but then Carter took the title 'Captain' and proclaimed 'Come my brave lads of Malden, I will be your leader for we will not starve', inciting unemployed cloth-workers to break into a grain store and removed the contents. A week later the government arrested Carter, put her on trial for sedition, and after the (all-male) jury convicted her, hanged her the next day.[10] Likewise, in Holland in 1652, 45-year-old Grietje Hendrickx was arrested, tried and sentenced for collecting stones in her apron, carrying them to the rioters and 'inciting bystanders to join in'. The following year, warrants went out for the arrest of two women who had led a riot:

> Griet Piet Scheer, aged 36, blond hair, thin face with blue eyes, fairly tall, slim figure, soberly dressed. She dresses at times in black and at others she wears a blue overall with red sleeves; she acted as captain. Alit Turfvolster, bearer of the flag, is as tall but somewhat stouter than the above-mentioned Griet; she sniffs somewhat through her nose, is brown of complexion with black hair and untidy clothes; she wears a bodice with a linen apron, and her age is 30.

According to Dutch court records, most of the women accused of leading street protests were, like those mentioned above, aged between thirty and forty-five.[11]

Accounts of revolts outside Europe also sometimes mention the participation of women, but only in subordinate roles. In India during the 1650s, when villagers resisted Mughal efforts to collect taxes, 'the women stood behind their husbands with spears and arrows. When the husband had shot off the matchlock his wife handed him the lance, while she reloaded the matchlock.' In China, some women doubtless participated in the popular rebellions of the late Ming period, but two considerations make it improbable that they took a prominent role like Ann Carter and Griet Piet Scheer: neither Chinese law nor practice had a concept similar to 'women can do no wrong'; and the practice of binding the feet and then secluding many women from puberty to menopause would have severely limited their ability to lead street protests.[12]

Clerics and Fools

Many European popular protests involved the clergy. Some became quite outspoken. A mid-seventeenth-century French catechism condemned as guilty of homicide those 'who failed to calm and disperse popular sedition when they have the power to do so, such as magistrates'. In Naples in 1647, priests assured insurgents that their struggle 'was just, because they were oppressed by excessive taxes and attacked and provoked by the Spaniards'.[13] Elsewhere, clerics often intervened as proxies in popular revolts. Throughout the Ottoman empire, local sheikhs (the heads of a Sufi or dervish lodge: Chapter 7) handled negotiations between the central government and a community either oppressed by taxes or by local officials, often securing redress of grievances before violence began. In China, Buddhist monks sometimes became spokesmen for the oppressed. In 1640, a monk in a Jiangnan town organized a grain strike (*da mi*) in which large crowds of peasants visited the houses of the rich asking for food: they spared houses that provided sustenance and burned those that refused. In general, however, Buddhist (and Daoist) clerics lacked the local authority wielded by their Christian counterparts, partly because most of them lived in temples largely isolated from the rest of the population, and also because the dominance of Confucian ethics undermined any claim to moral leadership made by others.[14]

A third proxy occasionally able to speak truth to power was the 'fool'. Most Islamic rulers grudgingly tolerated the criticisms and claims voiced by a *majdhūd* ('holy fool') – indeed Ottoman authorities may have initially overlooked the messianic claims of Shabbatai Zvi because they considered him a *majdhūd*. Holy fools were also common in Orthodox Christianity: they wore no clothes, draped themselves in chains and wore an iron hat, living in extreme poverty and begging for food. Although they normally spoke nonsense, they sometimes slipped sharp criticisms into their silliness, thus managing to confront even the tsar with unpalatable truths. In Portugal in 1637, those who protested against a new tax at Évora made a simpleton their leader: 'Manuelinho, secretary of the young people, ministers of divine justice' signed the manifestos posted in the streets against the 'tyrant Pharaoh'. That same year, when Archibald Armstrong, Charles I's Scottish fool, heard of the rioting in his native land provoked by the decision to impose a new Prayer Book on English lines he asked its author, Archbishop Laud, 'Who's the fool now?' Although Laud had Archie banished from the court and confiscated his fool's coat, the jester still had the last laugh. A courtier who encountered Archie without his fool's coat, and asked where it was, received the reply: 'Oh, my lord of Canterbury [Laud] has taken it from me, because he or some of the Scots bishops may have use for it themselves!'[15]

The Etiquette of Collective Violence

With or without proxies, most early modern rebels issued warnings before they resorted to violence. Posters would appear in the streets advising an individual whom the community identified as an oppressor to change his (or, less often, her) ways, and women would gather and kneel together in the open and noisily weep and wail in front of the house of an abusive landlord. If such shaming tactics failed to produce concessions, satirical songs might be composed and sung at night outside the offender's house. Artefacts might also be used to convey the warning: a cart left in the doorway (implying that another would soon carry away a coffin); a bonfire (suggesting that the owner's house would be next); or someone hanged in effigy from a gallows – the ultimate sign of disapproval short of violence. If all these coded messages failed to produce the desired changes, the aggrieved would graduate to destroying property, starting with distant assets such as vines, fruit trees, mills and storehouses, before moving on to stoning windows or smashing down the front door. After that, the terrified victims usually fled.

Those who ignored these warnings risked serious harm. In Barcelona, after the murder of the viceroy in June 1640, a royal judge watched in horror from a hiding place as rioters murdered every Castilian they could find 'without sparing the church, where they killed someone hiding beneath the altar, without seeing or knowing who it was until the blood running out from under the frontal revealed that some unfortunate was hiding there'. Angry Catalans surged through the streets of Barcelona again the following December, seeking out and murdering all suspected royalists: this time they disfigured their victims with repeated blows before hanging them from gallows in the city square. In southern France, William Beik has postulated the existence of a 'culture of retribution': once an angry crowd had decided that someone ' "had it coming to him", there was no such thing as excessive force'. In the province of Aquitaine, for example, at least thirty of the fifty known tax rebellions in the course of the seventeenth century involved the humiliation, execution and (often) mutilation of a tax collector.[16]

Anthropologist David Riches has noted that violence serves not only as a convenient and economical instrument to transform society, but also as an 'excellent communicative vehicle' with which to make symbolic statements.[17] In many parts of Europe, crowd violence often followed an etiquette that mirrored legal protocols. Protestors paraded their victim, often a tax collector, around the town with his hands bound and dressed only in a shirt, forcing him to make 'honourable amends' at each crossroads and square – just as happened to those convicted by the king's judges. Sometimes the crowd then set the victim free so he could warn others of the fate that awaited all tax collectors, but others they executed and quartered – again following the same measured ritual, often performed in the same places, as in state trials.

Nevertheless, European crowds rarely killed promiscuously. In Portugal, the conspirators who took over Lisbon on 1 December 1640 murdered Miguel de Vasconcelos, the hated agent of the Philip IV, and hurled him from a palace window into the square below where the crowd stripped his corpse, 'tore out his teeth, pulled out his moustache and beard and stabbed him repeatedly. They then cut off both his ears, which later the crowd displayed and offered for auction.' But Vasconcelos died alone: no other royal servant perished in the Restoration. The rioting in Istanbul that ended the life of Sultan Ibrahim in 1648 involved the death of relatively few except the ruler himself and his grand vizier. Admittedly, like Vasconcelos, the vizier met a spectacularly barbarous end – 'strangled and minced into mammock-pieces [shreds], one pulling out an eye, another cutting off an ear, a third a finger' – but he perished alone. In Naples the previous year, although several noblemen were butchered, and their mutilated bodies exposed naked on a special monument erected outside the rebel headquarters, these men had attempted to slay Masaniello and kill his followers. More typical was the rebels' protocol in torching houses: they followed a written list of targets, mostly the property of those involved in either imposing or collecting taxes.[18]

In China, too, most popular protestors initially displayed restraint. Many began with ritual wailing at local Confucian temples, or with public lamentations by aggrieved subjects who walked through the streets carrying placards that stated their petitions and grievances. In southern China, distressed villages also formed a covenant (known as a *gang*, or net) whose participants recorded their names in a register and made sacrifices to their ancestors before visiting the houses of the rich to beg for food. If such methods failed, crowds started to burn houses; but, as in Europe, they not only worked from a list of residences belonging to tax collectors but also notified the residents of adjacent houses in advance so that they could take steps to prevent the fire from spreading. In addition, they strictly forbade theft and often beat looters to death.

Those who protested against non-material grievances also often followed a distinctive etiquette and engaged in public displays. In England in 1641, opponents of the ecclesiastical innovations of the preceding decade publicized their contempt for certain items in original ways. The surplices now worn by ministers were ostentatiously shredded (often by women using knives and scissors) and put to a variety of profane uses: a bandage, a handkerchief, a shirt or (most eloquently) a sanitary towel. In London, four horsemen rode through the streets 'with the Book of Common Prayer in their hand, singing in derision thereof, and tearing it leaf by leaf, and putting every leaf to their posterior'.[19]

Place and Time

Popular resistance seems to have occurred with unusual frequently in regions where a reservoir of discontent could develop rapidly (cities and macroregions: Chapter 3). Some apparently sustained an intellectual tradition of resistance: Kingston, commanding a strategic bridge across the Thames, openly rejoiced over the murder of Charles I's favourite, Buckingham, in 1628, just as a generation later they would shelter both the Diggers (a group that cultivated common land and advocated sharing all things) and the Quakers. Areas protected by natural barriers such as marshes, forests or heaths could also become oases of insurrection in times of hardship. In Europe, the Austrian domains of the Schaunberg family, which could claim exemption from outside jurisdiction, lay at the heart of peasant revolts in 1620, 1626, 1632–33 and 1648, just as they had done in those of previous centuries. Likewise, men from St Keverne, a remote Cornish-speaking parish in England's deep southwest, led a major revolt in 1648, just as they had done in 1497, 1537 and 1548.

Frontier societies boasted larger oases of insurrection. The southern borderlands of Russia and the Polish-Lithuanian Commonwealth, where forest turned into steppe, offered a constant refuge to the oppressed and discontented, who periodically staged rebellions against the states to the north (most memorably in Ukraine after 1648, when Bohdan Khmelnytsky led a Cossack rebellion against its Polish overlords, and in 1670 when Stenka Razin defied Moscow: Chapter 6). In the Americas, Maroons (black slaves who escaped from European settlements) created fortified camps in several parts of Brazil, Central America and the Caribbean islands where jungle, canyons or swamps offered a measure of protection. From these refuges, often under the command of those who had been rulers before their abduction from Africa, they made common cause with Native Americans and welcomed any fugitive European servants and outlaws. Together they posed a constant threat to the colonies (especially by burning down plantations of sugar cane – a conveniently combustible crop). In China, in the 1620s and 1630s, rebels like Li

Zicheng found refuge in the deep forests between Shanxi and Henan provinces, just as Mao Zedong would do three centuries later; and in the 1640s the marshes around Liangshan mountain in Shandong province sheltered large bandit groups, just like those in *Water margin*, the popular novel of the day set in the twelfth century. The mere proximity of an oasis of insurrection could encourage resistance. As a group of Bengalis pointed out to their rulers, flight was always an option when one had 'A thousand countries to go to'.[20]

The calendar also influenced popular resistance. Revolts often began on market days – in 1641 the leading Irish conspirators planned to take Dublin Castle on a market day because large numbers of people would (like them) be entering the city the night before and so they would attract less suspicion – and on religious holidays, when the Church prohibited work, allowing people to throng the streets and taverns, talking and drinking, so that an insurrection could quickly gather momentum. Normandy erupted into violence on a festival in honour of the Virgin Mary in 1639; Barcelona on Corpus Christi, 1640; Palermo on Assumption Day, 1647. Revolutionary anniversaries could also provoke violence: Fermo in 1648 and Messina in 1674 both rebelled on the anniversary of Masaniello's revolt in Naples, 7 July.

Throughout the Muslim world religious holidays that involved public processions, such as Muharram, could also give rise to violence – particularly in areas like Mesopotamia that boasted both Sunni and Shiite populations, where supporters of one creed might disrupt the devotions of the other (a practice that continues to this day). In the Ottoman empire, the garrison of Istanbul became restless at the end of the month of Ramadan, when tradition demanded that they receive a bonus pay: if the treasury could not meet this obligation, or met it only in part, the troops might mutiny.

Weapons, Cadres and Emblems

A final common denominator of popular revolts was the transformative impact of groups already used to acting in unison, especially if they were familiar with weapons. In a society that lacked an effective police force, owning a weapon, especially a firearm, offered a measure of security against perceived threats: bandits, beggars, personal enemies and, in the countryside, natural predators. In some frontier zones of Europe and in Anglo-America, possession of firearms was seen as essential to survival. In France, the Peace of the Pyrenees in 1659 guaranteed the right of every inhabitant of Roussillon to bear arms; in Virginia, Governor Berkeley noted fearfully in 1676, the year of Bacon's rebellion, that his opponents were 'poor, indebted, discontented and armed'. In India, many seventeenth-century sources noted the abundance of 'labourers with their guns, swords and bucklers lying by them while they ploughed the ground', and the Mughals designated some areas *mawas* (rebellious lands) or *zor-talab* ('requiring coercion') because villagers there were armed and refused to part with their wealth, whether to the government or to bandits, 'without at least one fight'.[21]

Weapons became far more effective when in the hands of those familiar with their use, such as army veterans and outlaws. Whenever protestors forced open the gates of the local gaols they released into the crowd hundreds of men with experience of defying government, as well as links with other discontented groups still at large. Although some of the freed inmates immediately fled the country in search of safety, others seized the opportunity to inflict destruction on their oppressors. The participation of veterans also had a transformative effect on revolts, because they possessed not only familiarity with weapons but also discipline under fire and the ability to coordinate manoeuvres. In France, one contemporary claimed that the 8,000 armed peasant insurgents (the Croquants) who mobilized in Périgord in 1637 were 'mostly veterans from the most warlike provinces of the kingdom', serving under captains with extensive military experience, 'the best that one could find'. The Croquants certainly behaved like regular troops, posting sentries and issuing formal summons for towns in their path to surrender on pain of being 'ruined, razed and burnt to ashes'.[22] In Naples, Masaniello's regular drilling of his *ragazzi* – the 'boys' whom he had trained for a mock battle in the Piazza del Mercato – explains both the cohesion of his followers and his 'skill in digging trenches and keeping watch with sentries' (something that all commentators noted with surprise); and it was the ability of Gennaro Annese, an armourer by trade, to direct an effective bombardment of the Spanish fleet that kept the attackers at bay (Chapter 14). The Scots could not have prevailed against King Charles without the return of large numbers of veterans of continental armies with both equipment and expertise (Chapter 11).

In Asia, too, the participation of veterans secured the initial success of many revolts. At Shimabara (Japan) in 1638, 200 former samurai taught the rest of the rebels how to use firearms to defend Hara Castle against over 100,000 government troops. In China, deserters from the regular army (including Zhang Xianzhong, the future 'Great King of the West' in Sichuan) strengthened the bandits in the 1620s and 1630s; and after 1640 the influx of deserters from the Ming Army helps to explain Li Zicheng's ability to capture fortified towns. By 1644, Li's troops had reached such a high level of military effectiveness that they almost defeated the elite Ming troops of Wu Sangui at the Shanhai Pass, and only the intervention of fresh Manchu forces turned the tide.

Besides weapons and discipline, emblems and insignia could enhance the effectiveness of resistance to government. In Japan, the Catholic rebels at Shimabara in 1638 placed 'many small flags with red crosses around their parapets'; and the following year the Chinese insurgents in the Philippines also waved distinctive banners in defiance at the Spaniards.[23] In several parts of Europe, red now became the colour of revolution. In 1647, in Naples, the rebels 'unfurled the red flag [*lo stendado rosso*]' at their headquarters 'as a sign of war', and both there and in Palermo insurgents waved red flags as a sign of defiance. In 1647 the soldiers of the New Model Army in England tied a red ribbon around their left arms to show 'that we will defend the equity of our [cause] with our blood'. In France, the Ormée rebels of Bordeaux (1651–53) wore a *chapeau rouge*, and the Breton rebels in 1675 took their name from their *bonnets rouges*.[24]

Likenesses of many revolutionary leaders also circulated, to inspire their followers. Most leaders of the English Revolution – notably John Pym, Sir Thomas Fairfax and Oliver Cromwell – were frequently portrayed by both supporters and detractors in paintings, engravings, sculptures, medals and even artefacts. None, however, rivalled the posthumous fame of Masaniello of Naples. Although his reign lasted only nine days, the humble fisherman achieved an iconic status that anticipated that of Che Guevara in the twentieth century: artists captured his likeness in paintings, medals and wax statuettes (some for export); intellectuals composed epigrams extolling his achievements; plays about him were later published in England, Germany and the Dutch Republic. More than a century later, in England, one of the satirical prints denouncing a new tax on apples, entitled 'Mas-aniello or the Neapolitan insurrection', showed protestors in a marketplace surrounding a man standing on a barrel calling for resistance.[25]

In societies where many protagonists were illiterate, slogans and songs also played an important role in maintaining the coherence of rebellions. By far the commonest slogans in Western Europe were either economic or political: 'Long live the king and down with the evil government!' and 'Long the king and down with taxes!' Some protestors also composed special songs, including the 'Fadinger song' (Upper Austria, 1626) and the Cossacks' 'Victory march' (Ukraine, 1648), and in Ireland Gaelic bards composed and sang poems of freedom. Two seventeenth-century rebel songs enjoyed a long life. The melody of one of the many songs that commemorated Stenka Razin's revolt of 1670 entered the Western 'hit parade' nearly three centuries later as The Seekers' 'The carnival is over'; while the song of the *segadors*, with verses protesting the troops and policies imposed by the count-duke of Olivares and calling for armed defence of Catalan liberty, first heard in the summer of 1640, has become the hymn of Catalonia (albeit with some modernization of both words and tune).[26]

Concession or Repression?

Revolts placed governments everywhere in a quandary. Many ministers feared that a failure to respond swiftly and harshly would embolden others. In the words of a bishop in Spanish Peru in 1635 who sought to silence a critic of royal taxation: from 'small sparks are great fires easily set alight, and it is more prudent to remedy the damage in its beginning than to attempt to quench it when difficult or impossible'.[27] Although the bishop was correct, ignoring the pleas of desperate or intransigent subjects could also provoke disaster. In the 1640s the English parliamentary leader John Pym worried about the 'tumults and insurrections of the meaner sort of people' that would arise if they could not buy bread, because 'nothing is more sharp and pressing than necessity and want: what they cannot buy they will take'. A petition to Parliament from the porters of London warned that unless they received relief, their economic straits would 'force your petitioners to extremities, not fit to be named, and to make good that saying, that necessity has no law'. In 1648, in Italy, the people of Fermo rose up because they judged 'it better to die by the sword than to die of hunger'.[28] To avoid such dangers, many governments granted short-term concessions to protestors.

Caprice or scruples on the part of a ruler also sometimes brought relief. In China, each emperor normally forgave unpaid taxes to celebrate his accession or the birth of an heir. In 1641 Philip IV established a Committee on Conscience to examine whether any of the taxes he had imposed might have been unjust and therefore offended God, because 'I do not want to benefit from any tax that has the slightest suspicion of injustice'.[29] Some governments also established permanent machinery to consider individual appeals for tax relief. In China, magistrates regularly petitioned the central government for forgiveness or reduction of taxes, or at least for a delay in collection, because their district had suffered from some natural disaster. Not all did so in vain. The chief magistrate of Tancheng county, in Shandong province, petitioned the board of revenue for tax relief in the wake of a devastating earthquake in 1668: officials from the board made a personal inspection and, eighteen months later, local taxpayers received a 30 per cent reduction in their obligations (or, if they had already paid, a rebate). The board also reduced the county's compulsory labour services for the upkeep of roads and bridges. In the Ottoman empire, the sultan's council spent one day of each week hearing appeals from individuals and communities for a reduction in their tax assessment. Some cited a recent drought, flood or bandit attack, or general depopulation, to argue that the community could no longer meet its obligations; others claimed that they were wrongly included in a tax register (often because they were already on another); a few protested against extortion. Although the subsequent inquiry some-times discovered taxpayers falsely trying to reduce or escape their obliga-tions, normally the government agreed that an error existed in the registers, and therefore reduced or forgave the taxes specified in the thousands of petitions they considered.[30] In Castile, many taxpayers appealed to local judges against over-assessment, or pleaded misfortune or changed circum-stances to justify non-payment. Judges often reacted sympathetically, and if they did not (as in the Ottoman empire), taxpayers might send a petition directly to the crown protesting against their burdens – which the king's financial advisers often considered favourably.

Likewise, although the laws of Castile deemed resistance to authority as a capital offence, judges often hesitated to impose harsh penalties for fear of provoking a backlash. When the Council of Castile (in effect, the supreme court) heard evidence against 336 persons allegedly involved in nine separate anti-seigneurial riots between 1620 and 1685, it sentenced only ten to death and fewer than 100 to forced labour. In some other cases, violent resistance produced concessions. For example, in an area of France near to the war zone with Spain, in 1641 protestors burnt the offices of the king's tax collectors to the ground and ran his officials out of town to shouts of 'Thieves and taxmen: we'll kill you and exterminate all your kind so that no memory [of you] will remain.' Fearing reprisals, the magistrates sent agents to Paris to explain the hardship from which their town suffered and, on this occasion, the king acknowledged the justice of his subjects' complaints and the unreasonable behaviour of the tax collectors: a letter arrived from the agents that began 'Gentlemen: give thanks to God! No more taxes!'[31]

Such royal concessions received massive publicity – printed posters proclaiming the news went up everywhere – giving the impression that popular resistance worked; but revolt and resistance always involved high risks, because governments normally treated those who opposed them as traitors. When the Chinese community in the Philippines rebelled in 1639, the Spanish governor had all Chinese already in captivity shot in cold blood, murdered all Chinese servants in Christian households and set fire to Manila's Chinese suburb, incinerating all within it.[32] The insurgents would have fared no better in their native land, since both the Ming and Qing law codes included draconian penalties for all rebels and their families, and most Chinese uprisings therefore ended in mass executions. Rebels against the Romanovs in Russia met a similar fate. After the Moscow protests against currency debasement and high food prices in 1662, the tsar executed over 400 people,

> some beheaded, and [some] hanged. Others have had their feet and hands cut off and their tongues cut out of their throats. They have put 700 [persons] in chains, and these will be sent as soon as possible to Siberia with their wives and children, each one [of the men] with a brand mark burned into his left cheek ... Each of the young boys who were found among the rebels – twelve or fourteen years old – has had an ear cut off as a warning to others.[33]

In England, in 1685, the Catholic James II exacted a terrible price from the rebels who supported the attempt of his nephew, the Protestant duke of

Monmouth, to seize the throne. According to Monmouth's own account, he confronted the royal army with perhaps 7,500, of whom few escaped. 'The fight actually continued half an hour', the duke observed, with few losses on either side, but 'great slaughter was made in the pursuit'. Of the 1,300 men captured, the government butchered some immediately, executed almost 300 more (including the duke) and sentenced 850 to penal servitude in the American colonies.[34]

Even governments that promised pardon and concessions to rebels who surrendered might renege as soon as they regained the upper hand. The most notorious example occurred in Naples. When Gennaro Annese and the other leaders of the Serene Republic agreed to surrender, Philip IV's son Don John of Austria promised the city's inhabitants a 'general pardon for all crimes, whether committed through ignorance or malice, even if they involved treason, together with immunity from all excise duties', and the restoration of the traditional privileges. But when the king later asked a committee of theologians 'if we were obliged to uphold the pardons granted to the Neapolitans and respect the oaths sworn to respect the privileges?' their response was negative. The count of Oñate, who took over from Don John as viceroy, announced that his predecessor 'had acted according to the status he then held, but now it was up to him to proceed according to what he found'. He created a network of agents and spies to denounce enemies of Spain, and special courts to try them. Over the next three years, several hundred former rebels – called *Masanielos* in the Spanish sources – were executed, starting with Annese and several *capopopoli* of other towns in the kingdom. Thousands of others survived only because they fled abroad.[35]

Whether governments opted for concession or repression to restore order, most regarded rebellions as inconsequential unless or until members of the elite became involved. On hearing of the popular uprising in Évora in 1637, Olivares wrote dismissively: 'Normally we would take very little notice because we see popular tumults every day without any ill effects' – yet half the kingdom was soon in revolt. Three years later Cardinal Richelieu was equally dismissive when he first heard of the Catalan revolt, doing nothing because 'most disorders of this nature are normally just a brush fire [*un feu de paille*]' – yet the rebellion continued for almost two decades. More rashly, in 1648, Cardinal Mazarin dismissed the Day of the Barricades in Paris with exactly the same metaphor: 'The mild disorders that occurred in this city', he reassured a colleague, were 'just a brush fire which we extinguished as easily as it began' – yet the French capital had just experienced the worst disorders it had seen in sixty years, or would see again until the French Revolution.[36]

What Olivares, Richelieu and Mazarin failed to realize was that significant segments of the social elite seethed with resentment and would welcome any plausible excuse to 'turn petty rifts into major fights'. An official in the Papal States, forced to flee from rioters in 1648, had a better grasp of the nature of popular revolts: 'One must never underestimate the facility with which the common people [*la Plebe*] allow themselves to be persuaded by anyone who expresses concern for their fate', he warned the pope. In particular 'impassioned gentlemen are highly effective in making the people believe the impossible, especially when clothed in zeal for the public good'.[37] But in the mid-seventeenth century, 'impassioned gentlemen' were not the only ones who became 'highly effective in making the people believe the impossible'.

'People Who Hope Only For a Change': Aristocrats, Intellectuals, Clerics and 'Dirty People of No Name'[1]

In 1644 Nicholas Fouquet, later Louis XIV's chief fiscal officer but then his representative in Valence on the river Rhône, pondered the current unrest among the city's inhabitants. He concluded that, although the *origins* of the disorders he faced 'do indeed lie in the misery of the common people, their *progress* proceeds from the division that exists among the most powerful people, the ones who should oppose them'.[2] In the early modern world, three groups of 'powerful people' could manage to turn popular protests into something that threatened state breakdown – the nobility; literate lay men and women; and the clergy – and every major revolt in the mid-seventeenth century involved at least one of them. Sometimes, however, people from humble backgrounds also played important (albeit often ephemeral) roles, especially in urban revolts; and they, like their social betters, developed sophisticated theories and arguments to justify their opposition.

The Crisis of the Aristocracy

Although bitter rivals, Richelieu and Olivares agreed on at least one maxim of state. In 1624 the cardinal warned Louis XIII that 'keeping the nobles under the king's authority is the sole pivot around which the State turns', and that same year the count-duke lectured Philip IV on the necessity of keeping the nobles 'always reined in without letting any one of them grow too powerful'. Kings must 'under no circumstances allow noblemen, great or small, to make themselves popular'.[3] Over the next two decades both ministers followed their own advice ruthlessly. Aristocrats who openly challenged the government's policies, whether domestic (for example by defying a ban on duelling) or foreign (by plotting peace with the enemy), went to the scaffold, to prison or into exile.

Seventeenth-century Europe contained three distinct aristocratic zones. At one extreme stood areas where noble families made up a substantial proportion of the total population: Castile with 10 per cent, the Polish Commonwealth with 7 per cent, and Hungary with 5 per cent. By contrast, France, the British Isles, the Dutch Republic and the Scandinavian kingdoms belonged to a second zone, in which nobles were relatively scarce: 1 per cent or less of the total population. The rest of the continent fell somewhere in between these extremes, but almost everywhere the nobility expanded. Between 1644 and 1654, Queen Christina doubled the size of the Swedish aristocracy; between 1600 and 1640, Philip III and Philip IV almost tripled the number of nobles in the kingdom of Naples and almost doubled those in Castile; and James I and Charles I almost quadrupled the Irish peerage and more than doubled that of England.

Many of the new peers received their titles for the traditional reason – as a reward for outstanding service to the state – but many more became aristocrats either in return for money or as part of a strategy to empower one group (such as relatives of the favourite) at the expense of others. The new nobles therefore included many bankers, generals, lawyers and people with the same surname (Guzmán in Spain; Oxenstierna in Sweden; and so on). New or old, most peers had three ambitions: helping the king to govern; bringing to the king's attention the needs and interests of their family and followers; and preserving the liberties won by the blood of their ancestors in the service of the crown.

Domineering ministers like Richelieu, Olivares and Oxenstierna eventually alienated their fellow aristocrats by ignoring these ambitions. Their insistence that necessity knows no law resulted in repeated violations of aristocratic immunities by recruiters in search of troops, by tax collectors in search of funds and by commissioners demanding that nobles document their tax-exempt status. They also strove to restrict the king's attention solely to the views and interests of their own followers. Previously the economic health of each noble house had depended upon royal benevolence – the grant of a lucrative state office, or an edict that unilaterally reduced interest rates on their debts and offered protection against their creditors – but now it depended upon the benevolence of the favourite. At a time of economic adversity and heavy taxation many peers faced ruin if the favourite persuaded the king to withhold financial rewards.

The nobility had many ways of fighting back. In Castile, where the aristocracy had abandoned its right to attend the Cortes, rebellions by individual aristocrats came to nothing; but collective action could still succeed. After the surrender of Perpignan (Catalonia's second city) to the French in 1642, the grandees boycotted the court and made clear that they would stay away until His Majesty dismissed Olivares. As the aristocratic 'strike' entered its third week, Philip crumbled. In France, too, the aristocracy lacked a common forum in which to air their grievances – the crown did not convene the Estates-General between 1614 and 1789 – but discontented nobles found other ways to express their grievances. Some stated them in print: almost one-half of the authors of pamphlets published in France between 1610 and 1642 were aristocrats; while during the Fronde, several nobles maintained a stable of writers to propagate their views (Condé installed a printing press in his Paris mansion). Other nobles invoked (in the happy phrase of French historian Arlette Jouanna) their 'duty to rebel'.[4] Antoine Dupuy, lord of La Mothe la Forêt, who had recently retired from an active army career, agreed to lead the Croquants' rebellion in southwest France in 1637; two years later, in Normandy, the lord of Ponthébert became General Nu-pieds and commanded its Army of Suffering. Only one French noble, Condé, took the duty to rebel to the extremes seen in earlier centuries: he created a national following, sought to become chief minister, and when his attempt failed he entered the service of Spain, his country's arch-enemy, to further his cause.

Three considerations explain why Condé's extreme action remained virtually unique in seventeenth-century Europe. First, many attempts to create a national following foundered on an intense local hatred of the capital and the court. As an account of the Croquants' revolt noted, 'the very name "Parisian" excites such hatred and horror in everyone that just to say it is to risk being killed'. Local rebels, in France and elsewhere, normally 'resolved not to welcome any prince or lord fleeing the Court'.[5] Second, maintaining a rebel cause proved so expensive that most nobles lacked the means to sustain defiance for long: few could resist when the crown offered a settlement that restored their financial solvency. Third, some of Europe's great noble houses simply had too much to lose by outright defiance of the government. Duke John of Bragança boasted more power than any other Portuguese nobleman, if not more than any other aristocrat in Western Europe, and therefore stayed out of court politics because he had nothing more to gain. Instead he concentrated on conserving the assets, privileges and pre-eminences that he had inherited and only abandoned his prudent stance in 1640 when Olivares insisted that he raise and personally lead troops to fight in Catalonia. Fearing (probably correctly) that he might never be allowed to return, he threw in his lot with a group of conspirators who declared him king.

Some other European nobles also already exercised such extensive economic and political powers that they felt no duty to rebel. In Sweden, and also in Denmark until 1660, the aristocracy possessed vast landed domains and controlled the royal council, without whose approval the ruler could do little. The nobles of the Polish-Lithuanian Commonwealth acquired such a stranglehold on the federal parliament (Sejm) that a single veto could paralyze all business and force dissolution. In Russia, nobles used their prominent place in the Zemskii Sobor to agitate for legislation that granted them absolute authority over their serfs and shamelessly exploited the wave of popular disturbances that swept the empire in 1648–49 to secure the tsar's compliance (Chapter 6).

Outside Europe, by contrast, no group of hereditary nobles played an important role in the political upheavals of the mid-seventeenth century.

Some states, indeed, lacked any hereditary nobles: although the Mughal emperors granted fiefs (*jagirs*) to their leading followers, the grants never became hereditary; and although the Ottoman sultans also granted fiefs (*tımars*) to their cavalrymen, they never became an aristocracy in the European sense. Ming China boasted great landholding families, but most were Ming clansmen who remained staunchly loyal because they realized that, without imperial protection, they would lose everything (as happened in the 1640s, when rebel armies publicly humiliated and executed all members of the dynasty whom they encountered). The Tokugawa shoguns managed to control the 200 surviving *daimyō* families of Japan by demolishing all but one castle in each fief, demanding huge donations to their own construction projects and requiring all *daimyō* to spend half their time in attendance at Edo (and to leave their wives and children in the capital as a permanent guarantee of loyalty). When some discontented samurai conspired to overthrow the Tokugawa system after the death of Shogun Iemitsu in 1651, they received no *daimyō* support (Chapter 16).

Education and Revolution

In many parts of the mid-seventeenth century world, a second category of 'powerful people' (to use Fouquet's terminology) fostered political resistance: literate lay men and women. Ironically, the efforts by the state to generate highly educated officials also generated its most vocal critics and opponents. In China, the arithmetic was simple: 99 per cent of the 50,000 *shengyuan* (licentiates) who competed in each triennial provincial *juren* examination failed it, as did 90 per cent of the 15,000 men involved in each triennial metropolitan *jinshi* examination (Chapter 5). Although the overall population of China may have doubled under the Ming, if one includes all the men who prepared and sat for their examinations but failed, the student population probably increased twentyfold and by the 1620s some counties boasted over 1,000 frustrated scholars.

As the problems faced by the state multiplied, some started to criticize the government. One of the boldest protests led by scholars occurred in 1626 in the prosperous Jiangnan city of Suzhou, when the emperor's chief eunuch, Wei Zhongzian, ordered the arrest of one of his critics, Zhou Shunchang, a retired official who had displayed outstanding honesty and (according to one source) 'hated evil as a personal enemy'. Around 500 licentiates donned their formal attire and gathered in the courtyard of the local magistrate, begging him not to execute the warrant because it had been issued by Wei rather than by the emperor. Although the magistrate hesitated, the guards started to shackle Zhou anyway, and according to an eyewitness 'the affair got completely out of control, so that law and order could not be enforced'. The scholars beat one guardsman to death and chased away the rest. The disturbances lasted for three days. Wei later had five of the insurgent Suzhou scholars executed, degraded five more, and sentenced several others to hard labour.[6] Although these measures temporarily halted collective academic insurrections, individual intellectuals continued to criticize the policies pursued by the Ming government and denounced those who carried them out. Ai Nanying (1583–1646) achieved fame when his successful *juren* examination in 1624 was deemed to contain material critical of Wei Zhongzian, who decreed that Ai would have to wait nine years before attempting his *jinshi*. Ai responded by writing critical essays, letters and poems that proved so popular that 'Suzhou and Hangzhou booksellers paid him to come and write something – anything – they could publish'. Three editions of his collected works had appeared by the end of the century.[7]

As public order and economic conditions deteriorated, alienated intellectuals gained confidence by joining one of the 2,000 or so academies founded under the late Ming, of which the most famous was the Donglin Academy, where officials and candidates for civil service exams met other intellectuals to discuss current social issues. Wei suppressed the academy, but many of its alumni later joined the *Fu she* (Restoration Society, founded in 1629) or another scholarly society, where they debated not only literature, philosophy and history but also practical ways of ending government corruption and dealing with threats to the traditional order, both foreign and domestic. Although the leading lights of the academies and societies managed to introduce their ideas into the curriculum of the civil service examinations, they failed either to halt corruption or to influence the policies of the central government, and some therefore defected either to Li Zicheng or to the Qing (and, in some cases, to both: Chapter 5).

In Europe, the methods adopted by the state to produce highly educated officials likewise produced a pool of highly educated critics and opponents. By the 1620s, European men could study at almost 200 institutions of higher education, some of them surprisingly large. The student population of Naples University numbered about 5,000, and that of Salamanca University around 7,000. Almost 1,200 students graduated from Oxford and Cambridge each year during the reign of Charles I, and several hundred more studied law at the Inns of Court. In the 1650s, according to a Scottish visitor, James Fraser, 'the number of the Inns of Court students and Gentlemen are at present computed to a 1,000 or 1,200', while Cambridge 'could boast of 3,200 students and above'. Oxford ('much decayed of late') had somewhat fewer.[8]

In most European countries, the student body included many aristocrats. In Bavaria, almost one-fifth of those who matriculated at Ingolstadt University between 1600 and 1648 were nobles; and by 1620, one-half of the Protestant nobles of Lower Austria had acquired higher education. In England, according to James Fraser, 'all Gentlemen and a considerable number of ye higher

Nobility' were 'versed and accomplished here in rhetherick, logick, arithmetick, mathematicks, the French to argue, and Latin'.[9] In Spain, too, students from noble families filled the universities. By the 1630s, perhaps one young Englishman in forty and one young Castilian in twenty attended university – proportions that would remain unequalled until the later twentieth century. In addition, roughly one-third of all adult males in England acquired direct experience of law enforcement at some level, whether as magistrates, constables or jurors – a remarkably high participation ratio. Nevertheless, as in Ming China, no European government managed to employ all the alumni of its institutions of higher education. By the 1630s, perhaps 300 graduates left Oxford and Cambridge each year for a church living and another 200 went on to practice medicine or law – but this left some 700 others with no secure employment. The situation in Spain and other countries was similar: although many graduates entered the Church and a few became university professors, many more left their university without a job.

This overproduction of graduates terrified governments because, in the words of England's senior judge, a graduate of both Oxford and the Inns of Court, 'learning without a living doth but breed traitors'; and, he opined, 'we have more need of better livings for learned men, than of more learned men for these livings'. In Spain, the acerbic writer (and graduate) Francisco de Quevedo likewise argued that universities undermined rather than strengthened states: 'Armies, not universities, win and defend' states, he warned: 'battles bring kingdoms and crowns, but education gives you degrees and pompoms'. In Sweden, the chancellor of Uppsala University complained that 'there are more men of letters and learned fellows, especially in political matters, than means or jobs available to provide for them, and they grow desperate and impatient'. Emperor Ferdinand II (another graduate) blamed the revolt of Bohemia and Austria on the universities where his noble subjects had, he claimed, 'in their youth imbibed the spirit of rebellion and opposition to lawful authority'.[10]

These critics made a good point. In *Behemoth*, an examination of the causes of the English Civil War in dialogue form written in 1668, Thomas Hobbes claimed that Oxford and Cambridge 'have been to this nation as the wooden horse was to the Trojans' because 'out of the universities came all those preachers that taught' resistance. Hobbes did not limit his remarks to England. His dialogue boldly stated that 'The core of rebellion, as you have seen by this, and read of other rebellions, are the universities'.[11] Hobbes was right. In Bohemia, the rebel government included a dozen university graduates; in Sweden, two university-trained lawyers (the burgomaster and town secretary of Stockholm) and the historiographer royal led the opposition to Queen Christina in 1651; many of Philip IV's leading opponents were also university-trained lawyers or university-trained historians. Most striking of all, almost every one of Charles I's leading opponents had attended an institution of higher education. Of the first twenty-four graduates from Harvard College, no fewer than fourteen travelled to England to support Parliament; all the principal architects of the Scottish Revolution (Robert Baillie, Archibald Johnston of Wariston and Alexander Henderson) were graduates; in Ireland, about one-fifth of the Confederate General Assembly had studied at the Inns of Court (as had Sir Phelim O'Neill), and Lord Maguire (mastermind of the 1641 plot) had gone to Oxford. At least four-fifths of the English House of Commons elected in 1640 had studied at a university or the Inns of Court.

The Contentious Clergy

The clergy formed a third group of 'powerful people' with the capacity to transform popular revolts into revolutions, especially in Latin Christendom. The Scottish historian Sir James Balfour did not exaggerate when he denounced the clergy as 'the chiefest bellows that has blown this terrible fire' of civil war, because 'the best instruments, misapplied, doe greatest mischieffe and prove most dangerous to aney stait'.[12] Protestant clerics also acted as 'the chiefest bellows' of sedition elsewhere. In the Dutch Republic, Pastor Adriaan Smout of Amsterdam used his sermons to excoriate the city magistrates and the States-General for sending warships 'to assist the child of destruction, the child of Satan ... the king of France, Louis XIII'. Smout's oratory eventually secured the recall of the Dutch ships (shortly before the Amsterdam magistrates expelled Smout from their city in an attempt to silence him).[13]

Catholic clerics also fomented discontent. In 1637, the Jesuits of Évora (where they ran the university) openly supported the rebellion against Spain and toured adjacent areas spreading sedition. In the 1640s, almost eighty Portuguese preachers delivered sermons that endorsed the Restoration: some called on John IV to rebuild the kingdom of Portugal as Solomon had restored the Temple in Jerusalem, or to lead the Portuguese out of bondage as Moses had led the Hebrews out of Egypt; others warned their hearers that death would be preferable to 'returning under the vile yoke' of Spain, and that death in the Bragança cause would be equivalent to martyrdom.[14] In Catalonia, too, the clergy played a crucial role in legitimizing the revolt against Spain. In 1640 bishops excommunicated the king's soldiers after the latter set fire to churches in two villages that defied them; during the ensuing siege of Barcelona, clerics preached sermons, staged processions and heard confessions around the clock to rally resistance; and, as in Portugal, others assured their congregations that 'to die for the fatherland is to gain eternal life'. Seven years later, in Naples, dissident clerics 'animated the people so that they went freely to fight, and believed they would be martyrs and go to Paradise', and a group of priests took up arms and formed a regular militia company. At least three bishops of the kingdom were later charged with fomenting sedition, and at the provincial town of Nardò the government executed four cathedral canons for inciting the rebels.[15]

The Catholic clergy of Ireland proved the most bellicose of all. Although they apparently played little part in planning the 1641 uprising, many priests immediately lent their support. One urged the rebel troops on in battle ('Dear sons of St Patrick, strike hard the enemies of the holy faith'); another exhorted everyone to join the rebellion and called for extreme measures, 'assureing them that though the English did discharge musketts, and that some of them should be kill'd yet they should not feare, for such as soe died should be saints; and they should rush on with a multitude and kill all the Protestants'. In 1645 Conor O'Mahony, an Irish Jesuit teaching at Évora University, published a tract that congratulated his compatriots on the slaughter of 150,000 Protestant settlers and urged them to kill all the rest. After that they should replace King Charles with a native Irish monarch.[16]

The involvement of clerics in European revolts was significant in part because they constituted such a large proportion of the continent's public intellectuals: those who helped to shape popular opinion through the spoken and written word, through pulpit and sermon as well as pamphlet and book. In Spain, although clerics made up scarcely 5 per cent of the population, they wrote well over half of all works in Spanish published between 1500 and 1699. Likewise, clerics wrote almost half of all books published in the kingdom of Naples in the course of the seventeenth century, and over half the tracts that supported the Portuguese Restoration. Almost all Protestant ministers also boasted impressive academic credentials: every one of the 400 Scottish clerics involved in the revolt against Charles I had a university degree.

Three clerics actually led rebellions. Alexander Henderson became, in the unkind phrase of a Catholic adversary, 'the Scottish Pope', and such was his fame that Sir Anthony van Dyck, the king's painter, did a full-length portrait of him. Pau Claris orchestrated first defiance and then armed resistance by the Catalans until in 1641 he declared Catalonia to be an independent republic, with himself as its head. Giovanni Battista Rinuccini, papal nuncio to Ireland, in 1646 became president of the Supreme Council and thus chief executive of the Confederation of Kilkenny.

Clerical opposition also threatened the stability of some non-Christian states in the seventeenth century. In the 1630s the charismatic preacher Kadizade Mehmed provoked popular disorders, first in Istanbul and then throughout the Ottoman empire, with his call for a return to the beliefs and practices of Islam in the time of the Prophet Mohammed (Chapter 7). In Russia, a small group of literate clergymen and monks, protected by a few prominent lay patrons, articulated the cultural system of Old Believers in the 1650s and 1660s, creating a movement that challenged the right of the Romanov dynasty to rule (Chapter 6). In Ukraine, when Bohdan Khmelnytsky led the Cossacks in revolt against their Polish overlords in 1648, the Orthodox clergy offered enthusiastic support from their pulpits (Chapter 7). In India, the charismatic Guru Hargobind, the sixth Sikh Master, radicalized his followers and in the 1630s led them in battle against the forces of the Mughal emperor.

'Dirty People of No Name'

The political upheavals of the mid-seventeenth century empowered many others besides discontented members of the traditional elite. In East Asia, Nurhaci, the 'Great Ancestor' of the Qing dynasty, began his meteoric career as chief of a minor Manchurian clan that (by his own admission) initially possessed only thirteen suits of armour between them. The Ming loyalist leader Coxinga was the illegitimate son of a Japanese samurai's daughter and a Chinese pirate. Li Zicheng, who overthrew the Ming and proclaimed himself emperor of a new dynasty, had worked as a minor postal official. In Russia, Bohdan Khmelnytsky and Stenka Razin, leaders of the greatest Cossack rebellions of the early modern period, both came from poor families; Patriarch Nikon, who for a time enjoyed powers equal to those of the tsar of Russia, was the son of peasant.

Many European rebel leaders also came from obscure backgrounds – a fact that profoundly irritated the royalist Edward Hyde, earl of Clarendon. In his penetrating *History of the rebellion and civil wars in England,* and in other writings, Clarendon frequently condemned the king's opponents as 'inferior people who were notorious for faction and schism' and as 'dirty people of no name' – and indeed the Civil Wars catapulted many men from humble backgrounds to prominence.[17] Edward Sexby, once apprenticed to a grocer, in 1647 debated the rights of man with assurance and eloquence at Putney and later drew up *The principles, foundation and government of a republic* for the city of Bordeaux. Charismatic religious leaders of the period included George Fox, a former shepherd and apprentice shoemaker; John Bunyan, a tinker; and James Nayler, a yeoman. In Italy, many of the revolutionaries in 1647 also came from obscure backgrounds: Giuseppe d'Alesi of Palermo was a craftsman who had languished in prison until freed by rioters; Masaniello of Naples was an illiterate fishmonger.

Political upheaval and war also empowered some European women of humble birth. In England, they not only led street protests, like their continental sisters (Chapter 17), but also took an active role in the political process. In 1648–49, the Army Council granted two audiences to allow Elizabeth Poole (despite expulsion from her church for immorality and heresy) to explain her vision of what was best for the kingdom; and when they refused to heed her prophecies, she published them in pamphlets.

Mary Cary, who described herself as 'servant of the Gospel', also published several hundred pages of prophecy between 1647 and 1653, setting out God's plan for England and the world, based on twelve years' study of Scripture which started when she was fifteen. 'In a different world', it has been observed, 'such women might have become ministers'.[18] Some of their female Catholic contemporaries also achieved prominence. From 1643 to 1665, Philip IV of Spain wrote letters at least twice a month to Sor María de Ágreda, a nun who claimed prophetic powers, in which he solicited her advice on the problems he faced (as well as her prayers), making her the most powerful woman in the Spanish monarchy (Chapter 9).

Many of those empowered by the chaos of the seventeenth century, whatever their social background, were surprisingly young. Jack Goldstone has noted that 'the cohort that reached ages 26–35 during the 1630s was the largest youth cohort of the entire period 1500–1750'. Goldstone also noted that such a demographic structure is inherently unstable, not only because those who are young and single are the ones more likely to voice their discontent, but also because 'the participation of people in demonstrations or opposition movements depends to some extent on how great they perceive the support of that opposition movement to be'. Therefore, since 'the more timorous among the discontented may only join an opposition that appears widespread and successful', the growth of the 26–35 age cohort to unprecedented size in the 1630s probably increased the chances that people over the age of thirty-five would join the opposition, increasing the volatility of Stuart England.[19]

Age of the English population by percentage

Year	Total population	Age range				Total aged 15–29	Total aged 30+
		0–4	5–14	15–24	25–29		
1631	4,892,580	12.5	19.9	18.2	7.9	26.1	41.6
1641	5,091,725	11.8	20.5	17.3	8	25.3	42.4

Elsewhere, too, many of those who shaped revolutions were surprisingly youthful. Nathan of Gaza was twenty-two when he proclaimed Shabbatai Zvi to be the Messiah; Masaniello was twenty-seven when he became ruler of Naples, as was Johnston of Wariston when he drafted the National Covenant. The prince of Condé was twenty-eight when he tried to displace Mazarin as chief minister to Louis XIV; George Joyce was twenty-nine when he seized Charles I at Holdenby House; Dorgon and Wu Sangui were both thirty-two when in 1644 they made the agreement at the Shanhai Pass that settled the fate of China for almost three centuries. Sir Thomas Fairfax was thirty-three when he took command of the New Model Army in 1645 and thirty-eight when he resigned it. Coxinga was thirty-five when he led his great army of Ming loyalists up the Yangzi and almost took Nanjing in 1659. Lesser protagonists were also young. In Naples, the 'boys' who obeyed Masaniello were in their teens or twenties, just like the apprentices at the forefront of most London disturbances and those who took the lead in many popular revolts in France. As Christopher Hill observed about Revolutionary England in the 1640s and 1650s, it 'was a young man's world while it lasted'; but how long that world lasted, whether in England or elsewhere, depended in large measure upon the ability of the 'people who hope only for a change' to mobilize others with arguments that justified resistance.[20]

Justifying Disobedience

Religious, legal and historical texts that evoked a real or imagined Golden Age formed the commonest justification for disobedience in the seventeenth century. In the Ottoman empire, insurgents called for the abolition of all taxes imposed since the reign of Suleiman the Lawgiver; Kadizade Mehmed cited the Qur'an and the prophetic *hadith* to demand the abolition of all novelties; and the Sufi sheikh Niyāzī-i Mīşri compared the sultans of his day with Alexander the Great (advised by his 'sheikh' Aristotle). In China, too, protestors demanded a return to a Golden Age: in the 1640s, dissidents invoked the memory of a rebel hero from Fujian two centuries before, popularly known as the 'Pare-equal King' who had 'pared down master and serf, noble and menial, poor and rich, to make them equal'. Egalitarian ideas abounded in the popular culture of Ming China, notably two historical novels, *Romance of the three kingdoms* and *Water margin*, both proscribed by the government. *Romance*, which featured the evil power of eunuchs and their role in the downfall of the Han dynasty in the third century BC (an obvious parallel to the hated eunuchs of the Ming), became 'a veritable textbook on how regional militarists might bring the [ruling] dynasty to an end'.[21] *Water margin*, set in the mountains of Shandong in the early twelfth century, portrayed a Chinese Sherwood Forest peopled not only by heroic and selfless outlaws but also by itinerant monks, beggars, tricksters and experts in the martial arts. All of them opposed the corrupt and brutal government officials. During the late Ming period at least thirty editions of *Romance* appeared, while paintings and even printed playing cards popularized images of the protagonists. Nurhaci later claimed that he had learned about Chinese politics and military strategy from reading *Romance*, and his grandson ordered the book to be translated into Manchu and required all his followers to read it.

The Manchus also ransacked Chinese history for precedents to justify attacking the Ming and studied chronicles that described the rise and fall of earlier dynasties. In a letter written in 1621 to his Chinese neighbours, Nurhaci provided a list of unworthy rulers (going back to the eleventh century BC) who 'immersed themselves in liquor, women, and wealth and no longer troubled themselves about the country', and asserted that 'The [current] emperor of you Chinese does not rule fairly', because he allowed 'the eunuchs to take property' and persecuted 'those with property who are upright and honest'. The conclusion was obvious: '[Heaven] has given me the emperor's lands ... Heaven favours me'. His son Dorgon used similar rhetoric when he arrived in Beijing in 1644. His first proclamation contained the declaration: 'The empire is not an individual's private empire. Whosoever possesses virtue holds it. The army and people are not an individual's private army and people. Whosoever possesses virtue commands them. We now hold it.'[22] Nurhaci and Dorgon sought to justify their assault on the Ming with the concept of the Mandate of Heaven, found in both the *Classic of history* and the *Classic of songs* – texts by then 2,000 years old, with a status in East Asia equivalent to the Bible among Christians.

In Christian states, too, historical precedents served to justify popular rebellions. During the Irish rebellion of 1641, a priest in County Tyrone read from 'Hanmer's *Chronicles*, out of which [he] animated the rebels with the story of the Danes [in the eleventh century] who were discomfited by the Irish, though for the most part unarmed, and they paralleled that history with these times'. In Donegal, some of 'the rebells nowe expected the fulfilling of Columkill's Prophecie: which (as they did construe it to be) was that the Irish should conquer Ireland againe'. Charles I's Scottish opponents drew strength both from documents, such as the Declaration of Arbroath of 1320 that empowered the nobles of Scotland to protect their 'fundamental laws' against wayward kings, and from recent tradition, including the deposition of unsatisfactory rulers such as Mary Stuart (Charles's grandmother). His English opponents also sometimes cited the tumultuous history of their northern neighbour. Three days before his execution in January 1649, the High Court of Justice tactlessly reminded Charles of biblical and historical 'instances of kings being deposed and imprisoned by their subjects'.[23] Other opponents harped on the fact that England possessed its own distinctive laws and customs, sometimes going back to 1066, when foreign invaders imposed a 'Norman Yoke', and they urged True Englishmen to regain their ancient constitution and demand the abolition of any law or custom 'contrary to the Great Charter of England' (the Magna Carta of 1215).

In Spain, Catalan scholars found and published privileges granted by Carolingian emperors in the ninth century that granted Barcelona and its hinterland the right to self-government under loose Frankish protection, guaranteed the use of its own Visigothic laws and promised exemption from all future taxes. They therefore demanded that Philip IV do the same. Likewise, in France, the Croquants of Périgord in the 1630s demanded a return to 'the same state we enjoyed during the reign of Louis XII' (d. 1515), while the Nu-pieds of Normandy demanded respect for the charter granted to the duchy in 1315. In Italy, in 1647–48, the rebels of Palermo insisted on a return to the 'days of King Peter of Aragon' (d. 1285), and those of Naples wanted a restoration of the laws of John I (d. 1382), John II (d. 1435), and Charles V (d. 1558).

Apologists for rebellion regularly equated their leaders with biblical heroes and their opponents with biblical villains. Catholic preachers in Portugal after 1640 compared Philip IV with Saul or Herod and equated John IV with David or Christ; they also likened John IV's acclamation as king with that of King David and compared the sixty years of rule from Madrid with the 'Babylonish captivity' of the Jews. In Naples, sermons by preachers sympathetic to the revolt hailed the insurgent leaders as Daniel, David and Moses while comparing Viceroy Arcos with Nebuchadnezzar, Goliath and Pharaoh; royalists portrayed Viceroy Oñate (who restored order) as Gideon. Catalan pamphlets drew similarly invidious comparisons between Philip IV and those destroyed by God in the Old Testament.

In Protestant England, some compared John Felton (the discontented veteran who murdered the duke of Buckingham in 1628) with Phineas, Ehud and David. A generation later some compared the earl of Essex, commander of the parliamentary army, with John the Baptist, and many saw Oliver Cromwell as Gideon. In the Dutch Republic, many likened the kings of Spain to Pharaoh and equated the princes of Orange with Moses, Gideon, David and the Maccabees; and Joost van den Vondel's epic poem of 1612, *Passcha* (*Passover*), included a specific 'Comparison between the deliverance of the children of Israel from Egypt and the liberation of the United Provinces of the Netherlands from Spain'. In Scotland, the concept of a National Covenant as the foundation of resistance came directly from the Old Testament; and in 1638, as Archibald Johnston of Wariston watched his compatriots signing the document he had drafted, he discerned 'a very near parallel betwixt Israel and this church, the only two sworn nations to the Lord'. Like many of his compatriots, he saw the Scots as God's chosen people fighting Pharaoh.[24]

All over Europe, dissidents embraced not only the texts but also the tone of the Old Testament to justify their resistance. At regular intervals the English Parliament heard sermons that claimed scriptural warrant for extreme measures against the king and his supporters, citing texts such as 'Thou shalt smite them, and utterly destroy them; thou shalt make no covenant with them, nor show mercy unto them' (Deuteronomy 7:2); 'I will pour out my wrath on my enemies and avenge myself on my foes' (Isaiah 1:24); and 'Slay utterly old and young' (Ezekiel 9:2). During the trial of the earl of Strafford in 1641, a preacher reminded Parliament (which served as the earl's judge and jury) about the fate of Achan and Achitophel, those 'troublers of Israel' who had been rightly punished by death for giving their ruler false council. In other parts of the Protestant world, preachers imitated Jeremiah by threatening eternal damnation for rulers who deviated from God's commandments, and insisted (like Haggai) that since the end of the world was imminent, if rulers refused to impose the necessary reforms then their subjects must take over.[25]

Catholic propagandists also used Scripture to justify extreme violence. In France, pamphlets and pictures compared Mazarin with Goliath and his critics with David, and France under his rule with the lot of the Israelites under Pharaoh. After rebellion broke out in Ireland, a Franciscan exhorted his compatriots to 'fight to the end for our altars and hearths. We have no choice but to conquer or be conquered and either drive our enemies out of this land or be driven out ourselves. The country is too small for the English and the Irish.' A few years later, a Catholic commander informed his troops that 'You are the flower of Ulster … Maccabeans fighting against their enemy'; and (from the safety of Portugal) Father O'Mahony justified violent action against all non-Catholics in Ireland largely on the basis of Exodus 32, wherein Moses ordered the destruction of thousands of idolaters.[26]

In looking to the past for subversive precedents, European dissidents also scoured Roman and Greek texts. Scholars in Naples published classical works together with commentaries that unfavourably contrasted Spanish government with the supposed parity between nobles and citizens that had prevailed in the city's republican past. In England, according to Thomas Hobbes, 'As to rebellion, in particular against monarchy, one of the most frequent causes of it, is the reading of the books of policy, and histories, of the ancient Greeks and Romans'. He continued: 'From the reading, I say, of such books men have undertaken to kill their kings, because the Greek and Latin writers, in their books and discourses of policy, make it lawful and laudable for any man so to do – provided, before he do it, he call him a tyrant.' 'I cannot imagine how anything can be more prejudicial to a Monarchy', Hobbes concluded, 'than the allowing of such books to be publicly read'.[27]

Many rebels looked to the future as well as to the past, using prophecy, divination and portent to convince themselves and others of a favourable outcome to their resistance. In China, members of a popular religious sect known as the White Lotus Society had long predicted that a man named Li would one day be emperor, and as he strove to make this prediction come true, Li Zicheng consulted a medium (and when the prophet unwisely stated that 'Zicheng is not a true Son of Heaven' and predicted the imminent demise of his power, Li executed him). Like most Chinese, Li also placed great importance on portents. When a dust storm and yellow fog engulfed his temporary capital just after he proclaimed himself prince of Shun, Li panicked until his seers assured him that it was an auspicious sign because, when a new Chinese dynasty arose, the sun and moon temporarily lacked light.[28]

In Europe, too, rebels turned readily for both illumination and support to those who claimed a hotline to heaven. As the army of Philip IV sought to suppress the revolt of the Catalans in 1641, Sor Eufràsia Berenguer, a noblewoman of exemplary piety, experienced visions of Christ, the Virgin Mary and Saint Eulàlia (the patron saint of Barcelona) in the act of protecting the city. Her visions helped to animate the city's defenders: the last one 'occurred on the 22nd [of January 1641] and on the 26th came the victory on the hill of Montjuich'.[29] Some non-Catholics also consulted prophets. In 1638 many of Charles I's Scottish opponents drew comfort from 'the admirable speeches, exhortations, prayers [and] praises out of the mouth of a poor demoiselle, Margaret Mitchelson, who was transported in heavenly raptures and spoke strange things for the happy success of God's cause and Christ's crown in this kingdom, which was already enacted in heaven'. Margaret's ecstatic prophecies excited 'the astonishment of many thousand'; and some people who had previously harboured doubts 'were strongly confirmed and encouraged to add hand to this great work of God'.[30]

Some revolutionary prophets distributed their forecasts in print. William Lilly's *Prophecy of the white king* of 1644, which predicted the defeat and downfall of Charles I, sold 1,800 copies in the first three days and (according to an MP) his later writings 'kept up the spirits of the soldiers, the honest people of this realm, and many of us parliament men'. In 1650 George Foster, a former officer in the New Model Army, published his vision that God had chosen Sir Thomas Fairfax as his instrument to destroy Parliament and cut down 'all men and women, that he met with, that were higher than the middle sort, and raised up those that were lower than the middle sort and made them all equal'.[31] Some rebel leaders also exploited the prevailing millenarian climate to claim messianic powers for themselves. In Europe, Martin Laimbauer, a Protestant farmer who led a peasant revolt in Austria in 1635, sustained his cause for almost a year with his twin claims that the Apocalypse was imminent and that he was the Messiah. James Nayler, one of many Englishmen who claimed to be the Messiah during the 1640s and 1650s, gained such a following that at Bristol, England's second city, on Palm Sunday 1656 (a year in which many prophets predicted the world would end) he imitated Jesus and made a triumphal entry on a donkey while the people strewed palms before him. In Iran, in 1629, many Shiite Muslims hailed a rebellious provincial governor of Gilan as the Redeemer. In the Sunni Ottoman empire, Abaza Hasan (a rebellious provincial governor) in 1658, and the son of a Kurdish Sufi in 1667, both claimed to be the Mahdi (Chapter 7). In West Africa in the 1670s the charismatic Muslim sheikh Nasir al-Din 'claimed that he was sent by God' and 'preached penitence' and 'spoke only of the law of God and of welfare and freedom' (Chapter 15). Most spectacular of all, in 1665 many Jews hailed Shabbatai Zvi as both king of the world and Messiah (Chapter 7).

When they lacked sufficient precedents from Scripture, history and the classics, those in Europe 'who hope only for a change' deployed alternative strategies to justify resistance. Several rebels forged documents that seemed to justify their actions. In 1641, in Ulster, Sir Phelim O'Neill brandished 'a parchment or paper with a great seal affixed which he affirmed to be a warrant from the King's Majestie for what he did'. Although implausible (and untrue) many Protestant survivors testified that it had fooled them and others.[32] In 1647 the duke of Guise used an identical ruse to assure leaders of the Most Serene Republic of Naples that he possessed a letter from Louis XIV promising French support. Both Bohdan Khmelnytsky in Ukraine and Stenka Razin in Russia boasted to their Cossack followers that they possessed royal letters authorizing them to mobilize against their oppressors. Although each of these documents received wide credence, they were all almost certainly forgeries.

Thanks to their verbal diarrhoea, the Scottish Covenanters provided unusually detailed accounts of how they appropriated justifications for resistance invented by dissidents in other countries. In 1638, as the conflict with King Charles intensified, Wariston read histories of the successful Dutch Revolt against the king of Spain and 'studied all that week on *Althusii politica*' – a 1,000-page treatise written by Johannes Althusius, which claimed that a contract or covenant formed the basis of every human association (from families, through professional groups, towns and provinces up to states), and that the representatives of the lower associations could in certain circumstances resist a tyrannical superior. The following year, as he 'began to fall to the hypothesis of resistance in Scotland', Wariston 'epitomized Brutus his reasons' – a reference to a French Calvinist treatise, *Vindiciae contra tyrannos*, published sixty years before to justify armed resistance. Wariston's university preceptor, Robert Baillie, found justification for resistance in the writings of Martin Luther and other Protestants because they gave 'leave to subjects, in some cases, to defend themselves where the prince is absolute from any man, but not absolute from ties to the laws of church and state whereto he is sworn, which is the case of all Christian kings now'. Two weeks later, Baillie, Wariston and other radicals debated 'the lawfulness and necessity of defending ourselves in this case by arms'. Alexander Henderson, too, turned to modern Dutch writers when he considered what circumstances might justify opposing the king's commands 'and taking arms therefore': he borrowed from Hugo Grotius's *On the law of war and peace*, first published in 1625, the argument that 'the great force of necessity' might 'justify actions otherwise unwarrantable'. 'To sit still', Henderson continued, 'waiting for our own destruction' would be 'not only against religion, but [against] nature'. There could be, he concluded, 'no greater necessity' than the preservation of a country's religion and liberty because 'Necessity is a sovereignty, a law above all laws'.[33]

In England the gentleman scholar William Prynne revealed his preferred resistance theory in the title of his book: *The soveraigne power of parliaments and kingdomes, wherein the parliament's present necessary defensive armes against their sovereignes, and their armies in some cases, is copiously manifested to be just* (1643). It contained 200 closely printed pages of venomous attacks on King Charles, interspersed with quotations from the Bible, the classics and modern writers (Catholic as well as Protestant), followed by an appendix of foreign examples of resistance, deposition and regicide from ancient Israel to modern France; the full text of the Act of Abjuration by which the Dutch had renounced their allegiance to Philip II in 1581; and extracts in English from the *Vindiciae contra tyrannos* (which would soon appear in an English translation with the combative title, *A defence of liberty against tyrants*).[34]

Prynne believed that the core of England's political problems lay in its wayward monarch; therefore, its only chance of salvation lay in creating a republic. He had to hand two distinct republican visions: a state run by 'virtuous men' qualified to rule by their record of proven administrative competence, public service and legal expertise, such as the Dutch Republic; and an oligarchic state in which a few powerful families monopolized all power, as in Venice. Literature extolling both forms of government circulated widely in mid-seventeenth Europe, and their impact explains why, when Naples declared itself independent from Spain in 1647, it assumed the form 'Most Serene Republic', just like Venice, while the duke of Guise swore to defend the liberties of Naples 'just as the prince of Orange does in Holland'. Likewise some of the 1641 insurgents in Ireland claimed that it was their intention 'to have a free state of themselves as they had in Holland, and not to bee tyde unto any kinge or prince whatsoever'.[35]

The diffusion of such radical ideas in many parts of the world during the 1640s and 1650s was truly remarkable: never before had political news and ideas spread so far or so fast. Whereas earlier opposition movements had involved, at most, thousands of people, many of those in the mid-seventeenth century involved a million or more adherents. This transformation in scale rested upon two vital preconditions: both East Asia and Europe boasted not only a large number of eager readers, but also a large volume of printed material for them to read. Whether they appealed to Scripture or the classics, to ancient history or an ancient constitution – the 'people who hope only for change' attracted many more followers in the mid-seventeenth century than any of their predecessors because they managed to convey their arguments to audiences of unprecedented size.

'People of Heterodox Beliefs ... Who Will Join Up with Anyone Who Calls Them': Disseminating Revolution[1]

In *The tipping point* Malcolm Gladwell evaluated the impact of Paul Revere's ride through Massachusetts on the night of 18/19 April 1775 to spread the word that the following day British troops in Boston would try to arrest the leading American Patriots in Lexington and capture the weapons of the local militia in Concord. The ensuing hostilities began the American Revolutionary War. A critical element in Revere's success, according to Gladwell, was his status as a 'connector'. Revere's work as a silversmith and his frequent business travel had allowed him to create a wide network of casual acquaintances, from many different social groups, whose trust he had earned. As opposition to the British grew, Revere frequently carried messages between Patriot leaders, and so that night he knew where to find the boats and horses essential to his journey as well as where to find each Patriot leader – and how to avoid the British patrols – along the way. Revere's role as a connector enabled him to spread his news like a 'virus', and Gladwell hailed his ride as 'perhaps the most famous historical example of a word-of-mouth epidemic'.[2]

Several European observers in the mid-seventeenth century used similar medical metaphors to describe the remarkable speed with which revolts spread. The Spanish writer Francisco de Quevedo claimed in 1641 that revolts 'are the smallpox of kings: everyone gets them, and those who survive retain at least the marks of having had them'. A decade later, in his survey of the 'political uprisings of our times', the Italian historian Giovanni Battista Birago Avogadro declared: 'Popular uprisings are like contagious diseases, in which the deadly poison travels from one individual to another; and neither distance nor delay nor diversity of climate nor difference of life-styles can halt their dangerous effects.' In 1676 the governor of Barbados colony marvelled how the 'daily devastations of the Indians' had 'spread like a contagion over all the continent from New England ... to Maryland'. Nevertheless, as Hugh Trevor-Roper pointed out in his elegant 1959 essay that popularized the term General Crisis, although 'the universality of revolution owed something to mere contagion' nevertheless 'contagion implies receptivity: a healthy or inoculated body does not catch even a prevailing disease'.[3]

'Contagious Diseases' and Composite States

It is noteworthy that both Quevedo and Birago Avogadro drew their exam-ples of contagion from a type of polity that showed unusual receptivity: the composite state. More than half the rebellions that broke out in seventeenth-century Europe occurred in such entities, largely because their governments tried to impose similar policies on communities with different political, fiscal and cultural institutions and traditions – the Edict of Restitution in Germany; the Prayer Book in Scotland; the Union of Arms in the Spanish Monarchy and so on. Although these ambitious plans did not prosper, they provoked spirited resistance – in part because of the inflexibility of their proponents. When some German Catholics voiced fears about the risks of imposing the Edict of Restitution on all areas of Germany, Ferdinand II informed them that he was prepared to 'lose not only Austria but all his king-doms and provinces and whatever he has in this world, provided he save his soul, which he cannot do without the implementation of his Edict'. A decade later, Charles I likewise protested that 'So long as this Covenant is in force, I have no more power in Scotland than as a Duke of Venice, which I will rather die than suffer'; while Olivares exclaimed that 'if the Constitutions [of Catalonia] do not allow this, then the Devil take the Constitutions'.[4]

Not only did these proponents of uniformity fail to achieve their goals: they also failed to learn from their failure. Thus despite numerous warnings that imposing on a Prayer Book on Scotland would provoke opposition, Archbishop Laud made plans to impose it on Ireland too; and the miscar-riage of his attempt to subdue Scotland by force in 1639 did not stop Charles I from trying again in 1640 – nor yet, despite being soundly defeated, from contemplating a third attempt in 1641.

Such obstinacy was dangerous because rebellions that began on the periphery of each composite state often spread to other parts of the periphery. Bohemia in 1618 was the first to rebel, but almost all the other lands ruled by Ferdinand II – Hungary, Silesia, Moravia, Upper and Lower Austria – followed suit (Chapter 8). Two decades later in France, a judge commented that 'The news of the disorders that occurred in Lower Normandy' – the Nu-pieds revolt – 'redoubled the courage of the populace in Rouen', the duchy's capital, so that 'these disorders became the staple of popular conversation among the common people, who publicized them as heroic actions'. In 1647, on hearing that riots had occurred in Andalusia as well as in Sicily and Naples, a phlegmatic Spanish minister remarked that: 'In a Monarchy that comprises many kingdoms, widely separated, the first one that rebels takes a great risk because the rest can easily suppress it; but the second takes much less risk and from then onwards any others can try it without fear.' Seven years earlier, a French diplomat sent to liaise with the Catalan rebels believed that Portugal 'would never have dared revolt without the example of Catalonia, fearing that it would be rapidly overwhelmed if it joined in so dangerous a dance alone'.[5]

The Connectors

It is often difficult to reconstruct these dangerous dances because the dancers did their best to cover their tracks. Nevertheless connectors evidently spread sedition in Sicily, because the revolt of Palermo on 20 May 1647 was swiftly followed by urban uprisings in Trapani on the 25th, Cefalù and Marsala on the 27th, Castronuovo and Sanfilippo on the 29th, and so on. Likewise, in Russia, as soon as petitioners from provincial towns who had been in the capital in June 1648 returned home with news of the Muscovites' defiance of the tsar, local uprisings followed. Three years earlier, when James Howell sought to know 'upon whom to lay the blame' for the outbreak of the English Civil War, he argued (with an audacious mixture of metaphors), that the fire

> was first kindled in Scotland. The Puritans there were the womb of it; though I must tell you withall, the loins that begot this centaur were the Puritans here in England. If the flint and steel had not struck fire in England, the tinder had never took fire in Scotland, nor had the flame ever gone over into Ireland.[6]

Discontented subjects did not always require human connectors to 'kindle' their grievances, however: they could do it themselves. Philip IV's Italian subjects carefully monitored the progress of the Catalan revolt through letters, pamphlets and books. In Palermo, Vincenzo Auria (lawyer, poet and historian) reconstructed from the history books in his own library a full account of the earlier career of the unfortunate viceroy of Sicily, the marquis of Los Vélez, as viceroy of Navarre and Catalonia and ambassador in Rome, searching for a pattern of behaviour. In the Stuart monarchy, an Anglican bishop in Ireland complained in 1638 about the 'desperate example the contumacious Nonconformists [the Scottish Covenanters] have given both to England and to Ireland', and lamented that 'this contagion' had already begun to spread to Ulster. The following year, in the words of a professional 'letter-writer' (forerunner of newspaper reporters), 'The theatre for these kingdoms has now for a good while been chiefly placed at Edinburgh', so that others elsewhere would take 'what should be acted there' to 'frame the scene of their own interests accordingly'.[7]

No one watched Scottish events with greater attention than the Irish Catholics, who saw how 'the Scots, by pretending grievances and taking up arms to get them redressed, had not only gained divers privileges and immunities, but got £300,000 for their visit'. One Irish insurgent wanted to 'imitat Scotland, who gott a privilege by that course', in order to end 'the tyrannicall governement that was over them'; another boasted that 'The Scotts had theire willes by the force of armes and so would they heere in this kingdome'. When a Protestant pastor asked a leading Irish confederate: 'What? [Have you] made a Covenant amongst yow as the Scotts did?' "'Yea", said hee, "The Scotts have taught us our A.B.C."'[8]

Such contagion was not limited to composite states. For example, the various rebellions against Charles I attracted much attention in continental Europe. In 1648 one-third of the 'extraordinary issues' of the French *Gazette* focused exclusively on British affairs; between 1640 and 1660 some fifty German newspapers contained over 2,000 pages about events in Britain and Ireland, and German authors published more than 600 works on the subject; one-third of all pamphlets published in the Dutch Republic between 1640 and 1648 concerned English affairs; and Catalan insurgents published not only pamphlets with news of the parallel revolts against Charles I, but also Irish Catholic manifestos in Catalan translation. The initial success of the revolt of Naples against Philip IV attracted similar attention abroad. According to an ambassador, most Parisians believed 'that the Neapolitans have acted intelligently, and that in order to shake off oppression, their example should be followed'; and the crowds protesting against tax increases shouted 'Naples! Naples!' – a pointed reminder of the consequences of imposing unpopular taxes on a metropolis. In the Papal States, when a revolt broke out in Fermo in 1648, the first anniversary of Masaniello's revolt, many assumed that those who 'sacked and burned' the mansions of the wealthy merely imitated the 'example of the uprising of Naples'; and indeed several groups of revolutionaries crossed the border, encouraging insurgents in at least six other communities.[9]

The spate of rebellions in Europe also inspired malcontents overseas. In 1642 Don Guillén Lompart, an Irish protégé of Olivares who plotted to liberate Mexico 'from Spanish captivity' and declare himself 'king of New Spain', cited in his 'Declaration of Independence' the examples of others who 'have rebelled with good cause, having deliberated that it is better to die once for their restitution and liberty, than to live oppressed, tyrannized and violently subjected, as has been seen in the kingdoms of Portugal, Catalonia, Navarre and Biscay' – adding that 'in such remote and usurped kingdoms' as New Spain, abuses were 'far more widespread and grievous than over there' in Europe, so that the indigenous peoples oppressed by Philip IV 'not only can, but should rise up' against him. The following year, in Anglo-America, colonists in New England noted how their Native American neighbours took comfort from 'those sad distractions in England, which they have heard of, and by which they know we are hindered' from receiving protection. In 1647, Dutch visitors to Edo conducted an animated discussion with the shogun's leading advisers about the progress of the Dutch Republic's war with Spain and Portugal (revealing that the Japanese government had already learned a lot from other sources). Six years later, when a group of frustrated Portuguese colonists in Goa deposed the viceroy, they 'gave as justification that Portugal had done the same, and so had the people of England – while, near at hand', they added, 'Ceylon had done it'.[10]

Exporting Revolution

Many rebel leaders appealed for outside aid. In 1619–20 Frederick, the 'Winter King' of Bohemia, vainly requested military assistance from his fellow Protestants in Scandinavia, Britain and the Dutch Republic, as well as from the Ottoman sultan and his vassal the prince of Transylvania (only the last obliged); and in 1626 the rebels of Upper Austria asked Christian IV of Denmark, who had just invaded Germany, to send assistance (it never materialized). A decade later, Scottish opponents of Charles I mounted a successful diplomatic offensive to secure munitions from Denmark, the Dutch Republic and Sweden (albeit appeals to both the Catholic Louis XIII and the Protestant Swiss failed). The Portuguese also received favourable responses to their requests for aid: France, the Dutch Republic and eventually Britain all recognized the new regime and sent money, troops and warships. The Irish Catholic Confederation also gained diplomatic recognition (as well as munitions and funds) from Spain, France and the papacy, so that for the first and last time before the twentieth century, Ireland boasted its own diplomatic corps.

The Dutch supported several rebellions elsewhere as a matter of principle because, according to Lieuwe van Aitzema, the republic's official historian, 'the preservation of this state depended on the jealousy of its neighbours'. Its leaders therefore signed alliances 'with all the princes and potentates who opposed the tyranny and pretended Universal Monarchy of Spain': Catholic France and Venice, Protestant Denmark and Sweden, Orthodox Russia, Muslim Algiers and Tunis, and the Buddhist rulers of Sri Lanka. The republic also recognized a common interest (*gemeyn interesse*) with any group around the world that shared its 'enmity towards Spain'. As soon as news of Catalan revolt in 1640 arrived, the States-General established a special committee to coordinate support, and asked Cardinal Richelieu to facilitate contact between The Hague and Barcelona. The following year they also accepted the credentials presented by an ambassador sent by John IV of Portugal – thus recognizing the legitimacy of the 'Restoration' – and sent a fleet of twelve warships to protect Lisbon against the threat of a Spanish seaborne attack.[11] The republic also supported the enemies of Charles I. It allowed Scottish Covenanters to print pamphlets and purchase large quantities of arms and munitions; and a few months after the outbreak of civil war in England, a Dutch pamphleteer argued that 'we Netherlanders' must prevent 'suppressing of the Parliament' because 'if those that are on the king's side, together with him, get the upper hand' in England and Scotland, 'then shall they enter their action against us'.[12]

Charles I's opponents likewise sought to support and sometimes even foment rebellions elsewhere. In 1642 the London preacher John Goodwin assured his compatriots that successful opposition to the king would be 'cheering and refreshing' to 'your brethren in their several plantations in far countries [America]'; its 'heat and warmth' would 'pierce through many kingdoms great and large, as France, Germany, Bohemia, Hungary, Poland, Denmark, Sweden and many others'. Three years later, the Scots Parliament invited 'all Protestant potentates and republics to enter or join in the same or suchlike Solemn Covenant with the kingdoms of Great Britain, and so go on unanimously against the[ir] common enemy'. Most outspoken of all, in 1648 Hugh Peter delivered a sermon that claimed the New Model Army 'must root out monarchy, not only here but in France and other kingdoms round about'.[13] For a time, some in France speculated that 'the example of the neighbouring kingdom [England] would incite' critics of the monarchy to impose similar terms on the regency government, because 'Paris thinks itself no less than London'. Others asserted that 'In Paris they speak of nothing but republics and liberties'.[14]

The execution of Charles I in January 1649 changed everything. Admittedly the prolific French autodidact François Davant praised the regicides for reminding kings of the dangers of 'abusing their subjects', musing that 'troubled monarchies may give birth to republics' and predicting that France would be next; but few other Europeans agreed. Instead, in France, a spate of pamphlets denounced 'the most horrible and detestable parricide ever committed by Christians'; Pierre Corneille wrote a play sympathetic to monarchy; and four different French translations appeared almost immediately of *Eikon basilike* (allegedly written by Charles just before his execution: Chapter 12).[15] In the Dutch Republic, clerics condemned the regicide in their sermons; in Poland, the nobleman Albrycht Stanisław Radziwiłł included a detailed account of Charles's last hours in his memoirs, adding fervently 'let there be no similar examples [here]'; in Russia, Tsar Alexei expelled all English merchants as soon as he received news of the regicide. Shortly afterwards, in Sweden, Jakob de la Gardie warned the Council of State that some of his compatriots 'want to arrange things as they were in England some time past, making us all into pig's trotters'; and Queen Christina complained that 'neither king nor Parliament have their proper power, but the common man, the *canaille*, rules according to his fancy'.[16]

To stem this unfavourable tide of foreign opinion, the English Republic created new media to justify the new regime abroad. It produced a weekly newspaper in French, *Nouvelles ordinaires de Londres*; it maintained a Resident for the Parliaments of England and Scotland at Paris; and it appointed John Milton as secretary for foreign tongues. Milton began with a vitriolic counterblast to *The king's image*, provocatively entitled *Eikonoklastes*, and prepared numerous pamphlets and official publications specifically for foreign distribution. In 1654 his *Second defence of the English people* defiantly imagined that:

> From the pillars of Hercules [Cadiz in Spain] all the way to the farthest boundaries of [India], I seem to be leading home again everywhere in the world, after a vast space of time, Liberty herself, so long expelled and exiled ... I seem to introduce to the nations of the Earth a product from my own country: ... the renewed cultivation of freedom and civic life that I disseminate throughout cities, kingdoms and nations.

The success of his efforts can be measured by the prohibitions issued by several German states against owning or selling any work by Milton; and by Mazarin's denunciation of him as 'the most impudent and most wily apologist of the blackest of all parricides, by which the English nation has just been sullied'.[17]

In 1652 the English Republic retaliated by sending a team of agents, led by Edward Sexby, to the French port of Bordeaux, already in revolt against the central government (the Ormée: see Chapter 10). Sexby prepared two printed tracts that set out a blueprint for a republican form of government in the province of Guyenne (one of them clearly modelled on *An agreement of the people*, which he had drafted: see page 283), and the revolutionary government of Bordeaux sent delegates to London to secure English aid. Cromwell offered forty warships and 5,000 men in return for control of Bordeaux itself, leading a French pamphleteer to complain that English leaders saw themselves as 'so many Moses and Josuahs' who aspired to an 'empire of the universe'.[18]

A Public Sphere in the West?

The ability to export revolution reflected both the production of an unprecedented multitude of texts and the existence of an immense audience capable of receiving and understanding them. The Defenestration of Prague in 1618 – which almost everyone recognized as the harbinger of a major war (Chapter 8) – triggered a rapid expansion of a new medium: the newspaper. At least fifteen German cities published one by 1620, rising to thirty by 1640. By then, Hamburg published two newspapers, each with a run of 2,500–3,000 copies. Weekly newspapers were also common in the Dutch Republic. Every week Jan van Hilten sent twelve copies of the *Courante* he edited and printed in Amsterdam to a colleague in Leeuwarden, twenty-six copies to a bookseller in Nijmegen, and so on; and David Beck, a schoolmaster in The Hague, received a copy each week from his uncle in

Amsterdam and (according to his diary) sat down with friends 'until [they] had read the printed *Courante* in its entirety'. Many Dutch newspapers also boasted a foreign readership. Every Saturday between 1632 and his death in 1658 Michel le Blon sent Chancellor Oxenstierna of Sweden a package that contained that day's copies of the two newspapers printed in Amsterdam, the latest issues of newspapers printed in Arnhem and Antwerp, and of the *Gazette de France*, together with maps, military plans and a letter of advice 'containing political and cultural news which he based on reports from his various informers, on pamphlets and newspapers, and on rumours'.[19]

The audience for news comprised not only readers but also illiterate hearers. For example, late in 1659 General George Monck issued a pamphlet explaining his motives for leading an army from Scotland to restore parliamentary government in England, and calling for universal support. When a captain in the garrison of Leith, the port of Edinburgh, received a copy, he first read it himself, then discussed it with another officer and afterwards, since the pamphlet contained fewer than 1,000 words, he 'had it read to the soldiers' under his command. By this means, the same message reached hundreds of people, illiterate as well as literate, and if we imagine similar scenes throughout the army, the absence of effective resistance to Monck's march on London becomes more understandable (Chapter 12). The word had become mightier than the sword.[20]

Taken together, the combination of multiple media with a large audience created a popular public sphere: a series of arenas, at least partially free from government interference, where for the first time 'claims and counterclaims could be asserted and negotiated, and where the range of princely and imperial power could be questioned and contested'.[21] The destabilizing tendencies of this public sphere perplexed and frightened some contemporaries. 'It is strange to note how we have insensibly slid into this beginning of a civil war', an English MP lamented in 1642, through 'paper combats, by declarations, remonstrances, protestations, votes, messages, answers and replies'. Looking back just after the Restoration of Charles II in 1660, John Locke roundly condemned 'the scribbling of this age' and accorded 'the pens of Englishmen of as much guilt as their swords, judging that the issue of blood from whence such an inundation hath flowed had scarce been opened, or at least not so long unstopped had men been more sparing of their ink'. A decade later a royalist argued that nothing had 'hurt the late king [Charles I more] than the paper bullets of the press'[22] In Spain, a Catalan cleric loyal to Philip IV made exactly the same point: 'In this day and age, we fight more with books than with armies', he claimed, and he published a book crafted expressly to 'win back Catalonia in the same manner that it was lost'. Rebels throughout Europe seemed able to find a printing press with ease: the Ormée of Bordeaux issued tracts; the Nu-pieds of Normandy published manifestos; the Most Serene Republic of

Naples, 'considering the importance of works in print, and the way they are believed throughout the world', ordered printers to submit all future texts for their imprimatur or face a fine and confiscation of their presses.[23]

Many contemporaries blamed the emergence of Europe's first public sphere on education, and they made a valid point. A heightened appreciation of classical learning (the Renaissance), followed by the religious fervour that produced a more literate clergy and laity (the Reformation), had caused an educational revolution in Europe. Almost everywhere, schools sprang up to teach children to read, write and undertake simple arithmetical calculations, until by the 1640s half the parishes in some parts of England and Wales, three-quarters of the parishes of Lowland Scotland, and four-fifths of the parishes in and around Paris boasted their own schools (see also Chapter 21).

The resulting rise in literacy provoked stern countermeasures from some governments. In Spain, ministers called on Philip IV to 'close down some grammar schools newly founded in villages and small towns, because with the opportunity of having them so near, the peasants divert their sons from the jobs and occupations in which they were born and raised, and put them to study'. In France, Cardinal Richelieu wanted to close three-quarters of the *collèges de plein exercice* (schools that provided a general education in classical studies) because he, like Philip IV, reasoned that, if everyone received an education, 'the sons of the poor would desert the productive occupations of their parents for the comforts of office'. In 1659, the marquess of Newcastle, sometime preceptor to Charles II, warned his illustrious charge that 'there are too many grammar schools' in England: the country, he asserted, needed only enough schools 'to serve the church, and moderately the Law, for else they run out to idle and unnecessary people that become a factious burden to the Commonwealth'. Sir William Berkeley, the royalist governor of Virginia, agreed. 'Learning', he complained, 'has brought disobedience, and heresy, and sects into the world; and printing has divulged them and libels against the best government. God keep us from both!'[24]

Printing also served the semi-literate, primarily through the broadsheet: a single side of printed paper not unlike the front page of a modern newspaper, with a striking headline above a cartoon with an explanatory text beneath (often in rhyming verse, which made it easier for both readers and listeners to follow). To attract purchasers, the headlines always stressed either novelty or alarm, while the cartoons showed both ingenuity and ambiguity. France had no newspaper until 1631, when an official weekly *Gazette* began to appear; and although the government vetted its contents carefully, its feel-good diet whetted the appetite of readers for real political news, and after censorship collapsed in 1649, on some days no fewer than twelve new pamphlets appeared for sale in the streets of Paris. So many publications targeted Mazarin (hence their title: *Mazarinades*) that one critic jovially assured the cardinal that 'more attacks have been composed against you than against all the tyrants of Rome'. The 5,000 surviving *Mazarinades* fill more than 50,000 printed pages.[25]

The first printed Spanish weekly newspaper, the official *La Gaçeta Nueva*, began to appear only in 1661, but just as in France, failure to mention anything negative created a credibility gap, which a host of specialist writers filled by compiling manuscript *avisos* filled with news about assassinations and armed robberies; sodomy, rape and sexual promiscuity; political discontent, military defeats and rebellions. Discretion remained advisable, because several of those who criticized official policies in streets or taverns were never seen again; but anonymous manuscript newsletters avoided censorship and also appeared almost instantly. Some were the work of unemployed graduates, who could reproduce a few pages on demand almost like a modern photocopier, while others came from the pen of specialists who could reconstruct even a complex text (such as a play) from memory in a single night. As the situation of the Spanish monarchy worsened, Historiographer Royal José de Pellicer y Tovar used these resources to create a clandestine information network: he employed a team of scribes to write *avisos* to fellow scholars around the peninsula. Each one received a common core of news, together with additional items of local interest, and Pellicer expected full reports back to assuage what he called his 'sed de saber' ('thirst for knowledge') – and to pass on to other correspondents in his next bulletin. These *avisos* left no doubt about the perils facing the monarchy.

As in France, major changes occurred when rebellion put an end to censorship. Catalan pamphlet production, which averaged three items per year in 1620–34, and thirteen per year in 1635–39, soared to seventy in 1641, reflecting the decision of the rebel regime to spend 5 per cent of its total war budget on printing and distributing propaganda 'to inform all Catalans, men and women, old and young, of the true state of affairs, so that they can distinguish truth from lies'. Catalan printers published more in the 1640s than ever before – and more than in any later decade before the mid-nineteenth century.[26] Much the same happened in Portugal, where publications leaped from two in 1640 to 133 in 1641, and the 800 Portuguese works printed during the war with Spain (1640–68) exceeded the total of those printed during the rest of the century. In addition, the Portuguese published a gazette of their own modelled on the French precursor: the first newspaper to appear in that language. Like other rebel regimes, Philip IV's Iberian enemies used print to keep those living abroad abreast of its aspirations and achievements: many Catalan and Portuguese pamphlets were also published in France and the Dutch Republic, sometimes in translation, and a dozen or more printed justifications of the two rebellions circulated in Germany, sometimes translated into German.

The abolition of censorship in England had an even more dramatic impact on the ability to spread the contagion of revolution through print.

The year 1641 saw the publication of over 2,000 works in England, more than ever before, and in 1642 the number doubled – an annual total unequalled until the eighteenth century. Interested readers living in the English provinces had long been able to pay correspondents in London (rather like Pellicer in Madrid) to send them weekly manuscript news reports, but in the 1630s they had to pay £20 a year in return for one manuscript letter a week whereas a decade later a penny could buy thousands of words of printed news. Over sixty periodicals and newspapers came out in 1642 and seventy in 1648 (the highest number for any year of the seventeenth century), and each issue carried foreign and domestic news stories, including accounts of sermons and speeches taken down in shorthand by the first paid reporters in history. Each newspaper also had a party allegiance (to king or Parliament) that gave its reporting a distinctive spin, which (the marquess of Newcastle advised Charles II) 'overheat your people extremely, and do Your Majesty much harm . . . Every man is now is become a statesman, and it is merely with the weekly corantos [newspapers] both at home and abroad'. A Scottish visitor to England in 1657 agreed: 'There have been of late', he wrote, 'more good and more bad bookes printed and published in the English tongue than in all the vulgar languages of Europe'.[27]

A Public Sphere in China?

Just as no other seventeenth-century state became as politicized as England, no other seventeenth-century state saw as many people participate in political upheavals as China. Imperial subjects, from a wide range of backgrounds, reported and disseminated news throughout the empire in both speech and writing. According to Timothy Brook, 'More books were available, and more people read and owned more books, in the late Ming that at any earlier time in history, anywhere in the world'. This proliferation reflected, much as in Europe at the same time, the combination of a reading public of unprecedented size, with access to material of unprecedented quantity, and arenas with unprecedented freedom in which to discuss them.[28]

China, too, experienced an educational revolution in the sixteenth and early seventeenth centuries. Because Chinese is not an alphabetic language in which all words are composed from a relatively small number of characters, even functional literacy requires familiarity with several thousand characters, each one composed by several brushstrokes in a specific order from upper left to lower right. Nevertheless, schools to teach reading and writing abounded in late Ming China. At the beginning of the seventeenth century, a survey of 500 Chinese counties revealed almost 4,000 schools: one-quarter in the towns and the rest in the countryside.

China's educational revolution reflected two distinct stimuli. Some Confucian scholars stressed the need for introspection and intuition, believing that 'anyone could become a sage' and that moral principle might be found in the lives of 'ignorant men and women'. They therefore favoured education for all. Other scholars favoured the schooling system where boys learned to memorize and reproduce accurately the canon of classical texts on ethics and history necessary to climb the ladder of examination success, with all the social and economic advantages that success brought (Chapters 5 and 18). This process normally required years of classes that ran from dawn to dusk, with a short break for lunch, all year round (except for two weeks at New Year and a few holidays), because the canon required to pass even the *shengyuan* exam included 400,000 different characters of text, some of them archaic or arcane. Many students dropped out, but since even they acquired *some* reading skills, the functionally literate public of mid-seventeenth-century China far exceeded a million and may have exceeded 5 million. Put another way, perhaps 20 per cent of the adult male population of late Ming China boasted some educational attainments.

The existence of this huge potential readership fuelled a rapid expansion in printing. In the 1630s thirty-eight firms in Nanjing produced or sold books, with thirty-seven more in Suzhou, twenty-five in Hangzhou, and thirteen more in Beijing. Although some enterprises specialized in producing a few high-quality items in which calligraphy mattered almost as much as content, others shifted to a simpler artisanal style of cutting characters that reduced costs. The cumulative impact was remarkable: of 830 commercial works known to have been printed in Nanjing during the Ming era (1368–1644), over 750 appeared after 1573. The output of other centres seems to have increased at a similar rate: by the early seventeenth century, the printers of Suzhou employed 650 woodblock carvers.

Despite the complex calligraphy, Chinese printers enjoyed several advantages over their European colleagues. Whereas over fifty written languages were current in early modern Europe, all subjects of the Chinese emperor used the same script (even though they spoke many different tongues), so that a book published anywhere in China could be bought and read by millions of people – a market far larger than that enjoyed by any European printer. Moreover, the use of paper made from bamboo, rather than cloth fibres, reduced printing costs significantly; and the use of carved woodblocks (xylography) meant that Chinese booksellers could produce illustrated works without either a printing press or a stock of type – two expensive essentials for European printers who used movable type. In addition, Chinese publishers needed to print only enough copies to satisfy immediate demand, storing the blocks for future use: more copies from the existing blocks could be printed easily whenever required (no need to re-compose the text, as with movable type).

These various factors gave rise to a distinctive '*shengyuan* culture' in late Ming China, composed of satire and poetry, dictionaries and collected texts, how-to books (ranging from how to write letters to how to cure illnesses) and compendia of successful examination essays. For the first time in Chinese history, men below the official class participated in book culture and so created a public sphere: merchants published poetry as well as commercial manuals, and commoners published fiction. Some of these works became bestsellers, and a long-time European resident in China marvelled at 'the exceedingly large numbers of books in circulation here and the ridiculously low prices at which they are sold'.[29]

The last Ming emperors also resorted to print on an unprecedented scale. They not only issued innumerable posters for public display but also printed daily broadsheets known as *Dibao* (later the *Peking Gazette*) to inform all officials of imperial edicts and decrees, announce promotions and demotions, and provide news of domestic and foreign affairs. Manuscript copies also abounded, because regional officials hired scribes (many of them no doubt failed examination candidates) in the capital to make and mail copies of relevant entries in the *Gazette*. Some maintained a permanent news bureau where scribes copied unofficial as well as official news. Merchants produced commercial versions of these *Dibao*, often adding local news and gossip to the official pronouncements, and news entrepreneurs compiled and sold excerpts from the *Gazette* and other sources. The efficiency of this network is reflected in the memoirs Yao Tinglin, a minor official living in a small Jiangnan town. One day in 1644 he and 'other men of his family were drinking together when a friend rushed in in a panic, holding a "small gazetteer"' – that is, an unofficial news-sheet – 'that said that the troops of the rebel Li Zicheng had captured Peking ten days earlier and that the Chongzhen emperor had committed suicide'. Confirmation from the official *Peking Gazette* did not arrive until the following day.[30]

Yu Shenxing, a senior minister in Beijing, once complained about the fake news spread by 'news bureau entrepreneurs who are out for the most miniscule profits and give no thought to matters of national emergency'. He asked, like so many later politicians frustrated by journalists, 'Why aren't they strenuously prohibited?'[31] Even had Yu prevailed, closing down the news bureaux would not have prevented the diffusion of information, whether true or fake, because it also spread rapidly by word of mouth, via Ming China's excellent communications system. Travellers would find courier relay stations (in theory) no more than 25 miles apart on the extensive network of highways that connected all provincial and prefectural capitals, and postal stations (in theory) every 4 miles along the main roads of each county. This impressive infrastructure promoted social intercourse at many levels. In the words of the Jesuit Louis Le Comte in the 1680s: 'All of China is on the move: on the roads, on the highways, on the rivers, and along the coasts of the maritime provinces you see hordes of travellers.'[32] More specifically, it enabled tens of thousands of students to travel to take examinations in prefectural, provincial and (for the successful) metropolitan capitals. It also facilitated the journeys of thousands of government officials required to travel to distant postings, and of hundreds more sent on tours of inspection around the empire, to say nothing of itinerant merchants, peddlers carrying their wares between the different market towns in their area, and refugees hoping to find better conditions elsewhere. All these people wanted to hear news from home, and whatever their condition and wherever they went, travellers disseminated news of the outside world to entertain those with whom they stayed and those they met on the road, while their servants also exchanged news in their humble overnight lodgings.

In the 1620s Wei Zhongxian's persecution of Donglin alumni, and his subsequent fall from power (Chapter 5), offers an early snapshot of this developing 'public sphere'. Many intellectuals wrote private letters reporting each development that, together with the public edicts distributed by the courier and postal systems, excited public interest throughout China. Enterprising printers brought out compilations of personal accounts and official documents to satisfy public interest about what had happened and why; while the Suzhou rioters, punished so harshly for their support of the 'Donglin martyrs' (Chapter 18), became heroes of stage plays and popular literature, including four historical novels. The author of one novel assured readers that he had been at work for three years and 'based my book on what I read and heard', including scrutiny of a pile of copies of the *Peking Gazette* that stood 'more than three meters high', as well as 'several dozens of official documents and unofficial accounts'.[33]

John Dardess has argued that 'probably no earlier event in China's long history has available for modern retelling anything like the archive available for the Donglin affair'; but, just one generation later, those who survived the violent transition from Ming to Qing rule produced even more memoirs, almost 200 of which still survive. Among these, Grace Fong noted that Jiangnan produced 'a proportionately larger corpus of historical source materials' than any other area, reflecting the higher density of literate men and women living in the 'cultural and economic nexus of the Ming empire' who wanted to leave a written record of what they had seen and suffered before they died (many of them by their own hand).[34]

The diffusion of so much information about the common problems facing China allowed men and women in all regions to set their own experiences of adversity in a broader perspective and to develop comprehensive solutions. Huang Zongxi, a scholar whose father had been a Donglin martyr, probably exaggerated when he claimed in 1676 that in some areas of China, 'we find agricultural tenants, firewood gatherers, potters, brick burners, stone masons, and men from other humble walks of life attending public lectures and chanting classics'; but nevertheless several million imperial subjects took an active role in the Ming–Qing transition. The oldest state in the world had never seen anything like it – which is one reason why the transition claimed so many lives and lasted so long.[35]

A Public Sphere Elsewhere?

Although Islam is a 'religion of the book', and although Arabic is an alphabetic language, few parts of the vast Muslim world saw the emergence of anything resembling a 'public sphere' in the seventeenth century. West Africans, according to a French missionary, 'do not write: except for the *marabouts* [Sufi sheikhs] and some great lords, no one knows how to read or write'. A French merchant who lived in Senegal agreed: 'scarcely anyone, except those who want to be marabouts, study' – and even then, he added superciliously, 'they learn nothing except reading and writing. They devote themselves to no learned subject.'[36] Many other parts of the Islamic world probably resembled Senegal: literacy remained confined to the clergy and involved only religious learning.

India, by contrast, boasted both a large literate population and a rich literary culture. In the Mughal empire, an army of scribes 'copied and produced manuscripts in the hundreds of thousands' both in Persian and in the various languages of the subcontinent, some of them dealing with statecraft and politics; but their readership – and therefore their impact on the political life of the richest state on Earth – remains unknown. In South India, by contrast, 'everyday records were not penned on paper, but rather pressed into palm leaves, creating manuscripts that to survive had to be recopied each century'. Therefore most surviving Tamil documents from the period are poems, since only poems were deemed worthy of permanent preservation. In the Hindu states, intellectuals deemed most events insignificant and therefore compiled few written accounts.[37]

The intellectual life of the Ottoman empire was very different. The seventeenth century saw the proliferation of manuals of religious instruction known as *'ilm-i ḥāl*, aimed at women as well as men who could read Turkish, urging them to follow the Sharia and avoid 'blameworthy innovations' (such as sodomy, coffee and tobacco) if they wished to avoid adversity. One of the most popular authors of such tracts, Nushî, noted both the widespread literacy of Ottoman townsfolk, and their taste for Persian literature; he therefore stressed the need for proper religious education of the faithful in both Qur'an schools and households. His *Discourse on faith*, which began to circulate widely in the 1630s, even advocated an annual religious test for all Muslims aged seven or more on their knowledge of Islam, with rewards for those who knew their faith and punishment for those who did not.[38]

These works circulated in manuscript, because the Ottoman sultans allowed only two groups of their subjects to use printing presses: Orthodox Christians and Jews. In 1627 Patriarch Cyril Lukaris of Constantinople (born a Venetian subject in Crete and educated at Padua University) imported a printing press from England on which, with the aid of two Protestants, he produced editions of Patristic texts in Greek until jealous Catholics resident in the Ottoman capital persuaded the sultan that this was a seditious venture. He shut down the printing press, and later deposed and drowned Lukaris. This development did not affect the presses of Jewish printers in Istanbul and Thessalonica, which turned out Hebrew works in fascicles (rather than in completed books), allowing authors to receive comments that could be addressed in later segments. These printed works were distributed at synagogues on Shabbat, deposited in libraries (some of them public) and sent to notable scholars (some of whom made copies for use by their students), ensuring that news and ideas circulated far and fast. In the 1650s the Jewish community of the Anatolian port-city of Izmir, of which Shabbatai Zvi had been a member, began to print works not only in Hebrew but also in Spanish, including a new edition of Manasseh ben Israel's influential *Hope of Israel* (Chapter 7).

As in China, travellers within the Ottoman empire played an important role in spreading news and ideas. The central government tried to ensure that its senior officials rotated posts, so they would not put down roots in any area, and although the system did not always succeed, thousands of senior administrators, judges and soldiers travelled from one location to another at regular intervals. The career of Evliya Çelebi (1611–80?) offers an interesting example. Trained in Istanbul for a career in government service, he kept a detailed record of his assignments in Africa, Asia and Europe on military campaigns and on business, fiscal and diplomatic missions – during which he met and conversed with thousands of people. His *Book of travels* eventually filled ten volumes. Many other Muslims travelled around the empire to study with noted teachers. For example Sheikh Niyāzī-i Mīşri (1618–94), born in a small Anatolian town, went to Cairo (whose popular name at that time, 'Mīşr', he adopted), where he spent three years attending classes in the university attached to the al-Azhar mosque and residing in a Sufi lodge. Both there and in the numerous marketplaces and coffee houses of Cairo, Mīşri met and conversed with scholars from all over the Muslim world. Then he wandered through western Anatolia and the Balkans, attracting followers anxious to study with him. In the 1640s he went into exile, first on the island of Rhodes and then on Lesbos, after suggesting that Sultan Ibrahim, all his sons and his leading ministers were Jewish – a taint that (if true) rendered them unfit to rule over Muslims – and proposing replacement of the corrupted House of Osman with the Crimean khans. Despite his exile, Mīşri boasted many followers who read copies of his writings and formed a small Sufi order after his death. Although Niyāzī-i Mīşri never went to Mecca, many others did, for Islam expects every male Muslim to make a pilgrimage (*hajj*) there at least once. Along the way, as well as at their destination, pilgrims met people from other places with different experiences, skills and information, and expanded their mental horizons.[39]

The meteoric career of Shabbatai Zvi offers another demonstration of how far and how fast news and ideas travelled within the Muslim world. Although most rabbis and many Ottoman officials regarded him as a fraud, six months after Nathan of Gaza declared Shabbatai to be the Messiah in May 1665, the news had spread throughout the Jewish communities of North Africa from Cairo to Salé, on Morocco's Atlantic coast, and to Istanbul, whence it spread to Jewish communities in the Balkans, Hungary, Moldavia and the Crimea. Jewish printers in the Ottoman capital published two volumes of devotions composed by Nathan, one for nocturnal use and the other 'arranged to be said daily, brought from the Land of Zvi [Palestine]'. As soon as Shabbatai announced that he intended to travel to Istanbul to confront the sultan, thousands of Jews 'from Poland, the Crimea, Persia and Jerusalem, as well as from Turkey and the Frankish lands' converged on the Ottoman capital, and they were there to greet him when he arrived two months later (Chapter 7). Shabbatai's fame even reached the Americas: Jewish communities in the Caribbean islands expressed interest, and in Boston Increase Mather preached several sermons that drew attention to the 'constant reports' received 'that the Israelites were upon their journey towards Jerusalem, from sundry foreign parts in great multitudes'.[40]

This rapid diffusion of Shabbatai's message on four continents reflected not only its appeal at a time of millenarian excitement within both Judaism and Christianity, but also the impressive network of connectors who linked the Jewish communities of the eastern Mediterranean with the rest of the world. Shabbatai himself had lived in many cities of the Ottoman empire before 1665; his father had worked for the English merchants in Izmir; and his Polish-born wife had lived in Amsterdam, Venice and Livorno as well as in Egypt. Nathan of Gaza and the others who joined Shabbatai's entourage each boasted an extensive network of personal contacts whom they deluged with letters and (later) personal visits authenticating the Messiah's claims. In addition, Western merchants and diplomats resident in the Ottoman empire wrote detailed reports to their principals, spreading the news along Europe's Atlantic coast as far as Hamburg, where the rabbis inserted a blessing for Shabbatai into their prayers. In just eighteen months, Shabbatai and his network of connectors had turned the claim made by an obscure Jewish scholar in Hebron into a worldwide movement. Only news of his apostasy in September 1666 put an end to it.

The Rule of the Few

Despite the existence of extensive networks, new and old, for spreading the word about important events, most of the events themselves originated with a very small group of people who played a disproportionate role in 'turning the world upside down' (to use a popular phrase in revolutionary England). In 1640, a contemporary who watched the *segadors* rampage through the streets of Barcelona guessed that the hard core numbered no more than 500, while a mere thirty people maintained the clandestine Cloppenburg Press in London that produced and disseminated 'their own, patently incendiary version of what should come next'.[41] The following year, Lord Maguire planned to take Dublin Castle with fewer than 200 men (an enterprise that fewer than forty English officers accomplished in 1659); and about the same number enabled Sir Phelim O'Neill to capture almost all the strongholds of Ulster. In 1647 Masaniello began with no more than thirty *ragazzi*, many of them teenagers, when he turned a dispute over fruit excise in Naples into revolution; and Giuseppe d'Alesi had twelve co-conspirators when he seized control of Palermo. Both consolidated their authority with fewer than 500 men and boys. The following year Bohdan Khmelnytsky began his Cossack revolt with no more than 250 followers. Even successful revolutions might involve surprisingly few actors. In 1640 the coup d'état in Lisbon, a city of 175,000 people, that permanently restored Portuguese independence, involved at most forty noblemen with about 100 followers; twenty years later George Monck entered London, a city of perhaps 300,000 inhabitants, with fewer than 6,000 soldiers, exhausted after a 350-mile march in winter from the Scottish border. They nevertheless sufficed to end for ever Britain's sole republican experiment.

The explanation for such asymmetry – for how little things can make a big difference – lies in contingency, and especially in timing. In the words of a frustrated but perceptive French diplomat in London during the Civil War, 'affairs here change so fast that one no longer reckons time by months and weeks, but by hours and even by minutes'.[42] The same was true elsewhere. In Ireland, the Catholic rebellion gained unstoppable momentum when the O'Neills and their allies persuaded the castellans of half a dozen Ulster forts to admit them on the night of 22/23 October 1641 – only a few hours before a warning arrived from Dublin. Six years later, the duke of Arcos lost control of events in Naples in the few minutes it took for Masaniello and his *ragazzi* to win over the holiday crowds in the Piazza del Mercato. In each case, the government disposed of far superior resources right up until the tipping point, but the failure to deploy them in timely fashion proved fatal because of the speed with which the new information networks spread the contagion of revolutionary ideas – just as, a century later, the failure of British patrols to detain Paul Revere on his ride allowed him to spread the virus that would begin the American Revolution.

PART V
BEYOND THE CRISIS[1]

The political, social and economic upheavals known as the General Crisis largely ceased in the 1680s, yet global cooling continued for another generation. In the Paris region, the average monthly temperature fell below freezing eight times between 1691 and 1697, a phenomenon never again seen. In Finland, 'Long winters, rainy and cold summers, and early night frosts destroyed the harvest in both 1695 and 1696' causing 'a demographic catastrophe the scale of which is almost unique in European history', with population losses in some areas of over 40 per cent. In Boston, Massachusetts, Increase Mather complained in 1696 about the 'sore and long continued drought', adding ' 'Tis such a time of scarcity as the like never was these fifty years'; and John Tulley's *Almanac for the year of our Lord 1699* reported 'The seasons not as they used to be; the summers turn'd into winters, and the winters embittered with hardships. Which in the memory of man have not been known.' John Pike, who meticulously noted each winter's weather in Dover, New Hampshire, never recorded more than nineteen snowfalls per winter between 1682 and 1697, but over twenty – sometimes well over twenty – in seven winters between 1697 and 1709. The natural archive on climate indicates that average temperatures in 1687–1700 were 1.5°C lower than in the preceding decade, leading climatologists to christen this period the climax of the Little Ice Age.[2]

These oscillations may seem small, but since each change of 0.1°C in the global temperature advances or retards the ripening of crops by one day, the cooling of the 1690s delayed harvests by an average of two weeks in temperate zones, and by far more in sub-boreal regions. Sea temperatures around the Orkney Islands and Scandinavia in the 1690s were 5°C lower than today. In Ukraine in June 1695, after perhaps the worst winter in the past 500 years, it snowed as far south as Lviv; and a series of cool summers caused widespread crop failures. After a brief interlude of warmer weather, in 1708–9 Europe suffered what survivors would call the Great Winter. On the night of 5/6 January 1709, the temperature in Paris fell from 9°C to -9°C;

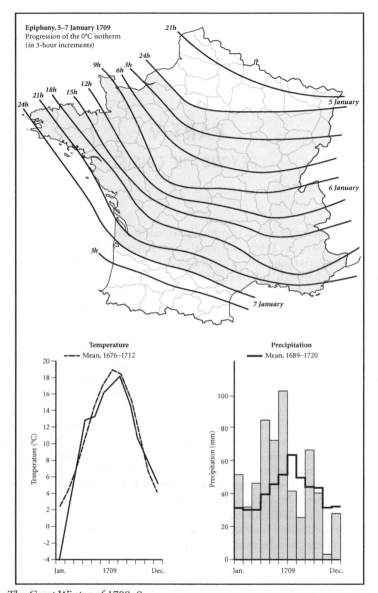

26. The Great Winter of 1708–9.
Dozens of people across France recorded the moment when rain turned to snow, and water turned to ice, at Epiphany 1709. Their attention reflected not only the rapid progress of Siberian air from Flanders to the Mediterranean – one of the last recorded 'extreme weather events' of the Little Ice Age – but also an increased awareness of climate change, and of the dangers that it posed. The year 1709 saw not only the coldest months recorded in the past 500 years but also an abnormally wet summer.

some towns on France's Atlantic coast received 60 centimetres of snow; on France's Mediterranean coast temperatures plunged to -11°C; at Venice, the rich went skating on the lagoon. At Versailles, the king's sister-in-law complained that 'It has never been so cold in the memory of man: no one remembers a winter like this.'[3] Temperatures in France stayed well below freezing for almost three weeks, making January 1709 the coldest month recorded in the past 500 years.

The underlying causes of this episode of global cooling remained the same: astronomers saw virtually no sunspots; El Niño episodes increased in frequency (1687–88, 1692, 1694–95 and 1697); volcanic activity peaked in 1693–94 and again in 1707–8 (with at least ten major eruptions, including Mount Fuji, which released perhaps 800 cubic metres of volcanic ash, some of which fell on Edo, 60 miles away). So did the consequences for humanity. In autumn 1690 Ottoman troops in the Balkans suffered from 'snow, rain and frost. The snow, being as high as the horses' chests, barred the roads, and the infantry could no longer move on; many animals dying, the officers were left to go on foot.' In 1691–92 drought produced widespread famine in China; and in New Spain hailstorms, a plague of locusts and heavy rains gave way to drought and early frosts that destroyed two maize harvests in a row and initiated a prolonged drought that lasted until 1697. Across the Atlantic, in Finland, some 500,000 people perished during the famine years of 1694, 1695 and 1696, and it took six decades for the country's population to recover. In France, between 1691 and 1701 climate change killed over a million people – a mortality, as Emmanuel Le Roy Ladurie observed, equal to France's losses during the Great War, but culled from a population that numbered only 20 million (not 40 million). At Versailles during the Great Winter 'every morning' there were 'people found frozen to death', and, in all, perhaps 600,000 more French men and women perished. In addition, those who survived these famines remained stunted for life, reaching an average height of only 1.617 metres – among the shortest Frenchmen ever recorded.[4]

As in the mid-seventeenth century, these episodes of climatic adversity occurred in wartime. Hostilities between Louis XIV of France and his enemies convulsed Western Europe from 1689 to 1697, and again from 1702 to 1713; the Great Northern War between Charles XII of Sweden and his enemies affected much of Eastern Europe from 1700 to 1721; the Qing emperor Kangxi led huge armies in the conquest of Inner Asia in the 1690s; and the Mughal emperor Aurangzeb campaigned ceaselessly against the Marathas and their allies in central India. All of these wars involved heavy taxation and caused widespread devastation.

It therefore seems astonishing that although the persistent global cooling caused misery and suffering on a scale that resembled the 1640s and 1650s, it was not accompanied by similar social and political upheavals. Admittedly, rioting broke out in several cities of New Spain, and in some regional capitals of the Ottoman empire, while in France the famine of 1709 that followed the Great Winter provoked almost 300 anti-tax revolts and even more bread riots. But none of these upheavals attracted the participation of 'people who hope only for a change' – the alienated aristocrats, intellectuals and clerics who had challenged and sometimes overthrown governments a generation before. In France, the unrest of 1709 remained unequalled until 1789 and there were no more Frondes. The fatal synergy had ended.

In his pioneering essay of 1959 on the General Crisis, Hugh Trevor-Roper used a flamboyant metaphor to describe the impact of the revolutions of the mid-seventeenth century. Thereafter, he wrote, 'Intellectually, politically, morally, we are in a new age, a new climate. It is as if a series of rainstorms has ended in one final thunderstorm which has cleared the air and changed, permanently, the temperature of Europe' (page xv). The chapters in Part II offer support for these claims regarding France, Spain, Britain, Germany and its neighbours; and make parallel assertions for some areas beyond Europe ignored by Trevor-Roper (China, Russia, Poland and the Ottoman empire). The chapters in Part III argue that Tokugawa Japan, Spanish Italy, Mughal India and its neighbours, as well as some parts of Africa and the Americas, managed to avoid a 'final thunderstorm', but still experienced an unpleasant 'series of rainstorms'.

The most important characteristic of the new age discerned by Trevor-Roper after the mid-century was nevertheless one that he overlooked: the 'series of rainstorms' often involved massive mortality. In China, the Ottoman empire, Russia and Europe, prolonged wars, famine and disease caused the death of hundreds of thousands of men, women and children. There were fewer people alive in the 1680s than in the 1640s, and therefore fewer mouths to feed when the harvest failed.

Yet depopulation alone cannot explain the lack of political upheavals in the late seventeenth century. Disasters, as Christof Mauch reminds us, often have a 'phoenix effect': some catastrophes 'improved emergency preparedness and spurred technological developments; they have also reduced the vulnerability of humans both in the emergent phase of natural catastrophes and during post-disaster recovery'.[5] Those who survived the mid-seventeenth century were no exception. Some survived by practising escapism (indulging in pursuits that dulled the senses amid encircling horror: Chapter 20). Others innovated (limiting the spread of plague through quarantine, and of smallpox through inoculation; planting crops with greater resistance to climate change; rebuilding towns in brick and stone to reduce the risk of catastrophic fires: Chapter 21). Others still developed new practical and scientific knowledge in the hope not only of repairing the damage done by past catastrophes but also of preparing for future ones – a legacy of the Global Crisis that led to a divergence between the West and the rest of the world (Chapter 22).

CHAPTER TWENTY

Escaping the Crisis

Getting Away From It All

Many of those who lived in the seventeenth century reacted to adversity and anxiety that they could neither explain nor avoid in much the same way as their descendants today: some killed themselves; others went to consult a therapist (to use today's nomenclature) or a cleric; many found solace in some absorbing pastime. All three categories are difficult to document because they left few traces in the surviving sources. Some of those who committed suicide appeared in court records (such as the findings of the juries convened by the coroners of England) or in chronicles (like the *Mingmo zhonglie jishi*, 'True record of Late Ming extreme loyalty', which honoured over 1,000 Chinese men and women who killed themselves rather than obey the Qing conquerors). A few left a note of explanation, like the desperate Chinese elite women who wrote *tibishi*, 'poems inscribed on walls', before killing themselves. The Scottish soldier, Patrick Gordon, recorded in his diary that he became 'careles of myself' after he returned to his camp after trying to rescue his captain in battle, only to receive a reproach for breaking ranks. This 'did so vexe me' that, with 'a desperate resolution', he rode between the two armies 'to seeke death ... swinging my pistol about my head [to] provoke any of them out to exchange bullets'. He was lucky to survive his suicide mission with only flesh wounds.[1]

Those who in adversity sought the advice of therapists and clerics left even fewer documentary traces. Richard Napier, an English country parson who gained national fame as a therapist, filled 15,000 folios with his notes on consultations with some 40,000 patients from 1597 to 1634. He deemed over 2,000 to be 'troubled', and over 150 'suicidal'. Twice as many women as men sought his help with psychological distress, of whom roughly half reported anxieties about courtship, marriage and bearing children. A quarter had recently been bereaved. One-third of his troubled patients were in their twenties and few were over sixty; most of them reported economic stress, chiefly through debt. Servants formed by far the largest occupational category.[2]

Napier determined that many of these patients suffered from 'melancholy', and the same diagnosis occurs in the casebooks of Theodore Turquet de Mayerne, the most famous (and best-paid) European physician of his day. His clients included Oliver Cromwell, the future lord protector, whom Turquet diagnosed in 1628 as 'valde melancholicus' ('exceedingly depressed'); and Princess Elizabeth, Charles I's youngest daughter, of whom Turquet wrote in 1650 that 'after the death of her father, she fell into a great sorrow, whereby all the other ailments from which she suffered were increased'. She died shortly afterwards.[3] Turquet and Cromwell had both read *The anatomy of melancholy*, a book by the Oxford academic Robert Burton that became a bestseller despite its enormous length (over 500,000 words), which argued that melancholy was 'a disease so grievous, so common' that 'in our miserable times' few 'feele not the smart of it'. He anatomized at great length the 'melancholy which goes and comes upon every smal occasion of sorrow, need, sicknesse, trouble, feare, griefe, passion, or perturbation of the minde, any manner of care, discontent, or thought, which causeth anguish, heavinesse and vexation of the spirits', concluding that 'from these melancholy dispositions, no man living is free'. Burton was no exception: in 1640 he hanged himself in his college rooms.[4]

Patients with untreated melancholy could endanger others as well as themselves. John Felton, the former army officer who stabbed to death the duke of Buckingham in 1628 was described by his brother as having 'a melancholy disposition', while his former neighbour remembered him as 'a melancholy man much given to reading of books'. Felton had no thought of escape: he had placed two statements justifying the deed inside his hat, in case he died in his attempt, and when instead he had the chance to flee amid the confusion, he announced 'I am the one' (thereby ensuring, as he must have anticipated, his arrest, torture and public execution).[5]

The chance survival of a dossier concerning someone afflicted by melancholy in Egypt in the 1690s demonstrates the existence of the same diagnosis in the Ottoman empire. A devout Sephardic Jew arrived in Egypt as the fiscal officer of a new provincial governor, leaving his wife behind him in Istanbul. 'From the day he arrived', the local rabbis noted, 'he was afflicted with various terrible illnesses so that he was falling apart'. The man consulted Jewish physicians and then, after they gave up on him, a Christian doctor who suggested that 'The illness had turned into *melancholia*' because the man's 'semen had built up and created an abscess in his body, and the vapors were rising to his head and reaching the heart'. The doctor predicted that if the official 'kept up in this way without discharging, the illness would overcome him'. Since Jewish doctrine prohibited masturbation, and the official's marriage contract prohibited bigamy, he asked the local rabbis for advice. They responded by interviewing both the Jewish and Christian doctors, and then 'conducted our own very thorough search in the books of the physicians to see whether an illness like this really exists in the world'. They eventually found that the *Canon of medicine* composed by the medieval Muslim physician and philosopher Avicenna (one of the authorities cited by Burton) contained a description that supported the diagnosis that sexual intercourse would solve the problem. Reasoning that even if they summoned the man's wife from Istanbul, the build-up of semen might kill him before she arrived, the rabbis allowed him to break his marriage oath and marry a second wife so he could ejaculate his way out of melancholy without committing the sin of Onan.[6]

Samuel Pepys committed that particular sin shamelessly on many occasions and in many places, including the Queen's Chapel during a Christmas Eve service – a feat that he recorded in the diary he kept for nine years. He also recorded sexual encounters with almost fifty women, several of them in 1665, when the departure of his wife and servants from London to avoid the Great Plague afforded him unusual freedom to sin. 'I have never lived so merrily', Pepys wrote at year's end, 'as I have done this plague-time.' Other seventeenth-century people seem to have sinned in order to avoid thinking about the disasters that surrounded them. In Germany, Elector Maximilian of Bavaria denounced in 1636 what he called the 'frivolous lifestyle' (*leichtfertige Leben*) that had developed among his subjects during 'the recent years of war'. He specified 'illegitimate pregnancies, especially in the countryside, among unmarried peasants and other common people' and claimed that 'the abominable vice of adultery' had become 'just as common as cursing and blasphemy among old as well as young people of both sexes'. Magistrates and ministers in neighbouring Protestant states agreed: 'All vices, and particularly swearing, have grown rampant because of the war', one lamented; 'Instead of making people more pious, the war made people nine times worse', echoed another. In Japan, four years after watching thousands die in Edo's catastrophic Meireki fire of 1657, Asai Ryōi published his *Tales of the floating world*, which called upon his readers to keep melancholy at bay by 'singing songs [and] drinking wine ... caring not a whit for the pauperism staring us in the face'.[7]

During the Ming–Qing transition in China, many discontented or disoriented intellectuals became monks. Some developed a religious vocation in response to 'the alienation from government affairs and feelings of despair, failure, worthlessness, and self-blame'; others did so to avoid compliance with the Qing edicts on tonsure and apparel, because Buddhist monks shaved their entire skull and wore traditional robes. They could thus avoid overt rejection of the new dynasty (a choice that normally ended in death: Chapter 5). Some cultivated elaborate gardens where they sought seclusion and composed poems, plays and prose; others succumbed to fatalism. In 1645 Yao Tinglin, a minor Chinese official living near Shanghai, sensed that with a new dynasty, a new dress code and a new social hierarchy, he had entered 'another world, with no restoring of the old order', and that he had been 'reborn in a new world' – a classic post-traumatic stress response. After failing both as a trader and as a farmer, as he entered his fourth decade in

1667, Yao wrote in his journal: 'I have the feeling that most of these forty years have been spent for nothing – that I have been through incredible hardship and yet have accomplished nothing so far.' He therefore resigned his official post, went back to the family village and opened a school where he worked as a teacher for the remaining three decades of his life.[8]

Some discontented and disoriented Europeans followed similar escapist strategies. In France, the chaos caused by the Fronde led a prominent critic of Mazarin, Robert Arnauld d'Andilly, who also looked after the gardens at the convent of Port Royal des Champs, near Paris, to publish a learned treatise on how to grow fruit trees, which offered an escape from insoluble political problems via raising, training and pruning trees to maximize their yields – an obvious metaphor for the peaceful pursuits that would restore general prosperity. Throughout Europe, the seventeenth century saw the rise of the geometrical garden, which not only offered a secluded place of escape but allowed those wearied or intimidated by the malign force of nature to tame it at the microcosmic level through obsessive trimming. In England, the disheartened royalist Izaak Walton wrote an overt invitation to escapism: *The compleat angler, or the contemplative man's recreation*, a conversation between a hunter, a falconer and a fisherman about pastimes that allowed the depressed to escape the fractured world around them. First published in 1653, it has seen over 400 editions.

Others tried to escape violence or vengeance by fleeing abroad. Thomas Hobbes left England for Paris in 1641, just before the Civil War, and did not return until it became clear that Parliament had won (and he had provided, in his *Leviathan*, a rationale for its rule that secured him a government pension). Many English royalists went into exile after the defeats of Marston Moor and Naseby and most of them remained in Europe until the Restoration in 1660 – when many republicans took their place. 'Bloodthirsty Mars' also forced many men and women to flee from Central Europe. Some were intellectuals at the height of their powers (like the musician Heinrich Schütz, the poet Martin Opitz, the mathematician and astronomer Johannes Kepler); others were country folk who could not protect their families, like the village shoemaker Hans Heberle, who fled with his family to Ulm thirty times during the Thirty Years' War. Some went into exile alone, like Hugo Grotius after the execution of Oldenbarnevelt in Holland; others moved as a group, like the thousands of *Masanielli* who in 1648 took refuge in Rome after the failure of the Neapolitan revolution, and the hundreds of *Malvizzi* from Messina thirty years later, who left for France on the fleet that evacuated the city's garrison (Chapter 14). In Eastern Europe, countless peasants fled misery at home by joining the Cossacks or the Tartars to the south, or (in the case of Russian peasants) by crossing the Urals into Siberia. In addition, several Polish, Transylvanian and Austrian nobles took out dual citizenship as a safeguard against potential catastrophe. Even Vasile Lupu (Basil the Wolf), prince of Moldavia, became a Polish subject when his daughter married the son of Bohdan Khmelnytsky, explicitly with the intention of creating a sanctuary in case his subjects should expel him. In China, some 60,000 Ming loyalists in 1661 followed their leader Coxinga to Taiwan, where they held out against the Qing for over two decades.

The violence of the mid-century even disheartened some of the victors. The spirits of the Manchu bannerman Dzengšeo sagged after he witnessed the death of comrades, and he recorded in his diary: 'In my heart I was frightened and, to keep myself safe, I pondered: "I have served on a military campaign for ten years, and have not lost my life in battle"' – but now he wondered if his luck would run out before he returned home. Dzengšeo was a senior officer, perhaps equivalent to a lieutenant colonel, normally attended by seven or eight servants, and his family had acquired a prestigious mansion in the Chinese capital; yet after rejoining his family in Beijing he confided to his journal: 'When I met my children and younger brothers I could not recognize them. Looking at them, the houses and the heated beds of the capital' seemed like 'a confused, hazy dream. The more I thought about it, the more I marvelled at what seemed like being born again' – another classic post-traumatic stress response. Neither wealth not victory brought him mental peace after the horrors he had seen.[9]

Keeping Score

Extreme swings of fortune led many seventeenth-century people to keep a minute and intimate record of everything that happened to them. Hugh Peter, a noted Protestant preacher in both England and New England, instructed his congregations that every day they should keep a journal and 'write down your sins on one side, and on the other side God's little mercies'. Nehemiah Wallington, a London craftsman who heard (and recorded) this exhortation, rejoiced that 'by God's mercy I practice it already', and from 1637 to 1654 he filled eight volumes with his nightly 'introspections' on public and private events. In 1660 Samuel Pepys began his celebrated diary as a spiritual journal and balance sheet, and at Whitsunday 1662, Isaac Newton (then a nineteen-year-old student at Cambridge) compiled a list of the forty-nine sins he could remember committing to that date (most of which involved disrespecting the Sabbath in some way, or beating people).[10] Devout Catholic contemporaries of Wallington, Pepys and Newton adopted similar strategies. In France, encouraged by Jansenist spiritual directors, some kept records of their good and bad deeds remarkably like those enjoined by Puritan preachers across the Channel; in Spain, Dr Gaspar Caldera de Heredia, a gentleman of Seville, kept a journal entitled *Political tariff* (subtitled 'A guide to life in our time'), full of reflections on where his own life had gone wrong because of 'the general corruption of morals that normally characterize a great empire'.[11]

In late Ming China, too, many worried individuals turned to introspection and self-criticism in the hope of averting disaster. Pre-printed ledgers of 'merit and demerit' (*gongguoge: gong*, meaning 'merits'; *guo* meaning 'faults') provided a calendar where their owners could record good and bad deeds, suggesting the appropriate score for each action and providing a space to enter a running tally. Thus someone who gave money to the poor, gaining (say) five merits, but also spread a slander about a neighbour, receiving (say) thirty demerits, would have to enter a score for that day of twenty-five demerits. In this way, Cynthia Brokaw has noted, each person's 'monthly balance helps him measure his moral progress, and at year's end his total score indicates whether he can expect good or bad fortune from the gods in the year ahead'. In the words of a prominent Ming intellectual: 'Every day know your errors; every day correct your faults.' Use of *gongguoge* proliferated in precisely those areas of China that experienced the greatest economic and social upheaval in the seventeenth century: Jiangnan, Fujian and Guangdong.[12]

The Psychoactive Revolution

David Courtwright has argued that the prevalence of melancholy in the seventeenth century caused the rapid spread of six substances that either stimulated or numbed the human senses, a phenomenon he called 'the psychoactive revolution'. Consumption of two (alcohol and opium) was already widespread, but the rest (coffee, tea, chocolate and tobacco) were new – and all seem to have been more potent then than they are today. Courtwright argued that consumption of these substances rose rapidly because they helped contemporaries to 'cope with lives lived on the verge of the unliveable' – in short, people 'who could use a smoke and a drink'.[13]

And drink they did. In 1632 an English envoy to King Christian IV, then aged fifty-five, primly remarked: 'Such is the life of that king: to drink all day and to lie with a whore every night.' Some years earlier, one of Christian's councillors recorded the court's drinking in his diary, grading the level of intoxication with one, two or three crosses: on one memorable night, he entered four crosses, followed by the prayer '*Libera nos domine*' ('God spare us'). Christian and his courtiers seem to have spent the equivalent of one month of each year dead drunk. Across the North Sea, the English consumed over 6 million barrels of beer every year – more than a pint a day for every man, woman and child – and Pepys visited well over 100 taverns in London in the 1660s where he drank both local ale and, thanks to the invention by English glassmakers of stronger bottle glass, more potent beers brewed in different parts of England. Pepys also relaxed in 'strong water houses' (another seventeenth-century invention) to consume Dutch gin, French brandy, Scots and Irish whisky and English rum. Protestants like

Christian IV and Samuel Pepys were not alone: some continental Catholics also often drank heavily in the seventeenth century. In the aphorism of Dr Caldera's *Political tariff*, 'wine has shipwrecked more boats than water'.[14]

The Mughal court in India also indulged in hard drinking. Alcohol abuse claimed one of Emperor Jahangir's uncles, both of his brothers and a nephew; and the future emperor was himself imprisoned at one point by his father in an effort to dry him out. Jahangir nevertheless celebrated the first New Year after his accession by decreeing that 'everyone could drink whatever intoxicants or exhilarants he wanted without prohibition or impediment', and his memoirs refer to prodigious drank at regular Thursday night parties at which he and his courtiers sampled 'here on Earth the future joys of paradise'. Jahangir also indulged in drugs – indeed he employed one steward for his wine and another for his opium.[15]

Visitors to other Muslim states in the seventeenth century also commented on the prodigious consumption of both opium and alcohol (a word derived from the Arabic *al-kohl*). Paul Rycaut, a long-time resident of the Ottoman empire, reported that it was normal for the sultan's subjects 'to be drunk, or intoxicate themselves with Aqua Vitae, opium, or any stupefying drugs'; and in Iran, although Shah Abbas II twice banned alcohol consumption and closed all taverns, in 1666, the year of the shah's death, his court consumed 145,000 litres of Shiraz wine. An English traveller in Iran also noted the widespread use of opium by soldiers about to fight, by messengers required to travel long distances at speed, and by ordinary people trying to stave off fatigue, boredom and stress. Some ingested it as pills, others as a cordial, and a few adventurous addicts inserted it as a suppository. A French visitor claimed that only one Persian in ten did not use opium, and that, thanks to the prevailing drug culture, it was not uncommon 'to come across individuals hallucinating in the streets – speaking with or laughing at angels'. It would be easy to write off such extreme descriptions as Western bias, but in the 1660s the Iranian mullah Qummi claimed that the Sufis 'eat hashish so as to quicken' the process of attaining God, and they dispensed it freely to their disciples. Qummi also argued that the drugs associated with the Sufi lifestyle attracted many who lacked a true vocation, so that 'when one asks them what sainthood means, they say it means being a bachelor and homeless' (apparently an early formulation of Dr Timothy Leary's call to 'turn on, tune in and drop out').[16]

The consumption of coffee, tea, chocolate and tobacco all increased dramatically during the seventeenth century. In the Muslim world, coffee had long served two purposes: a stimulus to sustain religious zealots, such as Sufis and dervishes, as they sang, danced and chanted; and a social lubricant consumed in coffee houses where men gathered to talk. For precisely those reasons, the Ottoman, Mughal and Safavid governments periodically persecuted and even executed consumers – although evidently without stemming demand (Chapters 6 and 13). Coffee houses spread to Christian Europe somewhat later – Venice from 1645; Oxford and London from 1652; Paris from 1672 – but then proliferated rapidly. London had over eighty by 1665, and in one of them Pepys 'did send for a cupp of tee (a China drink) of which I never had drank before' – but found that it cost twenty times more by the pound than coffee. Consumption of both beverages remained modest in northwest Europe until the custom of taking them with sugar gained ground after the 1680s, prompting the Jesuit poet Guillaume Massieu to devote dithyrambic verses to coffee, because 'thanks to this wonderful drink, you leap from your beds and hurry to do your duties, wishing the sun might rise earlier'. In particular, Massieu opined, before delivering a sermon preachers 'need to drink coffee because the liquid fortifies the weakened body, spreads a new vigour, a new life, in all actions, and gives the voice more strength'.[17]

Drinking chocolate, a practice which spread from Spain to France in the 1640s, and to England a decade later, was believed to have similar properties. In 1689, in Naples, Tommaso Strozzi, S.J., filled three volumes with Latin verses in its honour: *On the mind's beverage; or, the manufacture of chocolate*, described the origins of cacao in the Americas, the proper way to prepare a draught, and its therapeutic powers, which (he claimed) would cure diarrhoea (or constipation, depending on the dose), reduce fevers and stimulate the sexual appetite.[18]

The most popular and powerful stimulant of the seventeenth century was tobacco. Its reputation as a prophylactic against disease led Pepys 'to buy some roll-tobacco to smell and chew' during the London plague; but his most remarkable encounter with smoking came two years later when one of the horses pulling his coach became convulsed by shaking and seemed about to 'drop down dead' – until the coachman 'blew some tobacco [smoke or snuff] in his nose; upon which the horse sneezed, and by and by grows well and draws us the rest of our way'. When Pepys expressed surprise, the coachman observed phlegmatically: 'It's usual.' Seventeenth-century tobacco evidently contained far more powerful analgesic and psychotropic properties than any blend available today: apart from resurrecting half-dead horses, it eased pain, induced trances, suppressed hunger and staved off cold among humans. It also induced intoxication. In China in the 1620s the writer Yao Lü described how 'it can make one tipsy', and, three decades later, the German poet Jakob Balde entitled his satire on the abuse of tobacco: *Dry drunkenness (Die trückene Trünckenheit)*.[19]

The increase in tobacco consumption is staggering. England imported about thirty tons annually from its American colonies in the 1620s but over 5,000 tons by the end of the century. In China, Timothy Brook noted that 'Hundreds of poems on the subject of tobacco survive from the seventeenth and eighteenth centuries'. Even the elite of Qing China felt no shame about becoming 'tobacco's bondservant': indeed, claimed one writer, a true gentleman 'cannot do without it, however briefly, and to the end of their lives never tire of it' – no doubt because smoking offered solace to those who were depressed.[20] In the Ottoman empire, religious purists tried to eliminate tobacco because it was a stimulant, and therefore prohibited by the Qur'an, but (as elsewhere) it grew in popularity, as did the mixture of tobacco and opium known as *barsh*, widely smoked in coffee houses and marketplaces. In 1633 Sultan Murad outlawed the production, sale and consumption of tobacco, and had offenders summarily executed (Chapter 7), but still soldiers broke the taboo: according to Kâtib Çelebi, who served as a clerk on Murad's campaigns in Iraq, hungry and tired soldiers smoked in the latrines to avoid detection until the ban lapsed. In 1691, despite continuing clerical opposition, the government started to tax what they could not block.[21]

Peace Breaks Out

Each of these individual coping strategies to overcome the melancholy induced by the crisis of the mid-seventeenth century emerged independently of – often in spite of – the state. The same was true of several collective coping strategies. All of them depended on the restoration of peace and security, sometimes after decades of war.

Outside Europe, seventeenth-century peace normally broke out when one protagonist prevailed over its rivals by force of arms. Japan's Warring States Era ended when the armies of the Tokugawa Ieyasu defeated their remaining enemies in a series of battles around Osaka Castle, which they burnt down, and forced the opposing leaders to commit suicide; the Great Enterprise of the Qing ended when they had executed the last Ming pretender and compelled the last Ming loyalists to submit and shave their foreheads; the civil war among the four sons of the Mughal emperor Shah Jahan ended when Aurangzeb had defeated and executed virtually all his male relatives. In the Americas, wars between Europeans and native inhabitants, as well as among native inhabitants, normally ended with the death of most vanquished males and the enslavement of their families.

In Europe, by contrast, several savage and prolonged wars came to an end between 1648 and 1661 thanks to negotiated compromises born of exhaustion and popular anti-war sentiment. In the 1620s the French monk Emeric Crucé observed that 'for every two soldiers enriched by war you will find fifty who received only injuries or incurable diseases' – and such sentiments multiplied as war continued to ruin people and property. On each double-page spread of his manuscript 'Historical notes and meditations', which dealt with the campaigns of Charles I, Nehemiah Wallington wrote the heading: 'Of the bitternesse of warre and the miseries that war brings.'

One English pamphlet of 1642 warned that 'None knowes the misery of warre but those that see it'; another reminded readers of the 'manifold miseries' that civil war would bring, to judge 'by the examples of Germany, France, Ireland and other places'. Five years later, in Germany, a desperate entry in a peasant family's diary reads: 'We live like animals, eating bark and grass. No one could imagine that anything like this could happen to us. Many people say there is no God.'[22]

Anti-war sentiments also multiplied in art and literature. Many painters and engravers emphasized the arbitrary and catastrophic nature of war, especially after 1630, seeking not only to document but also to deter. The stark engravings of Jacques Callot's *Grandes misères de la guerre* (*Miseries of war*), each one with a caption condemning the barbarity of the soldiers, do not stand alone. In Germany, Hans Ulrich Franck produced a series of twenty-four engravings known as 'The theatre of war' with a title page on which curtains reveal a stage where an officer brandishes his weapon while trying to keep his balance upon Fortune's globe. The motto reads: 'O hark! Attend to the present; observe the future; and do not forget the end.' Heinrich Schütz, court musician of Electoral Saxony and perhaps the finest composer of his day, was compelled to arrange short choral pieces of religious music for only 'one, two, three or four voices with two violins, 'cello and organ', because the war left him neither choirs nor orchestras for anything grander. 'The times neither demand nor allow music on a big scale', he complained. 'It is now impossible to perform music on a large scale or with many choirs'. In the preface to his *Small spiritual concerti* of 1636, Schütz claimed that he no longer published so that his works could be performed, because few places boasted sufficient musicians, but so that he would not forget how to compose. Of his seventy surviving works, thirty were lamentations.[23]

Writers of both verse and prose cried out for peace with increased stridency as the wars continued. One of the moving hymns written by Lutheran pastor Paul Gerhardt began: 'Oh come on! Wake up, wake up you hard world: open your eyes before terror comes upon you in swift sudden surprise'. In Book II of *Paradise lost*, probably written in 1660, John Milton offered a powerful denunciation of the long-term as well as the short-term evils caused by war:

> Say they who counsel warr, we are decreed,
> Reserv'd and destin'd to Eternal woe;
> Whatever doing, what can we suffer more,
> What can we suffer worse?
>
> . . . And what peace can we return,
> But to our power hostility and hate,
> Untam'd reluctance, and revenge though slow,

Yet ever plotting how the Conqueror least
May reap his conquest, and may least rejoyce
In doing what we most in suffering feel?

In Germany, Hans Jakob Christoffel von Grimmelshausen published in 1668 *The adventures of a German simpleton* expressly to remind 'posterity about the terrible crimes that were committed in our German war'. It begins when the author, then a farm boy aged ten (and thus a 'simpleton' about the wider world), watches a group of 'iron men' (whom he later learns are cavalry troopers) torture the men and rape the women on his farm, and then take away all they can carry and burn the rest. The boy survives to narrate later outrages only because he feigns death when one of the 'iron men' shoots at him before riding away. The book proved an immediate bestseller.[24]

As Theodore K. Rabb observed in his landmark study, *The struggle for stability in early modern Europe*, by the mid-seventeenth century 'The shock of unbridled chaos, of a myriad of competing claims battling each other to extinction, made thoughtful men realize that these reckless assertions of private will were the surest route to disaster'.[25] Perhaps the most celebrated expression of this view appeared in Hobbes's *Leviathan* (1651), which enjoined obedience even to governments of dubious legitimacy, so long as they provided protection: 'If a monarch subdued by war render himself subject to the victor' (Hobbes hardly needed to remind his readers that Charles I had just done this), then 'his subjects are delivered from their former obligation, and become obliged to the victor'. A decade later, John Locke welcomed the Restoration of Charles II in part because 'I no sooner perceived myself in the world but I found myself in a storm, which hath lasted almost hitherto'. He therefore sought 'to endeavour [the] continuance' of the new regime 'by disposing men's minds to obedience to that government which brought with it quiet and settlement which our own giddy folly had put beyond the reach, not only of our contrivance, but hopes'.[26]

Similar views gained ground in other parts of Europe. In Catalonia, a French agent reported in 1644 that the local clergy 'say that since the obedience promised to France has no other foundation than the protection it promised, they are freed from their oath by the lack of such protection'; and five years later a Paris pamphleteer reminded his fellow subjects that monarchs 'owe us their protection just as we owe them our obedience'. Peder Schumacher, from Denmark, spent three years studying at Oxford, where he may have read the political works of Hobbes. He certainly witnessed the decline and fall of the English Republic in 1658–60 before he went to Paris and observed Louis XIV's measures to consolidate his personal power. On his return to Denmark, Schumacher applied these experiences when drafting the Royal Law of 1665 that conferred 'supreme power and authority' on the king.[27]

No More Wars

Other Europeans proposed mechanisms to limit and avoid future conflicts. In 1623 Emeric Crucé proposed the creation of a permanent international assembly of ambassadors to whom sovereigns would present their differences for adjudication, solemnly swearing to accept the majority decision (although, if they failed, other states might enforce a settlement through economic and even military sanctions). It would be a truly international body: although Crucé proposed Venice as its ideal location, he felt confident that 'navigation can overcome [the] difficulty' that delegates from Persia, China and the Americas would face in getting there. Seventy years later, William Penn published *An essay towards the present and future peace of Europe, by the establishment of an European Diet*, which suggested an international tribunal, very similar to Crucé's, to resolve international disputes peacefully, albeit only in Europe; and if any nation refuse to accept arbitration (or took up arms unilaterally) 'all other soveraignties, united as one strength, shall compel [its] submission'.[28]

These and other seventeenth-century writers who proposed universal solutions to what they perceived as a universal problem all wrote in wartime. Those who wrote just after a war, by contrast, tended to adopt a more limited and more pragmatic approach. In 1648 and again in 1652 the Estates of Hessen-Kassel submitted a formal petition asking the Supreme Court of the Holy Roman Empire to protect them with constitutional guarantees to safeguard them from 'being deprived of their liberties and from being led into one bloody massacre after another like innocent lambs'. When Hans Heberle, the long-suffering village shoemaker near Ulm, heard in 1667 that France had declared war on Spain again, he confided to his diary 'we beg God Almighty from the bottom of our hearts that he protect and shield our Germany and the whole Holy Roman Empire' from another incursion of 'foreign troops because we experienced and suffered enough in the Thirty Years' War'.[29] Perhaps the most striking expression of such 'never again' sentiments came in 1661, when Sweden's regency government discussed whether or not to continue the wars inherited from the late King Charles X. Gustav Bonde, the treasurer, reminded his colleagues that

> war makes the most serious inroads upon the resources of the crown, and entails impoverishment and ruin of the subject; and we have learned by experience that no war in past times has brought renown, profit or advantage to king and country, without also exacting large annual expenditure of our resources, and burdening the subject with

taxes and conscription ... It seems therefore necessary that we make up our minds to a period of peace, and lay aside all thought of war as long as peace is to be had.

Bonde conceded that recent wars had produced benefits for Sweden, but still concluded that his colleagues should not 'make plans upon the supposition that God will always arrange for a similar outcome to all our actions. We should remember rather that the issue of war is always uncertain, and is frequently most disastrous for those who believe themselves to have the justest cause.' Bonde's views prevailed – Sweden remained at peace for another decade – and although most states, including England and Russia as well as Sweden, would fight wars again, they did so far less frequently than in the preceding half-century. In England, for example, as soon as news arrived that Prince William of Orange had landed in southwest England with a powerful army in 1688, King James II and his ministers sent envoys to ask the prince to state his purpose, 'because they dread the thoughts of a war, for the bloudshed and all other evills that attend it alwaies'. This dread helps to explain why the Glorious Revolution involved no bloodshed in England.[30]

One of the 'evills' that attended the wars of the mid-seventeenth century was their duration. In Europe, even negotiations about ending hostilities could last for years – largely because neither side trusted the other. In summer 1643 the French and Spanish governments sent instructions to their plenipotentiaries at the Congress of Westphalia that virtually precluded a settlement: 'Experience has shown that [the other side] does not honour the treaties it makes' – indeed, both governments used identical words, and such incompatible views help to explain why it proved impossible to reconcile their outstanding differences at Westphalia (achieved only at the Peace of the Pyrenees sixteen years later).[31] Negotiations among the other participants in the Thirty Years' War dragged on for five years not only through lack of political trust but also through incompatible religious demands. The Protestant and Catholic German delegates finally agreed on a common formula in March 1648 – but only by agreeing to differ

in matters of religion, and in all other affairs, wherein the states of the empire cannot be considered as a single body, and when the Catholic states and the Lutheran states are divided into two parties; the difference shall be decided exclusively by amicable composition, without either side being coerced by a plurality of voices.

Many wept when the measure passed because it cleared the path to a final settlement hailed by later generations as 'the foremost bulwark of freedom and equality, built with so much blood'. For over a century, armed conflicts between Catholics and Protestants had provided an excuse for foreign states to intervene, and so civil wars often engendered foreign wars. Henceforth, no German ruler could exploit religious divisions to provoke or prolong a war – and neither could foreign rulers.[32]

This marked a major step towards restoring stability because, as Hobbes had forcefully stated in 1641: 'Experience teaches ... that the dispute for [precedence] betwene the spirituall and civill power has of late more than any other thing in the world bene the cause of civill warres in all places of Christendome'. Locke likewise complained that 'there hath been no design so wicked which hath not worn the visor of religion, nor rebellion which hath not' proclaimed 'a design either to supply the defects or correct the errors of religion'. He continued: 'almost all those tragical revolutions which have exercised Christendom these many years have turned upon this hinge':

> Hence have the cunning and malice of men taken occasion to pervert the doctrine of peace and charity into a perpetual foundation of war and contention, all those flames that have made such havoc and desolation in Europe, and have not been quenched but with the blood of so many millions, have been at first kindled with coals from the altar, and too much blown with the breath of those that attend the altar, who, forgetting their calling, which is to promote peace and meekness, have proved [to be] the trumpeters of strife and sounded a charge with a 'Curse ye Meros'.[33]

Although religious differences continued to affect European politics – some rulers expelled religious minorities, as the Catholic Louis XIV did with the French Huguenots while supporters hailed Protestant William III as Gideon and David when he went to war against Louis XIV – the views advanced by Locke steadily gained ground, and in 1689 he restated his condemnation of faith-based politics more forcefully in his Letter concerning toleration:

> Nobody, therefore, in fine, neither single persons nor churches, nay, nor even commonwealths, have any just title to invade the civil rights and worldly goods of each other upon pretence of Religion ... No peace and security ... can ever be established or preserved amongst men, so long as this opinion prevails: that dominion is founded upon grace, and that religion is to be propagated by force of arms.

According to the eminent religious historian Heinz Schilling, 'The end of confessional Europe, properly speaking, came around 1650' through 'the

internal dissolution of orthodoxy and through the state's deconfessionaliza-
tion of politics and society'; while, as his distinguished colleague Philip
Benedict has noted, 'religious conflicts spaced themselves out over time
with diminishing frequency'.[34]

Signing a peace treaty after a prolonged war often marked the beginning
of serious efforts to heal the wounds, end the fears and create a climate of
trust. To facilitate them, many authorities prohibited any discussion of the
recent contentious past. In 1648 the Peace of Westphalia forbade 'any person
to impugn in any place, in publick or in private, by preaching, teaching,
disputing, writing or consulting ... the present declaration or transaction;
or to render them doubtful'. In London, the group of scholars who would
later became the first fellows of the Royal Society had already made a similar
accommodation: from its 'first ground and foundation' in 1645, at their
weekly meetings members 'barred all discourses of divinity, state-affairs,
and of news ... confining ourselves to philosophical inquiries'.[35] Fifteen
years later, on his return to England, Charles II followed their wise example,
signing legislation that forbade the law courts to hear any suit arising from
things 'counselled, commanded, acted or done' during 'the late distractions'.
Furthermore, 'If any person or persons, within the space of three years next
ensuing, shall presume maliciously to call, or allege of, or object against any
other person or persons any name or names, or other words of reproach,
any way leading to revive the memory of the late differences, or the occasion thereof', the offender must pay a fine 'unto the party grieved'. The
impact of this policy of forced reconciliation may be seen in the careful
phrasing of petitions submitted by royalist veterans seeking compensation
for losses and injuries during the war: they studiously avoided terms like
'rebel' and 'rebellion'. Likewise, in his requests for government funding, the
scientist Samuel Hartlib always referred enigmatically to 'the troubles' when
explaining the delay in completing his project (Chapter 22). Thus did
England begin to heal itself after almost two decades of war.[36]

Such successes were remarkable because, as Sir John Plumb pointed out,
'by 1688 conspiracy and rebellion, treason and plot, were a part of the
history and experience of at least three generations of Englishmen'; and yet,
'by comparison, the political structure of eighteenth-century England
possesses adamantine strength and profound inertia'.[37] Plumb attributed the
growth of political stability to three structural changes in the wake of the
crisis of the mid-seventeenth century: a transition from population loss to
population growth; the resumption and diversification of economic activity;
and government decisions to invest both attention and resources in welfare
instead of (or as well as) warfare. His model works not only for England but
also for other states in the northern hemisphere where, taken together, the
same three changes ended the fatal synergy that both produced and
prolonged the crisis.

CHAPTER TWENTY-ONE

Warfare State or Welfare State?

The Phoenix Effect

Hans Jakob Christoffel von Grimmelshausen chose a striking image for the frontispiece of his 1668 novel, *The adventures of a German simpleton*: a phoenix, pointing to an open book that contains images of war. The verse below the engraving began:

Like a Phoenix I was born in the fire;

and continued with the question:

What often grieved me, and seldom brought luck?
What was it? I've written it down in this book.

Grimmelshausen's studied use of the past tense, and his image of a bird that rises from its own ashes, exuded confidence that the 'fire' was over; likewise the way his 'German simpleton' interspersed past events from 'our German war' with those from classical authors and the Bible implied that the age of wanton destruction of people and property had passed.[1]

To support Grimmelshausen's perception, historians can cite quantitative evidence of growth in toll records, tax returns, harvest yields and baptismal registers; but since almost all of them relate to individual communities, there is a risk that they may not be typical. By contrast the accounts of foreign travellers, though impressionistic, provide eyewitness observations that cover far larger areas. In 1663 Philip Skippon, the son of an English Civil War general and his German wife, toured Central Europe with his Cambridge tutor, Dr John Ray. The rapid postwar repopulation and reconstruction of Germany impressed both men. 'Since the instrument of peace [the Peace of Westphalia in 1648]', wrote Skippon, 'the people of this country have recruited themselves very much'; and both Skippon and Ray included numerous examples of urban renewal in their travelogues. At Mannheim, although 'the wars destroyed all the old town', they found 'the streets are designed to be uniform' with 'all the buildings alike in broad and straight streets'. Heidelberg, too, seemed 'populous, which is much considering the devastations made by the late wars in this country. The houses are most of timber, yet handsom and in good repair, which argues the inhabitants to be industrious and in a thriving condition'. Skippon found Vienna 'very populous' and its 'streets (except those at London) the most frequented we yet saw'; Ray considered it 'the most frequented and full of people that we have yet seen beyond the seas'.[2]

Urban regeneration was not universal, however. An anonymous Italian visitor to Germany in 1662 remarked that 'few towns have managed to recover from the damage sustained during the war, and many of the largest ones remain virtually depopulated'. The following year Ray considered Augsburg 'for the bigness, not very populous and [it] is, I believe, somewhat decayed, and short of what it hath been, both as to riches and multitude of inhabitants; which may be attributed to the losses and injuries it susteined in the late wars'. In 1671 the Paris physician Charles Patin still found at Höchst, just outside Frankfurt, many signs of 'the deplorable consequences of the war. This beautiful city' was now 'no more than a village'. Further east, between Jena and Leipzig, scene of several battles, Patin noted that the bodies of the 'nine or ten thousand men buried there still seems to provide manure for the fields' and predicted that 'all the surrounding towns will bear for a long time the sad traces of the war'. He concluded sadly: 'War spares nothing.'[3]

Most foreign visitors commented on the rapid recovery of Germany's rural economy. As he approached Munich in 1658, the itinerant Scots divinity student James Fraser admired the 'groves, gardins, parks, fertil cornfields and pretty brookes, fish ponds stored with carp and tinch and trout' all along his route – even though troops had repeatedly ravaged Bavaria during the second half of the Thirty Years' War. Further east, the following year Patrick Gordon (another itinerant Scot) praised the prosperity of Poland: they 'had abundance of all things – whereat I admired, considering how the countrey had been so often ruined by the enemyes, and no much better used by our owne soldiery'. He shrewdly observed that 'albeit many of their houses looke very waist lyke, as being destitute of hangings, standing beds, stooles or pictures ... yet there is superfluity of good, well-dressed [prepared] victualls and liquor'. A generation of war had evidently taught the Poles that it made sense to minimize possessions that could be burnt or taken as booty and maximize the portable necessities of life.[4]

In China, the nature of the surviving evidence complicates efforts to assess the devastation caused by the Ming–Qing transition. 'Official records of harvests, grain prices, rainfall, granary stocks, and the like were carefully kept, but no one counted or recorded the number of deaths from disasters'; so there are no estimates of human losses. Furthermore, virtually no Han Chinese artists included military themes in their repertory, while Manchu artists (who did) naturally eschewed scenes of the destruction wrought by their troops. By contrast, as Grace Fong has observed, 'Chinese poetry from its very beginning has given full expression to the tragedies of war': in the words of the poet Gui Zhuang, who lost one sister-in-law to soldiers and another to bandits, 'My grief has no outlet. I weep for her with poetry'.[5]

The mid-seventeenth-century crisis generated an outpouring of grief from men and women throughout China remarkable for its intensity and variety. One particularly moving poem came from the pen of the Shanghai poet and Ming minister Li Wen. He became prominent in the *Fu she* (Restoration Society) and remained in Beijing throughout 1644. When bandits murdered his father, Li offered his services to the Qing, and he served as Dorgon's secretary, drafting most of the regent's Chinese proclamations and public documents until 1646, when Li requested and received permission to visit his Jiangnan home. The desolation that he encountered on his journey south left him appalled, and he wrote a poem entitled 'On the road out: gazing in astonishment and seeing places which the bandits have destroyed':

> Stark are the thousand miles over which the bandits came,
> Seared are the many hills beneath the sun.
> Travellers gladly leave these hearths behind;
> Residents rely [only] on low walls.
> If half the Central Plain is like this,
> How can one escape wind and frost?[6]

Perhaps Li and other writers exaggerated. Poets continued for over a century to describe Yangzhou, whose sack in 1645 was the most brutal episode of the entire Ming–Qing transition (Chapter 5), as 'the weed-covered city' – yet the city's school reopened within a year and two years after that a dozen home-grown scholars took and passed the metropolitan examination, an unequivocal sign that some of the elite had survived the atrocities and accepted Qing rule. Within a decade Yangzhou boasted a new foundling home and several restored temples, and by 1664 more than one hundred gardens adorned the city's canals, where pleasure boats could be rented. When the Kangxi emperor visited two decades later, local merchants, officials, poets and artists of Yangzhou entertained him with opera, banquets and lantern shows. The weed-covered city had come a long way.

As in Europe, travel journals also illuminate the pace and extent of postwar recovery. Johan Nieuhof, secretary of a Dutch diplomatic mission to the Qing emperor in 1655–57, kept a detailed illustrated journal of what he saw as he travelled the 1,500 miles that separate Canton from Beijing. Although he included no images of a ruined city (perhaps because Qing officials forbade it), Nieuhof noted that one town after another south of the Yangzi had been 'totally ruined and sacked', starting with Canton, where the 'great fury' of the Tartars when they took it by storm left 'more than 80,000 people slain, not including those who later perished from hunger'. At Nanchang, a provincial capital, all but one of the 'rare buildings which had been formerly in this city were totally destroyed by the Tartars'; while Hukon (Jiangxi province), 'a very pleasant city full of industry before the destruction of China' now lay desolate.[7]

Nevertheless, like the poets who harped upon 'the weed-covered city', perhaps Nieuhof exaggerated. As in Germany, some urban centres 'totally ruined' by troops made a rapid recovery. Only six years after the sack, Nieuhof found Canton once move a thriving commercial centre. 'Although this city was lamentably laid waste', he wrote, 'it was in a few years after restored to its former lustre'. Moreover, Nieuhof noted that although 'It is a maxim among the Tartars, that such cities as revolt against them, and are subdued by force of arms', should be sacked, 'such as yield without any opposition, have no hurt done unto them'. Since almost every town north of the Yangzi had opened its gates to the advancing Qing, Nieuhof found bustling economic activity and abundant food almost everywhere, and the countryside alongside the Grand Canal 'so full of buildings as if it were all but one continued village'.[8]

Nevertheless, the mid-seventeenth-century crisis did not always produce a phoenix: sometimes it stifled rather than stimulated the survivors. To reprise three rural examples: in the Scottish borders, farmers never returned to cultivate the Pentland hills after global cooling and marauding troops ended the viability of their farms; in India, the cotton and cotton weavers of Gujarat never returned after famine and floods destroyed the market for their goods; and in China, sericulture vanished from the province of Shaanxi after the trauma of the Ming–Qing transition, despite a tradition that went back 2,000 years. The same was true of towns. For every Nanjing there was a Nanchang, for every Vienna there was a Höchst: places whose population would remain below – often far below – pre-war levels for the rest of the century and even beyond. Demographic recovery after the Global Crisis depended on a benign synergy of human and natural factors.

Be Fruitful and Multiply

Much of pre-industrial Europe saw a baby boom whenever wars ceased because (in the words of a French man of letters) some areas were 'so fertile that what war destroys in one year regenerates in two'; while others (in the words of his compatriot) 'resemble a fine fat bird: the more you pluck it, the more its feathers grow'. The later seventeenth century was no exception.[9] Abandoned farms on prime land normally recovered first and fastest, their natural fertility temporarily enhanced by enforced fallowing. As soon as it was safe to do so, unmarried or widowed survivors took advantage of the vacant lands and houses left by each catastrophe (whether human or natural in origin) to marry, move in and start a family. In Italy's Aosta Valley, war and plague killed 600 inhabitants of one village in 1630–31, leaving about

600 survivors; yet whereas the 1620s saw only five marriages a year, fourteen took place in 1630 and thirty-eight in 1631; and whereas the annual average number of baptisms in the 1620s had been twenty-four, the year 1630 saw forty-two, and each year of the next decade saw twenty-five. A generation half the size of its predecessor had thus produced more babies.

The return of security produced not only increases in the native population of war-ravaged villages on fertile land but also a flood of immigrants. The Thirty Years' War reduced by 80 per cent the population of the lands ruled by the Benedictine monastery of Ottobeuren in south Germany (a fief consisting of a market town, eighteen villages and scores of scattered hamlets), but as soon as peace returned thousands of travellers arrived 'from all over Germany and indeed from all over Europe'. Many of these homeless refugees settled in the half-empty villages of the Ottobeuren lands: almost half of all the marriages in the fief in the 1650s involved at least one immigrant.[10]

Urban populations could also rebound swiftly after a catastrophe. In Italy, marriages in the port-city of Genoa during each of the three years after the plague of 1656–57 were twice as numerous as in the three years before, because the catastrophe had left a glut of vacant houses and jobs: people could afford to marry younger. Elsewhere, in-migration proved critical. In the Ottoman empire, the population of the port-city of Izmir in Anatolia rose from perhaps 3,000 in 1603 to 40,000 in 1648 and over 100,000 in 1700, largely through migration – some from neighbouring villages, others from the Balkans (including the parents of Shabbatai Zvi), and others still from further afield, such as refugees from the massacres in the Polish-Lithuanian Commonwealth (including Sarah, the wife of Shabbatai). In Spain, the population of the port-city of Cadiz tripled from 7,000 in 1600 to 22,000 in 1650, and almost doubled again to 41,000 in 1700, thanks to strong in-migration from the surrounding villages, a strict quarantine that blocked the plague epidemic that devastated its economic rival Seville in 1649–50 and the city's growing trade with the Americas (largely at the expense of Seville). Port-cities seem to have recovered most rapidly: most others would not regain their pre-crisis populations until the nineteenth century.

Accommodating migrants sometimes required compromises. Venice lost almost 50,000 of its citizens during the plague of 1630–31, over half of them merchants and artisans belonging to the city's guilds. Previously, all guilds had admitted only native sons, but for three years after the plague they welcomed qualified immigrants in order to restore their numerical strength. In Brandenburg, Elector Frederick William issued a decree 'by virtue of princely power and sovereign authority' that 'all persons prepared to rebuild a devastated and abandoned peasant farm must without exception be granted six years free of taxes, rents, and military quartering' – a measure that provoked widespread protests from landowners who resented the unilateral declaration of a rent holiday and feared that foreign colonists would simply come, 'exhaust the soil' and then 'disappear into the dust'. Naturally the Elector ignored the protests. Most remarkable of all, the city-state of Lucca sought to replace its plague dead after 1631 by granting safe conduct and asylum to all outlaws from other states (provided they had not committed treason or some other heinous crime). Certificates of asylum issued over the next century expressly to restore the population to its pre-plague level fill two fat registers in the city archives.[11]

In many countries, returning veterans also helped many communities to recover after a war. Almost 5,000 infantry and 3,000 cavalry returned to Finland after the end of the Thirty Years' War, their pockets filled with the money provided by German taxpayers to pay their wage arrears; and even more veterans went home to Sweden, adding a substantial cohort of wealthy bachelors to communities where, for many years, women had far outnumbered men (Chapter 8). In England, the king's treasury paid out nearly £800,000 in wage arrears to the thousands of soldiers and sailors of the republic demobilized after the Restoration in 1660, and they returned to civilian life far richer than they had left it. In France, Spain, the Low Countries and Switzerland, too, the return of demobilized veterans no doubt injected wealth into countless communities

Nevertheless, steady and significant population growth required not only an increase in marriages, births and in-migration but also a reduction in deaths and a revival of prosperity. Although this combination prevailed in Japan and Mughal India after the 1650s, the population of most West European states only recovered after the famine of 1660–62. In China the Kangxi depression, caused by continuing war against Ming loyalists and the embargo on coastal trade, lasted until the 1680s. Even after that, the urban graveyard effect (Chapter 3) continued to take its toll. In northern Italy, famine, plague and war ravaged the duchy of Mantua in 1630–31, affecting city and countryside alike: three decades later, some English tourists declared Mantua to be 'a great city but not answerably populous, having not yet recovered it self of the losses it sustained when it was miserably sackt by the Emperor Ferdinand II's army in the year 1630'. By contrast, they noted that the surrounding country 'is very rich'. The tourists were correct: a census in 1676 revealed that the rural areas of the duchy had regained over 90 per cent of their pre-war population whereas the capital languished below 70 per cent. Similar asymmetries could be found elsewhere in Europe and they altered the balance of power between town and country. In 1600 the magistrates of the town of Memmingen in Germany contemptuously dismissed neighbouring peasants as bumpkins 'with neither enough cattle, grain nor other victuals or goods to justify the establishment of a single annual market'; but a century later, their descendants had to concede that the town and the surrounding villages 'are bound together, such that one always has need of the other, and in such a manner that each party must always uphold and maintain the motto "the one hand washes the other".[12]

The lack of parish registers makes it more difficult to establish whether a baby boom took place outside Europe in the seventeenth-century crisis, but in at least two areas this outcome seems unlikely. In West Central Africa and the lands around the Bight of Biafra, where between one-third and one-half of all slaves captured and deported were female, it is hard to see how communities could have maintained a stable population, let alone increased. Likewise, in those parts of Asia where female infanticide was the principal means of easing population pressure, births would have remained low for at least one generation after the crisis; so the notable population increase in early Qing China probably owed most to government measures to sponsor migration from poorer to richer lands. The new dynasty granted drifters (*liumin*) who had settled on abandoned lands a permanent title to them, regardless of the claims of others; lent seed corn to poor cultivators; created village pawnshops; and offered incentives (such as travel costs, start-up loans, freehold land, and farming stock and equipment) to any farmer prepared to migrate to the ravaged lands in the north and west. The number who took advantage of these schemes during the late seventeenth and eight-eenth centuries 'easily surpassed 10,000,000' and the total cultivated land in the empire, which had fallen to 67 million acres in 1645, climbed back to 90 million acres in 1661 and to 100 million acres in 1685 – although since over 191 million acres of Chinese soil had been under cultivation in 1600, the heavy footprint of the Global Crisis remained perceptible well into the eighteenth century.[13]

A Second Agricultural Revolution

A baby boom normally stimulates the agricultural sector, since every new mouth needs to be fed, encouraging farmers to improve the yield of tradi-tional crops and introduce new ones, and to invest in irrigation and drainage works. In Mughal India the versatile peasants of the Ganges Valley, who already cultivated over twenty different crops in the early seventeenth century, added maize and tobacco, two New World miracle crops, to their repertory; and the farmers of West Central Africa began to plant not only maize but also manioc (originally a Brazilian crop) as a safeguard against the failure of millet and sorghum harvests during drought. Manioc proved particularly valuable in wartime, since raiders might neglect the tubers below the ground. In the western Netherlands, and in eastern England, the disastrous harvests of the mid-seventeenth century encouraged the system-atic rotation of cereal with root vegetables (such as carrots and turnips), and the sowing of clover and other crops rich in nitrogen. In 1650 Samuel Hartlib, a refugee from the Thirty Years' War who settled in England, published *A discours of husbandrie used in Brabant and Flaunders*, later expanded to include information from New England and Ireland, showing how new methods of tillage and rotation made otherwise unproductive soils profitable; how to identify and to sow the strains best suited to each locality; and how to use chemical fertilizers to optimal effect. Some English historians have hailed these practical and theoretical efforts, which gathered momentum in the 1640s and 1650s, as a second agricultural revolution.[14]

Chinese farmers also began to cultivate maize, peanuts and sweet potatoes – all crops recently imported from the Americas that thrived in marginal soils, resisted both droughts and locusts, did not require transplanting like rice, and produced twice as much as other dry-land crops with far less labour input. According to a gazetteer from Jiangxi province, 'in general, maize is grown on the sunny side of the hills, sweet potatoes on the shady side', and maize 'provides half a year's food for the mountain dwellers'. Sucheta Mazumdar has hailed these improvements during the later seventeenth century as China's 'second agricultural revolution', one 'predicated on the maximum utilization of all crops and the development of complementary patterns of crop selection'. It allowed the average intake of adult males to rise above the vital threshold of 2,000 calories per day, and supplied more protein and vitamins – thereby both improving health and reducing hunger (Chapter 1).[15]

The Consumer Revolution

Domestic demand normally drove pre-industrial economies. Although the sight of stately East Indiamen returning to Europe's Atlantic ports filled with exotic goods from Asia captured the imagination of contemporaries, their cargoes amounted to (at most) 2 per cent of Europe's gross domestic product. Put another way, all the goods transported by sea from Asia to Europe in an entire year would today fit inside a single container ship. The trade goods exchanged with Europe were even less significant as a percentage of Asia's gross domestic product: as Jack Goldstone put it, 'The total volume of European trade was never more than just over 1 percent of China's economy'.[16] Even the intercontinental trade between Europe and the Americas reached significant volume and value only in the 1690s.

By contrast, the need to feed, clothe and house the baby boomers and their families after the crisis of the mid-seventeenth century stimulated every sector of the economy. In a bold study that focused on the production and consumption of households in Northwest Europe (above all the Dutch Republic) and North America, economic historian Jan de Vries perceived a Western version of Japan's Industrious Revolution, starting around 1650. According to de Vries:

A growing number of households acted to re-allocate their productive resources (which are chiefly the time of their members) in ways that increased both the supply of market-oriented, money-earning activities and the demand for goods offered in the marketplace. Increased production specialization in the household gives access to augmented consumption choices in the marketplace.

In short, the value as well as the volume of goods produced and consumed in the later seventeenth century increased because so many baby boomers wanted more than the bare necessities of life. To take the example of housing, after 1650 'brick construction replaced wood and lime; functional spaces became better defined, as drawing rooms and dining rooms appeared in middle class homes and distinct bed chambers came to be defined ... and these interior spaces came to be filled with more, and more specialized, furniture'. Similar diversification characterized the consumption, and therefore the production, of apparel, food and beverages. Craig Muldrew has uncovered a similar process in England: 'Before 1650, when food prices were rising', most labouring families had to work 'harder to merely maintain their living standards to avoid a fall towards poverty' but thereafter 'increasing food production and falling food prices' made it possible for labourers to earn more, and so to increase spending on their households. The mean total value of English labourers' households rose from £21 in 1600–49 to £33 in 1650–99 (far more than in the preceding and succeeding half-century). Moreover, since technology could not provide the water, illumination, heat and hygiene sought by wealthier consumers, the heightened demand for many of these goods also heightened the demand for domestic servants to operate them. By 1700, four-fifths of the English labouring population had been or were in service. 'Thus for most families children did not need to be supported after fourteen.'[17]

A burgeoning demand for comfort items also characterized late Ming China, but the trauma of the Ming–Qing transition brought it to an abrupt end in most areas. Suzhou, for example, with a population of 500,000, was in the early seventeenth century 'the most populous, and most prosperous, non-capital city on the face of the Earth'; but famines, plagues and protests under the late Ming undermined its prosperity, and its defiance of the Qing led to a sack that left only 'broken tiles and walls' within the city, while 'outside the city, four or five out of every ten houses had been destroyed'. Nevertheless, by 1676 four merchant guilds (including the printers and the drug merchants) as well as groups of foreign merchants had founded or reconstructed lodges in the city, and eleven more followed suit over the next two decades. Each lodge represented a major investment, and their multiplication reflected construction both of merchant ships in the city (a new development) as well as the quality goods to fill them. As he passed through

Nanjing in 1656, Johan Nieuhof noted that 'the shops of the chief citizens and merchants are filled with all manner of rich Chinese wares, as cottons, silk stuffs, porcelain, pearls, diamonds'; and on the Yangzi river 'the number of all manner of vessels is so great, that it seems as if all the shipping of the world were docked there'. The Qing compelled 47,000 men to carry out repairs to the Grand Canal, the longest artificial waterway in the world, which allowed the relatively rapid transport of goods and passengers between the Yangzi and Beijing, a distance of 1,000 miles.[18]

The recovery of another key element of infrastructure impressed foreign visitors: roads. Louis Le Comte, S.J., who travelled all over China in the 1680s, was deeply impressed by

> the care taken to make public highways accessible. They are about eighty feet wide, made of soil that is light and soon dries after rain. In some provinces you find (as with our [French] bridges) causeways for travellers on foot on the right and left hand, protected on both sides by an endless row of great trees, often enclosed by a wall eight or ten foot high on each side to keep passengers out of the fields.

Le Comte also noted with approval the post-stations and guardhouses built at regular intervals along every road, and the numerous arches which, with few exceptions, displayed

> a board on which characters that one can read a hundred paces away state the distance from the town one has just left and to the town to which the road leads. So guides are not necessary, and you always know where you are going, where you have been, how far you have travelled and how far you still have to go.[19]

Some of those highways remained in use in the twentieth century.

Communications infrastructure in Europe also improved dramatically after the troubles of the mid-century. When James Fraser arrived in London in 1657 he was amazed to find a postal network operated by 'a postmaster general and about 200 sub-postmasters', who conveyed letters at '120 miles in 24 houres'; and also 'hakeney [hackney] and stage coaches, wherin yow have yowr seat for a shilling the six miles, bravely sheltered from fowle ways and fowle wether'.[20] The following year, a coach managed to travel up the Great North Road between London and York (apparently the first wheeled transport to do so), covering 200 miles in four days; and five years later, an Act of Parliament gave magistrates along the Great North Road powers to erect tollgates and apply the proceeds to highway improvement. The English road network improved more rapidly after 1696 in response to a series of Turnpike Acts, which allowed magistrates to set up barriers where they could collect tolls from travellers, and use the money to appoint surveyors to improve highways. Almost forty turnpike trusts existed in England by 1720.

The later seventeenth century also saw important innovations in water transport throughout Western Europe. By 1665 the Dutch Republic boasted over 450 miles of inter-city transport canals, each of them with a towpath that permitted horse-pulled barges to convey passengers and freight quickly and cheaply between the principal cities of the coastal provinces, according to a regular schedule. Before long, fastidious Amsterdammers even sent their dirty laundry each day by barge to the cheaper and cleaner facilities in Haarlem, 15 miles away. In France, in 1667 a group of investors directed the efforts of up to 12,000 labourers on the 160 miles of the Royal Canal of Languedoc, linking the Mediterranean and the Atlantic, which involved creating an artificial lake near the summit, several aqueducts, and Europe's first tunnel constructed expressly for a canal. The first barges and almost 4,000 passengers paid to use the canal in 1682, its first year of operation, and the last barge sailed in 1989 when drought made it impassable.

Thanks in part to these improvements in transportation a consumer revolution took place in northwest Europe. Early in the eighteenth century, Daniel Defoe included in his *Compleat English tradesman* a chapter entitled 'Of the luxury and extravagancies of the age becoming virtues in commerce, and how they propagate the trade and manufactures of the whole nation', which argued that reintroduction of sumptuary laws would 'ruin thousands of families' because

> if a due calculation were made of all the several trades besides labouring, manufacturing, and handicraft business, which are supported in this nation merely by the sins of the people ... [namely by] the numberless gayeties of dress; as also by the gluttony, the drunkenness, and other exorbitances of life, it might remain a question, whether the necessary or the unnecessary were the greatest blessing to trade; and whether reforming our vices would not ruin the nation.[21]

A similar debate took place in China. The Confucian sage Mencius had pointed out the need for 'an intercommunication of the production of labour, and an interchange of men's services, so that one from his overplus may supply the deficiency of another'. A late Ming writer quoted this passage with approval, noting that although the frugality of a family 'can, perhaps, save it from becoming poor', the same was not true of larger communities because 'if a place is accustomed to extravagance, then the people there will find it easy to make a living, and if a place is accustomed to frugality, then the people there will find it difficult to make a living'. More specifically, when the rich 'are extravagant in meat and rice, farmers and cooks will share the profit; when they are extravagant in silk textiles, weavers and dealers will share the profit'. After the austerity of the Ming–Qing transition, China rediscovered the wisdom of Mencius: 'Spendthrifts may spend a million taels a day on foolish extravagances', wrote a government official around 1680, 'but the expended treasure circulates among the people through the hands of those who obtain it', whereas 'a miser who saves up a sizable amount of money will impoverish numerous families around him'.[22]

'Seeing Like a State'

In his paean of praise to conspicuous consumption, Defoe argued that 'exorbitances' of attire, food and drink had become 'so necessary to the support of the very government, as well as of commerce, that without the revenue now raised by them [through the excise], we can hardly see how the publick affairs could be supported'.[23] This was an extraordinary statement, considering that just one generation earlier excise taxes had precipitated major revolts in the Spanish monarchy, France and the Dutch Republic, and that even in England, enraged crowds had burned down excise offices and attacked excise collectors (Chapters 9, 10, 12, 14 and 17); and in fact Defoe exaggerated. The economic growth that he praised reflected a combination of private and public investment. Although even the costliest investments, such as improvements to roads and waterways, were effected by groups of private individuals known as undertakers, the state nevertheless played a crucial role by providing security and stability. In England, thanks to the new climate of business confidence created by the Glorious Revolution, individuals invested as much money in improving roads and rivers in the fifteen-year period 1695–1709 as in the eighty-five-year period 1604–88. Moreover, their investment substantially reduced transport costs, which benefited investors and producers alike, and thus stimulated further investment.

Increasing consumption made it possible for many governments to shift their fiscal base from direct to indirect taxes. In England, total revenues in 1650–59 exceeded £18 million, of which over one-third came from the assessment (a tax on landed property), whereas total revenues in 1680–89 fell below £15 million, of which only £530,000 came from the assessment. In France, although the total revenues of Louis XIV rose from 84 million livres in 1661 to 114 million in 1688, the yield of the taille (the principal tax on landed property) fell from 42 to 32 million livres. Statistics for other European states collected in the European State Finance Database (www. esfdb.org) show similar trends. These shifts – stable or rising state income, derived more from taxes on consumption than on production – benefited both richer taxpayers, because they spent proportionally less of their income on consumer goods, and the state, which gained a greater disposable income.

In his study, *Seeing like a state: how certain schemes to improve the human condition have failed*, James C. Scott examined the process by which modern governments used their tax revenues to impose order (something he called 'legibility') upon activities they wished to regulate. The process, he stressed, normally came at a cost because 'certain forms of knowledge and control require a narrowing of vision'.

> The great advantage of such tunnel vision is that it brings into sharp focus certain limited aspects of an otherwise far more complex and unwieldy reality. This very simplification, in turn, makes the phenomenon at the center of the field of vision more legible and hence more susceptible to careful measurement and calculation. Combined with similar observations, an overall, aggregate, synoptic view of a selective reality is achieved, making possible a high degree of schematic knowledge, control and manipulation.[24]

Scott traced both the growth of legibility and the narrowing of vision back to the mid-seventeenth century.

This chronology should not cause surprise. The trauma of the Global Crisis naturally made states want to 'know, control, and manipulate' both nature and society more effectively, in order to avoid any repetition. Japan led the way: the cadastral surveys carried out by the Tokugawa shoguns (Chapter 16), which enhanced the legibility of both agriculture and demography, proved crucial in enabling them to mitigate the effects of the Little Ice Age. In China, mindful of the role played by harvest failure in the fall of the previous dynasty, the Kangxi emperor regarded the systematic collection and collation of accurate information about the weather as crucial to political stability. Although provincial officials of the Ming had normally included a rain report in their regular accounts on local conditions, the central government paid little attention; the Kangxi emperor, by contrast, not only read the official weather reports but also requested reports from his bannermen and bondservants on droughts, floods and harvest projections (especially for Jiangnan, which supplied most of the capital's grain).

Kangxi's example inspired at least one minister in Europe. Sébastien le Prestre, marquis de Vauban (1633–1707), acknowledged that his ambitious plan to replace the complex tax structure of Louis XIV's France with a poll tax was 'more or less what they do in China', citing printed descriptions by the eyewitnesses Martino Martini (of which he owned a copy) and Louis Le Comte (Vauban attended the meeting of the Académie des Sciences that discussed Le Comte's observations). Vauban bombarded the king with more ambitious statistics, mostly derived from personal observation, often presenting his findings in visual form. After overseeing in person the construction or reconstruction of France's frontier fortresses, Vauban created a scale model of each so that the king could visualize the defences of his kingdom without leaving his palace – an early form of Google Earth. By 1700 almost 150 models existed, mostly on the same large scale of 1:600, with miniature walls, churches, houses and trees recreated in wood, silk, paper and sand. During a siege, daily updates from his field commanders were reproduced on the model, encouraging the king to micromanage operations via a stream of detailed instructions. It would be hard to find a better example of both the advantages and the perils of 'seeing like a state'.

Vauban also undertook speculative calculations (which he called *supputations*), based on his reading and travels, on such topics as how to improve the yield of crops (to feed the king's armies better), how to make trees grow taller and straighter (to provide better masts for the king's warships) and even how to get peasants to breed hogs so that in ten generations they would number over 3 million (to reduce the dependence of both peasants and soldiers on bread). He also devoted much attention to France's overseas colonies, especially to Canada, presenting the king with a memorandum in 1699 that suggested that the dispatch of more settlers every year would have a dramatic impact, because 'instead of the 13,000 to 14,000 souls that currently live in Canada, thirty years from now – that is, in about 1730 – there could be 100,000'. Moreover, if these men and women married and had four children each, Vauban predicted, the colony's population would double every generation so that 'in the year 2000 there might be 51 million people'.[25]

Such calculations were of course totally unrealistic – the product of the 'narrowing of vision' identified by James Scott. The idea that peasant farmers would keep their hogs alive so they could breed, while they themselves starved, betrays remarkable ignorance of French rural life. Vauban's addiction to geometry and calculations led him to overlook other critical factors. For example, although he cited 'the mortality of the year 1693 and the scarcity of food' as one reason for the decline in the number of Louis XIV's subjects, he 'considered climatic conditions as secondary', far less important than war and taxation. He even foolishly assured the king that 'dearth exists in the mind, not in reality'.[26]

The statistical approach pioneered by Vauban and other ministers of Louis XIV nevertheless found many admirers abroad. In a provocative article on the spread of 'political arithmetic' in seventeenth-century Europe, Jacob Soll has observed that 'European states shared not only complex economic, military, political, social and spiritual crises but also comparable responses to them' – and examples are not hard to find. Sir John Plumb underlined the envy of many of Charles II's English ministers for what they perceived as the 'systematic efficiency' of their French counterparts. Many royal officials became fellows of the Royal Society because they 'believed that the practical problems of life were best approached through knowledge' – although they stopped short of calculating the reproductive potential of hogs.[27]

The Containment of Disease

James Scott drew attention to another consequence of 'seeing like a state'. Starting in the 1650s, he observed, European statesmen seemed

> devoted to rationalizing and standardizing what was a social hiero-glyph into a legible and administratively more convenient format. The social simplifications thus introduced not only permitted a more finely tuned system of taxation and conscription but also greatly enhanced state capacity. They made possible quite discriminating interventions of every kind, such as public-health measures, political surveillance, and relief of the poor.[28]

Two public-health measures of the seventeenth century stand out, because they involved containment of the most lethal diseases of the age: bubonic plague and smallpox.

We now know that bubonic plague is a bacillus spread by infected fleas which moved first from rats (their normal 'hosts') to humans, and then among humans; but no one at the time (or, indeed, until the 1890s) knew that. Nevertheless, starting in Italy, an effective containment strategy emerged: creating a permanent health board, staffed by both doctors and magistrates, to prevent the movement of people and property in times of plague. At least in part thanks to the effective quarantine measures enforced by the health boards, significant areas of Europe became plague-free: Sicily after 1625, Scotland after 1649, Catalonia after 1651, and so on. In some areas, opposition limited or undermined the efficacy of quarantine (on the grounds that boarding up houses that contained suspected victims of the disease proved expensive and unpopular); and since the lethality of the disease remained high, wherever epidemics took hold they continued to cause high casualties. Plague afflicted the port-city of Marseilles in 1720–22, killing perhaps 100,000 people, because local merchants insisted on unload-ing the cargo of an infected vessel. 'In the end', as Kira Newman observed, 'opposition to plague policy died out only when plague itself did.'[29]

Elsewhere, governments had greater success in containing smallpox. The change began in northern China. Smallpox decimated the Manchus after they invaded the Ming empire, killing even the emperor in 1661, and his son and successor the Kangxi emperor, a smallpox survivor, took a keen interest in the disease. He discovered that Chinese doctors had developed a treatment called variolation: the deliberate administration of a mild form of the virus that most people managed to survive, thereby gaining life-long immunity because smallpox only strikes each victim once. Kangxi first tested the procedure 'on one or two people' (read: slaves) in the 1670s, and since it seemed to work, went on to inoculate his apprehensive family; then all the Banner troops on whom the safety of the Qing state depended; and finally more than half the population of Beijing. 'This is an extremely important thing, of which I am very proud', the emperor wrote in a testament for his children, because variolation 'has saved the health of millions of men' – and indeed, largely as a result of variolation, child mortality in China fell from 40 per cent to 10 per cent in the course of the eighteenth century. By any standard, this was a remarkable (and remarkably successful) public-health project.[30]

More successes in the field of public health could have occurred had medical practitioners in parts of seventeenth-century Europe (especially Britain) not insisted that physical illnesses arose from general and not specific causes, and deprecated any drug that did not produce some sort of excretion. This ruled out the bark of the cinchona tree (*Cinchona officinalis*), which contained quinine and so provided some protection against malaria. Jesuit missionaries in South America observed its therapeutic properties and in the 1640s began to send regular shipments to Rome where, after extensive tests, doctors used it to stabilize feverish malaria patients. Although some victims of the disease refused 'Jesuits' powder' – including Oliver Cromwell, whose death of malarial fever in 1658 doomed Britain's republican experiment – but shortly afterwards, the new treatment gained adherents. Robert Tabor carried out trials among malaria suffers in the Cambridge area until 1672, when he published his *Rational account of the cause and cure of agues*. His remedy cured Charles II, the queen regent of Spain, and the heir to the French throne. It also made him rich and earned him a knighthood. Such high-profile successes and such rewards encouraged seventeenth-century medical practitioners, both licensed and amateur, to examine and test new remedies, and in doing so they developed a sound methodology for experimental pharmacology. They also began to shift medical attention from applying remedies to individual patients according to their perceived 'humours' towards finding an efficient remedy for individual diseases.

Nourishing the People

After the 1650s, many governments also intervened to prevent their subjects from starvation in times of dearth. The Qing, for example, resuscitated the traditional network of state-run granaries in China 'with a consistency, intensity, and degree of centralization unknown in previous eras'.[31] Each county boasted a public reserve known as the 'ever-normal granary' (*changpingcang*), supplemented by community granaries (*shecang*) and charity granaries (*yicang*). Traditionally, each of them bought grain after harvest, when it was cheap, and sold it below the market price during late winter and spring to stabilize prices in good years and to prevent famine when necessary; but since the wars of the Late Ming left virtually no resources to devote to welfare, by the 1640s almost all granaries lay in disrepair.

In 1654 the Shunzhi emperor ordered officials to re-establish both county and community granaries with sufficient stocks to cope with any future famine relief; but it proved a slow process. After the drought-induced famine of 1691–92, the Kangxi issued a stream of legislation for more efficient storage, with dramatic results: whereas the target reserve for each province under the late Ming was 1,500 metric tons of rice, by the early eighteenth century at least twelve of China's provinces held a reserve of more than 50,000 tons each, while the state granaries in Yunnan, in the remote southwest, held 200,000 tons, and those of Gansu in the equally remote northwest held 272,000 tons (1 million bushels). According to Louis Le Comte, even though more than enough rice from Jiangnan arrived every year via the Grand Canal to feed the imperial household, 'they are so concerned about running short, that the granaries of Beijing always stock enough rice for three or four years ahead'.[32]

The Ottoman empire also maintained large granaries, especially in Istanbul and in Egypt, whose grain surplus fed not only the capital but also the pilgrims who flocked to the Holy Cities of Mecca and Medina. When drought in the African interior in 1694 meant that the Nile scarcely rose, leading to the worst famine of the century in Egypt, the sultan not only reduced or deferred taxes from affected areas (citing 'poor irrigation and drought') but also 'created a rationing system in which vouchers were distributed to peasants allowing them to claim an amount of food for themselves and their families. The vouchers could be bought and sold as currency'.[33]

In Europe, only the cities of Southern Europe managed to create permanent public granaries – although none of them could compare in size with those of Qing China. The Madrid granary normally handled 20,000 tons of wheat each year, but during the famine of 1630–31 it distributed over 40,000 tons. The grain silos of Naples could hold up to 13,000 tons, and although a visitor reported that they 'always store enough to provide the city seven years', he suspected 'there might be enough to supply for two or three years', but not more. The silos certainly did not suffice in 1647–48, a famine year, when the city revolted and troops loyal to Philip IV cut off its normal sources of supply, forcing the rebel regime to decree that bread would only be issued '*per cartella*' (that is, in exchange for a ration card), and that soldiers would be fed before any civilians.[34]

The states of Northern Europe lagged far behind. Proposals by James I in 1619 to build public granaries in England perished through the greed of grain producers who wanted to obtain the highest price possible for their crops and opposed any measure that might affect their profits. Six years later, similarly selfish motives defeated the efforts of his son, Charles I, to persuade the Scottish Parliament to

> mak, choise and designe suche placeis as thay sall think most fitte to be publict granaries whairin thair may be storehouses for preserving of all sortis of victuall and suche store of provisioun in the saidis store-houses as may prevent the extremitie of famine when bad yeiris sall happin, of the necessitie whairof we haif latelie had the experience.[35]

In the 1620s the Swedish government also called in vain on each parish to establish a granary, repeating the call more insistently in 1642 (after a run of disastrous harvests), and again in the 1690s (after more famine years); but on each occasion it failed. In France, Nicholas La Reynie, lieutenant general of police in Paris, also failed to overcome resistance to creating a municipal granary – but he eventually found another way to nourish the inhabitants of the French capital. Every winter he required the clergy to report the exact number of people in their parish who needed relief, and he then arranged the distribution of a wide variety of items by his troops. La Reynie did not intend to see bread riots in the capital on his watch.

Governments elsewhere also intervened to provide other forms of welfare to mitigate the harsh conditions of the seventeenth century. In Sweden, in 1633 the government established an orphanage in Stockholm, largely to accommodate the children of soldiers and sailors conscripted to fight in the continental war, because their mothers could not feed them; and in 1646 it created at Vadstena the earliest known home to care for injured and aged veterans. In Iceland, in 1651 the Danish crown established a hospital in each of the four quarters of the island, donated crown land for their construction, and introduced a special tax to support them, to be paid by everyone. The legal obligation was critical because, in the words of an English pamphlet of 1601: 'In this obdurate age of ours, neither godly perswasions of the pastors, or pitifull exclamations of the poore, can moove any to mercie unless there were a lawe made to compell them: whereby it appeareth, that most give to the poore rather by compulsion then of compassion.'[36]

The date of the pamphlet's publication is significant, because in that year the English Parliament passed the most comprehensive social legislation of the century, following a decade of global cooling. The Poor Law Act of 1601 required each parish in the kingdom to provide certain 'entitlements' (as we would call them today) in certain well-defined circumstances – old age, widowhood, illness, disability and unemployment – but only to those normally resident within the parish. Funding for the system came from a

tax on income from local property (primarily land and buildings), administered by 'overseers' chosen from local property owners, who had to present
their accounts annually for audit by the local magistrates. This devolution
of public welfare to parish level proved a stroke of genius for two reasons.
First, the involvement of magistrates (backed up, if necessary, by the royal
courts) ensured that all the rich contributed. Second, limiting benefits to
local residents meant that, in a crisis, the poor would stay in their parishes
where their entitlements were guaranteed instead of seeking relief in the
nearest town and overloading its resources (as happened elsewhere). The
system was by no means perfect, and even the great dearths of 1629–31 and
1647–49 failed to persuade magistrates to tax the rich to support the poor;
but during the famine of the 1690s government pressure ensured that 'virtually every parish in the country was part of a nationally co-ordinated relief
system' that brought benefits not only to the 10 per cent or so who needed
charity to survive at any given time, but also to the much larger group who
might require relief in some future crisis. The series of climate-induced
crises that afflicted England between the 1590s and the 1690s thus gave rise
to the world's first welfare state, and in doing so provided an essential
precondition for the first Industrial Revolution.[37]

Creative Destruction

The spate of urban fires in the seventeenth century also stimulated some
innovations. In Germany some medieval guilds existed specifically to share
fire risks, yet they could not cope with the proliferation of fires during
the Thirty Years' War; then in 1664, in Hamburg, regular guilds began to
issue fire-insurance contracts known as 'Prudence in everything' (*Alles
mit Bedacht*), which promised each member 1,000 thalers towards rebuilding
costs if his house burned down, in return for a 10-thaler annual premium;
and, as in other towns, the Hamburg magistrates also issued orders to
promote fire prevention and punish arson.[38] It seems surprising that England
did not immediately follow suit – almost 100 fires there in the second half of
the seventeenth century caused almost £1 million of damage – yet the only
remedy for most of those afflicted was to secure a 'charity brief' from a
sympathetic magistrate, authorizing them to collect money at the door of
their local church. Even the Great Fire of London in 1666 failed to effect
major changes. Admittedly, the following year, the London Rebuilding Act
required the city corporation to straighten and widen some streets and all
lanes, with compensation paid from a coal tax, and required future houses to
be built of brick and stone; and Sir Christopher Wren oversaw the rebuilding
of St Paul's Cathedral, fifty other churches, and numerous public buildings
in a uniform style. Fourteen years after the Great Fire of London, however,
Dr Nicholas Barbon (a physician turned property speculator) established

'The Insurance Office for Houses' with a capital fund of £30,000. In its first three years of operation the office insured 4,000 London houses, receiving £18,000 in premiums (charging twice as much for timber as for brick houses), and paying out some £7,000 in claims.[39] The profits of Barbon's venture led to the creation of other fire insurance societies, whose members affixed a distinctive fire-insurance mark to their property so that the company's private fire brigade could find them.

Some cities created a permanent public fire brigade. After the Meireki fire in Tokyo (1657), the shogun created an all-samurai fire service with four brigades at the centre, and special detachments assigned to bridges, granaries and other important structures; while London divided itself into four zones, each with 800 leather buckets, fifty ladders and twenty-four pickaxes. The basic technique of firefighters in both cities remained demolition, either tearing down or blowing up buildings to create a fire-break; but the frequency of major conflagrations prompted more sophisticated measures. Suzhou, the capital of Chinese silk production, created fire engines with forced water pumps mounted on wheeled vehicles, a technique also developed independently in Amsterdam, where the painter and entrepreneur Jan van der Heyden invented a suction-hose pump that he incorporated into fire engines equipped with leather hoses, each 50 feet long (which remains the standard length of fire hoses in Europe). He set up a factory and sold seventy of his engines to the city – one for each of its wards. The benefits were immediate and dramatic: between 1669 and 1673, eleven fires in Amsterdam caused a total of £100,000 of damage, but between 1682 and 1687 the forty fires extinguished by the new engines caused, in all, less than £2,000 of damage. Van der Heyden's invention soon spread – to Britain, Germany, Japan and Russia (after Tsar Peter the Great saw one during his visit to the Dutch Republic in 1698).

Other public initiatives also improved urban life. In 1658, James Fraser marvelled that in London he could 'walke as safe in the darkest night' as by day, 'for in the twailight every house hangs out a glasse lanthern and lighted candle over the door, so that the streets ar as a lamp of light'. He noted that 'it was to prevent picking of pockets in the dark that those lights were appointed', and now 'if any offer to wrong yow', he 'could not escap, not through the narrowest lane or alley, unseen'.[40] A decade later, the Amsterdam town council went one better. Van der Heyden had invented a lamp capable of burning the whole night (thanks to wicks of Cyprus cotton made in his lamp factory), and in 1669 the city accepted his bid to erect 1,800 lamps placed between 125 and 150 feet apart (which, the inventor calculated, combined maximum lighting efficiency with minimum cost). Within six months the system was in operation, serviced by 100 municipal lamplighters. Visitors immediately noted how disorder and crime fell, thanks to both the lamps and to a new neighbourhood watch comprised of 150 lightly armed citizens who patrolled the streets each night, with as many again in reserve. They summarily arrested anyone observed in anti-social behaviour: beating their wives or servants; engaged in rape, theft or blatant street prostitution; acting in a drunk or disorderly manner. By the end of the century, many cities in Germany and elsewhere in the Dutch Republic had emulated the Amsterdam system. Humans had for the first time tamed the night.

Non-Creative Destruction

In his influential analysis of what he called Creative Destruction, Joseph Schumpeter recognized that, in certain circumstances, exceptions existed to the process of 'incessantly destroying' old economic structures and then 'incessantly creating a new one' that he saw as central to economic growth. He wrote:

> Let us assume that there is a certain number of retailers in a neighborhood who try to improve their relative position by service and 'atmosphere' but avoid price competition and stick as to methods to the local tradition – a picture of stagnating routine. As others drift into the trade that quasi-equilibrium is indeed upset, but in a manner that does not benefit their customers. The economic space around each of the shops having been narrowed, their owners will no longer be able to make a living and they will try to mend the case by raising prices in tacit agreement. This will further reduce their sales and so, by successive pyramiding, a situation will evolve in which increasing potential supply will be attended by increasing instead of decreasing prices and by decreasing instead of increasing sales.

'Such cases do occur', Schumpeter noted, 'and it is right and proper to work them out. But as the practical instances usually given show, they are fringe-end cases to be found mainly in the sectors furthest removed from all that is most characteristic of capitalist activity. Moreover, they are transient by nature.'[41]

Schumpeter's assertion, made about the first half of the twentieth century (of which he had direct knowledge) is entirely incorrect for the second half of the seventeenth century, when the stagnating routine he described characterized almost every economic sector (not just 'fringe-end cases'). Moreover, instead of being transient, these exceptions persisted. Many survivors of the Global Crisis tried to 'narrow the economic space' expressly to preserve the existing equilibrium and to prevent anything likely to upset their comfortable ('stagnating', Schumpeter would say) routine. Indeed, the decision of most communities to enforce a low-pressure economic regime, as opposed to the high-pressure regime favoured in England and its neighbours, contributed to the Great Divergence discerned by many historians between Western Europe and the rest of the world.

The attitude of some European farmers towards alternative crops is instructive here. Thus although maize usually survives adverse weather that will ruin wheat and barley, and although it produces far better yields per acre, many farmers refused to cultivate it until virtually forced to do so by a crisis. In Spain, maize cultivation spread only after the famine of 1630–31 – and, even then only in some of the worst-affected areas. A comparison of demographic and crop records for Galicia reveals that those communities that grew maize increased far more, and far more rapidly, than those that remained dependent on cereals alone. This partial embrace of maize enabled Galicia to experience sustained demographic growth during the seventeenth century – yet elsewhere in Spain, tradition trumped expediency: farmers used maize (if they used it at all) mainly for fodder. Attitudes towards rice were much the same. Although farmers in the Tuscan village of Altopascio suffered from extreme climatic events in the mid-seventeenth century (unprecedented floods in 1654–56; drought in 1659), and saw deaths outnumber births one year in three, they still refused to diversify. Although they experimented with rice after one famine, they reverted to cereal cultivation as soon as possible because potential profits were higher. They introduced maize only after the catastrophic harvest that followed the Great Winter in 1710.[42]

A similar conservatism characterized many areas of Germany. For instance, although immediately after the Thirty Years' War ended in 1648 the authorities of Ottobeuren welcomed immigrants who wanted to settle down (page 452), but in the 1660s fewer than 20 per cent of all marriages involved an immigrant, and after the 1680s fewer than 10 per cent. The population of Ottobeuren therefore stood at only 72 per cent of the pre-war total in 1675 and still at only 92 per cent in 1707.[43]

The rural textile industry in the Black Forest region of Württemberg, centred on Calw, provides another striking example of stagnating routine. On the eve of the Thirty Years' War, villages in the area produced thousands of light woollen cloths (*Zeuge* or worsted), exporting them to Italy, Switzerland and Poland as well as other parts of Germany. Spinning, weaving and dyeing cloth employed up to half the families in some villages; but hostile troops occupied Württemberg for much of the war, burning Calw to the ground in 1634, cutting trade routes, increasing risks, and destroying several of its export markets. Once the war ended, the merchant-dyers formed an association and (supported by the local overlord) compelled the weavers' guilds to sign a permanent agreement with them which, they claimed, would guarantee a reasonable living for all. The contract contained three important restrictions: whereas previously each weaver had produced up to 200 cloths annually, henceforth the maximum would be fifty; each year the merchant-dyers' association unilaterally fixed a price and agreed not to purchase above or sell below that price; and weavers must swear not to sell to anyone outside the merchant-dyers' association, or to compete with one another for apprentices, spinners, or raw materials. Although these restrictions initially harmed only outsiders (consumers, employees, women, migrants, Jews), and protected the interests of native merchant-dyers, master-weavers, and male householders, they eventually undermined the competitiveness of Württemberg textiles.

This type of economic equilibrium appealed to guilds, which proliferated elsewhere in the wake of the crisis. Not far from Calw, on the Ottobeuren estates, before the war only potters and butchers had organized their own guilds, but between 1648 and 1700 rural guilds came to control virtually every sector of the economy. The activities of carters, blacksmiths, locksmiths, tailors, shoemakers, tanners, coopers, brewers, painters, gunsmiths, cutlers, rope-makers, bakers, millers, carpenters, cabinet-makers, glassmakers, masons, barbers and even bathhouse-keepers were regulated by guilds – and in every case the restrictive practices of the guilds guaranteed a living to all producers equally, and so facilitated a slow but sure recovery from the mid-century crisis. But in doing so they discouraged innovation and growth.

The development of book printing in a cluster of villages in Sibao, a mountainous area in Fujian province in southeast China, offers a final example of a low-pressure strategy for economic recovery. Devastation during the Ming–Qing transition left much abandoned land in the area until, with Qing encouragement, migrants 'who can farm mountain land [came] with their families and support[ed] themselves through their labour'. The sources referred to some of the newcomers as *pengmin* ('shed people', because they lived in flimsy shacks), which conveys their initial poverty, but some turned a profit by cultivating bamboo that others used to make paper. Starting in 1663, other Sibao families took advantage of this cheap local paper to cut woodblocks for books. 'They targeted primarily the largest textbook market: the students at the base of the educational pyramid' who were preparing to sit the civil-service examinations. By 1700 Sibao boasted over a dozen publishing houses, each distributing their products through peddlers who travelled from market to market selling their wares.[44]

The publishers of Sibao had thus managed to create something from nothing, but creativity soon gave way to conservatism. They now strove (like the worsted producers of Calw) to eliminate competition, claiming thereby to maintain a fair share of profits for all. To begin with, although all male heirs got an equal share of their father's goods upon his death, a practice that led to the division of woodblocks and other printing stock along with other assets, the sons and their families might choose to remain under the same roof and run the print shop as a cooperative family business. Within this labour pool, which might include seventy relatives, the household head divided responsibilities and scrutinized accounts to eliminate competition and conflict; and even if fragmentation took place, with some sons taking a share of the blocks and setting up on their own, families still minimized competition. At the end of each year the manager of each enterprise 'printed a sample cover page of each new work it planned to print in the coming year and posted these on the gate to its [premises]'. Should two enterprises propose the same title, village elders would intervene to broker an agreement or, if that proved impossible, to impose a settlement. In the words of Cynthia Brokaw, who uncovered the methods of the Sibao printers, 'these measures aimed to ensure some profit for even the smallest [print] shops' and to inhibit 'large concerns from monopolizing these inexhaustibly popular educational texts'. In this goal they succeeded: the households that founded the prosperous printing industry of rural Sibao after 1663 were still producing the same texts almost three centuries later.[45]

These low-pressure economic strategies observed in Calw, Ottobeuren and Sibao were probably the norm in many parts of the later seventeenth-century world. It seems unlikely that many regions adopted a high-pressure strategy that allowed migrants, labourers, women and Jews to participate in rural industries: such a system was flexible, adaptable and dynamic, but it also risked alienating vested interests and creating popular disorder. Even though opening up an enterprise to competition normally increased the size of the economic pie, the competitive melee normally left many with smaller slices. At least in the short term, it made good political and social sense to limit each enterprise to a closed corps of licensed experts, whose activities were micromanaged through their own professional associations, reinforced by state regulation, because vested interests remained content, minimizing the risk of disorder. Probably only regions with a sophisticated welfare system, such as England and the Dutch Republic, could take such risks.

Economic conservatism also involved risks, however. International teams of economic historians have published data from all around the eighteenth-century world on the number of 'baskets of basic goods' that unskilled labourers could purchase with their daily wages. Their research suggests that important disparities existed by 1700, if not before, and that those disparities increased over time. In London the minimum daily wage could buy four baskets of basic goods, and in both Amsterdam and Oxford it could buy three; but in Beijing, Canton, Suzhou, Shanghai, Edo, Kyoto, and Istanbul it could buy barely one. In addition, workers in the cities of Central and Southern Europe included in the survey had a welfare ratio no better than their comrades in Beijing; while no Asian city came close to achieving the welfare ratios already evident in parts of Northwest Europe. Put another way, the available data suggests that by 1700 (and probably earlier) a London labourer could support a family of four, and an Amsterdam or Oxford labourer could support a family of three, but a comrade in Asia or in South and East Europe could scarcely support himself. Moreover, in both England and Holland wage labourers formed perhaps one-half of the working population in some towns by the early eighteenth century, whereas in Jiangnan (and in many parts of continental Europe) they formed less than one-fifth of urban and less than one-tenth of rural populations. This disparity magnified the disparity in purchasing power.[46] Although each basket of basic goods included enough food to provide 1,940 calories per day, mainly from the cheapest available carbohydrate, London workers did not, of course, eat four times as many carbohydrates as before. Instead they ate more expensive foods, drank more expensive beverages and bought a wide range of nonfood items, fuelling the consumer revolution that so impressed Daniel Defoe. At least in Northwest Europe, it seems that the fatal synergy that produced the mid-seventeenth-century crisis had by 1700 given way to a benign synergy in which demand for food no longer exceeded supply, while warfare did not stifle welfare, enabling a return of political stability, demographic recovery and economic growth – thus allowing Grimmelshausen's phoenix to arise from the ashes. So why did it not arise everywhere?

CHAPTER TWENTY-TWO

The Great Divergence

The General Crisis, according to the eminent sinologist Samuel Adshead in 1970, 'marks the decisive point of divergence between the modern histories of Europe and China'. In a pioneering attempt to compare the experience of the two regions in the seventeenth century, Adshead examined economic, social and political data and concluded that 'European society emerged from this crisis reconstructed, more powerful and better integrated than before, while Chinese society remained relatively unchanged.' Thirty years later another eminent sinologist, Kenneth Pomeranz, challenged this interpretation in a powerful comparative study entitled *The Great Divergence* (a term that has subsequently gained widespread currency), and concluded that as late as 1750 little distinguished the economically advanced areas of China, such as Jiangnan, from the economically advanced areas of Europe, including England. Subsequent research by economic historians (notably on market integration and real wages) has generally supported Adshead, concluding that the start of the Great Divergence 'needs to be shifted back to the seventeenth century'. But what of other responses in the wake of the Global Crisis?[1]

Educate and Punish

In 1654, just after Louis XIV regained control of his capital, Abbot Jacques de Batencour, who claimed eighteen years' experience of teaching in primary schools, published *The parish school: or, how best to teach children in small schools*. The first section of his book compared running a school with running an army: the key to successful instruction was hierarchy and subordination. Effective classroom teachers needed 'observers' and 'admonishers' who noted the names of delinquent pupils to be punished; 'visitors' who followed students home to see how they and their families behaved after school hours (and later denounced any faults to the teacher); and 'repeaters' who recited the lesson and showed the alphabet to the youngest pupils. Then, after discussing how to teach children piety, reading, writing, arithmetic and the rudiments of Latin, Batencour turned to the subjects that evidently interested him most: how best to discipline and punish pupils. He stressed the need to segregate boys from girls, to prevent children from talking to each other and to administer punishments that would humiliate as well as hurt, because their effects lasted longer.[2]

The parish school attracted a wide audience and went through many editions. Louis XIV sent copies to Canada (Canadian libraries contain both extant copies of the first edition); French bishops commanded the schools of their diocese to follow its precepts. The book also set the agenda for the host of French charity schools founded during the later seventeenth century expressly 'to remedy the ignorance that prevails among the poor whose children, lacking money and unable to attend the parish schools, roam for the most part as vagabonds in the streets, without discipline and in total ignorance of the principles of religion'. Subsequent legislation included the same rationale, often word for word, and claimed that Louis XIV's initiative had not only educated the poor but also 'reformed the libertines whose excesses would have been a public scandal'.[3]

The Sun King was not the only European ruler to see basic education as an effective antidote to the disorders fomented by the General Crisis. In 1651 Duke Augustus of Brunswick-Wolfenbüttel issued a comprehensive ordinance that mandated universal primary education explicitly to avoid the next generation turning into savages ('*die Verwilderung der Jugend*'). 'Unfortunately', the duke thundered,

> the accursed recent war has destroyed (among other things) the education of the young, and unless it is remedied in time we can expect no end to misfortune and misery. Instead of learning honour, virtue and well-being, young people have grown up with the example and experience of barbarism, so that we can expect in future, indeed in just a few years, nothing except wicked and undisciplined subjects of the state, who will shun neither evil nor injustice, but will instead continue the destruction of the ruins that remain of the State that God has graciously saved from the searing flames of war.

To avert these dangers, the duke continued (anticipating Batencour), 'Young people must be carefully educated, with much wisdom and some severity, in order to turn them away from evil and attract them to the good'. Therefore, he decreed, 'all parents must send their children' to school, 'to study for as many years as it takes them to learn to understand the Catechism and to read printed texts'.[4] Similar concerns led other Protestant rulers to enjoin universal schooling: Sachsen-Gotha in 1642, Hannover in 1646 and Württemberg in 1649 in Germany; and, on a much grander scale, Scotland in 1646, where 160 out of 179 Lowland parishes boasted a school and a schoolmaster by the end of the century.

Many East Asian states likewise concluded that compulsory basic education would accelerate recovery from the crisis. In China, an imperial edict in 1652 called for the creation of a school in every village (whether by the community, a clan, a temple or a charity); and six years later another imperial edict imposed the national school curriculum on regions not inhabited by Han Chinese: henceforth local chieftains in all areas of the empire must complete a standard education before they could govern.[5] In Japan, the insistence of the Tokugawa rulers on conducting all the essential business of government in writing (Chapter 16) led to the creation of schools capable of creating cadres with the necessary reading, writing and arithmetical skills. A study of *daimyō* fiefs reveals that although only two boasted schools in the 1620s, and only eight by 1650, there were at least twenty by 1703.

The Crisis of the Universities

This post-crisis enthusiasm of governments for education had limits. In the Muslim world, *medreses* continued to concentrate on religious instruction (Chapter 19); and even the rulers of Japan, China and Europe saw little value in higher education. In Europe, the end of the Thirty Years' War in 1648 produced a swift recovery in the student population of most Germany universities, with matriculations almost resuming pre-war levels in the 1650s; but then, as Jonathan Israel has observed, 'a combination of social and especially cultural factors plunged Europe's universities into the deepest and most prolonged crisis in their history', because 'most universities not only ceased growing but steadily contracted'. Total student numbers 'fell uninterruptedly from the 1680s throughout the eighteenth century', and new foundations virtually ceased.[6]

The reasons for this crisis were no secret. When in 1680 the rector and senate of Heidelberg University examined why 'student numbers are constantly in decline', they compiled a list of damning reasons (some of which still sound familiar): 'there are not enough professors', especially in newer disciplines such as botany, anatomy and chemistry; in the traditional disciplines, instructors 'are careless in their lectures and public disputations'; 'discipline is either too strict or too soft'; 'fees and accommodation for students cost too much, and there are no scholarships'.[7] These problems were not confined to Heidelberg, or to Germany. All over Europe confessionalization and bureaucratization no longer stimulated university expansion: after the 1650s, church-building largely ceased, and theology began to lose its dominant place in intellectual life; and although states still sought highly educated officials and diplomats, most universities failed to provide 'useful' courses in established subjects like history, geography, philosophy and modern languages, and no courses at all in new subjects like physics, chemistry and biology.

Only a substantial injection of public funds could have reversed this trend, creating new teaching posts, academic facilities and scholarships;

but rulers remained reluctant, mindful of the prominence of highly educated scholars in recent rebellions (Chapter 18). In France, the government shut down the Huguenot academies that trained hundreds of young men in theology, philosophy and languages because, in the words of a Catholic bishop in 1659, 'the students here are serving their apprenticeship in rebellion and disobedience'. A Huguenot education, he asserted, was 'the source of all seditions'. Such anti-intellectualism was not confined to France. In the 1650s, many English radicals argued that in an ideal world 'children shall not be trained up only to book learning, and no other employment', because 'through idleness and exercised wit therein they spend their time to find out policies to advance themselves to be lords and masters above their labouring brethren', which 'occasions all the trouble in the world'. The Quakers, for their part, wanted to shut down all universities because they prepared ministers for the established church that they despised.[8]

Even had European universities received generous government support and general acclaim after 1650, it seems unlikely that they would have reformed their curricula to include new disciplines and methods, or even bought books that balanced authority with innovation. At Leiden University, violent disputes broke out between those who favoured an experimental approach in subjects such as astronomy, anatomy and botany, and those who saw no need to go beyond the published wisdom of Aristotle (who defined science as the contemplation and organization of eternal truths already discovered). To minimize the risk of intellectual innovation, the university's Curators (equivalent to the board of trustees at a modern North American university) ruled that students could enter the library on only two days per week, and forbade the librarian from purchasing any book that was not in Latin and on a traditional subject.[9]

In China, the Qing believed that the intellectual elite had played a critical role in fomenting the troubles faced by their predecessors, and therefore closed down all academies and learned societies. Moreover, although they reintroduced the national system of civil-service examinations, they made some important changes. First, Manchu subjects received encouragement to compete in the examinations, and permission to submit answers in either Manchu or Chinese. Second, because there were never enough successful Manchu candidates, the Qing regularly filled senior positions with men who lacked a *jinshi* degree but boasted other qualifications to govern. Finally, the new dynasty tolerated no irregularities in the examination process. In 1657 the Shunzhi emperor reacted savagely when evidence of cheating in the triennial *juren* examination came to light: he beheaded dozens of officials and examiners who had accepted bribes, and deported members of their families to serve as slaves in Manchuria. Every candidate had to take the exam again.[10]

The New Learning

Government hostility to institutions of higher education did not always manage to stifle intellectual speculation and innovation. In Europe, to quote Jonathan Israel again,

> down to around 1650, Western civilization was based on a largely shared core of faith, tradition and authority. By contrast, after 1650, everything, no matter how fundamental or deeply rooted, was questioned in the light of philosophical reason and frequently challenged or replaced by startlingly different concepts generated by the New Philosophy and what may still usefully be termed the Scientific Revolution.

The timeline proposed by John Aubrey, an early advocate of the New Philosophy in England, vindicated Israel's claim. 'Till about the year 1649', Aubrey wrote,

> 'twas held a strange presumption for a man to attempt an innovation in learning, and not to be good manners to be more knowing than his neighbours and forefathers. Even to attempt an improvement in husbandry, though it succeeded with profit, was look't upon with an ill eie ... 'Twas held a sinne to make a scrutinie into the waies of nature ... In those times, to have had an invention and enquiring witt was accounted affectation, which censure the famous Dr. William Harvey could not escape for his admirable discovery of the circulation of the blood. He told me himself that upon his publishing that booke, he fell in his practice extremely.

Aubrey condemned earlier scholarship as 'Paedantry', because 'criticall learning, mathematics and experimental philosophy was not known', and '*Things* were not then studied. My lord Bacon first led that dance.'[11]

'My lord Bacon' – Francis Bacon, Lord St Albans – sought 'to bring in industrious observations, grounded conclusions, and profitable inventions and discoveries', and to demonstrate how natural philosophy enhanced political power, by preparing a six-part project for mapping all human knowledge, which he called *The great instauration*. Bacon published the first part in 1605 as *The advancement of learning*, which boldly proposed that all knowledge could be organized according to the three intellectual faculties possessed by humans: memory (history), imagination (poetry) and reason (philosophy – including the sciences). This approach, he argued, both classified existing information and cleared the way for new discoveries. Bacon published the second part in 1620, entitled *Novum organum* (a play on Aristotle's *Organon* or 'logical works'), which proposed a method of scientific inquiry very different to that of Aristotle. Let 'the business be done as if by machinery', he argued, collecting and reviewing relevant cases in order to find 'certain and demonstrable knowledge'.[12]

Since neither Oxford nor Cambridge showed the slightest interest in the 'New Philosophy', their graduates possessed little or no scientific knowledge. The eminent mathematician John Wallis complained that when he was a student at Cambridge in the 1630s, the discipline was 'scarce looked upon as accademical studies, but rather mechanical, as the business of traders'. Therefore, 'amongst more than two hundred students' in his college, 'I do not know of any two (perhaps not any) who had more of mathematicks than I, (if so much), which was then but little; and but very few, in that whole university'. After the Civil War broke out in 1642, Wallis moved to London and attended the weekly meetings of 'divers worthy Persons, inquisitive into Natural Philosophy, and other parts of Humane Learning; and particularly of what hath been called the *New Philosophy or Experimental Philosophy*'. One of those 'worthy persons' was Robert Boyle, son of the richest man in Britain and the leading light of 'the invisible, or (as they term themselves) the philosophical college'. Boyle, like Wallis, applied himself to 'useful learning' – or, in his own words, to 'natural philosophy, the mechanics, and husbandry according to the principles of our new philosophical college, that values no knowledge, but as it hath a tendency to use'. To this end, Boyle would consult 'the meanest' person, provided 'he can but plead reason for his opinion'.[13]

The New Philosophy suffered from a serious handicap: few scholars outside Great Britain and its colonies read English. Bacon therefore had his *Advancement of science* translated into Latin so it 'will live, and be a citizen of the world, as English books are not', and sent complimentary copies to foreign scholars.[14] The recipients included Galileo Galilei of Florence, who in 1609 had created a telescope powerful enough to carry out systematic observations of the moon and Jupiter. The following year, Galileo published his findings in a short treatise in Latin (*Sidereus nuncius*, 'The starry messenger'), and sent copies of both his book and his telescope to foreign courts so that others could verify his claim – just as Bacon had suggested in his *Advancement*. William Harvey went one step further than Bacon: he did not even publish in England. His now famous book describing how blood circulates first appeared in Latin in Frankfurt, and he tucked away at the very end his affirmation that this proved Aristotle wrong. But if Harvey hoped that these stratagems would deflect criticism, he miscalculated: both his book and its argument were either ignored or condemned by his compatriots (and by most other Europeans) for two decades.

René Descartes, a French philosopher who moved to Holland to live as a private scholar, proved an exception. In 1637 his influential *Discourse on method for the correct use of one's reason and for seeking truth in knowledge*, praised Harvey's theory (although even he dared not name its author, citing merely 'an English doctor'); and, in an earlier work, Descartes approved Bacon's approach ('*la méthode de Verulamius*') because, although he had read all the classical authors, he had reached his conclusions by rigorous experiments that anyone could replicate. Descartes also praised Galileo's 'application of mathematical reasoning to physics' (again without naming the author) and argued that, since 'all the things which come within the scope of human knowledge are interconnected', if someone began with simple notions and proceeded step by step, 'it does not take great skill and capacity to find them'. Descartes thus exalted experiment over theory.[15]

Descartes's approach gained fame outside Europe. In Mughal India, Shah Jahan's eldest son Dara Shikoh translated from Sanskrit into Persian over fifty of the Upanishads, ancient Hindu philosophical texts, which he called *Sirr-i akbar* (*The great mystery*) and composed a philosophical work entitled *Majma-ul-Bahrain* (*The mingling of two oceans*), in which he stated that he 'became desirous of bringing in view all the heavenly books'. He therefore examined the Hebrew Bible and the Christian Gospels, as well as the Upanishads and the Qur'an, concluding that 'he did not find any difference, except verbal, in the way they all sought and comprehended Truth'. The prince also collected an impressive library and attracted to the holy city of Benares (Varanasi) a group of Sanskrit scholars who pursued *Navya Nayaya*, or 'new reason' (with the sense of evidence-based critical enquiry). The French traveller François Bernier, physician to Dara, lived in Benares for four years, where he translated works by Descartes and other French natural philosophers into Persian. One of Dara's protégé's composed two treatises that confronted the 'new reason' and Cartesian philosophy.[16]

In China, in 1608 (the year before Galileo turned his new telescope on the moon), the scholar-official Xie Zhaozhe published a 1,414-page treatise entitled *Wa za zu* (*Fivefold miscellany*), which anticipated Descartes by correcting classical writers (albeit followers of Confucius rather than of Aristotle) through systematic observation. For example, Xie dismissed the 'popular saying' that snowflakes at the winter solstice had five points because 'Every year, as the winter moves into spring, I have gathered snowflakes and looked at them. All are six-pointed.' Where Xie could not provide proof, he used his common sense: 'Since the conjunctions and the eclipses of the sun and the moon depend on their regular orbit, and they can be foreseen in numerical detail several tens of years in advance, it is not possible to escape them.' So 'is it not erroneous to point to them as portents of heaven?'[17]

Xie's experimental approach did not stand alone. The foundation charter of the *Fu She* (Restoration Society) in 1629 expressed the hope that it would revitalize 'the ancient learning and thus be of some use'; and in 1637 (as Descartes oversaw the publication of his *Discourse*), Chen Zilong, another scholar-official, published *The complete book of agricultural science*, which included material from many Western sources (including the translation of a treatise on hydraulic theory by a Jesuit). Chen justified the inclusion of foreign sources by twisting the traditional Confucian saying, 'If you have lost the proper way of doing things at court, search for it in the countryside': in China's present emergency, he asserted, 'If you have lost the proper way of doing things at court, search for it among the foreigners.' Two years later, Chen and two colleagues published *Select writings on statecraft from the Ming period*, a compendium of memoranda and advice submitted by over 400 officials from the fourteenth century to their own day, in the hope that they might suggest solutions to current Ming problems.[18]

Although Chen committed suicide rather than accept the Qing, some of his surviving colleagues promoted what became known as *kaozheng* (evidentiary research: knowledge that could be verified empirically) and *shixue* (practical knowledge). Gu Yanwu, from a gentry family in Jiangnan, epitomized the new approach. He memorized all the texts required to achieve *shengyuan* status and joined the Restoration Society, but

> rejected in the autumn triennial [*juren*] examination in 1639, I retired and read books. Realizing the many grievous problems with which the state was faced, I was ashamed of the meager resources which students of the Classics possessed to deal with these problems. Therefore, I read through the twenty-one dynastic histories as well as gazetteers from the whole empire. I read the collected literary works of the famous men of each period as well as memorials and documents. I noted down what I gained from my reading.

Initially, Gu used his practical knowledge to support the Southern Ming state, preparing memorials that showed how, four centuries before, the Southern Song had kept northern invaders at bay in similar circumstances; but after the Qing conquered Jiangnan he shaved his head and feigned obedience in order to travel all over the empire. He visited the Shanhai Pass, through which the conquerors had entered China, to try and understand the strategic geography of the area, and he interviewed veteran soldiers for their recollections of what had happened – and 'if what they told him was not in conformity with what was commonly accepted, he retired indoors to correlate their information with that contained in his books'. Everywhere Gu acquired and read rare books, and copied down (or made rubbings of) inscriptions, using them to verify and if necessary correct classical texts (which may have become corrupted) and historical chronicles (which may have been falsified). He also visited a host of friends and scholars – some lukewarm Qing subjects, others clandestine Ming loyalists – with whom he shared information. In the safety of their houses, Gu wrote books on history, archaeology and phonetics that exemplified the new scientific outlook.[19]

Gu survived because the conquerors initially encouraged such pragmatism. In 1652 an imperial decree ordered that henceforth only 'studies of principle, works on governance, and other books that contribute positively to learned affairs' should be printed; while the Kangxi emperor later claimed that 'since my childhood I have always tried to find out things for myself', adding 'If you want to really know something, you have to observe or experience it in person; if you claim to know something on the basis of hearsay, or on happening to see it in a book, you'll be a laughingstock to those who really know.' He advised his subjects to 'Keep an open mind and you'll learn things', and stressed the need to 'Ask questions about everything and investigate everything.' The emperor practised what he preached: he studied mathematics, surveying, music, mechanics and astronomy with the Jesuit missionaries resident at his court; he promoted a Jesuit to be director of the Imperial Bureau of Astronomy (which regulated the entire imperial calendar); and he allowed Western priests to teach his sons.[20]

Nothing Kangxi wrote or said could impress the scholars of Tokugawa Japan, who despised the Qing as barbarian descendants of the Mongols who had invaded the archipelago four centuries before – but they showed considerable interest in the epistemology of two other groups of foreigners. The professional translators (*tolken*: eventually about twenty families), who worked with the Dutch physicians and surgeons based on Dejima island at Nagasaki (Chapter 16), alerted the central government to the potential importance of Western knowledge, and in 1667 the shogun asked the Dutch East India Company to send him a physician with botanical and chemical experience. The Company selected Dr Willem ten Rhijne, a pupil of the Dutch scholar who had published Descartes's work. He arrived in Edo in 1675, but secured only one audience with the shogun before he returned the following year because (in the words of the neo-Confucian scholar and physician, Genshō Mukai, co-editor of *Western cosmography with critical commentaries*) Westerners 'are ingenious only in techniques that deal with appearance and utility' (a verdict that Descartes would no doubt have relished).[21]

The Tokugawa showed far more interest in a second group of refugees in Nagasaki: the Ming scholars who followed the path of evidentiary research, called in Japanese *jitsugaku*, or practical learning (from the word *jitsu*, 'practical'). In Japan, as in China, 'practical learning' involved four assumptions: first, a realization that the present was different from the past; second, prioritizing experience over theory; third, seeing knowledge as a process of continuous experience and re-evaluation; and finally, seeking knowledge of immediate utility. Also as in China, Japanese practitioners reacted to the seventeenth-century crisis by investigating how the world around them functioned, hoping to find some mechanism to escape from – or, at least, to mitigate – natural catastrophes. Kaibara Ekken, the son of a samurai, trained as a doctor in Nagasaki before becoming preceptor to a *daimyō* family. He wrote over one hundred treatises on topics that ranged from botany and medicine, through astronomy and topography, to ethics and education, many of them in the form of self-help manuals directed to non-expert readers, male and female. One of his works stressed that 'One should not blindly regard all one has heard as true and reject what others say merely because they disagree, nor be stubborn and refuse to admit mistakes'; and, just like his contemporary Robert Boyle, Kaibara Ekken assured his readers that 'I followed up on what the townspeople spoke of, salvaged what I could prove out of even the most insane utterances, and made enquiries of people of the most lowly station. I was always willing to inquire into the most mundane and everyday matters and give considera-tion to all opinions.'[22]

Several scholars who endorsed practical learning became counsellors to the leading *daimyō* of Japan, and through their written and oral advice helped to introduce practical measures designed to assist economic recovery, and to prepare better for any recurrence of crisis. Yamazaki Ansai, the son of a samurai who worked for Hoshina Masayuki (grandson of Tokugawa Ieyasu and a regent after the death of Shogun Iemitsu), studied and wrote on mathematics and science, and founded the Kimon School where he trained students to consult and learn from a wide variety of sources. Kumazawa Banzan, another scholar from a samurai family, served the Ikeda *daimyō* at Bizen and played an important role in developing relief strategies and reconstruction in the wake of destructive floods in the 1650s. The Chinese refugee scholar, Zhu Shunsui, also became prominent in *daimyō* service. Like other members of his family, Zhu studied for the civil-service examinations and read widely, and although he never became a minister of the Ming he refused to obey the head-shaving edict. He made several trips to both Japan and Annam to secure aid against the Manchus but, having failed, he settled in Nagasaki until in 1661, at the request of one of his Japanese disciples, he wrote a short tract explaining the fall of the Ming that reserved special blame for the 'empty' learning of scholar-officials who had succeeded only because their meaningless examination essays were judged according to form rather than content. He also argued that even the most martial samurai needed a literary education, and in 1665 another grandson of Ieyasu invited Zhu to serve as his adviser and teacher. In this role, Zhu stressed the need to provide everyone with a proper educa-tion through schools and public lectures, and he compiled an illustrated treatise on how one must learn to 'cope with concrete situations'.[23]

In many seventeenth-century states, educated men bombarded rulers with suggested remedies for the crisis – 'projectors' in Stuart England; *arbi-tristas* in Habsburg Spain; statecraft scholars in Ming China; memorialists in the Ottoman empire – but few achieved results. By contrast, in Japan men like Kaibara, Yamazaki, Kumazawa and Zhu managed to get many of their ideas implemented – but there were still limits. Suspicions about Kumazawa's ideas forced him to leave Ikeda service in 1657 and he opened a school in Kyoto, but after a few years the magistrates closed it down. He defiantly circulated the prospectus of a book to be called *Questions on the great learning*, which proposed reforms for the government of Japan – and spent the rest of his life in prison.[24]

The Thought Police

'If minds could be as easily controlled as tongues', Baruch Spinoza observed in his *Theological-political treatise* of 1670, 'every government would be secure in its rule, and need not resort to force; for every man would conduct himself as his rulers wished, and his views as to what is true or false, good or bad, fair or unfair, would be governed by their decision alone'. Spinoza predicted that 'utter failure will attend any attempt in a commonwealth to force men to speak only as prescribed by the sovereign' – but most seventeenth-century rulers aimed to achieve precisely that goal.[25] Spinoza, like Kumazawa, knew all about censorship. His father had fled from his native Portugal, where he had professed to be a Catholic to escape the Inquisition. The Spanish monarchy (of which Portugal then formed a part) boasted twenty-four tribunals of the Inquisition, stretching from Peru and Mexico in America, through the Canary Islands and Sicily, to Goa in India. In each one, inquisitors sought to control the practice, expression and circulation of heterodox ideas, and they acquitted few of those who came before them (mostly as the result of an anonymous denunciation). The Inquisitors of Coimbra, Portugal, who had jurisdiction over the village where Spinoza's father was born, tried almost 4,000 people between 1567 and 1631, and condemned over 250 of them to death.

Many Italian states also boasted Inquisitions. One of them, the Congregation of the Holy Office in Rome, handled public intellectuals who held views of which the Catholic Church disapproved, including Galileo, whose observations with telescopes suggested that the Earth revolved around the sun even though certain passages of Scripture asserted the contrary. When Jesuit astronomers corroborated Galileo's findings, the Inquisition agreed to tolerate *suggestions* that the solar system might be heliocentric but threatened to punish anyone who claimed it as a fact (unless they could prove it). Although Galileo respected the Inquisition's compromise while he searched for proof, he publicly ridiculed those (especially priests) who continued to claim that the sun circled the Earth. Many enemies therefore stood ready to denounce him when, in 1632, he published a new book that discussed the rotation of the sunspots he had observed as proof of the heliocentric theory. The Holy Office summoned Galileo to Rome, interrogated him, declared him guilty of heresy, banned his book and condemned him to life imprisonment. Although the pope grudgingly commuted the sentence to house arrest, he outlawed the publication of any of Galileo's books, past, present and future. In the phrase of a sympathizer, if Galileo requested a licence to print the Lord's Prayer, the pope would refuse to grant it.[26]

Galileo's fate stifled intellectual debate in many parts of Europe. Even Descartes, though living in the Dutch Republic, far beyond the reach of the Inquisition, was 'so astonished' by Galileo's fate 'that I have almost taken the decision to burn all my papers, or at least to let no-one see them'. He now abandoned the ambitious book he had just completed, called *The world*, because it asserted that the Earth moved round the sun, and 'although I think [the assertion] is founded on very sure and very clear proofs, I would nevertheless not want to sustain them against the authority of the Church'. Descartes continued: 'I desire to live in peace and to continue the life I have begun under the motto "to live well you must live unseen"'. His book would not appear until a decade after its author's death.[27]

Descartes was wise to seek anonymity. The publication in 1637 of his *Discourse on method*, in which 'I try to prove that the soul is distinct from the body' (Aristotle had argued that the soul was part of the body, and so of its 'essence'), triggered shouting matches at Dutch universities between supporters and critics of his views and denunciations by Calvinist ministers for dismissing the traditional proofs of God's existence. Some accused Descartes of seeking to corrupt ordinary people because he had published in French instead of Latin. Deeply demoralized by the attacks, in 1649 he boarded a ship for Sweden, where he died the following year.[28]

Spinoza also suffered for his beliefs. In 1656, when he was twenty-four, the Jewish community of his native Amsterdam issued a *cherem* ('ban'), which solemnly cursed him and forbade anyone to communicate with him, assist him or read his work, because of his 'evil opinions'; and four years later they denounced him to the magistrates as a menace to 'all piety and morals' and called for his expulsion from Amsterdam. Ejected by his own community, Spinoza Latinized his name (from Baruch to Benedictus) and left Amsterdam to live secluded in villages and small towns while he worked on his *Theological-political treatise*. The generally hostile reaction that greeted this book discouraged Spinoza from publishing anything more. Even the colleagues who published his last work posthumously identified its author by initials alone.[29]

Although the Dutch Republic's religious leaders persecuted both Descartes and Spinoza, the government left them in peace. In France, Louis XIV issued edicts that condemned individual works to be burnt; his police raided bookshops and searched travellers at the frontiers for forbidden publications; and those found guilty of importing or selling contraband books were imprisoned or sent to the galleys. When the French historiographer royal brought out an *Abridged chronology of the history of France*, which condemned the power exercised by favourites and mentioned over-taxation and financial corruption as a cause of the Fronde, he lost his pension; and when he refused to change his text for a second edition, he lost his job. Such treatment soon led to self-censorship. The *Pensées* of Blaise Pascal, for example, were heavily censored by their publisher: of his fifty essays on politics, only seventeen appeared in print during his lifetime.[30]

Government censorship elsewhere was even more brutal. In India, the spread of the 'new reason' and Cartesian philosophy virtually ceased after 1659, when in the succession war Aurangzeb defeated the forces of his brother Dara Shikoh, captured him, declared him an apostate from Islam and had him murdered. In China, although the Shunzhi emperor did his best to win the support of the Han Chinese literary elite, even authorizing the compilation of an official history of the previous dynasty, after his death in 1661 the Manchu regents for his young son immediately reversed this policy.

Singletons and Multiples

In the wake of the mid-seventeenth-century crisis, and largely as a result of it, the major states of both Europe and Asia thus witnessed an expansion of basic schooling combined with a limitation of higher education, and an efflorescence of new learning combined with increased censorship. Nevertheless, by 1700 the intellectual life of Western Europe already diverged from that of other areas. One key difference was well expressed by the Dutch inventor Jan van der Heyden (Chapter 21). His dedication to the second edition of his *Description of fire engines* in 1690 reminded his audience, the magistrates of Amsterdam, that:

> It is almost impossible to foresee and think through all that is necessary for the success [of any invention]. Small accidents often can spoil the whole result and demolish everything one believed unshakeable. Even the best-planned works are subject to endless chance and conflict, the more so when they are to be introduced for general use. So it happens, as has been accurately observed, that of a hundred inventions of which trials had been made (supposedly with good results and to which patents even had been granted) hardly one succeeds.

Van der Heyden knew whereof he spoke: he had been working (among other things) on improving fire hoses and fire engines for almost forty years, yet some of his prototypes 'caused more damage at fires than benefits'. He was probably unaware that the fire engine presented by Dutch envoys to the shogun of Japan in 1658 had malfunctioned and 'had been thrown into a pond nearby', but he certainly knew that repeated experiments and trials essential for scientific success required both copious funds and the free exchange of knowledge.[31]

In his study of India's 'new reason', Jonardon Ganeri astutely noted the absence of these two preconditions. Although practitioners found patrons, after the execution of Prince Dara in 1658, 'There were few institutions which brought together people of different intellectual persuasions and certainly nothing like the Royal Society.'[32] Likewise Qing China lacked learned societies, universities, museums and other institutions where scholars could meet together and freely present, discuss, re-evaluate and record their ideas. Although 'intermittent collaborations and occasional communications' took place between scholars, usually by letter, Mark Elvin has argued that 'the Chinese, in science, seem to have been loners in comparison with the Europeans'. They also lacked patrons. No doubt members of the Chinese elite 'placed a high value on objective natural knowledge', just like their European counterparts; but, as Harold J. Cook observed, most of them 'had almost nothing to do with government. The kind of knowledge they valued most highly could therefore hardly become dominant.' Mary Elizabeth Berry has likewise stressed that in Tokugawa Japan even the most original writers 'did not convert social knowledge into social science by analysing the effects of the data and systems they described. And they did not convert information into news by reporting on events and opinions.' In short, they 'never shifted register from observation to commentary'.[33]

Scholars throughout East and South Asia still made important scientific advances, of course, but those advances tended to be, in the phrase of the distinguished sociologist Robert K. Merton, 'singleton techniques', often discovered by chance; and while singletons can sometimes have a significant impact, further refinements and adaptations tend to be limited and soon run into diminishing returns. By contrast, Merton noted, 'once science has been institutionalized, and significant numbers of men are at work on scientific investigations, the same discoveries will be made more than once'. Merton called these 'multiples', and he noted that almost forty multiple discoveries occurred in the seventeenth century.[34]

Merton saw multiples as a critical and unique component of European thought, which he traced back to the research agenda established by Bacon, whose New Atlantis of 1626 described a 'college' (called Solomon's House), with a staff divided into observers, experimenters, compilers, interpreters and 'merchants of light' (those who travelled afar in order to bring back knowledge), who would collaborate to extend natural knowledge and apply its practical benefits.[35] By then, in several European cities, men interested in knowledge already met together in academies, and a few admitted women. The early academicians normally kept their activities secret. Even the noble members of the most famous Italian academy, the Lincei ('lynx-eyed') of Rome, at first used code names and wrote to each other in cypher; and

Wallis and his group of 'worthy persons, inquisitive into Natural Philosophy' in London significantly adopted the name the Invisible College.

Gradually, academies became less secretive. In 1649 some members of the Invisible College moved from London to Oxford, where they reinforced another group, the Experimental Philosophical Club, whose members came from diverse social, religious and political backgrounds (which they agreed never to discuss). They found a useful conduit for their discoveries in Samuel Hartlib, born in Poland of a German father and an English mother, who left home in 1628 to escape the continental wars and settled in London. Although Parliament rejected his proposal to establish an 'Office of Publike Addresse' to 'put in practice the Lord Verulam's designations *De augmentis scientiarum* [Bacon's *The advancement of science*]', Hartlib used the money they gave him to employ a team of translators and scribes who copied out descriptions of inventions, scientific developments and technological innovations. These he distributed to others who shared his belief that 'useful' knowledge could transform the world. Hartlib also received pamphlets and treatises from his correspondents, sent them to friends and colleagues for comment, and then had them printed.[36]

The chaos that preceded the Restoration of Charles II in 1660 (Chapter 12) destroyed the Hartlib circle and almost extinguished the Experimental Philosophical Club, whose members were 'scattered by the miserable distractions of that fatal year'; but soon after the king's return Wallis, Wilkins, Boyle and nine other natural philosophers reconvened to form 'a Colledge for the promoting of Physico-Mathematicall Experimentall Learning', which met in London every Wednesday. The college had grown to nearly 100 by 1662, when Charles II issued a charter that created the Royal Society for Promoting Natural Knowledge. Thomas Sprat, the Society's first historian, considered weekly meetings crucial because, 'in Assemblies, the Wits of most men are sharper, their Apprehensions readier, their thoughts fuller, than in their Closets'.[37]

Not all productive exchanges of scientific information took place in assemblies. Henry Oldenburg, like Hartlib a refugee from Germany who settled in England, stayed in touch with his friends abroad and added new ones whenever he travelled to Europe, creating (in effect) an enormous 'listserve'. After 1665, as founding secretary of the Royal Society, Oldenburg also solicited papers from his correspondents, which he sent to other scholars for what would now be called peer review before publishing them in the journal he edited: *Philosophical Transactions, giving some accompt of the present undertakings, studies and labours of the ingenious in many considerable parts of the world* – the world's oldest continuous scientific journal. Each issue had a print-run of 500 copies, with (for a while) additional volumes in Latin translation as well as a partial French edition for the convenience of intellectuals abroad.[38]

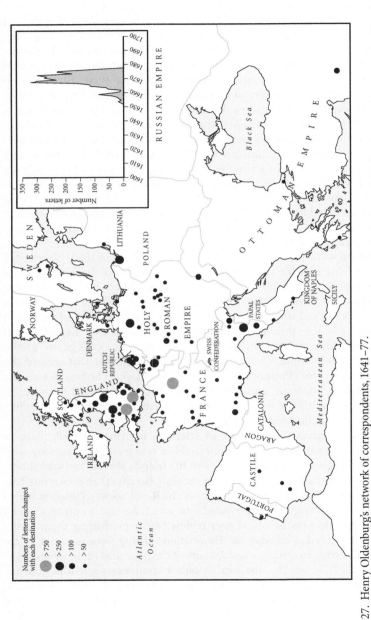

27. Henry Oldenburg's network of correspondents, 1641–77.
Over 3,000 of Oldenburg's letters are known for the years 1641–77, revealing that he had contacts all over Western Europe as well as in Scandinavia, Poland and the Ottoman empire. As founding secretary of the Royal Society of Great Britain, Oldenburg corresponded with far more scholars, and with scholars in far more countries, than any previous 'scientist'.

By 1700, similar scientific journals existed all across Europe: the *Journal des Sçavants* (Paris, from 1665, in French), the *Acta eruditorum* (Leipzig, from 1682, in Latin), the *Nouvelles de la République des Lettres* (Amsterdam, from 1684, in French), the *Monatgespräcke* (Leipzig, from 1688, in German), the *Boekzaal van Europe* (Rotterdam, from 1692, in Dutch), and many more. All of them reviewed and discussed books and ideas, and as such played a crucial role in disseminating scholarship, despite the obstacles posed by distance. In 1685 the *Journal des Sçavants* boasted that, whereas in the past it had proved difficult to secure copies of recent works published abroad, 'today, by means of the learned journals' readers are 'informed of everything that happens; and we learn each month what we only used to find out after some years'.[39]

As Alan Ross has noted, academicians were not the only ones to be informed of everything:

> European scholarship was a confusingly decentralized affair in the early modern period, focused less on institutions than the elusive 'Republic of Letters' ... Teachers, physicians, clerics, scribes, and printers – that is, members of those professions to which a Latin school education and some time at university provided access – all felt entitled to add their penny's-worth to the scholarly discussions of the day.

Christian Daum, rector of the Latin school at Zwickau in Saxony (where he had been both pupil and junior teacher), published over twenty books and owned 10,000, and received over 5,000 letters from almost 500 correspondents. He may have been 'a foot-soldier of the seventeenth-century Republic of Letters', serving in a relatively obscure outpost, but the fact that the Republic of Letters was 'constructed from below made it possible for individuals from relatively humble backgrounds and education to be at the centre of European intellectual life'.[40]

Personal contacts also advanced scientific knowledge. Many foreign scholars attended meetings of the Royal Society, and its fellows visited scholars on the continent. When two future fellows, Philip Skippon and his tutor John Ray, visited Naples they attended some of the weekly meetings of the Academici Investigantes, joining about sixty others to hear a paper that 'defended the Lord Verulam's [Bacon's] opinion' and to watch an 'experiment'. Everyone, they found, was 'well acquainted with writings of all the learned and ingenious men' of Europe, whether dead (such as Bacon, Harvey, Galileo and Descartes) or alive (the academics mentioned the work of Robert Boyle, Thomas Hobbes and Robert Hooke).[41]

The Republic of Letters also included practitioners who lived east of the Elbe and south of the Pyrenees. The Danzig brewer and astronomer Johannes Hevelius studied at Leiden, became a fellow of the Royal Society and welcomed Edmond Halley and other prominent scientists to the observatory attached to his house. In Spain, Miguel Marcelino Boix y Moliner asserted in a book entitled *Hippocrates illuminated* (1716) that 'the foreign doctors and philosophers of the last century' had only managed to 'make great advances' by plagiarizing their Spanish precursors. He singled out the work of 'Gideon' Harvey on the circulation of the blood, 'Renato' Descartes on philosophy, and Richard Morton on cinchona bark, all of whom (he claimed) had simply replicated earlier research by Spanish scholars – three little-known examples of contested multiples.[42]

The Limits of the Scientific Revolution

European scientists had less success when they tried to understand and explain the weather. Galileo had impressed on his illustrious pupil, Grand Duke Ferdinand of Tuscany, that instrumental observations and experiments could reveal the secrets of nature, and members of Florence's Accademia del Cimento (Experimental Academy) invented a rain-gauge to measure precipitation, an evaporimeter to measure humidity, a barometer to measure air pressure and a thermometer to measure atmospheric temperatures. In 1654 the grand duke established an international network of eleven observation stations, each equipped with identical instruments and protocols, to perform synchronized measurements of temperatures (and, in one station, also of atmospheric pressure) several times a day. An observer at each station recorded each reading on a standard sheet, and mailed copies to the grand duke. The network had assembled over 30,000 readings by 1667, when operations ceased under pressure from the papacy, which feared that the results would reinforce Galileo's dangerous idea that instrumental ideas were superior to the Bible in interpreting nature; and Grand Duke Ferdinand's death three years later ended any hope of processing the data.[43]

In England, Robert Hooke (curator of experiments of the Royal Society) proposed in 1663 a 'Method for making a history of the weather'. This would involve measurement in numerous stations of eight variables: half by standardized instruments (wind direction, temperature, humidity and air pressure) and the rest by observation (cloud cover, thunderstorms, 'any thing extraordinary in the tides', 'aches and distempers in the bodies of men', and 'what conveniences or inconveniences may happen in the year, in any kind, as by flouds, droughts, violent showers etc'.) Hooke designed a chart, divided into columns, on which a month's daily observations could be recorded at each station. He hoped to find 'in several parts of the world, but especially in distant parts of this kingdom, [people] that would undertake this work' of record-keeping. Here, however, lay the fatal flaw: making and distributing the delicate calibrated instruments, and paying the observers in each station, required money, and the Royal Society had none (its budget consisted of the annual dues from each fellow, many of whom failed to pay). The scheme therefore remained unrealized.[44]

Several Englishmen nevertheless kept a weather diary, including the London schoolmaster John Goad, who made detailed observations from 1652 to 1685, when he published many of his data, juxtaposed with planetary movements, and a tentative analysis of the results. Goad also exchanged information with the antiquarian and astrologer Elias Ashmole, who like him made a daily record of observed precipitation, wind direction and so on. These men knew what they were about. 'Casualty is inconsistent with science', Goad stated: since climate was not a matter of chance, he expected scientific observation to reveal patterns, and so permit prediction. John Locke, who kept a 'Register of the weather' throughout the year 1692, agreed:

> If such a register as this, or one that was better contriv'd with the help of some instruments that for exactness might be added, were kept in every county in England and so constantly published, many things related to the air, winds, health, fruitfulness, etc., might by a sagacious man be collected from them and several rules and observations concerning the extent of winds and rains etc be in time established, to the great advantage of mankind.

Robert Plot, director of experiments for the Oxford Philosophical Society, likewise hoped that the close study of the weather would empower scientists to

> learn to be forewarned certainly of divers emergencies (such as heats, colds, dearths, plagues, and other epidemical distempers) which are now unaccountable to us; and by their causes be instructed for prevention or remedies. Thence too we may hope to be informed how far the positions of the planets, in relation to one another and to the fixt stars, are concerned in the alterations of the weather, and in bringing and preventing diseases, or other calamities.[45]

Plot's rationale reveals two important limitations of the Scientific Revolution. First, the aim was admirable but unobtainable. Even in the twenty-first century, the wealth of meteorological data harvested via terrestrial and satellite observations does not suffice to forewarn us 'certainly of divers emergencies (such as heats [and] colds ...)', so that we can 'be instructed for prevention'. Even in August 2003 no one predicted – and, with the available technology, no one *could* have predicted – a heatwave in many parts of Europe of unprecedented intensity and duration, and yet it was 'the worst natural disaster in contemporary French history'. In several regions of France the normal mortality rate doubled; almost 20,000 people died prematurely in just two weeks.[46]

Plot's hope that systematic observation would show how far the positions of the planets affected the weather, diseases and other calamities revealed the second limitation of experimental philosophy: he still believed that occult forces shaped his environment. He was not alone. Bacon's last (and most popular) work, *Sylva sylvarum*, contained a chapter on telepathy, wart-charming and witchcraft; Harvey carried out experiments to see whether or not those who claimed to be witches had supernumerary nipples or familiars who performed supernatural tricks; and Boyle's last book (like Bacon's) examined the supernatural. Even Isaac Newton, knighted by the queen of England for his services to science, bought many books on magic, conducted alchemical experiments, and used the Book of Daniel to calculate the year in which the world would end (2060 AD).

The appearance of two comets in 1664–65 spawned over 100 publications filled with dire predictions that they would lead to war, plague, famine, drought, gales, floods, the death of princes, the downfall of governments and perhaps the end of the world. Two other brilliant comets in 1680 and 1682 had a similar impact. Almost 100 works discussing their significance appeared in Germany and German-speaking Switzerland, over thirty more in Spain, nineteen in France and the Netherlands, seventeen in England and its American colonies, and six in Italy. Even John Evelyn, a founding fellow of the Royal Society, wrote a detailed description of the comet in his diary because 'What this may portend (for it was very extraordinary) God only knows', and he prayed that 'God avert his judgements: we have had of late several comets, which though I believe appear from natural causes, and of themselves operate not, yet I cannot despise them. They may be warnings from God.'[47]

These reactions differed little from those of educated observers elsewhere in the world. In Sri Lanka, a chronicler associated the outbreak of rebellion against the king of Kandy with the 'fearful blazing star' of 1665. In India, 'during the whole duration of the comet' Aurangzeb, a devout Muslim, 'drank only a little water and ate a small quantity of millet bread', and also 'slept on the ground with only a tiger skin over him'. In 1680, in China, the Manchu bannerman Dzengšeo and his comrades fighting in Yunnan watched the progress of the comet with misgivings and argued 'that if it advanced towards the imperial palace it would be a bad omen'. In Massachusetts Increase Mather, president of Harvard College, delivered a sermon about the 1680 comet entitled 'Heaven's alarm to the world' warning 'that fearful sights and signs in heaven are the presages of great calamities at hand'. He later composed his comprehensive *Kometographia, or a discourse concerning comets*, which reminded his readers that the Thirty Years' War and the depopulation of the native population of New England had followed the comets of 1618, while the plague and fire that devastated London had followed that of 1664. It was only because 'we that live in America know but little of the great motions of

Europe, much less in Africa and Asia, until a long time afterwards', he insisted, that news of yet more catastrophic consequences had not reached Boston.[48]

Descartes took a different approach. In 1632 he expressed the hope that someone would publish in a single book all 'the observations of comets, with a table of the paths of each one', predicting that 'such a work would be of greater public utility than it might seem at first sight'; but (he continued with a sigh) 'I have no hope that anyone will do it' because 'I think it is a skill beyond the reach of the human mind'. Newton took up this challenge in the 1680s, carefully copying into his notebooks the descriptions he found in Aristotle, medieval chronicles and more modern accounts, as well as the observations made by his contemporaries – not only his compatriots Edmond Halley (who travelled to several European observatories to check their records) and John Flamsteed (the astronomer royal) but also the Jesuit Valentin Stansel from Brazil, the Harvard astronomer Thomas Brattle and his former schoolmate Arthur Storer, now a planter–slave-owner in Maryland, who transmitted outstanding observations of the 1682 comet. Newton incorporated these and other findings gleaned from informants all around the world in his *Mathematical principles* of 1687, of which the third book (entitled 'The System of the World') contained a long section about comets. Newton also used his newly invented mathematical technique, later known as calculus, to plot the course of the 1680 comet and concluded that it had come from outer space on a parabolic curve around the sun and would never return.[49]

Here Newton was wrong, as were other scientists who deduced that comets followed an elliptical orbit and so periodically returned. The French astronomer royal, Pierre Petit, argued (incorrectly) that the 1664 comet was the same as that of 1618, and would reappear in 1710; his successor Giandomenico Cassini deduced (also incorrectly) that the 1680 comet had previously appeared in 1577 and would therefore reappear in 1784. Edmond Halley would also have reached the wrong conclusion if he had decided to study the 1680 comet, but instead he focused on the one that appeared two years later – which happened to be the only short-period comet that is clear to the naked eye (others may be as bright, or brighter, but they appear only once in a millennium or more).

Like Newton, Halley combined observation with historical study, and his close reading of the accounts of all previous apparitions convinced him that the comets that appeared in 1531, 1607 and 1682 were one and the same, and that it must therefore circle the sun in a 75- or 76-year cycle. He then calculated its exact orbit, using Newton's physics to account for the effect of the gravitational pull of the planets that it passed. In 1705 Halley published a short pamphlet in Latin and English that set out his rationale about the orbit of comets in general, and the exact periodicity of the 1682 comet in particular. He then 'dare[d] venture to foretell' that the same comet 'will return again in the year 1758'. Although Halley did not live to see it, when a comet duly appeared in 1758, identical to the one described in 1682, contemporaries named it Halley's Comet in his honour. Halley's short pamphlet included another prediction: although no previous comet 'has threaten'd the Earth', he noted that the orbits of some comets came close enough to raise the possibility that one day a collision might occur, and he ended his pamphlet with the ominous words: 'What might be the consequences' of 'the shock of celestial bodies (which is by no means impossible to come to pass), I leave to be discuss'd by the studious of physical matters'.[50]

Halley's 1705 pamphlet epitomized the dazzling achievements (and self-confidence) of the Scientific Revolution. It provided irrefutable proof that comets, like planets, orbited the sun; it offered an impressive test of Newtonian physics; and it also made two accurate predictions: the first vindicated when the comet returned after exactly seventy-six years, the second when part of a comet struck Jupiter with dramatic results almost three centuries later. Yet no one at the time appreciated how dazzling Halley's achievement really was: even his first prediction could not be verified for seventy-six years.

One might therefore object that, in the scientific sphere as in the field of education, the Great Divergence between Europe and the rest of the world had not yet occurred; but this overlooks an important difference. In his introduction to the 2009 special issue of the *Journal of Interdisciplinary History*, dedicated to the General Crisis in Europe, Theodore K. Rabb noted 'the transformation of attitudes toward science', which 'in the decades after 1640 went from concerns about its bewildering, contested and controversial quest for knowledge to the acceptance of scientific method as the magisterial form of intellectual endeavour. To move from the condemnation of Galileo to the knighthood of Isaac Newton is to traverse a fundamental divide in European thought.' In another article, Rabb stressed that this divide lay not only in the advances made by scientists, but also in the political and social context in which their intellectual endeavour took place. In the first half of the seventeenth century, it

> corresponded closely with a growing malaise, as religious war, economic boom and bust, the spread of ever more virulent military conflicts, and the encroachment of centralizing governments on local autonomies caused widespread disruption and distress. In this context the unprovable demands that the scientists were putting forward for jettisoning old ideas seemed both a symptom of, and a stimulus to, the fires of doubt.

After 1650, by contrast, 'looking to restore a sense of confidence, Europe's elite found reassurance and a tangible certainty in the increasingly united claims for new truths about the physical world'. Rabb argued that, considered in isolation, 'the discoveries would not necessarily have gained wide acceptance. It was their role in restoring confidence in human capacity that gave them their magisterial status.' Useful knowledge had ceased to be threatening.[51]

Nevertheless this new confidence remained confined to a relatively small part of Europe. On their visit to Naples in 1663, Skippon and Ray heard complaints from the Academici Investigantes 'of the Inquisition, and their clergymen's opposition to the new philosophy; and of the difficulty they met with in getting books out of England, Holland etc'. Four years later, in Tuscany, papal pressure ended the grand duke's efforts to gather and study serial data on the climate. In England, Halley's doubts about the veracity of the Bible (for example, questioning whether the Creation had been a single act) cost him the chair of astronomy at Oxford University in 1691: the astronomer royal (one of his referees) warned that, if appointed, Halley would 'corrupt the youth of the university with his lewd discourse' and Newton and others duly prevented the election of such a subversive.[52] In Scotland, the High Court tried and executed a student at Edinburgh University for blasphemy in 1697.

It is here that Robert Merton's 'multiples' proved critical. In Europe, censorship in one country no longer affected scientific innovation elsewhere: despite the papal ban on holding heliocentric opinions, John Milton's *Paradise lost*, first published in London in 1667, contains an admirable summary of Galileo's *Dialogue of the two systems* as well as numerous references to telescopic observation, rotating sunspots and irregularities on the moon's surface. Likewise, although in 1663 the Roman Inquisition placed the works of Descartes on the Index of Prohibited Books, a few years later Louis XIV chose Jacques Rohault, France's leading Cartesian, as mathematics and philosophy tutor to his heir; and in the 1670s the young man attended open discussions of the heliocentric theory at the Academy of Sciences, and followed the rotation of the comets, stars and planets during his visits to the observatory created by his father. The Scientific Revolution had been nationalized.[53]

The situation in China was very different. In 1661 the Qing government set up a literary inquisition to review an unofficial *Ming history* published by a group of Han scholars in Jiangnan. Deeming it seditious, they executed some seventy men involved in the venture, including historians, printers, and even purchasers of the work; exiled their male relatives and condemned their female relatives to serve as slaves in Manchu households; confiscated all their property; and burned all copies and all blocks of the book. As in Europe, harsh penalties led to widespread self-censorship among survivors. A century later, the central government initiated a massive bibliographic project known as *The complete library of the four treasuries*, which aimed to secure a copy of all known books for the imperial library – but 'If the books contain language that is anti-dynastic', the emperor commanded, 'then the woodblocks and printed sheets must both be put to the flames. Heterodox opinions must be quashed, [so] that later generations may not be influenced.' The literary inquisition, which lasted for fifteen years, destroyed 70,000 printing blocks and incinerated almost 4,000 works, while authors and printers purged many other works themselves, re-carving some individual blocks and removing censored names and statements from others.[54]

The Qing continued to see intellectual innovation and learned discussion as potential threats, not potential assets. For them, 'new truths about the physical world' continued to seem (in Rabb's words) 'both a symptom of, and a stimulus to the fires of doubt'. Unlike rulers in northwest Europe, China's new masters denied their leading scholars both freedom of expression and freedom to exchange ideas. In the intellectual as in the economic sphere, reactions to the Global Crisis stimulated the Great Divergence.

Conclusion: The Crisis Anatomized

Winners and Losers

Crane Brinton's classic study *Anatomy of Revolution*, first published in 1938, sought 'uniformities' between the political upheavals in seventeenth-century England, eighteenth-century North America and France, and twentieth-century Russia. In the last chapter, Brinton asked 'What did these revolutions really change?', and he answered:

> Some institutions, some laws, even some human habits, they clearly changed in very important ways; other institutions, laws and habits they changed in the long run but slightly if at all. It may be that what they changed is more – or less – significant than what they did not change. But we cannot begin to decide the last matter until we have got the actual changes straight.

The preceding pages have attempted to get straight what changed and what stayed the same in the seventeenth-century world; it now remains to assess their relative significance and to ascertain the uniformities among the fifty or more revolutions and rebellions that occurred around the world between 1618 and 1688.[1]

At the individual level, the most significant change for most contemporaries was a sharp deterioration in the overall quality of life. During the Indian famine of 1631, English merchants living in Gujarat considered that 'The times here are so miserable that never in the memory of man [has] the like famine and mortality happened.' Ten years later, in China, according to the diary kept by a young scholar, 'Jiangnan has never experienced this kind of disaster'. In 1647 a Welsh historian opined: ' 'Tis tru we have had many such black days in England in former ages, but those parallel'd to the present are to the shadow of a mountain compar'd to the eclipse of the moon.' During the famine of 1649, a Scot lamented that 'The pryces of victuall and cornes of all sortes wer heigher than ever heirtofore aney[one] living could remember.'[2]

In a diary entry during the winter of 1660–61, the Essex clergyman Ralph Josselin cast doubt on such extreme verdicts, observing 'how apt wee are to account a harsh time the hardest wee ever felt and a mild the best'.[3]

Pessimists who claim that the hardships they face are the 'worst in living memory' can indeed be found in all times and places, but the surviving evidence from the mid-seventeenth century suggests that the lamentations of those alive then were justified: decreased solar energy, increased volcanic and El Niño activity, frequent wars and state breakdowns created several decades of unparalleled adversity.

Some groups suffered disproportionately. Slaves led the way. In China, in parts of Europe (notably Britain and Ireland), and above all in Africa, millions of men and women lost their liberty and often their lives when they became slaves; millions more in Russia and Eastern Europe lost their liberty when they became serfs. Women, whether free or unfree, also suffered disproportionately in most parts of the world. Many killed themselves after they had been raped and otherwise humiliated, or because they were destitute, or because they did not wish to survive the death or disappearance of their loved ones. Many survivors faced (in the memorable words of 'a poor woman with only a small field or so to her name' in Germany) a 'bitter living' (Chapter 4): she and her sisters had to work harder and longer just to stay alive.

Admittedly, some European women used the 'weapons of the weak' to retaliate against their oppressors. Female workers and servants abused by their employers could seek revenge not only through foot-dragging, pilfering and slander but also (in extreme cases) through arson and murder. Wives could plead with their abusive husbands in private, and also complain to their neighbours and to the courts; they could seek (or threaten) divorce; and they could threaten (or *in extremis* inflict) grievous bodily harm. In London, Elizabeth Pepys used all of these strategies in 1668 after she discovered her philandering husband Samuel making love to their sixteen-year-old servant. After tears and reproaches, she threatened to tell their neighbours, to leave him, even to join the Catholic Church. She also attacked Samuel with a pair of red-hot tongs and threatened to slit the servant's nose (a popular punishment for adultery); but Elizabeth's most effective weapon lay 'in matters of pleasure'. After three weeks of enforced celibacy, Pepys felt 'troubled to see how my wife is by this means likely for ever to have her hand over me, that I shall be for ever a slave to her'. Only English and Dutch women seem to have enjoyed this limited ability to retaliate, however. Their sisters in other areas of Europe might well ask, like Queen Christina of Sweden in the 1680s, 'What crime has the female sex committed to be condemned to the harsh necessity of being shut up all their days either as prisoners or slaves? I call nuns "prisoners" and wives "slaves".'[4]

Wars and revolutions killed and maimed women and men both directly through violence and brutality and indirectly through forced migration and destruction of property, especially in Western and Central Europe during the Thirty Years' War, in Eastern Europe and Russia during the Thirteen

Years' War, and in China during the Ming–Qing transition. For many people, the Global Crisis proved a terminal event, and it claimed the lives of so many millions that they amounted to a lost generation.

In some areas, a whole way of life disappeared. The Ming–Qing transition permanently destroyed sericulture in the province of Shaanxi, and the Gujarat famine and floods of 1628–31 did the same to one of India's premier cotton- and indigo-producing areas (Chapters 5 and 13). The plague epidemic that spread through Southern Europe in the decade after 1649, killing one-half of the inhabitants of Seville, Barcelona, Naples and other port-cities, set the seal on the decline of the Mediterranean as the heart of the European economy for ever. In many other areas, if the observations of Alex de Waal and Scott Cane concerning the effect of a prolonged 'hungry time' on farmers of marginal lands and on hunter-gatherers in the twentieth century (Chapters 3 and 15) also prevailed in the seventeenth century, then countless families and many communities must have crossed 'a threshold of awfulness' and perished, leaving no trace.

Of course, the turmoil did not only produce losers. In East Asia, both Nurhaci and Tokugawa Ieyasu were revered as gods soon after their deaths. Even today, about 130 Tōshōgū shrines in Japan still honour the divinity of the first shogun, issuing their own newsletter, making him by far the most successful denizen of the seventeenth-century world. The descendants of Alexei Romanov also prospered from the political, economic and social balance created by the crisis of 1648–49, cementing their control over an empire that expanded at the rate of 55 square miles a day – more than 20,000 square miles a year – for the next two centuries.

Many followers of these rulers also profited from the upheaval. Nurhaci and Ieyasu bequeathed to their numerous descendants a luxurious lifestyle that would endure for more than two centuries, and tens of thousands of Manchu bannermen and their families exchanged a precarious existence on the steppe for a life of plenty south of the Great Wall. Most of the military and civilian officials who swiftly transferred their allegiance from Ming to Qing also prospered: of 125 senior officials who received the ambiguous title *Er chen* ('ministers who served both dynasties'), 49 later became president or vice-president of a department of state. In Japan, the Tokugawa clansmen and most of their *daimyō* allies enjoyed more than two centuries of peace and plenty following the proclamation of the Genna Armistice in 1615. In Russia, the descendants of the nobles who in 1649 won control over their serfs through the *Ulozhenie* likewise maintained their gains for over two centuries. In India, the leaders who supported Aurangzeb when he challenged his father and brothers during the Mughal Civil War of 1657–59 became stakeholders in the richest state on Earth.

In Europe, among civilians, government office allowed Samuel Pepys to increase his personal fortune from £25, when he began to serve in 1660, to

£10,000 ten years later; Jean-Baptiste Colbert, who at school 'was so dull that he was always bottom of the class', died a millionaire and (thanks to Louis XIV) bequeathed a hereditary peerage to his son. Among soldiers, Sweden's commanders in Germany who survived the Thirty Years' War returned home with immense wealth: the castle of Skökloster near Stockholm testifies even today to the booty acquired by General Carl Gustav Wrangel; his colleague Hans Christoff Königsmarck, who began as a common soldier, died a nobleman with assets worth 2 million thalers. In England luck and good judgement during the Civil War allowed George Monck, the younger son of a squire (and lucky as a young man to escape hanging after he murdered a deputy sheriff), to become duke of Albemarle and commander-in-chief of England's armed forces in 1660, and to die with assets worth £60,000. Monck's followers also prospered. In return for facilitating the Restoration, the general insisted on full payment of the wage arrears of his men, and over the next two years the king's treasurers-at-war paid them £800,000.[5]

The balance sheet of many other states showed both losses and gains. Although Qing China and Romanov Russia topped the list of successful states, millions of their subjects lost either their lives or their freedom. In Ukraine, although Ruthenian culture flourished (and even spread to Russia), while serfdom disappeared, the name *Ruina* given by its historians testifies to the costs of the struggle to shake off Polish rule. Portugal exploited Spain's weakness to gain independence – the only entirely successful rebellion of the seventeenth century – but once again this success reflected immense material and personal sacrifices and led to the temporary loss of its colonies in Africa and Brazil. The Dutch Republic gained formal recognition as an independent state, and carved out a lucrative trading empire in south and southeast Asia at the expense of Portugal and independent rulers like the sultan of Mataram; but it lost its colonies in North and South America. Britain's brief republican experiment secured the Caribbean island of Jamaica, and commercial dominance in the North Atlantic, but the civil wars caused the premature death of perhaps 500,000 people in Britain and Ireland. Scotland and Ireland temporarily lost their independence. The weakness of the Ottoman empire allowed the Austrian Habsburgs to conquer most of Hungary, and the weakness of the Spanish Habsburgs allowed Louis XIV to advance the frontiers of France, but in both cases expansion cost hundreds of thousands of lives.

Other states suffered grave political losses in the mid-seventeenth century and gained little or nothing. The kingdom of Kongo in Africa, the Pequot of New England and the Wendat of New France all perished. The Polish-Lithuanian Commonwealth lost half its population, temporarily ceased to exist as an independent state and lost for ever its status as a great power. The Spanish monarchy, too, never recovered its political pre-eminence after the secession of Portugal and its overseas empire; and although Philip IV eventually overcame his rebels elsewhere, he did so only after making major concessions (in Catalonia, for example, he left the Constitutions intact and pardoned virtually all those who had defied him). In East Asia, the Shun dynasty founded by Li Zicheng in China disappeared without trace; and the fall of the Ming forced a reconstruction of Korean identity because it 'shattered the premise concerning the world order of which the Koreans felt they were a part' (just as it required Han Chinese intellectuals to refashion themselves).[6]

With the exception of Japan, New England and New France, the demographic balance of the seventeenth century was negative. Apart from the cases of drastic population loss already cited – China; Russia and Ukraine; the Polish-Lithuanian Commonwealth; most of Germany – Philip IV ruled far fewer subjects at his death in 1665 than at his accession four decades before. Famines, epidemics and civil war meant that Philip's son-in-law, Louis XIV, also ruled far fewer subjects when he began his personal rule in 1661 than at his accession eighteen years before, and the Little Ice Age combined with his repeated wars meant that he probably ruled fewer subjects at his death than when he began his personal rule.

In Search of Common Denominators

According to political scientist Mark Hagopian's book, *The phenomenon of revolution*, 'those seeking simplicity should study something else than the causes of revolution' because 'there is good reason to doubt the "completeness" that any explanation of revolution could possibly attain'. Francis Bacon agreed. In an essay entitled 'Of seditions and troubles', first published in 1612, he listed eleven 'causes and motives of sedition', namely:

> innovation in religion; taxes; alteration of lawes and customs; breaking of priviledges; general oppression; advancement of unworthy persons; strangers; dearths; disbanded soldiers; factions growne desperate; and whatsoever, in offending people, joyneth and knitteth them in a common cause.[7]

In his lectures to the Statistical Society of London in 1878 on 'The famines of the world: past and present', Cornelius Walford analysed just one of Bacon's categories: dearths. Walford discerned thirteen distinct precipitants of harvest failure, some natural (excessive rain, frosts, droughts, 'plagues of insects and vermin' and sunspot cycles), others human (war, 'defective agriculture', insufficient transport, legislative interference, currency manipulation, hoarding, and diverting grain from making bread to other purposes such as brewing or distilling). Walford also noted an 'enigma': that 'the very remedies which have been adopted to prevent, or to mitigate the severity of, these periodical visitations [of famine], have by some reflex action, apparently, either aided in producing them, or at least added very much to the severity of the results flowing from them' – results that might include rebellion and sometimes revolution. Nevertheless, after analysing nineteenth-century evidence from England and British India, Walford concluded that extreme climatic events normally played a greater role than human action in creating catastrophe.[8]

Does the seventeenth-century evidence support this analysis? Certainly, the major revolts almost all broke out in a period of unparalleled climatic adversity, notably when a 'blocked climate' produced either prolonged precipitation and cool weather or prolonged drought (1618–23, 1629–32, 1639–43, 1647–50, 1657–58 and 1694–96). Some areas suffered for longer: both Scotland (1637–49) and Java (1643–71) suffered the longest droughts in their recorded history. The century also saw a run of landmark winters, including some of the coldest months on record; two 'years without a summer' (1628 and 1675); and an unequalled series of extreme climatic events (the freezing over of both the Bosporus in 1620 and the Baltic in 1658); and the maximum advance of both the Alpine and Andean glaciers in the 1640s; Scandinavia experienced its coldest winter ever in 1641. These various climatic aberrations accompanied a major episode of global cooling in both northern and southern hemispheres that lasted at least two generations: something without parallel in the past 12,000 years. The famines caused by these climate changes caused what would today be called a humanitarian crisis in which millions of people died.

These same years of dearth also saw rebellions and revolutions, with two distinct peaks: Normandy, Catalonia, Portugal and its overseas empire, Mexico, Andalusia, Ireland, Scotland and England in 1639–42; Naples, Sicily, France, England (again), Russia, the Ottoman empire and Ukraine in 1647–48. Sometimes a link between rebellion and climate change is manifest. In Scotland, the summer of 1637 (when Charles I sought to impose his new liturgy) was the driest in two decades; the summer of 1638 (when he refused to make concessions to his Scottish opponents) was the driest in a century; the summer of 1639 (when he attempted to invade) was the coldest in more than a century. Government innovation and inflexibility at a time of such climatic adversity drove many Scots reluctantly into rebellion. The earl of Lothian, a prominent landowner, spoke for many when (having described how, in October 1637, 'the Earth has been iron in this land', ruining the harvest) he wrote 'I think I shall be forced this term to run away and let the creditors of the estate catch that catch may, for I cannot do impossibilities'. In the event, instead of running away his lordship signed the National Covenant and in 1640 led a regiment in the invasion of England, declaring that 'necessitie made us come from home' and 'in our laufull defence WE DARE DIE'.[9] In Ireland, too, the failed harvests of 1638–41 caused widespread hardship among the Catholic population, disposing many to support the rebellion that began in October 1641, when ice and snow covered many parts of the island. Then followed 'a more bitter winter than was of some years before or since seen in Ireland', which turned the brutal mistreatment of Protestant settlers by their Catholic neighbours into a massacre that would in turn provoke massive retaliation.[10] Likewise, in East Asia, repeated harvest failures caused by adverse weather in the early 1640s had two dramatic political effects. Famines and popular rebellions in Jiangnan fatally weakened the Ming as they struggled against the inroads of bandits from the northwest; drought and cold in Manchuria so reduced harvest yields that the Qing leaders concluded that invading China offered the only way to avoid starvation.

Climate-induced dearth contributed to many other rebellions. Perhaps, as Leon Trotsky wrote of the Russian Revolution of 1917, 'the mere existence of privations is not enough to cause an insurrection; if it were, the masses would always be in revolt' – but privations in the mid-seventeenth century apparently formed an exception. The revolts in Évora in 1637, Palermo in 1647, Fermo in 1648 and Andalusia in 1652 began in the same way as the greatest rebellion of the twentieth century in Petrograd in 1917: when adverse weather ruined a harvest and created a food shortage that brought hungry people onto the streets shouting 'bread'.[11]

In such a tense situation, even a small increase in government pressure could produce a disproportionate popular reaction. The revolt of Sicily in 1647 began when the government ended subsidies that had kept down the price of bread; the Naples revolution a month later began when the viceroy re-imposed an unpopular excise on fruit. In both cases, Philip IV overrode the misgivings of his ministers because he needed funds to pay for his wars – despite the fact that domestic rebellion opened a 'second front'. The same perverse logic prevailed in the French monarchy, where Louis XIII repeatedly raised taxes in times of high food prices so that his subjects had no money left to buy bread. 'Long live the king; death to taxes!' became the cry of rebellious subjects throughout Europe.

Governments could stimulate or spread insurrections by other means. Charles I's insistence on imposing a new liturgy on Scotland in 1637 inflamed and united his opponents as nothing else could have done. The desecration by royal troops of churches in villages that defied them had the same effect in Catalonia in 1640, and so did the decision of the Qing regent Dorgon to enforce the head-shaving edict on all males in China in 1645. The Scottish Revolution lasted a decade; the revolt of the Catalans lasted eighteen years; the resistance of Ming loyalists lasted thirty-eight years.

Government ineptitude could also encourage resistance. In Naples the inability of the *eletto del popolo* to settle a squabble over who should pay the fruit excise on the morning of 7 July 1647 allowed Masaniello and his *ragazzi* to galvanize irate bystanders into action. During the summer of 1648, revolts broke out in Moscow when the tsar refused to receive a supplication from his subjects that condemned corruption among his ministers, and in Paris when the queen regent botched an attempt to arrest her leading opponents as they left a service in Notre Dame Cathedral. In Barcelona in 1640, in Naples in 1647 and in Messina in 1674, rebellions began just after galley squadrons based in each port-city departed to fight elsewhere.

Violent opposition to governments in the mid-seventeenth century often began in a capital city – a circumstance that reflected the greater vulnerability of all urban areas to both climate change and government pressure. The major revolts against Charles I began in Edinburgh, Dublin and London; those against Philip IV in Barcelona, Lisbon, Palermo and Naples. Other insurrections that rocked and sometimes toppled seventeenth-century regimes also began in capital cities: Prague in 1618; Istanbul in 1622, 1648 and 1651; Manila in 1639; Paris in 1648; Moscow in 1648 and 1662; Tokyo in 1651.

Popular protests alone rarely toppled governments, however, and all the major rebellions of the mid-seventeenth century included members of the secular and, in most Christian and Muslim societies, also the clerical elite. Churchmen headed four rebel governments, at least for a time (Henderson in Scotland, Claris in Catalonia, Rinuccini in Ireland and Genoino in Naples); while throughout the French, Stuart and Spanish monarchies, clerics preached sermons and published propaganda in support of the rebel cause. In the Polish Commonwealth, the Ukrainian clergy threw its weight behind Khmelnytsky, and in the Ottoman empire the chief mufti played a pivotal role in legitimizing the deposition (and subsequent murder) of the sultan in 1622 and again in 1648.

Noblemen, too, took the lead in several European revolts – Condé and Longueville in France; Argyle and Hamilton in Scotland; Antrim and Maguire in Ireland; Essex and Manchester in England – and in all four countries virtually the entire nobility participated in the resulting civil war. In Portugal, Duke John of Bragança founded a new royal dynasty in 1640; in Castile, the duke of Medina Sidonia sought to become the head of an independent Andalusia in 1641; seven years later the duke of Guise established the short-lived Royal Republic of Naples.

Most of the remaining leaders of the major mid-seventeenth-century rebellions belonged to the intellectual elite. At least 80 per cent of the members of the English House of Commons between 1640 and 1642, and many English peers, had either studied law at the Inns of Court, or gone to university, or both. The Fronde in France began with the revolt of its senior judges. Those who had mastered China's national curriculum and started to climb its administrative ladder of success by passing state examinations took the lead both in paralyzing the Ming with factionalism and in opposing the Qing with suicidal energy.

Most insurgents in Europe claimed that they desired only the restoration of some past Golden Age. Rebels in Palermo and Naples demanded a return to the charters granted a century before; Catalans called for respect for their ancient Constitutions; the Portuguese wanted a return to the relationship with the king created at the Union of the Crowns in 1580 (and when they could not get it, a restoration of the constitutional situation that had prevailed *before* 1580). Initially, Charles I's enemies also called for a return to the past. In England, they demanded government by the crown-in-Parliament, as created by his predecessors; in Ireland, Catholics sought implementation of the Graces, which would end the recent trend in Protestant expansion at Catholic expense; in Scotland, Covenanters insisted on retaining their traditional liturgy. In France, judges wanted a return to the constitutional balance of power that they believed had prevailed in the Middle Ages; and nobles saw the liberties and franchises won by the blood of their ancestors in the service of the crown as their birthright, and to defend it they felt a duty to rebel. In Russia, the crowd wanted the tsar to accept their petitions as he and his predecessors had done before.

Rebels in other parts of the world also drew strength from precedents. In China, Li Zicheng, Zhang Xianzhong and the Qing, all of whom strove to replace the Ming dynasty, cited earlier examples (some of them two millennia earlier) of dynasties that had lost the Mandate of Heaven; and Wu Sangui would do the same in 1673 when he initiated the Revolt of the Three Feudatories against the Qing. In the Ottoman empire, Kadizade Mehmed and his followers called for a return to the political and religious conventions that had prevailed at the time of the Prophet Mohammed a millennium before. Many others, such as the Nu-Pieds of Normandy, demanded a return to a Golden Age when justice had prevailed. To quote Crain Brinton: 'Revolutions cannot do without the word "justice" and the sentiments it arouses.'[12]

Attempts to gain justice, at least in Europe, often attracted support from legal institutions of unquestioned legitimacy, such as the law courts or Parliament. To this end, rebel leaders in Scotland, Catalonia and Portugal immediately summoned representative assemblies to legitimize their challenge to established authority, as well as to enact appropriate policies and vote funds – thus creating an alternative government capable of winning widespread support both at home and abroad. In Ireland, since the Protestant-dominated Dublin government condemned the rebellion of 1641, Catholic leaders created their own General Assembly and Supreme Council at Kilkenny, which served for a decade as the government of an independent Ireland. In England, Parliament was already in lawful session when the king declared its members rebels, but both Houses continued to sit until, in January 1649, the surviving members of the Commons (the Rump) tried and executed him and then proclaimed England to be a republic with themselves as its sole sovereign body. The following year, in the Dutch Republic, the

States-General exploited the death of William II of Orange to gain control over the executive functions that he had exercised. In Ukraine, Hetman Bohdan Khmelnytsky from the first sought the approval of the assembly of Cossack freemen for his various actions, including a declaration of independence from the Polish Commonwealth and, later, a treaty of union with Russia that preserved most of the gains won by the initial revolt.

The unifying appeal of these aims helps to explain why so many seventeenth-century insurgencies lasted so long. The revolt of Bohemia against Habsburg authority in 1618 initiated a war that lasted thirty years. The revolt of Portugal against Habsburg authority in 1640 began a war that lasted twenty-eight years. The Cossacks' rejection of the authority of the Polish crown in Ukraine led to eighteen years of war.

Longevity changed the character of most rebellions. As John Wallis later observed about England: 'As is usual in such cases, the power of the sword frequently [passed] from hand to hand', because 'those who begin a war [are unable] to foresee where it wil end'.[13] None of the five MPs whom Charles I tried to arrest early in 1642 possessed military experience, and few had held executive office, so they gave way to those, like Oliver Cromwell, who could demonstrate an ability to win wars. In Naples, the lawyer-turned-priest Genoino replaced the illiterate demagogue Masaniello, only to lose his place to Gennaro Annese who possessed military experience. The rise of a second generation of more militant leaders, like Cromwell and Annese, helps to explain why revolutions became more violent the longer they lasted. At the same time, the experience of resistance habituated leaders to actions that previously would have seemed intolerable.

Rebellious regimes might also appeal for foreign aid, and in so doing fragment their domestic support. In Ireland, the Catholic Confederacy turned to their co-religionists in Europe, and although the papacy, France and Spain all provided valuable material assistance, each foreign power had its own agenda and did not scruple to create and exploit damaging domestic divisions in order to achieve them. In the Iberian Peninsula, Catalan opponents of Philip IV appealed for French assistance; and although French troops and military advisers helped to save Barcelona, Louis XIII demanded that the Catalan leaders abandon their resolve to become an independent republic and instead recognize him as their sovereign. Most spectacularly of all, in 1644 the Ming commander Wu Sangui appealed to his northern neighbours for military assistance against the bandits, and allowed the Manchu Grand Army to pass through the Great Wall to destroy them; but once this mission had been accomplished, the Manchus claimed that their victory conferred upon them the Mandate of Heaven to rule all China, which they did until 1911.

Within the composite states of Europe, opponents of the same ruler in one area often took active steps to encourage others to rebel. Immediately after his acclamation, King John IV of Portugal sent envoys to Barcelona to make common cause with the Catalan rebels; and somewhat later sent his principal advisor, the Jesuit António Vieira, to Rome to invite the pope to invest John's son as king of Naples (a papal fief). As soon as news arrived that protests in Palermo against excise duties in 1647 had secured their abolition, citizens of Naples put up 'pungent and bitter invectives' calling for 'a revolution like Palermo'; and as soon as the revolution began, 'some people from Palermo' urged the Neapolitans 'to demand everything, in the same way that had happened in Palermo'. In addition, in both kingdoms, revolt in the capital provoked copycat uprisings in numerous other towns (Chapter 14). The opponents of Charles I in different parts of his monarchy likewise created links across borders that aimed to improve their chances of success. Some Scottish ministers in northern Ireland found the earl of Strafford's religious policies so intolerable that in 1639 they chartered a vessel to take them to Massachusetts (John Winthrop had visited Ulster the previous year), but storms drove them back to their native land. They saw this as a divine sign that they should 'find an America in Scotland' and, once arrived there, joined the Covenanters' opposition to Charles I. In Russia, too, disorders spread throughout the empire largely because in June 1648 the capital was full of petitioners from provincial towns, and local uprisings followed as soon as the petitioners returned home with news of the Muscovites' apparently successful defiance of the tsar (Chapter 6).

If

Despite the unparalleled frequency of revolts in the mid-seventeenth century, it is possible to imagine a peaceful resolution to most of them. As Charles I reminded the Long Parliament in November 1640, while explaining how the Scots had managed to defeat his forces so swiftly: 'Men are so slow to believe that so great a sedition should be raised on so little ground'.[14] Accidents – totally unpredictable developments – could crucially affect the outbreak or outcome of a rebellion: the election by lot of two talented yet intransigent Catalan patriots, Pau Claris and Francesc de Tamarit, as the senior diputats of Catalonia, in 1638 (Chapter 9); the inter-regnum in the Polish-Lithuanian Commonwealth created by the death of King Władysław IV just after the Cossack rebels routed its field army in 1648 (Chapter 6); or the death from smallpox of William II of Orange without an adult heir just after he had defeated his domestic opponents in 1650 (Chapter 8).

Some accidents were more predictable – especially those that were caused by distance, which was (in Fernand Braudel's adage) 'the first enemy'. Philip IV's advisers hesitated to react immediately to the revolt of Naples 'because the state of affairs over there changes from one moment to the next, and what seems appropriate today might not be so tomorrow'; while his envoy to the Irish Catholic Confederation complained that distance constituted 'the greatest problem of my job' because 'I can neither send successive accounts of what is happening nor receive in good time the royal orders of Your Majesty'. Even within the Iberian Peninsula, as Sir John Elliott noted:

> The distance between Madrid and Barcelona meant that [the viceroy's] letters and those from Madrid never kept in step. While circumstances were changing from day to day in the Principality, Madrid was at least three days behind the news, and still legislating as if the situation was exactly the same as when the viceroy had written his last set of dispatches.[15]

Likewise, the central government in Madrid received the first reports of the Portuguese Revolution on 1 December 1640 just one week later, but refused to believe them. 'It is possible that a popular tumult might have produced a good deal of what we have heard', the Council of State informed Philip IV, 'but to proclaim a king the same day is not credible.' Philip did not sign letters warning ministers in Europe about 'the accident of Portugal' until 15 December; he did not instruct colonial administrators to take defensive measures until 27 December; he did not warn the treasure fleets coming from America to avoid Portuguese harbours until 5 January 1641; and he did not order the closing of all frontiers, both in the peninsula and in America, to commerce with the rebels until 10 January.[16]

Conversely, accidents could also unexpectedly derail rebellions. Lord Maguire's plot to seize Dublin Castle in 1641 failed because Owen Connolly, one of the conspirators, decided to betray his colleagues – but even then the magistrates 'gave at first very little credit to so improbable and broken a [story], delivered by an unknown, mean man, well advanced in his drink', and sent him away. Connolly only managed to sabotage the plot because he made a second attempt – this time successful – to betray his colleagues (albeit now too late to send a warning of the plot to Ulster, where it succeeded: see Chapter 11).[17] Likewise, ten years later, the samurai plot to seize Edo and destroy the Tokugawa regime came to light only because one of the conspirators became delirious and unwittingly shouted out the details (Chapter 16).

In each of these cases (and no doubt in many others) a minor rewrite of the historical record would produce a dramatically different outcome; and the same is true of natural disasters, such as earthquakes and volcanic eruptions, which occur with little or no warning: *if* the 1640s had not seen, at much the same time, the virtual disappearance of sunspots, much more volcanic activity, and double the number of El Niño episodes . . . Nevertheless, although contingency (like catastrophe) cannot be written out of history, when constructing 'What if?' scenarios, historians must always consider second-order (or reversionary) counterfactuals: the possibility that rewriting the *short-term* historical record, as in the examples above, might still not change the *long-term* outcome.

Reversionary counterfactuals take two forms: one positive (an accident could delay but not permanently divert a particular development) and the other negative (a development that was, so to say, an accident waiting to happen). Positive examples are relatively easy to find. From the human archive, twenty-two years after the death of William II and the Dutch Revolution that followed, his posthumous son William III recovered almost all of the traditional powers and influence of the princes of Orange; and Charles II regained virtually all of his father's powers eleven years after the regicide in 1649. Turning to the natural archive, since some parts of the planet could feed their inhabitants only in good years, even had fewer volcanic eruptions and El Niño episodes occurred in the 1640s, sooner or later bad years would have come and caused heavy mortality.

Pomponne de Bellièvre, a French diplomat in London, provided a good example of a negative reversionary counterfactual as he contemplated the situation in Ireland in 1648. He informed Mazarin that

> what surprises most of those who consider the affairs of that country [Ireland] is to see the people of the same country and the same religion, who know that the decision to exterminate them totally has been taken, so strongly divided by their private hatreds, so that zeal for their religion, the preservation of their country, and their own self-interest *does not suffice to make them abandon – at least for a while – the passions that incite them against each other.*[18]

The English conquest began the following year, and within three years Confederate Ireland was no more – but, in Bellièvre's view, even if the London government had delayed its campaign of repression, internal dissention still doomed the Catholic cause to ultimate defeat. Historian Julian Goodare has proposed a similar negative reversionary counterfactual for Scotland: given the character of both Charles I and the leading Covenanters, 'the Scottish crisis of 1637–38, with its momentous consequences for Britain, had been waiting to happen for some time; if the Prayer Book had not ignited it, something else soon would have done.'[19]

Many of Charles's fellow rulers – Dorgon, Alexei Romanov, Gustavus Adolphus and Christian IV – displayed similar inflexibility, and so did their principal ministers. None seemed prepared to contemplate alternatives to the policies they had adopted. In 1632 Thomas Wentworth, later earl of Strafford, informed a colleague 'Let the tempest be never so great, I will much rather put forth to sea, work forth the storm, or at least be found dead with the rudder in my hand' – an uncanny echo of the claim seven years earlier by the count-duke of Olivares that 'As the minister with paramount obligations, it is for me to die unprotesting, chained to my oar, until not a fragment is left in my hands.'[20] Yet, sooner or later, persevering with unpopular policies during the economic and social tempest caused by the Little Ice Age was bound to provoke resistance and rebellion.

The Two Worlds of Robinson Crusoe

Robinson Crusoe, perhaps the most famous fictional son of seventeenth-century Britain, grew up during the Civil War and left home in 1651, just after the execution of Charles I. After being marooned on a remote Caribbean island he returned to his native land in 1687, just in time to witness the flight of James II and the Glorious Revolution. Crusoe's 'Strange and surprizing adventures', first published in 1719, included not a word on these political changes: by contrast, Daniel Defoe, Crusoe's creator, repeatedly emphasized how the mental world in which his character had grown up differed from the mental world of his readers. For example, young Robinson kept a diary that initially resembled the spiritual journal and balance sheet maintained by many Puritans in the mid-seventeenth century (Chapter 20); but before long he filled it with balance sheets of profit and loss, reflecting the commercial outlook that had made England prosperous. Moreover, whereas England in the mid-seventeenth century had been riven by confessional strife, Crusoe despised religious intolerance. 'I allow'd liberty of conscience throughout my dominions' to Catholics, Protestants and pagans alike, and considered 'all the disputes, wranglings, strife and contention, which has happen'd in the world about religion . . . perfectly useless to us, as for ought I can yet see, they have been to all the rest of the world'. Crusoe favoured religious toleration because it facilitated international trade, which he pursued with great success and profit. He also successfully practised the new experimental philosophy (Chapter 22). He salvaged from his wrecked ship 'infinitely more than I knew what to do with', leading to the 'reflection, that all the good things of this world are no farther good to us than they are for our use'; and that, on the contrary, 'All I could make use of, was all that was valuable'. Crusoe also became a successful planter and soon found that his two most valuable assets were tools (the carpenter's chest he salvaged was 'much more valuable than a ship loading of gold would have been') and labour. Crusoe saved 'my man Friday', a fugitive from cannibals, and immediately set him to work on his colony (Crusoe's term), where the first English word Friday learned was 'Master'. Crusoe himself 'had never handled a tool in all my life', yet 'I improv'd myself in this time in all the mechanick exercises, which my necessities put me upon applying myself to'. A clearer example of experimental philosophy in action would be hard to find.[21]

The global climate had also changed by 1719. The frequency and violence both of volcanic eruptions and El Niño events diminished, an eleven-year sunspot cycle re-emerged, and surface temperatures on Earth began to rise. The benign climate, coinciding with a more systematic exploitation of the environment, permitted rapid population growth in more fertile areas. In China, the Kangxi emperor noted in 1716 that the population grew 'day after day', unlike the available arable land, and he complained – just like his predecessors a century before – about the increase in the number of 'unproductive consumers', singling out intellectuals (naturally), merchants and clerics. A few years later, a senior official in Fujian estimated that the population had doubled during the previous six decades. He too complained that 'while the population increases daily, the amount of land under cultivation does not'. The previous year, the central government had launched a drive to bring more land under cultivation because the 'population has increased of late, so how can [the people] obtain their livelihood? Land reclamation is the only solution.'[22] Thanks to such measures, by the mid-eighteenth century many parts of the globe boasted a denser population than ever before – but this time without a decline in life expectancy or standard of living. Equally important, the new equilibrium of population and resources made the demands of the fiscal-military state more bearable. The fatal synergy had ended.

Epilogue: 'It's the Climate, Stupid'[1]

Once upon a time, climate change was a hot topic. In 1979 the World Meteorological Organization (WMO), the United Nations Environment Programme, the National Science Foundation, the Ford Foundation and the Rockefeller Foundation paid for 250 historians, geographers, archaeologists and climatologists from thirty countries to share their expertise at the first International Conference on Climate and History, hosted by the Climatic Research Unit at the University of East Anglia (England) – a unit sponsored by (among others) British Petroleum and Royal Dutch Shell. That same year, the WMO created the World Climate Program, with a mandate to 'insert climatic considerations into the formulation of rational policy alternatives'; President Jimmy Carter created the Federal Emergency Management Agency (FEMA) to consolidate federal policies related to the management of civil emergencies, including climate-induced disasters; and the United States Congress invited a committee of scientists 'to assess the scientific basis for projection of possible future climatic changes resulting from man-made releases of carbon dioxide into the atmosphere'. The committee predicted that if airborne concentrations of carbon dioxide (CO_2) continued to increase, during the first half of the twenty-first century 'changes in global temperature of the order of 3°C will occur and ... will be accompanied by significant changes in regional climatic patterns'. It also warned that 'A wait-and-see policy might mean waiting until it is too late.' In response, Congress passed the Energy Security Act, which (among other things) ordered a Carbon Dioxide Assessment Committee (CDAC) to prepare a comprehensive survey of the 'projected impact, on the level of carbon dioxide in the atmosphere, of fossil fuel combustion'.[2]

These initiatives took place in the shadow of a world food crisis. The price of wheat tripled and that of rice quintupled between 1972 and 1974, a reflection of harvest failures in South Asia, North America, the Sahel and the Soviet Union, leading the United Nations to convene a World Food Conference that called on all countries to cooperate in the establishment of 'a world food security system which would ensure adequate availability of, and reasonable prices for, food at all times, irrespective of periodic fluctuations and vagaries of weather'.[3] Then, thanks to the Green Revolution (new high-yielding varieties of wheat, maize and rice, combined with increased investment in irrigation, fertilizers, pesticides and herbicides), food production dramatically increased. Famines virtually disappeared from the headlines, and concern about the vagaries of weather waned. The CDAC's report to Congress in 1983, entitled *Climate change*, categorically denied that 'the evidence at hand about CO_2-induced climate change would support steps to change current fuel-use patterns away from fossil fuels', and instead asserted that 'The direct effects of more CO_2 in the air are beneficial.'[4]

Climate change nevertheless included disturbing data. If the amount of CO_2 in the atmosphere continued to increase at the same rate, as the report predicted, then global temperatures would rise 'by about 1°C' as early as the year 2000, and 'in polar latitudes a doubling of the atmospheric CO_2 concentration would cause a 5 to 10°C warming'. This would increase droughts, and decrease 'yields of the three great American food crops [wheat, maize and soybeans] over the entire grain belt by 5 to 10%'. It would also cause sea levels worldwide to rise by 5 or 6 metres, so that even 'The old dream of a "Northwest passage" might become a reality.'[5] To avert alarm concerning these eerily accurate predictions, *Climate change* cited a few historical precedents to demonstrate how easily humans adapt to abrupt climate change: Europe in the fourteenth century (based entirely on Barbara Tuchman's *A distant mirror*); Dakota in the 1880s (citing *The Bad Lands cow boy* newspaper); and the Dust Bowl of the 1930s ('a natural experiment with results dramatized in John Steinbeck's *Grapes of wrath*'). Therefore, the report concluded cheerily: 'The safest prediction of any we shall make is: Farmers will adapt to a change in climate, exploiting it and making our preceding predictions too pessimistic.' Moreover, should local adaptation fall short, the CDAC argued that migration would solve all problems. Once again the report cited apparently reassuring precedents:

> People have moved from the seacoast to the prairie, from the snows to the Sun Belt. Not only have people moved, but they have taken with them their horses, dogs, children, technologies, crops, livestock, and hobbies. It is extraordinary how adaptable people can be in moving to drastically different climates.

In the unlikely event of a future climate-induced crisis, farmers would 'move as promptly as the Okies, saving themselves while abandoning the cropland to some other use'. Despite these spurious historical parallels, the CDAC report immediately became a foundational text for those who sought to deny global warming.[6]

Climate change failed to mention another precedent: the events of 1816, a year without a summer (the first since 1675, and the most recent). It occurred in the middle of a prolonged sunspot minimum amid major volcanic activity, which (as in the seventeenth century) both reduced average global temperatures by between 1°C and 2°C and caused extreme weather events. Intense cold prevailed from Finland to Morocco for most of the summer; rain fell on Ireland for 142 out of 153 days between May and September; grapes in French and Swiss vineyards ripened later than in any other year since continuous records began in 1437; the monsoon failed in India; and snow fell in Jiangnan and Taiwan. In America, north of a diagonal line stretching from British Columbia to Georgia, fronts of Arctic air produced temperature oscillations throughout the summer from 35°C to freezing in a single day, killing the crops: the price of wheat in New York City in 1816 would not be surpassed until 1973. The 'Yankee Chills', as survivors in North America called their miserable summer, produced massive migration from New England to the Midwest. 'The lands to the westward are luxuriant, and the climate mild and salubrious' crooned a land promoter, and from 1817 to 1820 the population of the State of Ohio rose by 50 per cent. Most newcomers were New Englanders fleeing the sudden climate change.[7]

If, two centuries later, the Yankee Chills (or any other natural disaster) should strike New England, flight to Ohio would bring little relief. As the 2011 version of the *State of Ohio homeland security strategic plan* points out: 'Getting food from farms to dinner tables involves a complex chain of events that could be interrupted at many different stages. Because food and agriculture are such vital industries to our state, Ohio must vigilantly protect animal, plant, and food supply chains' – but with over 11 million Ohioans, it is hard to see how the state could feed an additional 50 per cent in an emergency.[8] Admittedly, if the Chills killed *only* corn, or *only* affected New England, the transport and distribution infrastructure developed since 1816 could probably import sufficient emergency food rations to Ohio from unaffected (or less affected) areas; but this might prove impossible in the wake of a large-scale natural disaster, not least because, in the words of the current *State of Ohio emergency operations plan*, 'Manufacturing agencies within in the United States employ just-in-time inventory systems and do not stock large inventories, thus there may be a supply shortage nationwide for critical items.'[9] This is an understatement. Whereas once, as Stephen Carmel of Maersk Lines pointed out, 'We were self-sufficient in some but not all of what we needed, and we could trade the excess of what we made to fill the gaps . . . Now we are self-sufficient in nothing.' Therefore, 'As in any conveyor belt linking assembly lines, a disruption to any part of the system becomes a disruption to the whole system.' In addition, we are 'completely dependent on the uninterrupted flow of accurate information. Without it, trade simply will not happen' – and neither will relief efforts.[10]

The tragic experience of the Gulf Coast region after Hurricane Katrina struck in August 2005 highlighted the consequences of extreme weather when a society relies on just-in-time inventory systems and the uninterrupted flow of accurate information: almost 2,000 people killed; tens of thousands left without basic essentials for almost a week; over 1 million displaced; 92,000 square miles of land laid waste; and 200,000 homes destroyed. It was the largest and costliest natural disaster in the history of the United States, and it led to the largest domestic deployment of military forces since the Civil War. Yet, as a House of Representatives investigative committee reported, 'None of this had to happen. The potential effects of a Category 4 or 5 storm were predictable and were in fact predicted'; but 'despite years of recognition of the threat that was to materialize in Hurricane Katrina, no one – not the federal government, not the state government, and not the local government – seems to have planned for an evacuation of the city from flooding through breached levees'. The committee's report, entitled *A failure of initiative*, was especially scathing about the inability of the numerous teams of responders to contact one another: 'Catastrophic disasters may have some unpredictable consequences', they noted, 'but losing power and the dependent communications systems after a hurricane should not be one of them.' They cited the lament of the adjutant-general of the Mississippi National Guard that 'We've got runners running from commander to commander. In other words, we're going to the sound of gunfire, as we used to say in the Revolutionary War' (and also in the seventeenth century). The report concluded: 'We are left scratching our heads at the range of inefficiency and ineffectiveness that characterized government behavior right before and after this storm. But passivity did the most damage'. Its authors therefore wondered 'How can we set up a system to protect against passivity? Why do we repeatedly seem out of synch during disasters? Why do we continually seem to be one disaster behind?'[11]

A few months later, a United States Senate investigative committee reached similar conclusions in a report entitled *Hurricane Katrina: a nation still unprepared*. The 'still' referred to the inadequate responses to another catastrophic event revealed by the independent commission that studied the 9/11 attacks just four years earlier, and several members of Congress called for a similar independent commission into Katrina in order to learn from mistakes made – but the White House thwarted them.[12] So the United States remained 'one disaster behind': although the federal government improved its ability to mobilize and deliver massive quantities of supplies to assist state and local government in the days and weeks after a disaster (search and rescue, law enforcement, temporary shelter, emergency distribution of food, water and medicine), it did far less to help vulnerable localities prepare for long-term recovery (how to keep local government operating, rehouse displaced people, and prepare for the inevitable health problems, mental as well as physical).

When Hurricane Sandy struck New York and New Jersey in October 2012, therefore, entirely predictable infrastructure failures (loss of electrical power, closure of transportation systems, shortage of gasoline) once again afflicted tens of millions of people for days and sometimes weeks, and a smaller (but still substantial) number for months and sometimes years. For Adam Sobel, an earth scientist, the primary culprit for these defects was what *A failure of initiative* had termed 'passivity':

> Most of the damage could not have been prevented by any decision made or action taken in the days leading up to the storm. Most of it resulted from decisions made in the years, decades, and even centuries prior. Most was caused, perhaps unconsciously, by decisions to do nothing, or the absence of a decision to do something. These nondecisions were made not in response to scientific predictions, but despite them.[13]

The same passivity had characterized responses to many seventeenth-century disasters. To take one glaring example, plague epidemics in 1603, 1625 and 1636 had killed tens of thousands of Londoners, and when a new epidemic ravaged continental Europe in 1664–65 it was easy to anticipate the consequences if it reached the English capital. Nevertheless, neither local nor national government took appropriate action. Instead, when plague struck, the king and his court, many magistrates and almost all the rich fled. Parliament assembled in Oxford to debate appropriate measures, but no legislation passed because the peers demanded an exemption from restrictive measures such as quarantine and insisted that no plague hospitals be erected near their own homes. One may wonder why the central government did not act unilaterally to save its capital; but, as a contemporary pamphlet pointed out, 'their power was *limited* and they must proceed legally'. The rule of Oliver Cromwell and his army officers (which had ended only five years before) had left a bitter legacy, and Charles II dared not risk alienating his new subjects by imposing unpopular measures. The consequences of government nondecisions were therefore measured in the corpses of plague victims dumped daily into mass graves. In all, the epidemic killed 100,000 Londoners, one-quarter of the total population of the capital, plus 100,000 more victims elsewhere in England. Nevertheless, unlike Katrina and Sandy, this catastrophe proved to be a tipping point: the English government introduced and enforced stringent controls that ensured its citizens never again suffered a major plague epidemic (Chapter 21).[14]

Social psychologist Paul Slovik has argued that 'the ability to sense and avoid harmful environmental conditions is necessary for the survival of all living organisms', whereas 'humans have an additional capability that allows them to alter their environment as well as respond to it' – but only if they deploy two distinct skills: *learning processes* (the observation, measurement and classification of natural phenomena) and *learning steps* (the development of techniques, practices and instructions designed to reduce vulnerability in future hazards). In order to activate this additional capability, humans apparently need to experience natural disasters 'not only in magnitude but in frequency as well. Without repeated experiences, the process whereby managers evolve measures of coping with [disasters] does not take place.'[15] The National Hurricane Center, a division of the United States National Weather Service, confirmed this insight in the wake of the disastrous hurricane seasons of 2004 and 2005 (which included not only Katrina but seven of the nine costliest storm systems ever to strike the United States). Another 'disastrous loss of life is inevitable in the future', they concluded sadly, because the majority of those living in areas at risk have 'never experienced a direct hit by a major hurricane' and seemed incapable of envisaging what one is like, while the rest 'only remember the worst effects of a hurricane for about seven years.'[16] Adam Sobel reached a similar conclusion after Sandy:

> When a particular type of event has not happened before, predictions that the risk of that event is significant do not, historically, generate the collective will necessary for us to make investments in resiliency. This is true even when the science indicates quite clearly that the event is quite likely to happen eventually, and that the consequences of being unprepared for it will be severe. Just as the vulnerability of New Orleans was known for decades before Katrina, the vulnerability of New York City and the coastal areas around it was known for decades before Sandy.

In the words of Australian Professor of Public Ethics Clive Hamilton, 'Sometimes facing up to the truth is just too hard. When the facts are distressing it is easier to reframe or ignore them.'[17]

Perhaps cognitive dissonance explains why many societies fail 'to make investments in resiliency' to cope with our changing climate, despite the fact that the global temperature in 2016 was the warmest ever recorded, that it was the third consecutive year in which a new annual temperature record was set, and that the first sixteen years of the twenty-first century all ranked among the seventeen warmest on record; and despite the unprecedented extreme weather events – notably prolonged periods of unusual heat, heavy downpours and flash floods – that disrupted people's lives and damaged infrastructure. One may deny that the global climate is changing, but it is hard to deny that a heatwave in Europe in 2003 that lasted just two weeks caused the premature deaths of 70,000 people; that eighteen major flood events hit Texas, Louisiana, Oklahoma and Arkansas between March 2015 and August 2016; or that almost 18 inches of rain fell on Sacramento and other parts of California in January and February 2017, causing damage to roads, dams and other infrastructure that may cost $1 billion to repair.[18]

The scientific community is virtually unanimous on the reality of global warming: of 69,406 authors of recent peer-reviewed articles on the subject, only five rejected it. Likewise, only 2 per cent of the members of the American Association for the Advancement of Science, the world's largest multidisciplinary scientific professional society, denied that the Earth is warming (the same proportion who rejected evolution), a figure that fell to 1 per cent among earth scientists surveyed. By contrast, a 2016 Pew Research Center survey of adults living in the United States revealed that less than half believe 'that the Earth is warming mostly due to climate change', and scarcely a quarter believe that 'almost all climate scientists agree that human behaviour is mostly responsible for climate change'.[19]

These findings, and similar ones elsewhere, prompted a team of researchers to ask people in forty-seven different countries 'How serious do you consider global warming?' and to share their rationale.[20] The responses revealed three broad reasons for popular scepticism:

- A marked negative correlation between biblical fundamentalism and concern for the environment is evident, particularly among Christians in the United States, many of whom see natural disasters as divine punishments for sin, so that both preparation and mitigation are a waste of time and money (a peccatogenic outlook also common in the seventeenth century: see Chapter 1).[21]
- A negative correlation also exists between residence in regions highly exposed to natural disasters such as hurricanes and concern about global warming, either because hazard and disaster are accepted as aspects of daily life, or because it may be hard to admit that buying property in a location at high risk of damage from natural disasters such as hurricanes is foolish.
- Respondents in rich countries, and in countries with large carbon dioxide emission levels, showed less concern about global warming than those in poor countries or in countries with low emissions, no doubt because it is harder to accept global warming as a problem when it requires recognition that it is partly your fault and when mitigation requires major personal sacrifices.

All three reasons reflect the culture of passivity described in *A failure of initiative*, but they receive reinforcement from a small but powerful group of activists.

In his blueprint for scientific research, the *New Atlantis* in 1626, Francis Bacon extolled the 'merchants of light' who travelled afar to bring back scientific knowledge and then collaborated to debate its significance and apply its practical benefits (page 486). Almost all scientists today do precisely this, but on a few subjects they have to contend with 'merchants of doubt' who seek to discredit their research and promote distrust of their conclusions.[22] Over the past half-century, the merchants of doubt have vigorously challenged the scientific evidence for links between tobacco-smoking and lung cancer, between acid rain and environmental damage, and between chlorofluorocarbons (CFCs) and the depletion of the ozone layer, as well as between CO_2 emissions and climate change. In each case, they have found and funded contrarian experts (almost always from a scientific field not relevant to the environmental or public health issues in question) and insisted that journalists devote equal attention to the deniers, however few and however unqualified. Although the deniers eventually lose, they still cause immense harm. Thus, in 2006, after seven years of litigation, US Federal Judge Gladys Kessler determined that the tobacco industry had engaged in an 'unlawful conspiracy to deceive the American public about the health effects of smoking' since the 1950s. Moreover, because it 'consistently, repeatedly, and with enormous skill and sophistication, denied these facts to the public, to the Government, and to the public health community', it managed to delay the regulation of cigarettes by almost half a century, thereby causing 'a staggering number of deaths per year, an immeasurable amount of human suffering and economic loss'.[23]

In the case of climate change, the number of deaths, the amount of human suffering and the economic loss will be far higher. Extreme weather caused 91 per cent of almost 16,000 natural disasters recorded worldwide between 1980 and 2015. More specifically, in Europe, where extreme weather accounted for 92 per cent of all natural disasters reported in the same period, economic losses caused by weather-related events increased from a decadal average (adjusted for inflation) of over €7 billion in the 1980s, to over €13 billion in the 1990s, and over €14 billion in the 2000s.[24]

These alarming statistics have attracted the attention of the world's insurance companies. The International Association for the Study of Insurance Economics, also known as the Geneva Association, estimates that 'losses from weather events are growing at an annual 6 per cent, thus doubling every twelve years'; it predicts that 'rising loss trends will continue'; and it blames the loss trends on a synergy between climate change and human perversity.

The lack of preventive strategies (for instance, land zoning, building codes, and so on) in many countries' development planning results in increasing vulnerabilities and risks due to disasters and climate change. Further, ever more people and assets are concentrated in exposed (urban) areas such as coastal regions in low- and middle-income countries. At the same time, interconnected global supply and manufacturing chains are highly vulnerable to disaster-induced disruption. And, last but not least, climate change is believed to add to the increasing severity and frequency of extreme events.[25]

Amitav Ghosh has emphasized the special vulnerability of Asia. 'If we consider the location of those who are most at threat from the changes that are now under way across the planet', he wrote, 'the great majority of potential victims are in Asia.' Because of rising sea-levels, falling aquifers and desertification, 'the lives and livelihoods of half a billion people in South and Southeast Asia are at risk. Needless to add, the burden of those impacts will be borne largely by the region's poorest people, and among them, disproportionately by women' – and yet 'there exist very few polities or public institutions that are capable of implementing, or even contemplating, a managed retreat from vulnerable locations'.[26]

High-impact natural disasters will not be confined to South and Southeast Asia. A report published in 2007 by the military advisory board of the Center for Naval Analyses, a non-profit research organization based in Arlington, Virginia, presciently warned that 'Climate change acts as a threat multiplier for instability in some of the most volatile regions of the world. Projected climate change will seriously exacerbate already marginal living standards in many Asian, African, and Middle Eastern nations, causing widespread instability and the likelihood of failed states.' 'The major impact on Europe from climate change is likely to be migrations', they continued, 'now from the Maghreb (Northern Africa) and Turkey, and increasingly, as climate conditions worsen, from Africa.' Their prediction soon came true. In Syria, a country whose population had grown sevenfold in two generations, a multi-season, multi-year extreme drought, which began in the winter of 2006–7, reduced yields of wheat and barley by one-half and two-thirds respectively, and destroyed livestock herds. Over 1 million hungry and homeless people fled from rural areas to the cities, just as their ancestors had done when a similar prolonged drought struck in the seventeenth century (Chapter 7). There they raised the population of some cities by one-third, dramatically driving up food and housing prices, overstraining services such as hospitals and schools, increasing urban unemployment, economic dislocation and social unrest. Civil war broke out in 2012, resulting in over 1 million starving refugees crossing into Turkey and thence into Europe. By 2016, half the population of Syria had been displaced.[27]

What can be done to avoid such heart-rending scenarios? The Church of Jesus Christ of the Latter Day Saints currently enjoins its members to prepare for a sudden natural disaster by taking 'the amount of food you would need to purchase to feed your family for a day and multiply that by seven. That is how much food you would need for a one-week supply. Once you have a week's supply, you can gradually expand it to a month, and eventually three months.' More modestly, the State of Ohio recommends that each family should store 'enough food and water to last from several days up to two weeks'. In particular, since 'you can exist on very little food for a long time, but after a short time without adequate water, your body will not be able to function ... a family of 4 who wanted to keep a 1-week supply of water on hand would need to store 28 gallons'.[28] And after that? The devastation caused by Hurricanes Katrina and Sandy and countless other abrupt high-impact weather events lasted for months, and in a few areas for years, far exceeding the resilience of even the most forward-looking family or local government acting alone.

In December 2012, two months after Sandy, Mayor Michael Bloomberg of New York City created a Special Initiative for Rebuilding and Resiliency and tasked it with producing a plan for 'a stronger, more resilient New York'. Its report, 445 pages long, laid out over 250 initiatives to 'make our city even tougher', at a combined cost of almost $20 billion – but not alone. The report concluded: 'Given the important role played by the Federal government in flood risk assessment, flood insurance, and coastal protection measures, a clear Federal agenda for the City to pursue (in partnership with the State and the Congressional delegation) is critical to the successful implementation of the plan outlined in this report.' It added: 'While this list does not reflect all of New York City's needs from the Federal government, it does reflect a set of priorities that require immediate attention.'[29]

This recognition reflects the experience of other societies: preparing for, and coping with, a major weather-induced catastrophe requires resources that only a central government can command. The construction of the Thames Barrier in southeast England offers an instructive example. The river Thames has frequently burst its banks and flooded parts of London. In 1663 Samuel Pepys reported 'the greatest tide that ever was remembered in England to have been in this river: all White Hall having been drowned'. Proposals were made to erect a barrier to prevent the recurrence of similar catastrophes but the opposition of London merchants, whose trade would suffer if ships could not sail up the Thames, and disagreements among competing jurisdictions over the cost, thwarted them. Then, in 1953, a tidal surge in the North Sea flooded some 150,000 acres of eastern England and drowned more than 300 people. The government declared: 'We have had a sharp lesson, and we shall have only ourselves to blame if we fail to profit from it', and set up a committee to propose remedies. It, too, recommended the immediate construction in the Thames estuary of a 'suitable structure, capable of being closed', but opposition from shipping interests and cash-strapped local authorities again prevented action.[30]

In 1966 the government asked its chief scientific adviser, Hermann Bondi, to examine the matter afresh. A mathematician by training, Bondi devoted much attention to assessing risks; but he also consulted historical sources and found that the height of storm tides recorded at London Bridge had increased by more than 1 metre since 1791 (when records began). Although he could not identify the cause of this alarming development,

Bondi predicted that sea levels would continue to rise, increasing the probability of another 'major surge flood in London' that would deliver 'a knock-out blow to the nerve centre of the country'. He compared the likelihood of this with other low-probability/high-impact events, such as an asteroid or meteorite hitting central London, which would also cause immense damage; but concluded that the risk was remote and prevention almost impossible – whereas, given the rising level of the North Sea, another disastrous flood similar to, or worse than, that of 1953 was inevitable. He therefore unequivocally recommended the construction of a Thames Barrier.[31]

Although shipping interests and local authorities did their best to thwart this plan, too, in 1972 Parliament passed the Thames Barrier and Flood Protection Act and promised to fund Bondi's recommendation. By its completion in 1982, the barrier had cost £534 million – but the property it protects now exceeds £200 billion in value, and includes 40,000 commercial and industrial properties and 500,000 homes with over 1 million residents. If the Thames Barrier were not in place, and another flood were to 'drown' Whitehall, the heart of government today as in the time of Pepys, it would displace the 87,000 members of the central administration who work there. It would also flood the new Docklands economic development as well as sixteen hospitals, eight power stations and many fire stations, police stations, roads, railway lines, rail stations and underground stations, as well as the shops and suppliers needed to repair and replace items damaged in the flood. Londoners would therefore lose not only their homes and their jobs but also the essential means of response and recovery. In short, without the Thames Barrier, London would resemble New Orleans in 2005: vulnerable to a natural disaster that, like Katrina, is sooner or later inevitable. By 2016 the Thames Barrier had been activated to prevent flooding 176 times, 50 of them over the winter of 2013–14.[32]

Extreme weather is a great leveller. Despite all the obvious differences, humans in advanced societies have the same basic needs as humans elsewhere. We all need shelter, sufficient water and at least 2,000 calories a day; and we are all vulnerable during 'the hungry time' (the term used by the Aboriginal people of Western Australia for the season between the end of one annual cycle and the beginning of the next: Chapter 15), because if the power grid fails, very soon there will be no food on supermarket shelves and no water. The changing geographical distribution of the global population is increasing that vulnerability. In 1950, Europe had three times the population of Africa, but in 2016 the population of Africa was at least 50 per cent larger than that of Europe – a disparity that widens every year as the former grows and the latter declines. This shift increases the percentage of the global population that spends a high proportion of disposable income on basic needs such as water, food, energy and housing, often in areas where even central governments lack effective means of dealing with major disasters, making them more vulnerable to the effects of climate change. The Global Crisis of the seventeenth century prematurely ended the lives of *millions* of people. A natural catastrophe of similar proportions and duration today would prematurely end the lives of *billions* of people.

We face clear choices. As Britain's chief scientific adviser observed in 2004, summarizing the research of nearly ninety leading experts on the risks of flooding, 'We must either invest more in sustainable approaches to flood and coastal management or learn to live with increased flooding.' Anthony Zinni, former commander-in-chief of US Central Command (responsible for the Middle East, North Africa and Central Asia), made a similar point with characteristic bluntness in a 2007 interview: '"We will pay for this one way or another", he said. "We will pay to reduce greenhouse gas emissions today, and we'll have to take an economic hit of some kind. Or we will pay the price later in military terms. And that will involve human lives. There will be a human toll. There is no way out of this that does not have real costs attached to it".' In 2011, a study of the impact of climate change, based on thirty years of empirical data, prepared for New York State and published (ironically) just eleven months before Sandy struck, quantified the equation: 'There is an approximate 4-to-1 benefit-to-cost ratio of investing in protective measures to keep losses from disaster low.'[33] In short, we can pay to prepare, and commit substantial resources now, or we can incur far greater costs to repair at some future date.

Like Cassandra, historians who prophesy rarely receive much attention from their colleagues (or anyone else), and those who prophesy doom (whether or not they are historians) are normally dismissed as whiners – *hoggidiani*, to use the dismissive term in Secondo Lancellotti's 1623 bestseller *Nowadays* (page 299). Yet *hoggidiani* are not always wrong. As environmental historian Sam White has observed, 'Studying climate without considering the history of climate is like driving without a rearview mirror: it provides not just parables but also parallels about past climate change and its effects.' Earth scientists Tim O'Riordan and Tim Lenton concurred; because 'there is no single template for anticipating and adjusting' to weather-induced disasters, they argued that there is 'no substitute for good case history of successful practice'. We therefore 'need to follow examples of successful anticipation' and adjustment in the past 'in order to offer the best set of learning experience for others to follow'.[34]

The seventeenth-century Global Crisis offers two successful but very different learning experiences. Famine and unrest in Japan led Tokugawa Iemitsu and his advisers to create more granaries, upgrade the communications infrastructure, issue detailed economic legislation and avoid foreign wars in order to preserve sufficient reserves to cope with the consequences of extreme weather. But, as Ghosh notes, 'Climate change poses a powerful challenge to what is perhaps the single most important political conception of the modern era: the idea of freedom, which is central not only to contemporary politics but also to the humanities, the arts, and literature.'[35] The successful efforts of the Tokugawa to protect their subjects from starvation involved policies that few today would deem acceptable: they forbade freedom of speech, belief, assembly or movement; monopolized the possession and use of firearms; conducted constant surveillance and summarily executed offenders; and, when all else failed, they permitted *mabiki*, literally 'thinning out', but also the metaphor of choice for infanticide (Chapter 16). England followed a different strategy for adjusting to weather-induced disasters. Successive dearths in the 1590s, 1629–31, 1647–49 and the 1690s gradually forced reluctant property owners to accept the central government's argument that it was both cheaper and more efficient (as well as more humane) to support those who became old, widowed, ill, disabled or unemployed locally, thus creating the foundations of the world's first welfare state.

Nevertheless, not even England could cope with an abrupt change in the global climate, as George Gordon, Lord Byron, discovered in 1816 after he had fled the country amid accusations of incest, adultery, wife-beating and sodomy. He planned to relax in a villa near Lake Geneva with a former mistress, his personal physician John Polidori and a select group of close friends. Instead, the party spent a 'wet, ungenial summer' (Switzerland was one of the areas worst affected by global cooling), which forced Byron and his companions to spend almost all their time indoors. Among other recreations, they competed to see who could compose the most frightening story. Mary Wollstonecraft Shelley began work on *Frankenstein*, one of the first horror novels to become a bestseller; Polidori wrote *The vampyre*, the progenitor of the Dracula genre of fiction; Byron composed a poem that he called 'Darkness'. All three works reflected the disorientation and desperation that even a few weeks of sudden climate change can cause. As we debate whether it makes better sense today to invest more resources in mitigation or 'continually seem to be one disaster behind', we might re-read with profit Byron's poem because, unlike our ancestors in 1816 (and in the seventeenth century), we possess both the resources and the technology to choose whether to prepare today or repair tomorrow.

'Darkness' by Lord Byron[36]

I had a dream, which was not all a dream.
The bright sun was extinguish'd, and the stars
Did wander darkling in the eternal space,
Rayless, and pathless; and the icy Earth
Swung blind and blackening in the moonless air.

Morn came and went – and came, and brought no day,
And men forgot their passions in the dread
Of this their desolation; and all hearts
Were chill'd into a selfish prayer for light:
And they did live by watchfires – and the thrones,
The palaces of crowned kings – the huts,
The habitations of all things which dwell,
Were burnt for beacons; cities were consum'd,
And men were gather'd round their blazing homes . . .

And War, which for a moment was no more,
Did glut himself again: a meal was bought
With blood, and each sat sullenly apart
Gorging himself in gloom: no love was left.
All Earth was but one thought – and that was death,
Immediate and inglorious; and the pang
Of famine fed upon all entrails . . .

Chronology of the leading events of the Global Crisis, 1618–88

Year	Europe	Americas	Asia & Africa	Climate X = El Niño V = Major volcanic eruptions ** = Extreme climatic events	Year
1618	Bohemian Revolt; purges in Dutch Republic; deposition of Ottoman sultan Mustafa		Manchu leader Nurhaci declares war on Ming and invades Liaodong	X Three comets; Sunspot minimum	1618
				X	1619
1620	Ottoman-Polish war → 1621 War spreads to Germany → 1648			** X	1620
1621	Spanish–Dutch war resumes → 1648 *Kipper- und Wipperzeit* → 1623			X	1621
1622	Ottoman regicide		Revolt of Shah Jahan against Jahangir; revolt of Abaza Mehmed Pasha → 1628	V4	1622
1623	Deposition of Ottoman sultan Mustafa (again)				1623

1624	Danes invade Germany → 1629	Revolt against viceroy in Mexico; Dutch seize Salvador → 1625	Donglin crisis in China; Ottoman war with Iran → 1639		1624
1625	Act of Revocation (Scotland); Cossack revolt; French Huguenot revolt → 1629			V5; Sunspot maximum	1625
1626	Revolt in Upper Austria	Potosí flooded			1626
1627			Manchus raid Korea		1627
1628	War of Mantua begins → 1631		Gujarat famine → 1631	** Year without a summer	1628
1629	Edict of Restitution (Germany)		Manchus raid northern China		1629
1630	Plague in Italy; Cossack revolt; Swedes invade Germany → 1648/1654	Dutch capture Pernambuco → 1654		X / ** / V5, V4	1630
1631	Rioting in Istanbul; Kadizadeli movement gathers strength			V5	1631
1632	Russo-Polish war → 1634				1632
1633					1633
1634			Bandit armies invade Jiangnan		1634

Year	Europe	Americas	Asia & Africa	Climate X = El Niño V = Major volcanic eruptions ** = Extreme climatic events	Year
1635	France declares war on Spain → 1659			X **	1635
1636	Croquants' revolt (France) Revolt in Lower Austria		Hong Taiji proclaims Qing dynasty, raids northern China and invades Korea		1636
1637	Cossack revolt → 1638 Scottish Revolution → 1651 Évora & S. Portugal revolt → 1638	Pequot War (New England)	Revolt at Shimabara, Japan → 1638		1637
1638				V4 X	1638
1639	Nu-Pieds revolt (Normandy)		Chinese (Sangleys) revolt in Manila	X	1639
1640	Catalan revolt → 1659 Portugal declares independence: war with Spain →1668	'Beaver Wars' (Great Lakes)		** Two V4 & one V5	1640

1641	Irish rebellion → 1653 Conspiracy of Medina Sidonia (Andalusia)	'Panic in the Indies'	Revolt of Portuguese Asia → 1668; Dutch capture Angola	X ** One V4 & one V5	1641
1642	English Civil War → 1646		Qing raid northern China; Li Zicheng destroys Kaifeng	X	1642
1643	Sweden invades Denmark → 1645		Li Zicheng declares Shun Era		1643
1644			Li Zicheng takes and loses Beijing; Qing invade China, capture Beijing and occupy Central Plain	Weakest monsoon recorded in East Asia	1644
1645	Turco-Venetian war → 1669	Portuguese colonists in Brazil rebel → 1654	Qing invade South China; 'Southern Ming' resistance → 1662	Prolonged sunspot minimum (Maunder Minimum) begins → 1715	1645
1646			Macao revolt	V4 X	1646
1647	Naples revolt → 1648 Sicily revolt → 1648 'Putney Debates' (England)			**	1647

Year	Europe	Americas	Asia & Africa	Climate X = El Niño V = Major volcanic eruptions ** = Extreme climatic events	Year
1648	Fronde → 1653 Ukraine revolt → 1668 Híjar conspiracy (Madrid) Ottoman regicide and revolt Russian revolts → 1649 Second Civil War (England and Scotland) Succession crisis (Denmark)		Portuguese retake Angola	** X	1648
1649	English regicide; the Rump English army begins conquest of Ireland				1649
1650	Dutch regime change → 1672 Swedish crisis England invades and occupies Scotland → 1660			V4, V5 X	1650
1651	Bordeaux Ormée → 1653 Istanbul riots; murder of Valide Sultan		Yui conspiracy in Edo	X	1651
1652	Green Banner revolts (Andalusia)		Colombo revolt	X	1652
1653	Swiss Revolution Fall of Rump (England)		Goa revolt		1653

Year				Year
1654	Russo-Polish war → 1667 Last Swedish troops leave central Germany			1654
1655	Sweden invades Poland → 1661		V4	1655
1656	Istanbul riots			1656
1657		Revolt of Abaza Hasan Pasha → 1659		1657
1658	Sweden invades Denmark	Mughal Civil War → 1662	***	1658
1659	English Republic restored Spain and France make peace	Cape Colony at war → 1660	X	1659
1660	The 'Danish Revolution' 'Restoration' in Scotland, Ireland and England		One V6 and three V4 X	1660
1661		Qing order evacuation of all maritime areas → 1683	X	1661
1662	Moscow rebellion	Execution of last Ming claimant to Chinese throne		1662
1663	Ottoman–Habsburg war → 1664		V5	1663
1664			Comet	1664

Year	Europe	Americas	Asia & Africa	Climate X = El Niño V = Major volcanic eruptions ** = Extreme climatic events	Year
1665			Portuguese and allies destroy Kongo; Shabbatai Zvi declared Messiah	Comet X	1665
1666		Revolt of Laicacota (Peru)			1666
1667	Russia and Poland make peace French war against Spain → 1668			V5	1667
1668	Spain recognizes independence of Portugal and its empire				1668
1669	Venice and the Turks make peace		Dutch and allies defeat Makassar	X	1669
1670	Revolt of Stenka Razin → 1671				1670
1671			Portuguese and allies destroy Ndongo		1671

Year					Year
1672	Ottoman attacks on Poland → 1676 French attack Dutch Republic, leading to regime change and war →1678				1672
1673			'Revolt of the Three Feudatories' (China) → 1681	V5	1673
1674					1674
1675	'Red Caps' revolt (Brittany)			** Year without a Summer	1675
1676		King Philip's War (New England); Bacon's rebellion (Virginia)			1676
1677	Ottoman-Russian war → 1681				1677
1678					1678
1679					1679
1680		Pueblo revolt → 1692		V5 Comet	1680
1681			'Revolt of the Three Feudatories' ends		1681
1682	Moscow rebellion →1684			Last major comet of the century	1682

Year	Europe	Americas	Asia & Africa	Climate X = El Niño V = Major volcanic eruptions ** = Extreme climatic events	Year
1683	Ottomans war against Habsburgs and Poland (and Venice from 1684) → 1699		Qing troops take Taiwan and permit maritime trade again	**	1683
1684			First visit by a Qing ruler to southern China	X	1684
1685					1685
1686			Drought in India & Indonesia → 1688		1686
1687				X	1687
1688	'Glorious Revolution' (England) Louis XIV starts war with Britain, Dutch, Spain and Empire → 1697				1688

Sources

- Sunspot chronology from Usoshin, 'Reconstruction' modified by Vaquero, 'Revisited sunspot data'.
- El Niño chronology from Gergis and Fowler, 'A history of ENSO events'
- Volcanic chronology from the Smithsonian Institution's site http://www.volcano.si.edu/world/largeeruptions.cfm

I have omitted two other major eruptions in the Smithsonian series because of imprecise dating: one V6 between 1640 and 1680 and another V4 between 1580 and 1680.

Acknowledgements

In February 1998, I suddenly decided I wanted to write this book, and I immediately sent an email to Robert Baldock, then editing my *Grand strategy of Philip II* at Yale University Press, with the news:

> I always thought that the idea of my next book would come to me quite unexpectedly. Last night I awoke at 4 a.m. and realized that I wanted to write a book about the General Crisis of the seventeenth century – not a collection of essays (been there, done that) but an integrated narrative and analytical account of the first global crisis for which we possess adequate documentation for Asia, Africa, the Americas, and Europe. My account would adopt a Braudelian structure, examining long-term factors (climate above all), medium-run changes (economic fluctuations and so on) and 'events' (from the English Civil War and the crises in the French and Spanish Monarchies, through the murder of two Ottoman sultans, the civil wars in India and sub-Saharan Africa to the collapse of Ming China and the wars around the Great Lakes of North America). Besides examining each of the upheavals of the mid-century, the book would offer explanations of why such synchronic developments occur with so little warning and why they end. Although not the first 'general crisis' known to historians, it is the first one for which adequate data exists worldwide. Since it addresses issues of concern today – the impact of global climatic change and sharp economic recession on government and society – it should not lack interested readers.[1]

Although I had thus anticipated much of the structure of this book at the outset, I had no idea then how much reading and travel it would take to research and how much time it would take to write. I should have known better. In his introduction to *Science and civilization in China*, Joseph Needham warned that 'There is no substitute for actually seeing for oneself in the great museums of the world, and the great archaeological sites; there is no substitute for personal intercourse with the practising technicians themselves.' Needham followed this advice meticulously: he filled his magnificent multi-volume work with personal observations and testimony gathered during more than forty years of travel in China and around the world. He also selflessly furthered the work of students, and built on their achievements.[2] In researching my far more modest enterprise I have tried to follow Needham's precept and example. I have travelled extensively both to see for myself the places affected by the mid-seventeenth-century cataclysm and to consult 'the practising technicians' there who share my interest in them. Above all, having tried to further the research of students, I now build on their achievements: I am particularly indebted to Lee Smith, with whom I edited a collection of essays on the subject (*The General Crisis of the seventeenth century*, London 1978; revised edition 1997); and to Tonio Andrade, Derek Croxton, Matthew Keith, Pamela McVay, Andrew Mitchell, Sheilagh Ogilvie, William M. Reger IV and Nancy van Deusen, all of whom graciously shared with me the fruits of their research on the seventeenth century.

Unfortunately, I lack Joseph Needham's dedication to languages (he started to teach himself Chinese at age thirty-eight) and, in any case, as John Richards (another pioneer of Big History) observed, 'In the best of all worlds, the author would be proficient in a half-dozen more languages.'[3] I have therefore relied extensively on the skills of other colleagues and mentors. It is a particular pleasure to recall the assistance of Hayami Akira, who introduced me both to the wealth of surviving material on early modern Japanese history and to his extensive circle of erudite colleagues. I have learned more from him than I can say, and from Mary Elizabeth Berry, Phil Brown, Karen Gerhard, Iwao Seiichi, Ann Jannetta, Derek Massarella, Richard Smethurst, Julia Adeney Thomas and Ronald P. Toby, on the history of Japan. I am grateful for enlightenment to many other experts on seventeenth-century history: on China, to William S. Atwell, Cynthia Brokaw, Timothy Brook, Chen Ning Ning, Roger Des Forges, Nicola Di Cosmo, Kishimoto Mio, Joseph Needham, Evelyn S. and Thomas G. Rawski, Shiba Yoshinobu, Jonathan Spence, Lynn A. Struve, Joanna Waley-Cohen and Wang Jiafan; on Africa, to James de Vere Allen, John Lonsdale and Joseph C. Miller; and on South and Southeast Asia to Stephen Dale, Ashin Das Gupta, Michael Pearson, Anthony R. Reid, Niels Steensgaard, Sanjay Subrahmanyam and George Winius. Dauril Alden, Nicanor Domínguez, Shari Geistberg, Ross Hassig, Karen Kupperman, Carla Pestana and Stuart Schwartz guided my steps towards relevant sources on the Americas; while Paul Bushkovitch, Chester Dunning, Robert Frost, Josef Polišenský, Matthew Romaniello and Kira Stevens did the same for Russia and Eastern Europe; as did Nicholas Canny, David Cressy, Jane Ohlmeyer and Glyn Redworth for Britain and Ireland; and Günhan Börekçi, Jane Hathaway and Mircea Platon for the territories formerly comprising the Ottoman empire. Yet more of my debts are acknowledged with gratitude in individual chapters. On the big picture, I owe much to the advice and inspiration of Jonathan Clark, Karen Colvard, Robert Cowley, Jack A. Goldstone, Richard Grove, Joseph C. Miller, Paul

Monod, Ellen Mosley-Thompson, Kenneth Pomeranz, Nicholas Rodger, Lonnie Thompson and Joel Wallman.

For vital assistance in acquiring and interpreting foreign-language sources I thank Alison Anderson (German); Bethany Aram (Spanish); Maurizio Arfaioli (Italian); Günhan Börekçi (Turkish); Przemysław Gawron and Dariusz Kołodziejczyk (Polish); Ardis Grosjean-Dreisbach (Swedish); Mary Noll and Matthew Romaniello (Russian); Mircea Platon (Romanian); Taguchi Koijiro and Matthew Keith (Chinese and Japanese); and Věra Votrubová (Czech and German). I also thank Peter Davidson for graciously making available to me his transcript of Aberdeen University Library, Ms 2538, the 'Triennial travels' of James Fraser (an Aberdeen University student in the 1650s).

The International House of Japan provided a wonderful opportunity to familiarize myself with materials concerning the crisis in East Asia, by speaking with other scholars in residence and then following up their suggestions in its splendid library. I started writing this book in a room overlooking its magnificent samurai gardens in spring 2002, and completed the last phase of my research there on another visit in summer 2010. I thank the I-House community for their welcome, support and suggestions.

The cost of 'actually seeing for oneself', in East Asia and elsewhere, has increased prodigiously since Joseph Needham began his work, and my research would have been impossible without generous and prolonged financial support. I thank the Japan Society for the Promotion of Science, the Japan Academy (Nippon Gakushi-in), the Carnegie Trust for the Scottish Universities and the British Academy for grants; and the Harry Frank Guggenheim and the John Simon Guggenheim Foundations for fellowships. I also thank the eight universities where I have taught over the past four decades for their support of my research on this and earlier related projects: namely Cambridge, St Andrews, Keio, British Columbia, Illinois, Oxford, Yale and, above all, Ohio State, where almost all the final planning and writing of this book took place.

The sustained support of The Ohio State University's History Department and Mershon Center proved crucial in funding a course of lectures on the World Crisis, a succession of graduate research assistants, and prolonged periods of research abroad. The lectures given at the Mershon Center in 2001 by William Atwell, Paul Bushkovitch, Jack Goldstone, Richard Grove, Karen Kupperman, Anthony Reid, Stuart Schwartz and Joanna Waley-Cohen, and the discussions with them and my graduate seminar that followed, inspired and informed me as I started writing. Later on, their willingness to answer queries and pass on references expanded my horizons and saved me from many mistakes. The OSU graduate research assistants funded by the Mershon Center and the History Department helped me both to manage the material generated by this project and to maintain my enthusiasm. I thank Katherine Becker and Matthew Keith (both of whom also helped to organize the World Crisis lecture series), Günhan Börekçi, Andrew Mitchell and Leif Torkelsen. I also thank two other graduate students for their research assistance: Megan Wheeler in Oxford, and Taguchi Koijiro in Tokyo.

Thanks to all this I completed the first draft of this book in July 2007 and sent it to three expert readers: Paul Monod, Kenneth Pomeranz and Nicholas Rodger. They, as well as four other anonymous Yale University Press 'readers', provided me with magnificent advice, almost all of which I eventually followed. Other commitments prevented me from implementing this advice until 2010, when Rayne Allinson, Sandy Bolzenius, Kate Epstein and Mircea Platon (all at The Ohio State University) inspired me to get back to work on the project. I am immensely grateful to all four of them for their friendship, guidance and encouragement as I strove to implement their suggestions – but especially to Kate, who started to send me comments, encouragement, and references in May 2006 and has continued to do so ever since.

I have also been very fortunate in my editorial team at Yale University Press: Robert Baldock, my friend and publisher for forty years; Candida Brazil, Tami Halliday, Steve Kent, Rachael Lonsdale and Richard Mason. I am grateful to all of them for their expertise and their patience. Finally, I thank my four wonderful children – Susie and Ed, Richard and Jamie – and Alice Conklin for the love they have brought, and continue to bring, into my life.

Columbus–Oxford–Tokyo–Paris, 1998–2012

Additional Acknowledgements for the Abridged Edition

Several friends and colleagues supplied comments, corrections and suggestions after they read this book, completed in May 2012. I am deeply grateful to Lauren Benton, Mary Elizabeth Berry, Wolfgang Behringer, Michael Bevis, Sandy Bolzenius, Timothy Brook, John Brooke, Alice Conklin, Dagomar Degroot, Jan DeVries, Clem Dinsmore, Fabian Drixler, Sir John Elliott, Clara García Ayluardo, John-Paul Ghobral, Jack Goldstone, Gregory Hanlon, Antony Harper, Daniel Headrick, Emmanuel Le Roy Ladurie, Patrick Lenaghan, Scott Levi, Norman MacLeod, John McNeill, Franz Mauelshagen, Joseph Miller, Elizabeth Moyer, Angela Nisbet, Jürgen Osterhammel, Carla Pestana, Kenneth Pomeranz, Theodore K. Rabb, Eleanora Rohland, Hamish Scott, Curt Stager, Sanjay Subrahmanyam, Tale Sundlisæster, Julia Adeney Thomas, Edward Udovic, Cees van Lotringen, Jason Warren, Sam White and Ying Zhang. I also thank Richard Parker for the epigraph.

Kate Epstein, Ruth MacKay and Mircea Platon, as well as my editors Robert Baldock and Rachael Lonsdale, played a vital role in reducing my original typescript by half, a savage task I could not manage myself, and I am very grateful to them all.

Columbus, January 2017

Conventions

1. All dates are given according to the Gregorian calendar, even for European countries like Sweden and Russia that did not adopt it until later. I have also converted all dates from Chinese, Islamic, Japanese and Jewish systems of reckoning time to the Gregorian calendar. An exception occurs in Chapters 11 and 12, on the Stuart monarchy, where dates are given according to the Julian calendar (O.S.) because all historians of Britain and Ireland do so.

2. Where a recognized English version of a foreign placename exists, I have used it (thus Brussels, Lisbon, Moscow, Vienna); otherwise I have preferred the style used today in the place itself (thus Bratislava and not Pressburg or Pozsony; Lviv and not Lwów or Lemberg). Likewise, with personal names, where an established English usage exists I have adopted it (Gustavus Adolphus, Philip IV). In all other cases, I have used the style and title employed by the person concerned.

3. I have used Pinyin rather than Wade-Giles Romanization for Chinese (except when quoting from a work that used Wade-Giles); and I have given the family name first for all Japanese, Chinese and Turkish persons, past and present (thus Tokugawa Iemitsu, Li Zicheng and Kadizade Mehmed).

Note on Sources

Attempts to understand and explain the wave of political, social and economic upheavals that shook the seventeenth-century world began with the historians who lived through them. More than thirty accounts of individual rebellions appeared in print between 1643 and 1663; and, between them, *The history of the civil wars of these recent times* by Maiolino Bisaccione (1652), and *Memorable histories, containing the political uprisings of our times* by Giovanni Battista Birago Avogadro (1653), described upheavals in Brazil, Britain, Catalonia, France, Moscow, Naples, the Papal States, Poland, Portugal, Sicily, Switzerland and the Ottoman empire.[1]

A century later, Voltaire linked the political upheavals of Europe with those in Asia and Africa, postulating for the first time that a *global* crisis had occurred; but few comparative studies of the phenomena appeared until 1937, when Roger B. Merriman delivered a lecture (published the following year) entitled 'Six contemporaneous revolutions'. Merriman claimed that his subject 'has fascinated me for more than thirty years', and although he adopted a far narrower focus than Voltaire (looking only at Western Europe, and even then only at six 'anti-monarchical rebellions' during the 1640s), he considered their common 'parallels and philosophies', stressed the cross currents that linked them and compared them with the European revolutions of 1848. The comparative trail then went cold again until the 1950s, when the British historians Eric Hobsbawm and Hugh Trevor-Roper launched a debate on what they called 'The General Crisis of the seventeenth century' in a series of long articles in the journal *Past & Present: A Journal of Scientific History*. In 1965 the journal's editor, Trevor Aston, republished them, together with essays on England, France, Spain, Sweden and Ireland, in *Crisis in Europe, 1560–1660*. Five years later, Robert Forster and Jack Greene published the proceedings of a symposium on the subject, *Preconditions of revolution in early modern Europe*, with case studies of rebellions in the 1640s in France, England and the Spanish Monarchy. Much later, 'as a survivor' of 'those dramatic days in the 1950s and 1960s' John Elliott published an overview of both the debate and the protagonists: 'The General Crisis in retrospect'.[2]

Elliott himself had published an influential contribution to the debate entitled 'Revolution and continuity in early modern Europe', which sought common denominators for the events of the 1640s, and compared them with other waves of resistance in early modern Europe. In 1978, Geoffrey Parker and Lesley Smith reprinted this essay, together with others on the subject published since Aston's *Crisis in Europe,* in *The General Crisis of the seventeenth century*; and that same year a special issue of the journal *Revue d'histoire diplomatique* published studies of nine contemporaneous revolts (Portugal, Spain, England, Ireland, Spanish Italy, the Spanish Netherlands, Sweden, the Swiss Cantons, and Lübeck). The 1975 essay by Theodore Rabb, *The struggle for stability in early modern Europe,* looked not only at the crisis but also at its aftermath; and the following year Miroslav Hroch and Josef Petráň synthesized historical data and literature from Eastern Europe on the seventeenth-century crisis. In 1982, a special issue of the journal *Renaissance and Modern Studies* on '"Crisis" in Early Modern Europe' contained eight more case studies, and Perez Zagorin included over a dozen European upheavals in *Rebels and rulers, 1500–1800.*[3] In 1999 Francesco Benigno reviewed the literature on the European upheavals of the mid-seventeenth century, with special reference to the Fronde and the revolt of Naples, in *Specchi della rivoluzione* (translated with some modifications as *Mirrors of revolution: conflict and political identity in early modern Europe,* 2010).

All these works confined their attention to Europe – usually Western Europe. A global perspective re-emerged in 1989, at a panel on 'The General Crisis in East Asia' organized by William S. Atwell at the annual meeting of the American Association for Asian Studies, published the following year in *Modern Asian Studies* with essays by Atwell, Anthony Reid, John Richards and Niels Steensgaard.[4] In 1991, Jack Goldstone published a path-breaking study entitled *Revolution and rebellion in the early modern world,* which compared the contemporaneous state breakdowns in Stuart England, the Ottoman empire and Ming China. The following year, Ruggiero Romano's *Conjonctures opposées* presented data that linked the economic difficulties in seventeenth-century Europe with those in the Americas; and Sheilagh Ogilvie's essay 'Germany and the seventeenth-century crisis' provided the most satisfactory explanation to date of the crisis in Western Europe. In 2008–9, two of the leading United States historical journals – *American Historical Review* and *Journal of Interdisciplinary History* – devoted special issues to the subject. It is hard to disagree with Theodore K. Rabb's verdict, in his Introduction to the second collection: 'Like it or not, the Crisis seems here to stay.'[5]

Almost all these works on the General Crisis relied overwhelmingly upon written sources. This is understandable, since the quantity of surviving manuscript and printed material from the seventeenth century is almost overwhelming. A census of contemporary accounts of the Naples revolution of 1647–48 held just in the libraries and archives of the city of Naples revealed almost 300 manuscripts, most of them written by contemporaries and over twenty of them later printed. So many contemporaries described the rebellion of Fermo in central Italy that historians can follow its progress almost hour by hour; and the almost 400 extant accounts of the battle of Naseby in England (1645) enable a reconstruction of the action almost minute by minute. Despite this profusion, several additional sources only came to light relatively recently – in the 1930s the papers of Samuel Hartlib and his circle were discovered in a London solicitor's office; in the 1970s the Chinese government opened the archives of the victorious Qing to researchers – and many sources previously available only in the archives can now be consulted online.

Perhaps the most spectacular online source is the '1641 Depositions Project': some 8,000 statements, examinations and associated materials, taken down from over 3,000 Protestant survivors of the Irish rebellion in 1641 by a team of judges. They cover over 19,000 manuscript pages, subsequently bound in thirty-one fat volumes in the Manuscripts and Archives Research Library of Trinity College, Dublin, written in a variety of hands, full of abbreviations and idiosyncratic spelling and grammar; but they have now been scanned, transcribed and indexed so that it is possible to search for common denominators ('snow' and 'frost', 'rape' and 'murder') as well as for names. The data have also been plotted onto maps.[6]

The surviving written sources nevertheless remain incomplete because of destruction, both accidental and deliberate. The sources concerning the numerous urban revolts that rocked Romanov Russia in 1648 offer a telling example of accidental loss. Only the revolt of the small Siberian town of Tomsk has left a considerable documentary trace, because officials stationed east of the Urals reported to the Siberian Ministry in Moscow, whose archives remain largely intact, whereas their colleagues west of the Urals reported to other ministries in Moscow, most of whose archives were later either destroyed by fire or discarded. Historians can therefore reconstruct the minor troubles in Tomsk in considerable detail, whereas most of what we know about the major uprising in Moscow, which resulted in the murder of several ministers of the tsar and the burning of half his capital, comes from reports smuggled out of the country by foreigners (Chapter 6). Deliberate gaps also abound. In China, the victorious Qing announced in 1664 that 'should officials or commoners have in their homes books that record historical events of the late Ming, they should be sent in' to the nearest magistrate, who would destroy them. In Britain, in the 1630s, the growing tension between Charles I and his subjects led many protagonists to avoid committing their thoughts to paper. Some ceased to write at all (in 1639 an Irish nobleman wanted to visit a colleague because, he claimed, 'I have much to say to your lordship which I cannot trust to paper'); others burnt sensitive correspondence ('The more I think of the business of this letter [of yours and my reply]', wrote a Caroline courtier, the more he favoured 'burning them so soon as their business is answered and ended'). Twenty years later, in the turbulent months before the Restoration of Charles II, many protagonists again dared not commit their thoughts to paper: 'Many thinges I might write of', wrote an Engish Quaker, 'but not knowing in whose hands this might come I shall therefore be spareing.' As David Underdown, an eminent historian of seventeenth-century Britain, grimly observed: 'The first thing that disappears in a rebellion or Civil War is papers.'[7]

It is difficult (though not impossible) for historians to fill silences that have been deliberately created, but help is available in data sets concerning climate and population. Starting in the 1900s, scientists began to collect series of annual growth rings from two types of tree, those sensitive to changes in temperature and those sensitive to differences in precipitation, and arranged them in chronological sequences. By 1929, the archaeologist and astronomer Andrew E. Douglass had established a chronology for the Pueblos of the southwestern United States that ran back 1,200 years, and he speculated that the variations in ring width reflected not only growing conditions but also fluctuations in sunspot numbers. Douglass also established the Laboratory of Tree-Ring Research at the University of Arizona, which can now draw on almost 1,000 tree-ring chronologies from over 700 locations in North America alone. The thousands of chronologies from other regions archived within the International Tree-Ring Data Bank include many that cover the seventeenth century. For example, series of rings taken from over 300 living pine trees growing in the northern Cairngorms, in central Scotland, and almost 250 sub-fossil trees in adjacent Highland lakes, first published in January 2017, confirm that the area experienced some of the coldest years on record during the seventeenth century.[8]

In 1955 the meteorologist Marcel Garnier published a pioneering study that suggested that the date on which French communities began to harvest their grapes might also reflect the annual prevailing temperature during the growing season. Looking back in 2011, Emanuel Le Roy Ladurie recalled that he decided to become a historian of climate upon reading Garnier's work, and he published his path-breaking *History of the climate since the year 1000* in 1967, the first of a series of illuminating works on the subject.[9] Since then, climatologists have collected and published climatic observations from books and manuscripts around the world, and historians of climate (notably Hubert Lamb in the United Kingdom, Christian Pfister in Switzerland, Rudolf Brázdil in the Czech Republic, Mikami Takehiko in Japan, as well as Le Roy Ladurie in France, and their disciples) have studied the correlations and coincidences between these data and political and social upheavals.

Just after Garnier published his study on how to measure climate in the pre-instrumental era, a team of French demographers and archivists showed how to use parish registers of births, marriages and deaths to reconstruct family size and demographic trends before the era of national censuses. Thereafter, enthusiasts in many European countries transcribed entries in surviving parish registers to reconstruct family dossiers, from which it proved possible to track changes in rates of mortality and nuptiality, as well as in family size, in the duration of marriages and in life expectations. In the 1960s, economic historian Hayami Akira realized that the annual registers of believers kept by every Buddhist temple in Japan offered, in some ways, a superior source to parish registers because they were already organized by families, and he and a team of students began to transcribe and analyse them. In China, too, teams of historians painstakingly transcribed the detailed life records kept by the Qing dynasty of all its members, and calculated demographic trends.[10]

Since 2000, several path-breaking works have linked these serial data with specific historical events. Five studies stand out: Timothy Brook, *The troubled empire* (the history of China between the Mongol and the Manchu invasions) and *Vermeer's hat* (a set of connected histories from the seventeenth century); John Richards, *The unending frontier* (an environmental history of the early modern world); Sam White, *The climate of rebellion* (about the early modern Ottoman empire); and Bruce Campbell, *The great transition* (showing that in the fourteenth century climatic adversity and microbial mutation led to increasingly violent competition for resources, a wave of violent unrest and massive mortality).[11] Some journals include work by scholars who try to link climate and history. The interdisciplinary journal *Climatic Change* began publication in 1977. In 2012 the *Journal of Interdisciplinary History* published 'the first multi-proxy baseline synthesis of what is now known with some confidence about climate conditions across the Roman and post-Roman world from c. 100 BC to 800 AD'; and two years later it published a special feature on the Little Ice Age.[12] In 2015, the *William and Mary Quarterly* devoted a special forum to 'Climate and early American history', including a defiant challenge by its editor, Joyce Chaplin:

It remains to be seen how many other early Americanists might take up the invitation to consider climate as a central part of their field. Certainly, climate history – more than environmental history (broadly defined) – seems unignorable. Climate is the part of the environment that matters most to us now because changes in it can change everything else in the natural world, with terrible implications for the quality of human life.[13]

Historians of other regions of the early modern world have certainly taken up the invitation. Articles published from 2014 to 2016 include 'How climate change impacted the collapse of the Ming dynasty', 'Climate and the slave trade' and 'Jewish persecutions and weather shocks: 1100–1800'.[14] 'Like it or not', as Theodore Rabb might say, the role of climate change in inducing crises in human history is also 'here to stay'.

Currently the hottest topic concerning the Global Crisis is whether it gave rise to different economic paths in Europe and East Asia. An influential work by Kenneth Pomeranz, entitled *The Great Divergence*, published in 2000, argued that as late as 1750 little distinguished the economically advanced areas of China, such as Jiangnan, from the economically advanced areas of Europe, including England. A team led by Robert Allen later presented data from Chinese sources that suggested that in economic terms even Jiangnan lagged behind Northwest Europe by the 1730s, when comparable data first became available, and probably before. Subsequently Kent Deng and Patrick O'Brien suggested that a divergence between the two may have been 'visible and widening in the 1600s'; and a team of economic historians attempting to estimate changes in gross domestic product around the world since Roman times detected 'a large income gap between Asia and Western Europe at the onset of the Industrial Revolution'. The same team also detected a 'Little Divergence' between real wages in countries around the North Sea, which 'more or less doubled' between 1500 and 1800, and those in other parts of Europe which stagnated. A comparison of the economies of early modern Europe and India by Roman Studer has likewise suggested that the Great Divergence 'needs to be shifted back to the seventeenth century, if not earlier'. The problem with all these comparisons is that (in the words of Deng and O'Brien) 'the data published and potentially available for China (and probably for India and the Ottoman Empire) stand close to the unfounded guess work end of that spectrum'. It will be interesting to see how these 'potentially available' data affect the Great Divergence debate, and its links with the Global Crisis, once historians start to use them.[15]

Abbreviations Used in the Bibliography and Notes

AGI	Archivo General de Indias, Seville
IG	*Indiferente General* (*consultas* of the Council of the Indies)
Filipinas	(correspondence of the Council of the Indies with the Philippines)
México	(correspondence of the Council of the Indies with New Spain)
AGRB	Archives Générales du Royaume/Algemeen Rijksarchief, Brussels
SEG	*Secrétairerie d'État et de Guerre* (papers and correspondence of the Brussels government in Spanish)
AGS	Archivo General de Simancas
Estado	*Negociación de Estado* (papers of the Spanish Council of State)
GA	*Guerra Antigua* (papers of the Spanish Council of War)
SP	*Secretarías Provinciales* (papers of the Council of Italy), with *libro* or *legajo*
AHN	Archivo Histórico Nacional, Madrid
Consejos	*Consejos suprimidos* (papers of the Spanish Council of Castile)
Estado	*Consejo de Estado* (papers of the Spanish Council of State and of the Council of Portugal), with *libro* or *legajo*
AHR	*American Historical Review*
AM	Archivo Municipal
AMAE (M)	Archivo del Ministerio de Asuntos Exteriores, Madrid, *Manuscritos*
AMAE (P)	Archives du Ministère des Affaires Étrangères, Paris
CPA	*Correspondance Politique: Angleterre* (correspondence of the French government with diplomats in Britain)
CPE	*Correspondance Politique: Espagne* (correspondence of the French government with diplomats in Spain)

APW	Repgen, K., ed., *Acta Pacis Westphalicae*: series I, Instructions; series II, Correspondence (with multiple sub-series: II Abteilung B, die französischen Korrespondenzen; Abteiling C, die schwedische Korrespondenzen; and so on, each with multiple volumes); series III, Diaries
AUL	Aberdeen University Library
BL	British Library, London, Department of Western Manuscripts
Addl. Ms.	*Additional Manuscripts*
Eg. Ms.	*Egerton Manuscripts*
Harl. Ms.	*Harleian Manuscripts*
BNE	Biblioteca Nacional de España, Madrid, *Colección de manuscritos*
BNF	Bibliothèque Nationale de France, Paris, *Cabinet des manuscrits*
Ms. Esp.	*Manuscrit espagnol*
BNL	Biblioteca Nacional, Lisbon, Manuscripts
Bod.	Bodleian Library, Oxford, Department of Western Manuscripts
BR	Biblioteca Real, Madrid, *Colección de manuscritos*
BRB	Bibliothèque Royale Albert 1er, Brussels, *Collection des manuscrits*
CC	*Climatic Change*
CHC	*Cambridge history of China*, ed. D. Twitchett et al., 15 vols., some in two parts (Cambridge, 1978–2009)
CHJ	*Cambridge history of Japan*, ed. J. W. Hall et al., 6 vols. (Cambridge, 1988–99)
CODOIN	*Colección de documentos inéditos para la historia de España*, 112 vols. (Madrid, 1842–95)
CSPC	*Calendar of state papers, colonial series, America and West Indies*, ed. W. N. Sainsbury et al., 45 vols. (London, 1860–1994)
CSPD	*Calendar of state papers domestic series. Charles I*, ed. J. Bruce and W. D. Hamilton, 23 vols. (London, 1858–97)
CSPI	*Calendar of state papers relating to Ireland of the reign of Charles I*, ed. R. P. Mahaffy, 4 vols. (London, 1900–4)
CSPV	*Calendar of state papers and manuscripts relating to English affairs, existing in the archives and collections of Venice, and in other libraries of northern Italy, 1202–1674*, ed. H. F. Brown et al., 28 vols. (London 1864–1947)
EcHR	*Economic History Review*

HAG	Historical Archive, Goa,
Ms.	*Manuscripts* (correspondence of the viceroy of Portuguese India)
HJ	*Historical Journal*
HMC	Historical Manuscripts Commission
IVdeDJ	Instituto de Valencia de Don Juan, Madrid, Manuscript Collection (with envío and folio number)
JAS	*Journal of Asian Studies*
JIH	*Journal of Interdisciplinary History*
JMH	*Journal of Modern History*
JWH	*Journal of World History*
ODNB	*Oxford dictionary of national biography*, online version
PLP	*Proceedings in the opening session of the Long Parliament*, ed. M. Jansson, 7 vols. (Rochester, NY, 2002–7)
P&P	*Past & Present*
RAS	Riksarkivet, Stockholm
	Diplomatica: Muscovitica (correspondence from Swedish diplomats in Russia)
	Manuskriptsamlingen (manuscript collections)
RPCS	*Register of the Privy Council of Scotland*, 2nd series, ed. David Masson and P. Hume Brown, 6 vols. (Edinburgh, 1899–1908)
SCC	*Science and civilization in China*, ed. Joseph Needham and associates, 7 vols., several with multiple parts (Cambridge, 1954–2008)
SCJ	*Sixteenth Century Journal*
TCD	Trinity College Dublin
Ms.	*Manuscripts*: the 33 volumes of 'Depositions', now available in digitized form at http://1641.tcd.ie
TNA	The National Archives, Kew (formerly the Public Records Office)
SP	*State papers* (correspondence of the English secretaries of state)
WMQ	*William and Mary Quarterly*

Notes

Preface to the Abridged Edition

1. *Sunday Herald*, 24 Mar. 2013, review by Hugh MacDonald. Alas Ronnie Corbett died in March 2016.
2. Goldstone, 'Climate lessons', 37; Lonsdale, *Samuel Johnson*, I, 290, written c. 1780, just over a century after Milton's death.
3. Warde, 'Review article', 301, 297
4. Kenneth Pomeranz, John R. McNeill, and Jack Goldstone, '*Global Crisis:* a Forum', *Historically speaking*, XIV/5 (Nov. 2013), 29–39; Lauren Benton, Daniel Headrick, Joseph C. Miller, and Carla Gardina Pestana, 'Special Forum: the Afterlife of Geoffrey Parker's *Global Crisis*', *JWH*, XXVI (2015), 141–80 (based on a panel discussion at the 2014 annual meeting of the American Historical Association, organized by Pestana).

Prologue

1. White, 'The Real Little Ice Age', 328. In 2014, data presented in Neukom, 'Inter-hemispheric temperature variability', demonstrated for the first time that global cooling affected both northern and southern hemispheres in the seventeenth century.
2. Le Roy Ladurie, *Times of feast*, 119, 289; de Vries, 'Measuring the impact of climate', 23.
3. Fortey, 'Blind to the end'. See also Sobel, *Storm surge*, part II.
4. Trevor-Roper, 'The general crisis', 50.
5. Camuffo, 'The earliest temperature observations', 359; Rydval, 'Reconstructing'; Macadam, 'English weather', 234 (Fig. 8, based on data from the diary of Ralph Josselin); David, 'Patents', 265–6.
6. Zheng, 'How climate change', 173. White, 'The Real Little Ice Age', 348–51, evaluates the three current definitions of the term.
7. Le Roy Ladurie, *Times of feast*, 2. Presciently, Le Roy predicted that computers might overcome this challenge: ibid., 303.
8. BL *Harl. Ms.* 390/211, Mede to Sir Martin Stuteville, 24 Feb. 1627. Mede's 'star pupil' at Christ's College was John Milton.
9. Appleby, 'Epidemics and famine', 663.

Introduction

1. Burton, *Anatomy*, 3–4 ('Democritus to the reader'); *Calendar of the court minutes, directors of the East India Company to their agents in Surat*, Nov. 1644; Whitaker, *Ejrenopojos* (*The peacemaker*), 1–2, 9, a sermon on Haggai 2:7, 'And I will shake all nations'.
2. Elliott and La Peña, *Memoriales y cartas*, II, 276.
3. Porschnev, 'Les rapports', 160, quoting Johan Adler Salvius; Mentet de Salmonet, *Histoire*, ii; Parival, *Abrégé*, 'Au lecteur' and 477.
4. Ansaldo, *Peste*, 16 (Fra Francesco); BL *Addl. Ms.* 21,935/48 (Wallington); Mortimer, *Eyewitness accounts*, 185 (Minck).
5. Howell, *Epistolae Ho-Elianae*, Book III, Letter I, to Lord Dorset, '20 Jan. 1646' (but actually 1649).

6. Hobbes, *Leviathan*, 89.
7. Bisaccione, *Historia* (1652), 2; Rushworth, *Historical collections*, I (1659), preface. Howell, *Epistolae*, also began his collection of letters on historical matters with the outbreak of 'the wars of Germany' in 1618.
8. Piterberg, *Ottoman tragedy*, 1, quoting Kâtib Çelebi; Peterson, *Bitter gourd*, 35, quoting *Liudu wenjian lu* by Wu Yingji (1594–1645).
9. Hobbes, *Behemoth*, 1.
10. Kessler, *K'ang-hsi*, 131, quoting the personal report of Xizhu to the emperor in 1684.
11. Hugon, *Naples*, 16.
12. Johnson, *The works*, X, 365–6, 'Thoughts on the late transactions respecting Falkland's islands' (1771).
13. Gaddis, *Landscape*, 98–9; Gladwell, *The tipping point* (a term coined by sociologists in the 1960s). Lauren Benton prefers the term 'pivot' to 'tipping point' because the decisive shifts highlighted in Chapters 5 to 14 of this book 'are not moments of shift that only a climate-focused analysis can reveal' (Benton, 'After the crisis', 150–2). Perhaps: but I do not argue that the climate alone caused catastrophe.
14. Scott, *Weapons of the weak*, xvi–xvii.
15. TCD *Ms*. 833/228v, deposition of Rev. George Creighton, 15 Apr. 1643 (quoting Richard Plunkett).

Part I. The Placenta of the Crisis

1. Voltaire, *Essai*, II, 756–7.
2. Ibid., II, 756–7, 794, 806 and 941–7 (*Remarques pour servir de supplément à l'essai sur les moeurs*, 1763).
3. Cercas, *The anatomy*, 28, an analysis of the attempted military coup d'état in Spain on 23 Feb. 1981, argued that the simultaneous hostility to the government by various groups – politicians, journalists, bankers, businessmen, foreign governments, perhaps even the king – together formed 'the placenta of the coup, not the coup itself. The nuance is key in understanding the coup'. A parallel nuance is key in understanding the global crisis of the seventeenth century.

Chapter 1. The Little Ice Age

1. Special thanks for help in drafting this chapter to William S. Atwell, Rudolf Brázdil, Günhan Börekçi, John Brooke, Richard Grove, Karen Kupperman, Mikami Takehiko, David Parrott, Christian Pfister and José M. Vaquero.
2. Cysat, *Collectanea*, IV/2, 898.
3. Dunn, *The journal of John Winthrop*, 368, 384 and 387; Kupperman, 'The puzzle', 1,274, quoting Thomas Gorges.
4. Blair and Robertson, *The Philippine islands*, XXXV, 123 and 184, from reports on events in 1640–42 and 1643–44.
5. Odorico, *Conseils et mémoires de Synadinos*, 163, 169.
6. Naworth, *A new almanacke*, sig C2; BL *Harl. Ms*. 5,999/29v, 'Discourse' by Henry Jones and others, Nov. 1643.
7. Buisman, *Duizend jaar weer*, IV, 469–70 (quoting a burgher of Liège); Peters, *Ein Söldnerleben*, 166, diary entry at Neustadt-an-der-Saale, 7 Aug. 1640.
8. Vázquez de Espinosa, *Compendio*, 591–2, Bartolomé de Astete de Ulloa to Gerónimo López de Saavedra, Potosí, 17 Mar. 1626; Arzáns de Orsúa y Vela, *Historia*, II, 1–15; Vicuña Mackenna, *El clima de Chile*, 43–4, Dean Tomás de Santiago to the Tribunal of Lima, 23 June 1640 (a claim supported by tree-ring series: Morales, 'Precipitation changes', 659).
9. Bamford, *A royalist's notebook*, 120–1; Laing, *Letters and journals*, III, 62; Gilbert, *History of the Irish Confederation*, VI, 270–1.
10. Buisman, *Duizend jaar weer*, IV, 499–500; Labelle, *Dispersed*, 66.

11. Teodoreanu, 'Preliminary observations', 190, quoting Archdeacon Paul of Aleppo; Abbott, *Writings*, III, 225–8, Proclamation of 20 Mar. 1654.

12. Ådahl, *The sultan's procession*, 48, entry for 27 Feb. 1658; De Beer, *The diary of John Evelyn*, 388, 393; Van Aitzema, *Saken*, IX, book 38, 124–6.

13. Hoskins, 'Harvest fluctuations', 18; Goldie, *The entring book of Roger Morrice*, II, 450.

14. Teodoreanu, 'Preliminary observations', 191–2, quoting Philippe Le Masson du Pont in 1686.

15. Evliya Çelebi, *Seyahatname*, X, 508 (Egypt: I thank Jane Hathaway for both reference and translation.)

16. Silvestre de Sacy, *Lettres*, III, 321 and 345, Mme de Sévigné to Mme de Grignan, 28 June and 24 July 1675 ; *Da Qing shengzu [Kangxi] shilu* [Veritable records of the reign of the Wise Ancestor [Kangxi] of the Qing dynasty], 272.9b-10a, edict of Kangxi emperor, 26 May 1717 (graciously translated for me by Timothy Brook).

17. Smith, *The art of doing good*, 162 (Qi's diary); Wu, *Communication and imperial control*, 34–6 (Kangxi); Scott, *Love and protest*, 140 (folk song); Magisa, *Svcceso raro* ('implications' of the eruptions); Anon., *The way to get rain*, 3.

18. Wrightson and Walter, 'Dearth', 28–9. I salute Franz Mauelshagen for adding *peccatogenic* to our vocabulary.

19. AM Cádiz *Ms.* 26/127, Cabildo of 21 Aug. 1648, with a copy of a letter from Don Diego de Riaño y Gamboa, 1 Aug. 1648.

20. Haude, 'Religion', 541; Firth and Rait, *Acts and ordinances*, 26–7 and 1,070–2, 'Order for stage-playes to cease' (12 Sep. 1642 NS) and 'An ordinance for the utter suppression and abolishing of all stage-plays' (21 Feb. 1648 NS).

21. Pfister, 'Climatic extremes', 60–1 (praising Behringer) and 64 (a double histogram plotting the victims of witchhunts against cold weather); Balfour, *Historical works*, III, 436–7; Kuhn, *Soulstealers*, 229.

22. Porschnev, 'Les rapports', 160, quoting Johann Adler Salvius; BNE *Ms.* 2378/55, 'Reboluciones de Nápoles'; Bisaccione, *Historia*, 510.

23. BNE *Ms.* 2371/634, *Prognosticon* (1640); Naworth, *A new almanacke* (1642), sig. C2; Bournoutian, *The chronicle of Deacon Zak'aria*, 156.

24. [Voetius], *Brittish lightning*, sig. A3.

25. Kâtib Çelebi, *Fezleke-i Tarih*, II, 326.

26. Quenet, *Les tremblements*, 184–9 (on the earthquake of 1660) and 577 (table); Mallet and Mallet, *The earthquake catalogue*, plate IV.

27. Bainbridge, *An astronomicall description*, 30–1; Pacheco de Brito, *Discurso*, Sig. A11–11v

28. Roth, 'The Manchu–Chinese relationship', 7–8; Biot, 'Catalogue', 56–7; Brook, *The confusions*, 163–7; Moosvi, 'Science and superstition', 115, quoting the *Iqbalnama-i Jahangiri*.

29. Mather, *Kometographia*, 108–11.

30. Burton, *Anatomy* (1638 edn), Partition I, section I, subsection 1, 'Diseases in generall'.

31. Birago Avogadro, *Turbolenze di Europa*, 369; Riccioli, *Almagestum novum*, 96 (linking fewer sunspots with global warming).

32. Eddy, 'The "Maunder Minimum"', 268. Büntgen and Hellmann, 'The Little Ice Age', also reported that 'Warm (cold) summers generally matched episodes of high (low) solar activity'.

33. Amelang, *A journal*, 100 (chronicle of Andrés de la Vega); Yi, 'Meteor fallings', 205–6 and 217.

34. Magisa, *Svcceso raro*. Delfin, 'Geological, 14C and historical evidence', identified the volcano as Mt Parker; Atwell, 'Volcanism', 33, proposed that this was a Force Six eruption. Sigl, 'Timing', 547, found that all sixteen of the coldest decades during the past 2,500 years followed large volcanic eruptions.

35. AGI *Filipinas* 28/100, petition of Diego de Villatoro, 25 Aug. 1676, with further information in idem 28/90, memorial from Villatoro.

36. Song Yingxing, *Chinese technology*, 5. Zheng, 'How climate', 175–6, documents the impact of climate changes on crop production in twentieth-century China.

37. Costin, *Letopisețul Țării Moldovei*, 196–7 (I thank Mircea Platon for both reference and translation); Teodoreanu, 'Preliminary observations', 191.

38. Song Yingxing, *Chinese technology*, 4.
39. Kaplan, *The famine plot*, 62-3.
40. Ibid., 63; De Waal, *Famine that kills*, quoted in idem, 'A re-assessment', 470-1.
41. Calculations from Muldrew, *Food*, 130-1.
42. Lappalainen, 'Death and disease', 429
43. Ibid., 430; Smith, *The art of doing good*, 54, quoting Yang, 'A record of distributing padded jackets (*Shi mianao ji*)' (1612).
44. Komlos, 'An anthropometric history', 159; Ewert, 'Biological', 32 (especially Fig. 2); Hobbes, *Leviathan*, quoted on page xxi above.
45. Semedo, *Historica relatione*, 6-7, 13.
46. Andrews, *The colonial period*, 612-13, quoting Gorges (1611), the Virginia Company (1624) and many others; Canny, *The origins of empire*, 20, quoting Thomas Bowdler's commonplace book for 1635-36.
47. *Qing shilu*, VIII, ch. 86, 149, edict of the Yongzheng emperor 2 Nov. 1729 (translated by Ying Bao); Lenihan, 'War and population', 8, quoting Lawrence, *The interest of England*; Hull, *The economic writings*, I, 149-51, quoting Petty; Mortimer, *Eyewitness accounts*, 78, quoting Hans Conrad Lang; Antony and Christmann, *Johann Valentin Andreä*, 128; Arnauld, *Lettres*, II, 433, to the queen of Poland, 28 Jan. 1654.
48. Hayami and Tsubochi, *Economic and demographic developments*, 155; Lenihan, 'War and population', 20-1; Franz, *Der dreissigjährige Krieg*, 59; Jacquart, 'La Fronde', 283.

Chapter 2. The General Crisis

1. Data from Brecke, 'Violent conflicts'; Aho, *Religious mythology*, 195 (quoting Sorokin); Levy, *War*, 139-41.
2. St Vincent de Paul's correspondence (3,296 letters from 1607 to 1669: I thank Fr. Ed Udovic for sharing with me his calculation of word frequencies); Doglio, *Lettere di Fulvio Testi*, III, 204, to Francesco Montecuccoli, Jan. 1641; Gough, *The history of Myddle*, 71-2.
3. Carlton, *Going to the wars*, ch. 9; Gordon, *Diary*, I, 170; Zillhardt, *Der dreissigjährige Krieg*, 148, 225 and 158.
4. Hobbes, *Leviathan*, 88-9; Scott, *England's troubles*, 410, quoting Charles II's communications to the two Houses of the Convention Parliament in June-July 1660; Rodén, 'The crisis', 100, quoting Christina's 'Ouvrage de loisir'.
5. Struve, *Voices*, 2 and 48 (from 'Ten days in Yangzhou' by Wang Xiuchu); Meyer-Fong, *Building culture*, 11-12 (Weed-covered city), 150-2 (later memories by survivors) and 261-2 (assessment of Wang's account).
6. Helfferich, *The Thirty Years War*, 110, quoting von Guericke, *Die Belagerung, Eroberung und Zerstörung der Stadt Magdeburg*.
7. Sreenivasan, *The peasants*, 286; TCD *Ms.* 830/172, deposition of Christopher Cooe, Tuam, 21 Oct. 1645 (four years after the rape).
8. Quoted by Zysberg, 'Galley and hard labor convicts', 96.
9. Boyle, *A treatise*, I, 15; Behr, *Der verschantzte Turenne*, 'To the reader'. Both authors published in 1677.
10. Elliott and La Peña, *Memoriales y cartas*, I, 244, 'Resumen que hizo el rey don Felipe IV' (1627), discussed in Parker, *The military revolution*, 193 n. 3.
11. Botero, *Relatione*, fo. 19v; BNE *Ms.* 2362/61-2, marquis of Aytona to Philip IV, 28 Dec. 1630, copy; Davenant, *Essay*, 26-7.
12. Hellie, 'The costs'.
13. Stránský, *Respublica Bohemiae*, 495-6.
14. Glete, *War and the state*, 215.
15. Krüger, 'Dänische und schwedische Kriegsfinanzierung', 291, based on Oxenstierna's 'alternative' computation of the cost of the coming campaign, 8 Mar. 1633.
16. Salm, *Armeefinanzierung*, 163, quoting Hauptmann Holl in 1650; Di Cosmo, *The diary*, 48, 55.

17. Kishimoto, 'The Kangxi depression', 229, quotation Wei Jirui, secretary to the governor of Zhejiang, c. 1672.

18. Mut, *El principe*, 104; Seco Serrano, *Cartas*, I, 277–8, Sor María to Philip IV and reply, 1/12 June 1652, italics added.

19. Haboush, 'Constructing the center', 81–2, quoting a memorial by Hŏ Mok, government inspector, in 1660.

20. Foster, *The voyage of Thomas Best*, 176 (from the 'Journal' kept by Ralph Croft, who met the sultan in 1613); Reid, *Southeast Asia*, II, 257–8 (quoting Augustin de Beaulieu's account of his visit in 1619–22).

21. Brown, 'Tsar Aleksei', 140, quoting Gregorii Karpovich Kotoshikin, a Russian official who defected in 1664 and wrote a memoir about the government he had served.

22. McIlwain, *The political works of James I*, 307–8, speech to Parliament, 31 Mar. 1609; Forster, *The temper*, 9, funeral sermon for George of Hessen-Darmstadt in 1661; Cardin Le Bret, *De la souveraineté du roy*, 1 and 193–5; Christina, *Apologies*, 320.

23. Mormiche, *Devenir prince*, 231 (on reading Commynes's *Mémoires* of the reign of Louis XI), and passim. The models are currently displayed in the Musée de Plans-Relief, housed within the Hôtel des Invalides in Paris.

24. Mormiche, *Devenir prince*, 213, quoting the preceptor of Louis XIV's heir.

25. Ibid., 427, quoting the duke of Montausier.

26. Weiss, 'Die Vorgeschichte', 468, Frederick to Elizabeth, his wife, 19 Aug. 1619; Green, *The letters*, 70, Henrietta Maria to Charles, 11 May 1642 NS; Burnet, *The memoires*, 203, Charles to Hamilton, Dec. 1642; Halliwell, *Letters*, II, 383–4, Charles I to Prince Rupert, 31 July 1645.

27. Brown, 'Tsar Aleksei', 140, quoting Kotoshikin.

28. Elliott and de la Peña, *Memoriales y cartas*, I, 155, Olivares Memorial, 26 July 1626; Berwick y Alba, *Documentos escogidos*, 486, Philip IV to don Luis de Haro, 21 Oct. 1652.

29. Bergh, *Svenska riksrådets protokoll*, XIII, 17, debate on 21 Feb. 1649 OS; Christina, *Apologies*, 345.

30. Fernández Álvarez, *Corpus documental*, II, 104, Philip IV to don Gabriel Trejo Paniagua; Seco Serrano, *Cartas*, II, 42 and 48, Philip IV to Sor María, 11 Jan. and 19 Mar. 1656.

31. Jansson and Bidwell, *Proceedings in Parliament 1625*, 29, speech of Charles I to Parliament, 18 June 1625; Elliott, *The count-duke*, 378–9, Olivares's questions on 17 June 1629 and Philip's replies; Mormiche, *Devenir prince*, 306, Louis to the Princess Palatine, 23 Aug. 1693 (italics added).

32. Pursell, *Winter King*, 226, Rusdorf to Frederick V, 13 Aug. 1624, reporting the views of Sir Edward Conway; AHN *Estado* libro 714, unfol., *consulta* of the Council of State, 19 Oct. 1629; AGRB *SEG* 332/75, count of Oñate to the Cardinal-Infante, 8 Aug. 1634; Odhner, *Die Politik Schwedens*, 5, quoting Oxenstierna and Salvius.

33. RAS, *Manuskriptsamlingen* 68, Peter Loofeldt, 'Initiarum Monarchia Ruthenicae'; Scriba, 'The autobiography', 32; Morrice, *Entring book*, IV, 335.

34. Knowler, *Strafforde*, II, 243, Hopton to Wentworth, Madrid, 14 Nov. 1638 O. S.; *APW*, 2nd series, B IV, 579–81, Louis XIV [= Mazarin] to his diplomats at Münster, 14 Oct. 1646.

35. *APW*, 2nd series, B V, 1,151, Louis XIV to plenipotentiaries, 26 Apr. 1647, drafted by Mazarin; *CODOIN*, LXXXIII, 312–14, Peñaranda to the marquis of Caracena, governor of Lombardy, 27 June 1647; and 334, Peñaranda to the marquis of Castel Rodrigo, governor-general of the Spanish Netherlands, 12 July 1647.

36. Fletcher, 'Turco-Mongolian monarchic tradition', 238–9.

37. Anon. (probably James Howell), *The times dissected*, sig. A2; Bacon, *Essayes*, 46. This paragraph owes much to discussions with Daniel Nexon and Leif Torkelsen.

38. Burnet, *The memoires*, 55–6, Charles to Hamilton, 11 June 1638; Elliott, *The revolt*, 374–5, Olivares to Santa Coloma, viceroy of Catalonia, 7 Oct. 1639.

39. *ODNB*, s.v. Francis Russell, 4th earl of Bedford (1587–1641), by Conrad Russell, quoting Bedford's commonplace book.

40. Christina, *Apologies*, 142; Begley and Desai, *Shah Jahan Nama*, xvi n. 10; Wu, *Communication and imperial control*, 16.

41. Major, 'The crown', 639, quoting Father Joseph; Elliott, *The count-duke*, 42, exchange between Olivares and Uceda.
42. Ogilvie, 'Germany', 68.
43. Cressy, 'Conflict', 134, 137, 138, 139.
44. Ras, *Hikajat Bandjar*, 329.

Chapter 3. 'Hunger Is the Greatest Enemy'

1. Bacon, *Essayes*, 47 (first published in 1612).
2. Manuel de Melo, *Historia*, 25, 22; Bercé, 'Troubles frumentaires', 772, quoting 'rumore del popolo' in Fermo and neighbouring cities in 1648; Sibbald, *Provision for the poor*, 1. Italics added throughout.
3. Dodgshon, 'The *Little Ice Age*', 331, quoting the marquess of Lorne, whose family owned Kintyre.
4. *SCC*, IV.iii, 245, quoting Yan Sengfang, *Shang ting Chi*.
5. Jacquart, 'Paris', 108; McIlwain, *The political works of James I*, 343–4, speech to Star Chamber.
6. Jonson, *Epigrams*, CXXXIV, 'On the famous voyage' (1612); Howell, *Epistolae*, 25, to Captain Francis Bacon, Paris, 30 Mar. 1630; Dunstan, 'The Late Ming epidemics', 7, quoting Xie Zhaozhe, *Wu-za-zu* (1608).
7. Keene, 'Growth', 20–1.
8. Wrigley, 'Urban growth', 98–9 and 109; Davenant, *The first days*, 54–5 and 84.
9. Quotations from Evelyn, *Fumifugium*, 5–16.
10. Friedrichs, *The early modern city*, 276; Foster, *The voyage of Sir Henry Middleton*, 97–8 (Edmund Scott, 'Exact discourse').
11. Pepys, *Diary*, II, 128, entry for 30 June 1661.
12. Pepys, *Diary*, VII, 267–79, provides a memorable eyewitness account of the Great Fire. Mauelshagen, *Klimageschichte*, 127–9, notes that the summer of 1666 also saw devastating fires in a score of German cities, all associated with the unusual summer drought.
13. McClain, *Edo and Paris*, 106, quoting Asai Ryôi, *Musashi abumi*; Viallé and Blussé, *The Deshima registers*, XII, 294, (diary entries of Zacharias Wagenaer, 2–9 Mar. 1657), and 337 (12 Apr. 1658), final death toll.
14. Stoyle, '"Whole streets converted to ashes"', 141.
15. Elvin, 'Market towns', 446–7, quoting the gazetteer of Jiading county.
16. AMAE (M) *Ms.* 42/7, 11 and 15–16v, Chumacero to Philip IV, 6 Feb. and 20 May 1647.
17. Marks, *Tigers*, 112 (quotations, both from the Guangdong area in 1625–6).
18. Bray, *Technology and gender*, 228–9, quoting a gazetteer of 1660 about the town of Luzhou.
19. Kishimoto, *Shindai Chûgoku*, 29, writing of the early modern Chinese economy.
20. Van der Capellen, *Gedenkschriften*, I, 232. Studer, *The Great Divergence*, 113–15 maps and discusses the macroregions of Europe.
21. Lu, *Sources*, I, 216–18 (exclusion decrees); Tashiro, 'Tsushima', 97 n. 33 (order for an increase in silk imports via Korea and Tsushima); HAG *Ms.* 488/13v, viceroy to captain-general of Macao, 17 Apr. 1640 (on the loss of trade).
22. Kishimoto Mio, 'Comments' at a conference at International House of Japan, Tokyo, 16 July 2010.
23. Nakayama, 'On the fluctuation', 76–7 (quoting a contemporary journal).
24. Cervantes Saavedra, *Segunda parte de el ingenioso hidalgo Quijote*, ch. 20.
25. Weisser, *The peasants of the Montes*, 38–42.
26. Hindle, 'Exhortation', 119; idem, *On the parish*, 22–5; Barlow, *Barlow's journal*, 20–1.
27. Gutmann, *War and rural life*, 165 (St Truiden), 199 (Emael) and 84–6 (tithe receipts).
28. Smith, 'Benevolent societies', 325–7; idem, *The art of doing good*, 97–9, quoting Chen Longzheng.
29. Spence, *The death of Woman Wang*, 42, quoting the 1673 county gazetteer, and Huang Liuhong, *Fuhui quanshu [Complete book concerning happiness and benevolence]*.
30. Ono, *Enomoto Yazaemon*, 35; Mortimer, *Eyewitness accounts*, 77–8, quoting Hans Conrad Lang from near Ulm in 1635; Roupnel, *La ville*, 31–3 (quoting Girardot de Noseroy,

Histoire de dix ans de la Franche Comté de Bourgogne 1632-1642); Mortimer, *Eyewitness accounts*, 185, quoting Pastor Minck, from near Darmstadt, circa 1650, and 22-3, quoting Sebastian Bürster, a monk living near Überlingen, in 1647.

Chapter 4. Surviving in the Seventeenth Century

1. Special thanks to Matthew Connelly, Fabian Drixler, Gregory Hanlon, Eve Levin, Pamela McVay and Kenneth Pomeranz for help with framing this chapter; and to Peter Laslett and Tony Wrigley who first alerted me to the importance of demographic history. This chapter concentrates on demographic trends in Europe and Asia, which suffered the full force of the Little Ice Age and General Crisis; Chapter 15 examines the different demographic experience of Australia, Africa and America.
2. Jacquart, 'La Fronde', 283, on France; Telford, 'Fertility', 70-3, and Beattie, *Land*, 47 and 133 on Jiangnan.
3. John Donne, *Devotions*, XVII. 'Any man's death diminishes me, because I am involved in mankind, and therefore never send to know for whom the bell tolls; it tolls for thee.'
4. Sym, *Life's preservative against self-killing*, Preface and 124; Bosman, 'The judicial treatment', 15.
5. Fong, 'Writing from experience', 269-71.
6. Ho, 'Should we die?', 136, quoting Qian Xing, *Jiashen zhuanxin lu*; Spence, *The death of Woman Wang*, 9 and 14, from the memoirs of Magistrate Huang Liuhong.
7. Cherniavsky, 'The Old Believers', 21.
8. Kolff and van Santen, *De geschriften van Francisco Pelsaert*, 328.
9. Woods, *Death before birth*, 218, quoting Guillaume Mauqeste de la Motte, *Traité complète des accouchements naturels* (1722).
10. Nedham, *Medela medicinae*, 54.
11. Gatta, *Di una gravissima peste*, 6.
12. Moote, *The Great Plague*, 178 and 258-61.
13. Dunstan, 'The late Ming epidemics'; Odorico, *Conseils et mémoires*, 163-81.
14. Foster, *The English factories in India, 1630-1632*, 165-6, East India Company officials in Surat to their colleagues at Bantam, 8 Sep. 1631; and 178, same to London, 9 Dec. 1631.
15. Alfani, 'Plague', 18, 27.
16. Amelang, *A journal*, 40, 50, 58, 61, 68, 62 and 71.
17. Hastrup, *Nature*, 234-5; Galloway, 'Annual variations', 498-500; Patz, 'The effects', 282.
18. Rodger, *The command*, 73, quoting an officer who survived the Four Days' Battle, and 101, Commissioner Taylor to Samuel Pepys, Harwich, 4 Apr. 1667.
19. Schaufler, *Die Schlacht*, 7, quoting General Johann Werth.
20. Hutchinson, *Memoirs*, 208.
21. Ayala, *De iure*, 1.2 (Eng. Ed., I, 10-11).
22. Pascal, *Les pensées* ('Les hommes ne font jamais le mal si complètement et joyeusement que lorsqu'ils le font par conviction religieuse'); Ulbricht, 'The experience'.
23. Joshua 6: 21, 24; Carlton, *Going to the wars*, 178-9, quoting contemporary accounts of the siege of Basing House.
24. Abbott, *Writing and speeches*, II, 127, Cromwell to Speaker Lenthall, 17 Sep. 1649.
25. *HMC Ormonde*, n.s. II, 130-1, Lords Justices to Irish Commissioners, 7 June 1642.
26. Ogilvie, *A bitter living*, 1, quoting Catharina Schill, 'Eine arme Frau'.
27. Hufton, *The prospect before her*, 81.
28. Wrigley, *People, cities and wealth*, 240.
29. Capp, *When gossips meet*, 37.
30. Lee and Wang, *One quarter*, 10, 99.
31. Ko, *Teachers*, 261-3.
32. Cooper, *They came to Japan*, 58, quoting a letter from Cocks, Hirado, 10 Dec. 1614; Drixler, *Mabiki*, 110, quoting Manabe Genm'itsu, *Hokuchiku zakkō* (Kyoto, 1675).
33. Bray, *Technology and gender*, 289; Lee, Wang and Campbell, 'Infant', 405, 408, 410-11; Lee and Wang, *One quarter*, 47-51, 98, 107-8.

34. Levin, 'Infanticide' 219–21.

35. Patin, *Lettres*, III, 226, to André Falconet, 22 June 1660; Hanlon, *Human nature*, 119–20.

36. *Statutes of the realm*, 21 Jac. I c. 27; Wrightson, 'Infanticide'; Gowing, 'Secret births'.

37. Shakespeare, *Hamlet* (c. 1600), Act 3, Scene 1: 'Get thee to a nunnery: why wouldst thou be a breeder of sinners?'

38. 'Galarana Baratotti' [Elena Tarrabotti], *La simplicità ingannata* and *L'inferno monacale*, quoted in Dooley, *Italy in the Baroque*, 417.

39. Fildes, 'Maternal feelings', 143.

40. AHN *Consejos*, 41,391, n.p., 'Juez de comisión para la aberiguación de los fraudes que se an echo en el ospital de los niños expósitos' (I thank Fernando Bouza for bringing this volume to my attention).

41. Gray, *A good speed*, Sig. B3; Kupperman, *The Jamestown project*, 292, quoting Patrick Copland in 1622. See also Chapter 15.

42. For details, and for the term 'co-colonization', see Andrade, *How Taiwan became Chinese*.

43. Chen Zilong, 'The little cart', from Waley, *Translations*, 325 (I thank William S. Atwell for bringing this poem to my attention).

44. Özel, *The collapse*, 4, 176–80, and the relief map on 93. See also Chapter 7.

45. Ochoa, *Epistolario*, II, 64, Cristóbal Crespi, 16 May 1627.

46. Manuel de Melo, *Historia de los movimientos*, 38–40.

47. Schwartz, 'Silver, sugar and slaves', 10–11, quoting Fernando de Silva Solís; Donoghue, '"Out of the land of bondage"', 966.

48. T'ien, *Male anxiety*, 31. Ng, 'Ideology', 68, links an apparent prevalence of sodomy in Fujian to the surplus of bachelors.

49. Peters, *Ein Söldnerleben*; Helfferich, *The Thirty Years War*, 276–302.

50. Sen, *Poverty and famines*, 1; Kaplan, *The famine plot*, 2.

51. Bossuet, *Politics*, 65 (Book III, article iii).

Part II. Enduring the Crisis

1. Jia Yi, a Chinese political thinker of the second century BC, used as the epigraph to Elvin and Liu, *Sediments of time*.

2. Elliott, 'Revolution and continuity', 110. Other data quoted in Parker, 'La crisis de la década de 1590'.

3. Steensgaard, 'The seventeenth-century crisis', 33, originally written in 1970.

Chapter 5. The Great Enterprise in China, 1618–84

1. Special thanks for help in drafting this chapter to Tonio Andrade, William S. Atwell, Cynthia Brokaw, Timothy Brook, Roger Des Forges, Nicola Di Cosmo, Ann Jannetta, Kenneth Pomeranz, Evelyn Rawski, Christopher Reed, Lynn Struve, Kenneth Swope, Joanna Waley-Cohen and Ying Zhang in the United States; to Wang Jiafan and Chen Ning Ning in China; and to Hayami Akira, Iwai Shigeki, Kishimoto Mio, Mikami Takehiko, Shiba Yoshinobu and Yanagisawa Akira in Japan. I thank Taguchi Kojiro for research assistance with material in East Asian languages.

2. Struve, *Qing formation*, 334, quoting Xia, *Xingcun lu* (1645).

3. Brook, *The troubled empire*, 254–5.

4. Ibid., 243; *CHC*, IX, 40, quoting a Manchu document from 1615. Wang, 'Dessertification', Fig. 1, maps the limits of the summer monsoon and the different degrees of aridity of the steppe beyond.

5. Iwai, 'The collapse', 6, quoting Mao Yuanyi; Nakayama, 'On the fluctuations', 74.

6. Fei, *Negotiating urban space*, 43. The unusual name 'Single Whip' arose from the Chinese title *yitiao bianfa*, 'the conversion of tax assessments into a single item' – but because *bian* means not only 'convert' but also 'whip', 'single item' became 'single whip'.

7. Chan, *Glory and fall*, 189–97; Las Cortes, *Le voyage en Chine*, 183–8.

8. Dardess, *Blood and history*, 60, quoting Censor Zhou Zongjian.

9. Ibid., 163, quoting the 'Veritable records' of the Chongzhen reign.

10. Zhang, 'Politics and morality', 229–30.

11. Dardess, *Blood and history*, 166, quoting the *Mingshi*.

12. Zheng, 'How climate change', 172; Cheng and Lestz, *The search*, 5, quoting Song Yingxing, *Tiangong kaiwu*. The online supplement to Zheng's article, Fig. S1, provides six spectacular maps comparing drought-affected areas and regions of peasant uprisings from 1627 to 1643.

13. Wakeman, *The great enterprise*, 168–90.

14. Ko, 'The body as attire', 17, cites Hong Taiji's edicts.

15. Des Forges, *Cultural centrality*, 62, quoting the report of Wang Han, 'Sketches of a disaster', May 1640.

16. *SCC*, VI.ii, 64–70, discuss Xu Guangqi's *Nong zheng chuan shu*.

17. Atwell, 'East Asia', 6–7, quoting Chen Qide, *Zaihuang Zhishi* (*Record of disastrous famine*); Smith, *The art of doing good*, 137 and 153, quoting the diary of Lu Shiyi; and Will, 'Coming of age', 30, quoting Yao Tinglin, *Linian ji* (*Record of successive years*).

18. Qin Huitian, *Wuli tongkao* (1761).

19. Miller, *State versus gentry*, 140, quoting the foundation charter of the *Fu she*.

20. Cheng and Lestz, *The search*, 7, prints Li's proclamation to the citizens of Huangzhou.

21. Wakeman, *The great enterprise*, 306, quoting the Manchu proclamation to the Han Chinese.

22. Cheng and Lestz, *The search*, 25–7, Wu's letter to Dorgon, and Dorgon's reply.

23. Struve, *Voices*, 18–19, from 'Memoir of residing in Beijing' (1644) by Liu Shangyou.

24. Wakeman, *The great enterprise*, 316–17, Dorgon's edict of 5 June 1644.

25. Ibid., 418, quoting Song Quan.

26. Fu, *Ming-Ch'ing*, 99–101, quoting the bondservant leader Song Qi, and the gazetteer of Baoshan county (I thank Christopher Reed for this reference).

27. Mote, *Imperial China*, 828, quoting Dorgon. The term 'Southern Ming' only became common in the twentieth century. No Ming loyalist at the time used it, because it would have recognized the legitimacy of the Manchus' seizure of the north; and no Qing writer used it, because it would have accorded legitimacy to the resisters. I use it here for convenience.

28. Wakeman, *The great enterprise*, 456–7.

29. Ibid., 420–2, Dorgon's reluctant revocation of the edict on 25 June 1644.

30. Struve, *The Southern Ming*, 48, from the *Guoque* compiled by Tan Qian in the 1650s.

31. Kuhn, *Soulstealers*, 54, quoting Dorgon's outburst on 22 June and the decree of 8 July 1645. Cheng and Lestz, *The search*, 33–4, print the orders of the Board of Rites.

32. Martini, *Bellum*, 279.

33. Wakeman, 'Localism and loyalism', offers a detailed reconstruction of these events.

34. Will, 'Coming of age', 32–3, paraphrasing Yao's memoirs.

35. Kessler, *K'ang-hsi*, 14, quoting a memorial of 1649 (presumably echoing the proverb 'You can conquer China on horseback, but you cannot rule it on horseback').

36. Struve, *Voices*, 191, 194, 196. In 1646 a Southern Ming ruler bestowed on Zheng the imperial surname. Thereafter he was addressed as *guoxing ye* ('gentleman of the imperial surname') which Westerners rendered as 'Coxinga'.

37. Nakayama, 'On the fluctuations' 76–8.

38. Struve, *The Southern Ming*, 74 (for consistency, I have changed 'Ch'ing' to 'Qing' in this passage).

39. Wakeman, *Conflict and control*, 12, quoting one of those flogged and barred from taking exams.

40. Dott, *Identity reflections*, 177–8, quoting Kangxi's diary.

41. Nieuhof, *An embassy*, 48.

42. Waley-Cohen, *The culture of war*, 13; Will, 'Coming of age', 38–9, quoting from the final section of Yao's *Record*, explicitly comparing his experience under Ming and Qing.

43. Brook, *The troubled empire*, 242, 249.

Chapter 6. The 'Great Shaking'

1. Special thanks for help in drafting this chapter to Robert Frost, Przemysław Gawron, Dariusz Kołodziejczyk and James Lenaghan (on the Polish Commonwealth and Polish sources), and to Paul Bushkovitch, Irena Cherniakova, Chester Dunning, Mircea Platon, Matthew Romaniello, Mark Soderstrom and Kira Stevens (on Russia and Russian sources). I also thank Alison Anderson and Ardis Grosjean Dreisbach, respectively, for transcribing and translating German and Swedish documents, and Przemysław Gawron for translating Polish materials. Russia at this time used its own calendar, in which each year (counted from the Creation in 5508 BC) began on 1 September but in other respects followed the Julian calendar, which Sweden also observed. In this chapter all dates have been converted to the Gregorian calendar unless otherwise stated.

2. Baron, *Travels*, 120.

3. In this chapter, 'Ukraine' refers to the three palatinates of Kiev, Bratslaw and Volhynia (incorporated into Poland in 1569), plus the palatinate of Chernihiv (annexed from Russia at the Truce of Deulino in 1618 and added to the other three in 1635); and 'Ruthenian' refers to the Orthodox population of Ukraine.

4. RAS *Manuskriptsamlingen* 68, Peter Loofeldt, 'Initiarum Monarchiae Ruthenicae', 99, on 'zu Behauptunge ihren vermeinten kleinen Weldt'.

5. Kivelson, '"The Devil stole his mind"', 743.

6. Torke, *Die staatsbedingte Gesellschaft*, 219.

7. Gordon, *Diary*, II, 139.

8. Platonov, 'Novyi istochnik', 6–8: 'Kurtze vndt warhaftige Beschreibung desz gefährlichen Auffleutes des Gemeinen Pöbels Moscow'. Both this source (a manuscript from a private collection in Stockholm) and a dispatch from Ambassador Karl Anders Pommerenning mention a supplication on this day, yet no separate supplication has survived: only the one presented on 12 June 1648 NS. However, the 'Beschreibung' states that the supplication presented on both days was the same.

9. Romaniello, 'Moscow's lost petition', prints the only surviving text of this document, a Swedish translation of 'The supplication of the common man in Russia', sent by Pommerenning to Queen Christina together with his letter of 16 July 1648 NS.

10. Baron, *Travels*, 208 ('We must have you too'); RAS *Diplomatica: Muscovitica* 39, Pommerenning to Christina, 16 July 1648 (the musketeers' answer); Loewenson, 'Moscow rising', 153 (their number and pay arrears); Platonov, 'Novyi istochnik', 10 (the soothing words).

11. Loewenson, 'Moscow rising', 153–5; Platonov, 'Novyi istochnik', 13; RAS *Diplomatica: Muscovitica* 39, Pommerenning to Christina, 16 July 1648; RAS *Manuskriptsamlingen* 68, Peter Loofeldt, *Initiarum Monarchiae Ruthenicae*, 91.

12. Avrich, *Russian rebels*, 55; Platonov, 'Novyi istochnik', 19; Kivelson, 'The Devil', 747 (quoting a nobleman's servant).

13. Ellersieck, 'Russia', 89, citing Pommerenning to Christina, 6 July 1648, which Ellersieck decoded himself.

14. Torke, *Die staatsbedingte Gesellschaft*, 224–32, surveys the spread of unrest in 1648–9.

15. Hellie, *Enserfment*, 136, quoting later testimony from Patriarch Nikon.

16. Kivelson, 'The Devil', 752, quoting a memorandum later compiled by Prince Odoevskii, who headed the committee that drafted the code.

17. See http://pages.uoregon.edu/kimball/1649-Ulj.htm#ch11, 'Law code of the Assembly of the Land'.

18. RAS *Diplomatica: Muscovitica* 39, Pommerenning to Christina, Moscow, 17 Nov. 1649.

19. Plokhy, *The Cossacks*, 136, quoting a report by Stanislaw Koniecpolski to the Diet in 1631.

20. Sysyn, *Between Poland and the Ukraine*, 83, quoting Adam Kysil's second 'Discourse' on the Cossack problem in 1637; Plokhy, *The Cossacks*, 143, chronicle of Lviv.

21. Hannover, *Yaven Metzulah* (literally 'Deep mire', first published in 1653), 27–8.

22. Wrocław, Ossolineum, *Ms* 188/455v, 462, 463, 465v, 491, 499v, diary of Marcin Goliński of Kasimiersz (the Jewish quarter of Kraków); Beauplan, *Description*, 473–4 (on winters) and 471 (on locusts), reflecting seventeen years' living in Ukraine.

23. Wrocław, Ossolineum, *Ms* 188/516 (diary of Goliński); Namaczyńska, *Kronika*, 27–9.
24. Hrushevsky, *History*, VIII, 397, quoting a Russian source.
25. Ibid., VIII, 413 quoting a contemporary source.
26. Hannover, *Yaven Metzulah*, 50–77.
27. Bacon, 'The House of Hannover', 179–80 and 191; figures from Stampfer, 'What actually happened?'.
28. See Plokhy, *The Cossacks*, 220–35; Hrushevsky, *History*, VIII, 517–19; and Sysyn, 'Ukrainian-Polish relations', 63, 67 and 69–71.
29. Hrushevsky, *History*, VIII, 535, Khmelnytsky speech to Kysil, Feb. 1649.
30. RAS *Manuskriptsamlingen* 68, Loofeldt, *Initiarum Monarchiae Ruthenicae*, 93 and 97–9.
31. Ibid., 98–9.
32. Brown, 'Tsar Alexei', 124, order of Alexei to Prince Trubetskoi, late May 1654.
33. Rykaczewski, *Lettres de Pierre des Noyers*, 393, letter from Poznań, 8 Apr. 1658.
34. Davies, *Warfare*, 132.
35. Hellie, 'The costs', 64–6.
36. Crummey, 'The origins', 131, quoting Avraamii.
37. Michels, *At war*, 211.
38. Avrich, *Russian rebels*, 65, and Khodarkovsky, 'The Stepan Razin uprising', 8, both quote this 1667 document.
39. Avrich, *Russian rebels*, 76 and 78–9, quoting documents from 1670.
40. Details from ibid., 88–97, and Khodarkovsky, 'The Stepan Razin uprising', 14–18. No evidence exists that Razin possessed letters from Nikon.
41. Sysyn, 'The Khmelnytsky rising', 167; Davies, *Warfare*, 188.

Chapter 7. The 'Ottoman Tragedy', 1618–83

1. Special thanks for help in drafting this chapter to Günhan Börekçi, John Curry, Kaan Durukan, Suraiya Faroqui, Matt Goldish, Jane Hathaway, Colin Imber and Oktay Özel. I thank Allen Clarke for translating Arabic material for me, and Günhan Börekçi for analysing and translating Turkish sources, and for hosting me at the XI Congress of Social and Economic History of Turkey in Ankara in 2008, at which I learned so much.
2. Firpo, *Relazioni*, XIII, 170, Relazione of Lorenzo Bernardo, 1592; Sandys, *A relation of a Journey* (1615), 46.
3. Baer, 'Death in the hippodrome', 64.
4. Odorico, *Conseils et mémoires*, 163, 169 and 171.
5. Dankoff and Kim, *An Ottoman traveller*, 311–14 (Evliya Çelebi on Safed); Mikhail, *Nature and empire*, 23.
6. Özel, 'Banditry', 69.
7. Kâtib Çelebi applied this term to the deposition and murder of Osman: see Piterberg, *Ottoman tragedy*, 1.
8. Börekçi, 'Factions and favorites', 82–3, quoting Contarini, the Venetian agent in Istanbul, 3 Jan. and 18 Sep. 1604.
9. *Hasan Beyzade Tarihi*, 338–9; White, *The climate of rebellion*, 197–8, quoting Bostanzade Yahya.
10. *Peçevi Tarihi*, 385; *Topçular Kâtibi 'Abdulkadir*, 944–6, 985; *Hasan Beyzade Tarihi*, 375.
11. *Numarali Mühimme Defteri (H 1040/1630–1631)*, analysis of contents by Günhan Börekci.
12. Kâtib Çelebi, *Fezleke-i Tarih*, II, 326; Dujčev, *Avvisi*, 120–1, Turra to the pope, Ragusa, 9 July 1648, forwarding information sent from Istanbul on 12 June.
13. Monconys, *Journal*, I, 54, letter from Istanbul, 24 Aug. 1648.
14. Rolamb, 'Relation', 699.
15. Zilfi, 'Kadizadelis', 252.
16. Kunt, 'The Köprülü years', 100–15.
17. Scholem, *Sabbatai*, 136, quoting a letter of Rabbi Solomon Laniado in 1669, relating his meeting with Shabbatai four years before.
18. Scholem, *Sabbatai*, 435, quoting a letter written by Fr. La Croix.

19. The verdict of Scholem, *Sabbatai*, ix. My thanks to Matt Goldish and Benzion Chinn for helping me understand the Sabbatean phenomenon.
20. Finkel, *Osman's dream*, 276–7.

Chapter 8. Bloodlands: Germany and its Neighbours, 1618–88

1. Special thanks to Derek Croxton, Christopher Friedrichs and Paul Lockhart for valuable critiques of this chapter, and to Katherine Becker and Leif Torkelsen for help with some German and Scandinavian sources.
2. Dollinger, 'Kurfürst Maximilian', 298–9, Maximilian to his father (who had just abdicated), 21 June 1598.
3. Zillhardt, *Der dreissigjährige Krieg*, 93; Magen, *Reichsgräfliche Politik in Franken*, 190, statement of Hohenlohe's chancellor, June 1619.
4. Gindely, *Geschichte*, II, 164, Count Solms to Frederick of the Palatinate, Frankfurt, 28 Aug. 1619; Lee, *Dudley Carleton*, 270–1, letter of 18 Sep. 1619; Weiss, 'Die Vorgeschichte', 468, Frederick to Elizabeth, his wife, 19 Aug. 1619.
5. Reade, *Sidelights*, I, 388, Sir Edward Conway to Secretary of State Naunton, Nov. 1620.
6. Wilson, *The Thirty Years' War*, 353.
7. Turbolo, *Copia*, 6.
8. Mann, *Wallenstein*, 369, quoting the Bavarian council of war.
9. Bireley, *Religion and politics*, 54, Ferdinand's instructions to his representative at a meeting of the Electors at Mühlhausen, 4 Oct. 1627.
10. Parker, *The Thirty Years' War*, plate 7, reproduces the annotated copy of the edict (from Würzburg).
11. Bireley, *Religion and politics*, 125, eyewitness account from Regensburg by Kaspar Schoppe.
12. Wilson, *The Thirty Years War: a sourcebook*, 140–2, Treaty of Bärwalde, 23 Jan. 1631.
13. Suvanto, *Wallenstein*, 72, Questenberg to Wallenstein, 23 Apr. 1631.
14. Mortimer, *Eyewitness accounts*, 64–7 (Anna Wolff from Schwabach); Sreenivasan, *The peasants*, 282–6 (local parish registers); Helfferich, *The Thirty Years War*, 205–12 (Diary of Abbot Maurus Friesenegger of Andechs).
15. Bireley, *Religion and politics*, 214–17 (quoting Ludwig Crasius, S. J. in 1635); Bierther *Regensburger Reichstag*, 88 n. 69, Maximilian to his envoys, 27 Nov. 1640.
16. Odhner, *Die Politik Schwedens*, 163, Salvius to the Swedish regency council, 7 Sep. 1646.
17. *APW*, 2nd series, B II, 241, Mazarin to plenipotentiaries, 7 Apr. 1645; *ibid.*, B V, 1,151, Louis XIV to plenipotentiaries, 26 Apr. 1647, drafted by Mazarin.
18. *APW*, 1st series, I, 440–52, Instruction of Ferdinand III to Trauttmansdorff, Linz, 16 Oct. 1645, holograph (English translation in Helfferich, *The Thirty Years War*, 233–40).
19. *CODOIN*, LXXXIII 328 and 369, Peñaranda to Castel Rodrigo, 4 July and 2 Aug. 1647; Helfferich, *The Thirty Years War*, 250, quoting the diary of Clara Staiger.
20. Quotation from Roberts, 'Queen Christina', 198.
21. Data and quotations from Roberts, 'Queen Christina', 200, 201, 213 n. 62 and 217.
22. Details from Nordmann, 'La crise', 221–2.
23. Roberts, 'Queen Christina', 204, quoting Archbishop Lennaeus.
24. Roberts, *Sweden*, 101–5, prints the supplication of 8 Oct. 1650 OS, and on 105–8 the discussions of the noble Estate on 15 Oct.
25. Bergh, *Svenska riksrådets protokoll*, XV, 128, Jakob de la Gardie's speech to the council, 10 Oct. 1651.
26. Roberts, 'Queen Christina', 202 n. 26, quoting Christer Bonde, councillor of state, in 1655, with a reference to a similar remark by the queen.
27. Villstrand, 'Adaptation or protestation', 308–9.
28. Anon., *De na-ween vande vrede*, sig. A2v.
29. Poelhekke, *De vrede van Munster*, 256, 258, quoting a resolution of the States of Holland, 28 Feb. 1646, and the pamphlet *Ongeveynsden Nederlandtschen Patriot* (1647).
30. Details from Buisman, *Duizend jaar weer*, IV, 494–508 (Reijer Anslo, 'Op het regenachtige weer in het jaar 1648' quoted 494–5).

31. Israel, *The Dutch Republic and the Hispanic world*, 386 n. 31, quoting Antoine Brun to Philip IV, 25 Mar. 1650.
32. Meadows, *A narrative*, 33–5.
33. Roberts, *Sweden*, 171–2, debate of Charles X and the Swedish council, 23 July 1658.
34. Ekman, 'The Danish Royal Law', 102–7, prints a translation of several clauses.
35. Molesworth, *An account of Denmark*, 73, 74, 86.
36. Title from the evocative but misconceived pamphlet of Ergang, *The myth of the all-destructive fury of the Thirty Years War*.
37. Von Krusenstjern, *Selbstzeugnisse*, rubric B 8, notes the 'Schreibmotiv' whenever an author gave one. Melchior Brauch of Nuremberg, a Lutheran baker, stated explicitly that he wrote 'für mich' (idem, 57.)
38. Medick and Marschke, *Experiencing*, 80, quoting an anecdote from the evocatively titled *Sterbzeiten. Der dreissigjärigen Krieg in Herzogtum Westfalen. Eine Dokumentation*, ed. H. Conrad and G. Teske (Münster, 2000)
39. Outram, 'The socio-economic relations'; Lindegren, 'Men, money and means', 159; Medick and Marschke, *Experiencing*, vii.
40. Theibault, 'The demography', 12, 21.
41. Tacke, 'Mars, the enemy of art', 245–8, quoting from Sandrart, *Der teutsche Academie* (1675); Robisheaux, *Rural society*, 202.
42. Wedgwood, *Thirty Years War*, 526.

Chapter 9. The Agony of the Iberian Peninsula, 1618–89

1. Special thanks for help in drafting this chapter to James Amelang, Bethany Aram, Sir John Elliott, Xavier Gil, Alberto Marcos Martín, Andrew Mitchell, Martha Peach and Lorraine White.
2. IVdeDJ 82/444, duke of Sessa to Zúñiga, 28 Sep. 1600, minute; AGRB *SEG* 183/170v–171, Zúñiga to Juan de Ciriza, 7 Apr. 1619, copy.
3. Elliott, *The count-duke*, 231 and 236, Olivares to the count of Gondomar, 2 June and 3 July 1625.
4. Elliott, *The revolt*, 204 n. 2, Protonotorio Villanueva in Aug. 1626, on 'familiarizing the natives'.
5. AHN *Estado* libro 857/180–1, 'Papel que escrivió Su Magestad al Consejo Real', Sep. 1629.
6. Elliott, *The count-duke*, 365, records Olivares's prediction. Philip signed the Peace of the Pyrenees in 1659.
7. AHN *Estado* libro 857/180–183v, 'Papel que escribió Su Magestad', Sep. 1629; AHN *Estado* legajo 727/59, 'Orden de Su Magestad sobre su yda a Italia y Flandes', Oct. 1629.
8. AHN *Estado* libro 856, contains the proposals and 32 recommendations from the theologians, presented to the king on 23 Dec. 1629 (quotation from fo. 200).
9. BNL *Codex Ms.* 241/269–269v, Manuel de Faria e Sousa, 'Relação de Portugal'. I thank Lorraine White for this reference.
10. BNF *Ms. Esp.* 156/31–36v, *consulta* of the Council of State, 1631–32, copy.
11. Gelabert, *Castilla Convulsa*, 71–2, quoting Philip IV.
12. Quotations from Gelabert, *Castilla convulsa*, 53; Guiard Larrauri, *Historia*, II, 90 and 102–3 (from an anonymous 'Relación de lo suçedido en los alborotos' in Bilbao); and Elliott, 'El programa de Olivares', 434, royal apostil to a *consulta* on 4 Nov. 1632.
13. Elliott, *The count-duke*, 448, Olivares to the marquis of Aytona, 6 Oct. 1632; Subrahmanyam, *Explorations*, 129 and n. 78, Philip IV to Viceroy Linhares, 28 Feb. 1632.
14. Elliott, *The count-duke*, 482, Olivares to Pieter Roose, 29 Sep. 1634; Stradling, *Spain's struggle for Europe*, 116, Olivares' *voto* of 16 Jan. 1635.
15. Gelabert, *Castilla Convulsa*, 157.
16. AHN *Estado* libro 737/446–52, *consulta* of the Council of State, 16 Aug. 1624.
17. Valladares, *Epistolario*, 138 and 154, Olivares to Basto, 26 Nov. and 18 Dec. 1637; AGS *SP* libro 1536/3v–4, royal reply to a *consulta* of the Junta Grande de Portugal, 6 Nov. 1637.
18. Elliott, *The revolt*, 360, paper by Olivares, 12 Mar. 1639.

19. Ibid., 374–5, Olivares to Santa Coloma, 7 Oct. 1639, and 393, quoting Martí i Viladamor, *Noticia universal.*
20. Manuel de Melo, *Historia*, 25, 22.
21. Pasqual de Panno, *Motines*, 60 and 126; Simon i Tarrés, *Cròniques*, 253–4 (Relación of Judge Rubí de Marimon); Parets, *De los muchos sucesos*, I, 146–8 and 363–70.
22. BNE *Ms.* 2371/21, draft history of the year 1640 by Jerónimo de Mascarenhas; Corteguera, *For the common good*, 163, quoting the diary of Miquel Parets, and 165 n. 34, quoting a servant of Santa Coloma.
23. Rubí, *Les Corts*, 266, quoting Los Vélez.
24. Schaub, *Le Portugal*, 240, Vasconcelos to Soares, 30 Sep. 1640.
25. HAG *Ms.* 28/514v–515, viceroy of India to Philip IV, 2 Aug. 1641.
26. Loureiro de Souza, *Documentos*, 10–16, Câmara de Bahia to Philip IV, 22 Sep. and 13 Nov. 1640, and to John IV, 30 Apr. 1641 (I thank Thiago Krause for verifying these references for me.)
27. AGS *GA* 1331, n.p., *consulta* of the Junta de Ejecución, 7 Dec. 1640; TNA *SP* 94/42/73–4, Hopton to Windebank, 8 Dec. 1640.
28. Simon i Tarrés, *Els orígens*, 199 n. 81, quoting Albert Tormé i Lliori.
29. AMAE (P) *CPE: Supplément* 3/228v, account of the battle by Duplessis-Besançon.
30. AGS *GA* 1376, n.p., Olivares at the Junta Grande, 19 June 1641 (I thank to Lorraine White for this reference); AGI *México* 35/18, marquis of Caldereyta to Philip IV, 6 Dec. 1641, noting the decision to hold back the treasure in July 1640: the arrival of 750,000 ducats might have turned the tide in the peninsular war.
31. Hespanha, 'La "Restauraçao"', details the petitions.
32. BNE *Ms.* 2371/111–14, Nochera to Philip IV, 6 Nov. 1640, copy.
33. BNE *Ms.* 18,723 # 58, 'Copia del papel que dió a Su Magestad el duque de Medina Sidonia', 21 Sep. 1641.
34. Elliott, *The count-duke*, 611, Hopton to Vane, 3/13 April 1641.
35. AGS *Estado* 2667 n.p., and *Estado* 8341/3, *consultas* of the Council of State, 30 Jan. and 3 Feb. 1643; *CODOIN*, LIX, 304, Philip IV to Melo, 12 Feb. 1643, with a letter for Anne.
36. AGS *Estado* 2056 n.p., *consulta* of the Council of State 5 Dec. 1641, on Melo's letter of 11 Nov. (I thank Fernando González de León for this reference); AGS *Estado* 8341/3, *consulta* of the Council of State, 3 Feb. 1643, 'voto' of the count of Oñate.
37. AGS *Estado* 3848/154, *consulta* of the Junta of State, 23 Oct. 1643, *votos* of the counts of Monterrey and Oñate.
38. Gelabert, 'Alteraciones', 364, Chumacero to Philip IV, June 1645; AMAE (M) *Ms.* 39/218, same to same, 22 July 1645; AGS *GA* 3255, n.p., Haro to Gerónimo de Torre, 13 Feb. 1646 (two letters), and to Philip IV, 14 Feb. 1646, all from Cadiz.
39. BNE *VE* Ca 68–94, *Escrívense los sucessos de la Evropa desde abril de 46 hasta junio de 47 inclusive*; AMAE (M) *Ms.* 42/15–16v, Chumacero to Philip IV, 22 Oct. 1647, and fos. 45–8, *consulta*, 10 Sep. 1647; Thompson, 'Alteraciones granadinos', 799.
40. Morales Padrón, *Memorias de Sevilla*, 123–4; BNE *Ms.* 11,017/106–19, account of the troubles in Granada.
41. AGS *Estado* 2668, n.p., *consulta* of the Council of State, 4 July 1648; Seco Serrano, *Cartas*, I, 158–9, Philip IV to Sor María, 29 July 1648.
42. *CODOIN*, LXXXIV, 314–16, Peñaranda to Philip IV, 19 Aug. 1648.
43. AM Cadiz *Ms.* 26/161 and 168–74, resolutions of 26 Oct. and 24 Nov. 1648; Borja Paloma, *Historia crítica*, 297–9, quoting *Memorias sevillanas*; Morales Padrón, *Memorias de Sevilla*, 115–17.
44. Domínguez Ortiz, *Alteraciones andaluzas*, 86, *Real cédula*, 16 May 1652.
45. AGS *Estado Francia* K. 1618/C.5, Junta of State, 7 Jan. 1659; Seco Serrano, *Cartas*, II, 131, Philip IV to Sor María, July 1659.
46. Valladares, *La rebelión*, 204, 'Junta sobre materias de Inglaterra', 17 June 1665.
47. Espino López, *Catalunya*, 74 and 78, *consultas* of the Council of State, 4 Nov. 1687 and 13 Apr. 1688; Kamen, 'The decline of Castile', 63, quoting Ambassador Carlo Russini in 1695.
48. Sanz Camañes, 'Fronteras', 68.

49. Márquez Macías, 'Andaluces', quoting the marquis of Villars in 1681, who estimated that 6,000 people left for America each year.

50. Hoffman and Norberg, *Fiscal crises*, 176, estimate by I. A. A. Thompson.

51. Marcos Martín, '¿Fue la fiscalidad regia un factor de crisis?', 250–2.

52. BL *Eg. Ms.* 1820/340, Hopton to Secretary of State Coke, 6 Apr. 1634 NS. Brook quoted on page 121 above.

53. AGRB *SEG* 195/64, Philip IV to Infanta Isabella, his regent in the Netherlands, 9 Aug. 1626.

54. Monkhouse, *State papers*, III, 16, Hyde to Nicholas, Madrid, 14 Apr. 1650 NS; Alcalá-Zamora, 'Razón de Estado', 341, quoting the marques of Los Vélez, viceroy of Naples, to Carlos II, 11 Nov. 1678.

Chapter 10. France in Crisis, 1618–88

1. Special thanks for help in drafting this chapter to Robin Briggs, Laurence Brockliss, Oliver Herbert, David Parrott and Dale van Kley.

2. Moote, *The revolt*, 368.

3. *APW*, 1st series, I, 18–20, Louis XIII to Richelieu, 4 Aug. 1634.

4. Bercé, *Histoire des Croquants*, I, 414–19 discusses the fourteen surviving manifestos and the role of La Mothe.

5. Ibid., I, 443, quoting letters from royal ministers to Chancellor Séguier in June and July 1637. La Mothe survived in hiding until 1648 (ibid., I, 445).

6. Foisil, *La révolte*, 62, Bullion to Richelieu, 11 Oct 1639.

7. Ibid., 285, quoting Séguier's journal.

8. *APW*, 2nd series II, B IV, 241, Mazarin to d'Avaux, 20 July 1646.

9. Chéruel, *Histoire*, II, 497, quoting the *Journal manuscrit d'un bourgeois de Paris*; Descimon and Jouhaud, 'La Fronde en mouvement', 307, quote Richelieu's warning.

10. Motteville, *Mémoires*, II, 8 (the author accompanied the queen).

11. Le Boindre, *Débats*, 44, Broussel on 5 Feb. 1648; Chéruel, *Histoire*, II, 501–2, quoting Talon's speech and the Venetian ambassador in Paris.

12. Chéruel, *Lettres*, II, 917 and 948, Mazarin to Grimaldi, 4 July and 10 Sep. 1647; and 505, to Fontenay-Mareuil, 7 Oct. 1647.

13. Ranum, *Paris*, 283; Arnauld, *De la fréquente communion* (1643).

14. *CODOIN*, LXXXIV, 230–1, 234–5, Peñaranda to Don Luis de Haro and to Philip IV, 18 May 1648.

15. Chéruel, *Lettres*, III, 159–60, Mazarin to Chanut, 31 July 1648, and 173–81, Mazarin to Servien, 14 Aug. 1648.

16. Motteville, *Mémoires*, II, 98, 214.

17. Ibid., II, 355.

18. Parival, *Abrégé de l'histoire*, 480.

19. Carrier, *La Fronde*, I, no. 11, Robert Arnauld, *La vérité toute nue*.

20. Bonney, *Limits*, VIII, 336–7, on revoking the edicts of Parlement.

21. Carrier, *Le labyrinthe*, 150, Renaud de Sévigné to Christine de France, 19 July 1652; Arnauld, *Lettres*, II, 177, to the queen of Poland, 6 Sep. 1652; Jacquart, 'La Fronde', 279 quoting André d'Ormesson; *Mémoires de Mlle. de Montpensier*, II, 276, reporting the pessimism of Gaston, her father, at Easter 1655.

22. Robin Briggs, personal communication, May 2004.

23. Barnes, '"Playing the part"', 184, quoting René Le Voyer d'Argenson, *Annales de la Compagnie du Saint Sacrement* (manuscript completed in 1694).

24. Silvestre de Sacy, *Lettres*, III, 321 and 345, Mme de Sévigné to Mme de Grignan, 28 June and 24 July 1675; Le Roy Ladurie, *Histoire humaine*, 462–3 (Provence).

Chapter 11. The Stuart Monarchy

1. Special thanks for help in drafting this and the following chapter to Aidan Clarke, David Cressy, Richard Groening, Andrew Mackillop, Jane Ohlmeyer, Carla Pestana, Glyn

Redworth and John Walter. Britain in the seventeenth century used the Julian calendar ('Old Style'), running ten days earlier than the Gregorian calendar common on the continent, and I have used it in Chapters 11 and 12. Thus the Scottish Revolution began on 23 July 1637 in Edinburgh (and in this book), which in Paris, Rome and Madrid was 2 August.

2. Howell, *Cobbett's complete collection*, II, col. 114, speech by James I, 18 Apr. 1604; Davies, *A discovery*, 252

3. Thirsk, *The agrarian history*, IV, 582, quoting letters by local magistrates to the Privy Council.

4. James I, *His maiesties speech* (1607), sig. Cv; Cust, *Charles I*, 41.

5. Bamford, *A royalist's notebook*, 27–8 and 54.

6. Johnson et al., *Commons debates 1628*, II, 58–60, Sir Benjamin Rudyerd's speech on 22 Mar. 1628; Rushworth, *Historical collections*, I, 631–8, remonstrance presented on 17 June 1628.

7. Baker, 'The climate', 427 (from the diary of Richard Napier); Bamford, *A royalist's notebook*, 79; Wharton, *The history*, 47–9 and 51 (from Laud's *Diary*); TNA *SP* 16/282/134, a Latin poem on the 'most intense cold of January [1635] when the whole Thames froze over' (I thank David Cressy for this reference); *CSPV*, XXIV, 63, Anzolo Correr to Doge, 5 Sep. 1636.

8. Fincham, 'The judges' decision', 236, from Sir Roger Twysden's 'Remembrances'.

9. Clarendon, *The history*, 92.

10. TNA *SP* 16/527/103–7, draft proposal for a British Union of Arms (1627). Sir James Balfour explicitly linked the Union of Arms scheme with the Revocation: Haig, *Historical works*, II, 126 (quotations from ibid., 128, 134). Kishlansky, 'Charles I', 71, claimed that the Revocation was promulgated by Parliament, not by Prerogative, but this is misleading: it did not receive parliamentary approval until 1633.

11. Charles I, *A large declaration*, 16; Rushworth, *Historical collections*, II, 321, Charles's proclamation ordering Archbishops Laud and Spottiswoode to prepare a liturgy for Scotland, 19 Apr. 1636.

12. Donaldson, *The making*, 100 (proclamation authorizing the Prayer Book); Charles I, *A large declaration*, 18 (confirming that he had issued it by his own authority); *RPCS*, 2nd ser., VI, 448, Act of 13 June 1637 (placing non-compliers 'under pain of rebellion').

13. Rothes, *Relation*, 197 (use of proofs as wrapping paper); Braddick, *God's fury*, 3, quoting Montrose.

14. Charles I, *A large declaration*, 23; Bennett, *The Civil Wars*, 3, quoting the earl of Wemyss and Bishop Guthrie.

15. *RPCS*, 2nd ser., VI, 509–13, entries for 4, 5 and 9 Aug. 1637; Rothes, *Relation*, 2–5.

16. Dickinson and Donaldson, *Sourcebook*, III, 95–104, prints the Covenant, including the Oath and Subscription; italics added.

17. Paul, *Diary*, I, 322, 327–31 and 347, 27 Feb., Mar./Apr. and 4 May 1638.

18. Russell, *The fall*, 56, Hamilton to Charles, June 1638; Burnet, *The memoires*, 55–6 and 60–1, Charles to Hamilton, 11 and 25 June 1638; Hardwicke, *Miscellaneous state papers*, II, 118, Hamilton to Charles, Glasgow, 27 Nov. 1638.

19. New College, Oxford, *Ms.* 9502, Diary of Robert Woodford of Northampton, entry for 6 Apr. 1639; Wharton, *The history*, 56–7, Laud's diary entries for 14 Jan. and 27 Dec. 1639; Aston, 'The journal', 7; Bruce, *Letters and papers*, 238, Sir Edmund Verney to Ralph Verney, Newcastle, 19 May 1639.

20. Aston, 'The journal', 28; Paul, *The diary*, 58–62.

21. Adamson, 'England', 100.

22. Russell, *The fall*, 67 n. 135, Hamilton to Charles, 8 July 1639.

23. Adamson, *The noble revolt*, 23 and 17, quoting the king's speech on 5 May 1640.

24. *HMC third report*, 3, minutes taken by Vane at the meeting of the Committee of War, 5 May 1640.

25. *CSPD 1640*, 118 and 627, George Douglas to Roger Mowatt, 5 May 1640, and Strafford to Cottington, 24 Aug. 1640.

26. Haig, *Historical works*, II, 379, written in the 1650s; Adamson, *The noble revolt*, 47, Lord Savile to Lord Loudun, 8 July 1640.

27. Rushworth, *Historical collections*, III, 1,214–15, petition of the Yorkshire gentry, 28 July 1640.

28. Hardwicke, *Miscellaneous state papers*, II, 168–71, minutes of the council meeting, 2 Sep. 1640.

29. Gardiner, *Constitutional documents*, 134–5, petition of twelve peers, 28 Aug. 1640.

30. Hardwicke, *Miscellaneous state papers*, II, 182, Vane to Windebank, York, 18 Sep. 1640; Rushworth, *Historical collections*, III, 1,275, the king's speech on 24 Sep. 1640.

31. Ibid., 1,334 the king's speech on 3 Nov. 1640; *PLP*, I, 63, 65 and 69.

32. Clarendon, *The history*, 320–1, reporting his conversation with Essex on 26 Apr. 1641.

33. BL *Addl. Ms.* 21,935/138–9, Wallington's 'Historical notes and meditations' for 3 May 1641; Adamson, *The noble revolt*, 285–6, quoting Bishop Warner's diary, and 288–91, quoting Pym.

34. Cressy, 'The protestation', 266–7, 271, 273, quotations from John Turberville and Sir John Bramston.

35. Kilburn and Milton, 'The public context', 242, quoting Sir Philip Warwick.

36. *CSPD 1641–3*, 17, Henry Vane to Sir Thomas Roe, Whitehall, 18 June 1641; Fletcher, *The outbreak*, 192–9.

37. *Articles of the large treaty*, 48, treaty articles approved 7 Aug. 1641.

38. Perceval-Maxwell, 'Ulster 1641', 103, quoting the 'Desires concerning unity in religion' proposed by the Scottish commissioners.

39. TCD *Ms.* 839/135, deposition of Mulrany Carroll, County Donegal, 26 Apr. 1643.

40. Ohlmeyer, 'The "Antrim Plot" of 1641: a rejoinder', 434–7, printing Antrim's 'Information' in May 1650. The veracity of the earl's evidence, which deeply compromised Charles, has been hotly disputed ever since; but Lamont, 'Richard Baxter', 345–7, confirms Ohlmeyer's deduction that Antrim did not lie.

41. TCD *Ms.* 809/14, Examination of Owen Connolly, 22 Oct. 1641.

42. BL *Harl. Ms.* 5,999/29v and 27v, 'Discourse' by Henry Jones and the other judges who compiled the depositions, Nov. 1643.

43. TCD *Ms.* 837/12–13, deposition of Thomas Richardson, Co. Down, 13 June 1642. (Hickson, *Ireland*, I, 312, printed this document with the wrong date, the wrong profession – she read 'taylor' instead of 'saylor' – and several errors of transcription.)

44. Castlehaven, *Memoirs*, 28–9; *HMC Ormonde*, n.s, II, 251–2, Lords Justices to Charles I, 16 Mar. 1643.

45. TCD *Ms.* 830/41–2, deposition of Anthony Stephens, 25 June 1646, italics added.

46. TCD *Ms.* 809/8v and 10v, deposition of Archdeacon Robert Maxwell, 22 Aug. 1642, mentioning the figure '154,000' twice. Clarke, 'The "1641 massacres"', 41–2, charts the diffusion of Maxwell's false claim. According to Smythe, 'Towards a cultural geography', 81, 'geographic analysis of the data suggests a figure of about 4,200 murders and killings' of Protestants in October 1641.

47. Bray, *Diary*, IV, 97, Nicholas to Charles I, 25 Oct. 1641, with royal postscript dated 30th; TCD *Ms.* 835/158, deposition of John Right, Co. Fermanagh, 5 Jan. 1642. Hickson, *Ireland*, I, 114–15, published O'Neill's alleged royal commission, dated Edinburgh, 1 Oct. 1641, and the depositions she printed at 169–73 and 188–9 show how well the forgery worked.

48. Kenyon, *Stuart Constitution*, 228–40, the Remonstrance presented to the king on 11 Dec. 1641.

49. Details from letters calendared in *CSPD 1641–43*, 215–17.

50. Catherine Macaulay, *History*, III, 150, printed the queen's high words but gave no source – however, Henrietta Maria's threat five months later to enter a convent if her husband failed to do what she said (Green, *The letters*, 80, letter to Charles on 30 May 1642) lends plausibility to this earlier outburst.

51. *CSPD 1641–3*, 240–2, Thomas Wiseman and Robert Slingsby (one of the swordsmen who entered the Commons) to Pennington, 6 Jan. 1642; Cressy, *England on edge*, 393, quoting John Dillingham.

52. BL *Addl. Ms.* 21,935/162, Wallington; Cressy, *England on edge*, 396, quoting Ellis Coleman; and Groen van Prinsterer, *Archives*, 2nd series, IV, 7, Heenvliet to Orange, London, 19 Jan. 1642 OS, relating the queen's indiscreet comments at an audience the previous day.

53. Sharpe, *The personal rule*, 183; Scott and Bliss, *The works*, V, pt 2, 317–70, prints Laud's accounts and royal apostils (quotations from 319, 348 and 337).
54. The last sentence of James I, *His Maiesties speech* (1607), sig. Hv; Clarendon, *The history*, 567–8.
55. Sharpe, *The personal rule*, 930; Braddick, *God's fury*, 42 and 95.
56. Burnet, *The memoires*, 203, Charles to Hamilton, Dec. 1642.
57. Gardiner, *The Hamilton papers*, 6, Hamilton to Charles, 7 June 1638, italics added. On the need for a 'willingness to wink' among seventeenth-century rulers, see Chapter 2 above.

Chapter 12. Britain and Ireland from Civil War to Revolution, 1642–89

1. Rushworth, *Historical collections the fourth and last part*, II, 1,397, the 'charge' read by John Cook, 20 Jan. 1649.
2. Balfour, *Historical works*, III, 409; Gilbert, *History of the Irish Confederation*, VI, 270–1; Ó Siochrú and Ohlmeyer, eds., *Ireland, 1641*, 1.
3. Walter, *Understanding*, 201, the 'Humble petition' of Essex, 20 Jan. 1642 OS.
4. Kenyon, *Stuart Constitution*, 244–7 and 21–3, the Nineteen Propositions (1 June 1642) and the king's response (18 June 1642).
5. Ibid., 194, marginal comment of the king on a letter in Nov. 1641; Walter, *Understanding*, 129, instructions.
6. Symmons, *A vindication*, 241.
7. Cressy, *England on edge*, 313–14, quoting Thomas Knyvett to his mother, May 1642; Millstone, 'Seeing', 78 and 83–4 (quoting Parson John Rous).
8. Eisenstadt and Schluter, 'Early modernities', 25.
9. [Parker,] *The king's cabinet*, 7–8, Charles to Henrietta Maria, 5 Mar. 1645; 16, to Ormond, 27 Feb. 1645; 46–7 and 54–6 ('Annotations' arranged in six heads).
10. Woolrych, *Soldiers and statesmen*, 38; Firth, *The Clarke papers*, I, 425–6, 'Colonel Wogan's narrative'.
11. Rushworth, *Historical collections*, VI, 512, *The solemn engagement of the army*, 5 June 1647.
12. Ibid., VI, 564–70, 'A declaration, or representation from His Excellency Sir Thomas Fairfax and of the army under his command', 14 June 1647.
13. Gardiner, *Constitutional documents*, 316–26, 'The Heads of the Proposals ... to be tendered to the commissioners of Parliament residing with the army', debated by the General Council, 16–26 July 1647 OS; Firth, *Clarke Papers*, 1, 213, speech by William Allen; Woolrych, *Soldiers and statesmen*, 153–79.
14. Bamford, *A royalist's notebook*, 112.
15. Gardiner, *Constitutional documents*, 333–5, 'An agreement of the people'.
16. *Journals of the House of Lords*, IX, 529–31, 'A remonstrance from his excellency Sir Thomas Fairefax and his council of war', 14 Nov. 1647.
17. Gardiner, *Constitutional documents*, 247–52, 'The engagement between the king and the Scots', 26 Dec. 1647.
18. Howell, *Epistolae*, III, 26, letter of 10 Dec. 1647 to his nephew; Wildman, *Truths triumph*, 4–5 (published 1 Feb. 1648, describing his speech on 18 Jan.).
19. Rushworth, *Historical collections the fourth and last part*, II, 1,396–8 (the 'charge' read by John Cook); 1,406–14 (depositions); and 1,421 (sentence).
20. Firth and Rait, *Acts and ordinances*, II, 18–20, 'An Act for abolishing the kingly office in England and Ireland', 17 Mar. 1649; and 122, 'An act declaring and constituting the people of England to be a Commonwealth', 19 May 1649; Charles I, *Eikon basilike*, 183–4.
21. *Acts done and past*, 35–8, 'Proclamation of Charles the Second, king of Great Britain, France and Ireland', 5 Feb. 1649, and 'Act anent securing the covenant, religion and peace of the kingdom', 7 Feb. 1649.
22. Bremer, 'In defence of regicide', 103, quoting John Hull's journal.
23. Firth and Rait, *Acts and ordinances*, II, 325–9, 'An Act for subscribing the Engagement', 2 Jan. 1650; Worden, 'The politics of Marvell's Horatian Ode', 526–7, quoting Sir Henry Vane, Jr.

24. Balfour, *Historical works*, III, 409 (prices), 432-3 (weather) and 436-7 (witches); Larner, *Source book*, sub annis 1649-50.

25. Hobbes, *Leviathan*, 491, 484, and 154.

26. Abbott, *Writings and speeches*, II, 463 and 467, Cromwell to Lenthall, 4 and 8 Sep. 1651.

27. Ibid., II, 325, Cromwell to Lenthall, 4 Sep. 1650.

28. Scott, *Politics*, 201.

29. Firth and Rait, *Acts and ordinances*, II, 598-603, 'Act for the settlement of Ireland', 12 Aug. 1652. See Wells, 'English Law', for the work of the High Court of Justice for Ireland, and http://downsurvey.tcd.ie/1641-depositions.php for maps showing the dramatic change in Irish landholding in the 1650s.

30. Firth and Rait, *Acts and Ordinances*, II, 813-22, 'The government of the Commonwealth of England, Scotland and Ireland, and the Dominions thereunto belonging', 16 Dec. 1653. A few other states tolerated freedom of *private* worship but the Instrument of Government was the first to guarantee freedom of *public* worship.

31. Worden, 'Oliver Cromwell and the sin of Achan', 136-7, quoting 'A declaration of His Highness, inviting the people of England and Wales to a day of solemn fasting and humiliation' (Mar. 1656); De Beer, *The diary of John Evelyn*, 388, 393.

32. Rutt, *Diary of Thomas Burton*, III, 256, speech by Arthur Haselrig, one of the five members whom Charles I had tried to arrest in 1642.

33. Gardiner, *Constitutional documents*, 465, the Declaration of Breda, issued by Charles on 4 Apr. 1660.

34. Grey, *Debates*, VIII, 264, speech of Sir Henry Capel, 7 Jan. 1681.

35. Plumb, *The growth*, 60.

36. Kenyon, *Stuart Constitution*, 410-11, James II, 'Declaration of Indulgence', 27 Apr. 1688.

37. Williams, *Eighteenth-century Constitution*, 8-10, the 'invitation to William', 30 June 1688.

38. Williams, *Eighteenth-century Constitution*, 60 and 26, notes of discussions between William and the English peers in winter 1688-9, and 26-33, the Bill of Rights.

39. De Beer, *The diary of John Evelyn*, V, 288.

40. Cullen, *Famine*, 2, 10, and 49; Sinclair, *The statistical account*, XVII, 483, report from Insch (Aberdeenshire), and II, 551, report from Kilmuir (Skye).

41. Ingersoll, 'The lamp of experience', quoting Thomas Molyneux, MP; Anon., *An essay on government*, 95-6. (I thank Tom Ingersoll for this remarkable reference.)

42. Macaulay, *History*, II, 508-9. Macaulay wrote only about England: the Glorious Revolution caused the death of over 100,000 Scots and Irish, as well as prolonged material suffering: Carlton, *This seat of Mars*, 262.

Part III. Surviving the Crisis

1. Lancellotti, *L'hoggidi* (*hoggidiani* literally means 'people nowadays').

2. Green, *Spain*, IV, 6, Quevedo to Don Francisco de Oviedo, 21 Aug. 1645; Gracián, *El Criticón*, 3 parts published 1651, 1653 and 1657; Hobbes quoted on page xxi above.

Chapter 13. The Mughals and their Neighbours

1. Special thanks for help in drafting this chapter to Lisa Balabanlilar, Stephen Dale, Scott Levi, Sanjay Subrahmanyam, Tristan Mostert and Stephan van Galen. Although members of the dynasty never used the term 'Mughal', an Arabized word for 'Mongol' picked up by Europeans, it seems pedantic not to use it.

2. Dale, *The Muslim empires*, 100; Elliot and Dowson, *The history of India*, VII, 158.

3. Guha, *India*, I, 82 (Ovington in 1689); Sarkar, *Anecdotes of Aurangzeb*, 53, from Aurangzeb's advice to his son, c. 1695.

4. Temple, *Travels of Peter Mundy*, II, 42-4, 47-9, 55-6: entries for Nov.-Dec. 1630; Foster, *The English factories in India, 1630-1633*, 122, 165, 178 and 218-19, letters from East India Company officials in Surat to London, 31 Dec. 1630, 8 Sep. and 9 Dec. 1631, and 8 May 1632 OS.

5. Temple, *Travels of Peter Mundy*, II, 265, 275–6, entries for April–May 1633; HAG *Ms.* 1498/11–12, Viceroy Linhares to Philip IV, 10 Aug. 1631, copy.

6. Foster, *The English factories 1634–1636*, 64–5, East India Company officials in Surat to London, 29 Dec. 1634 OS; Temple, *Travels of Peter Mundy*, II, 265, 275–6, entries for April–May 1633.

7. Elliot and Dowson, *The history of India*, VII, 24–5, quoting Lahori's *Padshah Nama*; Begley and Desai, *Shah Jahan Nama*, 62.

8. Subrahmanyam, 'A tale of three empires', 73.

9. Foltz, 'The Mughal occupation', 57, quoting the *Tazkira-i Muqim Kahni*.

10. McChesney, *Waqf*, 141, quoting a *Manshur* (confirmation) issued by the ruler of Balkh in 1668–9.

11. Faruqui, 'Princes and power', 299, quoting the *Waqiat-i Alamgiri*; Elliot and Dowson, *The history of India*, VII, 178, quoting Muhammed Kazim, *Alamgir Nama*.

12. Foster, *The English factories, 1655–1660*, 263 and 256, letters from the English factors at Masulipatam, Mar. and Oct. 1659; and 210 and 310, letters from the English factors in Surat, Sep. and Oct. 1659 and Apr. 1660; Foster, *The English factories in India, 1661–1664*, 32, letters from the English factors at Madras, 28 Jan. 1661 (all dates OS).

13. Moinul Haq, *Khafi Khan's history*, 130–1; Singh, *Region and empire*, 116, (soup kitchens in Punjab 1658–60); Foster, *The English factories, 1661–1664*, 200, 321 and 329, letters from the English factors at Surat, 28 Jan., 4 Apr. and 26 Nov. 1664.

14. Bernier, *Travels*, 205, letter to Colbert; Moosvi, 'The Indian economic experience', 332.

15. Moosvi, 'Scarcities', a pioneering effort to remedy the 'neglect of the short-term fluctuations in the cycle of production and consumption', such as famine and climate, in the economic history of Mughal India.

16. Reid, 'The crisis', 219, quoting the report of a Dutch factor in the Philippines in 1699.

17. Dale, *The Muslim empires*, 218.

18. Floor, *The economy*, 61–2, quoting Jean Chardin; climatic details from Newman, *Safavid Iran*, 94–5 and 131–2, and from Matthee, *Politics of trade*, 175–7.

19. Subrahmanyam cited at page 306 above; Chardin, *Travels in Persia*, 130.

20. Musallam, *Sex and society in Islam*, 10 (*hadith*), 57–82 (pharmacy stocks and literature on abortion and contraception), and 118 (quotations from the treatises of Ibn Nujaim and Shawkani).

Chapter 14. Red Flag over Italy

1. Special thanks to Brian Pullan, who first introduced me to Masaniello in his lecture course at Cambridge University in 1965, and for commenting on this chapter, and to Mario Rizzo, who has shaped my interpretation of Spanish Italy ever since we first met in the Archivio di Stato in Naples in 1995. Early modern Italians used a twenty-four-hour clock, with each day beginning half an hour after sunset: thus, regardless of the season, sunset occurred every day at 'ore 2330'. When Italians wrote that something happened '*ad un'ora di notte* [at one o'clock at night]', they meant 'at 90 minutes after sunset'. In this chapter, I have converted the times given by Italian contemporaries into their modern equivalents.

2. Di Marzo, *Biblioteca storica*, III, 40–67 (Auria); ibid., IV, 64–6 and 70 (Rocco Pirri); AGS *SP* leg. 1444, n.p., Los Vélez to Philip IV, 23 May 1647.

3. Pocili, *Delle rivoluzioni*, 4–5; Di Marzo, *Biblioteca storica*, III, 68–71 (Auria).

4. Capaccio, *Il forastiero*, 703.

5. Details in Benigno, *L'ombra del re*, ch. 2, and Comparato, *Uffici*, 289–324.

6. Comparato, 'Toward the revolt', 291–2, quoting Francesco de Petri, *Responsa sive consilia* (Naples, 1634); Tutini, *Dell'origine e fundazione de' Seggi* (Naples, 1644).

7. Comparato, 'Toward the revolt', 280–1 and 315, quoting the 'Istoria' of Carlo Calà,); AGS *SP* libro 324/53, *consulta* of Council of Italy, 8 July 1647, noted that the tax on meat 'almost exceeds twice the sale price'.

8. Palermo, *Narrazioni*, 347, Medici to Grand Duke, 18 June and 25 June 1647.

9. Graniti, *Diario di Francesco Capecelatro*, I, 8–9 and 12–13, on the 'pessimo consiglio', 'gravissimo errore' and 'il secondo gravissimo errore che fece el duca d'Arcos'.

10. Howell, *An exact historie*, 13; Capograssi, 'La rivoluzione', 178, Andrea Rosso to Doge, 9 July 1647.

11. Graniti, *Diario di Francesco Capecelatro*, I, 15 and II, 67; BNE *Ms*. 2662/4v–5, 'Relación del tumulto'; *Cartas de algunos padres*, VII, 37–8, duchess of Arcos to her uncle, [15 July] 1647.

12. Capograssi, 'La revoluzione', 184, Rosso to Doge, and Palermo, *Narrazioni*, 387, Filomarino to Innocent X, both dated 16 July 1647.

13. Correra, 'Inedita relazione', 380; Capograssi, 'La revoluzione', 185, Rosso to Doge, 23 July 1647; Tontoli, *Il Mas'Aniello*, 154–5; BNE *Ms*. 2662/16v–17; Musi, *La rivolta*, 123–31.

14. Hugon, *Naples*, 95 and 100, quoting the Tuscan and French envoys.

15. AGS *SP* legajo 1444, n.p., *consulta* of 17 June 1647; AGS *SP* libro 218/72, *consulta* of 27 Aug. 1647, reviewing many letters about the revolt of Naples.

16. Seco Serrano, *Cartas*, I, 118, Philip IV to Sor María, 21 Aug. 1647.

17. Rovito, 'La rivoluzione', 414–17; Comparato, 'Toward the revolt', 312–15; and Musi, *La rivolta*, 138–43.

18. Conti, *Le leggi*, 52–3, edict of 25 Oct. 1647. On these complex events, see Villari, *Un sogno*, chs. 15–16.

19. Reinach, *Receuil*, X, 24, Mazarin to Duplessis-Besançon, 6 Apr. 1648, stressed his opposition to the idea of a republic in Naples. Ironically, the republic collapsed that same day.

20. Conti, *Le leggi*, 67–9 and 183–4, edicts of 4 Nov. and 17 Dec. 1647, exhorting the *regnicoli* to join the Republic. Several of Spadaro's paintings of the revolution now hang in the Certosa e Museo di San Martino and the Museo di Capodimonte, Naples.

21. Ibid., 150–2, edicts of 4–5 Dec. 1647; and 198–9, 211–13, 245, edicts of 23–24 Dec. 1647 (Guise's proclamation of himself *Duce*), 30 Dec. 1647 (a Constitution for the 'most serene and royal republic'), and 12 Jan. 1648.

22. Conti, *Le leggi*, 382, edict on the banks, 31 Mar. 1648 – four days before the Spanish troops re-entered the city.

23. Villari, *Un sogno*, ch. 18, follows the fate of the republican leaders.

24. AHN *Estado* libro 455, n.p. royal rescript to a *consulta* of 18 July 1648.

25. Ribot García, *La Monarquía*, 15.

26. Ribot García, *La revuelta*, 120, *consulta* of the Council of Italy, 9 Sep. 1669 and 124 n 272, biography of Hoyo.

27. Ibid., 166, viceroy to the queen regent of Spain, 28 Sep. 1672.

28. Ribot García, *La Monarquía*, 34, marquis of Astorga to the queen regent, 27 July and 5 Aug. 1674.

29. Ibid., 119, quoting Louis's *Mémoires*.

30. Ibid., 638.

31. Lottin, *Vie et mentalité*, ch. 4, 'Français malgré lui'.

Chapter 15. The Americas, Africa and Australia

1. Special thanks for help in drafting this chapter to Dauril Alden, Rayne Allinson, John Brooke, William Russell Coil, Ross Hassig, Karen Ordahl Kupperman, John Lamphear, Kathryn Magee Labelle, Joseph C. Miller, Margaret Newell, Carla Pestana, Jason Warren and Sam White. I also thank four of my honors students at OSU – Andrew Ashbrook, Nicole Emke, Matthew Schneider and Maria Widman – for drawing to my attention sources on New England, New France and southern Africa.

2. Kessell, *Kiva*, 170; Richter, 'War and culture', 537.

3. McNeill, *Mosquito empires*, 91.

4. Franklin, *Observations* (1751), paras 6–7.

5. *Winthrop papers*, III, 166, letter to Sir Nathaniel Rich, 22 May 1634; Anon., *New England's first fruits* (1643), 246.

6. Percy, 'A trewe relacyon' (written 1625, but describing 1609–10); Stahle, 'The lost colony', 567.

7. Kingsbury, *The records*, III, 485–90, Company to governor of Virginia, London, 25 July 1621; and IV, 73–4, George Sandys to his brother Samuel, Jamestown, 30 Mar. 1623.

8. Morton, *New English Canaan*, 23.

9. Starna, 'The Pequots', 44.

10. *Winthrop papers*, III, 149 and 167, letters to John Endicott, 3 Jan. 1634, and to Sir Nathaniel Rich, 22 May 1634; and 240, Williams to Winthrop, Providence [3 July 1637].

11. Dunn, *Journal*, 75 (6 Nov. 1634); Morton, *New-Englands memoriall*, 112; Grandjean, 'New world tempests', 77, 93–5.

12. Mason, *A brief history*, 17; Karr, "'Why should you be so furious?'", 907, quoting the Treaty of Hartford, 21 Sep. 1638.

13. Dunn, *Journal*, 181, 186, 256.

14. Pestana, *The English Atlantic*, 38, Freeholders of Barbados to the earl of Warwick, 1646.

15. Anon., *New England's first fruits*, 246.

16. *CSPC 1675–1676*, 365, Berkeley to Secretary Williamson, 1 Apr. 1676.

17. *Ibid.*, 366, Berkeley to Thomas Ludwell, his agent in London, 1 Apr. 1676.

18. Quotations from Slotkin, *The fatal environment*, 55–6, and Webb, *1676*, 411 and xvi. See also Warren, *Connecticut unscathed*, 4–5.

19. *Massachusetts Historical Society Collections*, 4th ser., IX (1871), 184–7, 'Declaration of Nathaniel Bacon in the name of the people of Virginia, July 30, 1676'; Webb, *1676*, 64–5 and 201–2 (for the 'rebellion of Ireland', see Chapter 11 above).

20. Van der Donck, *A description* , 184.

21. Bradford, *Of Plymouth plantation*, 206–7. Europeans rarely used the correct names of the Indian nations they encountered: *Huron* ('boar's bristle' in French) called themselves *Wendat*, 'islanders'; *Iroquois* (the opprobrious Huron term for them, meaning 'snakes') called themselves *Haudenosaunee*, 'builders of the longhouse'. *Mohawk* was an Algonquin insult meaning 'cannibals'; *Algonquin* meant 'allies'.

22. Thwaites, *The Jesuit relations*, XXIV, 295, Isaac Jogues, S. J., 'from the village of the Iroquois', 30 June 1643; Labelle, *Dispersed*, 63.

23. Galloway, *Choctaw genesis*, 347–8 and xiii (the 'black hole').

24. White, *The Middle Ground*.

25. Hemming, *Red gold*, 293, quoting two letters from São Luis: one by a Dutch official on 7 Apr. 1642, the other by the captain-general of Maranhão on 14 Mar. 1645.

26. Schwartz, 'Panic'; Álvarez de Toledo, 'Crisis', 272–4, quoting Palafox to Philip IV, 10 July 1641.

27. Schwartz, 'Panic', 220–1, quoting Mancera to Philip IV, 20 July 1642.

28. García Acosta, *Desastres agrícolas*, I, 178, quoting the *Historia de Nuevo León*.

29. Pérez de Ribas, *History*, 42; Scurlock, *From the Rio*, 24; Reff, 'Contact shock', 270.

30. García Acosta, *Desastres agrícolas*, I, 181, quoting López Cogolludo, *Historia de Yucatán*.

31. Schiebinger, *Plants and empire*, 1, quoting the commentary to plate 45 of Merian's *Metamorphosis insectorum surinamensium* (thus a book about caterpillars!). Schiebinger presents more contemporary reports of slave abortion and infanticide at 144–9.

32. Glave, *Trajinantes*, 199 n. 31, from a letter of 1671. Testimony concerning the revolt of Laicacota fills AGI *Escribanía de Cámara* 561–565. On the colder, drier climate, see Morales, 'Precipitation changes', 659, and Jomelli, 'Fluctuations', 269, 275–8

33. Scurlock, *From the Rio*, 26, 48 ('climate played a crucial role in the revolt'); Knaut, *The Pueblo revolt*, 14–15. Hämäläinen, *The Comanche*, 22–6, showed that, thanks to the Pueblo revolt, the Shoshone to the north acquired horses, guns and metal tools, moving, 'literally overnight, from the Stone Age to the Iron Age'.

34. Suárez, 'La "crisis" ', 317.

35. Thom, *Journal*, III, 195–7 (5–6 Apr. 1660). *Khoe* is the Nama word for 'person', and *Khoekhoen* (*Khoikhoi* in some earlier sources) is the Nama word for 'people'.

36. Tyson et al., 'The Little Ice Age', 125; Stager et al., 'Late holocene', 111, Fig 4; Grove, *Little Ice Ages*, 396–7.

37. Hair, *Barbot*, 76 and 83–4; Curtin, *Economic change, supplementary evidence*, 3, 5; Nicholson, 'Methodology'; and Martinson, *Natural climate variability*, 32–5 (by Nicholson).

38. Ritchie, 'Deux textes', 339 and 352, from Chambonneau's *Histoire de Tourbenan* (1678).
39. Hair, *Barbot*, 434; Thornton, 'Warfare', 15-16.
40. Harms, *River*, 33; Thornton, 'Warfare', 129, quoting Georg Oldendorp.
41. Fenske and Kala, 'Climate', 23-4, 31; Miller, 'The significance', 32.
42. Eltis and Richardson, *Atlas*, 192 and 194, 'Linguistic identifications of liberated Africans who embarked in Cameroon' and 'in the Sierra Leone region' in the early nineteenth century; 163 ('Gender and age of slaves carried from African regions to the Caribbean, 1545-1700'), and passim (quotations from ships' logs).
43. Grove, 'Revolutionary weather', 128.
44. Cook, 'Warm season temperatures', 84, Fig. 7A. Although on the previous page Cook and his co-authors declared that 'There is little indication for a "Little Ice Age" period of unusual cold', their graph shows a clear dip in 'warm-season temperature reconstruction' during the mid-seventeenth century. The same phenomenon appears in Pollack, 'Five centuries', 705, Fig. 4A, while his Fig. 4B shows an even sharper dip in tree-rings from New Zealand.
45. Cane, 'Australian aboriginal subsistence', 395-6; Connor, *The Australian frontier wars*, 2.
46. Cane, 'Australian aboriginal subsistence', 391 and 431 (quotations). The previous paragraphs rely heavily on Cane's research.

Chapter 16. Getting It Right

1. I thank Hayami Akira for guiding my steps through Japanese demographic, economic and social history ever since my first visit to Japan in 1983; Mary Elizabeth Berry for a trenchant critique of several drafts of this chapter; William S. Atwell, Fabian Drixler, Reinier Hesselink, Kishimoto Mio and Ronald P. Toby for valuable bibliographical advice; Matthew Keith and Taguchi Kojiro for assistance in researching, translating and interpreting Japanese materials; and the scholars who attended two seminars on this book organized by Professor Hayami at International House of Japan, Tokyo, in July 2010.
2. Data based on Hayami, *Economic history*, 36-40; and idem, *Population and family*, 10-11. Pomeranz, 'Weather, war, and welfare', 31, challenged my 'claim that Japan entered the crisis era "underpopulated"'. Now, the size of the population of early modern Japan is a highly contested field. Hayami Akira (the pioneer of Japanese historical demography) proposed a total population of 12 million (± 2 million) in 1600, whereas Saitō Osamu and others subsequently suggested much higher figures, ranging from 15 million to 17 million (Saitō Osamu, 'The frequency of famines'). Nevertheless, since almost all scholars agree that by 1700 the population of Tokugawa Japan had grown to some 30 million, even if the population doubled (as Saitō suggested) rather than almost tripled (as Hayami argued) such a development would have proved impossible without an inital population deficit.
3. Figures and quotation from Hayami, *Economic history*, 43, 163 and 218.
4. Totman, 'Tokugawa peasants', 465.
5. Ono, *Enomoto Yazaemon*, 137-8.
6. *Diaries kept by the heads of the Dutch factories*, VI, 87, entry for 15 July 1642, reporting news from the Japanese translators ('de tolcken') who worked for the Dutch factory at Nagasaki.
7. Geerts, 'The Arima rebellion', 57-61 and 96-8, Koekebacker to van Diemen, Hirado, 18 Jan. and 25 Mar. 1638 (translated directly from the Dutch text, since Geerts's English translation is sometimes unreliable); Elison, *Deus destroyed*, 220-1, quoting contemporary Christian circulars. The Messiah's original name was Masuda Shirō but, since he later took the name of the island of his birth, most sources call him Amakusa Shirō.
8. Hayami, *Population and family*, 6-8; idem, *Population, family and society*, 42-51 and 64-72. Hayami invented the now popular phrase 'Industrious Revolution' in a 1977 article.
9. Hayami, *Population and family*, 26-7.
10. Drixler, *Mabiki*, 61 (Sendai edict), 66 (etymology of *mabiki*), 81 (quotation), 105 (multiple births), and 99 (manuals). See also 101 for the laconic entries in a merchant's diary of how he had killed three of his children at birth.

11. Cooper, *They came to Japan*, 57, Cocks to the earl of Salisbury, 10 Dec. 1614, about the innovations just introduced by Tokugawa Ieyasu.

12. Tsunoda, *The sources*, 328–31, prints Hideyoshi's 'Sword collection edict' (1588) and 'Restrictions on change of status and residence' (1591). Berry, *Hideyoshi*, 102–10, expertly discusses these edicts, noting that both were 'absolutely without precedent in Japan'.

13. Yamamura, 'From coins to rice', 359 (Hideyoshi's instructions to his surveyors, 1594).

14. *CHJ*, IV, 196 (from Harold Bolitho's impressive chapter 'The Han'). Massive land redistribution also occurred in other states – after the Bohemian revolt in the 1620s (ch. 8), and in Ireland in the 1650s (ch. 12) – but it only occurred in exceptional circumstances and never became established government practice.

15. In fact *sankin* could mean either 'reporting for audience' or 'reporting for service', depending on the character used for *kin*: the Tokugawa scribes normally used the former, but meant the latter. Hall, *Feudal laws*, 293–7, prints the 1635 *Buke Sho-hatto* which codified these measures.

16. Ooms, *Tokugawa ideology*, 129 (on Suzuki's Hobbesian views) and 131, quoting Suzuki, *Banmin tokuyō [Right action for all]* (1652).

17. Ono, *Enomoto Yazaemon*, 35–6 and 137–8; Nagakura, 'Kan'ei no kikin', 75–8 (Minami-Otari village in modern Nagano prefecture).

18. Kuroita, *Shintei zōho kokushi taikei*, XL, 258, 269–71, Iemitsu's orders of 1, 2, 8, 17, 22, 24 Feb., and 25 May 1642. The Kan'ei era began in 1624 and ended in 1643.

19. Ibid., 279–81, 285 and 287–8, Iemitsu's orders of 28 June, 8 and 14 July 1642, and 20 and 21 Aug. (edict with 19 articles); *Diaries kept by the heads of the Dutch factories*, VI, 128–9, entry for 2 Sep. 1642 (reporting information received from Osaka); Nagakura, 'Kan'ei no kikin', 80–5 (legislation); Toyoda, *Aizu-Wakamatsu-shi, II*, 157–8 (the revolt of 1642).

20. Shively, 'Sumptuary regulation', 129 (no silk), 150–1 (laws for the *hatamoto*), and 152 (edict by Iemitsu); Kei'an Laws in Kodama and Ōishi, *Kinsei nōsei shiryōshû*, I, 35–40, summarized in Nakane and Ōishi, *Tokugawa Japan*, 39–42; other measures from Yamamoto, *Kan'ei jidai*, 199–203; and Sasaki, *Daimyō to hyakushō*, 233–9. In 1649, the Russian government issued far-reaching legislation immediately after a major crisis that likewise regulated (among other things) agrarian society: the *Ulozhenie* (Chapter 6 above).

21. Vlastos, *Peasant protests*, 38–9, Instruction of Hoshina Masayuki to his district magistrates; Howell, *Capitalism*, 33.

22. Berry, *Japan*, 32.

23. Viallé and Blussé, *The Deshima registers*, XII, 296 (3 Mar. 1657). Edo Castle, with a perimeter of perhaps 10 miles, covered a far larger area in the Tokugawa era than today. The five-storey donjon (*tenshudai*) was the tallest building in Japan until the Meireki fire destroyed it.

24. Hayami, *Economic history*, 169, tabulates the products listed in the 1637 *Kefukigusa*.

25. Ibid., 30–1. This paragraph rests on Hayami's insights, except for the '1945 tax rate' analogy, for which I thank Mary Elizabeth Berry.

26. Ooms, *Tokugawa ideology*, 297–8. *Deshima dagregisters*, XI (1641–50), 398, entry for 5 Aug. 1650 (arrival of news of the regicide at the Dutch factory in Nagasaki, and the decision of the city magistrates to forward it to Edo by express courier).

27. Ikegami, *Bonds of civility*, 307; Kornicki, *The book in Japan*, 63–5 (street vendors), and 324–52 (Chikamatsu's play, *Keisei Shimabara kaeru gassen*).

28. Berry, *Japan*, ch. 4, describes these rosters, known as *Mirrors*.

29. Data from Kornicki, *The book*, 20; Berry, *Japan*, 31; and Ikegami, *Bonds of civility*, 286. Compare, however, the 100 to 150 new titles printed each year in Japan, a country of perhaps 15 million, with the 2,000 new titles printed in England, a country of perhaps 5 million in 1642: page 279 above.

30. Nakane and Ōishi, *Tokugawa Japan*, 119; Ikegami, *Bonds of civility*, 173, 181–2.

31. Lane, *Images from the floating world*, 11, quoting from Asai, *Ukiyō monegatori [Tales of the floating world]* (1661).

32. Totman, 'Tokugawa peasants', 464–5 and 467.

Part IV. Confronting the Crisis

1. Hobbes, *On the citizen*, 29; Pascal, *Les pensées*, #451; Bacon, *Essayes*, 'Of seditions and troubles', 47.
2. Smith, *Nakahara*, 112, 115; Beik, 'The violence', 77–8, 92; Sibbald, *Provision for the poor*, 1–2; Angelozzi and Casanova, *La nobiltà*, 58; Carroll, 'Revenge', 107–15.
3. Des Forges, *Cultural centrality*, 176–7, quoting a memorial of Lü Kun to the Wanli emperor.

Chapter 17. 'Those Who Have No Means of Support'

1. Special thanks to John Walter for help in framing the argument of this chapter, and to him, Cynthia Brokaw, David Cressy, Stephen Dale, Kaan Durukan, Suraiya Faroqhi, Jane Hathaway and Sanjay Subrahmanyam for references.
2. Bamford, *A royalist's notebook*, 60.
3. Walter, 'Public transcripts', 128–9.
4. Scott, *Weapons of the weak*, xvi–xvii.
5. Gutiérrez Nieto, 'El campesinado', 70, quoting a manuscript treatise.
6. Bercé, *Histoire des Croquants*, II, 548 n. 44, count of Jonzac to Chancellor Séguier, 12 Dec. 1643; Bailey, 'Reading between the lines', 71, quoting Huang Liuhong's 1699 manual for magistrates, *A complete book concerning happiness*.
7. Scott, *Weapons of the weak*, 242–8.
8. Smythe, 'Towards a cultural geography', 90; Nicolas, *La rébellion*, 269–79.
9. Dekker, 'Women in revolt', 343–4; idem, *Holland in beroering*, 56–7; Walter, *Crowds*, 41, quoting William Lambarde's *Eirenarcha* (1619 edn) and a case from Star Chamber.
10. Walter, *Crowds*, 44, quoting *A briefe declaration concerning the state of the manufacture of woolls* (1629); Nicolas, *La rébellion*, 281, quoting rioters in 1694, 1699 and 1709; *ODNB* s.v. 'Ann Carter' by John Walter.
11. Dekker, 'Women in revolt', 351–2.
12. Khan, 'Muskets in the *mawas*', 93, quoting Manucci, *Storia do Mogor*. Chinese evidence is harder to interpret because officials normally named only rebel leaders – all male – and dismissed their followers as 'thugs' or 'wastrels' without specifying gender.
13. Pillorget, *Les mouvements*, 564, quoting a catechism from Avignon in 1633, reprinted at Aix-en-Provence, 1647; Hugon, *Naples*, 153–6, quoting the chronicle of Camillo Tutini, a cleric.
14. Wakeman, *The great enterprise*, I, 627, describing events in Wujiang, near Lake Tai.
15. Rushworth, *Historical collections*, II, 470–1; Rothes, *Relation*, 115, 208–9. On 'Archie the Fool', who outlived most other protagonists in the British civil wars, see Shannon, '"Uncouth language"'.
16. Simon i Tarrés, *Cròniques*, 269, account of Judge Ramon de Rubí, who only escaped death by disguising himself as a tonsured Jesuit; Beik, 'The violence', 77 and 87.
17. Riches, *The anthropology of violence*, 25.
18. Schaub, *Le Portugal*, 31–5; Pérez Samper, *Catalunya*, 243; Brennan, *Bargrave*, 82; Hugon, *Naples*, 303–8.
19. Walter, '"Abolishing superstition"', 90–2.
20. Wilson, '"A thousand countries to go to"', 84.
21. Nicolas, *La rébellion*, 412–13; Shy, *A people numerous and armed*, vii; Khan, 'Muskets in the *mawas*', 93, quoting Mundy, *Travels*, and Manucci, *Storia do Mogor*.
22. Bercé. *Histoire des Croquants*, I, 421–2.
23. Geerts, 'The Arima rebellion', 96–8, Koekebacker to Anthonio van Diemen, 25 Mar. 1638 (my translation from the Dutch original); Blair and Robertson, *Philippine islands*, XXIX, 220, report of Juan López, S.J.
24. Graniti, *Diario di Francesco Capecelatro*, II, 67; Anon., *The red-ribbon'd news*, 5; Kötting, *Die Ormée*, 111.

25. Hugon, *Naples*, 309–13, 328–57, on Masaniello; Stephens and George, *Catalogue*, IV, 240, #4014, 'Mas-aniello or the Neapolitan insurrection', Mar. 1763 (my thanks to Tom Ingersoll for this reference).

26. Heilingsetzer, *Der oberösterreichische Bauernkrieg*, 35–7; Hrushevsky, *History*, VIII, 450–1; Neumann, *Das Wort*, 214–18; and Lucas Val, 'Literatura i historia'.

27. Lorandi, *Spanish king of the Incas*, 23, quoting Bishop Juan de Vera of Cuzco in 1635.

28. Wood, 'Fear', 814; Hill, *The world*, 108, quoting *The mournfull cries* (and several other similar statements from 1648–9); Bercé, 'Troubles frumentaires', 772.

29. Cueto, *Quimeras y sueños*, 80–1, Philip IV's instructions to his *junta de conciencia*.

30. Spence, *The death of Woman Wang*, 13; Darling, *Revenue-raising and legitimacy*, 248–67; Faroqui, 'Political activity', 31–2; and Barkey, 'Rebellious alliances', 706.

31. Lorenzo Cadarso, *Los conflictos*, 178–9; Bercé, *Histoire des Croquants*, II, 597–9.

32. Blair and Robertson, *Philippines*, XXIX, 221–5, account of Juan López, S.J., on the brutal repression of the 1639 Sangley revolt.

33. RAS *Diplomatica: Muscovitica* 602, n.p., Adolph Ebbers to Charles XI of Sweden, Moscow, 21 Aug. 1662; and Gordon, *Diary*, II, 159–62.

34. Goldie, *The entring book of Roger Morrice*, III, 27–8; Clifton, *The last popular rebellion*, 231–41.

35. Hugon, *Naples*, 238–9 (quoting Don John on 21 Feb. 1648, and Philip IV and Oñate after the surrender on 6 April), 243–56 and 263–6.

36. AHN *Estado* libro 961/56–59v, Olivares, 'Relación' prepared for the duke of Bragança, Nov. 1637; AMAE (P) *CPE Supplément* 3/189–91, Duplessis-Besançon, 'Première négotiation des François en Cathalogne'. Chéruel, *Lettres*, III, 1,061, Mazarin to Ambassador Chanut in Stockholm, early Sep. 1648.

37. Bercé, 'Troubles frumentaires', 789, Giuseppe Caetano, governor of Perugia, to Cardinal Panzirolo, 14 July 1648.

Chapter 18. 'People Who Hope Only For a Change'

1. The title reprises the typology of rebellions by Lu Kun of China: pages 375–6 above.

2. Mandrou, 'Vingt ans après', 36, Fouquet to Chancellor Séguier, 1644, italics added.

3. Briggs, 'Richelieu and reform', 72, Richelieu paper of 1624; Elliott and La Peña, *Memoriales y cartas*, I, 55, 62, the Gran Memorial of 1624.

4. Jouanna, *Le devoir*, 268–70.

5. Bercé, *Histoire des Croquants*, II, 737, anonymous relation of the Croquants of Angoumois and Saintonge, 1636.

6. Details from Hucker, *Two studies*, 41–83; and Wakeman, *The great enterprise*, 109–10.

7. Brook, *Confusions of pleasure*, 1–4, quoting the *Shexian zhi [She County Gazetteer]* of 1609 compiled by Zhang Tao.

8. AUL *Ms.* 2538/21–2 and 37v, Fraser, 'Triennial travels'.

9. AUL *Ms.* 2538/21–2, Fraser, 'Triennial travels'.

10. Curtis, 'Alienated intellectuals', 299, quoting Lord Chancellor Ellesmere; Quevedo, *La fortuna con seso* (1632), I, 264; Roberts, 'Queen Christina', 217, quoting Magnus Gabriel de la Gardie in 1655; Zeman, 'Responses to Calvin', 45, quoting Ferdinand II.

11. Hobbes, *Behemoth*, 40, 70–1, 144, 147–8.

12. Balfour, *Historical works*, III, 426–7.

13. Groenhuis, *Predikanten*, 31–2, from a sermon in Jan. 1626.

14. Marques, *A parenética portuguesa e a Restauração*, I, 69, quoting Valenzuelo, *Portugal unido*.

15. Neumann, *Das Wort als Waffe*, 206 (quoting *Triomphos del Amor*, 1642); Capograssi, 'La revoluzione', 211, Rosso to Doge, Naples, 17 Sep. 1647.

16. Perceval-Maxwell, *The outbreak*, 231, quoting a Jesuit chronicle of the rebellion, Dec. 1641; TCD *Ms.* 817/37v, deposition of Rev. Thomas Fleetwood, Westmeath, 22 Mar. 1643; Ó hAnnracháin, '"Though hereticks and politicians"', 159–63, discussing O'Mahony, *Disputatio apologetica*.

17. Clarendon, *Brief view*, 319–20.

18. Firth, *Clarke papers*, II, 150–4 and 163–70 (the minutes of the council meetings at which Poole appeared); Poole, *A vision* (her own account); Brod, 'Politics and prophecy'; Davies, *Unbridled spirits*, 137–41; and *ODNB* s.v. Poole and Cary.

19. Goldstone, *Revolution*, 137–8; Adamson, 'England', 463.

20. Hill, *The world*, 366. Not all insurgent leaders were young: Pierre Broussel was seventy-three when his arrest provoked the Day of the Barricades in Paris; Giulio Genoino was over eighty when he led the Neapolitan Republic in 1647.

21. Wakeman, *The great enterprise*, I, 625–6, quoting a Jiangxi county gazetteer; Des Forges, 'Toward another Tang or Zhou?', 75.

22. Elliott, 'Whose empire?', 39, letters of Nurhaci to the Chinese inhabitants of Liaodong, late 1621; Wakeman, *The great enterprise*, 316–17, Dorgon's edict of 5 June 1644.

23. Hickson, *Ireland*, I, 194, deposition of Rev. John Kerdiff, Co. Tyrone, 28 Feb. 1642 citing Meredith Hanmer's *Chronicle of Ireland* (1571); TCD *Ms.* 839/134v, deposition of Mulrany Carroll, Co. Donegal, 26 Apr. 1643; Rutherford, *Lex, rex*, 449–53; Rushworth, *Historical collections the fourth and last part*, II, 1, 420–1, President John Bradshaw to Charles I, 27 Jan. 1649.

24. Schama, *The embarrassment of riches*, 113 (Vondel's *Passcha* of 1612); Groenhuis, *De predikanten*, 81 (Lydius, *Belgium Gloriosum*, 1667); Paul, *Diary*, I, 344 (Feb. 1638).

25. Trevor-Roper, 'The Fast Sermons', 280–1, quoting Samuel Fairclough, *Troublers of Israel* (with other texts).

26. Benigno, *Specchi*, 133, on France's 'Egyptian bondage' under Mazarin; Lenihan, *Confederate Catholics*, 73, quoting Fr Anthony Geoghegan, Sep. 1642; Casway, 'Gaelic Maccabeanism', 178, speech by Owen Roe O'Neill; O'Mahony, *Disputatio apologetica*.

27. Comparato, 'Barcelona y Nápoles'; Hobbes, *Leviathan*, 225–6, written just after the English regicide.

28. Parsons, *Peasant rebellions*, 189–99.

29. Mitchell, 'Religion, revolt', ch. 5, quoting the *Resumen de la vida de Sor Eufràsia Berenguer*.

30. Paul, *Diary*, I, 393, 395, 396 and 397 (Oct. and Nov. 1638).

31. Hill, *The world*, 90; Foster, *The sounding of the last trumpet*, 17–18.

32. TCD *Ms.* 835/158, deposition of John Right, Co. Fermanagh, 5 Jan. 1642. Hickson, *Ireland*, I, 114–15, prints O'Neill's alleged royal commission, dated Edinburgh, 1 Oct. 1641, and at 169–73 and 188–9, reveals how well the forgery worked.

33. Paul, *Diary*, I, 348 and 410–11 (19 May 1638 and 8–10 and 24 Feb. 1639); Laing, *Letters and journals*, I, 116–17, Baillie to William Spang, 12 Feb. 1639; Dunthorne, 'Resisting monarchy', 136–40, quoting Henderson's use of Grotius; Paul, *Diary*, I, 390 (Wariston's account of his discussion with Henderson and David Calderwood on foreign texts about resistance, 20 Sep. 1638).

34. Prynne, *The soveraigne power*, part IV, 153–99, is mostly a translation of the *Vindiciae*.

35. Hugon, *Naples*, 217–23; TCD *Ms.* 829/311, deposition of William Fytton, Limerick, 8 July 1643.

Chapter 19. 'People of Heterodox Beliefs . . . Who Will Join Up with Anyone Who Calls Them'

1. This chapter owes much to lively discussions with Cynthia Brokaw and David Cressy. As with chs. 17 and 18, the title incorporates the typology of rebellions suggested by Lü Kun of China: pages 375–6 above.

2. Gladwell, *The tipping point*, 30–4 and 57–60.

3. Quevedo, *La rebelión de Barcelona*, in *Obras*, I, 284; Birago Avogadro, *Turbolenze*, 369–70; *CSPC 1675–1676*, 368, Sir Jonathan Atkins to Secretary Williamson, 3 Apr. 1676; and Trevor-Roper, 'General Crisis', 61.

4. Ferdinand in 1630 quoted in Bireley, *Religion and politics*, 125, Charles I in 1638 quoted in Russell, *The fall*, 56–7 and Olivares in 1639 on page 208 above.

5. Foisil, *Révolte*, 231, quoting the memoirs of Bigot de Monville; *CODOIN*, LXXXIII, 313, count of Peñaranda, chief negotiator at Münster, to the marquis of Caracena, governor of

Milan, 27 June 1647; AMAE (P), *CPE Supplément* 3/240v–241, Duplessis-Besançon, 'Première négotiation des François en Cathalogne'.

6. [Howell], *A discourse* (1645), 15.

7. Di Marzo, *Bibliote storica*, III, 206–11 (citing books by Assarino, Birago Avogadro and Collurafi); *CSPI 1633–1647*, 182, Bishop Bramhall of Derry to Laud, 23 Feb. 1638 OS; Braddick, *God's fury*, 30, quoting John Castle to the earl of Bridgewater, 24 Oct. 1639.

8. Castlehaven, *Memoirs*, 13 (with corroborating statements at 14–16); TCD *Ms.* 834/18, deposition of Gerrard Colley, Co. Louth, 2 May 1642; TCD *Ms.* 828/194v, deposition of Thomas Dight, Co. Kerry, 24 May 1642, quoting an Irish priest; and TCD *Ms.* 833/228v, deposition of Rev. George Creighton, 15 Apr. 1643 (quoting Richard Plunkett).

9. Te Brake, *Shaping history*, 109–10, quoting Ambassador Nani to the doge and senate of Venice, Sep. 1647; page 234 above (the French echo); Bercé, 'Troubles frumentaires', 770 and 772 (Cardinal Montalto, 7, 8, and 17 July 1648); 775 (map of rebellious areas of the Papal States in 1648) and 779 ('i Masanielli').

10. Crewe, 'Brave New Spain', 77–8, quoting the paper that began 'Por quanto Dios Nuestro Señor compasivo de nuestrros duelos inhumanos', confiscated at Lompart's arrest; *Deshima dagregisters*, XI (1641–50), 263, account by the Dutch factor of an audience in Edo, 6 Jan. 1647; Winius, *Fatal history*, 141, count of Óbidos, Goa, 1653.

11. Van Aitzema, *Saken*, I, 146 (alliance with Tunis and Algiers because they all 'een machtigh vyandt hadden aen Spangien'), 905 (jealousy), and 1,103 (universal monarchy); Van de Haar, *De diplomatieke betrekkingen*, chs. 2–3; de Jong, 'Holland'.

12. [Voetius], *Brittish lightning*, sig. B.

13. Goodwin, *Anti-Cavalierisme*, 5 and 50, published in October 1642 (see also *ODNB*, s.v. John Goodwin); Young, 'The Scottish Parliament', 92, quoting instructions to Thomas Cunningham, Mar. 1645; Markham, *Anarchia anglicana*, part II, 49–50 (Hugh Peter's sermon).

14. Haan, 'The treatment', 30–1; Carrier, *Le labyrinthe*, 80, quoting Charles de Saumaise to Jacques Dupuy, 8 Sep. 1648; and 83, quoting Anon., *Epilogue, ou dernier appel du bon citoyen sur les misères publiques* (1649).

15. Carrier, *La Fronde*, I, no. 16, Davant, *Avis à la reine d'Angleterre et à la France* (1650), 3–6; Carrier, *Le labyrinthe*, 111–12, citing *Le Ti θείου de la maladie de l'état* (Paris, 1649); Corneille, *Pertharite, roy des Lombards*, first performed in 1651 ; Knachel, *England*, 66–70.

16. Van Aitzema, *Saken*, III (1645–57), 323 (news arrived in The Hague on 14 Feb. 1649); Radziwiłł, *Memoriale*, IV, 116–18 (written in Kraków on 18 Feb. 1649); Vernadsky, *Source book*, I, 246, decree of 1 June 1649 OS; Bergh, *Svenska riksrådets protokoll*, XIII, 17, minutes of de la Gardie's speech at the Council of State on 21 Feb. 1649 OS; Christina quoted in Roberts, 'Queen Christina', 196–7.

17. BL *Addl. Ms.* 4,200/14–70, letters from René Augier, resident for the Parliaments of England and Scotland in Paris, to Giles Greene in London, 1646–8; Milton, *Complete prose works*, VIII, 555–6, *Pro populo anglicano defensio secundo*, May 1654; Benigno, *Specchi*, 98.

18. Carrier, *La Fronde*, I no. 22, Anon., *Les cautelles de la paix* (May 1652), 17–18 ('l'empire de l'univers').

19. Van Groesen, M., '(No) news', 752; Keblusek, M., 'The business of news', 211–13.

20. Firth, *The Clarke papers*, IV, 231, Captain Newman of the Leith garrison to Monck, 31 Dec. 1659 OS.

21. Eisenstadt and Schluter, 'Early modernities', 25. Jürgen Habermas, who coined the term 'public sphere' in 1962, has insisted that it 'emerged only in competition with the literary public sphere of the late eighteenth century', and denied that the term can be properly applied to an earlier period. Nevertheless, from the 1640s onwards Western Europe witnessed both of the intersecting processes that Habermas considered essential ingredients of the public sphere: 'the communicative generation of legitimate power'; and, 'the manipulative deployment of media power to procure mass loyalty, consumer demand and "compliance" with systemic imperatives' (Calhoun, *Habermas*, 452, 464–5, from Habermas's response to his critics and his 'Concluding remarks'). See also the discussion in Dooley, 'News and doubts'; Condren, 'Public, private'; and Randall, 'Epistolary rhetoric'.

22. Whitelocke, *Memorials*, 176, speech on mobilizing an army, July 1642; Locke, *Political essays*, 5, from his 'First tract on government', written 1660 but never published; Raymond, *The invention*, 186, quoting Dudley, Lord North, in 1671.

23. Neumann, *Das Wort als Waffe*, 1, quoting Alexandre de Ros, *Cataluña desengañada* (Naples, 1646); Conti, *Le leggi*, 92–3, edict of 15 Nov. 1647.

24. Kagan, *Students*, 45; Brockliss, 'Richelieu', 245–6; Naudé, *Considerations*, 127–8; Newcastle, *Advice*, 20; Bremner, *Children and youth*, 90, quoting 'The report of Sir William Berkeley, governor of Virginia, on the state of free schools, learning and the ministry of the colony, 1671'.

25. Carrier, *La presse*, 56 and 58 (quoting the complaints) and 71 (calculation of total printed *Mazarinades*; another 800 exist only in manuscript).

26. Neumann, *Das Wort als Waffe*, 193.

27. Newcastle, *Advice*, 56 (see the 'young statesmen' reference on page 282 above); AUB *Ms.* 2538/44, 'Triennial travels' of James Fraser.

28. Brook, *The troubled empire*, 199. Wong, *China transformed*, 112–13 and 125–6, argued forcefully that 'Europe's public sphere' (which he defined as 'an arena in which politically engaged populations could express their claims against states') did not and could not exist in Qing China, mainly because of the lack of arenas 'in which reason could be heard and rationality could advance'. He ignored the freedom of debate that took place in many arenas during the Ming–Qing transition.

29. Gallagher, *China*, 21, quoting Mateo Ricci.

30. Will, 'Coming of Age', 31, quoting Yao's 'Record of successive years'.

31. Brook, *Confusions*, 171–2, quoting Grand Secretary Yu Shenxing (1545–1608).

32. Le Comte, *Nouveaux mémoires*, 498. Le Comte himself travelled thousands of miles around the Qing empire.

33. Wu, 'Corpses' 44, quoting Lu Yunlong, *Wei Zhongxian xiaosho chijianshu (Account to condemn the villainous Wei Zhongxian)*.

34. Dardess, *Blood and history*, 5; Fong, 'Writing from experience', 257–8.

35. Ho, *The ladder*, 199, paraphrasing the survey of Ming Confucian thought in Huang Zongxi, *Mingru xuean* (1676).

36. Labat, *Nouvelle relation*, II, 151 (adding that 'They use Arabic characters to write their own language'); Ritchie, 'Deux textes', 323–4, from Chambonneau's *Traité de l'origine des nègres* (1678).

37. Subrahmanyam, 'Hearing voices', 94–5; Ludden, *Peasant history*, 8.

38. Terzioğlu, 'Where 'İlm-i Ḥāl meets catechism', 89–100.

39. Evliya Çelebi, *Seyahatname [Book of travels]*; Terzioğlu, *Sufi and dissident*, 328–9, on the 'Jewish' slur against Ibrahim; 346–53, on the Crimean option (Mīşri met Crimean princes while exiled on the same island); and 464–90, on the Mīşri Order and other legacies.

40. Scholem, *Sabbatai*, 604 (quoting an Armenian living in Istanbul at the time), and 549 (quoting Mather).

41. Como, 'Secret printing', 78.

42. AMAE (P), *CPA* 54/101–7, M. de Bellièvre to Secretary of State Brienne, 31 Dec. 1646.

Part V. Beyond the Crisis

1. Special thanks to Derek Croxton, Kate Epstein, Jack Goldstone, Daniel Headrick, Paul Monod, Sheilagh Ogilvie and Kenneth Pomeranz for help in framing the final chapters of this book.

2. Lappalainen, 'Death and disease', 425; Wickman, '"Winters Embittered with Hardships"', 76–8, quoting Mather, Tulley, Pike and other beleaguered colonists. For an example of low temperatures from the natural archive, see Rydval, 'Reconstructing': according to tree-ring series in Scotland going back 800 years, the 1690s was 'the coldest decade in the record'.

3. Jaeglé, *Correspondance*, II, 11, duchess of Orléans to Amélie-Élizabeth of the Palatinate, Versailles, 17 Jan. 1709.

4. Teodoreanu, 'Preliminary observations', 189, quoting a Turkish chronicler; Le Roy Ladurie, *Les fluctuations*, 105–12, 114–15, 300–1; Jaeglé, *Correspondance*, II, 11, duchess of Orléans to Amélie-Élizabeth of the Palatinate, Versailles, 17 Jan. 1709.
5. Mauch and Pfister, *Natural disasters*, 6–7.

Chapter 20. Escaping the Crisis

1. Gordon, *Diary*, I, 259–60.
2. MacDonald, *Mystical bedlam*, 36, 38, 40–1, 55, 73.
3. Trevor-Roper, *Europe's physician*, 8 (Cromwell) and 363–4 (Princess Elizabeth).
4. Burton, *Anatomy*, 5 and 76 ('Democritus to the Reader') and second pagination 11 (Partition I, section I, subsection V: 'Melancholy in disposition'); Aubrey, *Brief lives*, s.v. 'Burton', reports his suicide.
5. *ODNB*, s.v. 'Felton' by Alastair Bellany, quoting from trial papers; *CSPD 1628–1629*, 343, examination of Elizabeth Josselyn, 3 Oct. 1628.
6. Goldish, *Jewish questions*, 131–3 (a responsum first published in Venice in 1697). Avicenna (Ibn Sina) discussed melancholy in part III of his treatise, dedicated to diseases of the brain.
7. Pepys, *Diary*, VI (1665), 342 (verdict on the plague year); Haude, 'Religion', 545–6, quoting Maximilian's directive (*Mandat*) of 20 Sep. 1636; the pastor of Hersbruck near Nuremberg; and Pastor Davis Wagentrotz of Brandenburg. Asai quoted page 371 above.
8. Struve, 'Dreaming', 159–60, part of a study of Xue Cai (1598–1665, *jinshi* 1631), one of at least 160 prominent literati known to have entered a monastery; Will, 'Coming of age', 33.
9. Di Cosmo, *The diary*, 46, 83, 87 (from 1682).
10. Seaver, *Wallington's world*, 11; Pepys, *Diary* (see, for example, his annual entries on 26 March, the anniversary of a dangerous but successful operation to remove a kidney stone); Westfall, 'Short-writing'.
11. Gallardo, *Ensayo*, II, col. 174, quoting Caldera's *Arancel político. Defensa del honor y práctica de la vida de nuestro siglo*.
12. Brokaw, *The ledgers*, 4; Fong, 'Reclaiming', 28, quoting Yuan Huang *Liming wen [Essay on determining fate]* (1601).
13. Courtwright, *Forces of habit*, 2, 59.
14. Lockhart, *Denmark*, 55, quoting the earl of Leicester in 1632; Larsen, *Eske Brock*, 36; Gallardo, *Ensayo*, II, col. 175. Muldrew, *Food*, 79–82, deduces that seventeenth-century beer was much more intoxicating than beer today. See also Withington, *Cultures*, 21–2.
15. Thackston, *Jahangirnama*, 320; Balabanlilar, *Imperial identity*, 91.
16. Rycaut, *The present state*, 114; Matthee, *The pursuit of pleasure*, 107 (quoting Thomas Herbert); Babayan, *Mystics*, 444–5 (quoting Jean Chardin and Rafael du Mans) and 446–7 (on Qummi). See also Matthee, 'Alcohol'.
17. Pepys, *Diary*, I, 253, entry for 25 Sep. 1660; Massieu, *Caffaeum* (c. 1700).
18. Haskell, *Loyola's bees*, 94, quoting Strozzi, *De mentis potu, sive de cocolatis opificio libri tres* (Naples, 1689).
19. Pepys, *Diary*, VI, 120 (7 June 1665), and VIII, 389–90 (18 Aug. 1667); Dikötter, '"Patient Zero"', 7, quoting Yao Lu; Balde, *Satira* (1657).
20. Brook, *Vermeer's hat*, 143–6.
21. Grehan, 'Smoking', 1,364–5 (quoting Kâtib Çelebi in 1653).
22. Crucé, *Nouveau Cynée*, 13; BL *Addl. Ms.* 21,935/78v–79, 88–92 (Wallington); Anon., *The victorious proceedings*, 2; Parker, *The manifold miseries*, 1; Kuczynski, *Geschichte*, 117, quoting a family Bible from Swabia in 1647.
23. Van Maarseveen, *Beelden*, and Bussmann and Schilling, *1648*, vol. II: *Art and culture* (on Callot and Franck); Raynor, *A social history of music*, 115 and 203–4, quoting Burckhart Grossman and Heinrich Schütz.
24. Rabb, *The struggle*, 119, quoting Gerhardt; Milton, *Paradise lost*, Book II, lines 160–4, 335–40; Grimmelshausen, *Der abentheurliche Simplicissimus Teutsch*, Book I, ch. iv.
25. Rabb, *The struggle*, 119, citing English and German texts.

26. Hobbes quoted page 287 above; Locke, *Political essays*, 7, from his 'First tract on government', written 1660 but never published.

27. AMAE (P) *CPE* 21/242–3v, Bishop Marca to Mazarin, Barcelona, 17 June 1644; Anon., *Avertissements aux rois*, 6; Schumacher's *Kongelov* discussed in Chapter 8 above.

28. Crucé, *Nouveau Cynée*; Hugo Grotius, *De jure belli ac pacis, libri tres* (Paris, 1625); William Penn, *An essay towards the present and future peace of Europe, by the establishment of an European Dyet, Parliament or Estates* (London, 1693).

29. Von Friedeberg, 'The making', 916, deposition before the Imperial Chamber Council in 1652; Zillhardt, *Der dreissigjährige Krieg*, 267.

30. Roberts, *Sweden*, 173–4, Gustav Bonde's memorial to the Council of the Realm, 26 June 1661; Morrice, *Entring book*, IV, 335.

31. Rohrschneider, *Der gescheiterte Frieden*, 81, 'La experiencia ha mostrado quan poco se puede fiar de las palabras y fee pública de Franceses en los tratados' (June 1643) and 'L'expérience nous fait cognoistre que les Espagnolz ne gardent leur traités' (Sep. 1643) – a stunning duplication of views, each composed in total secrecy

32. *Instrumentum Pacis Osnabrugensis*, V, 52 ('*sola amicabilis compositio lites dirimat non attenta votorum pluralitate*'); Heckel, 'Itio in partes', quoting a constitutional tract of 1722. On the tears, see Dickmann, *Westfälische Frieden*, 460.

33. Tuck, *Philosophy*, 319, quoting Hobbes to the duke of Devonshire, July 1641; Locke, *Political essays*, 40–1 (1660). Sermons preached on the text 'Curse ye Meroz' (Judges 5: 23) are discussed by Downs, 'The curse'.

34. Locke, *A letter*, 33; Schilling, 'Confessional Europe', 669; Benedict, 'Religion and politics', 133.

35. *Acta Pacis Westphalicae: supplementa electronica, IPO*, V:50; Wallis, *A defence of the Royal Society*, 7.

36. Kenyon, *Stuart Constitution*, 365–71, 'An Act of free and general pardon'; Stoyle, '"Memories of the maimed"'; Young, *Faith*, 54.

37. Plumb, *The growth*, xvi–xviii and 1. Scotland, Ireland and Anglo-America were, of course, a different story.

Chapter 21. Warfare State or Welfare State?

1. Grimmelshausen, *Der abentheurliche Simplicissimus Teutsch*, frontispiece. I thank Kenneth Pomeranz for many helpful comments on an earlier draft of this chapter.

2. Ray, *Observations*, 81–2 (Mannheim and Heidelberg) and 140 (Vienna); Skippon, *An account*, 432 (Mannheim), 439–40 (Heidelberg), and 476 (Vienna).

3. Meyer, 'Ein italienisches Urteil', 160–1; Ray, *Observations*, 109; Patin, *Relations* (1671), 144, 199–200, 212.

4. AUL *Ms* 2538, 'Triennial travels', III/7v (Munich) and 8 (Regensburg); Gordon, *Diary*, II, 8 and 36 (1659).

5. Li, *Fighting famine*, 9; Fong, 'Writing', 268–73.

6. Li Wen, 'On the road', graciously translated for me by Lynn Struve.

7. Nieuhof, *An embassy*, 3 (preface), 85 (Yangzhou), and 64 (Nanchang). The English edition claims that only 8,000 were slain at Canton, but Nieuhof's Dutch manuscript clearly states 80,000: Blussé and Falkenburg, *Johan Nieuhofs beelden*, 35 (also verified from BNF *Cartes et plans*, Ms. In-8o 17, fo. 26). Likewise the English edition claims that Hukon enjoyed a lively trade when the Dutch visited in 1656, but Nieuhof's original stated that this was 'voor de distructie van Chijna': Blussé and Falkenburg, *Johan Nieuhofs beelden*, 41.

8. Nieuhof, *An embassy*, 65 (Qing policy), 39 (Canton) and 84 (along the Grand Canal).

9. Brantôme, *Oeuvres*, VI, 326; La Noue, *Discours*, 160.

10. Sreenivasan, *The peasants*, 289–92 and 322–3 (all quotations).

11. *Venezia e la peste*, 98; Hagen, 'Seventeenth-century crisis', 325 (quoting the edict of 1661); Archivio di Stato, Lucca, *Anziani al tempo della libertà*, buste 707–708.

12. Ray, *Observations*, 221, and Skippon, *An account*, 550–1 (both in 1663); Sreenivasan, *The peasants*, 348, quoting statements by the magistrates of Memmingen in 1600 and 1702.

13. Rawski, 'The Qing formation', 217–18, notes the pawnshops; demographic data from Pomeranz, 'Is there an East Asian development path?', 325–6; acreage from Ho, *Studies*, 102 (but note that 1,161 million mou at 0.1647 acres per mou = 191.3 million acres and not 176 million acres as Ho stated).

14. Hartlib, *Samuel Hartlib his legacie*.

15. Ho, *Studies*, 146, quoting a gazetteer of 1760; Mazumdar, 'The impact', 69; Wong, *China transformed*, 28 (calculations from lineages in Tongcheng county).

16. Goldstone, *Revolution*, 372.

17. Quotations from de Vries, *The Industrious Revolution*, 10 and 128; Muldrew, *Food*, 17, 183, 200 and 218.

18. Marmé, 'Survival' (on Suzhou); Nieuhof, *An embassy*, 69 (shipping) and 75 (Nanjing).

19. Le Comte, *Nouveaux mémoires*, I, 118–20.

20. AUL *Ms*. 2538, 'Triennial travels', I/29.

21. Defoe, *The compleat English tradesman*, II, part 1, 99–102 and 107.

22. Yang, 'Economic Justification for Spending', 51, quoting an essay by Lu Chi of Shanghai, c. 1540, quoting Mencius; Kishimoto, 'Kangxi depression', 241–2, quoting an essay by Wei Shixiao, c. 1680.

23. Defoe, *A tour*, I, 'The author's preface' (published in 1724, after forty years of travels).

24. Scott, *Seeing like a state*, 11.

25. 'Moyen de retablir nos colonies de l'Amérique et de les accroître en peu de temps' (1699), in Vauban, *Les oisivitiés*, 539–73, quotation from 571. The actual population of Francophone Canada in 2000 was scarcely 7 million (and that of all Canada not even 31 million).

26. Virol, *Vauban*, 204 ('Du nombre d'hommes'), and 213 (dearth is 'dans l'opinion et non dans la réalité': from a mémoire of 1694).

27. Soll, 'Accounting', 237; Plumb, *The growth*, 11–13.

28. Scott, *Seeing like a state*, 3.

29. Newman, 'Shutt up', 829.

30. *SCC*, VI 6, 134–40 (Kangxi's personal testimony).

31. Li, *Fighting famine*, 167.

32. Le Comte, *Nouveaux mémoires*, 125 (writing of his personal inspection in the 1680s).

33. Mikhail, *Nature*, 216–17. Ibrahim, *Al-Azmat*, ch. 4, argued that the famine of 1694–5 was the worst of the seventeenth century.

34. Skippon, *An account*, 600, and Hugon, *Naples*, 75, 139, 141, on Naples.

35. Thirsk, *The agrarian history*, IV, 619; http://rps.ac.uk/search.php?a=fcf&fn=charlesi_ ms&id=25112&t=ms "Articles to be by oure chancellour to be proponed' to the Scottish Parliament, 1 Nov. 1625.

36. Anon., *An ease for overseers of the poore*, 22.

37. Hindle, *On the parish?*, 256; Solar, 'Poor relief', 4–6; Blaug, 'Poor Law Report', 229.

38. Allemeyer, '"Dass es wohl recht ein Feuer"', 218–20, noting orders from Braunschweig, 1647; Emden, 1666; Kirchward, 1673; Clausthal, 1687; Nürnberg, 1698.

39. Barbon, *A letter*, 1–2.

40. AUL *Ms*. 2538, 'Triennial travels', I/29.

41. Schumpeter, *Capitalism*, 85.

42. Marcos Martín, *España*, 462–3, 479–82; McArdle, *Altopascio*, 52–4 and 91.

43. Quotations from Sreenivasan, *The peasants*, 289–92, 322–3 and 326.

44. Brokaw, *Commerce*, 226 and 405.

45. Ibid., 179 (and all of her ch. 5 on 'Household division and competition').

46. Allen, 'Wages'; Deng and O'Brien, 'Establishing statistical foundations'; Bolt and van Zanten, 'The Maddison project'.

Chapter 22. The Great Divergence

1. Adshead, 'The XVIIth century General Crisis' 265, 251; Pomeranz, *The Great Divergence*, passim; Studer, *The Great Divergence*, 182; Deng and O'Brien, 'Establishing statistical

foundations', 630-1, 636, 638; Bolt and van Zanten, 'The Maddison project', 1075. I thank Kenneth Pomeranz and Ying Zhang for sharing with me their erudition and insights.

2. Batencour, *Instruction méthodique*, 32-46, 'De la justice du maistre'.

3. Jolibert, *L'enfance*, 18, *plaidoyer* before the Parlement of Paris, 25 Jan 1680; *Mélanges*, 7, Letter Patent of Louis XV, Sep. 1724, creating a charity school in Rouen.

4. Le Cam, 'Extirper la barbarie', 412-13, and idem, 'Die undeutlichen Grenzen', 50-1, quoting Duke Augustus of Brunswick-Wolfenbüttel's *Schulordnung* (1651) and *Allgemeine Landes-Ordnung* (1647), whose first two articles enjoined universal attendance at church and school – seen as the two pillars on which political stability rested.

5. Rawski, *Education*, 33-4 (on the 1652 schools edict) and 26 (citing an 'encyclopaedia' probably published in 1675-76); Herman, 'Empire' (on the 1658 edict on education for chieftains, the first of many).

6. Israel, *Radical enlightenment*, 128-9. Frijhoff, 'Surplus', 205-8, notes the recovery in the 1650s.

7. Hautz, *Geschichte*, II, 186-8, report of the rector and debate by the Senate, 25 Feb. and 5 Mar. 1680.

8. Garrisson, 'Les préludes', 13, bishop of Montauban to Mazarin, 8 July 1659, a few days after students from the Protestant Academy had unwisely invaded the neighbouring Jesuit College and beaten up its students; Dell, *Several sermons*, 612-13 and 644-7, two sermons delivered in 1652-53. See also Winstanley, *The law of freedom*, 68-9.

9. Verbeek, *Descartes and the Dutch*, 34-70. These measures remained in force until 1689.

10. Ho, *The ladder*, 191-2. Mote, *Imperial China*, 863-4, gives much higher figures for those involved.

11. Israel, *Radical enlightenment*, 3-4; Aubrey, *The natural history of Wiltshire*, 15 (the first paragraph of the Preface, written in the 1680s).

12. Hunter, *John Aubrey*, 41-2, quoting from notes made by Aubrey for his projected biography of Bacon; *ODNB*, s.v. 'Bacon'; Bacon's preface to the *Novum organum;* Bacon, *Works*, XIV, 120, Bacon to James I, 12 Oct. 1620.

13. Scriba, 'The autobiography', 26-9, and 39-40; Birch, *The works*, I, xxx-xxxv, Boyle to Isaac Marcombes, 22 Oct. 1646, and to Francis Tallent (his Cambridge tutor), 20 Feb. 1647.

14. Bacon, *Works*, XIV, 436, Bacon to Charles, prince of Wales, Oct. 1623, together with a Latin copy of *The advancement of learning*.

15. Descartes, *Discours*, 51 and 22. Descartes praised Bacon ('Verulamius': the Latin version of St Albans, Bacon's title) to Mersenne, 23 Dec. 1630 and 10 May 1632: *Oeuvres*, I, 195-6 and 251.

16. Ganeri, *The lost age*, discusses the 'new reason' and its practitioners (quotations from Dara Shikoh at 24-7).

17. Elvin, 'The man who saw dragons', provides a fascinating analysis of this work (quotations from 12 and 34).

18. Miller, *State versus gentry*, 140, quoting Zhang Pu's declaration; Chen's 'Rules of Compilation' *(fanli)* to the *Nongzheng quanshu* by Xu Guangqui (1562-1633), one of the most prominent Chinese Christian converts of his day; Atwell, 'Ming statecraft', 68-9, on the *Huang Mingjingshi wenbian* of 1639, inspired by a similar compilation of political tracts from earlier periods published in 1635: Chang Pu, *Memorials by famous officials through the ages*.

19. Details from Peterson, 'Ku Yen-wu' (quotations from 131 and 211). Gu completed his *Advantages and disadvantages of the provinces and prefectures of the empire* in 1662.

20. Struve, *The Ming-Qing conflict*, 30, quotes the edict; Spence, *Emperor*, 65-8, quotes Kangxi's own writings.

21. Based on Cook, *Matters*, ch. 4, 'Translating what works', quotation from 344-5 (Genshō Mukai, *Kenkon bensetsu*). Florentius Schuyle collated, translated and published Descartes's book as *De homine* in 1662.

22. Okada Takehiko, 'Practical learning', 270-1, quoting from the eight-volume complete works of Kaibara Ekken (1630-1714). This paragraph relies on the articles in de Bary and Bloom, *Principle and practicality*.

23. On Kumazawa (1619–91), and Zhu (1600–82), see McMullen, 'Kumazawa Banzan'; Ching, 'Chu Shun-Shui'; and idem, 'The practical learning'.

24. McMullen, 'Kumazawa Banzan'. Atwell, 'Ming observers', draws fascinating parallels between seventeenth-century *arbitristas* (from *arbitrio*, or 'remedy') in various countries.

25. Spinoza, *Tractatus theologico-politicus*, 291–2 (from the last chapter of the work, wherein 'It is shown that in a free commonwealth every man may think as he pleases and say what he thinks' – a quotation from Tacitus).

26. Details from Drake, *Galileo at work*, and Redondo, *Galileo*.

27. Descartes, *Oeuvres*, I, 270–1 and 285–6, to Mersenne, Nov. 1633 and Apr. 1634. *Le monde, ou traité de la lumière* only appeared in 1662.

28. Descartes made sure the first edition, published in Leiden, appeared anonymously; but the licence authorizing the French edition named him as its author. Descartes, *Oeuvres*, I, 338–41, Descartes to Mersenne, Mar. 1636, and 369, Descartes to someone involved in the licensing process, 27 Apr. 1637 (blaming Mersenne for naming him).

29. Spinoza, *Ethics*, 1–3 (from the introduction by Seymour Feldman).

30. Leffler, 'From humanist', 420, about François Eudes de Mézeray's *Abrégé chronologique de l'histoire de France*, 3 vols. (Paris, 1668).

31. Van der Heyden, *A description of fire engines* (2nd edn), 3; Viallé and Blussé, *The Deshima registers*, XII, 335–8 (on the ill-fated fire engine). See David, 'Patents', 276–8, on van der Heyden as a 'multiple' inventor.

32. Ganeri, *The lost age*, 248.

33. Elvin, 'The man who saw dragons', 22–3: Cook, *Matters*, 415; Berry, *Japan*, 51–2.

34. Merton, 'Singletons and multiples', 482–3.

35. Bacon, *New Atlantis. A worke unfinished*.

36. Hartlib, *Considerations*, 46–8. Details on earlier attempts from Blome, 'Office of Intelligence', and *ODNB*, s v. 'Samuel Hartlib'.

37. Sprat, *History*, 57–8; charters to the Royal Society, 15 July 1662 and 22 Apr. 1663; *ODNB*, s.v. 'Founder members of the Royal Society'.

38. *Philosophical Transactions* did not become the official journal of the Royal Society of London until 1752: Oldenburg took personal responsibility (as well as any profits) for his work as editor.

39. Bots, 'Le rôle des périodiques', 49, quoting Abbé Jean-Paul de la Roque, director of the *Journal des Sçavans*.

40. Ross, *Daum's boys*, 2–3, 6; idem, 'Pupils' choices', 314–15.

41. Skippon, *An account*, 607; Ray, *Observations*, 271–2.

42. Boix y Moliner, *Hippocrates aclarado, prólogo*, n.f. (the author claimed that Harvey had learned about the circulation of the blood from a commentary on Ecclesiastes by Padre Juan de Pineda in 1620; while Descartes had plagiarized a book by Dr Gómez Pereyra, published in 1554).

43. Camuffo, 'The earliest temperature observations'. By a cruel irony, the records of this precocious experiment were severely damaged by an extreme climatic event: the Florence floods in 1966. When eventually processed in the twenty-first century, the data revealed winters more than 1°C cooler than in the twentieth century.

44. Sprat, *History*, I, 173–9, prints Hooke's proposal; Fleming, *Historical perspectives*, 34–7, provides an overview of this and other systematic attempts to collect weather data.

45. Goad, *Astrometeorologica*; Baker, 'The climate', 428–32 (on Ashmole, Locke and Plot).

46. Keller, *Fatal isolation*, 2, 113, 162.

47. Robinson, *The great comet*, 120–6, and Álvarez de Miranda, 'Las controversías', list the works; De Beer, *The diary of John Evelyn*, IV, 235.

48. Lach and van Kley, *Asia*, III, 976; Tavernier, *Travels*, I, 309; Di Cosmo, *Diary of a Manchu soldier*, 57; Mather, *Heaven's alarm*; idem, *Kometographia*, 118, 124 and 107.

49. Descartes, *Oeuvres*, I, 251–2, Descartes to Mersenne, 10 May 1632; Schaffer, *The information order*, 36–44. See the sources listed in Book III, lemma IV, of Newton's *Principia mathematica*.

50. Halley, *A synopsis*, 19, 21–22, 24. It is possible that Halley knew that Cassini had seen and sketched something that *looked* like a comet hitting Jupiter in Dec. 1690: Peiser, *Natural catastrophes*, 7. Halley was also lucky because the 1682 comet was the last one visible to the naked eye to appear for sixty years.

51. Rabb, 'Introduction', 149; Rabb, 'The Scientific Revolution', 509. I thank Mircea Platon for alerting me to the second item.

52. Skippon, *An account*, 607; Ronan, *Edmond Halley*, 124, Astronomer Royal John Flamsteed to Isaac Newton in 1691.

53. Milton, *Paradise lost*, Book I, lines 287–91, and the beginning of Book VIII (I thank Kate Epstein for pointing out these references); Mormiche, *Devenir prince*, 338–9. The theologians of the Sorbonne protested at the discussion of Galileo's theory, until the Dauphin's religious preceptor Bossuet instructed them to desist. Compare Descartes's reaction when he read the sentence on Galileo: page 484 above.

54. Kessler, 'Chinese scholars', 181–4; Struve, *Ming–Qing conflict*, 30–2; Brook, 'Censorship', 177, quoting the Qianlong emperor's edict of 11 Dec. 1774.

Conclusion

1. Brinton, *Anatomy*, 237–8. Special thanks to Rayne Allinson, Kate Epstein and Ken Pomeranz for their trenchant criticism of earlier drafts of this chapter.

2. Foster, *The English factories in India, 1630–1633*, 218–19, letters from East India Company officials in Surat to London, 8 May 1632 OS; Smith, *The art of doing good*, 137, quoting Lu Shiyi's diary; Howell, *Epistolae*, III, 26, letter to his nephew, 10 Dec. 1647; Balfour, *Historical works*, III, 409.

3. Macfarlane, *The diary of Ralph Josselin*, 472, entry for 25 Nov. 1660 OS.

4. Pepys, *Diary*, VIII, 337–414; Cavaillé, 'Masculinité' para. 38, quoting Christina.

5. D'Aubert, *Colbert*, 23 (quoting Lefèvre d'Ormesson); *ODNB*, s.v. 'Monck'.

6. Haboush, 'Constructing the center', 51.

7. Hagopian, *The phenomenon*, 123; Bacon, *Essayes*, 46 (Bacon reprinted the essay, with certain changes, in the 1625 edition of his book, quoted here); Walford, 'The famines', part II, 79.

8. Walford, 'The famines', part II, 217.

9. Laing, *Correspondence*, I, 93–8, and 105, Lothian to his father, 19 Oct. 1637 and 8 Nov. 1640 (capitals in the original).

10. BL *Harl. Ms.* 5,999/29v, 'Discourse by Henry Jones, Nov. 1643 (more details on climate and catastrophe in both Scotland and Ireland in ch. 11 above).

11. Trotsky, *The history*, II, Introduction, and vol. I, ch. 7 ('Five Days').

12. Brinton, *Anatomy*, 35.

13. Scriba, 'The autobiography', 32.

14. Rushworth, *Historical Collections*, III, part 1, 11–12, the king's speech, 3 Nov. 1640. See also Chapter 11 above.

15. Braudel, *The Mediterranean*, I, part II, ch. 1, part 1; AGS *SP libro* 218/72, *consulta* of 27 Aug. 1647, reviewing many recent letters from Naples; AGS *Estado* 2566, n.p., Don Diego de la Torre to Philip IV, 18 Feb. 1648; Elliott, *The revolt*, 407.

16. AGI *IG* 435 *legajo* 10/258v–259v, Fernando Ruiz de Contreras orders the Casa de Contratación in Seville to warn the fleet about 'el accidente de Portugal', 5 Jan. 1641, minute, and *IG* 429 *legajo* 38/177–182v, Philip IV to the viceroy of Peru and others, 7 Jan. 1641, minute. AGI *IG* 761, n.p., *consulta* of 27 Dec. 1640, noting that the council had delayed advising the king on the correct steps to take 'until there should be more specific and general information'.

17. Gilbert, *History of the Irish Confederation*, I, 8–9; TCD *Ms.* 809/13v, examination of Connolly, 22 Oct. 1641.

18. AMAE (P) *CPA* 57/314–15, Bellièvre to Secretary of State Brienne, London, 13 Nov. 1648 (italics added).

19. Goodare, 'Debate: Charles I', 200–1.
20. Quotations from Elliott, 'The year of the three ambassadors', 181.
21. Defoe, *The life*, 286, 262, 152, 58, 79, 170. My reading of the book owes much to Hill, 'Robinson Crusoe'.
22. Will, 'Développement quantitatif ', 868, quoting the decrees of the Kiangxi emperor; Marks, *Tigers*, 291, quoting Han Liangfu in 1724 and the Yongzheng emperor in 1723.

Epilogue

1. Apologies to James Carville, author of the mantra of the successful Clinton–Gore presidential campaign of 1992, 'It's the economy, stupid'; and thanks to Oktay Özel who, at the panel on the Ottoman General Crisis at the XI International Congress of Social and Economic History of Turkey, Ankara, 18 June 2008, suggested that our title should be 'It's the climate, stupid'. Thanks also for helpful references and suggestions to Derrin Culp, Kate Epstein, Daniel Headrick, James Lenaghan, Ruth MacKay and Angela Nisbet; to Greg Wagman and a group of gifted Honors Students at Notre Dame University in 2007; and to those who attended a panel on *Global Crisis* organized by DePaul University in 2015.
2. Charney, *Carbon*, viii (quoting Verner Suomi), 1, 17; Nierenberg, *Changing climate*, 492–3, with the relevant section (Title VII subtitle B) of the Energy Security Act of 1980. I thank Christian Pfister and Martin Parry for sharing their recollections of the 1979 UEA conference. The text of Carter's Executive Order of 1979 creating FEMA is online at https://www.archives.gov/federal-register/codification/executive-order/12148.html
3. 'Universal declaration on the eradication of hunger and malnutrition', adopted by the World Food Conference on 16 November 1974, http://www.ohchr.org/EN/Professional Interest/Pages/EradicationOfHungerAndMalnutrition.aspx
4. Nierenberg, *Changing climate*, 4 (executive summary by William Nierenberg, a physicist) and 413 (by Paul E. Waggoner, a plant pathologist and agronomist). The motives of Nierenberg and his team in playing down the impact of CO_2 on climate change have been vigorously debated: see Oreskes and Conway, *Merchants of doubt*, 176–83, and Nierenberg, 'Early climate change consensus'.
5. Nierenberg, *Changing climate*, 387 and 404 (by Waggoner) and 49 (by Nierenberg). Thanks to global warming, in autumn 2016 the 68,000-ton *Crystal Serenity*, carrying 1,700 passengers and crew, traversed the Northwest Passage in a month.
6. Nierenberg, *Changing climate*, 384–7 and 409 (by Waggoner) and 53 (by Nierenberg). For the true costs of the abrupt climate changes on fourteenth-century Europe, see Campbell, *The great transition*; on the American West in the 1880s, see White, *'It's your misfortune and none of my own'*, 223–6; and on the 'Okies', see Gregory, *American exodus*, passim.
7. Quotations from Mussey, 'Yankee Chills', 442 and 449. On the eruption of Mount Tambora in 1815, apparently the most powerful eruption in the past 10,000 years, and its deleterious global consequences, see Wood, *Tambora*.
8. *State of Ohio homeland security strategic plan* (2011), 6, at http://www.ema.ohio.gov/Documents/Ohio_EOP/ESF_8_Tab_E_NonAcuteMF_ExerciseVersion_20161230.pdf, last accessed 13 Feb. 2017, when the 2011 version was still the most recent available.
9. *State of Ohio emergency operations plan*, Tab E to ESF-8 [=Emergency Support Function #8] of the Ohio E.O.P., NAMFIRP-14 Section H.2, promulgated by Governor John Kasich in Aug. 2013, last accessed 13 Feb. 2017, from a site that stated 'Download Full Plan (coming soon …)'. The only full plan available for download is the previous version, promulgated by Governor Ted Strickland in Jan. 2009.
10. Carmel, 'Globalization', 44, 53.
11. *A failure of initiative*, 123, 174 (quoting Major-General Harold A. Cross), 178 and 358–9 (conclusion).

12. See the appeals in *A failure of initiative*, 565–6, and *Hurricane Katrina*, 725–30. Bush had also tried to sabotage the 9/11 Commission, but without success: see May, 'When government writes history'.

13. Sobel, *Storm surge*, 279.

14. Cock, *Hygieine*, sig. B1v (italics in the original). The author had tried to put his case to a sub-committee of the London council, but was 'silenc'd'; now he tried again 'in paper'. Details on the catastrophe from Moote and Moote, *The great plague*.

15. Slovic, 'The perception of risk', 280; Slovic, *The perception of risk*, 8, introduction by R. W. Kates; Bankoff, *Cultures*, 3.

16. Blake, *The deadliest, costliest, and most intense United States tropical cyclones*, 5, 6, 25.

17. Sobel, *Storm surge*, 279; Hamilton, *Requiem for a species*, x.

18. NOAA Global analysis 2016; https://www.ncdc.noaa.gov/sotc/global/201613 (1998 is currently the eighth-warmest year on record). The World Meteorological Organization, which combines the NOAA data with reports from other national and regional temperature records, concurred: https://public.wmo.int/en/media/press-release/wmo-confirms-2016-hottest-year-record-about-11%C2%B0c-above-pre-industrial-era. Melillo, *Climate change impacts*, documents the increasing frequency and intensity of extreme events between 1991 and 2012 (see 36–7 for the increase in very heavy precipitation, especially in the US Midwest and northeast). The trend continues.

19. Powell, 'Climate scientists', 124; http://www.pewinternet.org/2015/07/23/an-elaboration-of-aaas-scientists-views/; http://www.pewinternet.org/2016/10/04/the-politics-of-climate/

20. Kvaløy, 'The publics' concern', based on data collected from the 2005–9 'World Values Survey', quotations from 11, 13–14 and 18.

21. A survey in 2014 revealed that 49 per cent of Americans 'reported that the severity of recent natural disasters is a sign of Biblical "end times"', a figure that rose to 61 per cent among those 'with a high school education or less': Jones, *Believers*, 23–4.

22. An epithet coined by Oreskes and Conway, *Merchants of doubt*, published in May 2010. Coincidentally, Hamilton's *Requiem for a species*, which reached the same conclusions, appeared one month earlier.

23. US District Court for the District of Columbia, Civil Action No. 99-2496, Amended Final Opinion by US District Court Judge Gladys Kessler, 17 Aug. 2006, 1, 3–4 and 259 (see also the evidence cited at 259–69), http://www.publichealthlawcenter.org/sites/default/files/resources/doj-final-opinion.pdf

24. Data from Geneva Association, *An integrated approach*, 5; and a report by the European Environment Agency, comprising thirty-three countries: http://www.eea.europa.eu/data-and-maps/indicators/direct-losses-from-weather-disasters-3/assessment, created Jan. 2017.

25. Geneva Association, *The insurance industry*, 42 and 62–3, and idem, *An integrated approach*, 6. The Geneva Association is a non-profit international insurance think tank comprising up to ninety chief executive officers from the world's leading insurance and re-insurance companies.

26. Ghosh, *The great derangement*, 87–8, 90, 54.

27. Center for Naval Analyses, *National security*, 6 and 27–8; Gleik, 'Water', 332 and fig. 3 (population growth from 3 million in 1950 to 21 million in 2007), 334 (crop yields).

28. http://lds.about.com/od/preparednessfoodstorage/p/foodstorage.htm; http://ship.oh.networkofcare.org/ph/library/article.aspx?hwid=tf6355

29. *A stronger, more resilient New York*, 6, 39, 412–13.

30. Pepys, *Diary*, IV, 323–4 (entry for 7 Dec. 1663); *Hansard House of Commons debate*, 19 Feb. 1953, speech by Sir David Maxwell Fyfe.

31. TNA *HLG* 145/151, 'The London flood barrier: the Bondi report and branch comments' (a document only declassified in 1999). The risk of a meteorite or asteroid strike is no longer considered so remote, and scientific teams around the world constantly monitor the situation, ready to take appropriate action if necessary: see Drube, 'The NEOTωIST mission', and http://www.dlr.de/pf/en/desktopdefault.aspx?tabid-174/319_read-507/ (I thank Tale Sundlisaeter for these references).

32. The floods of 1953 also affected the Netherlands, leading to thirteen major projects to protect the Low Lands from another catastrophe, including the 5-mile Oosterscheldekering (Eastern Scheldt Storm Surge Barrier), completed in 1986 – like the Thames Barrier, more than three decades after the initial catastrophe. It was fully activated against predicted storm surges twenty-five times between 1986 and 2014.
33. King, *Foresight, flood and coastal defence*; Center for Naval Analysis Corporation, *National security*, 31; Rosenzweig, *Responding to climate change*, 353.
34. Comment by Sam White at a workshop on climate change held at the Mershon Center, The Ohio State University, 28 Apr. 2015; O'Riordan and Lenton, 'Tackling tipping points', 23, 26, 27.
35. Ghosh, *The great derangement*, 119.
36. Written in July 1816, and first published five months later: Coleridge, *The works of Lord Byron . . . Poetry*, IV, 42–5.

Acknowledgements

1. Email to Robert Baldock, 21 Feb. 1998.
2. Needham, *SCC*, I, preface.
3. Richards, *The unending frontier*, 3.

Note on Sources

1. Bisaccione, *Historia*; Birago Avogadro, *Delle historie memorabili*.
2. Voltaire, *Essai* (see page 1 above); Merriman, *Six contemporaneous revolutions*; Aston, *Europe in crisis*; and Elliott, 'The General Crisis'.
3. *Revue d'histoire diplomatique*, XCII (1978), 5–232; Hroch and Petráň, *Das 17 Jahrhundert – Krise der Feudalgesellschaft?* (first published in Czech in 1976); *Renaissance and Modern Studies*, XVI (1982), 1–107.
4. *Modern Asian Studies*, XXIV (1990), 625–97, mostly reprinted in the 1997 edition of Parker and Smith, *General Crisis*.
5. Goldstone, *Revolution*; Romano, *Conjonctures opposés*; Ogilvie, 'Germany'; *AHR*, CXIII (2008), 1,029–99 ('The General Crisis of the seventeenth century revisited'); *JIH*, XL (2009), 145–303 ('The crisis of the seventeenth century: interdisciplinary perspectives') – Rabb quoted from 150.
6. The depositions are available at http://1641.tcd.ie; the maps at http://downsurvey.tcd. ie/1641-depositions.php. See also the essays in Darcy, *The 1641 depositions*. Other important seventeenth-century documentary series now available online include the Hartlib papers at Sheffield University (25,000 manuscripts), online at https://hridigital.shef.ac.uk/hartlib/; and the papers of the English secretaries of state from 1509 to 1782, scanned from various British archives (over 1 million manuscript images for the seventeenth century): 'State Papers Online' provides to institutional subscribers (only) the entire archives of the Tudor and Stuart governments.
7. Struve, *The Ming–Qing conflict*, 32; Ohlmeyer, 'The Antrim Plot', 912, Lord Antrim to Hamilton, 13 July 1639; Scott and Bliss, *The works*, VII, 211, Laud to Wentworth, 30 Nov. 1635; Peters, 'The Quakers', 98, letter from William Caton in late 1659; Amussen and Kishlansky, *Political culture*, 262–3, quoting Underdown.
8. Douglass, 'The secret of the southwest'; Stahle, 'Tree-ring reconstructed megadroughts'; Rydval, 'Reconstructing'.
9. Le Roy Ladurie, 'Naissance', pays tribute to Garnier's influence.
10. Fleury and Henri, *Des registres paroissiaux*; Wrigley and Schofield, *The population history*; Hayami, *The historical demography*.
11. See also the monographs of Raphael, *Climate*, and Ellenblum, *The collapse*, on climate change in the medieval eastern Mediterranean.

12. McCormick; 'Climate change'; Kelly and Ó Gráda, 'The waning'; White, 'The real Little Ice Age'; and Büntgen and Hellmann, 'The Little Ice Age'.

13. Chaplin, 'Ogres and omnivores', 32. The forum may be found in *WMQ*, LXXII (2015), 25–158, ending with a review article entitled 'It's the climate, stupid' – fame indeed.

14. Zheng, 'How climate change' (2014); Fenske and Kala, 'Climate' (2015); Anderson, 'Jewish persecutions' (2016).

15. Allen, 'Wages'; Deng and O'Brien, 'Establishing statistical foundations', 1075 and 1057 (quote); Bolt and van Zanten, 'The Maddison project', 631; Studer, *The Great Divergence reconsidered*, 182.

Bibliography

A. Printed primary sources cited in the notes

Abbott, W. C., ed., *Writings and speeches of Oliver Cromwell*, 3 vols. (Cambridge, MA, 1937–47)

Acta Pacis Westphalicae: supplementa electronica, 1, 'Die westfälischen Friedensverträge vom 24. Oktober 1648. Texte und Übersetzungen', http://www.pax-westphalica.de/ipmipo/

Acts done and past in the second session of the second Triennall Parliament of our soveraign lord Charles . . . and in the first Parliament of our soveraign lord Charles the II (Edinburgh, 1649)

Ådahl, K., ed., *The sultan's procession: the Swedish embassy to Sultan Mehmed IV in 1657–1658 and the Rålamb painting* (Istanbul, 2006)

Amelang, J., ed., *A journal of the plague year: the diary of the Barcelona tanner, Miquel Parets, 1651* (Oxford, 1991)

Anon., *An ease for overseers of the poore abstracted from the statutes, allowed by practise, and now reduced into forme, as a necessarie directorie for imploying, releeuing, and ordering of the poore* (Cambridge, 1601)

Anon., *A true and strange relation of fire, which by an eruption brake forth out of the bowels of the earth* (London, 1639)

Anon., *The times dissected, or a learned discourse of several occurrences very worthy of observation to deter evill men and encourage good* (London, 1642)

Anon., *The victorious proceedings of the Protestants of Ireland* (London, 1642)

Anon., *Nicandro o antídoto contra las calumnias que la ignorancia y envidia ha esparcido por deslucir y manchar las heroicas y inmortales acciones del conde-duque de Olivares después de su retiro* (Madrid, 1643)

Anon., *New England's first fruits in respect of the progress of learning, in the Colledge at Cambridge in Massachusetts-Bay* (London, 1643), in *Collections of the Massachusetts Historical Society for the year 1792*, I (Cambridge and Boston, 1792), 242–50

Anon., *Escrívense los sucessos de la Evropa desde abril de 46 hasta junio de 47 inclusive* (Madrid, 1647)

Anon., *The red-ribbon'd news from the army* (London, 1647)

Anon., *The way to get rain; by way of question and answer. Shewing the true cause both of too much want, and too much abundance of raine* (London, 1649)

Anon., *Le Ti θεῖου de la maladie de l'état* (Paris, 1649)

Anon., *Avertissements aux rois et aux princes pour la traité de la paix et le sujet de la mort du roi de la Grande Bretagne* (Paris, 1649)

Anon., *Avis à la reine d'Angleterre et à la France* (Paris, 1650)

Anon., *De na-ween vande vrede. Ofte ontdeckinge vande kommerlijcke ghelegentheydt onses lieven vaderlants: . . . met de remedien daer teghen* (Amsterdam, 1650)

Anon., *Les cautelles de la paix* (Paris, 1652)

Anon., *A letter of the officers of the army of Scotland, under the commander in chief there, to the officers of the army of England* (London, 1659)

Anon., *An essay upon government adopted by the Americans. Wherein the lawfulness of revolutions are demonstrated in a chain of consequences from the fundamental principles of society* (Philadelphia, 1775)

Antony, P. and H. Christmann, eds., *Johann Valentin Andreä: ein schwäbisher Pfarrer im dreissig-jährigen Krieg* (Hildesheim, 1970)

Arnauld, Angélique, *Lettres de la révérende mère Marie Angélique Arnauld*, 3 vols. (Utrecht, 1741–2)

Arnauld, Antoine, *De la fréquente communion où les sentimens des pères, des papes et des conciles, touchant l'usage des sacremens de pénitence et d'Eucharistie, sont fidèlement exposez* (Paris, 1643)

Arnauld d'Andilly, Robert, *La vérité toute nue* (Paris, 1652)

Articles of the large treaty concerning the establishing of the peace betwixt the kings Majesty, and his people of Scotland, and betwixt the two kingdomes: agreed upon by the Scottish, and English Commissioners in the city of Westminster the 7th day of August. 1641: allowed and published for the use of the kingdome of Scotland (London, 1641)

Arzáns de Orsúa y Vela, Bartolomé, *Historia de la Villa Imperial de Potosí,* 3 vols., ed. L. Hanke and G. Mendoza (Providence, 1965)

Aston, John, 'The journal of John Aston, 1639', in *Six North Country diaries* (Edinburgh, 1910: Surtees Society, CXVIII), 1–34

Aubrey, John, *Brief lives chiefly of contemporaries set down by John Aubrey between the years 1669 and 1696,* ed. A. Clark (Oxford, 1898)

Ayala, Baltasar de, *De iure et officiis bellicis et disciplina militari libri III* (Douai, 1582, ed. J. Westlake, 2 vols., Washington, DC, 1912)

Bacon, Francis, *The proficience and advancement of learning, divine and humane* (London, 1605)

— *Novum organum Francisci de Verulamio Instauratio Magna* (London, 1620; English edn, *The new organon,* by L. Jardine and M. Silverthorne, Cambridge, 2000)

— *The essayes or counsels, civill and morall* (1625; ed. M. Kiernan, Oxford, 2000)

— *New Atlantis. A worke unfinished appended to: Sylva Sylvarum, or, a natural historie in ten centuries* (London, 1626)

— *The works of Francis Bacon, baron of Verulam, Viscount St. Albans, and lord high chancellor of England,* ed. J. Spedding, R. L. Ellis and D. D. Heath, 14 vols. (London, 1861–79)

Baily [or Bayly], Charles, *A true & faithful warning unto the people and inhabitants of Bristol* (London, 1663)

Bainbridge, John, *An astronomicall description of the late comet from the 18. of November 1618. to the 16. of December following, with certain morall prognosticks or applications drawne from the comets motion* (London, 1619)

Balde, Jakob, *Satira contra abusum tabaci* (Nuremberg, 1657; trans. into German as *Die trückene Trünckenheit,* 1658)

Balfour, Sir James, *Historical works: see* Haig

Bamford, F. ed., *A royalist's notebook; the commonplace book of Sir John Oglander* (London, 1936)

Baratotti, Galarana [Elena Tarrabotti], *la simplicità ingannata o la tirannia paterna* (Leiden, 1654; ed. F. Medioli, Turin, 1989)

Barbon, Nicholas, *A letter to a gentleman in the country, giving an account of the two Insurance Offices; the Fire-Office and friendly-society* (London, 1685)

Barbot, Jean: *see* Hair et al.

Barlow, Edward, *Barlow's journal of his life at sea in king's ships, East & West Indiamen & other merchantmen from 1659 to 1703,* 2 vols., ed. B. Lubbock (London, 1934)

Baron, S. H., *The travels of Olearius in seventeenth-century Russia* (Stanford, 1967)

[Batencour, Jacques de], *Instruction méthodique pour l'école paroissiale, dressée en faveur des petites écoles, dividée en quatre parties* (Paris, 1669; a re-edition of the 1654 original: *L'escole paroissiale ou la manière de bien instruire les enfans dans les petits escoles*)

Beauplan, Guillaume Le Vasseur, sieur de, *A description of Ukraine* (Rouen, 1651: English trans., New York, 1959)

Begley, W. E. and Z. A. Desai, *The Shah Jahan Nama of 'Inayat Khan: an abridged history of the Mughal emperor Shah Jahan, compiled by his royal librarian* (Delhi, 1990)

Behr, Johann Heinrich, *Der Verschantzte Turenne oder grundliche Alt- und Neue Kriegsbaukunst* (Frankfurt, 1677)

Bergh, S., ed., *Svenska riksrådets protokoll, 1621–1658*, 17 vols. (Stockholm, 1878–1925)

Bernier, François, *Travels in the Mogul Empire, AD 1656–1668* (Westminster, 1891)

Berwick y Alba, Duchess of, *Documentos escogidos de la casa de Alba* (Madrid, 1891)

Birago Avogadro, Giovanni Battista, *Delle historie memorabili che contiene le sollevationi di stato di nostri tempi* (Venice, 1653; reissued as *Turbolenze di Europa dall'anno 1640 sino al 1650*, Venice, 1654)

Birch, Thomas, *The works of the honourable Robert Boyle*, 6 vols. (London, 1772)

Bisaccione, M., *Historia delle guerre civili di questi ultimi tempi, cioè di Inghilterra, Catalogna, Portogallo, Palermo, Napoli, Fermo, Moldavia, Polonia, Svizzera, Francia, Turco* (1st edn, Venice, 1652; 4th edn, 'ricorretta et in molte parti accresciuta', Venice, 1655)

Blair, E. H. and J. A. Robertson, *The Philippine islands*, 55 vols. (Cleveland, 1905–11)

Blussé, L. and R. Falkenburg, *Johan Nieuhofs beelden van een Chinareis 1655–1657* (Middelburg, 1987)

Boix y Moliner, Miguel Marcelino, *Hippocrates aclarado y sistema de Galeno impugnado, por estar fundado sobre dos aphorismos de Hippocrates no bien entendidos, que son el tercero, y veinte y dos del primer libro* (Madrid, 1716)

Bossuet, Jean-Bénigne, *Politics drawn from the very words of Holy Scripture* (1679; Paris, 1709; English trans. Cambridge, 1990)

Botero, Giovanni, *Relatione della Republica Venitiana* (Venice, 1605)

Bournoutian, G., *The chronicle of Deacon Zak'aria of K'anak'er (Zak'areay Sarkawagi Patmagrut'iwn)* (Costa Mesa, CA, 2004)

Boyle, Robert: *see* Birch

Boyle, Roger, earl of Orrery, *A treatise on the art of war*, 2 vols. (London, 1677)

Bradford. William, *Of Plymouth plantation, 1620–1647* (1650; ed. S. E. Morison, New York, 2002)

Brantôme, Pierre de Bourdeille, seigneur de, *Oeuvres*, VI (Paris, 1787)

Bray, W., ed., *Diary and correspondence of John Evelyn, F. S., to which is subjoined the private correspondence between King Charles I. and Sir Edward Nicholas*, new edn, 4 vols. (London, 1887)

Bremner, R. H., ed., *Children and youth in America: a documentary history*, I (Cambridge, MA, 1970)

Brennan, M. G., ed., *The travel diary of Robert Bargrave: Levant merchant (1647–1656)* (London, 1999: Hakluyt Society, 3rd series, III)

Brittish lightning: see Voetius, Gisbertius

Bruce, John, ed., *Letters and papers of the Verney family down to the end of the year 1639* (London, 1853: Camden Society, LVI)

Burnet, Gilbert, *The memoires of the lives and actions of James and William, dukes of Hamilton* (London, 1677)

Burton, Robert, *The anatomy of melancholy, what it is: with all the kinds, causes, symptomes, prognostickes, and several cures of it; in three partitions with their severall sections, members & subsections. Philosophically, medicinally, historically, opened & cut up* (1621; 4th edn, Oxford, 1638)

Calendar of the court minutes etc. of the East India Company 1644–1649, ed. E. B. Sainsbury (Oxford, 1912)

Calendar of state papers, colonial series, America and West Indies, 1675–1676, ed. W. N. Sainsbury (London, 1893)

Calendar of state papers, domestic series. Charles I, ed. J. Bruce and W. D. Hamilton, 23 vols. (London, 1858–97)

Calendar of state papers and manuscripts relating to English affairs, existing in the archives and collections of Venice, and in other libraries of northern Italy, 1202–1674, ed. H. F. Brown et al., 28 vols. (London, 1864–1947)

Calendar of state papers relating to Ireland of the reign of Charles I, ed. R. P. Mahaffy, 4 vols. (London, 1900–4)

Capograssi, A., 'La rivoluzione di Masaniello vista dal residente veneto a Napoli', *Archivio storico per le province napolitane*, NS, XXXIII (1952), 167–235

Cardin Le Bret, P., *De la souveraineté du roy* (Paris, 1632)

Carrier, H., *La Fronde: contestation démocratique et misère paysanne. 52 Mazarinades*, 2 vols. (Paris, 1982)

Cartas de algunos padres de la Compañía de Jesús sobre los sucesos de la monarquía entre 1634 y 1648, ed. Pascual de Gayangos, 7 vols. (Madrid, 1861–65: *Memorial Histórico Español*, XIV–XIX)

Castlehaven, James Tuchet, earl of, *Memoirs of the Irish wars* (1684; Delmar, NY, 1974)

Cayet, P. V., *Chronologie novenaire* (Paris, 1608; ed. Michaud and Poujoulat, Paris, 1838)

Cervantes Saavedra, Miguel de, *Segunda parte de el ingenioso hidalgo, Don Quijote de la Mancha* (Madrid, 1615)

Chardin, Jean, *Sir John Chardin's travels in Persia* (Paris, 1676; London, 1724; New York, 2010)

Charles I, king of Great Britain, *A large declaration concerning the late tumults in Scotland* (London, 1639)

— *Eikon basilike* (London, 1648/9; ed. J. Daems and H. F. Nelson, Peterborough, 2005)

Chen Zilong, 'The little cart', in A. Waley, *Translations from the Chinese* (New York, 1941)

Cheng Pei-kai and Michael Lestz, with J.D. Spence, *The search for modern China: a documentary collection* (New York, 1999)

Chéruel, Adolphe and Georges Avenel, eds., *Lettres du Cardinal Mazarin pendant son ministère*, 7 vols. (Paris, 1872–93)

Christina, queen of Sweden, *Apologies*, ed. J.-F. de Raymond (Paris, 1994)

Churchill, A. and J., eds., *A collection of voyages and travels: some now first printed from original manuscripts, others now first published in English in six volumes* (London, 1732)

Clarendon, Edward Hyde, earl of, *The history of the rebellion and civil wars in England, begun in the year 1641*, ed. W. D. Macray, 6 vols. (Oxford, 1888)

— *Brief view and survey of the dangerous and pernicious errors to church and state in Mr. Hobbes's book entitled 'Leviathan'* (Oxford, 1676)

Cock, Thomas, *Hygieine, or, a plain and practical discourse upon the first of the six non-naturals, viz. air with cautionary rules and directions for the preservation of people in this time of sickness, very necessary for the gentry and citizens that are now in the country to peruse before they come into London* (London, 1665)

Coleridge, E. H., ed., *The works of Lord Byron. A new, revised and enlarged edition, with illustrations. Poetry*, IV (London, 1922)

Collurafi, A., *Tumultazioni delle plebe di Palermo* (Palermo, 1661)

Conti, V., *Le leggi di una rivoluzione. I bandi della repubblica napoletana dall'ottubre 1647 all'aprile 1648* (Naples, 1983)

Cooper, M., ed., *They came to Japan: an anthology of European reports on Japan, 1543–1640* (Berkeley, 1965)

Corneille, Pierre, *Pertharite, roy des Lombards, tragédie* (Paris, 1654)

Correra, L., 'Inedita relazione dei tumulti napoletani del 1647', *Archivio storico per le province napolitane*, XV (1890), 353–87

Costin, Miron, *Letopisețul Țării Moldovei* (Bucharest, 1975)

Crucé, Émeric, *Nouveau Cynée ou discours d'estat représentant les occasions et moyens d'establir une paix générale et la liberté de commerce pour tout le monde* (Paris, 1623)

Cysat, Renward, *Collectanea Chronica und denkwürdige Sachen pro Chronica Lucernensi et Helvetiae*, IV.2, ed. J. Schmid (Luzern, 1969)

Dankoff, R., and S. Kim, eds., *An Ottoman traveller: selections from the* Book of travels *of Evliya Çelebi* (London, 2010)

Davenant, Charles, *An essay upon the ways and means of supplying the war* (London, 1695)

[Davenant, William], *The first days entertainment at Rutland-House* (London, 1656)

Davies, John, *A discovery of the true causes why Ireland was never entirely subdued* (1612; 3rd edn, London, 1666)

De Beer, E. S., ed., *The diary of John Evelyn*, 6 vols. (Oxford, 2000)

Defoe, Daniel, *The life and strange surprizing adventures of Robinson Crusoe, of York, mariner, who lived eight and twenty years all alone in an un-inhabited island on the coast of America, near the mouth of the great river of Oroonoque; having been cast on shore by shipwreck, wherein all the men perished but himself*, 4th edn (London, 1719)

— *A journey thro' the whole island of Great Britain*, 4th edn, 4 vols. (1724; London, 1748)

— *The compleat English tradesman*, vol. II, in two parts (London, 1727)

Dell, William, *Several sermons and discourses* (London, 1652; reprinted 1709)

Descartes, René, *Discours de la méthode: pour bien conduire sa raison et chercher la vérité dans les sciences* (Leiden, 1637; bilingual English and French edn by G. Heffernan, Notre Dame, 1994)

— *Oeuvres de Descartes*, ed. C. Adam and P. Tannery, 12 vols. (Paris, 1897–1913)

Di Cosmo, N., ed., *The diary of a Manchu soldier in seventeenth-century China: 'My service in the army' by Dzengšeo* (New York, 2006)

Di Marzo, G., ed., *Biblioteca storica e letteraria di Sicilia. Diari dell città di Palermo dal secolo XVI al XIX*, 28 vols. (Palermo, 1869–86)

Diaries kept by the head of the Dutch factory in Japan, 1633–49, 11 vols. (Tokyo, 1974–2012)

Dickinson, W. C. and G. Donaldson, eds., *A sourcebook of Scottish history*, III (Edinburgh, 1961)

Doglio, M. L., ed., *Lettere di Fulvio Testi*, 3 vols. (Bari, 1967)

Donaldson, G., *The making of the Scottish Prayer Book of 1637* (Edinburgh, 1954)

Donne, John, *Devotions upon emergent occasions* (London, 1623)

Dooley, B., ed., *Italy in the Baroque: selected readings* (New York, 1995)

Dujčev, I., *Avvisi di Ragusa. Documenti sull'Impero Turco nel secolo XVII e sulla guerra di Candia* (Rome, 1935: Orientalia Christiana Analecta, CI)

Dunn, R. S., J. Savage and L. Yeandle, eds., *The journal of John Winthrop, 1630–1649* (Cambridge, MA, 1996)

Duplessis, Armand-Jean, cardinal-duke of Richelieu: *see* Richelieu

Elliot, H. M. and J. Dowson, *The history of India as told by its own historians: the Muhammadan period*, 8 vols. (London, 1867–77)

Elliott, J. H. and J. F. de la Peña, eds., *Memoriales y cartas del conde-duque de Olivares*, 2 vols (Madrid, 1978–81)

Evelyn, John, *Diary*: *see* De Beer

— *Fumifugium: or the inconveniency of the aer and smoke of London dissipated* (London, 1661)

Evliya Çelebi, *Seyahatname*, 15 vols. (Istanbul, 1969–71); *see also* Dankoff

Fernández Álvarez, M., *Corpus documental de Carlos V*, 5 vols. (Salamanca, 1973–81)

Fincham, Kenneth, 'The judges' decision on Ship Money in February 1637: the reaction of Kent', *Bulletin of the Institute of Historical Research*, LVII (1984), 230–7

Firpo, Luigi, ed., *Relazioni di ambasciatori veneti al Senato, tratte dalle migliori edizioni disponibili e ordinate cronologicamente*, 14 vols. (Turin, 1965–96)

Firth, Charles H., ed., *The Clarke papers: selections from the papers of William Clarke*, 4 vols. (London, 1891–1901: Camden Society, vols. XLIX, LIV, LXI, LXII)

Firth, Charles H. and Robert S. Rait, *Acts and ordinances of the Interregnum, 1642–1660*, 2 vols. (London, 1911)

Fleming, David H., ed., *Diary of Sir Archibald Johnston of Wariston*, II (Edinburgh, 1919: Scottish History Society, XVIII); *see also* Paul

Foster, George, *The sounding of the last trumpet or, severall visions declaring the universall overturning and rooting up of all earthly powers in England, with many other things foretold, which shall come to passe in this year, 1650* (London, 1650)

Foster, William, ed., *The English factories in India, 1630–1633* (Oxford, 1910)

— *The English factories in India, 1634–1636* (Oxford, 1911)

— *The English factories in India, 1637–1641* (Oxford, 1912)

— *The English factories in India, 1655–1660* (Oxford, 1921)

— *The English factories in India, 1661–1664* (Oxford, 1923)

— *The voyage of Thomas Best to the East Indies, 1612–1614* (London, 1934: Hakluyt Society, 2nd series, LXXV)

— *The voyage of Sir Henry Middleton to the Moluccas, 1604–1606* (London, 1943: Hakluyt Society, 2nd series, LXXXVIII)

Franklin, Benjamin, *Observations concerning the increase of mankind and the peopling of countries &c.* (Philadelphia, 1751)

Gallagher, L., ed., *China in the sixteenth century: the journals of Matthew Ricci, 1583–1610* (New York, 1951)

Gallardo, B. J., *Ensayo de una biblioteca española de libros raros y curiosos*, 2 vols. (Madrid, 1863–89)

García Acosta, V., J. M. Pérez Zevallos and A. Molina del Villar, eds., *Desastres agrícolas en México. Catálogo histórico. I. Época prehispánica y colonial, 958–1822* (Mexico, 2003)

Gardiner, S. R., ed., *The Hamilton papers* (London, 1880: Camden Society, new series, XXVII)

— *Constitutional documents of the Puritan Revolution* (London, 1906)

Gatta, G., *Di una gravissima peste che nella passata primavera e estate dell'anno 1656 depopulò la città di Napoli* (Naples, 1659)

Geerts, A. J. M., 'The Arima rebellion and the conduct of Koekebacker', *Transactions of the Asiatic Society of Japan*, XI (1883), 51–116

Gilbert, J. T., ed., *History of the Irish Confederation and the war in Ireland by Richard Bellings*, 7 vols. (Dublin, 1882–91)

Giraffi, Alessandro, *Le rivoluzioni di Napoli* (Venice, 1647; *see also* Howell, James, *An exact historie*)

Girard, *Le voyage en Chine*: *see* Las Cortes

Goad, J., *Astrometeorologia, or aphorisms and discourses of the bodies celestial, their nature and influence* (London, 1686)

Goldie, M., ed., *The entring book of Roger Morrice*, 7 vols. (Woodbridge, 2007)

Goodwin, John, *Anti-Cavalierisme, or, truth pleading as well the necessity as the lawfulness of this present war* (London, 1642)

Gordon of Auchleuchries, Patrick, *Diary of General Patrick Gordon of Auchleuchries 1635–99*, ed. D. Fedosov, 6 vols. (Aberdeen, 2009–16)

Gough, R., *The history of Myddle* (written 1700–2), ed. D. Hey (Harmondsworth, 1981)

Gracián, Baltasar, S.J., *El criticón*, 3 vols. (Zaragoza, 1651, 1653, 1657)

Graniti, A., ed., *Diario di Francesco Capecelatro contenente la storia delle cose avvenute nel reame di Napoli negli anni 1647–50*, 3 vols. (Naples, 1850–4)

Gray, Robert, *A good speed to Virginia* (London, 1609)

Green, M. A. E., ed., *Letters of Queen Henrietta Maria including her private correspondence with Charles the First* (London, 1857)

Grey, Anchitell, *Debates of the House of Commons, from the year 1667 to the year 1694*, 10 vols. (London, 1763)

Grimmelshausen, Hans Jacob Christoph von, *Der abentheuerliche Simplicissimus Teutsch* (1669; Tübingen, 1967)

Groen van Prinsterer, Guillaume, ed., *Archives ou correspondance de la maison d'Orange-Nassau*, 2nd series, 5 vols. (Utrecht, 1857–61)

Guha, J. P., ed., *India in the seventeenth century*, 2 vols. (New Delhi, 1984)

Haig, J., ed., *The historical works of Sir James Balfour of Denmylne and Kinnaird*, 3 vols. (Edinburgh, 1825)

Hair, P. E. H., A. Jones and R. Law, eds., *Barbot on Guinea: the writings of Jean Barbot on West Africa, 1678–1712*, 2 vols. (London, 1992: Hakluyt Society, 2nd series, CLXXV–CLXXVI)

Hall, J. C., ed., *Feudal laws of Japan* (Washington, DC, 1979)

Halley, Edmond, *A synopsis of the astronomy of comets* (Oxford, 1705)

Halliwell, J. O., ed., *Letters of the kings of England*, 2 vols. (London, 1846)

Hannover, Nathan, *Yaven Metzulah* (Venice, 1653: trans. A. J. Mesch as *Abyss of despair*, New York, 1950)

Hardwicke, Philip Yorke, earl of, ed., *Miscellaneous state papers from 1501 to 1726*, 2 vols. (London, 1778)

Hartlib, Samuel, *Considerations tending to the happy accomplishment of England's Reformation in church and state. Humbly presented to the piety and wisdome of the High and Honourable Court of Parliament* (London, 1647)

— *Samuel Hartlib his legacie; or, an enlargement of the discourse of husbandry used in Brabant and Flaunders* (London, 1650, incorporating the 'Discourse' compiled by Sir Richard Weston)

Hasan Beyzade Tarihi, ed. N. Aykut (Istanbul, 1980)

Heberle, Hans: *see* Zillhardt

Helfferich, T., *The Thirty Years War: a documentary history* (Indianapolis, 2009)

Hickson, Mary, *Ireland in the seventeenth century, or the Irish massacres of 1641–2, their causes and results*, 2 vols. (London, 1884)

Historical Manuscript Commission, *Third report* (London, 1872)

Hobbes, Thomas, *On the citizen* (*De cive*, 1641; English edn, ed. and trans. R. Tuck and M. Silverthorne, Cambridge, 1998)

— *Leviathan, or the matter, forme, and power of a common-wealth, ecclesiasticall and civill* (London, 1651; ed. R. Tuck, Cambridge, 1996)

— *Behemoth or The Long Parliament* (written 1668, first published 1679; 2nd edn, ed. F. Tönnies, London, 1969)

Howell, James, *A discourse discovering some mysteries of our new state . . . shewing the rise and progresse of England's unhappinesse, ab anno illo infortunato 1641* (Oxford, 1645)

— *Epistolae ho-elianae or familiar letters* (London, 1650; ed. J. Jacobs, London, 1890)

— *An exact historie of the late revolutions in Naples and of their monstruous successes not to be parallel'd by any antient or modern history* (2nd edn, London, 1664: an English translation of A. Giraffi, *Le revolutioni di Napoli*)

Howell, Thomas B., *Cobbett's complete collection of state trials*, 33 vols. (London, 1809–26)

Huang Liu-hung, *A complete book concerning happiness and benevolence: Fu-hui ch'üan-shu, a manual for local magistrates in seventeenth-century China* (1699; trans. and ed. Djang Chu, Tucson, 1984)

Hull, C. H., ed., *The economic writings of Sir William Petty*, 2 vols. (Cambridge, 1899)

Hutchinson, Lucy, *Memoirs of the life of Colonel Hutchinson, governor of Nottingham* (written c. 1665; 10th edn, London, 1863)

Hyde, Edward: *see* Clarendon

Israel, Menassah ben, *Esperança de Israel* (Amsterdam, 1650)

Jacobs, *Epistolae ho-elianae: see* Howell

Jaeglé, E., *Correspondance de Madame, duchesse d'Orléans*, 2 vols. (Paris, 1880)

Jahangir, *Memoirs: see* Thackston

James I, king of Great Britain, *His Maiesties speech to both the houses of Parliament, in his Highnesse great chamber at Whitehall* (London, 1607)

James I: *see also* McIlwain

Jansen, Cornelius, *Augustinus, seu doctrina S. Augustini de humanae naturae sanitate, aegritu-dine, medicina, adversus Pelagianos et Massilienses*, 3 vols. (Leuven, 1640; 2nd edn, Paris, 1641)

Jansson, M. and W. B. Bidwell, eds., *Proceedings in Parliament, 1625* (New Haven, 1987)

Johnson, R. C., M. F. Keeler, M. Jansson and W. B. Bidwell, eds., *Commons debates 1628*, 3 vols. (New Haven and London, 1977)

Johnson, Samuel, *The works of Samuel Johnson. X. Political writings*, ed. D. J. Greene (New Haven and London, 1977)

Jonson, Ben, *Epigrams* (London, 1612)

Josselin, Ralph, *see* Macfarlane

Journals of the House of Lords, 39 vols. (London, 1767–1830)

Kâtib Çelebi, *Fezleke-i Tarih*, 2 vols. (Istanbul, 1870)

Kenyon, J. P., ed., *The Stuart Constitution, 1603–1688: documents and commentary* (Cambridge, 1966)

Kepler, Johannes, *Prognosticum astrologicum auff das Jahr . . . 1618* (Linz, 1618; reprinted in V. Bialas and H. Grüssing, eds., *Johannes Kepler Gesammelte Werke*, XI part 2, Munich, 1993)

Kingsbury, S. M., *The records of the Virginia Company of London*, 4 vols. (Washington, DC, 1906–35)

Knowler, W., ed., *The earl of Strafforde's letters and dispatches*, 2 vols. (London, 1739)

Kodama Kōta and Ōishi Shinzaburo, *Kinsei nōsei shiryōshû. I. Edo bakufu horei* (Tokyo, 1966)

Kolff, D. H. A., and H. W. van Santen, eds., *De geschriften van Francisco Pelsaert over Mughal Indië, 1627. Kroniek en Remonstrantie* (The Hague, 1979: Werken uitgegeven door de Linschoten Vereeniging, LXXXI)

Kuroita Katsumi, ed., *Shintei zôho kokushi taikei*, XL (Tokyo, 1964)

La Noue, François, *Discours politiques et militaires* (Basel, 1587)

Labat, Jean-Baptiste, *Nouvelle relation de l'Afrique occidentale: contenant une description exacte du Sénégal et des païs situés entre le Cap-Blanc et la rivière de Serrelionne*, 5 vols. (Paris, 1728)

Laing, D., ed, *The letters and journals of Robert Baillie, 1637–62*, 3 vols. (Edinburgh, 1841)

— *Correspondence of Sir Robert Kerr, first earl of Ancram, and his son William, third earl of Lothian*, 2 vols. (Edinburgh, 1875: Roxburghe Club, C)

Lancellotti, Secondo, *L'hoggidì, overo il mondo non peggiore ne più calamitoso del passato* (Venice, 1623)

Larner, C. J., C. J. H. Lee and H. V. McLachlan, *A source book of Scottish witchcraft* (Glasgow, 1977)

Larsen, C., ed., *Eske Brock medt egen handt: Eske Broks dagbøger 1604–1622* (Copenhagen, 2005)

Las Cortes, Adriano de, *Le voyage en Chine d'Adriano de las Cortes, S. J. (1625)*, ed. M. Girard (Paris, 2001)

Laud, William: *see* Scott and Bliss, *and* Wharton

Lawrence, Richard, *The interest of England in the Irish transplantation stated* (London, 1655)

Le Boindre, J., *Débats du Parlement de Paris pendant la minorité de Louis XIV*, ed. R. Descimon and O. Ranum (Paris, 1997)

Le Comte, Louis, *Nouveaux mémoires sur l'état présent de la Chine*, 2 vols. (Paris, 1691–7; English trans., London, 1698)

Lee, M., ed., *Dudley Carleton to John Chamberlain, 1603–1624. Jacobean letters* (New Brunswick, 1972)

Leslie, John, earl of Rothes: *see* Rothes

Locke, John, *A letter concerning toleration* (1689, ed. J. H. Tully, Indianapolis, 1983)

— *Political essays*, ed. M. Goldie (Cambridge, 1997)

Loofeldt, Peter, 'Initiarum Monarchiae Ruthenicae', RAS *Manuskriptsamlingen*, 68

Loubère, S. de la, *A new historical relation of the kingdom of Siam* (Paris, 1693)

Loureiro de Souza, A., ed., *Documentos históricos do Arquivo Municipal. Cartas do Senado 1638–1673* (Salvador, 1951)

Lu, J. D., *Sources of Japanese history*, I (New York, 1974)

Macfarlane, A., ed., *The diary of Ralph Josselin, 1616–1683* (London, 1976)

McIlwain, C. H., ed., *The political works of James I* (Cambridge, MA, 1918)

Magisa, Raymundo, *Svcceso raro de tres volcanes, dos de fuego y uno de agua, que reventaron a 4 de enero de este año de 1641 a un mismo tiempo en diferentes partes de islas Filipinas* (Manila, 1641)

Manucci, Niccolo, *Storia do Mogor, or Mogul India, 1653–1708*, ed. W. Irving, 4 vols. (London, 1906)

Manuel de Melo, Francisco, *Historia de los movimientos, separación y guerra de Cataluña* (Lisbon, 1645; Madrid, 1912 edn.)

Markham, Clement ['Thodorus Verax'], *Anarchia anglicana, or the history of independency, the second part* (London, 1649)

Martí i Viladamor, Francesc, *Noticia Universal de Cataluña* (Barcelona, 1640)

Mason, John, *A brief history of the Pequot War: especially of the memorable taking of their fort at Mistick in Connecticut in 1637* (Boston, 1736)

Massieu, Guillaume, *Caffaeum* (c. 1700, reprinted in Latin with French translation in *Etrennes à tous les amateurs de café, pour tous les temps, ou manuel de l'amateur de café*, I, Paris, 1790, 81–109)

Mather, Increase, *Heaven's alarm to the world, or, a sermon wherein is shewed that fearful sights and signs in heaven are the presages of great calamities at hand* (Boston, 1681)

— *Kometographia, or a discourse concerning comets, wherein the nature of blazing stars is enquired into, with an historical account of all comets which have appeared since the beginning of the world* (Boston, 1683)

Meadows, Philip, *A narrative of the principal actions occurring in the wars betwixt Sueden and Denmark before and after the Roschild treaty* (London, 1677)

Medick, H., and B. Marschke, eds., *Experiencing the Thirty Years War. A brief history with documents* (Boston, 2013)

Mélanges de notices diverses sur les frères des écoles chrétiennes (Paris, 1818)

Mentet de Salmonet, Robert [Robert Menteith], *Histoire des troubles de la Grande Bretagne* (Paris, 1649)

Meyer, A. O., 'Ein italienisches Urteil über Deutschland und Frankreich um 1600', *Quellen und Forschungen aus italienischen Archiven und Bibliotheken*, IX (1906), 155–69

Milton, John, *Paradise lost* (2nd edn, revised and augmented, 1674; ed. B. K. Lewalski, Oxford, 2007)

— *Complete prose works of John Milton*, VII (revised edn, New Haven, 1980)

— *Complete prose works of John Milton*, VIII (New York, 1933)

Moderate Intelligencer, ed. John Dillingham (a weekly newspaper, London, 1645–49)

Molesworth, Robert, *An account of Denmark as it was in the year 1692* (London, 1694)

Monconys, Balthasar de, *Journal des voyages de monsieur de Monconys*, 3 parts (Lyon, 1666)

Monkhouse, Thomas, ed., *State papers collected by Edward, earl of Clarendon*, 3 vols. (Oxford, 1767–86)

Montpensier, Mlle de, *Mémoires*, 2 vols. (Paris, 1728)

Morales Padrón, F., *Memorias de Sevilla (1600–78)* (Córdoba, 1981)

Morrice, Roger: *see* Goldie

Morton, Nathaniel, *The New-England's memoriall: or, a briefe relation of the most memorable and remarkable passages of the providence of God, manifested to the planters of New-England in America* (1669; Plymouth, 1826)

Morton, Thomas, *New English Canaan, or New Canaan containing an abstract of New England* (London, 1637)

Motteville, Mme de, *Mémoires*, 2 vols. (Paris, 1904)

Mundy, Peter: *see* Temple, R. C.

Mut, Vicente, *El príncipe en la guerra, y en la paz, copiado de la vida del emperador Iustiniano* (Madrid, 1640)

Naudé, Gabriel, *Considérations politiques sur les coups d'état* (Rome 1639; English edn: *Political considerations upon refin'd politicks and the master-strokes of state*, London, 1711)

Naworth, George [Sir George Wharton], *A new almanacke and prognostication for the yeere of our Lord and Saviour Iesus Christ, 1642* (London, 1642)

Nedham, Marchamont, *Medela medicinae. A plea for the free profession and a renovation of the art of physic . . . tending to the rescue of mankind from the tyranny of diseases* (London, 1665)

Newton, Isaac, *Philosophiae naturalis principia mathematica* (London, 1687)

Nieuhof (or Nieuhoff), Johan, *Het gezantschap der Nederlandtsche Oost-Indische Compagnie aan de tartarischen cham* (1655–57; Amsterdam, 1665; English translation, *An embassy from the East-India Company of the United Provinces, to the Grand Tartar Cham, emperor of China*, London, 1673); *see also* Blussé and Falkenburg

North, Sir Dudley, *Discourses upon trade, principally directed to the cases of the interest, coynage, clipping, increase of money* (London, 1691)

Numarali Mühimme Defteri (H. 1040/1630–1631) (Ankara, 2001)

O'Mahony, Conor, *Disputatio apologetica de iure regni Hiberniae pro Catholicis Hibernis adversus haereticos Anglos* (Lisbon, 1645; reprinted Dublin, c. 1828)

Ochoa, E. de, ed., *Epistolario español: cartas de personajes varios*, 2 vols (Madrid, 1870)

Odorico, P., ed., *Conseils et mémoires de Synadinos, prêtre de Serrès en Macédonie (17e siècle)* (Paris, 1996)

Ono Mizuo, ed., *Enomoto Yazaemon Oboegaki: kinsei shoki shōnin no kiroku* (Tokyo, 2001)

Pacheco de Britto, Mendo, *Discurso em os dous phaenominos aereos do anno de 1618* (Lisbon, 1619)

Palermo, F., *Narrazioni e documenti sulla storia del Regno di Napoli dall'anno 1522 al 1667* (Florence, 1846: *Archivio storico italiano*, IX)

Parets, M., *De los muchos sucesos dignos de memoria que han ocurrido en Barcelona y otros lugares de Cataluña: crónica escrita por Miguel Parets entre los años 1626 a 1660*, 6 vols. (Madrid, 1851–1948: *Memorial Histórico Español*, XX–XXV)

Parival, J.-N. de, *Abrégé de l'histoire de ce siècle de fer, contenant les misères et calamitez des derniers temps, avec leurs causes et pretextes* (1653; 2nd edn, Brussels, 1655)

Parker, Henry, *The manifold miseries of civil warre and discord in a kingdome: by the examples of Germany, France, Ireland, and other places* (London, 1642)

[Parker, Henry], *The king's cabinet opened: or, certain packets of secret letters & papers written in the king's own hand and taken in his cabinet at Nasby-Field, June 14, 1645* (London, 1645)

Pascal, Blaise, *Les Pensées sur la religion et sur quelques autres sujets* (Paris, 1670; English edn trans. W. F. Trotter)

Pasqual de Panno, Francisco, *Motines de Cataluña*, ed. I. Juncosa and J. Vidal (Barcelona, 1993)

Patin, Charles, *Relations historiques et curieuses de voyages, en Allemagne, Angleterre, Hollande, Bohême, Suisse, &c.*, 2nd edn (Lyon, 1676)

Patin, Guy, *Lettres*, ed. J. H. Reveillé-Parise, 3 vols. (Paris, 1846)

Paul, G. M., ed., *The diary of Archibald Johnston Lord Wariston 1639* (Edinburgh, 1896: Scottish Historical Society, XXVI)

— *Diary of Sir Archibald Johnston of Wariston. I: 1632–7* (Edinburgh, 1911: Scottish Historical Society, LXI): *see also* Fleming

Peçevi Tarihi, ed. B. S. Baykal, 2 vols. (Ankara, 1982)

Pelsaert, Francisco: *see* Kolff and van Santen

Pepys, Samuel, *The diary of Samuel Pepys*, ed. R. C. Latham and W. Matthews, 11 vols. (London, 2000)

Percy, George, 'A trewe relacyon of the procedeings and ocurrentes of momente which have hapned in Virginia', *Tyler's Quarterly Historical and Genealogical Magazine*, III (1922), 259–82

Petty, Sir William: *see* Hull, C. H.

Pérez de Ribas, Antonio, *History of the triumphs of our holy faith among the most barbarous and fierce peoples of the new world* (1645: ed. D. T. Reff, Tucson, 1999)

Peters, J., *Ein Söldnerleben im dreissigjährigen Krieg. Eine Quelle zur Sozialgeschichte* (Berlin, 1993)

Petitot, A. and M. Monmerque, eds., *Mémoires du duc de Guise*, 2 vols. (Paris, 1826: *Collection des mémoires relatifs à l'histoire de France*, LV–LVI)

Pocili, A. [pseudonym for Placido Reina], *Delle rivoluzioni della città di Palermo avvenute l'anno 1648* [sic] (Verona, 1649)

Poole, E., *A vision wherein is manifested the disease and cure of the kingdom* (London, 1648)

— *Another alarum for war* (London, 1649)

Prynne, William, *The soveraigne power of parliaments and kingdomes: divided into foure parts* (London, 1643)

Qing shilu [Veritable records of the Ming], VIII (Beijing, 1985)

Quevedo, Francisco de, *La fortuna con seso y la hora de todos, fantasia moral* (1632) in *Obras de Francisco de Quevedo Villegas*, 3 vols. (Antwerp, 1699)

— *La rebelión de Barcelona* (1641), in A. Fernández-Guerra y Orbe, ed., *Obras de Don Francisco de Quevedo y Villegas*, I (Madrid, 1876: Biblioteca de Autores Españoles, XXIII)

Radziwiłł, Albrycht Stanisław, *Memoriale rerum gestarum in Polonia, 1632–1656*, ed. A. Przyboś and R. Żelewski, 4 vols. (Wrocław, 1968–74)

Ras, J. J., ed., *Hikajat Bandjar: a study in Malay historiography* (The Hague, 1968)

Ray, John, *Observations topographical, moral, and physiological, made in a journey through part of the Low-countries, Germany, Italy, and France* (London, 1673)

Register of the Privy Council of Scotland, 2nd series, ed. David Masson and P. Hume Brown, 6 vols. (Edinburgh, 1899–1908)

Reinach, J., ed., *Recueil des instructions données aux ambassadeurs et ministres de France. X. Naples et Parme* (Paris, 1913)

Repgen, K., et al., ed., *Acta Pacis Westphalicae*, Series I. *Instruktionen. Frankreich, Schweden, Kaiser*, I (Münster, 1962)

— *Acta Pacis Westphalicae*, Series II, Part B. *Die französischen Korrespondenzen*, 7 vols. to date (Münster, 1979–2011)

Riccioli, Giovanni Battista, *Almagestum novum, astronomiam veterem novamque* (Bologna, 1651)

Richelieu, Armand-Jean Duplessis, cardinal-duke of, *Testament politique*, ed. L. André (Paris, 1947)

Ritchie, C. I. A., 'Deux textes sur le Sénégal (1673–1677), *Bulletin de l'Institut fondamental de l'Afrique noire*, XXX, série B (1968), 289–353

Roberts, M., *Sweden as a great power 1611–1697: government, society, foreign policy* (London, 1968)

Rolamb, Nicholas, 'A relation of a journey to Constantinople', in Churchill, *A collection of voyages and travels*, V, 669–716

Romaniello, M. P., 'Moscow's lost petition to the tsar, 2 June 1648', *Russian History*, XLI (2014), 119–25

Rothes, John Leslie, earl of, *A relation of proceedings concerning the affairs of the Kirk of Scotland, from August 1637 to July 1638*, ed. J. Nairne (Edinburgh, 1830: Bannatyne Club)

Rushworth, John, *Historical collections or private passages of state, weighty matters in law, remarkable proceedings in five parliaments, beginning the sixteenth year of King James, anno 1618*, 6 vols. (London, 1659–1722)

Rutherford, Samuel, *Lex, rex: the law and the prince. A dispute for the just prerogative of king and people* (London, 1644)

Rutt, J. T., ed., *Diary of Thomas Burton, esq., member in the Parliaments of Oliver and Richard Cromwell from 1656 to 1659*, 4 vols. (London, 1828)

Rycaut, Paul, *The present state of the Ottoman empire* (London, 1668: facsimile edn, New York, 1971)

Sandys, George, *A relation of a iourney begun an. dom. 1610* (London, 1615)

Sarkar, J., ed., *Anecdotes of Aurangzeb* (English trans. of Ahkam-i Alamgiri, ascribed to Hamid-ud-din Khan Bahadur (3rd edn, Calcutta, 1949)

Scott, W. and J. Bliss, eds., *The works of William Laud*, 7 vols. (Oxford, 1847–60)

Scriba, C. J., 'The Autobiography of John Wallis, F.R.S.', *Notes and records of the Royal Society of London*, XXV (1970), 17–46

Seco Serrano, C., ed., *Cartas de Sor María de Jesús de Ágreda*, 2 vols. (Madrid, 1958)

Semedo, Alvaro, S.J., *Historica relatione del gran regno della Cina* (Rome, 1653)

Sévigné, marquise de: *see* Silvestre de Sacy

Shakespeare, William, *The tragical history of Hamlet, prince of Denmark*, (c. 1600; London, 1623)

Sibbald, Sir Robert, *Provision for the poor in time of dearth and scarcity* (Edinburgh, 1699)

Silvestre de Sacy, S. U., ed., *Lettres de Marie de Rabutin-Chantal, marquise de Sévigné, à sa fille et à ses amis*, 11 vols. (Paris, 1861)

Simon i Tarrés, A., ed., *Cròniques de la Guerra dels Segadors* (Barcelona, 2003)

Sinclair, John, *The statistical account of Scotland, drawn up from the communications of the different parishes*, 21 vols. (Edinburgh 1791–99)

Skippon, Philip, *An account of a journey through the Low Countries, Germany and France* (1663), in Churchill, *A collection of voyages and travels*, VI, 359–736

Song Yingxing, *Tiangong kaiwu* (1637, translated as *Chinese technology in the seventeenth century*, New York, 1966)

Song Zhenghai, *Zhongguo Gudai Zhong de Ziran Zaihai he Yichang Nianbiao* (Guangzhou, 1992)

Sorbière, Samuel, *A voyage to England: containing many things relating to the state of learning and religion, and other curiosities of that kingdom* (1663; London, 1709)

Spinoza, Baruch, *Tractatus theologico-politicus* (1670; trans. S. Shirley, Leiden, 1991)

— *The ethics: treatise on the emendation of the intellect and selected letters* (1677; trans. S. Shirley, ed. S. Feldman, 2nd edn, Indianapolis, 1992)

Sprat, Thomas, *The history of the Royal-Society of London for the improving of natural knowledge* (London, 1667)

Stephens, F. G., and M. D. George, *Catalogue of prints and drawings in the British Museum. Division I. Political and personal satires*, 11 vols. (London, 1870–1954)

Struve, L. A., ed., *Voices from the Ming-Qing cataclysm: China in tigers' jaws* (New Haven, 1993)

Struve, L. A., *The Ming-Qing conflict, 1619–1683. A historiography and source guide* (Ann Arbor, 1998)

Sym, John, *Life's preservative against self-killing* (London, 1637)

Symmons, Edward, *A vindication of King Charles or, a loyal subject's duty manifested in vindicating his soveraigne from those aspersions cast upon him by certaine persons, in a scandalous libel, entitled, The kings cabinet opened* (London, 1647)

Tarrabotti, Elena, *L'inferno monacale*, ed. F. Medioli (Turin, 1989); *see also* Baratotti

Tavernier, Jean-Baptiste, *Travels in India*, ed. V. Ball and W. Crooke, 2 vols. (Oxford, 1925)

Temple, R. C., ed., *The travels of Peter Mundy in Europe and Asia, 1608–1667*, 5 vols. in 6 parts (London 1907–36: Hakluyt Society, 2nd series, XVII, XXXV, XLV, XLVI, LV, LXXVIII)

Thackston, W. M., ed., *The Jahangirnama: Memoirs of Jahangir, emperor of India* (Oxford, 1999)

Thom, H. B., ed., *Journal of Jan van Riebeeck*, 3 vols. (Cape Town/Amsterdam, 1952–8)

Thwaites, R. G., ed., *The Jesuit relations and allied documents: travels and explorations of the Jesuit missionaries in New France 1610–1791*, 71 vols. (Cleveland, 1898–1901)

Topçular Kâtibi 'Abdulkâdir (Kadri) Efendi Tarihi, ed. Z. Yilmazer (Ankara, 2003)

Toyoda Takeshi, ed., *Aizu-Wakamatsu-shi, II: Kizukareta Aizu Han* (Aizu, 1965)

Turbolo, Giovanni Donato, *Copia di quattro discorsi* (Naples, 1629)

Valladares, R., *Epistolario de Olivares y el conde de Basto (Portugal 1637–1638)* (Badajoz, 1998)

Van Aitzema, Lieuwe, *Saken van staet en Oorlogh, in ende omtrent de Vereenigde Nederlanden*, 6 vols. (The Hague, 1669–72)

Van der Capellen, Adrian, *Gedenkschriften, 1621–54*, 2 vols. (Utrecht, 1777)

Van der Donck, Adriaen, *A description of the New Netherlands* (1653: Madison, 2003: Wisconsin Historical Society Digital Library and Archives)

Van der Heyden, Jan, *A description of fire engines with water hoses and the method of fighting fires now used in Amsterdam* (1690; English edn by L. Multhauf, 1996)

Vauban, Sébastien Le Prestre, marquis de, *Les oisivités de monsieur de Vauban*, ed. M. Virol (Paris, 2007)

Vázquez de Espinosa, Antonio, *Compendio y descripción de las Indias Occidentales*, ed. C. U. Clark (Washington, DC, 1948: Smithsonian Miscellaneous Collections, CVIII)

Vernadsky, George, ed., *A source book for Russian history from early times to 1917* (New Haven, 1972)

Viallé, C. and L. Blussé, *The Deshima registers. XI: 1641–1650* (Leiden, 2001: Intercontinenta, XXIII)

— *The Deshima registers. XII: 1650–1660* (Leiden, 2005: Intercontinenta, XXV)

— *The Deshima registers. XIII: 1660–1670* (Leiden, 2010: Intercontinenta, XXVII)

[Voetius, Gisbertius; 'G. L. V.'], *Brittish lightning, or, suddaine tumults in England, Scotland and Ireland to warne the United Provinces to understand the dangers and the causes thereof* (Amsterdam, 1643)

Voltaire, F. M. A. de, *Essai sur les mœurs et l'esprit des nations et sur les principaux faits de l'histoire depuis Charlemagne jusqu'à Louis XIII* (1741–42; first published 1756; Paris, 1963)

Von Sandrart, Joachim, *Der Teutschen Academie, Zweyter und letzter Haupt-Teil, von der edlen Bau-Bild und Mahlerey-Künsten*, 3 vols. (Nuremberg, 1679)

Wallington, Nehemiah: *see* Seaver *and* Webb

Wallis, John, *A defence of the Royal Society, and the Philosophical Transactions* (London, 1678)

— 'Autobiography': *see* Scriba

Walton, Izaak, *The compleat angler, or the contemplative man's recreation* (1653; 2nd edn, London, 1655)

Wariston, Archibald Johnston of: *see* Paul *and* Fleming

Wharton, Henry, ed., *The history of the troubles and tryal of the Most Reverend Father in God and blessed martyr, William Laud, lord arch-bishop of Canterbury, wrote by himself during his imprisonment in the Tower; to which is prefixed the diary of his own life*, 2 vols. (London, 1695–1700)

Whitaker, Jeremiah, *Ejrenopojos. Christ the settlement of vnsettled times, in a sermon preached before the honourable House of Commons at their publicke fast in Margarets Church at Westminster the 25 day of January* (London, 1642)

Whitelocke, Bulstrode, *Memorials of the English affairs* (1682; reprinted in 4 vols., Oxford, 1853)

Wildman, John, *Truths triumph, or treachery anatomized* (London, 1648)

Williams, E. N., *The eighteenth-century constitution, 1688–1815: documents and commentary* (Cambridge, 1960)

Wilson, P. H., *The Thirty Years War: a sourcebook* (Basingstoke, 2010)

Winstanley, Jerrard, *The law of freedom in a platform, or true magistracy restored* (London, 1652)

Winthrop papers, 3 vols. (Boston, 1929–43)

Winthrop, John: *see* Dunn

Zhongguo Jin-wubai-nian Hanlao Fenbu Tu-ji (Beijing, 1981)

Zillhardt, G., ed., *Der dreissigjährige Krieg in zeitgenössischer Darstellung. Hans Heberles 'Zeytregister' (1618–1672)* (Ulm, 1975)

B. Secondary sources cited in the notes

A failure of initiative. Final report of the Select Bipartisan Committee to investigate the preparation for and response to Hurricane Katrina (Washington, DC, 2006)

A stronger, more resilient New York (New York, 2013)

Adamson, J. S. A., 'England without Cromwell: what if Charles I had avoided the Civil War?', in N. Ferguson, ed., *Virtual history: alternatives and counterfactuals* (London, 1997), 91–123

— *The noble revolt: the overthrow of Charles I* (London, 2007)

Adshead, S. A. M., 'The XVIIth century General Crisis in China', *France-Asie/Asia*, XXIV/3–4 (1970), 251–65

Aho, J. A., *Religious mythology and the art of war: comparative religious symbolisms of military violence* (Westport, CT, 1981)

Alcalá-Zamora y Queipo de Llano, J., 'Razón de estado y geoestrategia en la política italiana de Carlos II: Florencia y los presidios, 1677–81', *Boletín de la Real Academia de Historia*, CLXXIII (1976), 297–358

Alfani, G., 'Plague in seventeenth-century Europe and the decline of Italy: an epidemiological hypothesis' (IGIER Working Paper n. 377, accessed Feb. 2011)

Allemeyer, M. L., '"Dass es wohl recht ein Feuer vom Herrn zu nennen gewesen …", Zur Wahrnehmung, Deutung und Verarbeitung von Stadtbränden in norddeutschen Schriften des 17. Jahrhunderts', in Jakubowski-Tiessen and Lehmann, *Um Himmels Willen*, 201–34

Allen, R. C., et al., 'Wages, prices, and living standards in China, 1738–1925: in comparison with Europe, Japan, and India', *EcHR*, 2nd series, LXIV/S1 (2011), 8–38

Álvarez de Miranda, P., 'Las controversias sobre los cometas de 1680 y 1682 en España', *Dieciocho; Hispanic Enlightenment*, XX, Extra; Anéjo 1 (1997), 21–52

Álvarez de Toledo, C., 'Crisis, reforma y rebelión en el mundo hispánico: el caso Escalona', in G. Parker, ed., *La crisis de la Monarquía de Felipe IV* (Barcelona, 2006), 255–86

Anderson, R. W., et al., 'Jewish persecutions and weather shocks: 1100–1800', *Economic Journal* (2016; online publication date: 1 Sep, 2016).

Andrade, T., *How Taiwan became Chinese: Dutch, Spanish, and Han colonization in the seventeenth century* (New York, 2007)

Andrews, C. M., *The colonial period of American history. I. The settlements* (New Haven, 1934)

Angelozzi, G., and C. Casanova, *La nobiltà disciplinata. Violenze nobiliare, procedure di giustizia e scienza cavalleresca a Bologna nel XVII secolo* (Bologna, 2003)

Ansaldo, M., *Peste, fame, guerra: cronache di vita Valdostana del secolo XVII* (Aosta, 1976)

Appleby, A. B., 'Epidemics and famine in the Little Ice Age', *JIH*, X (1980), 643–63

Atwell, W. S., 'Ming statecraft scholarship and some common themes in the political writings of Ch'en Tzu-lung (1608–47) and Ogyū Sorai (1666–1728)', in Yue-him Tam, ed., *Sino-Japanese cultural interchange: the economic and intellectual aspects. Papers of the International Symposium on Sino-Japanese cultural interchange*, III (Hong Kong, 1985), 61–85

— 'Ming observers of Ming decline: some Chinese views on the "17th-century crisis" in comparative perspective', *Journal of the Royal Asiatic Society of Great Britain and Ireland*, CXX (1988), 316–48

— 'East Asia and the "World Crisis" of the mid-seventeenth century' (lecture at the Mershon Center, Ohio State University, 2001)

— 'Volcanism and short-term climatic change in East Asian and world history, c. 1200–1699', *JWH*, XII (2001), 29–98

— 'Another look at silver imports into China, c. 1635–1644', *JWH*, XVI (2006), 467–89

Avrich, P., *Russian rebels 1600–1800* (London, 1972)

Babayan, K., *Mystics, monarchs and messiahs: cultural landscapes of early modern Iran* (Cambridge, MA, 2002)

Bacon, G., '"The House of Hannover": Gezeirot Tah in modern Jewish historical writing', *Jewish History*, XVII (2003), 179–206

Baer, M. D., 'Death in the hippodrome: sexual politics and legal culture in the reign of Mehmet IV', *P&P*, CCX (2011), 61–91

Bailey, C. D. A., 'Reading between the lines: the representation and containment of disorder in Late Ming and Early Qing legal texts', *Ming Studies*, LIX (2009), 56–86

Baker, J. N. L., 'The climate of England in the seventeenth century', *Quarterly Journal of the Royal Meteorological Society*, LVIII (1932), 421–39

Balabanlilar, L., *Imperial identity in the Mughal empire: memory and dynastic politics in early modern South and Central Asia* (London, 2012)

Bankoff, G., *Cultures of disaster: society and natural hazard in the Philippines* (London, 2003)

Barkey, K., 'Rebellious alliances: the state and peasant unrest in early seventeenth-century France and the Ottoman empire', *American Sinological Review*, LVI (1991), 699–715

Barnes, A. E., '"Playing the part of angels": the Company of the Holy Sacrament and the struggle for stability in early modern France', in Benedict and Gutmann, *Early modern Europe*, 168–96

Beattie, H., *Land and lineage in China: a study of T'ung-ch'eng county, Anhwei, in the Ming and Ch'ing dynasties* (Cambridge, 1979)

Behringer, W., H. Lehmann and C. Pfister, eds., *Kulturelle Konsequenzen der 'Kleinen Eiszeit'* (Göttingen, 2005)

Beik, W., 'The violence of the French crowd from Charivari to Revolution', *P&P*, CXCVII (2007), 75–110

Benedict, P., 'Religion and politics in the European struggle for stability, 1500–1700', in Benedict and Gutmann, *Early modern Europe*, 120–38

Benedict, P. and M. P. Gutmann, eds., *Early modern Europe: from crisis to stability* (Newark, DE, 2005)

Benigno, F., *L'ombra del re. Ministri e lotta politica nella Spagna del Seicento* (Venice, 1992)

— *Specchi della rivoluzione. Conflitto e identità politica nell'Europa moderna* (Rome, 1999; translated with some modifications as *Mirrors of revolution: conflict and political identity in Early Modern Europe*, Turnhout, 2010)

Bennett, M., *The Civil Wars experienced: Britain and Ireland 1638–1661* (London, 2000)

Benton, L., 'After the crisis: the politics of a global pivot', *JWH*, XXVI (2015), 149–55

Bercé, Y.-M., 'Troubles frumentaires et pouvoir contralisateur: l'émeute de Fermo dans les Marches (1648)', *Mélanges d'archéologie et d'histoire de l'École française de Rome*, LXXIII (1961), 471–505, and LXXIV (1962), 759–803

— *Histoire des Croquants. Étude des soulèvements populaires au XVIIe siècle dans le sud-ouest de la France*, 2 vols. (Geneva, 1974)

Bergin, J., and L. W. B. Brockliss, eds., *Richelieu and his age* (Oxford, 1992)

Berry, M. E., *Hideyoshi* (Cambridge, MA, 1982)

— 'Public life in authoritarian Japan', *Daedalus*, CXXVII/3 (1998), 133–65

— *Japan in print: information and nation in the early modern period* (Berkeley, 2006)

Bierther, K., *Der Regensburger Reichstag von 1640/1641* (Kallmünz, 1971)

Biot, E., 'Catalogue des comètes observées en Chine depuis l'an 1230 jusqu'à l'an 1640 de notre ère', *Connaissance des Temps* (Paris, 1843), 44–59

Bireley, R., *Religion and politics in the age of the Counter-Reformation: Emperor Ferdinand II, William Lamormaini, S. J., and the formation of imperial policy* (Chapel Hill, NC, 1981)

Blake, E. S., E. N. Rappaport and C. W. Landsea, *The deadliest, costliest, and most intense United States tropical cyclones from 1851 to 2006 (and other frequently requested hurricane facts)* (Miami, 2007: NOAA Technical Memorandum NWS TPC-5)

Blaug, M., 'The Poor Law Report re-examined', *Journal of Economic History*, XXIV (1964), 229–45

Blome, A., 'Offices of intelligence and expanding social spaces', in Dooley, *The dissemination*, 207–22

Bolt, J., and J. L. van Zanten, 'The Maddison project: collaborative research on historical national accounts', *EcHR*, 2nd series, LXVII (2014), 627–51

Bonney, R. J., *Political change in France under Richelieu and Mazarin, 1624–1661* (Oxford, 1978)

— *The limits to absolutism in ancien régime France* (Aldershot, 1995)

Bonney, R. J., ed., *The rise of the fiscal state in Europe, c. 1200–1815* (Oxford, 1999)

Börekçi, G., 'Factions and favorites at the courts of Sultan Ahmed I (r. 1603–17) and his immediate predecessors' (Ohio State University, PhD thesis, 2010)

Borja Palomo, F., *Historia crítica de las riadas o grandes avenidas del Guadalquivir en Sevilla* (Seville, 2001)

Bosman, M., 'The judicial treatment of suicide in Amsterdam', in J. R. Watt, ed., *From sin to insanity: suicide in early modern Europe* (Ithaca, NY, 2004), 9–23

Bots, H., 'Le rôle des périodiques néérlandais pour la diffusion du livre (1684–1747)', in C. Berkvens-Stevelinck, et al., eds., *Le magasin de l'univers: the Dutch Republic as the centre of the European book trade* (Leiden, 1992), 49–70

Bottigheimer, K. S., *English money and Irish land: the 'adventurers' in the Cromwellian settlement of Ireland* (Oxford, 1971)

Braddick, M. J., *God's fury, England's fire: a new history of the English Civil Wars* (London, 2008)

Braddick, M. J., and J. Walter, eds., *Negotiating power in early modern society: order, hierarchy, and subordination in Britain and Ireland* (Cambridge, 2001)

Braudel, F., *The Mediterranean and the Mediterranean world in the age of Philip II*, 2 vols. (London, 1972–3)

Bray, F., *Technology and gender: fabrics of power in late imperial China* (Berkeley, 1997)

Brecke, P., 'Violent conflicts 1400 AD to the present in different regions of the world' (paper prepared for the 1999 meeting of the Peace Science Society International Conference, Ann Arbor, Michigan)

Bremer, F. J., 'In defence of regicide: John Cotton on the execution of Charles I', *WMQ*, 3rd series, XXXVII (1980), 103–24

Briggs, R., 'Richelieu and reform: rhetoric and political reality', in Bergin and Brockliss, eds., *Richelieu and his age*, 71–97

Brinton, C., *Anatomy of revolution* (1938; 2nd edn, New York, 1965)

Brockliss, L., 'Richelieu, education and the state', in Bergin and Brockliss, eds., *Richelieu and his age*, 237–72

Brod, M., 'Politics and prophecy in seventeenth-century England: the case of Elizabeth Poole', *Albion*, XXXI (1995), 395–413

Brokaw, C. J., *The ledgers of merit and demerit: social change and moral order in late Imperial China* (Princeton, 1991)

— *Commerce in culture: the Sibao book trade 1663–1946* (Cambridge, MA, 2007)

Brook, T., 'Censorship in eighteenth-century China: a view from the book trade', *Canadian Journal of History*, XXII (1988), 177–96

— *The confusions of pleasure: commerce and culture in Ming China* (Berkeley, 1998)

— *Vermeer's hat: the seventeenth century and the dawn of the global world* (New York, 2008)

— *The troubled empire: China in the Yuan and Ming dynasties* (Cambridge, MA, 2010)

Brooke, J. L., *Climate change and the course of global history: a rough journey* (Cambridge, 2014)

Brooks, G. E., *Landlords and strangers: ecology, society and trade in western Africa, 1000–1630* (Boulder, CO, 1993)

Brown, P. B., 'Tsar Aleksei Mikhailovich: Muscovite military command style and legacy to Russian military history', in Lohr and Poe, *The military and society in Russia*, 119–45

Buisman, J., *Duizend jaar weer, wind en water in de Lage Landen. IV. 1575–1675* (Franeker, 2000)

Büntgen, U., and L. Hellmann, 'The Little Ice Age in scientific perspective: cold spells and caveats', *JIH*, XLIV (2014), 353–68

Bussmann, K. and H. Schilling, eds., *1648: War and peace in Europe*, 3 vols. (Münster, 1998)

Calhoun, C., ed., *Habermas and the public sphere* (Boston, 1992)

Cambridge history of China: see Twitchett

Campbell, B. M. S., *The great transition: climate, disease and society in the late medieval world* (Cambridge, 2016)

Camuffo, D., and C. Bertolin, 'The earliest temperature observations in the world: the Medici network (1654–1670)', *CC*, CXI (2012), 335–62

Cane, S., 'Australian aboriginal subsistence in the Western Desert', *Human Ecology*, XV (1987), 391–434

Canny, N., ed., *The origins of empire: British overseas enterprise to the close of the seventeenth century* (Oxford, 1998: *The Oxford history of the British Empire*, I)

Capp, B., *When gossips meet: women, family and neighbourhood in early modern England* (Oxford, 2003)

Carlton, C., *Going to the wars: the experience of the British Civil Wars 1638–1651* (London, 1992)
— *This seat of Mars: war and the British Isles, 1485–1746* (New Haven and London, 2011)

Carmel, S. M., 'Globalization, security, and economic well-being', *Naval War College Review*, LXI (2013), 41–55

Carrier, H., *La presse de la Fronde (1648–53): les Mazarinades* (Geneva, 1989)
— *Le labyrinthe de l'etat. Essai sur le débat politique en France au temps de la Fronde (1648–1653)* (Paris, 2004)

Carroll, S., 'Revenge and reconciliation in early modern Italy', *P&P*, CCXXXIII (2016), 101–42

Castellanos, J. L., ed., *Homenaje a don Antonio Domínguez Ortiz*, 3 vols. (Granada, 2008)

Casway, J., 'Gaelic Maccabeanism: the politics of reconciliation', in Ohlmeyer, *Political thought*, 176–88

Cavaillé, J.-P., 'Masculinité et libertinage dans la figure et les écrits de Christine de Suède', *Les Dossiers du Grihl* (online), 2010–01 | 2010, accessed 4 Mar. 2012. URL: http://dossiersgrihl.revues.org/3965; DOI: 10.4000/ dossiersgrihl.3965

Center for Naval Analysis Corporation, *National security and the threat of climate change* (Washington, DC, 2007)

Cercas, J., *The anatomy of a moment: thirty-five minutes in history and imagination* (New York, 2011)

Chaplin, J. E., 'Ogres and omnivores: early American historians and climate history', *WMQ*, LXXII (2015), 25–32

Charney, J., et al., *Carbon dioxide and climate: a scientific assessment. Report of an ad hoc study group on carbon dioxide and climate, Woods Hole, Massachusetts, July 23–27, 1979, to the Climate Research Board, Assembly of Mathematical and Physical Sciences, National Research Council* (Washington, DC, 1979)

Cherniavsky, M., 'The Old Believers and the new religion', *Slavic Review*, XXV (1966), 1–39

Chéruel, A., *Histoire de France pendant la minorité de Louis XIV*, 2 vols. (Paris, 1879)

Ching, J., 'Chu Shun-Shui, 1600–82: a Chinese Confucian scholar in Tokugawa Japan', *Monumenta Nipponica*, XXX/2 (1975), 177–191
— 'The practical learning of Chu Shun-Shui (1600–82)', in de Bary and Bloom, eds., *Principle and practicality*, 189–229

Clarke, A., 'The "1641 massacres"', in Ó Siochrú and Ohlmeyer, eds., *Ireland, 1641*, 37–41

Clifton, R., *The last popular rebellion: the Western Rising of 1685* (London, 1984)

Como, D. R., 'Secret printing, the crisis of 1640, and the origins of Civil War radicalism', *P&P*, CXCVI (2007), 37–82

Comparato, V. I., *Uffici e società a Napoli (1600–1647). Aspetti dell'ideologia del magistrate nell'età moderna* (Florence, 1974)
— 'Toward the revolt of 1647', in A. Calabria and J. A. Marino, eds., *Good government in Spanish Naples* (New York, 1990), 275–316

— 'Barcelona y Nápoles en la búsqueda de un modelo político: analogías, diferencias, contactos', *Pedralbes: revista d'historia moderna*, XVIII (1998), 439–52

Condren, C., 'Public, private and the idea of the "public sphere" in early-modern England', *Intellectual History Review*, XIX (2009), 15–28

Connor, J., *The Australian frontier wars, 1788–1838* (Sydney, 2005)

Cook, E. R., B. M. Buckley, R. D. D'Arrigo and M. J. Peterson, 'Warm-season temperatures since 1600 BC reconstructed from Tasmanian tree rings and their relationship to large-scale sea surface temperature anomalies', *Climatic Dynamics*, XVI (2000), 79–86

Cook, H. J., *Matters of exchange: commerce, medicine and science in the Dutch Golden Age* (New Haven and London, 2007)

Corteguera, L. R., *For the common good: popular politics in Barcelona 1580–1640* (Ithaca, NY, 2002)

Courtwright, D. T., *Forces of habit: drugs and the making of the modern world* (Cambridge, MA, 2001)

Cressy, D., 'Conflict, consensus and the willingness to wink: the erosion of community in Charles I's England', *Huntington Library Quarterly*, LXI (1999–2000), 131–49

— 'The Protestation protested, 1641 and 1642', *HJ*, XLV (2002), 251–79

— *England on edge: crisis and revolution 1640–1642* (Oxford, 2006)

Crewe, R. D., 'Brave New Spain: an Irishman's independence plot in seventeenth-century Mexico', *P&P*, CCVII (2010), 53–97

Crummey, R. O., 'The origins of the Old Believers' cultural system: the works of Avraamii', *Forschungen zur osteuropäischen Geschichte*, L (1995), 121–38

Cueto, R., *Quimeras y sueños. Los profetas y la monarquía católica de Felipe IV* (Valladolid, 1994)

Cullen, K. J., *Famine in Scotland: the 'Ill Years' of the 1690s* (Edinburgh, 2010)

Curtin, P. D., *Economic change in Precolonial Africa: Senegambia in the era of the slave trade*, 2 vols. (Madison, 1975)

Curtis, M. H., 'The alienated intellectuals of early Stuart England', *P&P*, XXIII (1962), 25–43 (reprinted in Aston, ed., *Crisis in Europe*, 295–316)

Cust, R., *Charles I: a political life* (Harlow, 2005)

Dale, S. F., *The Muslim empires of the Ottomans, Safavids and Mughals* (Cambridge, 2010)

Darcy, E., A. Margey and E. Murphy, *The 1641 depositions and the Irish rebellion* (London, 2012)

Dardess, J. W., *Blood and history in China: the Donglin faction and its repression, 1620–1627* (Honolulu, 2002)

Darling, L. T., 'Ottoman politics through British eyes: Paul Rycaut's *The present state of the Ottoman empire*', *JWH*, I (1994), 71–97

— *Revenue-raising and legitimacy: tax collection and finance administration in the Ottoman empire, 1560–1660* (Leiden, 1996)

David, K., 'Patents and patentees in the Dutch Republic between c. 1580 and 1720', *History and technology* XVI (2000), 263–83

Davies, B. L., *Warfare, state and society on the Black Sea steppe, 1500–1700* (London, 2007)

Davies, S., *Unbridled spirits: women of the English Revolution, 1640–1660* (London, 1998)

De Bary, W. T. and I. Bloom, eds., *Principles and practicality: essays in neo-Confucianism and practical learning* (New York, 1979)

De Jong, M., 'Holland en de Portuguese Restauratie van 1640', *Tijdschrift voor Geschiedenis*, LV (1940), 225–53

De Vries, J., 'Measuring the impact of climate on history: the search for appropriate methodologies', *JIH*, X (1980), reprinted in R. I. Rotberg and T. K. Rabb, eds., *Climate and history: studies in interdisciplinary history* (Princeton, 1981), 19–50

— *The Industrious Revolution: consumer behavior and the household economy, 1650 to the Present* (Cambridge, 2008)

De Waal, A., 'A re-assessment of entitlement theory in the light of recent famines in Africa', *Development and Change*, XXI (1990), 469–90

Dekker, R., *Holland in beroering. Oproeren in de 17e en 18e eeuw* (Baarn, 1982)

— 'Women in revolt: popular revolt and its social bias in Holland in the seventeenth and eighteenth centuries', *Theory and Society*, XVI (1987), 337–62

Delfin, F. G., et al., 'Geological, 14C and historical evidence for a 17th-century eruption of Parker volcano, Mindanao, Philippines', *Journal of the Geological Society of the Philippines*, LII (1997), 25–42

Deng, K., and P. O'Brien, 'Establishing statistical foundations of a chronology for the great divergence: a survey and critique of the primary sources for the construction of relative wage levels for Ming–Qing China', *EcHR*, 2nd series, LXIX (2016), 1057–82

Des Forges, R. V., *Cultural centrality and political change in Chinese history: northeast Henan in the fall of the Ming* (Stanford, 2003)

— 'Toward another Tang or Zhou? Views from the Central Plain in the Shunzhi reign', in Struve, ed., *Time, temporality and change*, 73–112

Descimon, R. and C. Jouhaud, 'La Fronde en mouvement: le développement de la crise politique entre 1648 en 1652', *XVIIe siècle*, CXLV (1984), 305–22

Diamond, J., *Collapse: how societies choose to fail or succeed* (London, 2005)

Dickmann, F., *Der westfälische Frieden* (Münster, 1959)

Dikötter, F., '"Patient Zero": China and the myth of the "Opium Plague"' (inaugural lecture, SOAS, London, 2003)

Dodgshon, R. A., 'The *Little Ice Age* in the Scottish Highlands and Islands: documenting its human impact', *Scottish Geographical Journal*, CXXI (2005), 321–37

Dollinger, H.,'Kurfürst Maximilian I. von Bayern und Justus Lipsius', *Archiv für Kulturgeschichte*, XLVI (1964), 227–308

Domínguez Ortiz, A., *Alteraciones andaluzas* (Madrid, 1973)

Donoghue, J., '"Out of the land of bondage": the English Revolution and the Atlantic origins of abolition', *AHR*, CXV (2010), 943–74

Dooley, B., 'News and doubt in early modern culture: or, are we having a public sphere yet?', in B. Dooley and S. Baron, eds., *The politics of information in early modern Europe* (London, 2001), 275–90

Dooley, B., ed., *The dissemination of news and the emergence of contemporaneity in early modern Europe* (Aldershot, 2010)

Douglass, A. E., 'The secret of the southwest solved by talkative tree rings', *The National Geographic Magazine*, LVI (1929), 736–70

Downs, J. S., 'The curse of Meroz and the English Civil War', *HJ*, LVII (2014) 343–68

Drake, S., *Galileo at work: his scientific biography* (Mineola, NY, 1995)

Drixler, F., *Mabiki: infanticide and population growth in eastern Japan, 1660–1950* (Berkeley, 2013)

Drube, L., et al., 'The NEOTωIST mission (Near-Earth Object Transfer of angular momentum spin test)', *Acta astronautica*, CXXVII (2016), 103–11

Duccini, H.,'Regard sur la littérature pamphlétaire en France au XVIIe siècle', *Revue Historique*, CCLX (1978), 313–39

Dunstan, H., 'The late Ming epidemics: a preliminary survey', *Ch'ing-shih wen- t'i*, III.3 (Nov. 1975), 1–59

Dunthorne, H., 'Resisting monarchy: the Netherlands as Britain's school of revolution in the late sixteenth and seventeenth centuries', in R. Oresko, G. C. Gibbs and H. M. Scott, eds., *Royal and republican sovereignty in early modern Europe: Essays in memory of Ragnhild Hatton* (Cambridge, 1997), 125–48

Eddy, J. A., 'The "Maunder Minimum": sunspots and climate in the reign of Louis XIV', in Parker and Smith, eds., *The General Crisis*, 264–97

Eisenstadt, S. N. and W. Schluter, eds.,'Introduction: paths to early modernities: a comparative view', *Daedalus*, CXXVII.3 (1998), 1–18

Ekman, E., 'The Danish Royal Law of 1665', *JMH*, XXVII (1959), 102–7

Elison, G., *Deus destroyed: the image of Christianity in early modern Japan* (Cambridge, MA, 1973)

Ellersieck, H. E., 'Russia under Aleksei Mikhailovich and Fedor Alekseevich, 1645–1682: the Scandinavian sources' (UCLA PhD thesis, 1955)

Elliott, J. H., *The revolt of the Catalans: a study in the decline of Spain* (Cambridge, 1963)

— 'The year of the three ambassadors', in H. Lloyd-Jones et al., eds., *History and imagination: essays in honour of H. R. Trevor-Roper* (London, 1981), 165–81

—'El programa de Olivares y los movimientos de 1640', in J. M. Jover Zamora, ed., *La España de Felipe IV* (Madrid, 1982: Historia de España Espasa-Calpe, XXV), 335–62

— *The count-duke of Olivares: the statesman in an age of decline* (New Haven and London, 1986)

— 'Revolution and continuity in early modern Europe', in Parker and Smith, *The General Crisis*, 108–27

— 'The General Crisis in retrospect: a debate without end', in Benedict and Gutmann, *Early Modern Europe*, 31–51

— *Empires of the Atlantic world: Britain and Spain in America 1492–1830* (New Haven and London, 2006)

Elliott, M. C., 'Whose empire shall it be? Manchu figurations of historical process in the early seventeenth century', in Struve, ed., *Time, temporality and imperial transition*, 31–72

Eltis, D. and D. Richardson, *Atlas of the transatlantic slave trade* (New Haven and London, 2010)

Elvin, M., 'Market towns and waterways: the county of Shanghai from 1480 to 1910', in G. W. Skinner, ed., *The city in late imperial China* (Stanford, 1977), 441–73

— 'The man who saw dragons: science and styles of thinking in Xie Zhaozhe's Fivefold Miscellany', *Journal of the Oriental Society of Australia*, XXV–XXVI (1993–4), 1–41

Elvin, M., and Ts'ui-Jung Liu, eds, *Sediments of time: environment and society in Chinese history* (Cambridge, 1998)

Ergang, R., *The myth of the all-destructive fury of the Thirty Years War* (Pocono Pines, PA, 1956)

Espino López, A., *Catalunya durante el reinado de Carlos II. Política y guerra en la frontera catalana, 1679–1697* (Bellaterra, 1999)

Ewert, H. C., 'Biological standard of living in a cold century', *Francia*, XXXI/2 (2004), 25–53

Faruqui, M. D., 'Princes and power in the Mughal empire, 1569–1657' (Duke University, PhD thesis, 2002)

Fei Si-yen, *Negotiating urban space: urbanization and Late Ming Nanjing* (Cambridge, MA, 2009)

Fenske, J. and N. Kala, 'Climate and the slave trade', *Journal of Development Economics,* CXII (2015), 19–32

Fildes, V., 'Maternal feelings re-assessed: child abandonment and neglect in London and Westminster, 1550–1800', in Fildes, ed., *Women as mothers in pre-industrial England: essays in memory of Dorothy McLaren* (London, 1990), 139–78

Finkel, C., *Osman's dream: the story of the Ottoman empire* (London, 2005)

Fleming, J. R., *Historical perspectives on climate change* (Oxford, 1998)

Fletcher, A., *The outbreak of the English Civil War* (London, 1981)

Fletcher, J., 'Turco-Mongolian monarchic tradition in the Ottoman empire', *Harvard Ukrainian Studies*, III–IV/1 (1979–80), 236–51

Fleury, M. and L. Henri, *Des registres paroissiaux à l'histoire de la population: manuel de dépouillement et d'exploitation de l'état civil ancien* (Paris, 1956)

Floor, W., *The economy of Safavid Persia* (Wiesbaden, 2000)

Foisil, M., *La révolte des Nu-pieds et les révoltes normandes de 1639* (Paris, 1970)

Foltz, R., 'The Mughal occupation of Balkh, 1646–1647', *Journal of Islamic Studies*, VII (1996), 49–61

Fong, G. S., 'Reclaiming subjectivity in a time of loss: Ye Shaoyuan (1589–1648) and autobiographical writing in the Ming–Qing transition', *Ming Studies*, LIX (2009), 21–41

— 'Writing from experience: personal records of war and disorder in Jiangnan during the Ming–Qing transition', in N. Di Cosmo, ed., *Military culture in Imperial China* (Cambridge MA, 2009), 257–77

Forster, L. W., *The temper of 17th-century German literature* (London, 1952)

Forster, R. and J. P. Greene, eds., *Preconditions of revolution in early modern Europe* (Baltimore, 1970)

Fortey, R. A., 'Blind to the end', *New York Times*, 26 Dec. 2005, online at http://www.nytimes.com/2005/12/26/opinion/26ihtedfortey.html

Franz, G., *Der dreissigjährige Krieg und das deutsche Volk* (1940; 4th edn, Stuttgart, 1979)

Friedrichs, C. R., *The early modern city, 1450–1750* (London, 1995)

Frijhoff, W., 'Surplus ou déficit. Hypothèses sur le nombre réel des étudiants en Allemagne à l'époque moderne (1576–1815)', *Francia*, VII (1979), 173–218

Fu I-ling, *Ming-Ch'ing Nung-Ts'un she-hui ching-chi* (Beijing, 1961)

Gaddis, J. L., *The landscape of memory: how historians map the past* (Oxford, 2002)

Galloway, P., 'Annual variations in deaths by age, deaths by cause, prices and weather in London, 1670 to 1830', *Population Studies*, XXXIX (1985), 487–505

Galloway, P. K., *Choctaw genesis, 1500–1700* (Lincoln, NE, 1995)

Ganeri, J., *The lost age of reason: philosophy in early modern India, 1450–1700* (Oxford, 2011)

Garnier, E., 'Calamitosa tempora, pestis, fames. Climat et santé entre les XVIIe et XIXe siècles', online at http://hal.archives-ouvertes.fr/docs/00/59/51/45/PDF/6-JSE-2009-Garnier-Manuscrit-2009-03-09.pdf

Garrisson, C., 'Les préludes de la Révocation à Montauban (1659–1661)', *Société de l'histoire du Protestantisme français: bulletin historique et littéraire*, XLII (1893), 7–22

Gelabert González, J. E., '*Alteraciones* y Alteraciones (1643–1652)', in Castellanos, *Homenaje*, II, 355–78

— *Castilla convulsa (1631–52)* (Madrid, 2001)

Geneva Association, *The insurance industry and climate change – contribution to the global debate* (Geneva, 2009: Geneva Reports, Risk and Insurance Research, II)

— *An integrated approach to managing extreme events and climate risks: towards a concerted public-private approach* (Zurich, 2016)

Ghosh, A., *The great derangement: climate change and the unthinkable* (Chicago, 2016)

Gladwell, M., *The tipping point: how little things can make a big difference* (New York, 2000)

Gindely, A., *Geschichte des dreissigjährigen Krieges*, 4 vols. (Prague, 1869–80)

Glave, L. M., *Trajinantes. Caminos indígenas en la sociedad colonial, siglos XVI/XVII* (Lima, 1989)

Gleik, P. H., 'Water, drought, climate change, and conflict in Syria', *Weather, Climate, and Society*, VI (2014), 331–40

Glete, J., *War and the state in early modern Europe: Spain, the Dutch Republic and Sweden as fiscal-military states, 1500–1660* (London, 2002)

Goldish, M., *Jewish questions: responsa on Sephardic life in the early modern period* (Princeton, 2008)

Goldstone, J. A., *Revolution and rebellion in the early modern world* (Berkeley and Los Angeles, 1991)

— 'Climate lessons from history', *Historically Speaking*, XIV/5 (2013), 35–7

Gommans, J., *Mughal warfare* (London, 2002)

Gowing, L., 'Secret births and infanticide in seventeenth-century England', *P&P*, CLVI (1997), 87–115

Grandjean, K., 'New world tempests: environment, scarcity and the coming of the Pequot War', *WMQ*, LXVIII (2011), 75–100

Gregory, J. N., *American exodus: the dust bowl migration and Okie culture in California* (Oxford, 1989)

Grehan, J., 'Smoking and "early modern" sociability: the great tobacco debate in the Ottoman Middle East (seventeenth to eighteenth centuries)', *AHR*, CXI (2006), 1,352–77

Groenhuis, G., *De predikanten. De sociale positie van de gereformeerde predikanten in de Republiek der Verenigde Nederlanden voor c. 1700* (Groningen, 1977)

Grove, J. M., *Little Ice Ages, ancient and modern*, 2nd edn, 2 vols. (London, 2004)

Grove, R., 'Revolutionary weather: the climatic and economic crisis of 1788–1795 and the discovery of El Niño', in T. Sharratt, T. Griffiths and L. Robin, eds., *A change in the weather: climate and culture in Australia* (Sydney, 2005), 128–39

Guiard Larrauri, T., *Historia de la noble villa de Bilbao*, II (Bilbao, 1906)

Gutiérrez Nieto, J. I., 'El campesinado', in J. Alcalá-Zamora, ed., *La vida cotidiana en la España de Velázquez* (2nd edn, Madrid, 1999), 43–70

Gutmann, M. P., *War and rural life in the early modern Low Countries* (Princeton, 1980)

Haan, R. L., 'The treatment of England and English affairs in the Dutch pamphlet literature, 1640–1660' (University of Michigan PhD thesis, 1959)

Haboush, JaHyun Kim, 'Constructing the center: the ritual controversy and the search for a new identity in seventeenth-century Korea', in JaHyun Kim Haboush and Martina Deuchler, eds., *Culture and the state in late Chosôn Korea* (Cambridge, MA, 1999), 46–90, 240–9

Hagen, W. W., 'Seventeenth-century crisis in Brandenburg: the Thirty Years' War, the destabilization of serfdom, and the rise of absolutism', *AHR*, XCIV (1989), 302–35

Hagopian, M. N., *The phenomenon of revolution* (New York, 1974)

Hämäläinen, P., *The Comanche empire* (New Haven and London, 2008)

Hamilton. C., *Requiem for a species: why we resist the truth about climate change* (London, 2010)

Hanlon, G., *Human nature in rural Tuscany: an early modern history* (New York 2007)

Harms, R. W., *River of wealth, river of sorrow: the central Zaire basin in the era of the slave and ivory trade, 1500–1891* (New Haven and London, 1981)

Harrell, S., ed., *Chinese historical micro-demography* (Berkeley, 1995)

Haskell, Y. A., *Loyola's bees: ideology and industry in Jesuit Latin didactic poetry* (Oxford, 2003)

Hastrup, K., *Nature and policy in Iceland 1400–1800: an anthropological analysis of history and mentality* (Oxford, 1990)

Haude, S., 'Religion während des dreissigjährigen Krieges (1618–1648)', in G. Litz, H. Munzert and R. Liebenberg, eds., *Frömmigkeit, Theologie, Frömmigkeitstheologie/Contributions to European church history. Festschrift für Berndt Hamm zum 60. Geburtstag* (Leiden, 2005), 537–53

Hautz, J. F., *Geschichte der Universität Heidelberg*, 2 vols. (Mannheim, 1862–4)

Hayami, A., *The historical demography of pre-modern Japan* (Tokyo, 2001)

— *Population, family and society in pre-modern Japan* (Folkestone, 2009)

— *Population and family in early-modern central Japan* (Kyoto, 2010)

Hayami, A., O. Saitō and R. P. Toby, *The economic history of Japan, 1600–1990. I. Emergence of economic society in Japan, 1600–1859* (Oxford, 2004)

Hayami, A. and Y. Tsubochi, eds., *Economic and demographic developments in rice-producing societies: some aspects of East Asian economic history, 1500–1900* (Leuven, 1990)

Heilingsetzer, G., *Der oberösterreichische Bauernkrieg 1626* (Vienna, 1976: Militärhistorische Schriftenreihe, XXXII)

Hellie, R., *Enserfment and military change in Muscovy* (Chicago, 1971)

— 'The costs of Muscovite military defence', in Lohr and Poe, eds., *The military and society*, 41–66

Hemming, J., *Red gold: the conquest of the Brazilian Indians* (London, 1978)

Herman, J. E., 'Empire in the southwest: early Qing reforms to the native chieftain system', *JAS*, LVI (1997), 47–74

Hespanha, A. M., 'La "Restauração" portuguesa en los capítulos de las cartas de Lisboa de 1641', in J. H. Elliott et al., *1640: la monarquía hispánica en crisis* (Barcelona, 1992), 123–68

Hill, C., *The world turned upside down: radical ideas during the English Revolution* (2nd edn, London, 1972)

— 'Robinson Crusoe', *History Workshop Journal*, X (1980), 7–24

Hindle, S., 'Exhortation and enlightenment: negotiating inequality in English rural communities, 1550–1650', in Braddick and Walter, eds., *Negotiating power*, 102–22

— *On the parish? The micro-politics of poor relief in rural England, c. 1550–c. 1750* (Oxford, 2004)

Ho, K.-P., 'Should we die as martyrs to the Ming cause? Scholar-officials' views on martyrdom during the Ming–Qing transition', *Oriens Extremus*, XXXVII (1994), 123–57

Ho, P.-T., *Studies in the population history of China* (revised edn, Cambridge, MA, 1967)

— *The ladder of success in Imperial China: aspects of social mobility, 1368–1911* (2nd edn, New York, 1980)

Hoffman, P. T. and K. Norberg, eds., *Fiscal crises, liberty, and representative government, 1450–1789* (Stanford, 1994)

Hoskins, W. G., 'Harvest fluctuations and English economic history, 1620–1759', *Agricultural History Review*, XVI (1968), 15–31

Howell, D. L., *Capitalism from within: economy, society and the state in a Japanese fishery* (Berkeley, 1995)

Hroch, M. and J. Petráň, *Das 17 Jahrhundert – Krise der Feudalgesellschaft?* (Hamburg, 1981; original Czech edn 1976)

Hrushevsky, M., *History of Ukraine-Rus'*, VIII (1913–22; English translation, Edmonton, 2002)

Hucker, C. O., ed., *Two studies on Ming history* (Ann Arbor, 1971: Michigan papers in Chinese Studies, XII)

Hufton, O., *The prospect before her: a history of women in Western Europe 1500–1800* (New York, 1995)

Hugon, A., *Naples insurgée 1647–1648. De l'événement à la mémoire* (Rennes, 2010)

Hunter, M., *John Aubrey and the realm of learning* (London, 1975)

Hurricane Katrina: a nation still unprepared. Special report of the Committee on Homeland Security and Governmental Affairs, United States Senate (Washington, DC, 2006)

Hütterroth, W.-D., 'Ecology of the Ottoman lands', in S. Faroqhi, ed., *The Cambridge history of Turkey, III: the later Ottoman empire, 1603–1839* (Cambridge, 2006), 18–43

Ibrahim, Nasir Ahmad, *Al-Azmat al-ijtima 'iyya fi misr fi al-qarn al-sabi' 'ashar* (Cairo, 1998)

Ikegami, E., *Bonds of civility: aesthetic networks and the political origins of Japanese culture* (Cambridge, 2005)

Ingersoll, T. N., 'The lamp of experience and the shadow of Oliver: history and politics in 1776' (unpublished manuscript)

Israel, J. I., *The Dutch Republic and the Hispanic world, 1606–1661* (Oxford, 1982)

— *Radical Enlightenment: philosophy and the making of modernity, 1650–1750* (Oxford, 2001)

Iwai Shigeki, 'The collapse of the Ming and the rise of the Qing in the seventeenth-century general crisis' (conference paper at I-House, Tokyo, July 2010)

Jacquart, J., 'La Fronde des princes dans la région parisienne et ses consequences matérielles', *Revue d'histoire moderne et contemporaine*, VII (1960), 257–90

— 'Paris: first metropolis of the early modern period', in P. Clark and B. Lepetit, eds., *Capital cities and their hinterlands in early modern Europe* (Aldershot, 1996), 105–18

Jakubowski-Tiessen, M. and Harmut Lehmann, eds., *Um Himmels Willen. Religion in Katastrophezeiten* (Göttingen, 2003)

Janku, A., '"Heaven-sent disasters" in late imperial China: the scope of the state and beyond', in Mauch and Pfister, eds., *Natural disasters*, 233–64

Jolibert, B., *L'enfance au 17e siècle* (Paris, 1981)

Jomelli, V., et al., 'Fluctuations of glaciers in the tropical Andes over the last millennium and palaeoclimatic implications: a review', *Palaeogeography, Palaeoclimatology, Palaeoecology*, CCLXXXI (2009), 269–282

Jones, R. P., D. Cox and J. Navarro-Rivera, *Believers, sympathizers, and skeptics: why Americans are conflicted about climate change, environmental policy, and science. Findings from the PRRI/AAR religion, values and climate change survey* (Washington, DC, 2014)

Jouanna, A., *Le devoir de révolte. La noblesse française et la gestation de l'état moderne (1559–1661)* (Paris, 1989)

Kagan, R. L., *Students and society in early modern Spain* (Baltimore, 1974)

Kamen, H., 'The decline of Castile: the last crisis', *EcHR*, 2nd series, XVII (1964–65), 63–76

Kaplan, S. L., *The famine plot persuasion in eighteenth-century France* (Philadelphia, 1982: *Transactions of the American Philosophical Society*, LXXII, part 3)

Karr, R. D., '"Why should you be so furious?" The violence of the Pequot War', *Journal of American History*, LXXXV (1998), 876–909

Keblusek, M., 'The business of news: Michel le Blon and the transmission of political information to Sweden in the 1630s', *Scandinavian Journal of History*, XXVIII (2003), 205–13

Keene, D., 'Growth, modernization and control: the transformation of London's landscape, c. 1500–c. 1760', in P. Clark and R Gillespie, eds., *Two capitals. London and Dublin 1500–1840* (Oxford, 2001), 7–37

Keller, R. C., *Fatal isolation: the devastating Paris heatwave of 2003* (Chicago and London, 2015)

Kelly, M., and Ó Gráda, C., 'The waning of the Little Ice Age: climate change in early modern Europe', *JIH*, XLIV (2014), 301–25 and XLV (2014), 57–68

Kessell, J., *Kiva, cross and crown: the Pecos Indians and New Mexico, 1540 to 1840* (Washington, DC, 1979)

Kessler, L. D., 'Chinese scholars and the early Manchu State', *Harvard Journal of Asiatic Studies*, XXXI (1971), 179–200

— *K'ang-hsi and the consolidation of Ch'ing rule 1661–1684* (Chicago, 1976)

Khan, I. A., 'Muskets in the *mawas*: instruments of peasant resistance', in Panikkar, ed., *The making of history*, 81–103

Khodarkovsky, M., 'The Stepan Razin uprising: was it a "peasant war"?', *Jahrbücher für Geschichte Osteuropas*, XLII (1994), 1–19

Kilburn, T. and A. Milton, 'The public context of the trial and execution of Strafford', in J. F. Merritt, ed., *The political world of Thomas Wentworth, earl of Strafford, 1621–1641* (Cambridge, 1996), 230–51

King, D., *Foresight, flood and coastal defence project* (London, 2004), online at http://www.publications.parliament.uk/pa/cm200304/cmselect/cmenvfru/558/4051202.htm

Kishimoto Mio, 'The Kangxi depression and early Qing local markets', *Modern China*, X (1984), 227–56

— *Shindai Chûgoku no buka to keizai hendô* (Tokyo, 1997)

Kishlansky, M. A., 'Charles I: a case of mistaken identity', *P&P*, CLXXXIX (2005), 41–80 and CCV (2009), 175–237

Kivelson, V. A., '"The Devil stole his mind": the tsar and the 1648 Moscow uprising', *AHR*, XCVIII (1993), 733–56

Knachel, P. A., *England and the Fronde: the impact of the English Civil War and revolution in France, 1649–58* (Ithaca, NY, 1967)

Knaut, A. L., *The Pueblo revolt of 1680. Conquest and resistance in seventeenth-century New Mexico* (Norman, OK, 1995)

Ko, D., *Teachers of the inner chambers: women and culture in seventeenth-century China* (Stanford, 1994)

— 'The body as attire: the shifting meanings of footbinding in seventeenth-century China', *Journal of Women's History*, VIII/4 (1997), 8–27

Komlos, J., et al., 'An anthropometric history of early-modern France', *European Review of Economic History*, VII (2003), 159–89

Kornicki, P., *The book in Japan: a cultural history from the beginnings to the nineteenth century* (Honolulu, 2001)

Kötting, H., *Die Ormée (1651–3). Gestaltende Kräfte und Personenverbindungen der Bordelaiser Fronde* (Münster, 1983: Schriftenreihe der Vereinigung zur Erforschung der neueren Geschichte, XIV)

Krüger, K., 'Dänische und schwedische Kriegsfinanzierung im dreissigjährigen Krieg bis 1635', in Repgen, ed., *Krieg und Politik*, 275–98

Kuczynski, J., *Geschichte des Alltags des deutschen Volkes. I: 1600–50* (Berlin, 1981)

Kuhn, P. A., *Soulstealers: the Chinese sorcery scare of 1768* (Cambridge, 1990)

Kunt, I. M., *The Köprülü years, 1656–1661* (Princeton, 1971)

Kupperman, K. O., 'The puzzle of the American climate in the early colonial period', *AHR*, LXXXVIII (1982), 1,262–89

Kupperman, K. O., *The Jamestown project* (Cambridge, MA, 2007)

Labelle, K. M., *Dispersed but not destroyed: a history of the seventeenth-century Wendat people* (Vancouver, 2013)

Lach, D. F. and E. J. van Kley, *Asia in the making of Europe*, 3 vols (Chicago, 1965–93)

Lachiver, M., *Les années de misère: la famine au temps du Grand Roi 1680–1720* (Paris, 1991)

Lamont, W., 'Richard Baxter, "popery" and the origins of the English Civil War', *History*, LXXXVII (2002), 336–52

Lane, R., *Images from the floating world: the Japanese print, including an illustrated dictionary of ukiyo-e* (New York, 1982)

Lappalainen, M., 'Death and disease during the Great Finnish Famine 1695–1697', *Scandinavian Journal of History*, XXXIX (2014), 425–47

Le Cam, J.-L., 'Extirper la barbarie. La reconstruction de l'Allemagne protestante par l'école et l'église au sortir de la Guerre de Trente Ans', in F. Pernot and V. Toureille, eds., *Lendemains de guerre* (Brussels, 2010), 407–14

Le Cam, J.-L., 'Über die undeutlichen instutionellen Grenzen der Elementarbildung. Das Beispiel des Herzogtums Braunschweig-Wolfenbüttel im 17. Jahrhundert', in A. Hanschmidt and H.-U. Musolff, eds., *Elementarbildung und Berufsausbildungs, 1450–1750* (Cologne, 2005: Beiträge zur historischen Bildungsforschung, XXXI), 47–72

Le Roy Ladurie, E., *Times of feast, times of famine: a history of climate since the year 1000* (London, 1973: original French edn, Paris, 1967)

— *Historie humaine et comparée du climat*, 3 vols. (Paris, 2004–9)

— 'Naissance de l'histoire du climat' (paper delivered at the 'Climate and History' conference at the Deutsches Historisches Institut, Paris, 3 Sep. 2011)

Le Roy Ladurie, E., D. Rousseau and A. Vasak, *Les fluctuations du climat: de l'an mil à aujourd'hui* (Paris, 2011)

Lee, J. Z. and F. Wang, *One quarter of humanity: Malthusian mythology and Chinese realities, 1700–2000* (Cambridge, 1999)

Lee, J. Z., F. Wang and C. Campbell, 'Infant and child mortality among the Qing nobility: implications for two types of positive check', *Population Studies*, XLVIII (1994), 395–411

Leffler, P. K., 'From Humanist to Enlightenment historiography: a case study of François Eudes de Mézeray', *French Historical Studies*, X (1978), 416–38

Lenihan, P., 'War and population, 1649–1652', *Irish Economic and Social History*, XXIV (1999), 1–21

— *Confederate Catholics at war, 1641–1649* (Dublin, 2001)

Levin, E., 'Infanticide in pre-Petrine Russia', *Jahrbücher für Geschichte Osteuropas*, XXXIV (1986), 215–24

Levy, J. S., *War in the modern great power system, 1495–1975* (Lexington, MA, 1983)

Li, L., *Fighting famine in north China: state, market and environmental decline, 1690s–1990s* (Stanford, 2007)

Lindegren, J., 'Men, money and means', in P. Contamine, ed., *War and competition between states* (Oxford, 2000), 129–62

Lockhart, P. D., *Denmark in the Thirty Years War 1618–1648* (Cranbury, NJ, 1996)

Loewenson, 'The Moscow rising of 1648', *Slavonic and East European Review*, XXVII (1948–9), 146–56

Lohr, E. and M. Poe, *The military and society in Russia, 1450–1917* (Leiden, 2002)

Lorandi, A. M., *Spanish king of the Incas: the epic life of Pedro Bohorques* (Pittsburgh, 2005)

Lorenzo Cardoso, P. L., *Los conflictos sociales en Castilla (siglos XVI–XVII)* (Madrid, 1996)

Lottin, A., *Vie et mentalité d'un Lillois sous Louis XIV* (Lille, 1968)

Lucas Val, N. de, 'Literatura i historia. Identitats collectives i visions de "l'altre" al segle XVII', *Manuscrits*, XXIV (2006), 167–92

Ludden, D., *Peasant history in south India* (Princeton, 1985)

Macadam, J., 'English weather: the seventeenth-century diary of Ralph Josselin', *JIH*, XLIII (2012), 221–246

McArdle, F., *Altopascio: a study in Tuscan rural society, 1587–1784* (Cambridge, 1978)

Macaulay, C., *The history of England from the accession of James I. to that of the Brunswick line*, 8 vols. (London, 1763–83)

Macaulay, T. B., *History of England since the accession of James the Second*, 5 vols. (New York, 1848)

McChesney, R. D., *Waqf in Central Asia: four hundred years in the history of a Muslim shrine, 1480–1889* (Princeton, 1991)

McCormick, M., et al., 'Climate change during and after the Roman empire: reconstructing the past from scientific and historical evidence', *JIH*, XLIII (2012), 169–220

MacDonald, M., *Mystical bedlam: madness, anxiety and healing in seventeenth-century England* (Cambridge, 1981)

McClain, J. L., J. M. Merriman and Kaoru Ugawa, *Edo and Paris: urban life and the state in the early modern era* (Ithaca, NY, 1994)

McMullen, I. J., 'Kumazawa Banzan and "Jitsugaku": toward pragmatic action', in de Bary and Bloom, eds., *Principle and practicality*, 337–73

McNeill, J. R., *Mosquito empires: ecology and war in the greater Caribbean, 1620–1914* (Cambridge, 2010)

Magen, F., *Reichsgräfliche Politik in Franken. Zur Reichspolitik der Grafen von Hohenlohe zur Vorabend und Beginn des dreissigjährigen Krieges* (Schwäbische Hall, 1975)

Major, J. R., 'The crown and the aristocracy in Renaissance France', *AHR*, LXIX (1964), 631–45

Makey, W., *The Church of the Covenant, 1637–1651* (Edinburgh, 1979)

Mallet, R. and J. W. Mallet, *The earthquake catalogue of the British Association with the discussion, curves and maps etc* (London, 1858)

Mandrou, R., 'Vingt ans après, ou une direction de recherches fécondes: les révoltes populaires en France au XVIIe siècle', *Revue Historique*, CCXLII (1969), 29–40

Manley, G., 'Central England temperatures: monthly means 1659 to 1973', *Quarterly Journal of the Royal Meteorological Society*, C (1974), 389–405

Mann, G., *Wallenstein* (Frankfurt, 1971)

Mantran, R., *Istanbul dans la seconde moitié du XVIIe siècle* (Paris, 1962)

Marcos Martín, A., *España en los siglos XVI, XVII y XVIII. Economía y sociedad* (Barcelona, 2000)

— '¿Fue la fiscalidad regia un factor de crisis en la Castilla del siglo XVII?', in G. Parker, *La crisis de la Monarquía de Felipe IV* (Barcelona, 2006), 173–253

Marks, R. B., *Tigers, rice, silk and silt: environment and economy in late imperial South China* (Cambridge, 1998)

Marmé, M., 'Survival through transformation: how China's Suzhou-centred world economy weathered the general crisis of the seventeenth century', *Social History*, XXII/2 (2007), 144–65

Marques, J. F., *A parenética portuguesa e a Restauração 1640–1668. A revolta e a mentalidade*, 2 vols (Porto, 1989)

Márquez Macías, R., 'Andaluces en América. Recuerdos y añoranzas', *Trocadero*, XXI–XXII (2009–10), 9–20

Martinson, D. G., ed., *Natural climate variability on decade-to-century time scales* (Washington, DC, 1995)

Matthee, R. P., *The politics of trade in Safavid Iran: silk for silver 1600–1730* (Cambridge, 1999)

— *The pursuit of pleasure: drugs and stimulants in Iranian history, 1500–1900* (Princeton, 2005)

— 'Alcohol in the Islamic Middle East: ambivalence and ambiguity', in Withington, ed., *Cultures*, 100–25

Mauch, C. and C. Pfister, eds., *Natural disasters, cultural responses: case studies towards a global environmental history* (Lanham, MD, 2009)

Mauelshagen, F., *Klimageschichte der Neuzeit, 1500–1900* (Darmstadt, 2010)

May, E. R., 'When government writes history: the 9/11 Commission report', *New Republic*, CCXXXII/4 (25 May 2005), 30–5

Mazumdar, S., 'The impact of New World food crops on the diet and economy of China and India, 1600–1900', in R. Grew, ed., *Food in global history* (Boulder, CO, 2000), 58–78

Meinert, C., ed., *Climate and culture: nature, environment and culture in East Asia: the challenge of climate change* (Leiden, 2013)

Melillo, J. M., T. Richmond and G. W. Yohe, eds., *Climate change impacts in the United States: the third National Climate Assessment* (Washington, DC, 2014)

Merriman, R. B., *Six contemporaneous revolutions* (Oxford, 1938)

Merton, R. K., 'Singletons and multiples in scientific discovery', *Proceedings of the American Philosophical Society*, CV (1961), 470–86

Meyer-Fong, T., *Building culture in early Qing Yangzhou* (Stanford, 2003)

Michels, G. B., *At war with the church: religious dissent in seventeenth-century Russia* (Stanford, 1999)

Mikhail, A., *Nature and empire in Ottoman Egypt: an environmental history* (Cambridge, 2011)

Miller, H., *State versus gentry in Late Ming Dynasty China, 1572–1644* (London, 2008)

Miller, J. C., 'The significance of drought, disease and famine in the agriculturally marginal zones of West-Central Africa', *Journal of African History*, XXIII (1982), 17–61

Millstone, N., 'Seeing like a statesman in early Stuart England', *P&P*, CCXXIII (2014), 77–127

Mitchell, A. J., 'Religion, revolt, and the creation of regional identity in Catalonia, 1640–1643' (Ohio State University PhD thesis, 2005)

Moosvi, S., 'Scarcities, prices and exploitation: the agrarian crisis, 1658–70', *Studies in History*, I (1985), 45–55

— 'Science and superstition under Akbar and Jahangir: the observation of astronomical phenomena', in I. Habib, ed., *Akbar and his India* (Delhi, 1997), 109–20

— 'The Indian economic experience 1600–1900: a quantitative study', in Panikkar, ed., *The making of history*, 328–57

Moote, A. L., *The revolt of the judges: the Parlement of Paris and the Fronde, 1642–1652* (Princeton, 1971)

Moote, A. L. and D. C. Moote, *The Great Plague: the story of London's most deadly year* (Baltimore, 2004)

Morales, M. S., et al., 'Precipitation changes in the South America Altiplano since 1300 AD reconstructed by tree rings', *Climate of the Past* VIII (2012), 653–66

Mormiche, P., *Devenir prince. L'école du pouvoir en France, XVIIe–XVIIIe siècles* (Paris, 2009)

Morrill, J. S., *Revolt in the provinces: the people of England and the tragedies of war, 1630–1648* (2nd edn, London, 1999)

Mortimer, G., *Eyewitness accounts of the Thirty Years War 1618–1648* (London, 2002)

Mote, F. W. *Imperial China, 900–1800* (Cambridge, MA, 1999)

Muldrew, C., *Food, energy and the creation of industriousness. Work and material culture in agrarian England, 1550–1780* (Cambridge, 2011)

Musallam, B. F., *Sex and society in Islam: birth control before the nineteenth century* (Cambridge, 1983)

Musi, A., *La rivolta di Masaniello. Nella scena politica barroca* (2nd edn, Naples, 2002)

Mussey, B., 'Yankee chills, Ohio fever', *New England Quarterly*, XXII (1949), 435–451

Nadal, J., 'La población española durante los siglos XVI, XVII y XVIII. Un balance a escala regional', in V. Pérez Moreda and D. S. Reher, eds., *Demografía histórica de España* (Madrid, 1988), 39–54

Nagakura Tamotsu, 'Kan'ei no kikin to bakufu no taio', in Kodama Kota et al., eds., *Edo jidai no kikin* (Tokyo, 1982), 75–85

Nakane, C. and S. Ōishi, eds., *Tokugawa Japan: the social and economic antecedents of modern Japan* (Tokyo, 1990)

Nakayama, M., 'On the fluctuation of the price of rice in the Chiang-nan region during the first half of the Ch'ing period (1644–1795)', *Memoirs of the Research Department of the Toyo Bunko*, XXXVII (1979), 55–90

Namaczyńska, S., *Kronika klęsk elemntarnych w Polsce i w krajack sąsiednich w latach 1648–1696* (Lwów, 1937)

Naquin, S., *Peking: temples and city life, 1400–1900* (Berkeley, 2000)

Needham, J., et al., *Science and civilization in China*, 7 vols. in 27 parts to date (1954–)

Neukom, R., et al., 'Inter-hemispheric temperature variability over the past millennium', *Nature Climate Change*, IV (May 2014), 362–7

Neumann, K., *Das Wort als Waffe: politische Propaganda im Aufstand der Katalanen 1640–1652* (Herbolzheim, 2003)

Newman, A. J., *Safavid Iran: rebirth of a Persian empire* (London, 2006)

Newman, K. L. S., 'Shutt up: bubonic plague and quarantine in early modern England', *Journal of Social History*, XLV (2012), 809–34

Ng, V., 'Ideology and sexuality. Rape laws in Qing China', *JAS*, XLVI (1987), 57–70

Nicholson, S. E., 'The methodology of historical climate reconstruction and its application to Africa', *Journal of African History*, XX (1979), 31–49

Nicolas, J., *La rébellion française. Mouvements populaires et conscience sociale 1661–1789* (Paris, 2002)

Nierenberg, N., et al., 'Early climate change consensus at the National Academy: the origins and making of *Changing Climate*', *Historical Studies in the Natural Sciences*, XL (2010), 318–49

Nierenberg, W. A., et al., *Changing climate: report of the Carbon Dioxide Assessment Committee* (Washington, DC, 1983)

Nordmann, C., 'La crise de la Suède au temps de Christine et de Charles X Gustave (1644–1660)', *Revue d'histoire diplomatique*, XCII (1978), 210–32

Ó hAnnracháin, T., '"Though hereticks and politicians should misinterpret their goode zeal": political ideology and Catholicism in early modern Ireland', in J. H. Ohlmeyer, ed., *Political thought in seventeenth-century Ireland: kingdom or colony* (Cambridge, 2000), 155–75

O'Riordan, T., and T. Lenton, 'Tackling tipping points', *British Academy Review*, XVIII (2011), 21–7

Ó Siochrú, M., and J. H. Ohlmeyer, eds., *Ireland, 1641. Contexts and reactions* (Manchester, 2013)

Odhner, C. T., *Die Politik Schwedens im westfälischen Friedenscongress und die Gründung der schwedischen Herrschaft in Deutschland* (Gotha, 1877)

Ogilvie, S. C., 'Germany and the seventeenth-century crisis', in Parker and Smith, eds., *The General Crisis*, 57–86

— *A bitter living: women, market and social capital in early modern Germany* (Oxford, 2003)

Ohlmeyer, J. H., 'The Antrim Plot of 1641 – a myth?', *HJ*, XXXV (1992), 905–19

Okada Takehiko, 'Practical learning in the Chu Hsi School: Yamazaki Ansai and Kaibara Ekken', in de Bary and Bloom, eds., *Principles and practicality*, 231–305

Ooms, H., *Tokugawa ideology: Early constructs, 1570–1680* (Princeton, 1985)

Oreskes, N., and E. M. Conway, *Merchants of doubt: how a handful of scientists obscured the truth on issues from tobacco smoke to global warming* (New York, 2010)

Outram, Q., 'The socio-economic relations of warfare and the military mortality crises of the Thirty Years' War', *Medical History*, XLV (2001), 151–84

Özel, O., 'Banditry, state and economy: on the financial impact of the *Celali* movement in Ottoman Anatolia', *Proceedings of the IXth International Congress of Economic and Social History of Turkey* (Ankara, 2007), 65–74

— *The collapse of rural order in Ottoman Anatolia. Amasya 1576–1643* (Leiden, 2016)

Özvar, E., 'Fiscal crisis of the Ottoman empire in the seventeenth century?' (unpublished paper at the XI Congress of Social and Economic History of Turkey, Ankara, 2008)

Panikkar, K. N., T. J. Byres, and U. Patnaik, eds., *The making of history: essays presented to Irfan Habib* (London, 1985)

Parenti, C., *Tropic of chaos: climate change and the new geography of violence* (New York, 2011)

Parker, G., ed., *The Thirty Years' War* (2nd edn, London, 1997)

Parker, G., *The military revolution. Military innovation and the rise of the West, 1500–1800* (3rd edn, Cambridge, 2008)

— 'La crisis de la década de 1590 reconsiderada: Felipe II, sus enemigos y el cambio climático', in A. Marcos Marín, ed., *Libro Homenaje para José Luis Rodríguez de Diego* (Valladolid, 2011), 643–70

Parker, G. and L. M. Smith, *The General Crisis of the seventeenth century* (2nd edn, London, 1997)

Parsons, J. B., *The peasant rebellions of the late Ming dynasty* (Tucson, AZ, 1970)

Patz, J. A., et al., 'The effects of changing weather on publlic health', *Annual Review of Public Health*, XXI (2000), 271–307

Perceval-Maxwell, M., *The outbreak of the Irish rebellion of 1641* (Quebec, 1994)

— 'Ulster 1641 in the context of political developments in the three kingdoms', in B. MacCuarta, ed., *Ulster 1641: aspects of the rising* (2nd edn, Belfast, 1997), 93–106

Peiser, B. J., *Natural catastrophes during Bronze Age civilizations* (Oxford, 1998)

Pérez Samper, M. A., *Catalunya i Portugal el 1640: dos poles en una cruilla* (Barcelona, 1992)

Pestana, C. G., *The English Atlantic in an age of revolution, 1640–1661* (Cambridge, MA, 2004)

Peters, K., 'The Quakers and the politics of the army in the crisis of 1659', *P&P* 231 (2016), 97–128

Peterson, W. J., 'The life of Ku Yen-wu (1613–1682)', *Harvard Journal of Asiatic Studies*, XXVIII (1968), 114–56, and XXIX (1969), 201–47

— *Bitter gourd: Fang I-chih and the impetus for intellectual change* (New Haven and London, 1979)

Pfister, C., 'Climatic extremes, recurrent crises and witch hunts: strategies of European societies in coping with exogenous shocks in the late sixteenth and early seventeenth centuries', *Medieval History Journal*, X/1-2 (2007), 33–73

Pillorget, R., *Les mouvements insurrectionnels de Provence entre 1596 et 1715* (Paris, 1975)

Piterberg, G., *An Ottoman tragedy: history and historiography at play* (Berkeley, 2003)

Platonov, S. F., 'Novyi istochnik istochnik dlia istorii Moskovskikh volnenii', *Chteniiia v imperatorskom obshchestve istorii i drevnostei Rossiiskikh Moskovskom universitete* (1893/1), 3–19

Plokhy, S., *The Cossacks and religion in early modern Ukraine* (Oxford, 2001)

Plumb, J. H., *The growth of political stability in England, 1675–1725* (London, 1967)

Poelhekke, J. J., *De vrede van Munster* (The Hague, 1948)

Pollack, H. N., S. Huang and J. E. Smerdon, 'Five centuries of climate change in Australia: the view from under-ground', *Journal of Quaternary Science*, XXI (2006), 701–6

Pomeranz, K. W., *The Great Divergence: China, Europe, and the making of the modern world economy* (Princeton, 2000)

— 'Is there an East Asian development path? Long-term comparisons, constraints, and continuities', *Journal of the Economic and Social History of the Orient*, XLIV (2001), 322–62

— 'Weather, war, and welfare: persistence and change in Geoffrey Parker's *Global Crisis*', *Historically Speaking*, XIV/5 (Nov. 2013), 30–3

Porschnev, B. F., 'Les rapports politiques de l'Europe occidentale et de l'Europe orientale à l'époque de la Guerre de Trente Ans', in *Rapports du XIe Congrès des Sciences Historiques* (Stockholm, 1960), IV, 136–63

Powell, J. L., 'Climate scientists virtually unanimous: anthropogenic global warming is true', *Bulletin of Science, Technology and Society*, XXXV/5-6 (2015), 121–4

Pursell, B., *The Winter King: Frederick V of the Palatinate and the coming of the Thirty Years War* (Aldershot, 2002)

Quenet, G., *Les tremblements de terre aux XVIIe et XVIIIe siècles. La naissance d'un risque* (Seyssel, 2005)

Rabb, T. K., *The struggle for stability in early modern Europe* (Oxford, 1975)

—'The Scientific Revolution and the problem of periodization', *European Review*, XV (2007), 503–12

— 'Introduction: the persistence of the "Crisis"', *JIH*, XL (2009), 145–50

Randall, D., 'Epistolary rhetoric, the newspaper, and the public sphere', *P&P*, CXCVIII (2008), 3–32

Ranum, O., *Paris in the age of absolutism: an essay* (2nd edn, University Park, PA, 2002)

Rawski, E. S., *Education and popular literacy in Ch'ing China* (Ann Arbor, 1979)

— 'The Qing formation and the early-modern period', in Struve, ed., *Qing formation*, 207–41

Raymond, J., *The invention of the newspaper: English newsbooks 1641–1649* (Oxford, 1996)

Raynor, H., *A social history of music* (London, 1972)

Reade, H. G. R., *Sidelights on the Thirty Years' War*, 3 vols. (London, 1924)

Redondo, P., *Galileo: heretic* (Princeton, 1987)

Reff, D. T., 'Contact shock in northwestern New Spain, 1518–1764', in J. W. Verano and D. H. Uberlaker, eds., *Disease and demography in the Americas* (Washington, DC, 1992), 265–76

Reid, A. R., 'The crisis of the seventeenth century in southeast Asia', in Parker and Smith, eds., *The General Crisis*, 206–34

Ribot García, L. A., *La revuelta antiespañola de Mesina. Causas y antecedentes (1591–1674)* (Valladolid 1982)

— *La Monarquía de España y la guerra de Mesina (1674–1678)* (Madrid, 2002)

Richards, J., *The unending frontier: an environmental history of the early modern world* (Berkeley, 2003)

Riches, D., *The anthropology of violence* (Oxford, 1986)

Richter, D. K., 'War and culture: the Iroquois experience', *WMQ*, 3rd series, XL (1983), 528–59

Roberts, M., 'Queen Christina and the General Crisis of the seventeenth century', in Aston, ed., *Crisis in Europe*, 195–221

Robinson, J. H., *The great comet of 1680: a study in the history of rationalism* (Northfield, MN, 1916)

Robisheaux, T., *Rural society and the search for order in early modern Germany* (Cambridge, 1989)

Rodén, M. L., 'The crisis of the seventeenth century: the Nordic perspective', in Benedict and Gutmann, eds., *Early modern Europe*, 100–19

Rodger, N. A. M., *The command of the ocean: a naval history of Britain, 1649–1815* (London, 2004)

Rohrschneider, M., *Der gescheiterte Frieden von Münster: Spaniens Ringen mit Frankreich auf dem westfälischen Friedenskongress (1643–1649)* (Münster, 2007: Schriftenreihe der Vereinigung zur Erforschung der neueren Geschichte, XXX)

Romano, R., *Conjonctures opposées. La crise du 17e siècle en Europe et en Amérique latine* (Geneva, 1992)

Ronan, C. A., *Edmond Halley: genius in eclipse* (New York, 1969)

Rosenzweig, C., ed., *Responding to climate change in New York State: The ClimAID integrated assessment for effective climate change adaptation. Final report* (Albany, NY, 2011)

Ross, A. S., 'Pupils' choices and social mobility after the Thirty Years' War: a quantitative study', *HJ*, LVII (2014), 311–41

— *Daum's boys: schools and the Republic of Letters in early modern Germany* (Manchester, 2015)

Roth, G., 'The Manchu-Chinese relationship, 1618–1636', in Spence and Wills, eds., *From Ming to Ch'ing*, 4–38

Roupnel, G., *La ville et la campagne au XVIIe siècle: étude sur les populations du pays dijonnais* (2nd edn, Paris, 1955)

Rovito, P. L., 'Le rivoluzione constitutionale di Napoli (1647–48)', *Rivista storica italiana*, XCVII (1986), 367–462

Russell, C., *The fall of the British monarchies, 1637–1642* (Oxford, 1991)

Rydval, M., et al., 'Reconstructing 800 years in Scotland from tree rings', *Climate Dynamics*, XLVII (2017), 1–24

Saitō Osamu, 'The frequency of famines as demographic correctives in the Japanese past', in T. Dyson and C. Ó Gráda, eds., *Famine demography: perspectives from the past and present* (Oxford: Oxford University Press, 2002), 218–39

Salm, H., *Armeefinanzierung im dreissigjährigen Krieg. Die niederrheinisch-westfälische Reichskreis, 1635–1650* (Münster, 1990: Schriftenreihe der Vereinigung zur Erforschung der neueren Geschichte, XVI)

Sanz Camañes, P., 'Fronteras, poder y milicia en la España moderna. Consecuencias de la administración militar en las poblaciones de la frontera catalano-aragonesa durante la Guerra de Secesión catalana (1640-1652)', *Manuscrits*, XXVI (2008), 53–77

Sasaki Junnosuke, *Daimyō to hyakushō* (Tokyo, 1966)

Schaffer, S., *The information order of Isaac Newton's* Principia Mathematica (Uppsala, 2008)

Schama, S. M., *The embarrassment of riches: an interpretation of Dutch culture in the Golden Age* (New York, 1987)

Schaub, J. F., *Le Portugal au temps du comte-duc d'Olivares (1621–1640). Le conflit de juridictions comme exercice de la politique* (Madrid, 2002)

Schaufler, H. H., *Die Schlacht bei Freiburg im Breisgau* (Freiburg, 1979)

Schiebinger, L., *Plants and empire: colonial bioprospecting in the Atlantic world* (Cambridge, MA, 2004)

Schilling, H., 'Confessional Europe', in T. A. Brady, H. A. Oberman and J. D. Tracy, eds., *Handbook of European history, 1400–1600*, 2 vols. (Cambridge, 1995), II, 641–81

Scholem, G., *Sabbatai Sevi: the mystical Messiah, 1626–1676* (Princeton, 1973)

Schumpeter, J. A., *Capitalism, socialism and democracy* (1942; ed. R. Swedberg, London, 2003)

Schwartz, S. B., 'Panic in the Indies: the Portuguese threat to the Spanish empire', in Thomas and de Groof, eds., *Rebelión y resistencia*, 205–17

— 'Silver, sugar and slaves: how the empire restored Portugal' (EUI conference paper, 2003)

Scott, D., *Politics and war in the three Stuart kingdoms, 1637–1649* (Basingstoke, 2004)

Scott, J., *Love and protest: Chinese poems from the sixth century AD to the seventeenth century AD* (London, 1972)

— *England's troubles: seventeenth-century English political instability in European context* (Cambridge, 2000)

Scott, J. C., *Weapons of the weak: everyday forms of peasant resistance* (New Haven, 1985)

— *Seeing like a state: how certain schemes to improve the human condition have failed* (New Haven, 1998)

Scurlock, D., *From the Rio to the Sierra. An environmental history of the middle Rio Grande basin* (Ft Collins, CO, 1998)

Seaver, P. S., *Wallington's world: a Puritan artisan in seventeenth-century London* (Stanford, 1985)

Sen, A. K., *Poverty and famines: an essay on entitlement and deprivation* (Oxford, 1981)

Shannon, A., '"Uncouth language to a prince's ears": Archibald Armstrong, court jester, and early Stuart politics', *SCJ*, XLII (2011), 99–112

Sharpe, K., *The personal rule of Charles I* (New Haven, 1992)

Shively, D. H., 'Sumptuary regulation and status in early Tokugawa Japan', *Harvard Journal of Asiatic Studies*, XXV (1964–5), 123–64

Shy, J., *A people numerous and armed: reflections on the military struggle for American independence* (2nd edn, Oxford, 1990)

Sigl, M., et al., 'Timing and climate forcing of volcanic eruptions for the past 2,500 years', *Nature*, DXXIII (2015), 543–9

Singh, C., *Region and empire: Panjab in the seventeenth century* (Delhi, 1991)

Slotkin, R. *The fatal environment: the myth of the frontier in the age of industrialization, 1800–1890* (2nd edn., Norman, OK, 1998)

Slovic, P., 'The perception of risk', *Science*, NS, CCXXXVI (1987), 280–5

— *The perception of risk* (London, 2000)

Smith, J. H., 'Benevolent societies: the reshaping of charity during the late Ming and early Ch'ing', *JAS*, XLVI (1987), 309–37

— *The art of doing good: charity in late Ming China* (Berkeley, 2009)

Smith, T. C., *Nakahara: Family farming and population in a Japanese village, 1717–1830* (Stanford, 1977)

Smythe, W. J., 'Towards a cultural geography', in Ó Siochrú and Ohlmeyer, eds., *Ireland, 1641*, 71–94

Sobel, A., *Storm surge: Hurricane Sandy, our changing climate, and extreme weather of the past and future* (London, 2014)

Solar, P. M., 'Poor relief and English economic development before the Industrial Revolution', *EcHR*, 2nd series, XLVIII (1995), 1–22

Soll, J., 'Accounting for government: Holland and the rise of political economy in seventeenth-century Europe', *JIH*, XL (2009), 215–38

Spence, J. D., *Emperor of China: self-portrait of K'ang-hsi* (New York, 1974)

— *The death of Woman Wang: rural life in China in the seventeenth century* (London, 1978)

Spence, J. D. and J. E. Wills, eds., *From Ming to Ch'ing: conquest, region, and continuity in seventeenth-century China* (New Haven, 1979)

Sreenivasan, G. P., *The peasants of Ottobeuren, 1487–1726: a rural society in early modern Europe* (Cambridge, 2004)

Stager, J. C., et al., 'Late Holocene precipitation variability in the summer rainfall region of South Africa', *Quaternary Science Reviews*, LXVII (2013), 105–20

Stahle, D. W., et al., 'The lost colony and Jamestown droughts', *Science*, CCLXXX (1998), 564–7

— 'Tree-ring reconstructed megadroughts over North America since A. D. 1300', *Climatic Change*, LXXXIII (2007), 133–49

Stampfer, S., 'What actually happened to the Jews of Ukraine in 1648?', *Jewish History*, XVII/2 (2003), 207–27

Starna, W. A., 'The Pequots in the early seventeenth century', in L. M. Hauptman and J. D. Wherry, eds., *The Pequots in southern New England: the fall and rise of an American Indian nation* (Norman, OK, 1990), 33–47

Steensgaard, N., 'The seventeenth-century crisis', in Parker and Smith, *The General Crisis*, 32–56

Stoyle, M. J., '"Whole streets converted to ashes": property destruction in Exeter during the English Civil War', *Southern History*, XVI (1994), 62–81

— '"Memories of the maimed": the testimony of Charles I's former soldiers, 1660–1730', *History*, LXXXVIII (2003), 204–26

Studer, R., *The great divergence reconsidered: Europe, India, and the rise to global economic power* (CUP, 2015)

Struve, L. A., *The Southern Ming, 1644–1662* (New Haven, 1984)

Struve, L. A., ed., *The Qing formation in world-historical time* (Cambridge, MA, 2004)

— *Time, temporality and change of empire: East Asia from Ming to Qing* (Honolulu, 2005)

— 'Dreaming and self-search during the Ming collapse: the *Xue Xiemeng Biji*, 1642–1646', *T'oung Pao*, XCIII (2007), 159–92

Suárez, M., 'La "crisis del siglo XVII" en la región andina', in M. Burga, ed., *Historia de América Andina. Formación y apogeo del sistema colonial*, II (Quito, 2000), 289–317

Subrahmanyam, S., 'Hearing voices: vignettes of early modernity in South Asia, 1400–1750', *Daedalus*, CXXVII.3 (1998), 75–104

— *Explorations in connected history: Mughals and Franks* (Oxford, 2005)

— 'A tale of three empires: Mughals, Ottomans and Habsburgs in a comparative context', *Common Knowledge*, XII.1 (2006), 66–92

Suvanto, P., *Wallenstein und seine Anhänger am Wiener Hof zur Zeit des zweiten Generalats, 1631–1634* (Helsinki, 1963)

Symcox, G., ed., *War, diplomacy and imperialism, 1618–1763* (London, 1974)

Sysyn, F. E., 'Ukrainian-Polish relations in the seventeenth century: the role of national consciousness and national conflict in the Khmelnytsky movement', in P. J. Potichnyj, ed., *Poland and Ukraine: past and present* (Edmonton, 1980), 58–82

— *Between Poland and the Ukraine: the dilemma of Adam Kysil, 1600–1653* (Cambridge, MA, 1985)

— 'The Khmel'nyts'kyi uprising: a characterization of the Ukrainian revolt', *Jewish History*, XVII (2003), 115–39

Tacke, A., 'Mars, the enemy of art: Sandrart's *Teutsche Academie* and the impact of war on art and artists', in Bussmann and Schilling, eds., *1648*, II, 245–52

Tashiro Kazui, 'Tsushima Han's Korean trade, 1684–1710', *Acta Asiatica*, XXX (1976), 85–105

Te Brake, W., *Shaping history: ordinary people in European politics 1500–1700* (Berkeley, 1998)

Telford, T. A., 'Fertility and population growth in the lineages of Tongcheng county, 1520–1661', in Harrell, ed., *Chinese historical micro-demography*, 48–93

Teodoreanu, E., 'Preliminary observations on the Little Ice Age in Romania', *Present Environment and Sustainable Development*, V (2011), 187–94

Terzioğlu, D., 'Sufi and dissident in the Ottoman empire: Niyāzī-i Mīṣri (1618–94)' (Harvard University PhD thesis, 1999)

— '"Where 'Ilm-i Ḥāl meets catechism: Islamic manuals of religious instruction in the Ottoman empire in the age of confessionalization", *P&P*, CCXX (2013), 79–114

Theibault, J. C., 'The demography of the Thirty Years War revisited: Günther Franz and his critics', *German History*, XV (1997), 1–21

Thirsk, J., *The agrarian history of England and Wales. V. 1640–1700*, 2 vols (Cambridge, 1985)

Thomas, W. and B. de Groof, eds., *Rebelión y resistencia de el mundo hispánico del siglo XVII* (Louvain, 1992)

Thompson, I. A. A., 'Alteraciones granadinos: el motín de 1648 a la luz de un nuevo testimonio presencial', in Castellanos, *Homenaje*, II, 799–812

Thornton, J. K., 'Warfare, slave trading and European influence: Atlantic Africa 1450–1800', in J. M. Black, ed., *War in the early modern world, 1450–1815* (London, 1999), 129–46

Tong, J. W., *Disorder under heaven: collective violence in the Ming dynasty* (Stanford, 1991)

Torke, H. J., *Die staatsbedingte Gesellschaft im Moskauer Reich. Zar und Zemlja in der altrussischen Herrschaftsverfassung, 1613–1689* (Leiden, 1974)

Totman, C., 'Tokugawa peasants: win, lose or draw?', *Monumenta Nipponica*, XLI (1986), 457–76

Trevor-Roper, H. R., 'The general crisis of the seventeenth century', *P&P*, XVI (1959), 31–64

— 'The Fast Sermons of the Long Parliament', in idem, *Religion, the reformation and social change* (2nd edn, London, 1972), 273–316

— *Europe's physician: the various life of Sir Theodore de Mayerne* (New Haven and London, 2006)

Trotsky, L., *The history of the Russian Revolution*, 3 vols. (1930; English edn, 1932)

Tsunoda, R., W. T. de Bary and D. Keene, *The sources of Japanese tradition*, 2 vols. (New York, 1958)

Twitchett, D. and F. W. Mote, eds., *The Cambridge history of China, VIII, Part 2: the Ming* (Cambridge, 1998)

Tyson, P. D., W. Karlén, K. Holmgren and G. A. Heiss, 'The Little Ice Age and medieval warming in South Africa', *South African Journal of Science* XCVI (2000), 121–26

Ulbricht, O., 'The experience of violence during the Thirty Years' War; a look at the civilian victims', in Canning, J., H. Lehmann and J. Winter, eds., *Power, violence and mass death in pre-modern and modern times* (Aldershot: Ashgate, 2004), 97–127

Urban, H., 'Druck und Drücke des Restitutionsedikt von 1629', *Archiv für Geschichte des Buchwesens*, XIV (1974), 609–54

Valladares, R., *La rebelión de Portugal. Guerra, conflicto y poderes en la Monarquía Hispánica (1640–1680)* (Valladolid, 1998)

Van de Haar, C., *De diplomatieke betrekkingen tussen de Republiek en Portugal, 1640–1661* (Groningen, 1961)

Van Groesen, M., '(No) news from the Western Front: the weekly press of the Low Countries and the making of Atlantic news', *SCJ*, XLIV (2013), 739–60

Van Maarseveen, M. P., et al., eds., *Beelden van een strijd. Oorlog en kunst vóór de vrede van Munster, 1621–1648* (Delft, 1998)

Vaquero, J. M., et al., 'Revisited sunspot data: a new scenario for the onset of the Maunder Minimum', *Astrophysical Journal, Letters*, DCCXXXI (2011) L 24

Venezia e la peste, 1348–1797 (Venice, 1980)

Verbeek, T., *Descartes and the Dutch: early reactions to Cartesian philosophy, 1637–50* (Carbondale, IL, 1992)

Vicuña Mackenna, B., *El clima de Chile. Ensayo histórico* (Buenos Aires, 1970)

Villari, R., *Un sogno di libertà. Napoli nel declino di un impero, 1585–1648* (Milan, 2012; expanded version of Villari, *The revolt of Naples,* Cambridge, 1993; original Italian edn, 1967)

Villstrand, N. E., 'Adaptation or protestation: local community facing the conscription of infantry for the Swedish armed forces 1620–79', in L. Jespersen, ed., *A revolution from above? The power state of sixteenth- and seventeenth-century Scandinavia* (Odense, 2000), 249–313

Virol, M., 'Connaître et accroître les peuples du royaume: Vauban et la population', *Population*, LVI (2001), 845–75

Vlastos, S., *Peasant protests and uprisings in Tokugawa Japan* (Berkeley, 1986)

Von Friedeburg, R., 'The making of patriots: love of Fatherland and negotiating monarchy in seventeenth-century Germany', *JMH*, LXXVII (2005), 881–916

Von Glahn, R., *Fountain of fortune: money and monetary policy in China, 1000–1700* (Berkeley, 1996)

Wakeman, F. C., *The Great Enterprise: the Manchu reconstruction of imperial order in 17th century China* (Berkeley, 1985)

Wakeman, F. C., 'Localism and loyalism during the Ch'ing conquest of Kiangnan', in F. C. Wakeman and C. Grant, eds., *Conflict and control in Late Imperial China* (Berkeley, 1975) , 44–85

Waley-Cohen, J., *The culture of war in China: empire and the military under the Qing dynasty* (London, 2006)

Walford, C., 'The famines of the world: past and present', *Journal of the Statistical Society of London*, XLI (1878), 433–535, and XLII (1879), 79–275

Walter, J., *Understanding popular violence in early modern England: the Colchester plunderers* (Cambridge, 1999)

— '"Abolishing superstition with sedition"? The politics of popular iconoclasm in England, 1640–2', *P&P*, CLXXXIII (2004), 79–123

— 'Public transcripts, popular agency and the politics of subsistence in early modern England', in Braddick and Walter, eds., *Negotiating power*, 123–48

— *Crowds and popular politics in early modern England* (Manchester, 2006)

Wang X. M., et al., 'Desertification and the rise and collapse of China's historical dynasties', *Human Ecology*, XXXVIII (2010), 157–72

Warde, P., 'Review article: global crisis or global evidence', *P&P*, CCXXVIII (2015), 287–301

Warren, J. W., *Connecticut unscathed: victory in the Great Narragansett War, 1675–1676* (Norman, OK, 2014)

Webb, S. S., *1676: the end of American independence* (New York, 1984)

Webster, J. B., ed., *Chronology, migration and droughts in interlacustrine Africa* (London, 1979)

Wedgwood, C. V., *The Thirty Years War* (London, 1938)

Weiss, J. G., 'Die Vorgeschichte des böhmischen Abenteuers Friedrichs V. von der Pfalz', *Zeitschrift für die Geschichte des Oberrheins*, new series, LIII (1940), 383–492

Weisser, M. R., *The peasants of the Montes: the roots of rural rebellion in Spain* (Chicago, 1976)

Wells, J., 'English law, Irish trials and Cromwellian state building in the 1650s', *P&P*, CCXXVII (2015), 77–119

White, R. *'It's your misfortune and none of my own': a new history of the American West* (Norman, OK, 1991)

— *The middle ground: Indians, empires and republics in the Great Lakes region 1650–1800* (Cambridge, 1991)

White, S., *The climate of rebellion in the early modern Ottoman empire* (Cambridge, 2011)

— 'The real Little Ice Age', *JIH*, XLIV (2014), 327–52

Wickman, T., '"Winters embittered with hardships": severe cold, Wabanaki power, and English adjustments, 1690–1710', *WMQ*, LXXII (2015), 57–98

Wigley, T. M. L., M. Ingram and G. Farmer, eds., *Climate and history: studies in past climates and their impact on man* (Cambridge, 1981)

Will, P.-E., 'Développement quantitatif et développement qualitatif en Chine à la fin de l'époque impériale', *Annales HSS*, XLIX (1994), 863–902

— 'Coming of age in Shanghai during the Ming-Qing transition: Yao Tinglin's (1628–after 1697) *Record of the successive years*', *Gu jin lung heng*, XLIV (2000), 15–38

Wilson, J. E., '"A thousand countries to go to": peasants and rulers in late eighteenth-century Bengal', *P&P*, CLXXXIX (2005), 81–109

Wilson, P. H., *The Thirty Years War: Europe's tragedy* (Cambridge, MA, 2009)

Winius, G., *The fatal history of Portuguese Ceylon: transition to Dutch rule* (Cambridge, MA, 1971)

Withington, P., ed., *Cultures of intoxication, P&P*, CCXXII (2014), Supplement IX

Wong, R. Bin, *China transformed: historical change and the limits of European experience* (Ithaca, NY, 1997)

Wood, A., 'Fear, hatred and the hidden injuries of class in early modern England', *Journal of Social History*, XXIX (2006), 803–26

Wood, G. D., *Tambora: the eruption that changed the world* (Princeton, 2014)

Woods, R., *Death before birth: fetal health and mortality in historical perspective* (Oxford, 2009)

Woolrych, A., *Soldiers and statesmen: the General Council of the Army and its debates, 1647–1648* (Oxford, 1987)

Worden, B., 'The politics of Marvell's Horatian Ode', *HJ*, XXVII (1984), 525–47

— 'Oliver Cromwell and the sin of Achan', in D. Beales and G. Best, eds., *History, society and the churches: essays in honour of Owen Chadwick* (Cambridge, 1985), 125–45

Wrightson, K. E., 'Infanticide in earlier seventeenth-century England', *Local Population Studies*, XV (1975), 10–22

Wrightson, K., and J. Walter, 'Dearth and the social order in early modern England', *P&P*, LXXI (1976), 22–42

Wrigley, E. A., *People, cities and wealth: the transformation of traditional society* (Oxford, 1987)

— 'Urban growth in early modern England: food, fuel and transport', *P&P*, CCXXV (2015), 79–112

Wrigley, E. A., and R. S. Schofield, *The population history of England, 1541–1871: a reconstruction* (2nd edn, Cambridge, 1989)

Wu, H. L., 'Corpses on display: representations of torture and pain in the Wei Zhongxian novels', *Ming Studies*, LIX (2009), 42–55

Wu, S., *Communication and imperial control in China: evolution of the palace memorial system, 1693–1735* (Cambridge, MA, 1970)

Yamamoto Hirofumi, *Kan'ei jidai* (Tokyo, 1989)

Yamamura, K., 'From coins to rice: hypotheses on the *Kandaka* and *Kokudaka* systems', *Journal of Japanese Studies*, XIV (1988), 341–67

Yang, Lien-Sheng, 'Economic justification for spending – an uncommon idea in traditional China', *Harvard Journal of Asiatic Studies*, XX/1 (1957), 36–52

Yi Tae-jin, 'Meteor fallings and other natural phenomena between 1500–1750, as recorded in the Annals of the Chosŏn dynasty (Korea)', *Celestial mechanics and dynamical astronomy*, LXIX (1998), 199–220

Young, J. R., *The Scottish Parliament 1639–1661: a political and constitutional analysis* (Edinburgh, 1996)

Young, J. T., *Faith, medical alchemy and natural philosophy: Johann Moriaen, reformed intelligencer, and the Hartlib circle* (Aldershot, 1998)

Zagorin, P., *Rebels and rulers 1500–1660*, 2 vols. (Cambridge, 1982)

Zeman, J. K., 'Responses to Calvin and Calvinism among the Czech Brethren (1540–1605)', *American Society for Reformation Research. Occasional papers*, I (1977), 41–52

Zhang Ying, 'Politics and morality during the Ming-Qing dynastic transition' (University of Michigan PhD thesis, 2010)

Zheng, Jingyun, et al., 'How climate change impacted the collapse of the Ming Dynasty', *CC*, CXXVII (2014), 169–82, and electronic supplementary material

Zilfi, M., 'The Kadizadelis: discordant revivalism in seventeenth-century Istanbul', *Journal of Near Eastern Studies*, XLV (1986), 251–69

Zysberg, A., 'Galley and hard labor convicts in France (1550–1850)', in P. Spierenburg, ed., *The emergence of carceral institutions: prisons, galleys and lunatic asylums 1550–1900* (Rotterdam, 1984), 78–124

Index